Handbook of Synthetic Photochemistry

Edited by
Angelo Albini and Maurizio Fagnoni

Further Reading

S. Ma (Ed.)

Handbook of Cyclization Reactions

2 Volume Set

2010

ISBN: 978-3-527-32088-2

B. Wardle

Principles and Applications of Photochemistry

2010

ISBN: 978-0-470-01493-6

G.I. Likhtenshtein

Stilbenes

Applications in Chemistry, Life Sciences and Materials Science

2010

ISBN: 978-3-527-32388-3

M. Bandini, A. Ronchi-Umani (Eds.)

Catalytic Asymmetric Friedel-Crafts Alkylations

2009

ISBN: 978-3-527-32380-7

N. Mizuno (Ed.)

Modern Heterogeneous Oxidation Catalysis

Design, Reactions and Characterization

2009

ISBN: 978-3-527-31859-9

E.M. Carreira, L. Kvaerno

Classics in Stereoselective Synthesis

2009

ISBN: 978-3-527-32452-1 (Hardcover)

ISBN: 978-3-527-29966-9 (Softcover)

Handbook of Synthetic Photochemistry

Edited by
Angelo Albini and Maurizio Fagnoni

WILEY-VCH Verlag GmbH & Co. KGaA

The Editors

Prof. Dr. Angelo Albini
University of Pavia
Department of Organic Chemistry
Via Taramelli 10
27100 Pavia
Italy

Prof. Dr. Maurizio Fagnoni
Department of Organic Chemistry
University of Pavia
Via Tamarelli 10
27100 Pavia
Italy

All books published by Wiley-VCH are carefully produced. Nevertheless, authors, editors, and publisher do not warrant the information contained in these books, including this book, to be free of errors. Readers are advised to keep in mind that statements, data, illustrations, procedural details or other items may inadvertently be inaccurate.

Library of Congress Card No.: applied for

British Library Cataloguing-in-Publication Data
A catalogue record for this book is available from the British Library.

Bibliographic information published by the Deutsche Nationalbibliothek
The Deutsche Nationalbibliothek lists this publication in the Deutsche Nationalbibliografie; detailed bibliographic data are available on the Internet at http://dnb.d-nb.de

© 2010 WILEY-VCH Verlag GmbH & Co. KGaA, Weinheim

All rights reserved (including those of translation into other languages). No part of this book may be reproduced in any form – by photoprinting, microfilm, or any other means – nor transmitted or translated into a machine language without written permission from the publishers. Registered names, trademarks, etc. used in this book, even when not specifically marked as such, are not to be considered unprotected by law.

Printed in Great Britain
Printed on acid-free paper

Cover Design   Adam-Design, Weinheim
Typesetting   Thomson Digital, Noida, India
Printing and Binding   T.J. International Ltd, Padstow, Cornwall

ISBN: 978-3-527-32391-3

Foreword

From its origin over a century ago, organic photochemistry has undergone a transformation from an area of science populated by a few specialized organic and physical chemists to a field that now attracts the interest of members of the broad synthetic organic chemistry community. Along the way, the basic chemical and physical foundations of the science were developed and the full synthetic potential of photochemical reactions of organic substrates has been realized.

The science of organic photochemistry can be traced back to observations made in the nineteenth century, which showed that ultraviolet irradiation of certain organic substances leads to formation of products that have unique and sometimes highly strained structures. An example of this is found in studies in the early 1800s, which demonstrated that irradiation of the naturally occurring, cross-conjugated cyclohexadienone, α-santonin, in the crystal state induces a deep-seated, multistepped rearrangement reaction. It is fair to conclude that at that time observations like this could only have been attributed to the magic of Nature, since little if anything was known about the fundamental principles of the light absorption process and the relationships between structures and decay pathways of electronic excited states.

The science of organic photochemistry experienced a significant transformation in the middle part of the twentieth century when it began to attract the interest of organic chemists, who were skilled in the use of valence bond theory, and physical chemists, who were able to probe and theoretically analyze the properties of electronic excited states. These efforts led to a basic mechanistic framework for understanding and predicting how electronic excited states of organic substrates undergo reactions to form products. Clear examples of the insight provided by organic chemists during this era are found in ground-breaking investigations performed independently by Zimmerman and Chapman that probed the photochemistry of simple, cross-conjugated cyclohexadienones. The realization that these processes could be described by utilizing Lewis electron-dot-line structures of excited states and reactive intermediates brought organic photochemistry into the intellectual sphere of organic chemists, who already had learned the benefits of writing arrow-pushing mechanisms for ground-state reactions.

Another important contribution to the field of organic photochemistry arose from investigations of excited state redox processes in the latter part of the twentieth

Handbook of Synthetic Photochemistry. Edited by Angelo Albini and Maurizio Fagnoni
Copyright © 2010 WILEY-VCH Verlag GmbH & Co. KGaA, Weinheim
ISBN: 978-3-527-32391-3

century. These efforts showed that when the oxidation and reduction potentials and excited state energies of interacting electron donors and acceptors are appropriate, thermodynamically and kinetically favorable excited state single electron transfer (SET) will take place to produce ion radical intermediates. This phenomenon expanded the vista of organic photochemistry, since it enabled the unique and predictable reactivity profiles of charged radicals to be included in the concept library used to design new photochemical transformations. Many examples of the exceptional impact that SET has had on the field of organic photochemistry came from the pioneering work of Arnold and a cadre of other organic chemists who developed synthetic applicable SET photochemical processes.

It is clear that studies in the area of organic photochemistry have led to the discovery of a large number of novel reactions, and that some of these processes meet the high standards needed for use as preparative methodologies. The compilation in this Handbook, which begins with a useful chapter describing practical experimental methods used in photochemistry, reviews several of the more synthetically prominent photochemical reactions of organic substrates.

There is no doubt that the field of organic photochemistry was subjected to intense scrutiny in the latter half of the twentieth century, and that efforts during this period led to a firm understanding of basic photochemical principles and to the discovery of a wealth of highly unique chemical reactions. Moreover, during this period members of the synthetic organic chemistry community recognized that several photochemical processes could be applied as key steps in routes for the construction of complex target molecules. It is likely that activity in the area of organic photochemistry will not diminish in the twenty first century where it will used in finding matchless solutions to challenging chemical problems. Thus, rather than being caused by the need to prepare sophisticated organic substances made by Nature, problems in the new century are likely to revolve about the search for green methods for promoting chemical reactions and for processes that can be performed in confined spaces (e.g., cells), defined patterns (e.g., lithography), and precisely controlled time domains (e.g., triggers). Organic photochemistry is uniquely applicable to these types of challenges and, as a result, it should continue to be an interesting area in which to work.

Patrick S. Mariano
Department of Chemistry and Chemical Biology
University of New Mexico
Albuquerque, NM, USA

Contents

Foreword V
Preface XV
List of Contributors XVII

1 **Photochemical Methods** 1
 Angelo Albini and Luca Germani
1.1 Photochemical Methods 1
1.1.1 Photochemistry and Organic Synthesis 1
1.2 Irradiation Apparatus 2
1.2.1 General 2
1.2.2 Low-Pressure Mercury Arcs 3
1.2.3 Medium- and High-Pressure Mercury Arcs 7
1.2.4 Other Light Sources 9
1.3 Further Experimental Parameters 11
1.3.1 Concentration and Scale 11
1.3.2 Effect of Impurities, Oxygen, and Temperature 15
1.3.3 Safety 17
1.3.4 Planning a Photochemical Synthesis 17
1.4 Photochemical Steps in Synthetic Planning 19
 References 22

2 **Carbon–Carbon Bond Formation by the Photoelimination of Small Molecules in Solution and in Crystals** 25
 Saori Shiraki and Miguel A. Garcia-Garibay
2.1 Introduction 25
2.1.1 Synthesis of Unstable Molecules 27
2.2 Photochemical C–C Bond Formation in Solution 30
2.2.1 Concerted Reactions 30
2.2.2 Photoelimination of N_2 31
2.2.2.1 Synthesis of Three-Membered Rings 31

Handbook of Synthetic Photochemistry. Edited by Angelo Albini and Maurizio Fagnoni
Copyright © 2010 WILEY-VCH Verlag GmbH & Co. KGaA, Weinheim
ISBN: 978-3-527-32391-3

2.2.2.2	Synthesis of Cyclobutanes and Polycyclic Compounds	33
2.2.3	Photoelimination of CO from Ketones in Solution	35
2.2.4	Photoelimination of CO_2 from Lactones	39
2.2.5	Photoelimination of Sulfur from Sulfides, Sulfoxides, and Sulfones	40
2.3	Reactions in the Solid State	41
2.3.1	Reactivity and Stability in the Solid State	42
2.3.2	Restricting the Fate of the Radical Intermediates in Solids	43
2.3.3	Crystalline Diacyl Peroxides	44
2.3.4	Decarbonylation of Crystalline Ketones	50
2.3.4.1	Early Observations	50
2.3.4.2	Reactivity and Stability	51
2.3.4.3	The RSE $>$ 11 kcal mol^{-1} Condition	53
2.3.4.4	Scope of the Reaction	55
2.3.4.5	Reaction Enantiospecificity	56
2.3.4.6	Synthesis of Natural Products	57
2.3.4.7	Quenching Effects	57
2.3.4.8	Reaction Scale and Experimental Conditions	59
2.4	Concluding Remarks	60
	References	60

3 Intermolecular Addition Reactions onto C–C Multiple Bonds 67
Valentina Dichiarante and Maurizio Fagnoni

3.1	Introduction	67
3.1.1	Scope and Mechanism	68
3.2	Addition to C–C Double Bonds	69
3.2.1	H–C Addition (Hydroalkylation Reactions)	69
3.2.1.1	Addition of Alkanes	70
3.2.1.2	Addition of Alcohols (Hydrohydroxymethylation), Ethers, and (2-substituted) 1,3-Dioxolane(s)	71
3.2.1.3	Addition of Amines (Hydroaminomethylation) or Amides	72
3.2.1.4	Hydrofluoromethylation	74
3.2.1.5	Addition of Nitriles or Ketones	75
3.2.1.6	Hydroacylation and Hydrocarboxylation Reactions	75
3.2.1.7	Hydroarylation (Photo-EOCAS)	76
3.2.2	H–N Addition (Hydroamination)	76
3.2.3	H–P Addition	78
3.2.4	H–O Addition	80
3.2.5	H–S Addition	82
3.2.6	Addition of X-Y Reagents to Alkenes	83
3.2.6.1	Halogenation	84
3.2.6.2	Addition with the Formation of C–C Bonds	84
3.3	Addition to C–C Triple Bonds	86
3.3.1	Hydroalkylation Reactions	87
3.3.2	Addition of X–Y Reagents	87

3.4	Concluding Remarks 88	
	References 89	
4	**Formation of a Three-Membered Ring** 95	
	Takashi Tsuno	
4.1	Introduction 95	
4.2	Di-π-Methane Rearrangement 96	
4.2.1	Di-π-Methane Rearrangement of Barrelene, Benzobarrelene, Dibenzobarrelene, and Related Derivatives 96	
4.2.2	Di-π-Methane Rearrangement of Acyclic Systems 100	
4.2.3	Di-π-Methane Rearrangement in Natural Compounds 103	
4.3	Oxa-di-π-Methane Rearrangement and Related Rearrangements 105	
4.3.1	Oxa-di-π-Methane Rearrangement of β,γ-Unsaturated Ketones and Aldehydes 105	
4.3.2	Aza-di-π-Methane Rearrangement 109	
4.3.3	Synthetic Applications of Oxa-di-π-Methane Rearrangement 110	
4.4	[2+1] Cycloaddition of Alkenes with Carbenes 111	
4.4.1	Intramolecular [2+1] Cycloaddition 111	
4.4.2	Novel Triplet Sensitizers for the Generation of Carbenes 111	
4.4.3	Metal-Catalyzed Cyclopropanation-Supported Photochemistry 112	
4.4.4	Novel Precursors of Carbenes 113	
4.5	Formation of a Cyclopropane via Intramolecular Hydrogen Abstraction 114	
4.5.1	Formation of Cyclopropanol via Intramolecular β-Hydrogen Abstraction 115	
4.5.2	Formation of Cyclopropane Ring via Intramolecular γ-Hydrogen Abstraction 117	
4.6	[3+2] Cycloaddition of Arenes with Alkenes 119	
4.6.1	Intermolecular [3+2] Cycloaddition 119	
4.6.2	Intramolecular [3+2] Cycloaddition 119	
4.6.3	Application of the Photochemical [3+2] Cycloaddition in the Synthesis of Natural Products 122	
4.7	Photochemical Synthesis of Three-Membered Heterocycles 123	
4.7.1	Epoxides 123	
4.7.2	Aziridines 123	
	References 126	
5	**Formation of a Four-Membered Ring** 137	
	Norbert Hoffmann	
5.1	Introduction 137	
5.2	[2+2]-Photocycloaddition of Nonconjugated Alkenes 137	
5.3	[2+2]-Photocycloaddition of Aromatic Compounds 144	
5.4	Photochemical Electrocyclic Reactions 150	
5.5	Intramolecular γ-Hydrogen Abstraction (Yang Reaction) 153	

5.6	Metal-Catalyzed Reactions	*156*
5.7	Other Methods	*157*
5.8	Concluding Remarks	*160*
	References	*160*

6 Formation of a Four-Membered Ring: From a Conjugate Alkene *171*
Jörg P. Hehn, Christiane Müller, and Thorsten Bach

6.1	Introduction *171*	
6.2	[2+2]-Photocycloaddition of Enones (Substrate Type A1) *173*	
6.2.1	Cyclopentenones *173*	
6.2.2	Cyclohexenones *177*	
6.2.3	*para*-Quinones and Related Substrates *181*	
6.3	[2+2]-Photocycloaddition of Vinylogous Amides and Esters (Substrate Classes A2 and A3) *182*	
6.3.1	Endocyclic Heteroatom Q in β-Position (Substrate Class A2) *183*	
6.3.1.1	4-Hetero-2-Cyclopentenones *183*	
6.3.1.2	4-Hetero-2-Cyclohexenones *185*	
6.3.2	Exocyclic Heteroatom Q in β-Position (Substrate Class A3) *186*	
6.4	[2+2]-Photocycloadditon of α,β-Unsaturated Carboxylic Acid Derivatives (Substrate Classes A4, A5, and A6) *189*	
6.4.1	No Further Heteroatom Q in β-Position (Substrate Class A4) *189*	
6.4.1.1	α,β-Unsaturated Lactones *189*	
6.4.1.2	α,β-Unsaturated Lactams *192*	
6.4.1.3	Coumarins *193*	
6.4.1.4	Quinolones *194*	
6.4.1.5	Maleic Anhydride and Derivatives *196*	
6.4.1.6	Sulfur Compounds *197*	
6.4.2	Endocyclic Heteroatom Q in β-Position (Substrate Class A5) *198*	
6.4.2.1	1,3-Dioxin-4-Ones *198*	
6.4.2.2	4-Pyrimidinones *200*	
6.4.3	Exocyclic Heteroatom Q in β-Position (Substrate Class A6) *201*	
6.4.3.1	Lactones *202*	
6.4.3.2	Lactams *203*	
6.5	Concluding Remarks *205*	
	References *205*	

7 Formation of a Four-Membered Ring: Oxetanes *217*
Manabu Abe

7.1	Introduction *217*
7.2	The Generally Accepted Mechanism of the Paternò–Büchi Reaction *220*
7.3	Regioselective and Site-Selective Syntheses of Oxetanes *221*
7.4	Stereoselective Syntheses of Oxetanes *226*
7.5	Concluding Remarks *233*
	References *233*

8	**Formation of a Five-Membered Ring** 241
	Ganesh Pandey and Smita R. Gadre
8.1	Introduction 241
8.2	Formation of Five-Membered Rings: Intramolecular δ-H Abstraction 241
8.2.1	Formation of Cyclopentanol Ring System 242
8.2.1.1	Synthesis of Indanols 243
8.2.2	Synthesis of Tetrahydrofuranols 247
8.2.2.1	Formation of Benzofuranols 250
8.2.3	Synthesis of Pyrrolidine Derivatives 250
8.3	Formation of Five-Membered Rings via [3+2]-Cycloadditions 254
8.3.1	Photofragmentation of Oxiranes to Carbonyl Ylides: Synthesis of Tetrahydrofurans 254
8.3.2	Generation of Azomethine Ylides by the Photolysis of Aziridines: Synthesis of the Pyrrolidine Framework 258
8.3.3	Vinyl Cyclopropane to Cyclopentene Rearrangement 260
8.4	Photochemical Electrocyclization Reactions: Synthesis of Fused, Five-Membered Ring Compounds 261
8.5	Photoinduced Electron Transfer-Mediated Cyclizations: Synthesis of Five-Membered Carbocyclic and Heterocyclic Ring Systems 266
8.5.1	Radical Cation-Mediated Carbon–Carbon Bond Formation 266
8.5.2	Radical Anion-Mediated Cyclizations 272
8.5.3	Intramolecular Trapping of Radical Cations by Nucleophiles 276
	References 279
9	**Formation of Six-Membered (and Larger) Rings** 287
	Julia Pérez-Prieto and Miguel Angel Miranda
9.1	Introduction 287
9.2	Photoelectron Transfer-Initiated Cyclizations 287
9.2.1	Phthalimides as Electron Acceptors 287
9.2.2	Aromatic Ketones as Electron Acceptors 291
9.2.3	Chloroacetamides as Electron Acceptors 292
9.2.4	Electron-Deficient Aromatic Compounds as Electron Acceptors 293
9.3	Photoinduced 6π-Electrocyclization 295
9.3.1	Stilbene-Like Photocyclization 295
9.3.2	Vinyl-Biphenyls Photocyclization 299
9.3.3	Anilides and Enamides Photocyclization 299
9.4	Photocycloaddition Reactions 300
9.4.1	Photochemical Diels–Alder Reaction 301
9.4.2	Photoenolization/Diels–Alder Reaction 302
9.4.3	[4+4]-Photocycloaddition 302
9.4.4	Transition Metal Template-Controlled Reactions 304
9.5	Remote Intramolecular Hydrogen Abstraction 307
9.6	Ring Contraction and Ring Enlargement 308
9.7	Other Reactions 311

9.7.1	Intramolecular [2+2]-Cycloadditions *311*
9.7.2	Photocyclization of Cinnamylanilides *311*
9.7.3	Photocycloaddition of Aromatic Compounds *311*
9.8	Concluding Remarks *313*
	References *313*

10	**Aromatic and Heteroaromatic Substitution by $S_{RN}1$ and S_N1 Reactions** *319*
	Alicia B. Peñéñory and Juan E. Argüello
10.1	Introduction *319*
10.2	General Mechanistic Features *320*
10.2.1	$S_{RN}1$ Mechanism *320*
10.2.2	S_N1 Mechanism *322*
10.3	Carbon–Carbon Bond Formation *323*
10.3.1	Carbanions from Ketones, Esters, Acids, Amides, and Imides as the Nucleophiles *323*
10.3.2	Alkenes, Alkynes, Enols, and Vinyl Amines as the Nucleophiles *326*
10.3.3	Aryl Alkoxide and Aryl Amide Anions as the Nucleophiles *329*
10.3.4	Cyanide Ions as the Nucleophile *331*
10.4	Carbon–Heteroatom Bond Formation *332*
10.4.1	Tin Nucleophiles *332*
10.4.2	Sulfur Nucleophiles *333*
10.5	Synthesis of Bi-, Tri-, and Polyaryls *334*
10.5.1	Consecutive $S_{RN}1$–Pd(0)-Catalyzed Crosscoupling Reactions *334*
10.5.2	Photo-S_N1 as an Alternative to Metal Catalysis *336*
10.6	Synthesis of Carbocycles and Heterocycles *338*
10.6.1	Carbocycles *338*
10.6.2	Nitrogen Heterocycles *341*
10.6.3	Oxygen Heterocycles *344*
10.6.4	Sulfur Heterocycles *346*
	References *346*

11	**Singlet Oxygen as a Reagent in Organic Synthesis** *353*
	Matibur Zamadar and Alexander Greer
11.1	Introduction *353*
11.2	Dioxetanes *354*
11.2.1	Background Information *354*
11.2.2	Adamantyl-Substituted Alkenes *355*
11.2.3	Alkoxy-Substituted Alkenes *356*
11.2.4	Phenyl- or Methyl-Substituted Alkenes *357*
11.2.4.1	Diphenylindene Photooxidation *357*
11.2.4.2	Electron-Transfer Photooxidation *357*
11.2.5	Summary *358*
11.3	Endoperoxides *358*
11.3.1	Background Information *358*

11.3.2	Arenes	359
11.3.2.1	Benzenes	359
11.3.2.2	Naphthalenes	361
11.3.2.3	Anthracenes, Polyacenes, and Carbon Nanotubes	362
11.3.3	Electron-Transfer Photooxidation	364
11.3.4	Conjugated Dienes	364
11.3.4.1	Acyclic Dienes	364
11.3.4.2	Cyclopentadienes and Cyclohexadienes	364
11.3.4.3	Heterocycles and Cyclohexatriene	365
11.3.5	Summary	368
11.4	Allylic Hydroperoxides	368
11.4.1	Background Information	368
11.4.2	Simple Alkenes	368
11.4.3	"Ene" Reactions Confined in Zeolites	370
11.4.4	Summary	370
11.5	Tandem Singlet Oxygen Reactions	371
11.5.1	Background Information	371
11.5.2	Bisperoxides	371
11.5.2.1	Phenyl-Substituted Alkenes	371
11.5.2.2	Cyclic Alkenes	372
11.5.3	Rearrangement to a Hemiketal Hydroperoxide	374
11.5.4	Rearrangements to Spiro Compounds	374
11.5.5	Summary	376
11.6	Concluding Remarks	377
	References	377

12 **Synthesis of Heteroaromatics via Rearrangement Reactions** 387
*Nicolò Vivona, Silvestre Buscemi, Ivana Pibiri,
Antonio Palumbo Piccionello, and Andrea Pace*

12.1	Introduction	387
12.2	Synthesis of Five-Membered Rings with One Heteroatom	388
12.2.1	Pyrroles	388
12.2.2	Furans	391
12.2.3	Thiophenes	392
12.3	Synthesis of Five-Membered Rings with Two Heteroatoms	393
12.3.1	Pyrazoles	393
12.3.2	Imidazoles	394
12.3.3	Oxazoles	398
12.3.4	Thiazoles	400
12.4	Synthesis of Five-Membered Rings with Three Heteroatoms	402
12.4.1	Oxadiazoles	402
12.4.2	Triazoles	404
12.4.3	Thiadiazoles	405
12.5	Synthesis of Six-Membered Rings	406
12.6	Synthesis of Seven-Membered Rings	406

12.6.1	Azepines	*406*
12.6.2	Diazepines	*407*
12.6.3	Oxazepines	*409*
12.7	Concluding Remarks	*410*
	References	*411*

13 Photolabile Protecting Groups in Organic Synthesis *417*
Christian G. Bochet and Aurélien Blanc

13.1	Introduction	*417*
13.2	Photolabile Protecting Groups	*418*
13.2.1	Ortho-Nitrobenzyl Alcohol Derivatives	*418*
13.2.2	Benzyl Alcohol Derivatives	*421*
13.2.3	Other Types of Protecting Group	*423*
13.2.3.1	Norrish Type II	*423*
13.2.3.2	Norrish Type I	*424*
13.2.3.3	Thioketals	*424*
13.2.3.4	Silicon Ethers	*424*
13.2.4	Z/E Photoisomerization	*425*
13.2.4.1	Cinnamyl Esters	*425*
13.2.5	Phenacyl Derivatives	*426*
13.2.5.1	Mechanism	*426*
13.2.6	Benzoin Derivatives	*428*
13.2.6.1	Mechanism	*428*
13.2.7	Indolines	*429*
13.3	Chromatic Orthogonality	*430*
13.4	Two-Photons Absorption	*431*
13.5	Concluding Remarks	*432*
13.6	Appendix	*433*
	References	*439*

Index *449*

Preface

Practitioners of organic photochemistry feel that this science has a great potential for synthesis. Indeed, nowadays many reactions are known that lead to useful transformations and have been exploited as key steps in complex synthetic plans. These achievements attract the interest of synthetic chemists. However, photochemical methods are probably less often adopted than they may be, and are still less familiar to the broad chemical community than other methods. In the present handbook it has been attempted to offer an easy approach to the use of photochemical methods in synthesis. Thus, rather than discussing the chemistry of the various chromophores, as usual in photochemistry books, reactions have been grouped according to the molecular transformation involved and care has been given that experimental aspects (much less elaborate with many other methods) are clearly presented. We are convinced that a more general application in nonspecialized laboratories will lead to the discovery of new applications and even new reactions.

It was chosen to have a multiauthor book because this allows a breadth of approaches that could not otherwise be reached. The distinguished photochemists who accepted to participate in this project patiently tolerated the long work required to avoid the risk of discontinuity. We thank them heartily and any deficiency in the book is certainly not their fault. Thanks go to Dr. Heike Nöthe, a friendly and capable help during both the preparation and production phases and to Davide Ravelli and Matteo Albini for the pictures.

Angelo Albini and Maurizio Fagnoni

List of Contributors

Manabu Abe
Hiroshima University (HIRODAI)
Graduate School of Science
Department of Chemistry
1-3-1 Kagamiyama
Higashi-Hiroshima
Hiroshima 739-8526
Japan

Angelo Albini
University of Pavia
Department of Organic Chemistry
Via Taramelli 10
27100 Pavia
Italy

Juan E. Argüello
Universidad Nacional de Córdoba
Facultad de Ciencias Químicas
INFIQC – Dpto Química Orgánica
Ciudad Universitaria
5000 Córdoba
Argentina

Thorsten Bach
Technische Universität München
Lehrstuhl für Organische Chemie I
85747 Garching
Germany

Aurélien Blanc
University of Strasbourg
Institut de Chimie, UMR 7177/CNRS
Laboratoire de Synthèse et Réactivité
Organiques
4 rue Blaise Pascal
67000 Strasbourg
France

Christian G. Bochet
University of Fribourg
Department of Chemistry
9 Ch. du Musee
1700 Fribourg
Switzerland

Silvestre Buscemi
Università degli Studi di Palermo
Dipartimento di Chimica Organica
"E. Paternò"
Viale delle Scienze, Parco d'Orleans II
Edificio 17
90128 Palermo
Italy

Valentina Dichiarante
University of Pavia
Department of Organic Chemistry
Via Taramelli 10
27100 Pavia
Italy

Maurizio Fagnoni
University of Pavia
Department of Organic Chemistry
Via Taramelli 10
27100 Pavia
Italy

Smita R. Gadre
National Chemical Laboratory
Division of Organic Chemistry
Dr. Homi Bhabha Road
Pune 411008
India

Miguel A. Garcia-Garibay
University of California
Department of Chemistry and Biochemistry
Los Angeles, CA 90095
USA

Luca Germani
University of Pavia
Department of Organic Chemistry
Via Taramelli 10
27100 Pavia
Italy

Alexander Greer
Graduate Center and
The City University of New York (CUNY)
Brooklyn College
Department of Chemistry
Brooklyn, NY 11210
USA

Jörg P. Hehn
Technische Universität München
Lehrstuhl für Organische Chemie I
85747 Garching
Germany

Norbert Hoffmann
Université de Reims Champagne-Ardenne, CNRS
Institut de Chimie Moléculaire de Reims, UMR 6229
Groupe de Photochimie, UFR Sciences
B.P. 1039
51687 Reims
France

Patrick S. Mariano
University of New Mexico
Department of Chemistry and Chemical Biology
Albuquerque, NM 87131
USA

Miguel Angel Miranda
Universidad Politécnica de Valencia
Instituto de Tecnología
Química UPV-CSIC
Departamento de Química
Camino de Vera sn
46071 Valencia
Spain

Christiane Müller
Technische Universität München
Lehrstuhl für Organische Chemie I
85747 Garching
Germany

Andrea Pace
Università degli Studi di Palermo
Dipartimento di Chimica Organica
"E. Paternò"
Viale delle Scienze, Parco d'Orleans II
Edificio 17
90128 Palermo
Italy

Ganesh Pandey
National Chemical Laboratory
Division of Organic Chemistry
Dr. Homi Bhabha Road
Pune 411008
India

Alicia B. Peñéñory
Universidad Nacional de Córdoba
Facultad de Ciencias Químicas
INFIQC – Dpto Química Orgánica
Ciudad Universitaria
5000 Córdoba
Argentina

Julia Pérez-Prieto
Universidad de Valencia
Instituto Ciencia Molecular
Polígono La Coma sn
46980 Paterna, Valencia
Spain

Ivana Pibiri
Università degli Studi di Palermo
Dipartimento di Chimica Organica
"E. Paternò"
Viale delle Scienze, Parco d'Orleans II
Edificio 17
90128 Palermo
Italy

Antonio Palumbo Piccionello
Università degli Studi di Palermo
Dipartimento di Chimica Organica
"E. Paternò"
Viale delle Scienze, Parco d'Orleans II
Edificio 17
90128 Palermo
Italy

Saori Shiraki
University of California
Department of Chemistry and
Biochemistry
Los Angeles, CA 90095
USA

Takashi Tsuno
Nihon University
College of Industrial Technology
Department of Applied Molecular
Chemistry
Narashino, Chiba 275-8575
Japan

Nicolò Vivona
Università degli Studi di Palermo
Dipartimento di Chimica Organica
"E. Paternò"
Viale delle Scienze, Parco d'Orleans II
Edificio 17
90128 Palermo
Italy

Matibur Zamadar
Graduate Center and
The City University of New York
(CUNY)
Brooklyn College
Department of Chemistry
Brooklyn, NY 11210
USA

1
Photochemical Methods
Angelo Albini and Luca Germani

1.1
Photochemical Methods

1.1.1
Photochemistry and Organic Synthesis

A cursory look to the literature shows that only about 1% of the published papers classed as organic syntheses by *Chemical Abstracts* involve a photochemical step. On the other hand, in photochemistry courses it is often stated that excitation by light multiplies by 3 the accessible reaction paths, because the chemistry of the excited singlet and triplet states are added to that of the ground state. It thus appears that photochemical reactions are less used as they may be. As it has been again recently remarked, this limited diffusion may be due to ill-founded prejudices [1].

Two conditions should be verified in order that the potential of photochemical reactions is more extensively exploited. These are:

- That the knowledge of the main classes of such reactions is more largely diffused among synthetic practitioners, so that a photochemical step comes more often into consideration when discussing a synthetic plan.

- That the prejudice that photochemical reactions are mostly unselective, experimentally cumbersome and at any rate difficult to generalize is overcome, so that there is no hesitation in considering the introduction of a photochemical step on the basis of the analogy with known examples, just as one would do with a thermal reaction.

The connection between synthesis and photochemistry is vital. As long as photochemistry is felt as a "sanctuary" of the small group of "professional" photochemists, many synthetic perspectives will be ignored, and this is a negative impact also on mechanistic photochemistry that loses part of its interest. As a matter of fact, this remark is not new. In a talk in Leipzig in 1908, Hans Stobbe, a pioneer of photochemistry (well known for his innovative studies on the photochromism of fulgides), stressed the importance of devising new photochemical applications in organic chemistry [2]. "Then probably..." he hoped "...organic chemists would become interested and take into account the effect of light on their experiments. Known photoreaction would become better known and new photoreactive compounds will be looked for. Final products and intermediates would be isolated, their structure demonstrated and on the basis of the chemical structure the process will be understood. In this way the physical chemist would always have in his hands a wealth of material for his favorite studies of kinetics and for investigating the relation between radiation and chemical energy."

Stobbe's wish has been only partially fulfilled in the century which has elapsed in the meantime. Whilst many photochemical reactions have been discovered, certainly many more wait to be uncovered, and it still holds true that more photochemistry carried out by synthetic chemists would contribute to the growth of photochemistry as a whole. This Handbook represents a modest attempt to contribute towards this aim and to foster the synthetic use of photochemistry. The presentation is referred to the small-scale laboratory synthesis of fine chemicals. In this aspect, the photochemical literature does not differ from the large majority of published synthetic work, most of which is carried out on the 100 mg scale for exploratory studies. However, there is no reason to think that a photochemical reaction is unfit for scaling up. As will shown below, an increase up to the 100 g scale can be obtained in the laboratory by simple arrangements. Furthermore, while the presently running industrial applications are limited in number, they are nonetheless rather important [3]. Some of these are well established, an example being the synthesis of vitamin D_3 which has been produced at the several tons level each year for several decades, and for which dedicated plants continue to be built. This indeed demonstrates that photochemical syntheses are commercially viable.

1.2
Irradiation Apparatus

1.2.1
General

As the name implies, photochemical reactions result from the absorbance of light by the starting reagent. Conditions for a successful course of the photoreaction are that:

- There is good matching between the emission of the light source and the absorption by the reagent; that is, the wavelength emitted by the lamp falls within the absorption band of the reagent.
- Nothing interferes with the photons before they reach the target molecule; for example, the wall of the vessel and the solvent are transparent to λ_{ex}.
- Nothing interferes with the electronically excited states and quench them before they react (see Scheme 1.1).

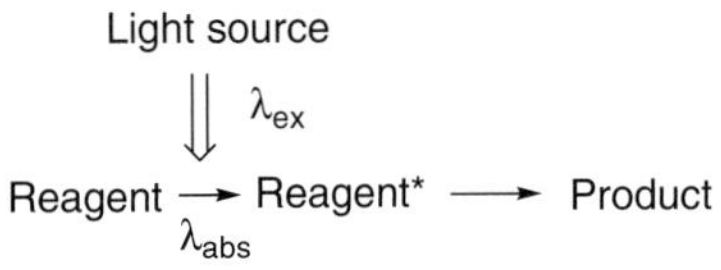

Light source
$$\downarrow \lambda_{ex}$$
Reagent $\longrightarrow$ Reagent* $\longrightarrow$ Product
λ_{abs}

Scheme 1.1

In other cases, rather than irradiating the reagent ("direct" excitation), a photosensitizer or photocatalyst is irradiated and activates the reagent by some mechanism (energy transfer, a redox step, hydrogen abstraction). In this case, the above conditions apply to the sensitizer.

Today, there are several companies which supply lamps as well as complete photochemical reactors (lamp + power supply + reaction flask with accessories, e.g. for gas inlet). However, the complete set may be rather expensive and not necessarily provide the most convenient solution. The most widely used light sources are mercury vapor arcs, both in photochemistry and in indoor and outdoor illumination, and which are classed according to the operating pressure. It is important that the wattage on the lamp label is not confused with the amount of light emitted. The efficiency of conversion into light is low, and the lamp output is dispersed over a range of wavelengths and towards all directions; thus, only a part of the light emitted (in turn, a fraction of the electrical power dissipated) is absorbed. Therefore, it is important to take care of the geometry of the lamp/reaction vessel system as well as of the wavelength matching between lamp emission and reagent absorption, because these factors are at least as important as the lamp power in determining how many molecules of the reagent will be excited. The quantum yield then indicates the fraction of excited states that reacts [Φ = (molecules reacted)/(photons absorbed)], provided that no competitive quenching occurs. The main characteristics of lamps used for photochemical synthesis are presented in the following sections.

1.2.2
Low-Pressure Mercury Arcs

The most widely used lamps are low-pressure (10^{-5} atm under operating conditions) Hg arcs, of 6–16 W, that are often identified as germicidal lamps or mercury resonance lamps. These are supplied as quartz (or rather "fused silica," a synthetic amorphous SiO_2) tubes of various lengths, typically 20–60 cm (although lamps >1 m long are available), and with 1.0–2.4 cm diameter (see Figure 1.1). In these lamps, >80% of the emission occurs at 254 nm (and a fraction at 185 nm, a wavelength to which the common "quartz" is not transparent and thus is available only if a high-purity "UV-grade" quartz is used).

Under these conditions, the excitation of most classes of organic compounds (including many solvents!) is ensured. It must be taken into account that, given the large size of the lamp, the amount of photons emitted per surface unity is low. Therefore, these lamps are most useful for external irradiation by using (quartz!) tubes for the irradiated solutions. The heating under operating conditions is modest.

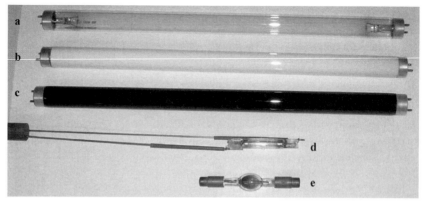

Figure 1.1 Lamps used for photochemical syntheses. (a) Low-pressure mercury arc; (b, c) phosphor-coated lamps, emission centered at 305 and 370 nm; (d, e) medium- and high-pressure mercury arcs, respectively.

Multilamp apparatus are commercially available where between eight and 12 lamps are arranged in a circular fashion (40–60 cm diameter), with room inside to accommodate the vessel in which the solution to be irradiated is contained. These were initially marketed by the Southern New England Ultraviolet Co. under the name of "Rayonet," now the name is often used for similar devices by other companies. These units are fitted with a fan which maintains the temperature below 40 °C; otherwise, this might increase in such a confined space (see Figure 1.2).

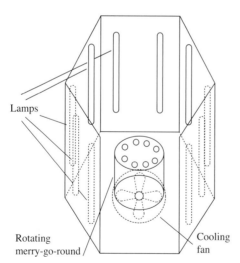

Figure 1.2 Multilamp apparatus fitted with low-pressure mercury lamps and a rotating "merry-go-round" that ensures the uniform illumination of several test tubes. Alternatively, test tubes or other vessel(s) containing the solution to be irradiated can be accommodated.

However, anybody can build an "amateur" version of the irradiation apparatus, simply by placing one to three pairs of lamps (with each pair mounted on a normal lamp holder for household "fluorescent" lamps) around a small space where two to four test tubes or a single cylindrical vessel of larger diameter can be placed. This home-made apparatus can be easily installed (but well separated from the laboratory, in order to maintain appropriate safety precautions, or better still under a ventilated hood to remove ozone; see below and Figure 1.3). In order to maximize the fraction of light absorbed, it is convenient that the tubes are as long as the lamps, or even slightly shorter. The manufacturers can provide lamps of this type in different shapes (U-shaped, coiled) with a more concentrated emission; this makes their use possible in different set-ups, an example being an immersion well apparatus with internal irradiation (see Figure 1.4).

Figure 1.3 Two pairs of lamps used for external irradiation. In the arrangement shown, only a small fraction of the light flux is used.

Low-pressure mercury arcs are manufactured for much more widespread use than preparative photochemistry, and therefore are by far the cheapest light source (particularly if buying them from companies selling optical components can be avoided). Furthermore, these lamps are long-lived (>10 000 h, depending on how they are used), consume less energy, and require only an inexpensive transformer and a starter for operation.

Whilst there is no doubt that these are the most convenient sources, their geometric optimization is difficult and part of the lamp emission may be lost. In the external irradiation set-up, the most convenient choice is to use tubes which are about the same length as the lamps, and contain 20 ml of solution. In fact, this set-up works very well for small-scale photochemical syntheses, with irradiated volumes in the region of 100 ml distributed in a number of quartz test tubes or in a single cylindrical vessel. This set-up is also convenient for optimizing the reactions, since results can easily be compared under different conditions but constant irradiation when placing different solutions in the tubes. One available accessory for these multilamp apparatuses is a "rotating merry-go-round"; this can hold several tubes and ensures equivalent irradiation in all positions (see Figure 1.2).

The lamp emission can be changed by means of a coating made from a phosphor (or a combination of phosphors) that absorbs the almost monochromatic Hg radiation and emits a range of longer wavelengths. Phosphor-coated lamps maintain the same advantages of "quartz" lamps (including price, due to their large-scale manufacture for different uses, including household illumination), and are available in a variety of wavelength ranges. Apart from "fluorescent" lamps for household illumination, which emit over most of the visible (and are useful for dye-photosensitized irradiations), lamps with emission centered at 305, 350 and 370 nm (half-height width 20–40 nm; the last one is known as "Wood" or "black light" lamp) that are most useful for photochemical applications are commercially available (as well as lamps with the emission centered at various wavelengths in the visible). Except for the 305 nm type, Pyrex glassware (transparence limit 300–310 nm, but take into account that the transparency changes somewhat with the use) [4] can be used with phosphor-coated lamps, as there is no emission below that wavelength. The phosphor coating does not alter the electrical characteristics, and these lamps can be interchanged with germicidal lamps in all of the settings mentioned above. Having available three to four pairs each of 254, 305, and 350 (or 370) nm lamps, as well as lamps emitting in the visible range, allows one to carry out any type of small-scale photochemical reaction with negligible financial investment.

One subcategory of low-pressure lamp that might become more important in the future is the electrodeless discharge lamp, which is energized by an external field. These lamps comprise a quartz tube that has been evacuated, leaving behind a small pressure of argon and mercury or other metal or metal halide. Emission is obtained by placing the lamp in a microwave field, for example. Whilst these lamps are available commercially, they may also be built in-house rather easily [5].

1.2.3
Medium- and High-Pressure Mercury Arcs

Medium-pressure (sometimes dubbed "high pressure," 1–10 atm) mercury arcs are available in different types, ranging from 100 to 1000 W. They are supplied as small ampoules (from 3 to 15 cm in length, depending on the power; see Figure 1.1). The emission consists of a range of lines (the most prominent are those at 313, 366, 405, and 550 nm) over a continuum, while the 254 nm line is strongly diminished. The emission from these lamps is at least 10-fold stronger than that of low-pressure arcs, and occurs over a much smaller surface. In contrast to the previous type, these sources develop a considerable amount of heat, and require several minutes to achieve their optimal temperature, where the emission reaches full intensity. Cooling is required, but running tap water is normally sufficient to maintain the temperature at about 20 °C. Due to these characteristics, these lamps are typically used in an immersion well apparatus with circulating water. If the cooling well is made from Pyrex, the (small) fraction of emission below 300 nm is lost, which may make a difference (see below and Figure 1.4a). The most powerful lamps require a forced circulation for cooling. A suitable power supply is also required for operation, the lifetime is limited, and overall the system is considerably more expensive than the low-pressure lamps. There may also be some concern regarding safety aspects; it is suggested that the reactor is provided with a switch that will cut the power supply in case of an increase in temperature.

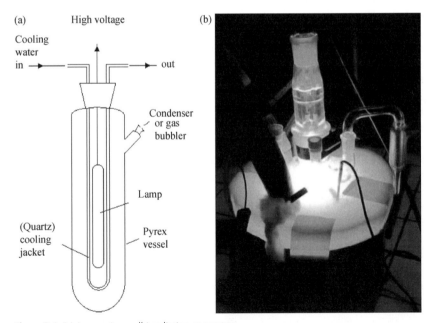

Figure 1.4 (a) Immersion well irradiation apparatus; (b) a refrigerated apparatus for conducting reactions at low temperature.

These compact and rather powerful sources are convenient for internal irradiation of volumes of between 100 and 1000 ml, where the emission in any direction is exploited (obviously within the range of absorbed λ); and are well suited for preparative irradiations up to the gram scale. The apparatus can be easily adapted to low-temperature experiments (e.g., at −80 °C) by circulating a refrigerant liquid through the lamp jacket (in this case, the lamp must be ignited outside and placed in position when lit, otherwise it will not function) and adding an external cooling bath (see Figure 1.4b) [6]. Lamps doped with different metals are also available; these yield an emission which is richer in some regions of spectrum, and may be better suited to particular photoreactions, although they are generally more expensive.

High-pressure (or "very high" pressure, 200 atm; see Figure 1.1) arcs, ranging from 150 to 1000 W and above, operate at higher temperatures. In this case, the contribution of the continuum is much more important than that at a lower pressure, although the maxima may still be distinguished. The optimal temperature requires several minutes before it is reached, and must be maintained by appropriate cooling. These Hg-lamps are the most powerful and the smallest sources, with a distance between the electrodes of only a few millimeters. In view of the severe operating conditions, such lamps are used in explosion-proof cases (finned for cooling, unless forced cooling is required) that are fitted with mirror and lenses. In this way a collimated emission is obtained, typically 5 cm in diameter, and the lamp is mounted on a optical bench where other optical components can be added (see Figure 1.5).

By inserting either an interference filter or a colored filter, it is possible to select a more or less extended region of the spectrum; likewise, by adding an optical fiber it is possible to direct the beam where desired. This set-up best exploits the characteristics of these powerful lamps, and offers an excellent choice for the irradiation of small surfaces. Consequently, spectrophotometric cuvettes or cylindrical cuvettes are used for the irradiation, which involves small volumes. Such restrictions, as well as the high price and short lifetime of the lamp and its accessories, favors the use of these arcs for kinetics studies and quantum yield measurements, rather than for preparative photochemistry.

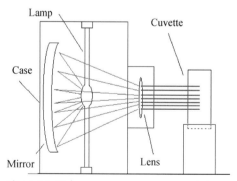

Figure 1.5 High-pressure mercury arc mounted on an optical bench.

1.2.4
Other Light Sources

Several other sources are available commercially, such as high-pressure xenon arcs (that mimic solar emission), mercury–xenon or antimony–xenon arcs (which are richer in the UV) or sodium arcs (dominated by the strong yellow-green emission of sodium metal). The latter type may be useful for dye-sensitized photo-oxidation, but are relatively expensive and short-lived; light-emitting diodes (LEDs) are more convenient in this application (see below). Tungsten (incandescent) and tungsten–halogen lamps evolve large amounts of heat (the former type is quite inefficient, and will likely be banned from commercial application; the latter is more efficient). These lamps may be used for photoinitiated chain reactions (where $\Phi \gg 1$; see Chapter 10), but emit poorly in the UV range and are rarely used for proper photochemical reactions ($\Phi \leq 1$).

Lasers have been shown to serve as an efficient light source for some reactions. In particular, the 308 nm emission of the XeCl "excimer" laser is a convenient source that, unlike arc lamps, is monochromatic and does not emit heat. A commercial 3 kW laser of this type (60 cm long, vertically mounted) has been used to build a falling film reactor capable of converting 10 g or more in 10–20 h [7]. At least at present, however, these light sources are rather expensive and require considerable care for their maintenance; consequently, they cannot be considered for adoption by an organic photochemistry laboratory requiring a versatile tool for preparative applications.

Substitute systems that may in time become a convenient alternative are the rapidly developing LEDs. These are semiconductors that emit an incoherent electroluminescence over a short wavelength range. These small devices are fitted with optics that shape the emission (parallel or with an angle), and are available in a large variety of "monochromatic" emitting types (actually over a narrow range, typically 20 nm) for almost any wavelength. The conversion of absorbed power (which is in the hundred mW range) into light depends on the wavelength, and LEDs that are reasonably effective light sources are available over the whole of the visible spectrum. LEDs emitting in the UV-A, down to 320–310 nm are also available, although their emission is much less intense (one-tenth or less compared to those emitting in the visible), and they are more expensive. The advantages are that such sources are cheap and long-lived ($>10^4$ h), and although a single LED is too weak (more powerful sources are expected in a few years) a set of perhaps 20 LEDs would be adequate. Mounting these light sources on a cylindrical surface, at the center of which test tubes containing the solution can be accommodated, might represent an efficient approach to carrying out photochemical reactions with small volumes (2–30 ml) and will most likely become increasingly common [8].

Having available a certain number of LED sets might also be an alternative to having the above-mentioned sets of phosphor-coated lamps. Whilst the former set-up would be more expensive than the latter, it would be much more versatile, because in the case of LEDs a much wider choice of λ is available (unfortunately, this is not as extensive as might be wished in the UV, which is the most interesting region for synthetic photochemistry). The precise choice of λ_{ex} may be advantageous if a specific class of reactions is under investigation, and where fine-tuning of the wavelength is important (typically, to avoid the primary product undergoing further unwanted photochemical reactions). Another area where LEDs are convenient is that of dye-sensitized reactions (which in practice are oxidations), as lamps emitting in the visible spectrum are already more powerful and are undergoing a more rapid development (in view of their use for illumination purposes). Moreover, further improvements are expected in this area.

Solar light is an obvious and ecologically convenient alternative, its main limitations being a low density and scarce UV component. Nevertheless, placing a solution on the window sill on a sunny day represents a very convenient way of carrying out a photochemical reaction on the 100 mg scale, even with compounds of which the absorption is limited to UV-A (see Figure 1.6a). The obvious lack of reproducibility according to season and weather, may make this not the preferred choice. However, for any compound absorbing at 330 nm or at a longer wavelength, there is no reason why, in an exploratory test, a one- or two-day exposure to the sun should not be substituted for an overnight exposure to a multilamp apparatus, with significant energy savings.

Given the importance of using alternative energy sources, some attention has been paid also to organic syntheses by solar light on a (relatively) large scale. For this, a variety of apparatus have been built, which generally circulate the solution through tubes exposed to the sun by means of a pump, with the addition of a heat exchanger in order to avoid overheating (see Figure 1.6b). Exposure to the sun can be achieved either by using a simple flat-bed arrangement, or concentrating the solar emission by locating the tubes in the axis of parabolic mirrors. Sophisticated versions of this apparatus are equipped with a mechanical system that allows the sun to be tracked during the day. In this way, some reactions have been scaled up to 500 g [9].

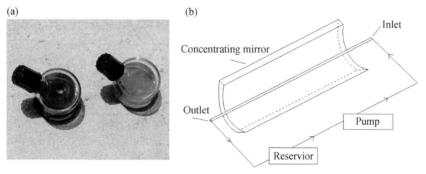

Figure 1.6 (a) Cylindrical vessels exposed to the sun on a window sill; (b) moderately concentrated solar light. The solution is circulated through the tube placed in the axis of a parabolic mirror.

1.3
Further Experimental Parameters

1.3.1
Concentration and Scale

As mentioned above, small-scale photoreactions are quite often carried out in quartz or Pyrex tubes, by external irradiation. However, this is certainly not an optimal solution for maximizing the exploitation of the emitted radiation. Internal irradiation is obviously better from the geometric point of view, but (relatively) large-scale preparations must take into account all of these factors and achieve optimal light and mass transfer. These elements are not taken into account in exploratory studies or small-scale syntheses, just as is the case for thermal reactions, where the optimization is considered at a later stage; the essential requirement is that the explorative study is carried out under conditions where occurrence of the reaction is not prevented. Thus, it is important that the source is matched with the reagent absorption, the vessel is of the correct material, and the solvent does not absorb competitively (unless it acts also as the sensitizer). Figure 1.7 and Table 1.1 may help in this choice, in conjunction with the UV spectra of all of the compounds used (it is recommended that the spectra are measured on the actual samples used, in comparison with those taken from the literature, in order to check for absorption by impurities).

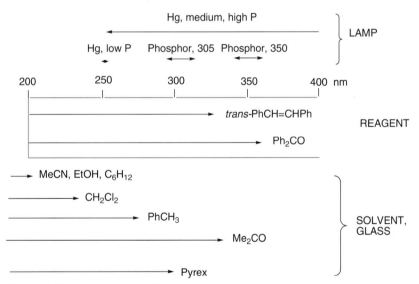

Figure 1.7 Checking that the conditions for a successful photochemical reaction are met. To use this system: (1) Insert into the frame the range of active wavelengths (up to the longest wavelength where the reagent absorbs significantly; two representative examples are shown). (2) Check whether this fits with the lamp chosen, the solvent and the material from which the reaction vessel, cooling well, etc. are constructed.

Table 1.1 Choosing a solvent with reference to the reagent irradiated.

Solvent	λ_{lim} [a]	Reagent	λ [b]
Acetone	329	Aniline	308
Acetonitrile	190	2-Cyclohexenone	310
Benzene	280	Stilbene	333
Cyclohexane	205	Benzophenone	360
Dichloromethane	232	1,4-Naphthoquinone	385
Diethyl ether	215	Uracil	285
N,N-Dimethylformamide	270	Phenanthrene	345
Dimethylsulfoxide	262	Anthracene	378
Ethanol	205	Pyrrole	238
Pyridine	305		
Pyrex, Vicor	λ_{lim} [c] ca. 300		

[a] Limiting wavelength; the wavelength at which a 1 cm layer of the solvent absorbs 90% of the light impinging; use only when the reagent absorbs above this value.
[b] The longest wavelength at which the reagent has absorbance $A = 1$ at a 0.01 M concentration.
[c] Wavelength at which a 1 mm layer of the glass absorbs 90% of the light.

How light is absorbed is also important. Beer's law states that the absorbance of a solution is $A = \varepsilon bc$, where c is the molar concentration and b the optical path. The value of ε_{max} ranges from about 10 for a "forbidden" band (e.g., aliphatic ketones at 280 nm), to 10^4 and more for "allowed" transitions. When considering a solution at a preparatively sensible concentration such as 0.1 M, this means that when using a tube with an internal diameter of 1 cm or an immersion well apparatus with a 1 cm path between the cooling well and the outer wall, $A = 1$ when irradiating on a maximum, and $\varepsilon_{\text{max}} = 10$. Since $A = \log(1/T)$, this means that 10% of the light flux is transmitted and 90% is absorbed. On the other hand, when $\varepsilon_{\text{max}} = 10^4$ the absorbance $A = 10^3$; that is, 99.9% of the light is absorbed. Thus, there is only a 10% difference in the overall number of photons absorbed in the two cases. However, in the latter case 90% of the photons are absorbed in the first 0.1 mm layer, and a further 9% in the following 0.9 mm, while the remaining part of the solution is excluded from activation. Obviously, when considering irradiation in a part of the spectrum different from the maximum, a more equilibrated absorption results; however, the problem remains that homogeneous activation is difficult to attain at concentrations useful for synthesis.

This has some negative implications. The first implication is that it lengthens the irradiation time; mechanistic studies obviate to this limitation simply by using a low concentration, but this obviously does not apply to a preparative irradiation. A very effective mixing would help, but this is difficult to obtain in a simple laboratory set-up, such as the usual cylindrical immersion well apparatus. On the contrary, attaining a satisfactory mixing is one of the main points of attention for chemical engineers when designing a large-scale reactor (the "mass transfer" problem; see above) [10]. The second implication is that, if the product of the photochemical reaction absorbs in the same wavelength range as the starting material (which is a common

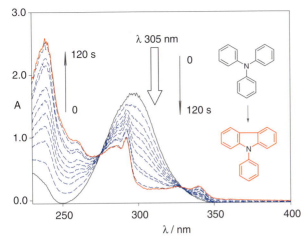

Figure 1.8 Photochemical conversion of triphenylamine into N-phenylcarbazole. At 305 nm the photoproduct has only a modest absorption, and irradiation at that wavelength leads to a regular conversion that can be continued up to completion. This would not be the case if the irradiation were carried out at a shorter wavelength (because the inner filter effect by the photoproduct would progressively slow down the reaction), or if using a solvent that absorbs competitively, such as acetone.

occurrence), it may hinder completing the photoreaction when it is photostable (the so-called "inner filter" effect), or it may lead to a mixture of primary and secondary photoproducts when the product itself is photoreactive.

The desirable situation where it is possible to irradiate the reagent at a wavelength where the product absorbs much less is shown in Figure 1.8. Under this condition, the conversion takes place at a regular pace and one arrives at a quantitative conversion, or close to it. When this does not apply, it may be difficult to bring the reaction to completion and/or to avoid the formation of secondary photoproducts. These are the actual limitations to the synthetic use of photochemical reactions, and they involve a general concentration limit (a clean reaction is rarely obtained above a 0.01–0.1 M concentration). Although there is no simple and general way to overcome this negative aspect, several methods are available that allow many cases to perform better.

Significant – and sometimes spectacular – improvements may be obtained by circulating the solution by means of a peristaltic pump; this is particularly the case when exposing a thin layer to the lamp at any time, rather than stirring the whole of the solution and thus having a longer optical path. "Falling film photoreactors" are commercially available in which a solution from a reservoir is sprayed onto the top of a (cooled) lamp, using a peristaltic pump, although a cheap, home-made device has been shown to be almost equally effective. For this, good results have been obtained by coiling a plastic tubing (that is transparent to the light used. and preferably made from fluorinated polyolefins for purpose of chemical resistance) around the cooling well of a medium-pressure lamp, and circulating the solution. In this way, considerable volumes (up to several liters) can be irradiated in a limited

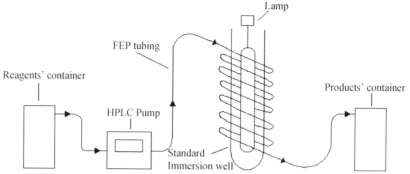

Figure 1.9 The circulation of a solution around an immersion well fitted with a medium-pressure mercury lamp.

overall time, with a better conversion and yield than when irradiating in a static reactor (see Figure 1.9). In this case, a tubing with an internal diameter sufficiently wide (e.g., >1 mm) so as to avoid any build-up of pressure must be used. Under such conditions, the tubing may be used to connect several lamps in series, so that the irradiated volume and amount of reagent transformed reach industrially significant levels. In fact, when using such an inexpensive set-up with a single lamp, a daily production on the order of 500 g has been achieved [11].

An alternative approach is that of *process intensification through miniaturization*, which involves the application of microreactors to photochemical reactions. For this, the reactors were created by engraving thin grooves on a variety of surfaces; the grooves are then covered by a transparent plate, which enables the solution to be circulated through the resultant channels. The short optical path so formed may lead to partial absorption, but also ensures homogeneous excitation – a condition that is difficult to achieve otherwise (as indicated above). "Home-made" continuous flow microreactors have been described in the literature, and shown to have significant advantages in terms of the photochemical reactions carried out (see Figure 1.10) [12].

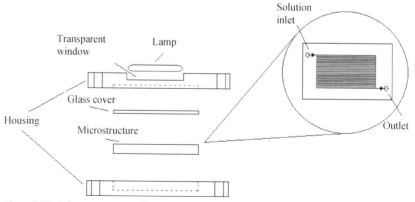

Figure 1.10 A flow microreactor for photochemical reactions.

Another way to minimize secondary photoreactions is to use lamps with a narrow λ range emission, as this may match the reagent absorption better than that of the primary product. Low-pressure Hg arcs with almost exclusive emission at 254 nm can be advantageous in this sense, and even better LEDs that are available for various λ ranges. The merging of these two ideas – that is, illuminating planar microreactors with a set of LEDs mounted on a flat support – may combine the advantage of both choices and thus best exploit the spatially limited output of these light sources. Indeed, this may be a future solution to this problem [13].

The above considerations regarding concentrated solutions imply that a solid-state photochemical reaction is practicable only when the product formed does not hinder the penetration of light down to the inner molecules of the reagent. In the opposite case, the crystals may undergo a conspicuous change in appearance when exposed to light; however, dissolving the crystals and analyzing the solution shows that the photodecomposition is next to negligible, because only the first layer has reacted. Nonetheless, a considerable number of crystal-phase photochemical reactions have been reported, and advantage has been taken of the potential of this method for selective reactions, at least in some cases rationalized on the basis of the crystal structure [14], and including enantioselective processes by using chiral auxiliaries. The irradiation of a thin layer of small-dimension crystals, either on a glass plate or on the internal walls of a rotating tube, may produce good results; for example, a solution can be slowly evaporated in a rotating tube held horizontally, and the thin coat obtained then irradiated [14]. An alternative, which has been shown to be convenient in some cases, is to irradiate a microcrystals suspension obtained by adding a solution of the reagent in a water-miscible solvent (e.g., acetone) to neat water, while stirring [15]. Irradiation with the reagent adsorbed onto a variety of matrices has likewise been reported [16].

1.3.2
Effect of Impurities, Oxygen, and Temperature

The lifetime of the excited state is short, in the order of $\tau < 1\,\mu s$ for triplets, and $\tau < 1\,ns$ for singlets. Thus, in order that a chemical reaction competes with an unproductive physical decay, it must be quite fast. This has both positive and negative implications. On the plus side, it makes the effect of impurities relatively small, such that the extensive purification of reagents and solvents is not generally required, provided that neither absorbing impurities (this can easily be monitored with UV spectroscopy) nor highly reactive molecules (e.g., alkenes as impurities in alkanes, which react easily with ketones) are present. The exception here is oxygen, which quenches excited states at a diffusion-controlled rate. In many air-equilibrated organic solvents the amount of oxygen present in solution is in the range 0.002 to 0.003 M (0.0005 M in water), which is high enough to quench more than 90% of long-lived triplets. The effect is much smaller with singlets or with short-lived triplets.

The effect can easily be eliminated by flushing the solution with an inert gas for some minutes, so that the amount of oxygen dissolved drops by approximately three orders of magnitude (this is frequently carried out also with radical reactions). At any rate, it is advisable to carry out at least the first tests under "deoxygenated" conditions (or rather with a low oxygen concentration) because O_2 may react also with any further intermediates that are often formed along the reaction path, such as radicals or radical ions. To summarize, taking into account that oxygen interferes only with some families of photochemical reactions – and likewise that moisture is a problem only in some cases – the very fact that excited states are highly reactive and short-lived often makes photochemical reactions easier to carry out than "typical" procedures of advanced organic synthesis. In fact, photochemistry requires nothing like the exhaustive dehydration and purification of solvents needed by most reactions via enolates or transition-metal catalysis. On the negative side is the fact that, in bimolecular reactions, a large trap concentration is required to capture the short-lived excited state, often a great excess with respect to the reagent.

Again, for the same reasons, photochemical reactions generally are little affected by changes in temperature, or at least are reactions of the excited states (though there are exceptions). However, the course of the overall process which occurs on irradiation may change dramatically, generally due to an effect on the system before or after the actual photochemical step [17]. For example, when the reagent is liable to some equilibrium, the temperature may affect distribution between the forms, so that a different reagent would be excited at a different temperature. (In the same way, a compound may be hydrogen-bonded in certain solvents, such that the excitation produces a different excited state that results in a photochemistry which differs from that seen in a non-hydrogen-bonding solvent.) On the other hand, photochemical reactions do create (in most cases) a ground-state product (i.e., the lowest-lying species at that configuration), although this need not be a stable particle in absolute terms. Indeed, the photochemical reaction quite often produces a highly reactive species (a radical, an ion, a nitrene, etc.), the fate of which depends more on the conditions than that of the excited state, and results in an overall dependence on the temperature. In fact, the "cold" generation of highly reactive intermediates is an important advantage of photochemical reactions. For example, the generation of alkyl radicals by thermal hydrogen abstraction from α-substituted aldehydes leads unavoidably to decarbonylation. However, the highly favored hydrogen abstraction by an excited photocatalyst is also effective at low (e.g., $-80\,°C$) temperatures, when decarbonylation is slowed down, so that different end-products are obtained at different temperatures (Scheme 1.2) [6].

In fact, this characteristic has even more general implications. Photochemical reactions tend to be (relatively) independent of the conditions (except when the nature of the excited state changes, see above), and thus the primary photoproduct can be created under different conditions (temperature, viscosity, proticity, etc.), a fact that provides such methods with an unparalleled versatility.

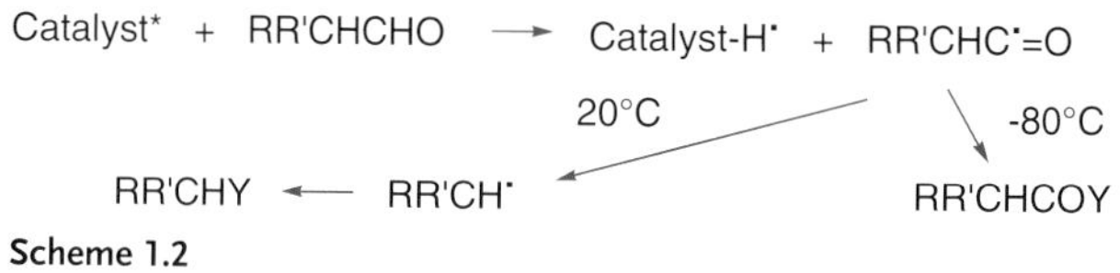

Catalyst* + RR'CHCHO → Catalyst-H• + RR'CHC•=O

20°C → RR'CH• → RR'CHY

−80°C → RR'CHCOY

Scheme 1.2

1.3.3
Safety

Intense light sources (including solar light) may be dangerous for the skin (UV-B, 280–320 nm causes erythema, while UV-C below 280 nm is genotoxic) and the eyes. An annoying *conjunctivitis* may result from exposure to short-wavelength light, with the onset of pain occurring some hours after the exposure, without warning. In general, it is difficult to avoid indirect or reflected radiation, the only practical approach being *always* to wear protective glasses (specially designed to block UV light) when working close to an irradiation apparatus. It is advisable to switch off the lamps before approaching the apparatus, even when wearing such glasses. This may be problematic with high-pressure, but not low-pressure, arcs. A photochemical safety cabinet which switches off automatically when opened is available commercially, but careful behavior is probably sufficient. Attention must also be paid to the formation of toxic ozone, and irradiations should always be carried out under a fume hood. With high-power lamps, nitrogen flushing in the close vicinity of the arc should be contemplated so as to avoid ozone building up to dangerous concentrations. Obviously, mercury-polluted exhausted lamps should be disposed of correctly and safely.

1.3.4
Planning a Photochemical Synthesis

To summarize, it may be concluded that with the range of lamps available, the most convenient choice would be two pairs (each of 15 W) of low-pressure arcs with their emission centered at 254, 305, 350 and 370 nm; an alternative (which is likely to become more common in the future) would be a set of LEDs. Quartz tubes would be a sensible acquisition for any synthetic laboratory as they allows an exploration of the viability of photochemical steps. The limited precautions and safety requirements involved with photochemistry make it a much more easily used method than many of the reactions that involve thermal/oxygen/moisture-labile and/or toxic/flammable reagents, or a delicate catalyst that are routinely considered and carried out in synthetic laboratories. With photochemistry, an experiment can be easily conducted, with failure by beginners more likely due to naïve oversight (see below) than to an inadequate experimental capability. An experiment with a time-honored method, such as the exposure to solar light, may also be appropriate, at least as a first indication, for reagents absorbing in the near UV spectrum. There are also not negligible advantages in this case of using a freely available source requiring no investment at all!

Some of the key parameters useful for planning a photochemical reaction are summarized in Table 1.1 and Figure 1.1. Thus, a lamp must be chosen that emits where the putative photoreactive molecule absorbs (check the absorption spectrum). A check must also be made that neither the material from which the apparatus is built nor the solvent absorb at that wavelength. In practice, with a low-pressure mercury arc, a quartz apparatus and a solvent chosen among alkanes, alcohols, ethers or

Table 1.2 Some key indications for planning a photochemical reaction.

Lamp	Irradiation (nm)	Einstein (min^{-1} cm^{-2})	Volume irradiated (ml)	Time taken to convert a 5×10^{-2} M solution ($\Phi = 1$) (h)
Low-pressure Hg	ext, 254[a]	$\approx 4 \times 10^{-6}$	10–100	0.25
Phosphor-coated	ext, 305[a]	$\approx 1 \times 10^{-6}$	10–100	1
	ext, 350[a]	$\approx 8 \times 10^{-7}$	10–100	1.2
Medium-pressure Hg	imm, 300–400[b]	$\approx 10^{-5}$	80–1000	0.1
LED	ext, 310[c]	$\approx 2 \times 10^{-8}$	1–10	40
LED	ext, 400[c]	$\approx 1 \times 10^{-6}$	1–10	1
Solar light	ext, 330–400	$\approx 10^{-7}$	Any	10

[a] In a multilamp apparatus fitted with six 15 W lamps.
[b] For the range indicated, the lamp emits also in the visible.
[c] Six LEDs circularly placed around the test tube containing the solution.

simple haloalkanes, should be used. Acetonitrile is a common choice as solvent because it adds to the overall transparency in the UV its low chemical reactivity. A wider choice is possible when utilizing the UV-A spectrum.

When planning the scale of the experiment, the data listed in Table 1.2 might be of value. The quantities indicated are: (i) the light flux (in Einstein, moles of photons) on a 1 cm^2 surface close to the lamp; and (ii) the time required for converting a 5×10^{-2} M solution by using that lamp, assuming that all of the flux is absorbed, the quantum yield is unitary, and that there is no inner filter effect (in other words, the minimal time required for converting the above amount).

Clearly, the amount of reagent converted depends on the volume that can be illuminated, and thus on the surface of the lamp. Further consider that in an immersion well apparatus, all of the flux is absorbed, whereas with an external illumination the fraction absorbed will vary considerably. Phosphor-coated lamps and LEDs, for example, may have the same flux density (at $\lambda > 360$ nm), but the former lamp is larger and a much larger overall flux is emitted. The problem then is to place the vessel with the solution in such a way that a large fraction of the flux is absorbed. This target is achieved in a multilamp apparatus with mirror walls, where volumes of 100 ml or more can be effectively irradiated (i.e., according to Figure 1.2 rather than to Figure 1.3). In contrast, while the flux from a LED is readily directed towards the sample, the total emission is weak and only a few milliliters of solution will be irradiated by using about ten LEDs (which is still useful for explorative reactions).

It can be seen that amounts of up to a few grams can be converted in 5–10 h in the case of complete absorption and unitary quantum yield (ignoring any internal filter effect). Lower quantum yields or an incomplete absorption proportionally lengthens the time required for the transformation. It is most likely safe to say that exploratory reactions on the 100 mg scale can be carried out in a reasonable time, provided that $\Phi \geq 0.1$–0.05. Even reactions with a quantum yield at the lower limit or below may be interesting for a preparation if the reaction is clean. Given the minimal safety precautions required, leaving the reaction for a few hours or even overnight in a tube exposed to the light of a low-pressure Hg arc may often be a convenient choice as a test experiment. Larger amounts require additional planning, however, the best approach probably being to use a flow system that, as shown above, can even be implemented by using a home-made apparatus and subsequently developed to the kilogram scale.

1.4
Photochemical Steps in Synthetic Planning

A number of excellent textbooks of organic photochemistry are available, and the reader is directed to these to acquire an appropriate introduction to the field of organic chemistry (for example, see [18]). Presentations of the synthetic aspects of organic photochemistry are likewise available in books and reviews [19–21], arranged according to the chemical function reacting or to the type (mechanism) of the reaction involved.

In this handbook, it was deemed more appropriate to present the reactions according to the transformation occurring, in the way that an appreciated treatise such as Theilheimer does (albeit more simply). This should help when considering a photochemical alternative if planning a synthesis. Unfortunately, this choice has the drawback that very little is said regarding the general features of any photochemical reaction, let alone the detailed mechanism. Moreover, a single reaction may be mentioned in different chapters if it leads to different synthetic targets, whereas reactions that have nothing in common are dealt with in the same chapter when their synthetic targets are the same.

No attention is given to the "mechanistic" importance of a reaction; rather, an attempt has been made to concentrate on reactions that have an actual (potential) synthetic role. This is not always an obvious selection, because photochemistry has not been sufficiently used in such syntheses, and mechanistic studies are not necessarily a reliable guide towards this aim. As an example, hydrogen abstraction by ketones (Scheme 1.3), which probably is the most thoroughly studied photochemical reaction, is not mechanistically discussed. Neither is presented the resultant photoreduction of ketones (Scheme 1.3, path *a*), because this will hardly ever become a sensible synthetic alternative for the reduction. However, other reactions arising from the same primary photoprocess, namely bimolecular reduction (path *b*) and

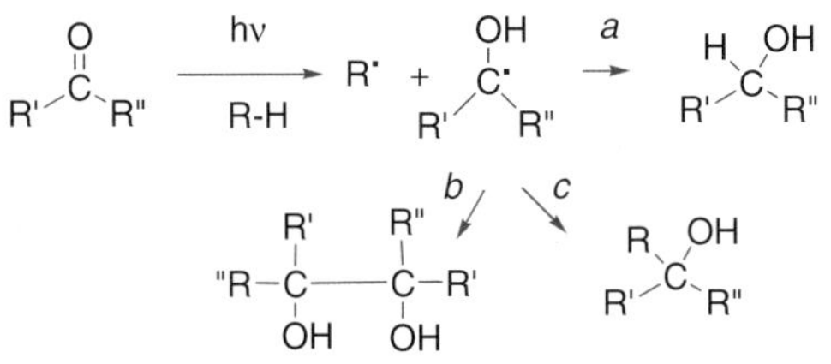

Scheme 1.3

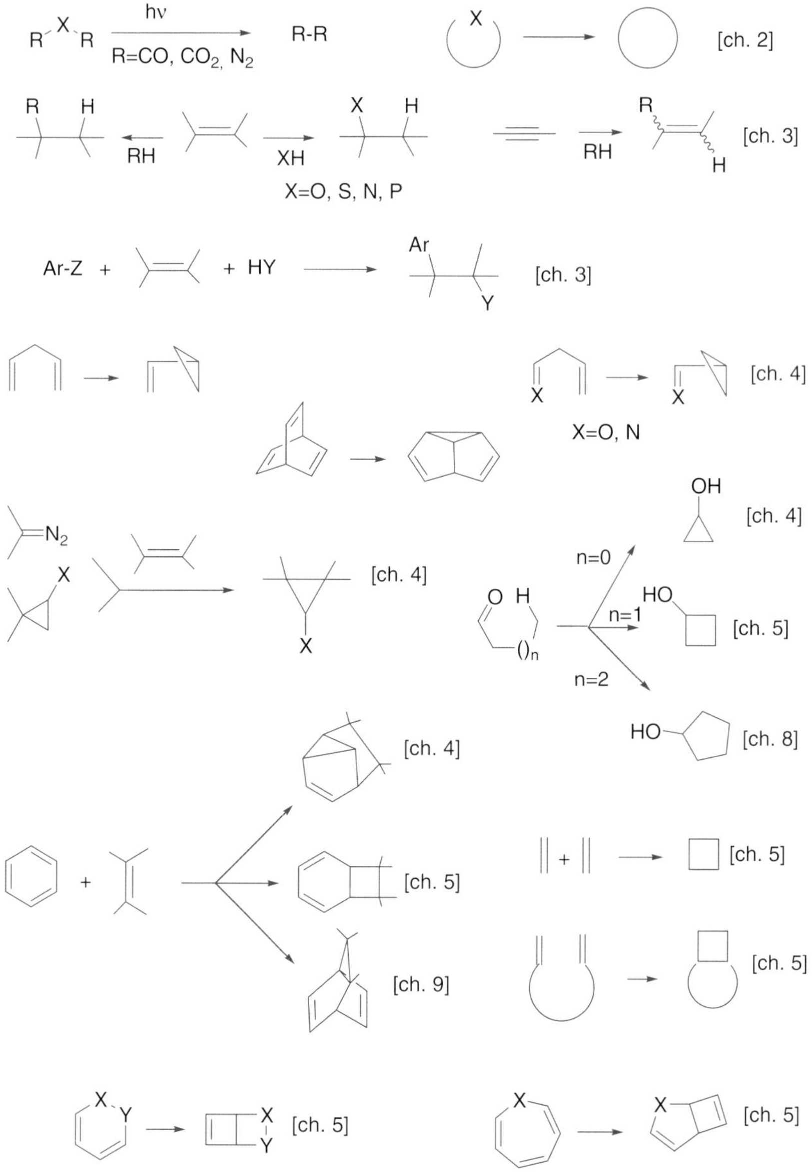

Scheme 1.4 The chapter number where each reaction is discussed is shown in square brackets.

coupling (path c), are presented in the appropriate chapter according to the characteristics of the (carbon–carbon) bond formed (in a open-chain compound, forming a ring, of which dimension).

The general schemes for some of the main photochemical reactions are reported in Scheme 1.4 (the relevant chapter number is shown in square brackets).

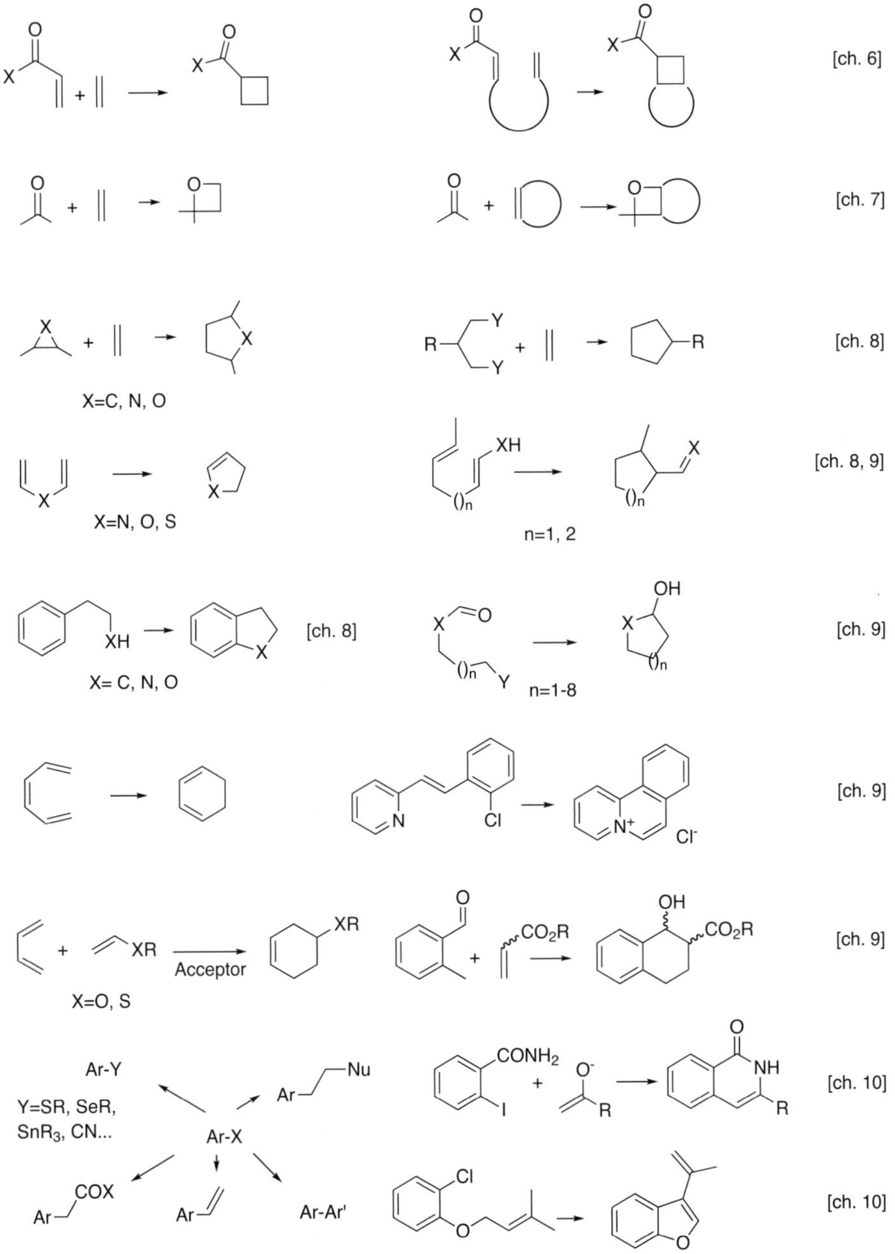

Scheme 1.4 (*Continued*)

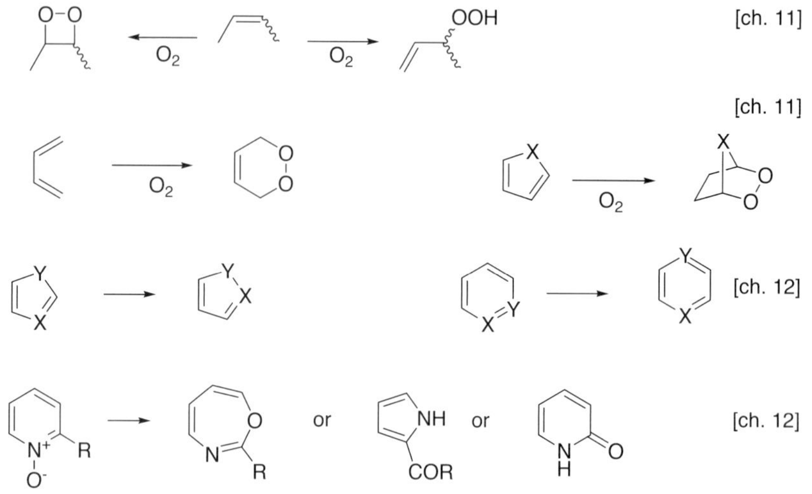

Scheme 1.4 (Continued)

References

1 Ciana, C.L. and Bochet, C. (2007) Clean and easy photochemistry. *Chimia*, **61**, 650–654.
2 Stobbe, H. (1908) Die Photochemie organischen Verbindungen. *Zeitschrift Elektrochemie*, **33**, 473–483.
3 (a) Pfoertner, K.H. (1990) Photochemistry in industrial synthesis. *Journal of Photochemistry and Photobiology, A: Chemistry*, **51**, 81–86; (b) Braun, A.M. (1997) New potentials for photochemical technology. *Speciality Chemicals*, **17**, 21–22, 24–25; (c) Pape, M. (1975) Industrial applications of photochemistry. *Pure and Applied Chemistry*, **41**, 535–558; (d) Roloff, A., Meier, K., and Riediker, M. (1986) Synthetic and metal organic photochemistry in industry. *Pure and Applied Chemistry*, **58**, 1267–1272.
4 Morrison, H. and Malski, R. (1974) Changes in ultraviolet transmission of Corning Code 9700 glass tubing. *Photochemistry and Photobiology*, **19**, 85.
5 Cirkva, V., Vlkova, L., Relich, S., and Hajek, M. (2006) Preparation of the electrodeless discharge lamps for photochemical applications. *Journal of Photochemistry and Photobiology A: Chemistry*, **179**, 229–233.
6 Esposti, S., Dondi, D., Fagnoni, M., and Albini, A. (2007) Acylation of electrophilic olefins through decatungstate-photocatalyzed activation of aldehydes. *Angewandte Chemie, International Edition*, **46**, 2531–2534.
7 Griesbeck, A.G., Maptue, N., Bondock, S., and Oelgemöller, M. (2003) The excimer radiation system: a powerful tool for preparative organic photochemistry. A technical note. *Photochemical and Photobiological Sciences*, **2**, 450–451.
8 (a) Lapkin, A.A., Boddu, V.M., Aliev, G.N., Goller, B., Polisski, S., and Kovalev, D. (2008) Photo-oxidation by singlet oxygen generated on nanoporous silicon in a LED-powered reactor. *Chemical Engineering Journal*, **136**, 331–336; (b) Ghosh, J.P., Langford, C.H., and Achari, G. (2008) Characterization of an LED based photoreactor to degrade 4-chlorophenol in an aqueous medium using coumarin (C-343) sensitized TiO_2. *Journal of Physical Chemistry A*, **112** (41), 10310–10314;

(c) Imperato, G. and König, B. (2008) Acceleration of Suzuki-Miyaura- and Stille-type coupling reactions by irradiation with near-UV-A light. *ChemSusChem*, **1**, 993–996.

9 (a) Schiel, C., Oelgemöller, M., Ortner, J., and Mattay, J. (2001) Green photochemistry: the solar-chemical 'Photo–Friedel–Crafts acylation' of quinones. *Green Chemistry*, **3**, 224–228; (b) Esser, P., Pohlmann, B., and Scharf, H.D. (2003) The photochemical synthesis of fine chemicals with sun light. *Angewandte Chemie, International Edition in English*, **33**, 2009–2023; (c) Oelgemöller, M., Healy, N., de Oliveira, L., Jung, C., and Mattay, J. (2001) Green photochemistry: solar-chemical synthesis of juglone with medium concentrated sunlight. *Green Chemistry*, **8**, 831–834.

10 Ballari, M.M., Brandi, R., Alfano, O., and Cassano, A. (2008) Mass transfer limitations in photocatalytic reactors employing titanium dioxide suspension. I. concentration profiles in the bulk. *Chemical Engineering Journal*, **136**, 50–65.

11 Hook, B.D.A., Dohle, W., Hirst, P.R., Pickworth, M., Berry, M.B., and Booker-Milburn, K.I. (2005) A practical flow reactor for continuous organic photochemistry. *Journal of Organic Chemistry*, **70**, 7558–7564.

12 (a) Maeda, H., Mukae, H., and Mizuno, K. (2005) Enhanced efficiency and regioselectivity of intramolecular $(2\pi + 2\pi)$ photocycloaddition of 1-cyanonaphthalene derivatives using microreactors. *Chemistry Letters*, **34**, 66–67; (b) Coyle, E.E. and Oelgemöller, M. (2008) Micro-photochemistry: photochemistry in microstructured reactors. The new photochemistry of the future? *Photochemical & Photobiological Sciences*, **7**, 1313–1322; (c) Fukuyama, T., Hino, Y., Kamata, N., and Ryu, I. (2004) Quick execution of [2 + 2] type photochemical cycloaddition reaction by continuous flow system using a glass-made microreactor. *Chemistry Letters*, **33**, 1430–1431;

(d) Meyer, S., Tietze, D., Rau, S., Schäfer, B., and Kreisel, G. (2006) Photosensitized oxidation of citronellol in microreactors. *Journal of Photochemistry and Photobiology A: Chemistry*, **186**, 248–253; (e) Gorges, R., Meyer, S., and Kreisel, G. (2004) Photocatalysis in microreactors. *Journal of Photochemistry and Photobiology A: Chemistry*, **167**, 95–99; (f) Wootton, R.C.R., Fortt, R., and de Mello, A.J. (2002) A microfabricated nanoreactor for safe, continuous generation and use of singlet oxygen. *Organic Process Research & Development*, **6**, 187–189.

13 (a) Meyer, S., Tietze, D., Rau, S., Schäfer, B., and Kreisel, G. (2007) Photosensitized oxidation of citronellol in microreactors. *Journal of Photochemistry and Photobiology A: Chemistry*, **186**, 248–253; (b) Lapkin, A.A., Boddu, V.M., Aliev, G.N., Goller, B., Polosski, S., and Kovalev, D. (2008) Photo-oxidation by singlet oxygen generated on nanoporous silicon in a LED-powered reactor. *Chemical Engineering Journal*, **136**, 331–336.

14 (a) Cohen, M.C. (1987) Solid phase photochemical reactions. *Tetrahedron*, **43**, 1211–1224; (b) Williams, J.R. and Abdel-Magid, A. (1981) Photolysis of 3-oxo-$\Delta^{5(10)}$-steroids in alcoholic solvents and in the solid phase. *Tetrahedron*, **27**, 1675–1677.

15 Kuzmanich, G., Natarajan, A., Chin, K.K., Veerman, M., Mortko, C.J., and Garcia-Garibay, M.A. (2008) Solid state photodecarbonylation of diphenyl-cyclopropenone: a quantum chain process made possible by ultrafast energy transfer. *Journal of the American Chemical Society*, **130**, 1140–1141.

16 Johnston, L.J. (1991) Phototrans-formations of organic molecules adsorbed on silica and alumina, in *Photochemistry in Confined Media* (ed. V. Ramamurthy), Ch. 8, VCH Publishers, New York.

17 (a) Heredia-Moya, J. and Kirk, K.L. (2007) Photochemical Schliemann reaction in ionic liquids. *Journal of Fluorine Chemistry*, **128**, 674–678; (b) Literík, J., Relich, S.,

Kulhánek, P., and Klán, P. (2003) Temperature dependent photochemical cleavage of 2,5-dimethylphenacyl esters. *Molecular Diversity*, **7**, 265–271; (c) Warrener, R.N., Pitt, I.G., and Russell, R.A. (1993) The photochemistry of isobenzofuran. I. Structure of the dimers resulting from ultraviolet irradiation of isobenzofuran in acetone and ether solution. *Australian Journal of Chemistry*, **46**, 1515–1534.

18 (a) Horspool, W.H. and Armesto, D. (1992) *Organic Photochemistry: A Comprehensive Treatment*, Ellis Horwood, London; (b) Turro, N.J., Ramamurthy, V., and Scaiano, J.C. (2009) *Modern Molecular Photochemistry of Organic Molecules*, Palgrave Macmillan, Basingstoke; (c) Klan, P. and Wirz, J. (2009) *Photochemistry of Organic Compounds*, Wiley-Blackwell, Weinheim.

19 (a) Müller, E. (ed.) (1975) *Methoden der Organischen Chemie (Houben-Weil). Photochemie*, Vol. 1 & 2, Georg Thieme V., Stuttgart; (b) Horspool, W.H. and Lenci, F. (eds) (2004) *Handbook of Organic Photochemistry and Photobiology*, 2nd edn, CRC Press; (c) Horspool, W.M. (ed.) (1984) *Synthetic Organic Photochemistry*, Plenum, New York; (d) Ninomiya, I. and Naito, T. (eds) (1989) *Photochemical Synthesis*, Academic Press, London.

20 (a) Ramamurthy, V. and Schanze, K.S. (2008) Organic molecular photochemistry, in *Molecular and Supramolecular Photochemistry*, Vol. 3 (eds V. Ramamurthy and K.S. Schanze), Marcel Dekker, New York; (b) Griesbeck, A.G. and Mattay, J. (eds) (2005) *Molecular and Supramolecular Photochemistry*, Vol. 12 (eds V. Ramamurthy and K.S. Schanze), Marcel Dekker, New York; (c) Mattay, J. and Griesbeck, A. (eds) (1994) *Photochemical Key Steps in Organic Synthesis*, VCH, Weinheim.

21 (a) Hoffmann, N. (2008) Photochemical reactions as key steps in organic synthesis. *Chemical Reviews*, **108**, 1052–1103; (b) Fagnoni, M., Dondi, D., Ravelli, D., and Albini, A. (2007) Photocatalysis for carbon-carbon bond formation. *Chemical Reviews*, **107**, 2725–2756.

2
Carbon–Carbon Bond Formation by the Photoelimination of Small Molecules in Solution and in Crystals
Saori Shiraki and Miguel A. Garcia-Garibay

2.1
Introduction

The formation of C—C bonds is arguably the most important reaction in organic synthesis. While many strategies have been developed to link organic fragments with a wide range of thermal processes, photochemical syntheses based on the elimination of a small molecule, "X" (Scheme 2.1), offer significant advantages when the desired structures are highly strained or hindered. In this chapter, we will describe examples of the generation of C—C bonds through the photoinduced extrusion of small molecules such as N_2, CO_2, CO, and SO_2, as well as the formal extrusion of S atoms. After describing the most general mechanistic aspects covering these reactions and the current scope and limitations, we will illustrate examples involving these reactions in solution. Following that, we will describe a strategy based on the use of crystalline solids that helps to control the fate of the reaction intermediates. Finally, we will illustrate the use of crystalline ketones (X=CO) as one of the most promising general methods for the synthesis of compounds with all-carbon adjacent stereogenic quaternary centers.

The photoelimination of a small stable molecule X accompanied by the formation of a new C—C bond could, in principle, take place in a concerted manner through cyclic transition states with simultaneous bond-breaking and bond-making steps that follow trajectories determined by orbital symmetry considerations. However, as it pertains to photodecarbonylation reactions, theoretical analysis on the formation of hydrogen and CO from formaldehyde suggests that the concerted process is significantly less likely than the stepwise cleavage of the H—C(O)—H bonds [1]. Similar conclusions have been made for the elimination of N_2 and H_2 from simple diazene (HN=NH) [2]. In fact, evidence for concerted reactions is available only in a handful of cases (*vide infra*). Most examples involve a four-step process that starts with electronic excitation (step 1) and is followed by an α-cleavage reaction (step 2) to give a primary biradical or radical pair (**BR-1**). Subsequent dissociation of the small molecule X (step 3) results in the formation of the secondary biradical or radical pair (**BR-2**), which is followed by radical–radical combination to give the desired product (step 4).

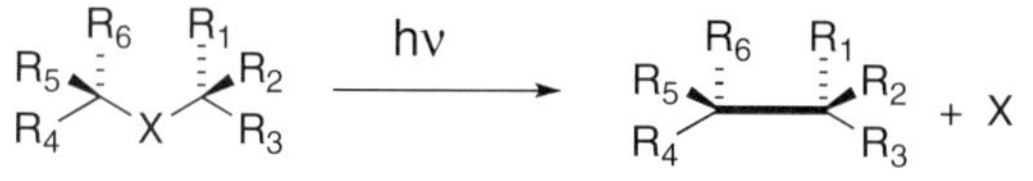

2 Carbon–Carbon Bond Formation by the Photoelimination of Small Molecules

Small molecule "X" eliminated (functional group)		
N_2 (diazenes)	CO_2 (diacylperoxides & esters)	CO (ketones)
S (sulfides)	SO_2 (sulfones)	

Scheme 2.1

As qualitatively implied by the energetic ordering of the cyclic reactant and the intermediates in Figure 2.1, the photoinduced elimination of X is ideal for the synthesis of structures that have a higher energy content than the starting material. Highly endothermic reactions are made possible by the large energy input that occurs upon absorption of a photon. It is also implied in the figure that photoelimination reactions will be more efficient if α-cleavage and each subsequent step is highly exothermic and, as expected by the Hammond postulate, very fast. As an experimental guideline to establish the enthalpic feasibility of a photoelimination reaction, one can determine the excitation energy from the absorption spectrum of the reactant and use quantum mechanical methods to estimate the energy content of the **R, BR-1, BR-2,** and **P**. An important aspect of the excited states and radical intermediates involved in photoelimination reactions is that they are open shell structures with singly occupied orbitals, which may exist in singlet or triplet states. Photodissociation may start with α-cleavage from either **S1** or **T1**, and the subsequent radical pairs are

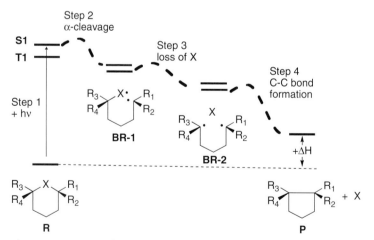

Figure 2.1 Energetics of the four steps involved in the photoelimination of small molecule X and the formation of a new C—C bond.

formed in the corresponding multiplicity (^{1}BR-1 or ^{3}BR-1 and ^{1}BR-2 or ^{3}BR-2). However, bond formation in the final step requires the radical pair to exist in the singlet state, and there are cases where spin inversion may be the rate-limiting step [3]. For the reaction to be as selective and efficient as possible, steps 1–4 should be faster than the time it takes for the two radical centers to lose their stereochemical information and to separate. For that reason, photoelimination reactions in solution will be more efficient when they proceed very rapidly along the singlet manifold, for example, $R + h\nu \to S1 \to {}^1BR_1 \to {}^1BR_2 + X \to P$. In contrast, it has been speculated that photoelimination reactions in crystals may be more efficient along the triplet surface [4]. This argument suggests that preferred sequence in solids involve the triple state: $R + h\nu \to S1 \to T1 \to {}^3BR_1 \to {}^3BR_2 + X \to {}^1BR_2 + X \to P$.

With synthetic applications in mind, one can recognize that photoelimination reactions will have some potential advantages and some obvious disadvantages. As mentioned above, one of most important advantages of radical–radical C–C bond-making reactions is that they are strongly exothermic (from the excited state reactant), and have relatively small activation energies. Hence, it is possible to make bonds between carbons bearing high steric hindrance and to prepare highly strained chemical structures. The obvious disadvantages here are that radicals in solution tend to diffuse apart, lose their stereochemical information rather quickly, and have many undesirable competing pathways. In fact, photochemical C–C bond-forming reactions going through radical intermediates are generally useful only when the fate of the radicals is controlled by: (i) a suitable molecular scaffold; or (ii) by limiting the rotational and translational diffusion of the radical pair within the rigid environment, such the crystalline solid state (Scheme 2.2).

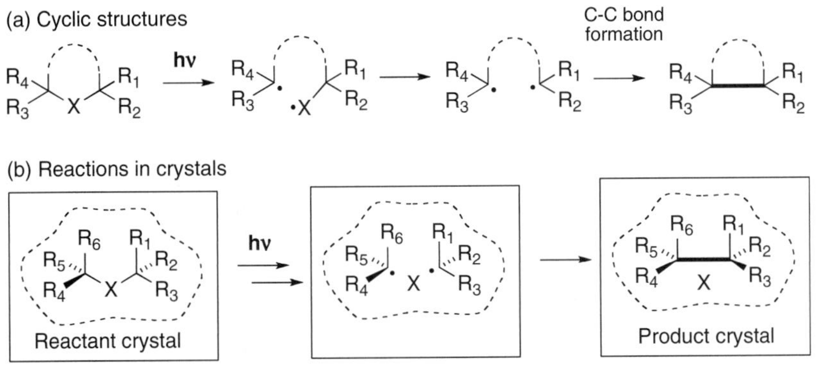

Scheme 2.2

2.1.1
Synthesis of Unstable Molecules

Given that the absorption of photons with wavelengths in the range of $\lambda \approx 250–300$ nm results in the injection of about 115–95 kcal mol^{-1}, it is not surprising that one of

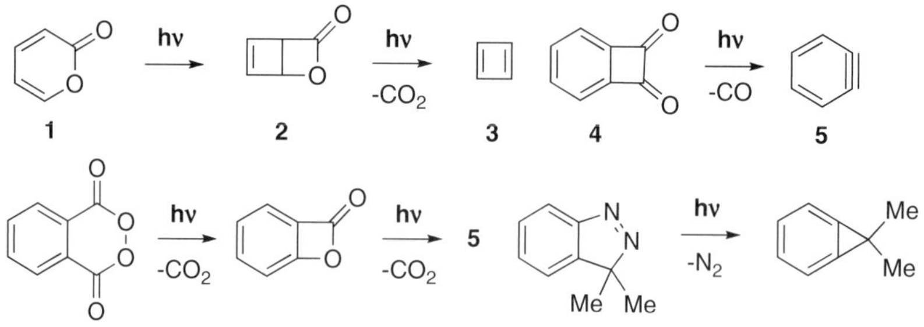

Scheme 2.3

the most successful applications for the photochemical synthesis of C–C bonds is in the construction of high-energy structures that are not readily available by other means. Some examples to illustrate this application include the syntheses of cyclobutadiene **3** by sequential cyclization of pyranone **1**, followed by photodecarboxylation of bicyclic lactone **2** reported by Cram [5], the preparation of benzyne **5** by the photodecarbonylation of benzocyclobutanone **4** [6] or by the double decarboxylation of diacyl peroxide **6** [7, 8], and the synthesis of benzocyclopropane **9** by the denitrogenation of indazole **8** (Scheme 2.3) [9]. Because of the intrinsic instability of these structures, reactions of this type must be carried out in low temperature matrices for the spectroscopic characterization of the products by UV-Vis and IR methods. An alternative and elegant strategy pioneered by Cram involves the use of carcerands and other supramolecular capsules to carry out these types of reaction in solution at ambient temperatures, with the synthesis of cyclobutadiene [5] and benzyne [10] being the best known examples in this class. More recently, the photochemical elimination of small molecules within crystalline substrates has been used to characterize the resulting reactive intermediates by single crystal X-ray diffraction (XRD) [11]. The use of single crystals as an ideal media to investigate the structure and reactivity of several radical pairs by taking advantage of single crystal XRD and spectroscopic techniques such as electron paramagnetic resonance (EPR) and Fourier transform infrared (FTIR) spectroscopy was originally pioneered by McBride *et al.* [12].

The relative generality and utility of the C–C bond-forming reaction by the photoelimination of small molecules can be illustrated also with the synthesis of *para*-[2,2]-cyclophane **14** from precursors **10–13** (Scheme 2.4). The initial interest in cyclophanes can be traced back to the 1950s, due to their significant strain and expected electronic interaction between the stacked aromatic rings [13]. The cyclophane precursors **10–13** were readily constructed by double cyclization reactions of the α,α′-phenylene halides (dibromides or dichlorides) with a suitable carbon-, carboxylate-, or sulfur nucleophiles ($^-$Y-CH$_2$C$_6$H$_4$CH$_2$-Y$^-$) in overall yields ranging from 22% to 97% [14–17]. The yields of the photochemical double elimination reactions varied widely, depending on the molecules being extruded. While the photodecarbonylation of **10** and photodecarboxylation of **11** gave good to moderate yields of 97% and 70%, respectively, the photoextrusion of sulfur from **12**

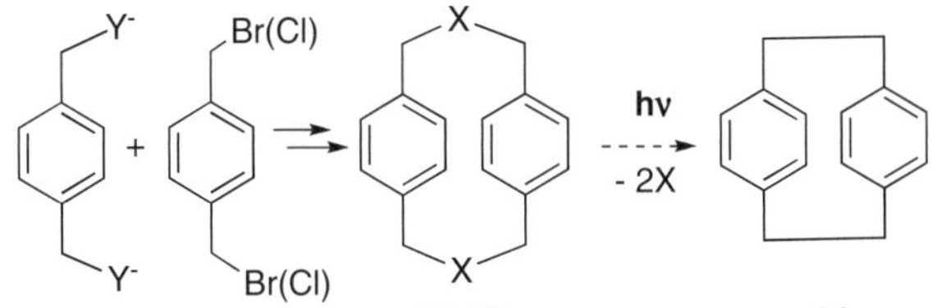

Compound	X	Yield
10	CO	97
11	CO_2	70
12	S	55 (85)
13	SO_2	16

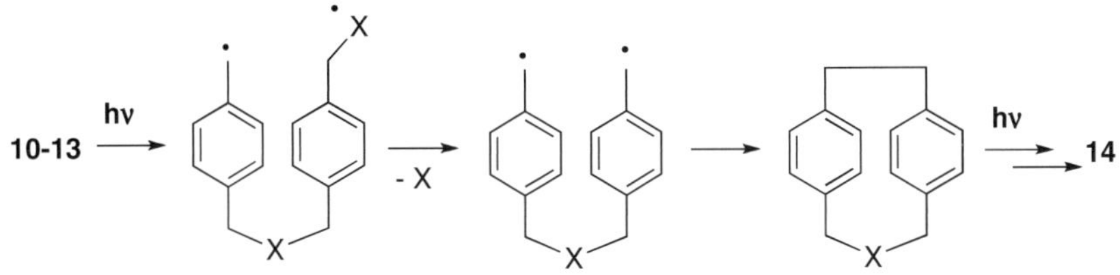

Scheme 2.4

occurs at 55% in the presence of triethyl phosphite, presumably in the form of the phosphorus–sulfur adduct. The elimination of sulfur dioxide from **13** (16%) proceeded in lower yields. The irradiation of **10** in a 1.1 mM benzene solution for 20 min gave **14** in 10% yield, and the monodecarbonylation product in 21% yield. In contrast, the irradiation of a 0.73 mM solution for 4 h gave the cyclophane in 97% yield, with no mono-decarbonylation product. The photoinduced decarboxylation of **11** was successful both in methanol and 1,2-dimethoxyethane, with yields approaching 70%. The irradiation of bis(sulfide) **12** in the presence of triethyl phosphite in benzene gave the cyclophane product in 55% yield (85% by photoelimination of sulfur), presumably in the form of a phosphorothioic acid. The lowest yields were obtained by the irradiation of bis-sulfone **13**, which gave the desulfonated product in only 16% yield when irradiations were carried out by direct irradiation in benzene. However, reactions carried out in acetone acting as a solvent and as a triplet sensitizer yielded side reactions, including polymerization, to yield the cyclophane **14** in less than 2% yield. Step-wise extrusion is likely in all of these cases, and has been confirmed by the isolation and characterization of the single-elimination product in the case of diketone **10**. The spin multiplicities of the reaction have not been studied in detail, except in the case of *bis*-sulfone **13**, which Givens and Olsen demonstrated to react in good yields only from the singlet excited state. It is important to note that the photoelimination of CO from suitable ketones is by far the best understood and most reliable reaction. This is followed by the elimination of CO_2 and then by photoextrusion of the SO_2 from sulfones (or other sulfur groups), which have been studied and applied to a significantly lesser extent. It is interesting that the synthesis of *para*-[2,2]-cyclophane **14** has not been attempted by the elimination of N_2 from the corresponding diazene derivative. However, as illustrated below, diazenes are very useful and reliable intermediates for synthetic applications involving the photoinduced elimination of N_2 followed by C–C bond formation.

2.2
Photochemical C—C Bond Formation in Solution

2.2.1
Concerted Reactions

One of the primary goals of all C—C bond-forming reactions is to control the stereoselectivity of the product when either or both of the bonding atoms are stereogenic. For the photoelimination of small molecules, the stereochemical integrity of the two carbon atoms associated in the bond-breaking and bond-making steps is generally compromised for reactions that take place by radical intermediates. While examples are very limited, exceptions exists for concerted reactions with transition states defined by orbital symmetry considerations and for reactions that occur with a memory of chirality.

The number of photoelimination reactions that occur in a concerted manner is very small. One of the best-characterized examples does not lead to the formation of a C—C bond between the two α-carbons. Rather, it involves the stereospecific photoelimination of CO from 3,5-cycloheptadienone **15** to 1,3,5-octatriene **16** (Scheme 2.5) [18]. The reaction occurs in a stereospecific manner with the relative stereochemistry of the two α-carbons determining the configuration of the resulting 1,5-double bonds. Another interesting and more relevant example involves the photodecarboxylation of enantiomerically pure aromatic esters derived from (+)- or (−)-2-methylbutyric acid and 2,4,6-trimethylphenol **17**, which occurs with 100% retention of configuration in the product **18**. [19] While this is a promising lead that

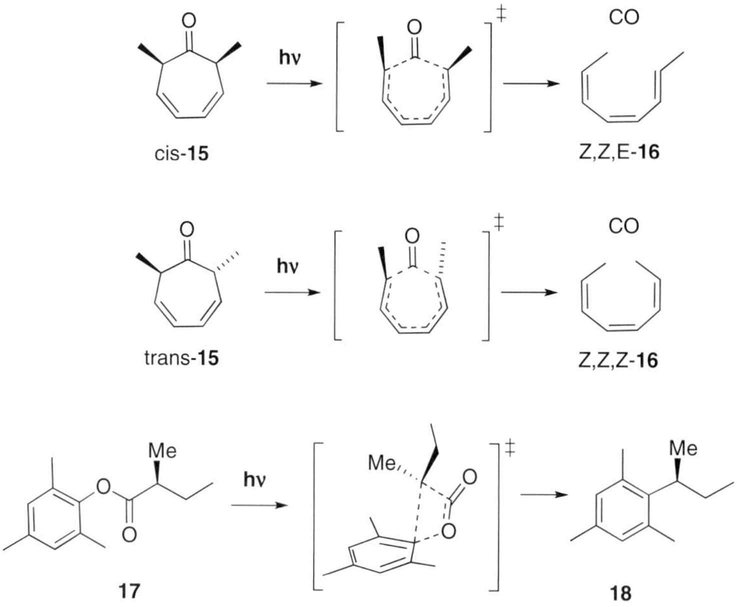

Scheme 2.5

2.2 Photochemical C–C Bond Formation in Solution

deserves further analysis, the generality of the reaction at this time appears to be limited to compounds that block the more common photo-Fries rearrangement by bearing substituents in the two *ortho-* and *para-* positions.

2.2.2
Photoelimination of N_2

2.2.2.1 Synthesis of Three-Membered Rings

One of the most synthetically useful photoelimination reactions is the photo-denitrogenation of pyrazolines for the preparation of substituted cyclopropanes. The reaction is particularly valuable through the strategy illustrated in Scheme 2.6. It involves the conversion of an activated olefin into the corresponding pyrazoline by a thermal dipolar cycloaddition, followed by the photochemical elimination of N_2 to generate the three-membered ring. While the regioselectivity of the cycloaddition reaction is not relevant in the overall reaction scheme [20], a suprafacial addition ultimately leads to the stereospecific formation of the three-membered ring.

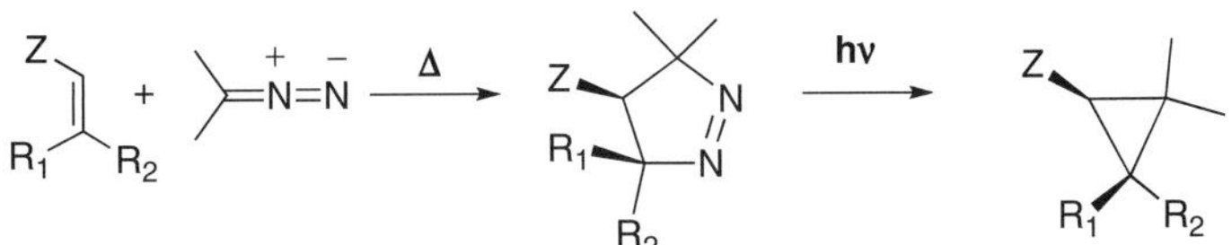

Scheme 2.6

Several interesting applications of the above reaction sequence are illustrated in Scheme 2.7, including a general strategy for the synthesis of several naturally occurring cyclopropyl-containing amino acids in good to excellent yields [21]. Among these, compound **19** (R = Cbz and Boc) was prepared in 100% yield and photochemically transformed into compound **20** in quantitative yield. The latter was used for the total synthesis of (−)-allo-coronamic acid, (−)-allo-norcoronamic acid, (−)-(Z)-2,3-methanohomoserine, and (−)-(Z)-2,3-methanomethionine [22]. Similarly, the photochemical reaction of pyrazoline **21** led to the key cyclopropane intermediate **22** for the synthesis of (±)-carnosadine, a marine natural product isolated from the red alga *Grateloupia carnosa* [21a]. Pyrazoline **23** led to the efficient synthesis of methanoglutamic acid **25**. Closely related examples include the enantiospecific synthesis of several cyclopropyl phosphonate derivatives **26** used in agricultural chemistry [23].

Other examples involving intermolecular dipolar cycloadditions include the total synthesis of the cyclocolorene from precursor **28** [24], prostratin and DPP from **30** [25], and the marine diterpene halimedatrial from diazolactone **31** (Scheme 2.8) [26]. It is noteworthy that compound **28** was obtained by activating one of the α,β-double bonds of the otherwise fully conjugated tropone with an $η_4$-iron tricarbonyl complex. Two intramolecular versions of the dipolar cycloaddition, followed by photodenitrogenation of the corresponding pyrazolines, are illustrated in Scheme 2.9. In the formal synthesis of longifolene by Schultz and Pulg, the *in situ* generation of diazoalkane **33** led to the formation of the tricyclic pyrazoline **34**, which underwent

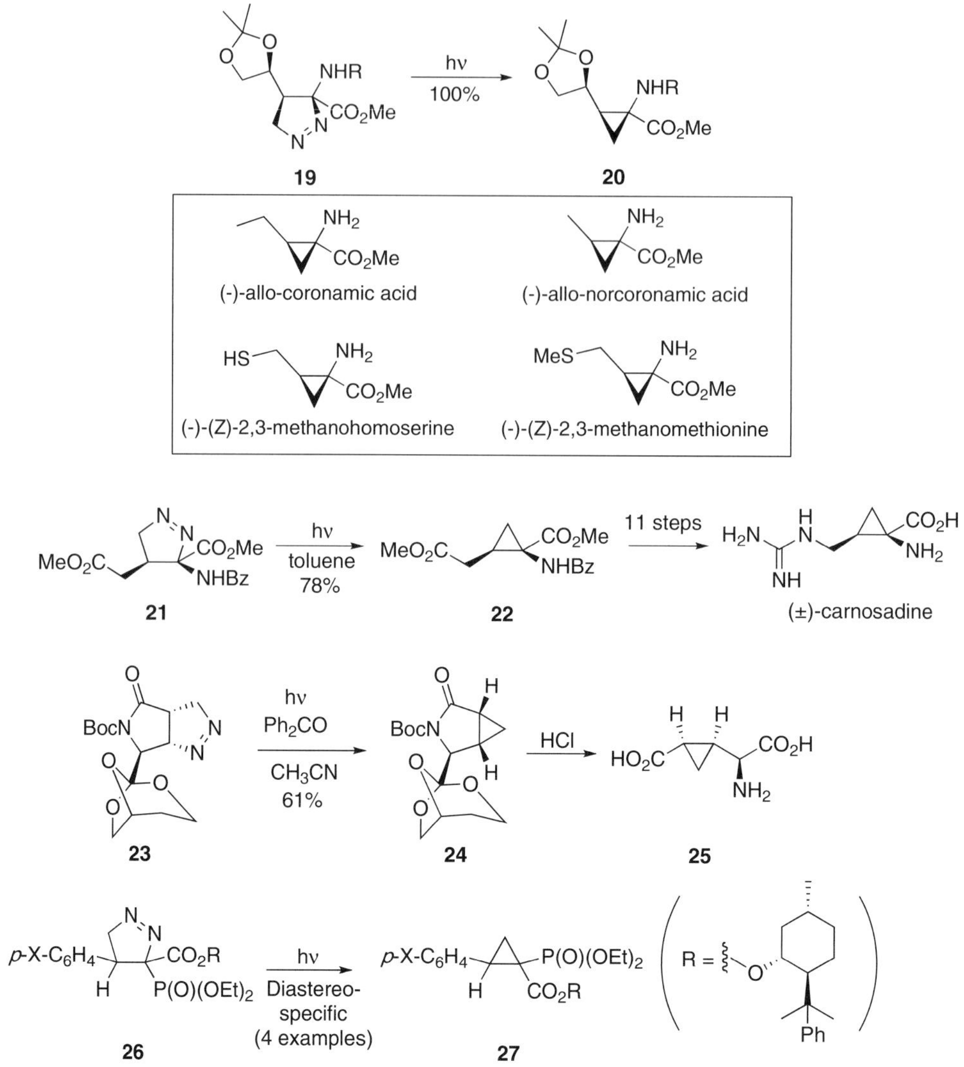

Scheme 2.7

nitrogen elimination to give the three-membered ring in **35** [27]. The thermal rearrangement of **35** to **36** had been formulated as the key component in their strategy wherein an intramolecular 1,4-cycloaddition of a carbene intermediate to the diene to yield the desired tricyclic structure. The tricyclic keto ester **36** had been previously used as an intermediate in the total synthesis of longifolene. In the second example, Ohfune et al. [28] reported the generation of diazoalkane **37** followed by an intramolecular dipolar cycloaddition to yield azolactam **38**, which gave the strained cyclopropyl intermediate **39** upon photoinduced loss of N_2. Compound **39** was transformed in six steps into the desired mGluR2 antagonist (+)-LY354740.

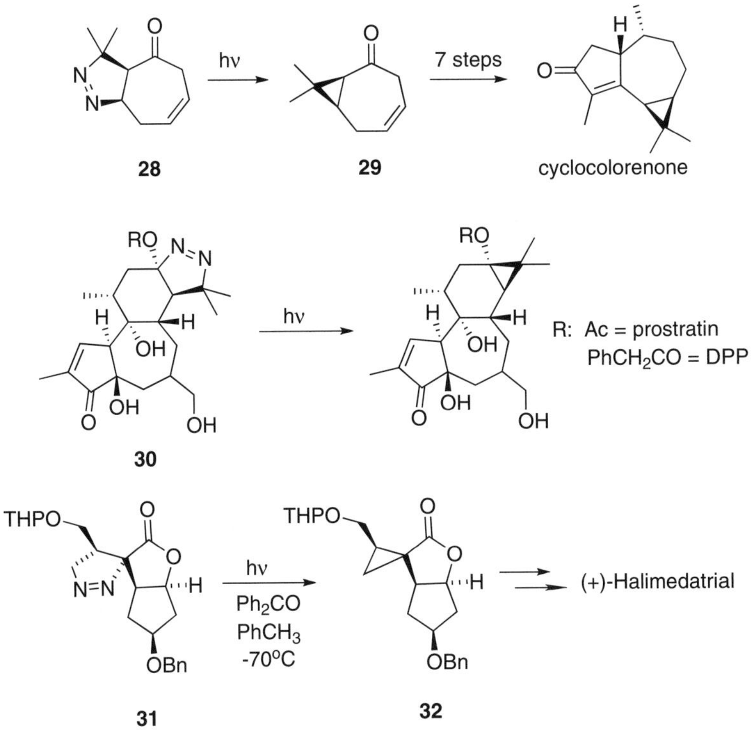

Scheme 2.8

2.2.2.2 Synthesis of Cyclobutanes and Polycyclic Compounds

In a process analogous to the C—C bond formation that occurs upon ring contraction of the five-membered ring pyrazolines to the three-membered ring cyclopropanes, the photoinduced denitrogenation of tetrahydropyridazines leads to the formation of four-membered rings. As illustrated in Scheme 2.10, the reaction is most useful with bicyclic structures such as diazabicyclo[2.2.2]octane and diazabicyclo[2.2.1]heptane. As shown in the scheme, the corresponding azo compounds are generally prepared from a suitable diene by a Diels–Alder reaction with diethyl diazene-1,2-dicarboxylate (step i) followed by decarboxylation (step ii) and, if desired, hydrogenation of the remaining olefin (step iii).

The subject of significant early inquiry, the denitrogenation of cyclic alzoalkanes, has been shown to proceed in a stepwise manner through radical intermediates with stereospecificity and chemical yields that depend on the precursor and reaction conditions [29]. A few representative examples in Scheme 2.11 include the synthesis of tricyclic housane **41** [30] and ladderane **43** [31] from diazenes **40** and **42**, respectively. These compounds were obtained in about 20% yield with inversion of configuration by direct irradiation in ether at low temperature. Notably, the more substituted and more thermally stable tricyclic diazene **44** reacted in ether at ambient temperature to give compound **45** with retention of configuration in essentially

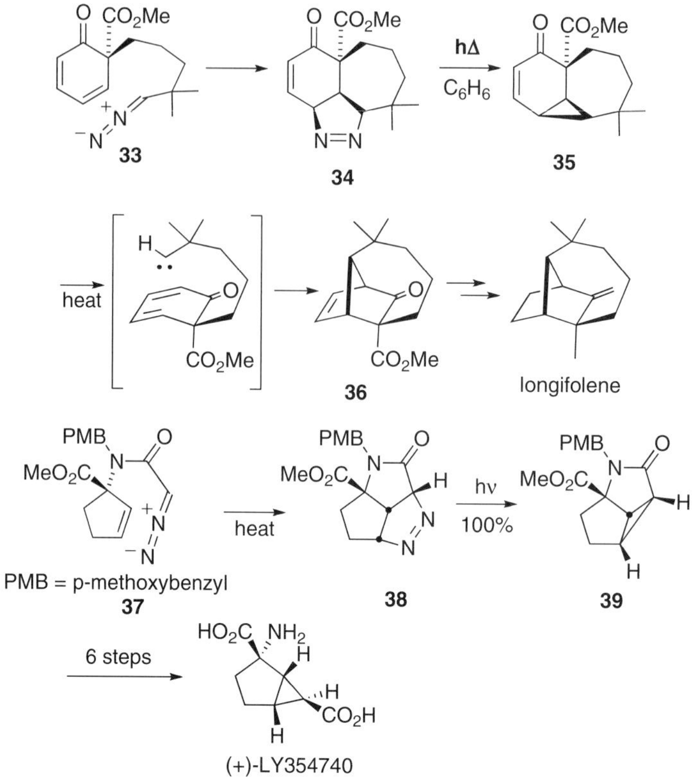

Scheme 2.9

quantitative yield (>95%) [32]. Other interesting examples include the *syn*-spirocycloheptadienyl derivative **46**, which was reported to give the corresponding housane **47** in 76% yield with retention of the diazene stereochemistry upon irradiation in acetonitrile at −78 °C [33]. The last two examples illustrate the sensitivity of the reaction to the substrate and the need to optimize reactions conditions. While structurally analogous, the formation of quadricyclene **49** and prismane **51** occurs very differently. While **49** is obtained in up to 90% yield upon direct or sensitized irradiation in ether at ambient temperature [34], the best yields of prismane **51** were obtained when irradiations were carried out in toluene at 30 °C [35].

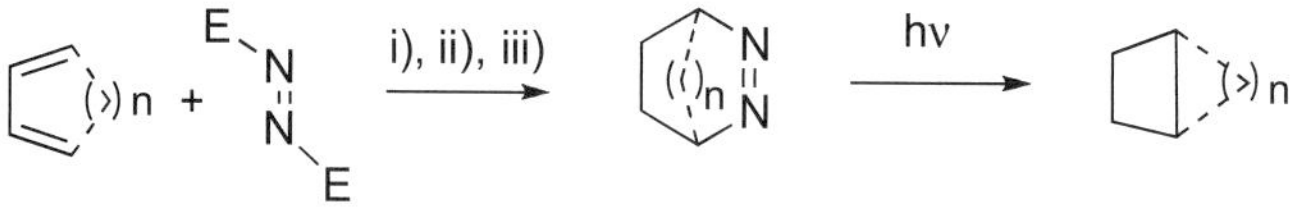

i) Diels-Alder reaction, ii) decarboxylation, iii) hydrogenation

Scheme 2.10

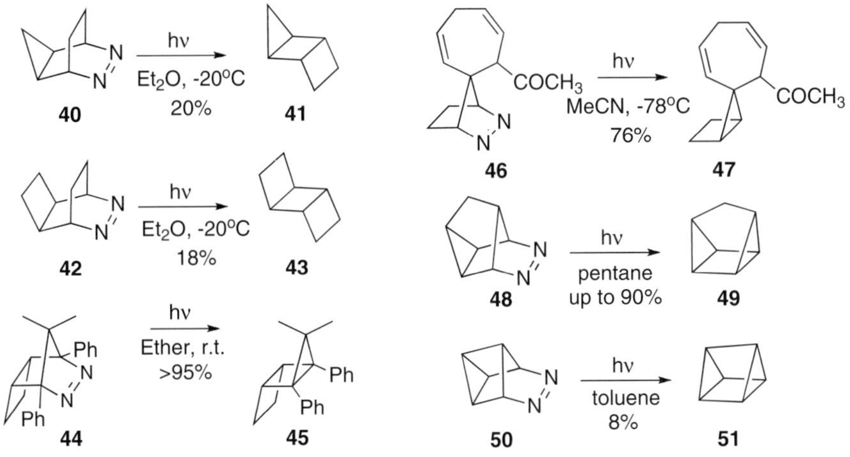

Scheme 2.11

An application of the N$_2$ photoelimination reaction was recently reported in the total synthesis of (±)-pentacycloanemmoxic acid (Scheme 2.12) [36]. It is interesting that all four-membered rings in the ladderane structure were obtained by photochemical reactions. The key diazene intermediate **52** was prepared with a protocol analogous to that illustrated in Scheme 2.10, and led to the formation of **53** upon irradiation in acetic acid and water. It is worth mentioning that rings 3 and 4 were obtained by a 2π + 2π cycloaddition involving cyclopentenone, while rings 1 and 2 were formed by the elimination of N$_2$ from the diazabicyclo[2.2.2]octane **52**, and cyclobutane ring 5 was formed by a photoinduced Wolff rearrangement of a diazoketone. As the reaction of **52** proceeded in a very low yield (6%), a subsequent enantioselective synthesis was based on a more efficient strategy based on iterative 2π + 2π cycloadditions and Wolf rearrangements [37].

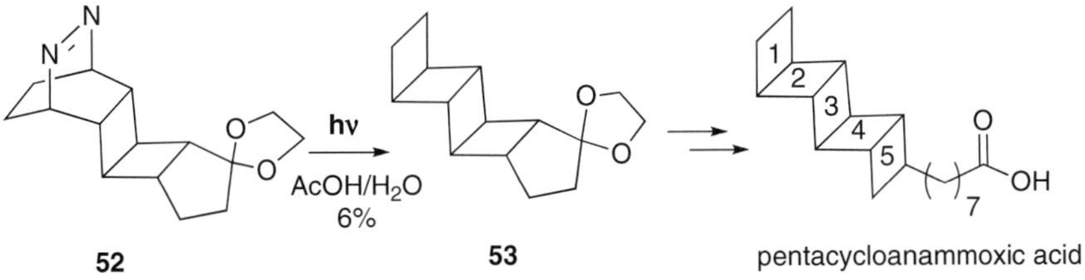

Scheme 2.12

2.2.3
Photoelimination of CO from Ketones in Solution

The two steps involved in the photoinduced decarbonylation of ketones in solution are among the most studied elementary processes in excited state and radical chemistry. and many reviews have dealt with various aspects of the

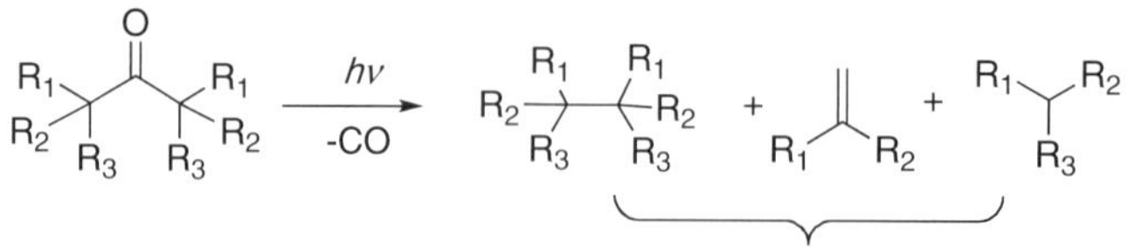

Scheme 2.13

Diethyl ketone	2%
Diisopropy ketone	51%
Ditertbutyl ketone	>99%

reaction [38, 39]. It is well known that the α-cleavage step, or Norrish type-I reaction, takes places from both singlet and triplet n,π* excited states to give an alkyl–acyl radical pair of the corresponding multiplicity. As aliphatic ketones have a relatively slow rate of intersystem crossing from the singlet excited state to the triplet state, α-cleavage often occurs from both singlet and triplet manifolds. As expected for dissociative processes, the rate and efficiency of the excited state reaction and the rate of decarbonylation from the acyl radical depend on the bond dissociation energy of α-bonds. This is nicely illustrated by examples in Scheme 2.13 [40]. An increase in the chemical yield of photodecarbonylation products occurs with the number of α-methyl substituents as the reaction proceeds through primary (diethyl ketone), secondary (isopropyl ketone), and tertiary (*tert*-butyl ketone) radicals. It is interesting to point out that the reaction of *tert*-butyl ketone occurs with similar efficiency from the singlet (43%) and triplet states (57%).

Synthetic applications of the reaction are somewhat limited as the highly reactive biradicals and radical pairs tend to undergo reactions that compete with C–C bond formation. As in previous cases, the reaction may have synthetic value for the synthesis of strained structures involving small rings. For example, the preparation of the simplest [2]-ladderane **55** by photodecarbonylation of bicyclo[3.2.0]heptan-3-one **54** gave the bicyclic structure in 5% yield with a ring-opened 1,5-heptadiene being the dominant product (Scheme 2.14) [41].

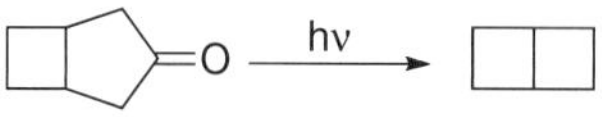

Scheme 2.14

Studies by Lee-Ruff and coworkers indicate that the photodecarbonylation of substituted cyclobutanones in solution has some synthetic potential in the preparation of three-membered rings. As indicated in the box of Scheme 2.15, the reaction starts by formation of a 1,4-acyl-alkyl biradical, which may undergo: (i) the desired decarbonylation to give the 1,3-dialkyl analogue; (ii) a cycloelimination reaction to give the corresponding alkene and ketene components; or (iii) form a ring-expanded oxacarbene intermediate (the latter is typically observed in polar protic media) [42]. In summary, 2- and 3-alkyl-substituted cyclobutanones provide cyclopropanes in low to moderate yields, with diastereoselectivities that are probably determined by

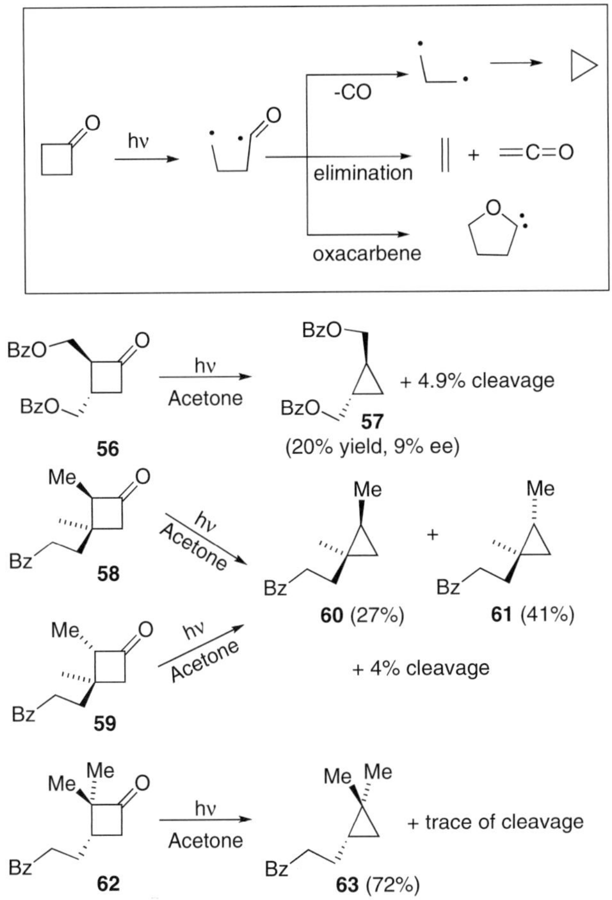

2.2 Photochemical C–C Bond Formation in Solution | 37

Scheme 2.15

substituents in the configurationally uncompromised 3-position. Among several examples studied by Lee-Ruff and coworkers, it was shown that the optically pure chiral *trans*-2,3-disubstituted cyclobutanone **56** reacts under triplet-sensitized irradiation in acetone to yield cyclopropane **57** in a significantly diastereospecific manner, with a 79% enantiomeric excess (ee), but in a relative low yield (20%) [42]. While the authors preferred a mechanism that compromises the configuration of C2 at the stage of the acyl–alkyl biradical, it is also possible that the observed enantioselectivity is the result of an equilibrium to give the more stable *trans*-product determined by the substituent at C3. The latter suggestion is supported by results with epimeric cyclobutanones **58** and **59**, which give an identical product mixture of cyclopropanes **60** and **61** [43]. The best results in these studies were obtained with the optically pure α,α-dimethyl cyclobutanone **62**, which gave exclusively cyclopropane **63** in 72% yield. It is worth noting that direct irradiation in these examples was reported to yield primarily the cycloelimination products, which suggested that the

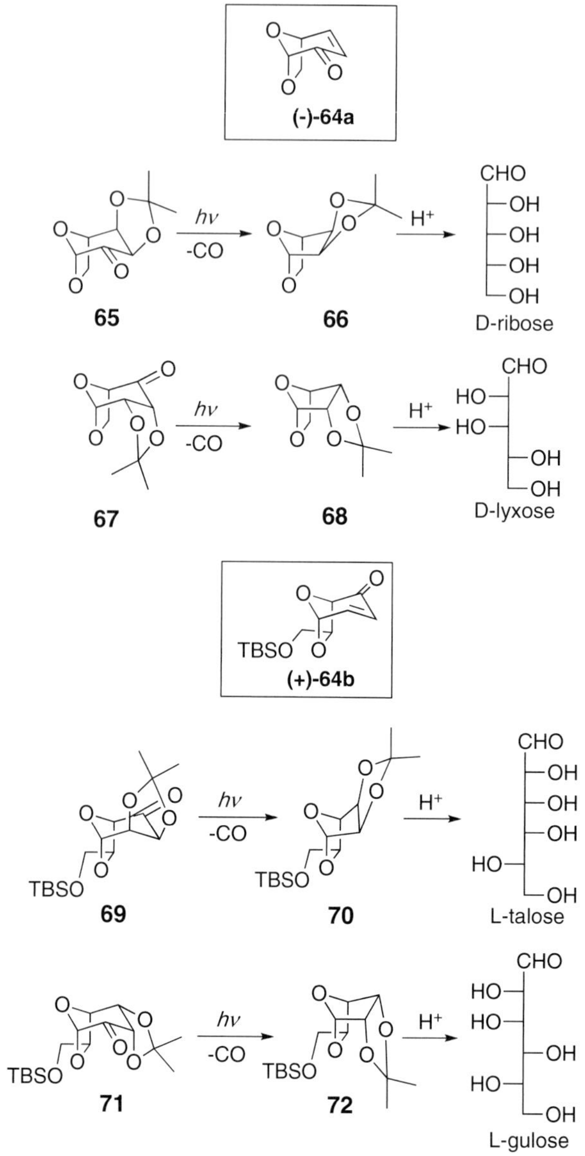

Scheme 2.16

longer-lived triplet acyl-alkyl biradical provides an opportunity for decarbonylation to take place. The authors also showed the α,α-dichlorocyclobutanones to yield exclusively the elimination products [43].

An interesting set of examples illustrating the use of the reaction with substrates possessing limited options to compromise the configuration of the ketone α-centers involves the photochemistry of ketones **65–71** (Scheme 2.16) [44]. Compounds **65** and **67** were prepared from the optically active α,β-unsaturated (−)-**64a** and compounds **69** and **71** from analogue (+)-**64b**. All four ketones were shown to photodecarbonylate to give their corresponding products by enantiospecific radical–radical combination reactions in reasonable to low (41% to 11%) chemical yields. The utility of this reaction for a novel strategy for carbohydrate synthesis was nicely demonstrated with reactions leading to the synthesis of D-ribose and D-lyxose, as well as L-talose and L-gulose.

2.2.4
Photoelimination of CO_2 from Lactones

The photodecarboxylation of lactones has not been as widely used as the photodecarbonylation of ketones [45]. In an interesting application to the synthesis of epimaalienone **76** [46], Green et al. took advantage of an observation regarding the direct photodecarboxylation of dihydro santonin (Scheme 2.17, **73** R = H) to the cyclopropane derivative **75** in good yield by exposure to sunlight after about 10 days [47, 48]. Epimaalienone was later used for the synthesis of α-cyperone and β-cyperone, which differ only by the position of the B-ring double bond. In a subsequent example by Murai et al., epimaalienone **76** was used for the synthesis of Aubergenone, a stress metabolite found in diseased eggplants [48].

While the reaction of **73** is stereoselective, it has been shown that the photodecarboxylation of simple γ-butyrolactones gives product selectivities that depend on the nature of the substrate. From the phenyl-substituted lactones **77–80**, only **79** gives rise to phenyl cyclopropane in low yields (Scheme 2.18) [49]. Diphenyl-substituted lactones cis-**80** and trans-**80** give identical mixtures of cis- and trans-1,2-diphenylcyclopropane in 50–60% yield. Cyano-lactone **81** was the most efficient of the set, giving cis- and trans-1-cyano-2-phenylcyclopropanes in 44% and 48% yields. The simple trend that surfaces from this small number of examples (i.e., **73** and **77–81**) is that radical-stabilizing substituents alpha to the lactone carbonyl may be important for the reaction to take place.

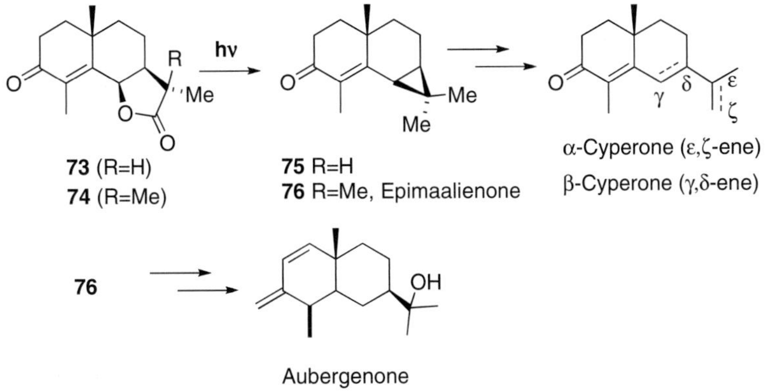

73 (R=H)
74 (R=Me)

75 R=H
76 R=Me, Epimaalienone

α-Cyperone (ε,ζ-ene)
β-Cyperone (γ,δ-ene)

76 → Aubergenone

Scheme 2.17

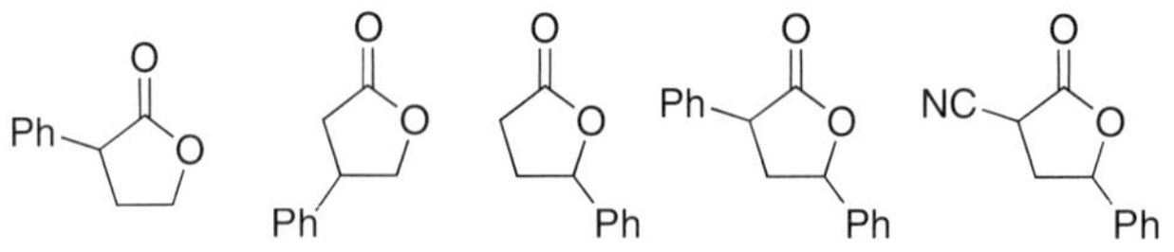

77 **78** **79** **80** (cis and trans) **81**

Scheme 2.18

2.2.5
Photoelimination of Sulfur from Sulfides, Sulfoxides, and Sulfones

The formation of C—C bonds by the photochemical extrusion of sulfur from cyclic and bridged disulfides was originally suggested by Corey and Block [50]. As illustrated in Scheme 2.19 with 1,5-cyclooctadiene **82**, the formation of bridged disulfides includes the reaction of sulfur dichloride with a nonconjugated diene to form the bridged dihalosulfide **83**, followed by reductive elimination of the halogen substituents to form the bridged sulfide **84** [51]. The reaction is thought to proceed by homolysis of an α-bond in **84**, followed by removal of the intermediate sulfur radical by a trivalent phosphorous species such as trialkylphosphites, which are used as solvents, with the reaction culminating with the radical-radical C—C bond formation to yield **85**. While yields in the original study were relatively low, they correlated with the relative stability of the intermediate biradical or radical pair. It is, therefore, not surprising that the reaction has been most successfully used for the synthesis of several cyclophanes where the intermediate biradicals are benzylic. As illustrated in Scheme 2.20, the photoelimination of S to form cylcophanes has been remarkably robust over a wide range of structural variations.

The photochemical elimination of SO from sulfoxides was recently reviewed and analyzed by Jenks et al. [52]. In general, the photoelimination of SO is rare, perhaps associated to the fact that SO, like O_2, has a ground-state triplet. A few known examples involving formal chelotropic reactions resulted in the formation of conjugated systems rather than C—C bond formation. While α-cleavage is a relatively common reaction, the cleavage of sulfinyl radicals RSO• to form SO is highly endothermic [53]. The most common process after α-cleavage is bond formation at the oxygen atom to form the corresponding sulfenates, RS-OR′. In contrast to sulfoxides, there are several examples of photochemical SO_2 elimination from sulfones, suggesting that further analysis may be warranted. The chemical efficiency of the photochemical reaction varies from excellent to low. As illustrated in

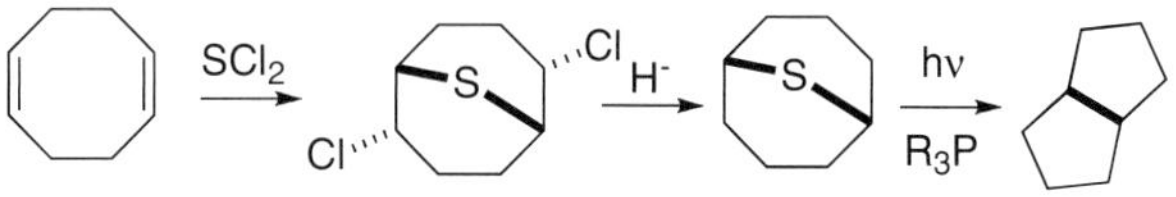

82 → SCl₂ → **83** → H⁻ → **84** → hv, R₃P → **85** (ca. 20%)

Scheme 2.19

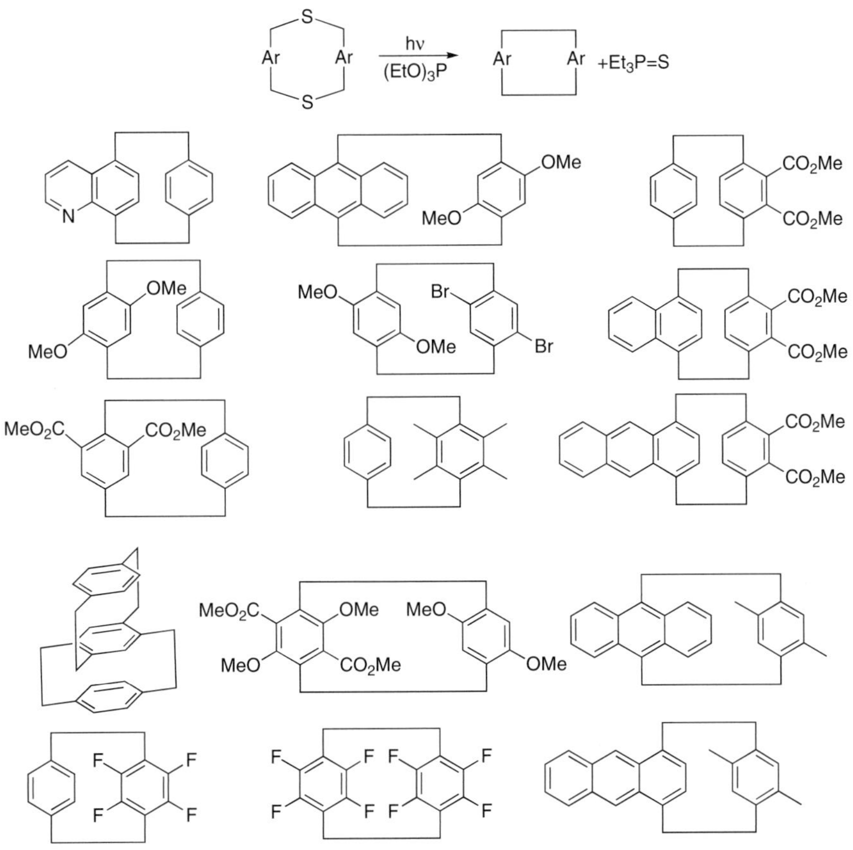

Scheme 2.20

Scheme 2.21, simple cyclic sulfones such as **86** [54] and **88** [55] give the photoelimination and C—C bond-formation products **87** and **89**, respectively, in yields that vary from 88% to 95%. In contrast, when the sulfone moiety is part of a lactam (**90**), the formation of β-lactam **91** occurs in only 10% yield.

Another interesting example of sulfone photochemistry pertains to diyne **92**, which was investigated as an early approach for the synthesis of the highly unstable nine-membered ene-diyne ring of neocarzinostatin Chrom A:1 by Wender et al. [56]. The benzophenone-sensitized desulfonation of **92** resulted in the formation of the desired, but highly unstable, compound **93** in 9–15% yield (Scheme 2.22).

2.3
Reactions in the Solid State

Many of the above-described examples highlight the limitations of radical reactions in solution. However, the potential of a highly exothermic and diffusion-controlled C—C

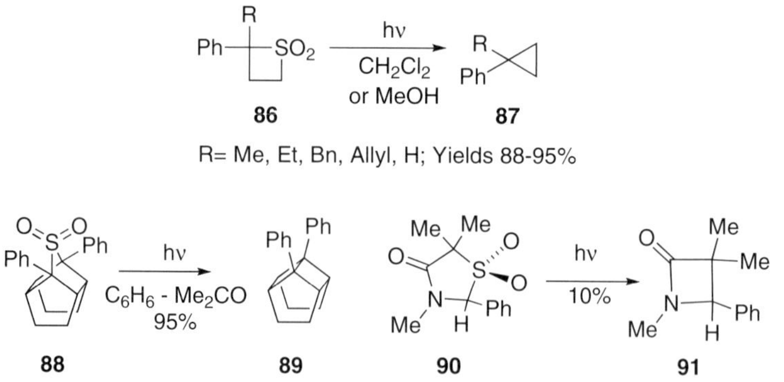

Scheme 2.21

R = Me, Et, Bn, Allyl, H; Yields 88–95%

bond-forming reaction is evident from the number of challenging structures synthesized so far. Given that most radical reactions are limited (or strongly affected) by molecular dynamics in the reaction media, a very simple approach to control their reactivity is to limit their conformational motions, whole-body rotation, and translational diffusion by taking advantage of the solid state.

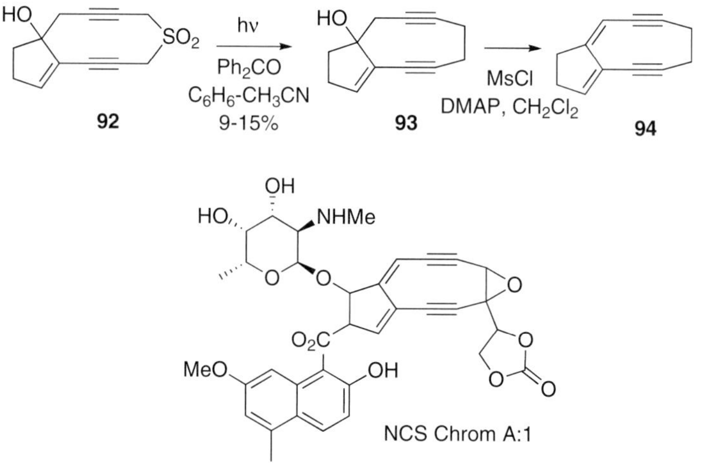

Scheme 2.22

2.3.1
Reactivity and Stability in the Solid State

While photochemical reactions in solids are somewhat counterintuitive and relatively rare in organic synthesis, recent examples have suggested that reactions in crystals may be as reliable and efficient as their solution counterparts [57]. As it pertains to the photoelimination of small molecules and C—C bond-forming reactions, it is essential to recognize that product formation relies on the sequential cleavage of two sigma

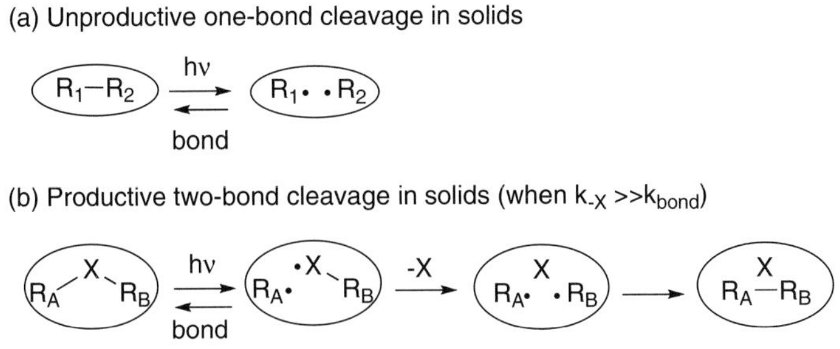

2.3 Reactions in the Solid State

(a) Unproductive one-bond cleavage in solids

$$R_1\text{-}R_2 \underset{\text{bond}}{\overset{h\nu}{\rightleftharpoons}} R_1\cdot\cdot R_2$$

(b) Productive two-bond cleavage in solids (when $k_{\text{-}X} \gg k_{\text{bond}}$)

$$R_A\text{-}X\text{-}R_B \underset{\text{bond}}{\overset{h\nu}{\rightleftharpoons}} R_A\cdot\;\cdot X\text{-}R_B \overset{\text{-}X}{\longrightarrow} R_A\cdot\;\cdot R_B \longrightarrow R_A\text{-}R_B$$

Scheme 2.23

bonds (Scheme 2.23). While cleavage reactions in the solid state are expected to be reversible and unproductive by virtue of the cage effect that prevents the two radicals (R_1 and R_2) from escaping, this analysis will only apply to one-bond cleavage reactions. In the case of two-bond cleavage reactions, reactions in crystals will proceed to the desired secondary radical pair when their structures make the second bond-cleavage (loss of X) faster than recombination to the starting material. Reliable reactions have been demonstrated for many crystalline diacyl peroxides ($X=CO_2$), and aliphatic ketones ($X=CO$).

2.3.2
Restricting the Fate of the Radical Intermediates in Solids

An illustration of the effect of the medium on the dynamics of a radical pair $A^\bullet\,B^\bullet$, generated from an optically pure precursor ($A^*\text{-}X\text{-}B^*$), is summarized in Scheme 2.24. It is assumed in the scheme that the relative orientation of the initial (geminate) radical pair ($A^\bullet\,B^\bullet$) is the same as that in the starting material and that, depending on the fluidity of the environment, the two radicals may rotate and escape from the original solvent cage to become freely diffusing radicals. As shown in the bottom part of Scheme 2.24, the higher the fluidity of the medium, the more the structural

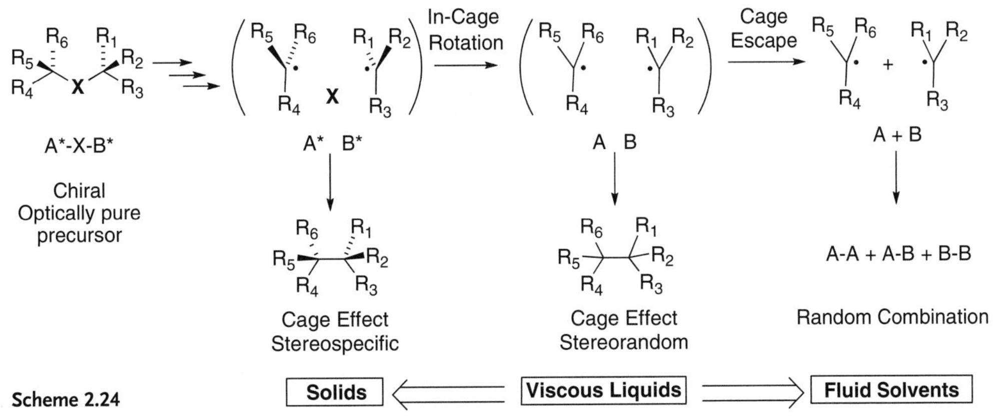

Scheme 2.24

information of the starting material is likely to be lost. It can be seen, from right-to-left, that:

- Free radicals formed in fluid solvents lose the stereochemical and constitutional information of the precursor. Bond-forming reactions occur not only between radicals A$^{\bullet}$ and B$^{\bullet}$, but also between pairs A$^{\bullet}$ A$^{\bullet}$ and B$^{\bullet}$ B$^{\bullet}$, in a statistical manner.

- Radical pairs held together in a viscous solvent cage may be capable of rotating and exposing their two enantiotropic faces, losing their configuration to give a mixture of all possible stereoisomers of the pair A$^{\bullet}$ B$^{\bullet}$.

- In contrast, radical pairs generated in highly rigid media are unable to diffuse apart or rotate, such that the bond-forming reaction may be able to preserve the stereochemistry of the starting material in the structure of the final product.

As illustrated in the scheme, the conditions leading to random radical combinations, a nonstereospecific cage effect with stereorandom bond formation, and a stereospecific cage effect with retention of configuration, can be observed in fluid solutions, in viscous liquids, and in the solid state.

It should be clear from the relationship of molecular dynamics and viscosity, depicted in Scheme 2.23, that sequential photoelimination and C—C bond formation will be synthetically more valuable when carried out in the solid state. An advantage of using pure solid precursors comes from the green chemical perspective of reactions without solvents, and from the absence of competing reagents and adventitious radical traps, such as molecular oxygen. While photochemical reaction in pure crystalline compounds and amorphous solids have not been exploited as much as reactions in solution, the examples in this section unveil a very simple and promising approach to the construction of challenging C—C bonds.

2.3.3
Crystalline Diacyl Peroxides

The first set of C—C bond-forming reactions to be discussed in this section involves the preparation of the seemingly simple but remarkably challenging CH$_2$—CH$_2$— bonds of nonfunctionalized, long-chain alkanes. The reaction involves the double

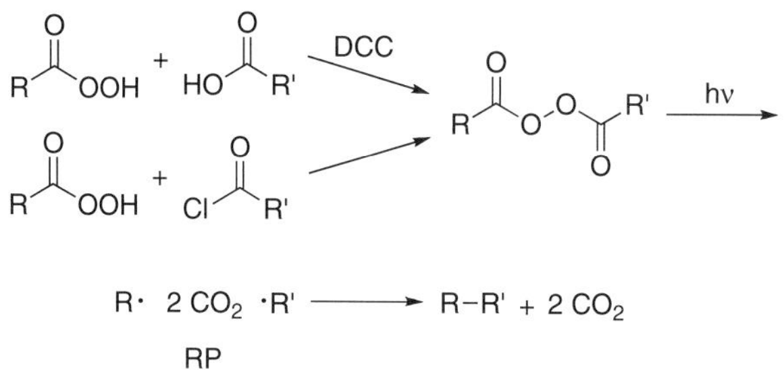

Scheme 2.25

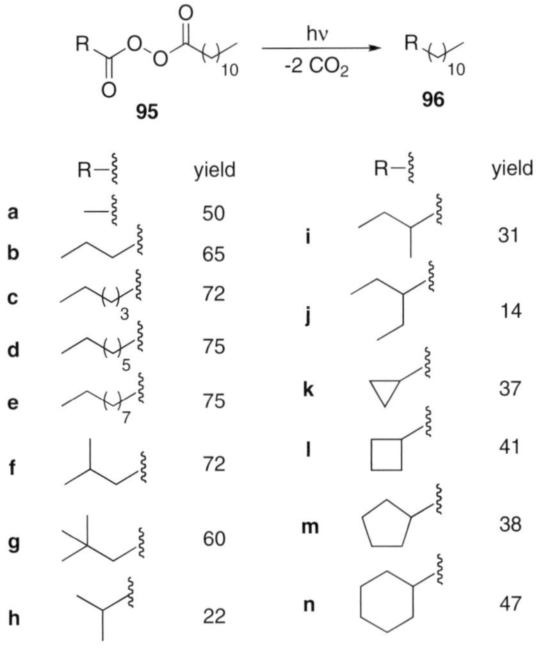

Scheme 2.26

R–C(O)–O–O–C(O)–R′ (95) where R′ = structure 10

$\xrightarrow{h\nu,\ -2\,CO_2}$ R–R′ (96)

	R–	yield		R–	yield
a	–CH₃	50	i	isopropyl	31
b	ethyl	65	j	sec-butyl	14
c	–(CH₂)₃–	72	k	cyclopropyl	37
d	–(CH₂)₅–	75	l	cyclobutyl	41
e	–(CH₂)₇–	75	m	cyclopentyl	38
f	isobutyl	72	n	cyclohexyl	47
g	neopentyl	60			
h	tert-butyl	22			

decarboxylation of diacyl peroxides to generate the intermediate radical pair (RP). Diacyl peroxides are efficiently synthesized by the N,N'-dicyclohexylcarbodiimide (DCC) coupling of peracids with carboxylic acids or acyl chlorides (Scheme 2.25). The photolysis of the resulting diacyl peroxide at 254 nm gives rise to the extrusion of two carbon dioxide molecules and the formation of an alkane.

In one of the earliest and most thorough studies of this reaction, Feldhues and Shäfer analyzed a wide range of diacyl peroxides, covering more than two dozen compounds. Representative examples are illustrated in Schemes 2.26 to 2.29 [58, 59]. Most reactions were carried at low temperatures with neat solid or oily compounds until the starting material had been completely consumed (up to 5 days). The product yields in the schemes represent the reaction selectivity, and are an indication of alternative reaction pathways, such as radical disproportionation and others. It was shown that the photolytic decarboxylation reaction tolerates a wide

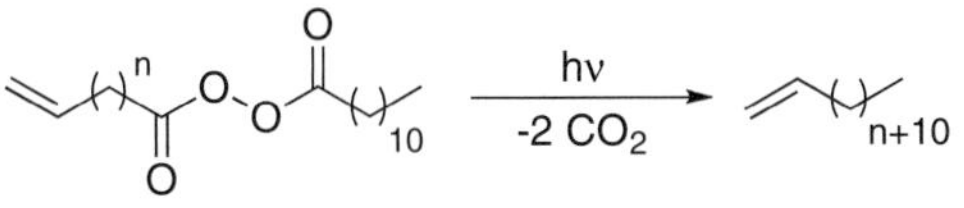

Scheme 2.27

CH₂=CH–(CH₂)ₙ–C(O)–O–O–C(O)–R′ $\xrightarrow{h\nu,\ -2\,CO_2}$ CH₂=CH–(CH₂)_{n+10}

	n	product	yield
97	2	100	56
98	4	101	48
99	8	102	64

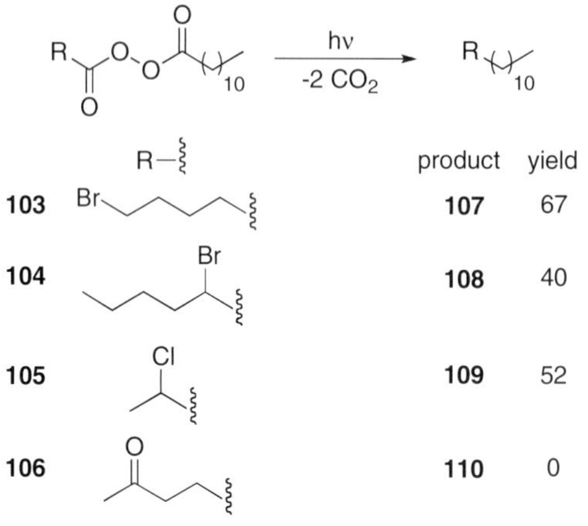

Scheme 2.28

	R–	product	yield
103	Br-(CH₂)₄–	107	67
104	CH₃CH₂CH₂CH(Br)CH₂–	108	40
105	(CH₃)₂CH(Cl)– (2-chloropropyl)	109	52
106	CH₃C(O)CH₂CH₂–	110	0

range of substitution. For unbranched alkyl groups, the longer chain peroxides led to higher yields (**95a–e**), perhaps reflecting an increase in the melting point in the case of crystalline materials and the glass transition point in the case of amorphous solids. The β-branched peroxides (**f, g**) gave significantly higher yields than the α-branched peroxides (**h–j**). The latter gave a greater amount of disproportionation products, as expected from the stability of the corresponding alkenes. While cycloalkyl peroxides (**k–n**) also decarboxylated upon UV irradiation, the corresponding C–C products were obtained in relatively low yields (37–47%). Diacyl peroxides with alkenyl (Scheme 2.27) and halogen (Scheme 2.28) groups also gave decarboxylation products in moderate yields. Alkenyl diacyl peroxide **98** is of particular interest, as photolysis in a low-viscosity solvent (such as pentane) [60] leads to more than 50% cyclization of the hexenyl radical intermediate, while only minimal (1%) cyclization takes place upon photolysis in the solid state. One functional group that gave none of the C–C coupling product was the keto derivative **106**, which underwent several competing reactions. The efficiency of this method in the coupling of alkyl groups was used to elongate the alkyl chain of dodecanoic acid by two or three carbons (Scheme 2.29). Lomolder and Schafer also demonstrated the importance of a low temperature to restrict molecular motion of the radical intermediate, as well as the retention of stereocenters (Scheme 2.30) [61]. The photolysis of **115** and **116** led to the corresponding decarboxylated product at −60 °C with 89–95.6% and 95.6% diastereomeric excess (de), respectively. When the temperature of the reaction was

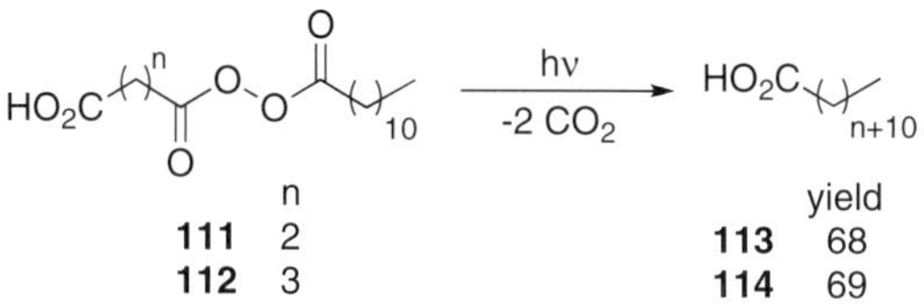

	n		yield
111	2	113	68
112	3	114	69

Scheme 2.29

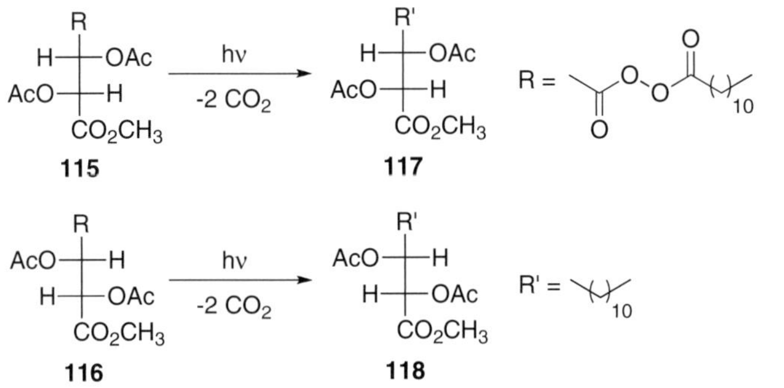

2.3 Reactions in the Solid State | 47

Scheme 2.30

	T (°C)	yield	% de
115	-60	44-64	89-95.6
115	-20	25	35.7
116	-60	48	95.6
116	5	17	3.4

raised, the diastereoselectivity decreased to 35.7% at −20 °C for **115**, and to 3.4% at 5 °C for **116**.

The decarboxylation of diacyl peroxides was also applied in the synthesis of cyclopentadecanone, a compound with a musky odor of interest in the perfume industry (Scheme 2.31) [62]. The photolysis of tetraacyl diperoxides **119** led to the extrusion of four molecules of CO_2 to give **121** in 73% yield. When decarboxylation occurred thermally, the product was obtained in only 41% yield. While the diacyl peroxide containing a keto group (**106**) was shown to give side products, the ketal-protected diperoxide **120** decarboxylated efficiently to give **122** in 65% yield.

Included in Scheme 2.32 are a series of bis(amino acids) prepared by Spantulescu et al. [63] by the decarboxylation of diacyl peroxides derived from combinations of protected aspartic and glutamic acids. In all cases, the reactions gave yields ranging from 52% to 66%. To prevent the formation of side products, the irradiation was performed on neat samples at low temperatures (−78 °C or −96 °C).

Jain and Vederas utilized the reaction to synthesize protected cyclopropyl amino acids **136–140** (Scheme 2.33) [64]. The decarboxylation of the corresponding cyclopropyl precursors **131–135** gave the products in 52–58% yield. These authors also explored the diastereoselectivities of the reaction in their synthesis of two compounds

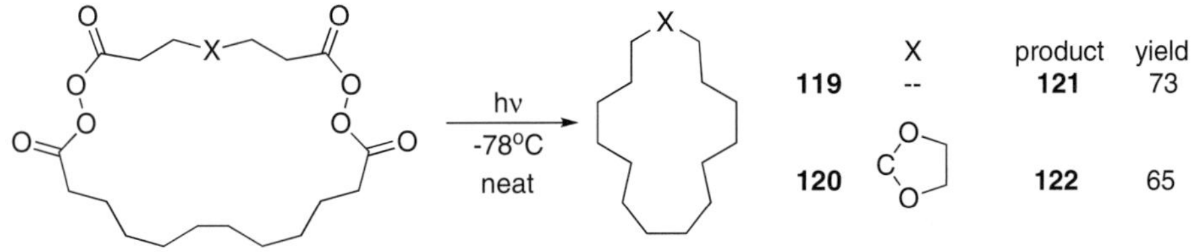

Scheme 2.31

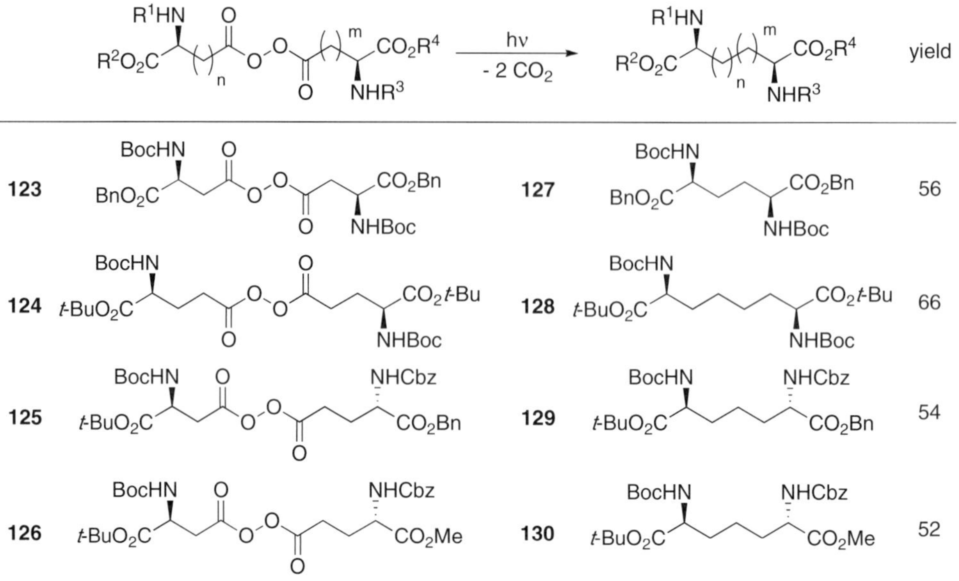

2 Carbon–Carbon Bond Formation by the Photoelimination of Small Molecules

Scheme 2.32

#	Starting material	#	Product	yield
123	BocHN, BnO$_2$C, CO$_2$Bn, NHBoc	127	BocHN, BnO$_2$C, CO$_2$Bn, NHBoc	56
124	BocHN, t-BuO$_2$C, CO$_2$t-Bu, NHBoc	128	BocHN, t-BuO$_2$C, CO$_2$t-Bu, NHBoc	66
125	BocHN, t-BuO$_2$C, NHCbz, CO$_2$Bn	129	BocHN, t-BuO$_2$C, NHCbz, CO$_2$Bn	54
126	BocHN, t-BuO$_2$C, NHCbz, CO$_2$Me	130	BocHN, t-BuO$_2$C, NHCbz, CO$_2$Me	52

related to the natural products belactosin A and hypoglycin A (Scheme 2.34). Belactosin A is a metabolite from *Streptomyces* species that exhibits antitumor activity. The irradiation of precursor **141** gave 47% of **142**, with high diastereoselectivity (≥95 : 5). In addition, 41% of the monodecarboxylated product **143** was also recovered. Diacyl peroxide **144** decarboxylated to give 24% of protected hypoglycin A **145**, also with high diastereoselectivity (dr ≥ 95 : 5). Hypoglycin A was isolated from *Blighia sapida* (Jamaican ackee tree), and causes Jamaican vomiting sickness.

The decarboxylation of diacyl peroxides was also used in the synthesis of a nonapeptide **148** and 5-methyl L-arginine **151** (Scheme 2.35) [65, 66]. The peptide **148** is an analogue of an oxytocin receptor antagonist. The photolysis of precursor **146** gave 34% of product **147**, which was reprotected with Fmoc and then incorporated into the oligopeptide. 5-Methyl arginine **151** was synthesized as a probe to study the nitric oxide synthase (NOS) active site and mechanism. The photolysis of **149** gave **150**

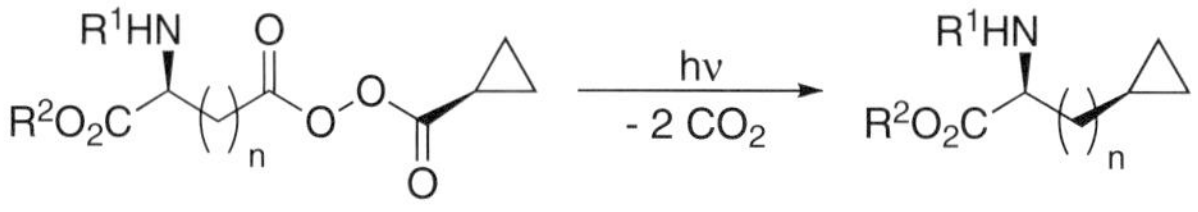

	R^1	R^2	n	product	yield
131	Boc	t-Bu	1	136	58
132	Boc	t-Bu	2	137	58
133	Boc	Bn	2	138	53
134	Cbz	Me	1	139	57
135	Boc	Bn	1	140	52

Scheme 2.33

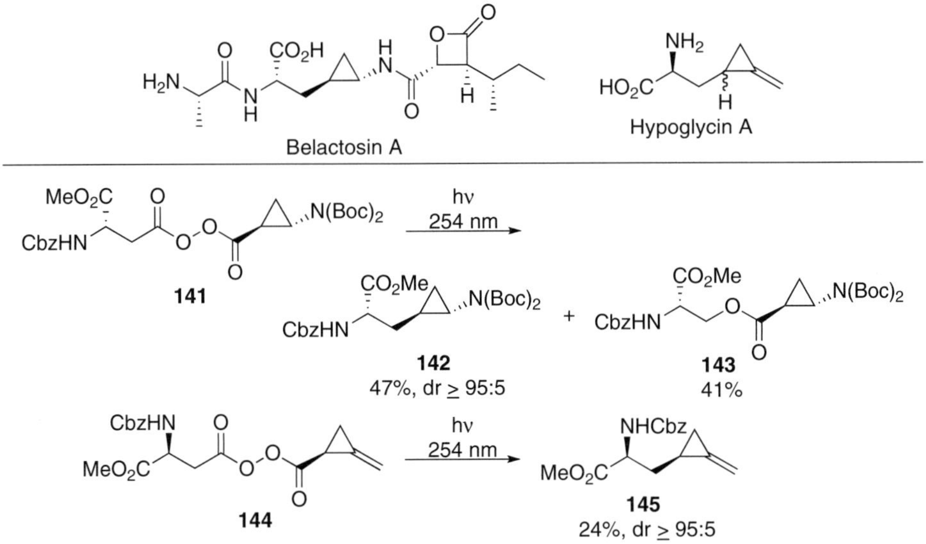

2.3 Reactions in the Solid State

Scheme 2.34

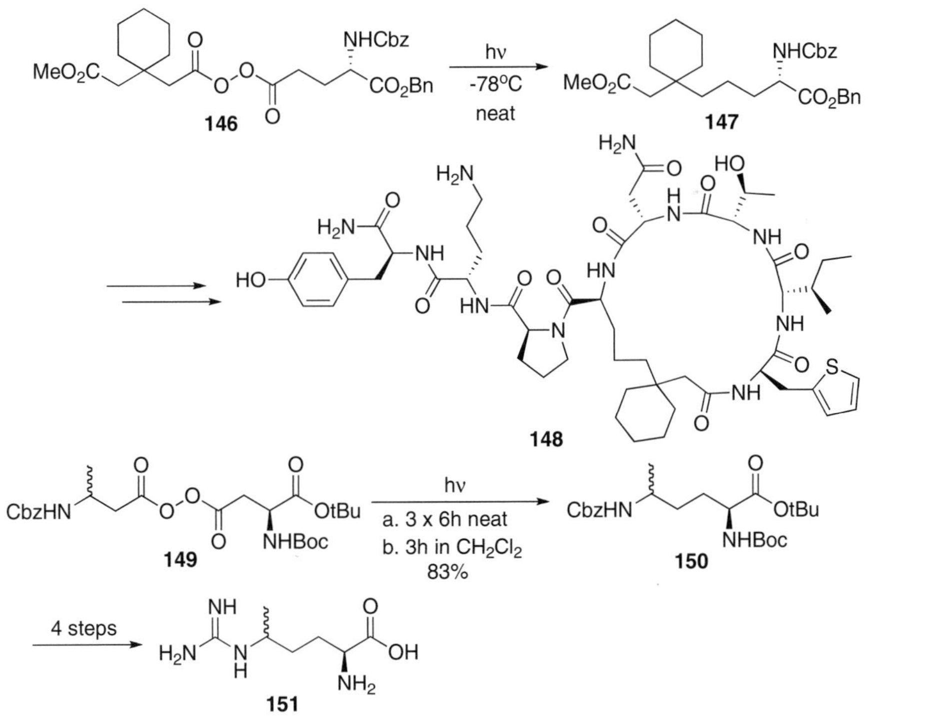

Scheme 2.35

in high yield (83%) after irradiation of the solid sample three times, followed by irradiation in dichloromethane to remove any remaining starting material. The protected ornithine was converted to the methyl arginine in four steps.

2.3.4
Decarbonylation of Crystalline Ketones

With a host of strategies available for stereoselective synthesis and subsequent derivatization, ketones represent an ideal organic functionality to relay the synthesis of other challenging structural motifs. With that in mind, recent studies have addressed the use of hexasubstituted ketones as a general starting point for the synthesis of structures with adjacent stereogenic quaternary centers [85]. As indicated in Scheme 2.36, the suggested strategy begins with the introduction of structural complexity at the two α-carbons; this is followed by crystallization of the resulting ketone, and is completed by a solid-state photodecarbonylation to form the target compound. The premise of the first step relies on well-developed carbonyl chemistry procedures for the diastereo- and enantioselective addition of substituents at the two α-carbons. The crystallization step helps assure the purity of the starting material, and also sets the stage for the solid-state photoreaction, which helps control the reactivity of the radical pair intermediate and assures stereospecificity of the bond-forming reaction.

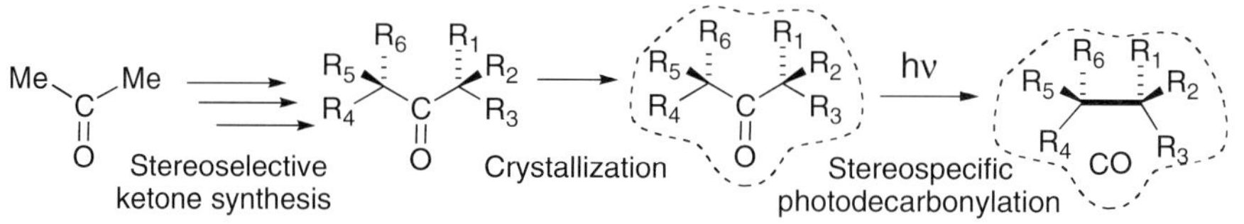

Scheme 2.36

2.3.4.1 Early Observations

The first reports on the photochemical decarbonylation of ketones in the crystalline state were documented during the early 1960s by Quinkert et al., while exploring the effects of viscosity on the recombination of radical pairs. Examples included acyclic structures such 1,1,3-tiphenylpropanone **152** (Scheme 2.37) [67] and cyclic structures such cis- and trans-diphenylindanones **157** and **158** (Scheme 2.38) [68].

The key observation in the case of **152** is that photolysis in benzene conforms to the expected α-cleavage and decarbonylation reactions to form diphenylmethyl (**A$^\bullet$**) and benzyl radicals (**B$^\bullet$**), which are free to diffuse apart. The statistical combination of free radicals **A$^\bullet$** and **B$^\bullet$** gives a 1 : 2 : 1 mixture of products **154**, **155**, and **156**. In contrast, photochemical excitation in the crystalline phase led to the exclusive formation of **153** by combination of the *geminate* radical pair **A$^{\bullet\bullet}$B** with a 100% cage effect.

The experimental observations in the case of 1,3-diphenyl-2-indanones **157** and **158** demonstrate convincingly that the stereochemical integrity of the two radical centers is maintained to high levels when reactions are carried out in the solid state. Crystals

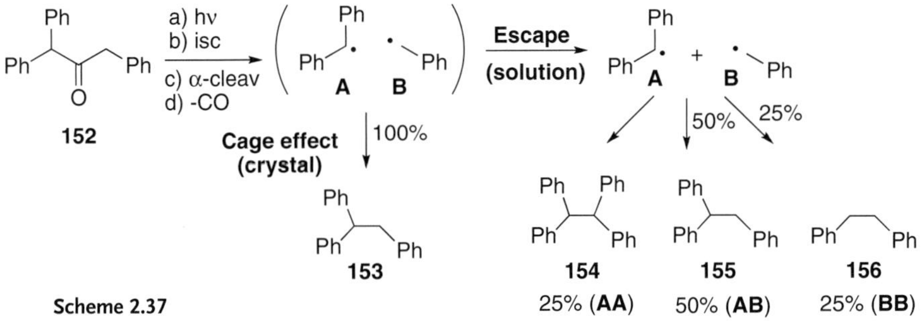

Scheme 2.37

of the *cis*-ketones and *trans*-ketones give respectively the *cis*- and *trans*-products in yields that vary from 86% to 95%. By comparison, reactions carried out in solution (not shown) gave product mixtures that favored the more stable *trans*-diphenylbenzocyclobutane isomers. The photochemical excitation of *cis*-157 and *trans*-157 in solution gave the same product mixture *trans*-159:*cis*-159 = 89:11. Similarly, solutions *cis*-158 and *trans*-158 favored the *trans*-isomer by 70–90%.

2.3.4.2 Reactivity and Stability

While the synthetic potential of the solid-state decarbonylation as a C–C bond-forming reaction was strongly hinted by the examples in Schemes 2.37 and 2.38, common wisdom indicates that most crystalline ketones should be photostable. It was not until the mid-1990s that a systematic effort was launched to address the factors that control the reactivity and stability of crystalline ketones. Having recognized that sequential bond-cleavage reactions were required for the solid-state reaction to occur, Choi et al. postulated that reactivity would require radical-stabilizing substituents in the two α-positions [69]. Their hypothesis was tested by measuring the relative rates of a set of cyclohexanones with increasing substitution

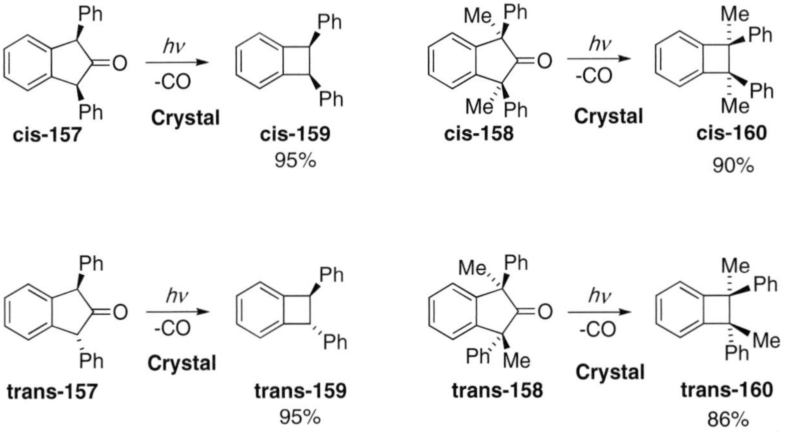

Scheme 2.38

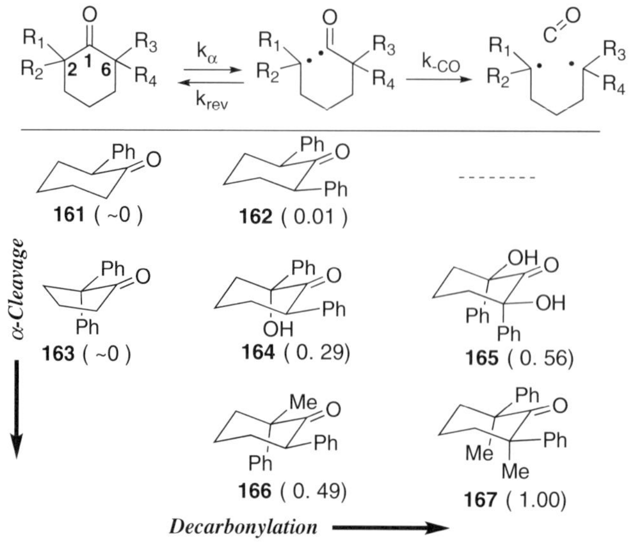

Scheme 2.39

161 (~0), **162** (0.01)

163 (~0), **164** (0.29), **165** (0.56)

166 (0.49), **167** (1.00)

α-Cleavage ↓   Decarbonylation →

at the two α-carbons C1 and C6. It was expected that radical-stabilizing substituents at C2 would increase the efficiency of α-cleavage, while radical-stabilizing substituents at C6 would increase the efficiency of decarbonylation. Their experimental observations, summarized in Scheme 2.39 (with unpublished data from ketone **163** included), confirmed that reactivity in crystals depends on the combined effects of all the α-substituents. Ketones **161** and **163** may undergo α-cleavage, but the decarbonylation is too slow to compete with the reversible formation of the C1–C2 bond. In contrast, the phenyl groups at C2 and C6 in ketone **162** make it possible for the two cleavage reactions to take place, albeit with a relatively low efficiency. It can be seen from the results with ketones **164–167** that additional substituents augment the reaction efficiency. It was pointed out by the authors that reactions of **165** were as efficient in the solid state as in solution, which suggested that the forward reactions in the solid state were faster than any reversible pathway. It was also noted that the reactions in crystals were chemoselective and stereoselective, yielding only the cycloalkane products with the stereochemistry of the starting ketone.

The need for radical-stabilizing substituents at both α-carbons and the generation and reactivity of the 1,4-biradical was further explored with cyclopentanones **168–170**, all of which were shown to give complex product mixtures in solution (Scheme 2.40) [70]. Irradiation experiments carried out over extended periods of time with fine powders of **168** and **169** resulted in no reaction, which suggested that alkyl substituents do not have the radical-stabilizing abilities required to facilitate the decarbonylation reaction. It was suggested that α-cleavage in **169** should be relatively easy, but that decarbonylation does not take place. In particular, the generation of a relatively stable diphenyl methyl radical in **169** lowers the bond dissociation energy of the corresponding α-bond, but the formation of a tertiary radical upon the loss of CO is unfavorable, such that the ketone is regenerated. In contrast, diphenyl substituents

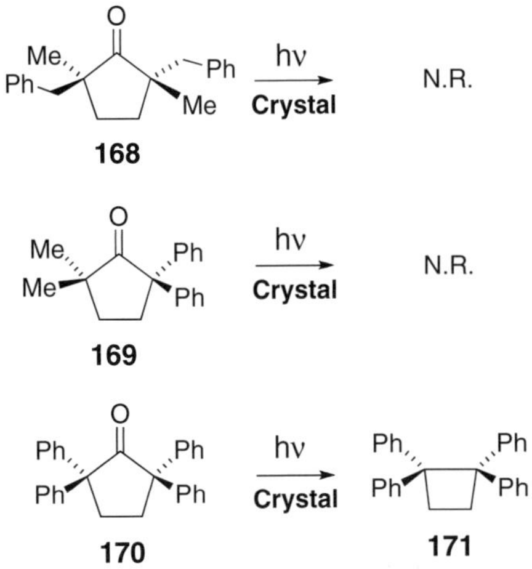

2.3 Reactions in the Solid State

[Structures 168, 169, 170 with hv/Crystal conditions; 168 and 169 give N.R., 170 gives 171 (tetraphenyl-cyclobutane)]

* All ketones are photoreactive in solution

Scheme 2.40

at both α-carbons in the case of **170** make the two-bond cleavage reactions highly efficient to yield tetraphenyl-cyclobutanone **171** as the only product, and in quantitative yield.

2.3.4.3 The RSE > 11 kcal mol^{-1} Condition

With an emerging correlation between the solid-state reactivity and the radical stabilizing energy (RSE) of the α-substituents, Campos et al. suggested that solid-state reactivity may be predictable from the RSE values of the substituents [71]. Taking the reaction of acetone as a reference, and assuming that reactions in crystals must be thermoneutral or exothermic, these authors suggested that substituents with RSE 11 kcal mol^{-1} on both α-carbons should make the reaction thermodynamically possible. As indicated in Scheme 2.41, the proposed RSE value derives from the

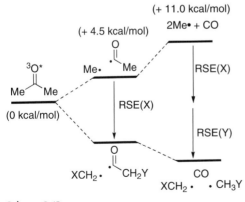

Scheme 2.41

minimum stabilization energy required to make the decarbonylation step of an α-substituted acetyl radical thermoneutral. This suggestion is based on the reasonable assumption that α-cleavage will occur at the weakest α-bond of the triplet ketone (i.e., XCH_2–CO in Scheme 2.41), so that decarbonylation will depend on the RSE of the second substituent (Y).

To test this hypothesis, experimental observations and computational studies were carried out with crystals of diadamantyl-1,3-acetonedicarboxylate with a varying extents of α-methyl substitution, **172a–f**. The corresponding structures are shown in Scheme 2.42, along with a plot that indicates the energetics of the α-cleavage and decarbonylation steps for each compound and acetone, with the assumption that the excited state energy for compounds **172a–f** is the same as that of acetone. The key experimental observation was that dicarboxyketones **172a** and **172b**, lacking at least one methyl group at each α-carbon are photostable, while compounds **172d–f** with one or two methyl groups per α-carbon are reactive in the solid state. All compounds were reactive in solution. It was also shown that the solid-state reaction efficiency increased with the number of methyl groups, with **172f** reacting about twice as fast as **172e**, which had a similar relation with **172d**. Quantum mechanical density functional theory (DFT) calculations indicated that the RSE of an α-ester with no methyl groups was only 8.9 kcal mol^{-1}, which would make the loss of CO endothermic and the reaction for **172a** and **172b** energetically forbidden. In contrast, the calculated RSE values for one α-ester and one or two methyl groups were 11.0 and 19.8 kcal mol^{-1}, respectively, which helped to account for the reactivity trends observed with **172d–f**. Support for the 11 kcal mol^{-1} RSE requirement was also obtained from unpublished observations with di-*tert*-butyl-ketone and hexaalkyl ketone analogues. These compounds are expected to have an RSE of 9.3 kcal mol^{-1} and, as predicted, have been shown to be stable in the solid state.

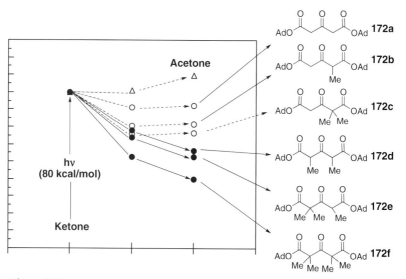

Scheme 2.42

2.3.4.4 Scope of the Reaction

Studies reported to date have suggested that the scope of the solid-state photodecarbonylation reaction for linear ketones, cyclopentanones, cyclohexanones, and cycloheptanones may include a wide range of substituents and structural variations, as long as the substituents have values of RSE $\gg$ 11 kcal mol^{-1} (Schemes 2.37–2.42). It is significant that reaction selectivities and chemical yields in crystals are consistently high when carried out with fine powders or nanocrystals suspended in water. Additional examples of highly efficient reactions in Scheme 2.43 include a series of aryl-substituted dicumyl ketones **173**, all of which react to give dicumenes **174** in excellent yields [72]. Studies with dithiophenyl ketone **175** showed that the reaction proceeds well with heteroaromatic compounds, although the formation of **176** in good yields requires irradiation in nanocrystalline suspensions [73]. Similarly, the irradiation of *trans*-dialkenyl cyclohexanone **177** in the solid state produced the cyclopentane **178** in quantitative yield, confirming the enabling effect of the β,γ-double bonds, the stereospecificity of the reaction and, as in the other examples, the lack of numerous byproducts observed in solution [74]. The solid-state photo-

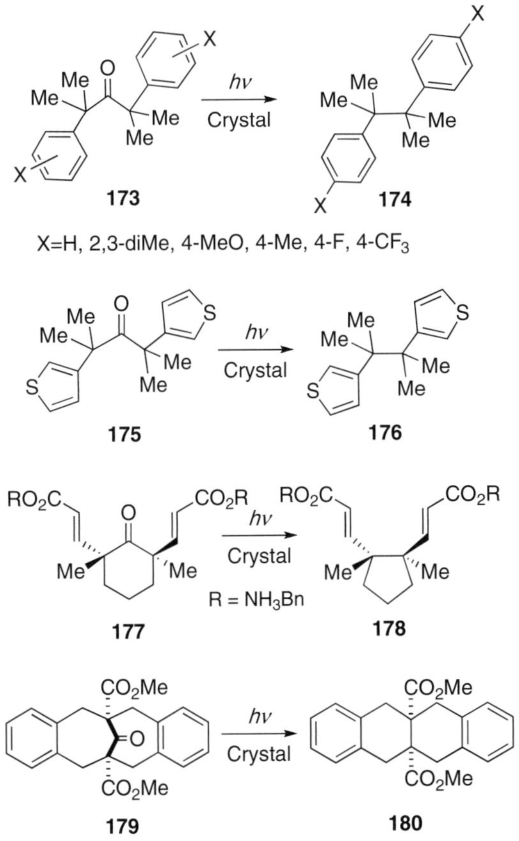

Scheme 2.43

reaction of bicyclic ketodiester **179** is particularly interesting because the C—C bond-formation product **180** is not observed at all in solution, where only disproportionation takes place, but is formed quantitatively in the solid state [75].

The examples in Scheme 2.43 include ketones with π-stabilizing substituents in the two α-positions, such as phenyl (**173**), thiophenyl (**175**), alkenyl (**177**), and ester carbonyls (**179**). Other examples of reaction-enabling substituents that support the generality of the solid-state reaction include the α,α-dialkoxy groups of bisketal **181**, which has an aromatic π-system out of alignment to stabilize the radical intermediates (Scheme 2.44) [76]. The spiro-cyclobutanone **183** [77] and diphenyl cyclopropenone **185** [78] serve to illustrate an alternative strategy to enable the solid-state reaction based on the release of strain energy. Compound **183** reacts by a double decarbonylation process to give bis-cyclohexylidene **184**, and cyclopropenone **185** reacts with an unprecedented quantum chain transformation to generate diphenylacetylene **186**.

2.3.4.5 Reaction Enantiospecificity

Many of the above-described examples highlight the diastereoselectivity of the solid-state reaction. Cyclic ketones with *cis*- or *trans*-α,α′-substituents give the ring-contraction product with the same relative stereochemistry. Examples include 2-indanones **157** and **159** (Scheme 2.38), cyclohexanones **162** and **164–167** in Scheme 2.39, and ketones **177** and **179** in Scheme 2.43. The diastereospecificity and enantiospecificity of linear ketones was also tested with the pentasubstituted acetone derivative **187** in its racemic and enantiomerically pure forms (Scheme 2.45). Notably, both samples reacted efficiently and in a completely diastereospecific manner when reactions were carried out to 60% conversion. Sample of (+)-(2R,4S)-**187** gave the corresponding combination product in 100% ee and

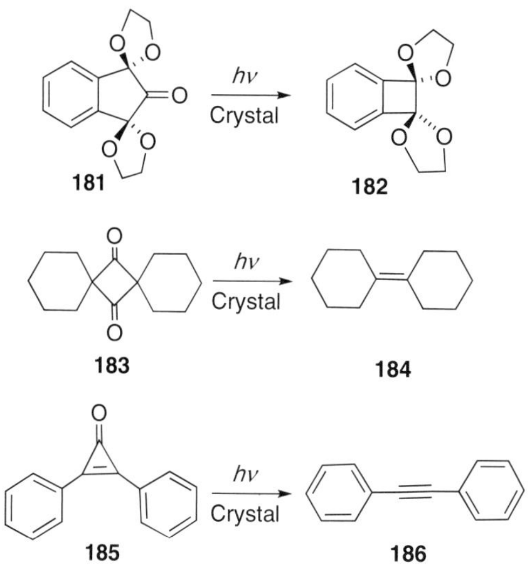

Scheme 2.44

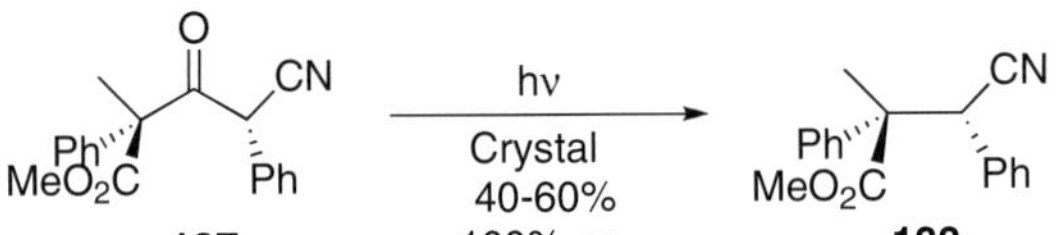

Scheme 2.45

187 → (hv, Crystal, 40-60%, 100% ee) → 188

>95% de, suggesting a small amount of inversion at one carbon center, but not at both [79].

2.3.4.6 Synthesis of Natural Products

The utility of the solid-state photodecarbonylation of crystalline ketones was recently demonstrated in the total syntheses of two natural products, where the key step is the solid-state reaction. The first example involves the synthesis of the sesquiterpene (±)-herbertenolide [80] by the solid-state decarbonylation of cyclohexanone 189, followed by cyclization of the photoproduct 190 (Scheme 2.46). With precursor 189 obtained by simple methods and a solid-state reaction carried out to 76% conversion, herbertenolide was obtained in good overall yield in a record number of steps from commercial starting materials. With a similar synthetic strategy, samples of the natural product (±)-α-cuparenone were obtained in about 60% overall yield from 191 by a very succinct procedure that included four simple steps and a solid-state reaction at −20 °C [81].

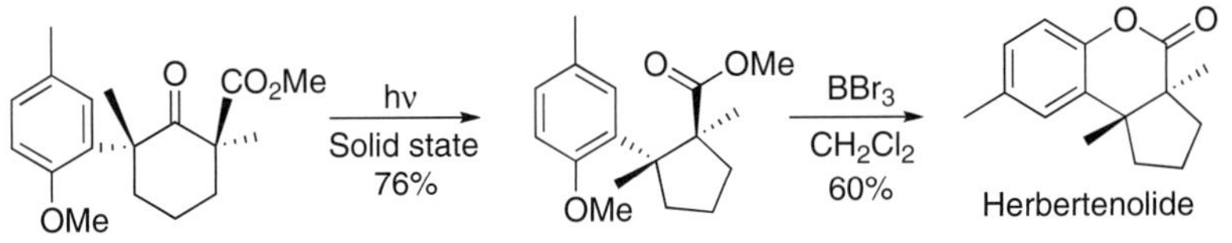

Scheme 2.46

189 → (hv, Solid state, 76%) → 190 → (BBr$_3$, CH$_2$Cl$_2$, 60%) → Herbertenolide

The enantioselective syntheses of (R)-α-cuparenone and (S)-α-cuparenone, both of which are natural products from different sources, were also completed using the solid-state photodecarbonylation of diasteromerically pure difluorodioxaborinane ketones 192 and 194 (Scheme 2.47). The latter were prepared in two steps from 191, and irradiated as nanocrystalline suspensions to optimize the chemical yields of the transformation. The photoreaction of the optically pure ketones was 100% stereoselective with an isolated yield of 80%. The two natural products were obtained by simple acid removal of the chiral auxiliary.

2.3.4.7 Quenching Effects

As is true for all chemical processes, the photodecarbonylation of ketones is limited by structural effects known to result in fast competing pathways, and by structural attributes that give rise to quenching interactions. Whilst one might have predicted, based on the RSE values of the α-substituents, that the three ketones in Scheme 2.48 should have reacted in the solid state by loss of CO and radical combination, the

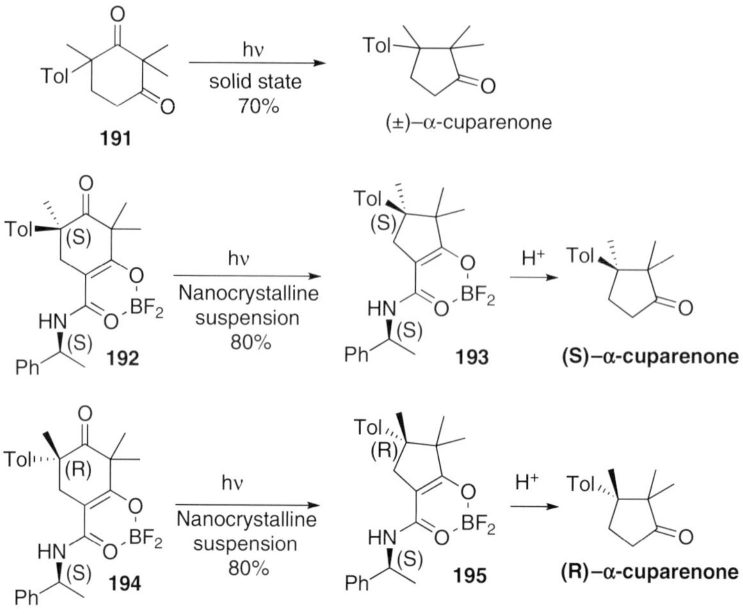

2 Carbon–Carbon Bond Formation by the Photoelimination of Small Molecules

Scheme 2.47

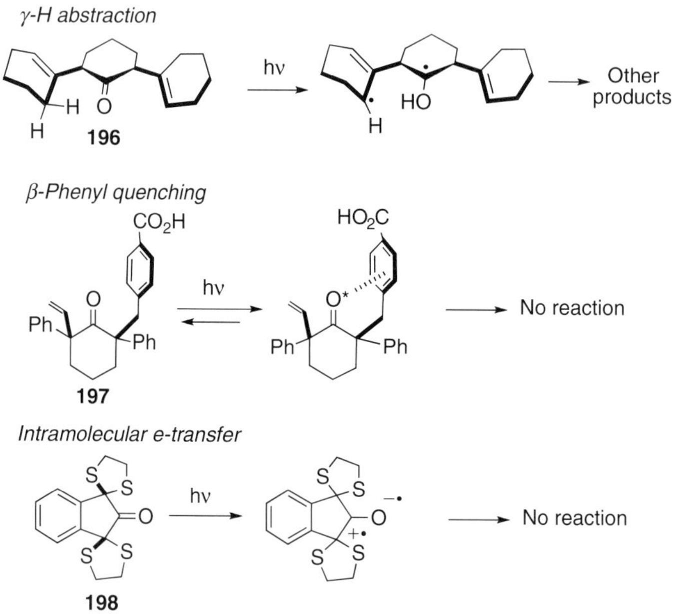

γ-H abstraction

β-Phenyl quenching

Intramolecular e-transfer

Scheme 2.48

experimental results indicate otherwise. Crystalline samples of 2,6-di-cyclohexenyl-cyclohexanone **196** react exclusively by intramolecular γ-hydrogen abstraction, indicating that the latter reaction is much faster than α-cleavage [82]. In fact, single crystal XRD studies confirmed that one allylic γ-hydrogen is ideally positioned for a very fast reaction [82]. Complete solid-state photostability rather than side reactions were observed with samples of ketones **197** and **198** [76, 83]. These represent two examples of ultrafast electron-transfer processes that are capable of quenching the excited state prior to the slower α-cleavage reaction. Ketone **197** is one of several cases of the well-known β-phenyl quenching effect, which depends on conformations that position the face of the phenyl group near the photochemically active carbonyl n-orbital. It is thought that the stability of ketone **198** originates from a very fast intramolecular electron transfer from one of the electron-rich sulfides to the ketone carbonyl. As the efficient deactivation of the excited states of **196** and **197** are conformationally controlled, it may be possible to predict their solid-state stability if the ground-state conformation, which is generally the one used by molecules in crystals, can be reliably determined using quantum mechanical methods.

2.3.4.8 Reaction Scale and Experimental Conditions

Solid-state photoreactions can be performed on very small scales (ca. <1–5 mg) for analytical and exploratory purposes, by exposing to light a fine powder spread between two microscope slides. In order to prevent sample melting it is convenient to immerse the sealed sample in a cold bath while it is exposed to a suitable light source. For the detailed factors affecting the progress of a solid-state reaction, the reader is referred elsewhere [84, 85]. In general terms, photochemical experiments involving crystalline ketones should be carried out with light sources that emit in the 300–350 nm range. Useful sources are Hg arc lamps (Hanovia) with a Pyrex ($\lambda > 290$ nm) filter, reactors with $\lambda = 300$ nm light tubes (Rayonete, Luz Chem, etc.), and even the UV-A and UV-B portion of the spectrum of solar light [74]. The extent of reaction can be monitored using gas chromatography (GC), thin-layer chromatography (TLC), or ^1H nuclear magnetic resonance (NMR). Reaction progress can be monitored also *in situ* by monitoring disappearance of the ketone carbonyl using FTIR spectroscopy. Larger samples can be analyzed also using solid-state ^{13}C cross-polarization magnetic-angle spinning (CPMAS) NMR. There are cases where the reaction stops at low conversions due to impurities that act as filters of the incident light. However, this problem can be circumvented by taking advantage of samples prepared as nanocrystalline suspensions. The latter can be obtained by the so-called "re-precipitation method" [86], which involves dissolution of the starting material in the smallest possible amount of a good, water-miscible solvent such as acetone, acetonitrile, or tetrahydrofuran (THF). The concentrated solution is quickly added via a syringe into an excess volume of water with (or without) a small amount of a surfactant, well below the critical micelle concentration (CMC). The surfactant helps to prevent agglomeration of the nanocrystals, and aids the formation of a stable suspension. The use of nanocrystalline suspensions makes photochemical reactions of crystalline solids as simple as reactions in solution; it also makes it possible to use large-scale reactors, as shown recently by Veerman *et al.* [86]. It should be stressed that, in order for reactions in crystals to proceed with high selectivity and high chemical yields, it is advantageous to use crystalline solids. While reactions in amorphous solids, glasses, and polymeric matrices tend to give a radical-combination with results that imply a large cage effect, they tend to give varying amounts of combination and disproportionation products. At the same time, the stereoselectivity of the combination pathway tends to be much lower than that observed in crystalline solids. It should be also mentioned that the good stereoselectivities observed with crystalline solids depend on the degree or order in the medium, and are largely lost when the crystals begin to melt [81, 82].

2.4
Concluding Remarks

The photolysis of diazoalkane, sulfides, sulfones, esters, or ketones leads to the excited state. The increase in energy accompanying this excitation can be utilized to cleave the α bonds surrounding a small molecule. This leads to an extrusion of the small molecule (nitrogen, sulfur, sulfur dioxide, carbon dioxide, or carbon monoxide), leaving a biradical or radical pair which, upon recombination, will form a new carbon–carbon bond. Although some stereochemical information may be lost during the radical intermediate stage, this can be mitigated by the use of rigid media such as crystals. The utility of the photochemical extrusion of small molecules has been demonstrated by the syntheses of natural products such as aubergenone and α-cuperanone.

References

1 Halevi, E.A. (1992) *Orbital Symmetry and Reaction Mechanism*, Springer-Verlag, Berlin.

2 Michl, J. and Bonacic-Koutecky, V. (1990) *Electronic Aspects of Organic Photochemistry*, Wiley-Interscience, New York.

3 (a) Weir, D. and Scaiano, J.C. (1985) Lifetimes of the biradicals produced in the Norrish type I reaction of cycloalkanones. *Chemical Physics Letters*, **118**, 526–529; (b) Doubleday, C. Jr., Turro, N.J., and Wang, J.-F. (1989) Dynamics of flexible triplet biradicals. *Accounts of Chemical Research*, **22**, 199–205; (c) Zimmt, M.B., Doubleday, C., Gould, I.R., and Turro, N.J. (1985) Nanosecond flash photolysis studies of intersystem crossing rate constants in biradicals: structural effects brought about by spin-orbit coupling. *Journal of the American Chemical Society*, **107**, 6724–6726; (d) Tsentalovich, Y.P., Morozova, O.B., Avdievich, N.I., Ananchenko, G.S., Yurkovskaya, A.V., Ball, J.D., and Forbes, M.D.E. (1997) Influence of molecular structure on the rate of intersystem crossing in flexible biradicals. *Journal of Physical Chemistry A*, **101**, 8809–8816.

4 Garcia-Garibay, M.A. and Campos, L.M. (2004) Photochemical decarbonylation of ketones: recent advances and reactions in crystalline solids, in *CRC Handbook of Organic Photochemistry and Photobiology*, 2nd edn (eds W. Horspool and F. Lenci) CRC Press, Boca Raton, pp. 48/1–48/41.

5 Cram, D.J., Tanner, M.E., and Thomas, R. (1991) The taming of cyclobutadiene. *Angewandte Chemie, International Edition*, **30**, 1024–1027.

6 Chapman, O.L., Mattes, K., McIntosh, C.L., Pacansky, J., Calder, G.V., and Orr, G. (1972) Benzyne. *Journal of the American Chemical Society*, **95**, 6134–6135.

7 Chapman, O.L., McIntosh, C.L., Pacansky, J., Calder, G.V., and Orr, G. (1973) Benzopropiolactone. *Journal of the American Chemical Society*, **95**, 4061–4062.

8 Jones, M.J. and DeCamp, M.R. (1971) Photochemically generated benzyne. *Journal of Organic Chemistry*, **36**, 1536–1539.

9 Fehr, O.C., Grapenthin, O., Kilian, J., and Kirmse, W. (1995) Photochemistry of 3H-indazoles in protic media. Benzyl cations via protonation of 2-methylene-3,5-cyclohexadienylidenes. *Tetrahedron Letters*, **36**, 5887–5890.

10 Warmuth, R. (1997) o-Benzyne: strained alkyne or cumulene? NMR characterization in a molecular container. *Angewandte Chemie, International Edition*, **36**, 1347–1350.

11 (a) Kawano, M., Hirai, K., Tomioka, H., and Ohashi, Y. (2007) Structure determination of triplet diphenylcarbenes by *in situ* X-ray crystallographic analysis. *Journal of the American Chemical Society*, **129**, 2383–2391; (b) Kawano, M., Kobayashi, Y., Ozeki, T., and Fujita, M. (2006) Observation of a coordinatively unsaturated transition-metal complex *in situ* generated within a self-assembled cage. *Journal of the American Chemical Society*, **128**, 6558–6559.

12 (a) McBride, J.M., Vary, M.W., and Whitsel, B.L. (1978) EPR studies of radical pairs. The benzoyloxy dilemma, in *ACS Symposium Series, Organic Free Radicals*, Vol. 69, pp. 208–223; (b) Walter, D.W. and McBride, J.M. (1981) Neophyl rearrangements in crystalline bis(3,3,3-triphenylpropanoyl)peroxide. 2. Structural studies by x-ray and EPF. Mechanistic role of inert molecules. *Journal of the American Chemical Society*, **103**, 7074–7084; (c) Hollingsworth, M.D. and McBride, J.M. (1990) Photochemical mechanism in single crystals: FTIR studies of diacyl peroxides. *Advances in Photochemistry*, **15**, 279–379; (d) Jaffe, A.B., Malament, D.S., Slisz, E.P., and McBride, J.M. (1972) Solvent steric effects. III. Molecular and crystal structures of azobisisobutyronitrile and azobis-3-cyano-3-pentane. Structural deuterium isotope effect. *Journal of the American Chemical Society*, **94**, 8515–8521; (e) Skinner, K.J., Blaskiewicz, R.J., and McBride, J.M. (1972) Solvent steric effects. IV. Solid-state photolysis of azobis-3-phenyl-3-pentane. Crystal-lattice control of the stereochemistry of radical disproportionation. *Israel Journal of Chemistry*, **10**, 457–470; (f) Karch, N.J., Koh, E.T., Whitsel, B.L., and McBride, J.M. (1975) X-ray and electron paramagnetic resonance structural investigation of oxygen discrimination during the collapse of methyl-benzoyloxy radical pairs in crystalline acetyl benzoyl peroxide. *Journal of the American Chemical Society*, **97**, 6729–6743; (g) Vary, M.W. and McBride, J.M. (1979) Single crystal EPR studies of radical pairs in dibenzoyl peroxide. *Molecular Crystals and Liquid Crystals*, **52**, 437–448.

13 Cram, D.J. and Cram, J.M. (1971) Cyclophane chemistry: bent and battered benzene rings. *Accounts of Chemical Research*, **4**, 204–213.

14 Shinmyozu, T., Nogita, R., Akita, M., and Lim, C. (2004) Photochemical routes to cyclophanes involving decarbonylation reactions and related process, in *CRC Handbook of Photochemistry and Photobiology*, 2nd edn (eds W. Horspool and F. Lenci) CRC Press, Boca Raton, pp. 51/1–51/6.

15 Kaplan, M.L. and Truesdale, E.A. (1976) [2.2]Paracyclophane by photoextrusion of carbon dioxide from a cyclic diester. *Tetrahedron Letters*, **17**, 3665–3666.

16 Iwata, M. and Kuzuhara, H. (1985) Functionalized dithia(2,5)pyridinophanes

as vitamin B6 analogues. Synthesis, properties, and catalytic activity for racemization reaction. *Bulletin of the Chemical Society of Japan*, **58**, 2502–2514.

17 Givens, R.S. and Olsen, R.J. (1979) Photoextrusion of sulfur dioxide: general route to [2.2]cyclophanes. *Journal of Organic Chemistry*, **44**, 1608–1613.

18 Shuster, D.I. and Wang, L. (1983) Photochemistry of ketones in solution. 70. Stereospecific photodecarbonylation of cis- and trans-2,7-dimethyl-3,5-cycloheptadienone: observation of an apparent symmetry-forbidden concerted cheletropic fragmentation. *Journal of the American Chemical Society*, **105**, 2900–2901.

19 Mori, T., Weiss, R.G., and Inoue, Y. (2004) Mediation of conformationally controlled photodecarboxylations of chiral and cyclic aryl esters by substrate structure, temperature, pressure, and medium constraints. *Journal of the American Chemical Society*, **126**, 8961–8975.

20 The regioselectivity of addition in reactions involving diazomethane and R-monosusbtituted olefins is controlled by the HOMO of diazomethane giving the 3-pyrazoline product, regardless of the electron-donating or -withdrawing nature of R Please see: Houk, K.N. and Yamaguchi, K. (1984) Theory of 1,3-dipolar cycloadditions, in *1,3-Dipolar Cycloaddition Chemistry*, Vol. 2 (ed. A. Padwa), John Wiley & Sons, New York, pp. 407–450.

21 (a) Wakamiya, T., Oda, Y., Fujita, H., and Shiba, T. (1986) Synthesis and stereochemistry of carnosadine, a new cyclopropyl amino acid from red alga *Grateloupia carnosa*. *Tetrahedron Letters*, **27**, 2143–2144; (b) Oba, M., Nishiyama, N., and Nishiyama, K. (2003) Synthesis and reactions of a novel 3,4-didehydro-pyroglutamate derivative. *Chemical Communications*, 776–777.

22 Jimenez, J.M., Rife, J., and Ortuno, R.M. (1996) Enantioselective total syntheses of cyclopropane amino acids: natural products and protein methanologs. *Tetrahedron: Asymmetry*, **7**, 537–558.

23 Takagi, R., Nakamura, M., Hashizume, M., Kojima, S., and Ohkata, K. (2001) Stereoselective cyclopropanation of 3-aryl-2-phosphonoacrylates induced by the (−)-8-phenylmethyl group as a chiral auxiliary. *Tetrahedron Letters*, **42**, 5891–5895.

24 Saha, M., Bagby, B., and Nicholas, K.M. (1986) Colbalt-mediated propargylation/annelation: total synthesis of (±)-cyclocolorenone. *Tetrahedron Letters*, **27**, 915–918.

25 Wender, P.A., Kee, J.M., and Warrington, J.M. (2008) Practical synthesis of prostratin, DPP, and their analogs, adjuvant leads against latent HIV. *Science*, **320**, 649–652.

26 Nagaoka, H., Miyaoka, H., and Yamada, Y. (1990) Total synthesis of (+)-halimedatrial: the absolute configuration of halimedatrial. *Tetrahedron Letters*, **31**, 1573–1576.

27 Schultz, A.G. and Pulg, S. (1985) The intramolecular diene-carbene cycloaddition equivalence and an enantioselective Birch reduction-alkylation by the chiral auxiliary approach. Total synthesis of (±)- and (−)-longifolene. *Journal of Organic Chemistry*, **50**, 915–916.

28 Ohfune, Y., Demura, T., Iwama, S., Matsuda, H., Namba, K., Shimamoto, K., and Shinada, T. (2003) Stereocontrolled synthesis of a potent agonist of a group II metabotropic glutamate receptors, (+)-LY354740, and its related derivatives. *Tetrahedron Letters*, **44**, 5431–5434.

29 Adam, W. and Sahin, C. (1995) Photochemical decomposition of cyclic azoalkanes, in *CRC Handbook of Organic Photochemistry and Photobiology* (eds W.M. Horspool and P.-S. Song), CRC Press, Boca Raton, pp. 937–953.

30 Tanida, H., Teratake, S., Hata, Y., and Watanabe, M. (1969) Photochemical and thermal decompositions of anti-6,7-diazatricyclo[3.2.2.02,4]non-6-ene evidence for mechanistic difference

between the reactions. *Tetrahedron Letters*, **10**, 5345–5347.
31. Tanida, H., Teratake, S., Hata, Y., and Watanabe, M. (1969) Photochemical decomposition of anti-7,8-diazatricyclo [4.2.2.02,5]dec-7-ene. *Tetrahedron Letters*, **10**, 5341–5343.
32. Paquette, L.A. and Leichter, L.M. (1971) Pyrolysis of anti-tricyclo[3.2.0.02,4] heptanes. The role of 2,4 substitution in the operation of σ bond assisted cyclobutane fragmentations. *Journal of the American Chemical Society*, **93**, 4922–4924.
33. Carroll, G.L., Harrison, R., Gerken, J.B., and Little, R.D. (2003) Coping with substituent effects in divinylcyclopropyl dizaene rearrangements. *Tetrahedron Letters*, **44**, 2109–2112.
34. Turro, N.J., Cherry, W.R., Mirbach, M.F., and Mirbach, M.J. (1977) Energy acquisition, storage, and release. Photochemistry of cyclic azoalkanes as alternate entries to the energy surfaces interconnecting norbornadine and quadricyclene. *Journal of the American Chemical Society*, **99**, 7388–7390.
35. Katz, T.J. and Acton, N. (1973) Synthesis of prismane. *Journal of the American Chemical Society*, **95**, 2738–2739.
36. Mascitti, V. and Corey, E.J. (2004) Total synthesis of (±)-pentacycloanammoxic acid. *Journal of the American Chemical Society*, **126**, 15664–15665.
37. Mascitti, V. and Corey, E.J. (2006) Enantio-selective synthesis of pentacyclo-anammoxic acid. *Journal of the American Chemical Society*, **128**, 3118–3119.
38. Bohne, C. (1995) Norrish type I processes of ketones: selected examples and synthetic applications, in *CRC Handbook of Organic Photochemistry and Photobiology* (eds W.M. Horspool and P.-.S. Song), CRC Press, Boca Raton, pp. 423–429.
39. Weiss, D. (1981) The Norrish type-I reaction in cycloalkene photochemistry, in *Organic Photochemistry*, Vol. 5 (ed. A. Padwa), Marcel Dekker, New York, pp. 347–420.
40. (a) Yang, N.C., Feit, E.D., Hui, M.H., Turro, N.J., and Dalton, J.C. (1970) Photochemistry of di-*tert*-butyl ketone and structural effects on the rate and efficiency of intersystem crossing of aliphatic ketones. *Journal of the American Chemical Society*, **92**, 6974–6976; (b) Abuin, E.B., Encina, M.V., and Lissi, E.A. (1973) The photolysis of 3-pentanone. *Journal of Photochemistry*, **1**, 387–396; (c) Encinas, M.V., Lissi, E.A., and Scaiano, J.C. (1980) Photochemistry of aliphatic ketones in polar solvents. *Journal of Physical Chemistry*, **84**, 948–951.
41. (a) Nouri, D.H. and Tantillo, D.J. (2006) They came from the deep: syntheses, applications, and biology of ladderanes. *Current Organic Chemistry*, **10**, 2055–2074; (b) Cremer, S. and Srinivasan, R. (1960) A photochemical synthesis of [2.2.0]-hexane. *Tetrahedron Letters*, **1** (42), 24–27.
42. Ramnauth, J. and Lee-Ruff, E. (1997) Photodecarbonylation of chiral cyclobutanones. *Canadian Journal of Chemistry*, **75**, 518–522.
43. Ramnauth, J. and Lee-Ruff, E. (2001) Photochemical preparation of cyclopropanes from cyclobutanones. *Canadian Journal of Chemistry*, **79**, 114–120.
44. Kadota, K. and Ogasawara, K. (2001) A carbohydrate synthesis employing photochemical decarbonylation. *Tetrahedron Letters*, **42**, 8661–8664.
45. Budac, D. and Wan, P. (1992) Photodecarboxylation: mechanism and synthetic utility. *Journal of Photochemistry and Photobiology A*, **67**, 135–166.
46. Green, A.E., Muller, J.-C., and Ourisson, G. (1971) A new photochemical approach to dimethylcyclopropanes. Synthesis of epimaalienone and conversion to α- and β-cyperones. *Tetrahedron Letters*, **12**, 4147–4149.
47. Perol, G.W. and Ourisson, G. (1969) Photolyse des dihydro-1,2-(6β,11α et 11β)-santonines. Une photosensibilisation intramoleculaire stereospecifique. *Tetrahedron Letters*, **10**, 3871–3873.

48 Murai, A., Abiko, A., Ono, M., and Masamune, T. (1982) Synthesis of aubergenone, a sesquiterpenoid phytoalexin from diseased eggplants. *Bulletin of the Chemical Society of Japan*, **55**, 1191–1194.

49 Givens, R.S. and Oettle, W.F. (1972) Mechanistic studies in organic photochemistry. VI. Photodecarboxylation of benzyl esters. *Journal of Organic Chemistry*, **37**, 4325–4334.

50 Corey, E.J. and Block, E. (1969) Sulfur-bridged carbocycles. II. Extrusion of the sulfur bridge. *Journal of Organic Chemistry*, **34**, 1233–1240.

51 (a) Weil, E.D., Smith, K.J., and Gruber, R.J. (1966) Transannular addition of sulfur dichloride to cyclooctadienes. *Journal of Organic Chemistry*, **31**, 1669–1679; (b) Corey, E.J. and Block, E. (1966) New synthetic approaches to symmetrical sulfur-bridged carbocycles. *Journal of Organic Chemistry*, **31**, 1663–1668.

52 Jenks, W.S., Gregory, D.D., Guo, Y., Lee, W., and Troy, T. (1997) The photochemistry of sulfoxides and related compounds, in *Molecular and Supramolecular Photochemistry* (eds V. Ramamurthy and K.S. Schanze), Marcel Dekker, New York, pp. 1–56.

53 The loss of SO from CH_3SO is estimated to be endothermic by 50 kcal mol^{-1}: (a) Chatgilialoglu, C. (1988) *The Chemistry of Sulfones and Sulfoxides* (eds S. Patai, Z. Rappoport and C.J.M. Stirling), John Wiley & Sons Ltd, New York, pp. 1081–1087; (b) Benson, S.W. (1978) Thermochemistry and kinetics of sulfur-containing molecules and radicals. *Chemical Reviews*, **78**, 23–35.

54 Finlay, J.D., Smith, D.J.H., and Durst, T. (1978) The photolysis of 2-phenylthietane 1,1-dioxides: preparation of substituted cyclopropanes. *Synthesis*, 579–581.

55 Sponholtz, W.R. III, Trujillo, H.A., and Gribble, G.W. (2000) Tandem 1,3-dipolar cycloaddition of mesoionic 2,4-diphenyl-1,3-oxathiolium-5-olate with cycloocta-1,5-diene. A new synthesis of the tetracyclo[6.2.0.04,10.05.9]decane ring system. *Tetrahedron Letters*, **41**, 1687–1690.

56 Wender, P.A., Harmata, M., Jeffrey, D., Mukai, C., and Suffert, J. (1988) Studies on DNA-active agents: the synthesis of the parent carbocyclic subunit of neocarzinostatin chromophore A. *Tetrahedron Letters*, **29**, 909–912.

57 (a) Quinkert, G., Palmowski, J., Lorenza, H.P., Wiersdorff, W.W., and Finke, M. (1971) Non-cheletropic photodecarbonylation of 1,3-diphenyl-substituted 2-indanone derivatives. *Angewandte Chemie, International Edition*, **10**, 199–1199; (b) Quinkert, G., Tabata, T., Hickmann, E.A.J., and Dobrat, W. (1971) Increase in selectivity of photoelimination products on replacing the liquid phase by the crystalline phase. *Angewandte Chemie, International Edition*, **10**, 199–200.

58 Feldhues, M. and Schafer, H.J. (1985) Selective mixed coupling of carboxylic acids (I). Electrolysis, thermolysis and photolysis of unsymmetrical diacyl peroxides with acyclic and cyclic alkyl groups. *Tetrahedron*, **41**, 4195–4212.

59 Feldhues, M. and Schafer, H.J. (1985) Selective mixed coupling of carboxylic acids (II). Photolysis of unsymmetrical diacylperoxides with alkenyl-, halo-, keto-, carboxyl-groups and a chiral α-carbon. Comparison with the mixed Kolbe electrolysis. *Tetrahedron*, **41**, 4213–4235.

60 Sheldon, R.A. and Kochi, J.K. (1970) Pair production and cage reactions of alkyl radicals in solution. *Journal of the American Chemical Society*, **92**, 4395–4404.

61 Lomolder, R. and Schafer, H.J. (1987) Low-temperature photolysis of peracetylated dodecanoyl peroxides of tartaric acid and D-gluconic acid in the solid state- a diastereoselective radical coupling. *Angewandte Chemie, International Edition*, **26**, 1253–1254.

62 Feldhues, M. and Schafer, H.J. (1986) Selective mixed coupling of carboxylic acids- III. Synthesis of cyclopentadecanone from cyclic tetraacyl diperoxides. *Tetrahedron*, **42**, 1285–1290.

63 Spantulescu, M.D., Jain, R.P., Derksen, D.J., and Vederas, J.C. (2003) Photolysis of diacyl peroxides: a radical-based approach for the synthesis of functionalized amino acids. *Organic Letters*, **5**, 2963–2965.

64 Jain, R.P. and Vederas, J.C. (2003) Synthesis of β-cyclopropylalanines by photolysis of diacyl peroxides. *Organic Letters*, **5**, 4669–4672.

65 Stymiest, J.L., Mitchell, B.F., Wong, S., and Vederas, J.C. (2005) Synthesis of oxytocin analogues with replacement of sulfur by carbon gives potent antagonists with increase stability. *Journal of Organic Chemistry*, **70**, 7799–7809.

66 Martin, N.I., Woodward, J.J., Beeson, W.T., and Marletta, M.A. (2007) Design and synthesis of C5 methylated L-arginine analogues as active site probes for nitric oxide synthase. *Journal of the American Chemical Society*, **129**, 12563–12570.

67 Quinkert, G., Opitz, K., Wiersdorff, W.W., and Weinlich, J. (1963) Lichtinduzierte Decarbonylierung gelöster Ketone bei Raumtemperatur. *Tetrahedron Letters*, **4**, 1863–1868.

68 (a) Quinkert, G., Cimbollek, G., and Buhr, G. (1966) Die photolyse konfigurations-isomerer cyclobutanon-derivate. *Tetra-hedron Letters*, **38**, 4573–4578; (b) Quinkert, G., Opitz, K., Wiersdorff, W.W., and Finke, M. (1966) Light-induced reactions. IV. Stereospecific adduct formations by benzocyclobutenes. *Justus Liebigs Annalen der Chemie*, **693**, 44–75; (c) Quinkert, G., Wiersdorff, W.W., Finke, M., and Opitz, K. (1966) Uber das tetraphenyl-o-xylylen. *Tetrahedron Letters*, **7**, 2193–2200.

69 (a) Choi, T., Cizmeciyan, D., Kahn, S.I., and Garcia-Garibay, M.A. (1995) An efficient solid-to-solid reaction via a steady-state phase-separation mechanism. *Journal of the American Chemical Society*, **117**, 12893–12894; (b) Choi, T., Peterfy, K., and Garcia-Garibay, M.A. (1996) Molecular control of solid state reactivity by decarbonylation of crystalline ketones. *Journal of the American Chemical Society*, **118**, 12477–12478.

70 Peterfy, K. and Garcia-Garibay, M.A. (1998) Generation and reactivity of a triplet 1,4-biradical conformationally trapped in a crystalline cyclopentanone. *Journal of the American Chemical Society*, **120**, 4540–4541.

71 Campos, L.M., Dang, H., Ng, D., Yang, Z., Martinez, H.L., and Garcia-Garibay, M.A. (2002) Engineering reactions in crystalline solids: predicting photochemical decarbonylation from calculated thermochemical parameters. *Journal of Organic Chemistry*, **67**, 3749–3754.

72 Resendiz, M. and Garcia-Garibay, M.A. (2005) Hammet analysis of photo-decarbonylation in crystalline 1,3-diarylacetones. *Organic Letters*, **7**, 371–374.

73 Resendiz, M.J.E., Taing, J., and Garcia-Garibay, M.A. (2007) Photodecarbonylation of 1,3-dithiophenyl propanone: using nanocrystals to overcome the filtering effect of highly absorbing trace impurities. *Organic Letters*, **9**, 4351–4354.

74 Mortko, C.J. and Garcia-Garibay, M.A. (2005) Green chemistry strategies using crystal-to-crystal photoreactions: stereoselective synthesis and decarbonylation of trans-α,α′-dialkenoylcyclohexanones. *Journal of the American Chemical Society*, **127**, 7994–7995.

75 Choe, T., Khan, S.I., and Garcia-Garibay, M.A. (2006) Combination vs. disproportionation in dialkyl biradicals. Selectivity reversal in a crystalline solid. *Photochemistry & Photobiological Sciences*, **5**, 449–451.

76 Ng, D., Yang, Z., and Garcia-Garibay, M.A. (2002) Engineering reactions in crystals: *gem*-dialkoxy substitution enables the photodecarbonylation of crystalline 2-indanone. *Tetrahedron Letters*, **43**, 7063–7066.

77 Krapcho, A.P. and Abegaz, B. (1974) Photochemistry of dispiro-1,3-cyclobutanediones in methylene chloride and

methanol solutions. *Journal of Organic Chemistry*, **39**, 2251–2255.
78 Kuzmanich, G., Natarajan, A., Chin, K.K., Veerman, M., Mortko, C.J., and Garcia-Garibay, M.A. (2008) Solid state photodecarbonylation of diphenylcyclopropenone: a quantum chain process made possible by ultrafast energy transfer. *Journal of the American Chemical Society*, **130**, 1140–1141.
79 Ellison, M.E., Ng, D., Dang, H., and Garcia-Garibay, M.A. (2003) Enantiospecific synthesis of vicinal stereogenic tertiary and quaternary centers by combination of configurationally-trapped radical pairs in crystalline solids. *Organic Letters*, **5**, 2531–2534.
80 Ng, D., Yang, Z., and Garcia-Garibay, M.A. (2004) Total synthesis of (±)-herbertenolide by stereospecific formation of vicinal quaternary centers in a crystalline ketone. *Organic Letters*, **6**, 645–647.
81 Natarajan, A., Ng, D., Yang, Z., and Garcia-Garibay, M.A. (2007) Parallel syntheses of (+)- and (−)-α-cuparenone by radical combination in crystalline solids. *Angewandte Chemie, International Edition*, **46**, 6485–6487.
82 Mortko, C.J., Dang, H., Campos, L.M., and Garcia-Garibay, M.A. (2003) H-Abstraction prevails over α-cleavage in the solution and solid state photochemistry of cis-2,6-di(1-cyclohexenyl)cyclohexanone. *Tetrahedron Letters*, **44**, 6133–6136.
83 Ng, D., Yang, Z., and Garcia-Garibay, M.A. (2001) Engineering reactions in crystals: suppression of photodecarbonylation by intramolecular β-phenyl quenching. *Tetrahedron Letters*, **42**, 9113–9116.
84 Keating, A.E. and Garcia-Garibay, M.A. (1998) Photochemical solid-to-solid reactions, in *Organic and Inorganic Photochemistry*, Vol. 2 (eds V. Ramamurthy and K. Schanze), Marcel Dekker, New York, pp. 195–248.
85 Mortko, C.J. and Garcia-Garibay, M.A. (2006) Engineering stereospecific reactions in crystals: synthesis of compounds with adjacent stereogenic quaternary centers by photodecarbonylation of crystalline ketones, in *Topics in Stereochemistry*, Vol. 25 (eds S.E. Denmark and J.S. Siegel), John Wiley & Sons, Hoboken, NJ, pp. 205–253.
86 Veerman, M., Resendiz, M.J.E., and Garcia-Garibay, M.A. (2006) Large-scale photochemical reactions of nanocrystalline suspensions: a promising green chemistry method. *Organic Letters*, **8**, 2615–2617.

3
Intermolecular Addition Reactions onto C−C Multiple Bonds
Valentina Dichiarante and Maurizio Fagnoni

3.1
Introduction

Addition reactions across a double or a triple bond to form 1,2-difunctionalized products are an important tool in organic synthesis [1, 2]. Many of these reactions conform to the postulates of "green chemistry," an important characteristic nowadays, because in such processes the atom economy or atom efficiency can be 100%. The character of alkenes, nonconjugated and bearing an electron-donating substituent (e.g., a methoxy group), is nucleophilic and accordingly they react with electrophiles; reactions such as halogenation, hydrohalogenation, hydration are well known examples of this class. When the additive is not sufficiently electrophilic (e.g., water), activation of the olefin is required, such as the addition of an acidic catalyst (most often protonation) or coordination to a metal salt [3], as in the Pd(II)-mediated addition of water to ethylene in the synthesis of acetaldehyde via the Wacker process.

In contrast, electrophilic olefins such as α,β-unsaturated ketones, esters and nitriles undergo nucleophilic addition by stabilized C-nucleophile (Michael addition) or 1,4-conjugate addition of organometals or organocuprates [1]. One of the drawbacks of these reactions is that 1,2-addition of the nucleophile can efficiently compete. Free-radical addition is an alternative, mild approach for arriving at 1,4-conjugate addition reactions exclusively [4]. In the case of the latter process, alkyl iodides or bromides are largely used as radical precursors, and the reaction mainly makes use of highly toxic tin derivatives as chain carriers, although organoboranes may be valid substitutes. Within this context, photochemical reactions may also have a role [5], since photoaddition to olefins can be carried out under environmentally friendly conditions, where organic molecules can be "activated" by a clean reagent such as light, as illustrated in the following sections.

3.1.1
Scope and Mechanism

The aim of this chapter is not to offer an exhaustive coverage of the subject, but rather to provide familiarization with the idea that applying a photochemical method to organic synthesis is, in most cases, advantageous. However, this possibility has not yet been greatly exploited. Accordingly, some significant examples (when possible, those reported in the past decade) have been selected with particular regard to the simplicity of the procedure and the high product yields obtained. When convenient, reference is made to reviews or book chapters rather than to the primary literature. Those reactions which are significant only from a mechanistic point of view, but which lead to a mixture of products, are not considered at this point; neither are the photopolymerization of alkenes and Bu_3SnH-photoinitiated reactions. The *intra*molecular version of the reaction is likewise not described, as this involves the formation of a ring; this topic is described in Chapters 5 to 9.

Two main approaches are used for the photochemical functionalization of alkenes. For the first approach, a C, P, S-based radical R· is generated photochemically from a precursor R–H(Y) by: (i) photohomolysis of a R–Y bond (Y = halogen, hydrogen, PhS, PhSe, etc.); (ii) homolytic hydrogen abstraction from R–H by a suitable photomediator (or photocatalyst) **P**; and (iii) via a photoinduced electron transfer (PET) oxidation of R–H(Y) followed by the loss of an electrofugal group $H^+(Y^+)$ (or respectively, reduction followed by the loss of the nucleofugal group Y^-; see further below and Scheme 3.1a) [6]. The radical thus formed, whether electrophilic or nucleophilic, attacks the C–C multiple bond to form the corresponding addition product.

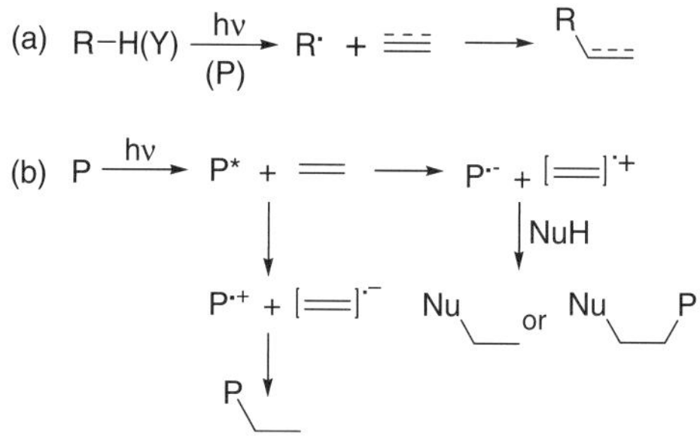

Scheme 3.1

In the second approach, activation of the olefin takes place (Scheme 3.1b). Thus, an excited sensitizer (or a photocatalyst) **P*** is able either to oxidize or to reduce a double bond by a PET process. The radical cation of the olefin is trapped by a nucleophile NuH; alternatively, a three-component reaction occurs if **P** is also incorporated into the end product. In rare instances, **P*** reduces the olefin to the corresponding radical anion and functionalization of **P** results.

3.2
Addition to C—C Double Bonds

The reactions discussed in this section are summarized in Scheme 3.2.

1,2-Difunctionalized olefins are always obtained, but in most cases a hydrogen atom is one of the olefin substituents (H—Y addition). Thus, in the first part of this section, attention will be focused on H—C, H—N, H—O, H—P, and H—S addition. Some sparse examples of X—Y addition reactions (where X is different from hydrogen) are also mentioned briefly at the end of this chapter.

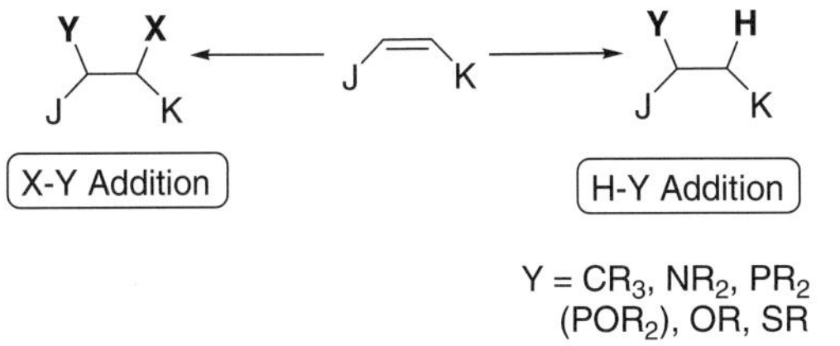

$Y = CR_3, NR_2, PR_2$
$(POR_2), OR, SR$

Scheme 3.2

3.2.1
H—C Addition (Hydroalkylation Reactions)

Photohydroalkylations are in most cases carbon-centered radical conjugate additions onto electron-deficient olefins [7]. Scheme 3.3 summarizes in detail the pathways for the photogeneration of radicals from R—H(Y) **1**. In path *a*, a photocatalyst **P** (when excited) cleaves homolytically a suitable C—H bond, and the resultant radical adds to the olefin **2** to form the adduct radical **3**. α,β-Unsaturated nitriles, ketones, and esters

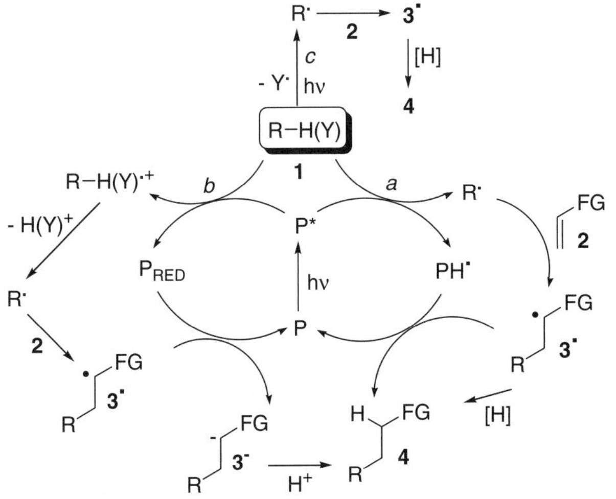

Scheme 3.3

are mainly used as the olefins. The functionalized olefin **4** can be formed via a further hydrogen abstraction step from the solvent or a suitable added hydrogen donor. If the reduced photocatalyst PH· is able to donate back the hydrogen, **P** is regenerated to close the catalytic cycle. Aromatic ketones (e.g., benzophenone) or decatungstate salts [e.g., $(n\text{-}Bu_4N)_4W_{10}O_{32}$; TBADT] are typically used in this role. This reaction is suitable for the efficient generation of (substituted) alkyl radicals directly from hydrogen donors such as alcohols, ethers, alkanes, aldehydes, and amides [7]. In an alternative route (path b), **P*** oxidizes **1** to the corresponding radical cation (R–H (Y)$^+$) with a concomitant reduction of the photocatalyst (P_{RED}, left part of Scheme 3.3). The loss of a proton or of an electrofugal group (Y$^+$) from this radical cation leads again to an alkyl radical. In this case, radical **3**• (formed by the addition of R• onto **2**) undergoes reduction by P_{RED}, and the anion **3**$^-$ thus formed affords adduct **3** when protonated by the reaction medium. Aromatic esters, nitriles and ketones are well-suited as the oxidizing agents, whereas oxidizable substrates are amines, stannanes and 3,3-dialkyldioxolanes [6, 7]. Use of the energy of radiation for cleaving homolytically a C–Y bond (path c) is less common.

3.2.1.1 Addition of Alkanes

Alkyl radicals can be generated from tetramethylammonium phenyltrialkylborate salts (used as electron donors) in the PET reaction with excited benzophenone in the presence of a 10-fold excess of an electron-poor olefin in an acetonitrile/benzene (1 : 5) solution [8]. The advantages of this method are that more than one equivalent of alkyl radicals is obtained per molecule of the borate derivative, and that part of the benzophenone is usually recovered after the reaction [8].

Reductive addition to electron-deficient olefins can be also performed by using nontoxic (diacyloxyiodo)arenes (**5**, Scheme 3.4) [9]. In this case, the photocleavage of an iodo–oxygen bond forms an alkylcarboxyl radical that in turn decarboxylates to yield the desired alkyl radical. As an example, phenyl vinyl sulfones are easily alkylated (e.g., by the addition of adamantane) in good yields, provided that a good hydrogen donor such as 1,4-cyclohexadiene is present in solution.

The generation of alkyl radicals by direct and selective activation under mild conditions of C–H bonds in alkanes is clearly desirable, and still considered as the

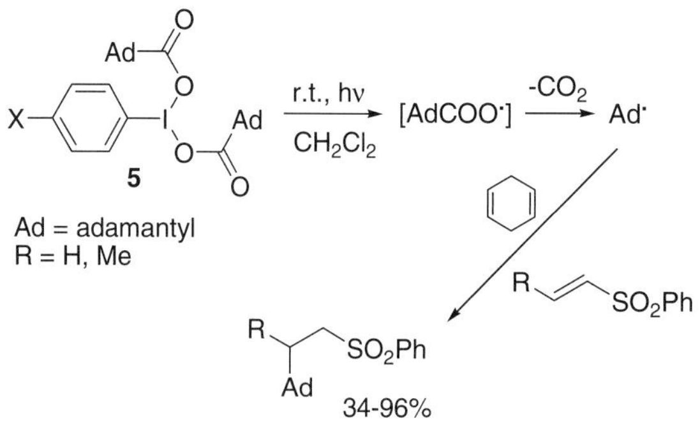

Scheme 3.4

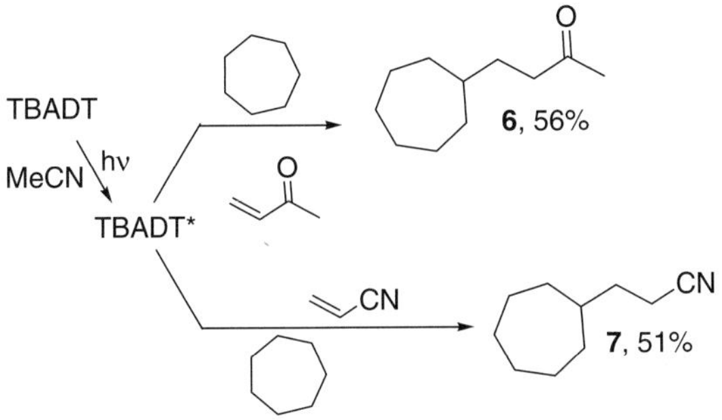

Scheme 3.5

"Holy Grail" of chemistry. Photocatalysis (by hydrogen abstraction) by using benzophenone or some polyoxoanions (notably TBADT) as the photocatalyst appears to be a promising method in this respect [7]. Two examples are shown in Scheme 3.5 that involve cycloheptane as the hydrogen donor. In both cases, excited TBADT is able to cleave homolytically the C—H bond of the cycloalkane, with the resultant cycloheptyl radical being used for the synthesis of β-cycloalkyl derivatives such as 4-cycloheptyl-2-butanone **6** (56% yield, from 3-buten-2-one) [10] and 3-cycloheptylpropanenitrile **7** (51% yield, from acrylonitrile) [11]. This simple alkylation procedure is environmentally benign, since TBADT is used in small amounts (2% mol. equiv) and is easily eliminated after the reaction.

Solar light irradiation can be used for the formation of C—C bonds, as demonstrated in the benzylation of electron-poor olefins via TiO_2 photocatalysis [7]. A substituted benzyl(trimethyl)silane is well-suited for this reaction, and acts as fragmentable electron donor that generates a benzyl radical upon desilylation of the corresponding radical cation. A typical example is the irradiation of a MeCN solution containing 4-methoxybenzyl(trimethyl)silane, maleic anhydride and TiO_2, that was carried out in a solar platform (in Almeria, Spain). A multigram synthesis of the benzylated product **8** was obtained after 10 h irradiation (Scheme 3.6) [12]. As both the solvent and semiconductor can be easily recovered and reused, this preparation conforms to the "green chemistry" postulates.

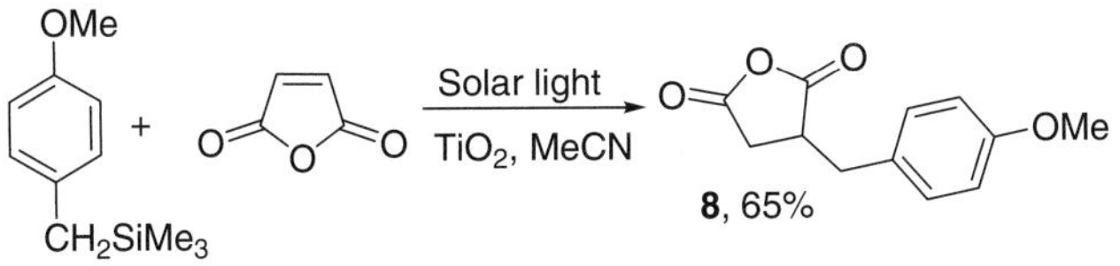

Scheme 3.6

3.2.1.2 Addition of Alcohols (Hydrohydroxymethylation), Ethers, and (2-substituted) 1,3-Dioxolane(s)

α-Hydroxy(alkoxy)alkyl radicals are likewise easily obtained through hydrogen abstraction from the corresponding alcohols, ethers, or (2-substituted) 1,3-dioxolanes [7]. Here, aliphatic (e.g., acetone) or aromatic (benzophenone, aceto-

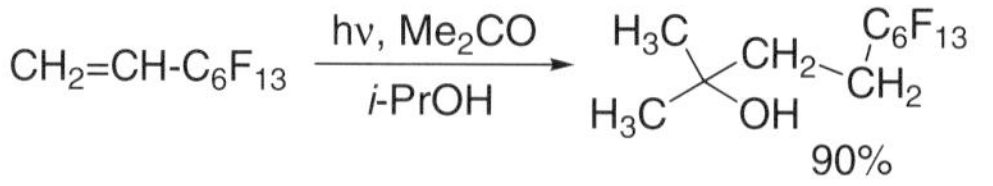

Scheme 3.7

$$CH_2=CH-C_6F_{13} \xrightarrow[i\text{-PrOH}]{h\nu,\ Me_2CO} (H_3C)_2C(OH)-CH_2-CH_2-C_6F_{13} \quad 90\%$$

phenone) ketones are largely used as the photomediators. The reactions were mainly carried out by using the hydrogen donor as the reaction medium, with several examples having been reported [7]. The mild conditions used allow the alkylation of labile substrates such as unsaturated aldehydes. In the latter case, the solvent/reagent could in part be replaced by water; this in turn allows the use of a water-soluble photocatalyst, such as disodium benzophenondisulfonate, that is easily eliminated on completion of the reaction. γ-Lactols (and γ-lactones from these, via mild oxidation) are thus accessible through a 1,4 attack of the α-hydroxyradical [13].

Of note, perfluoroalkylethylenes were also easily alkylated when irradiated in an acetone/i-PrOH mixture [14], yielding end products in a completely regioselective and almost completely chemoselective fashion, with preparative yields ranging from 71% to 92%. Scheme 3.7.

Although ethers have been used less frequently than alcohols, it has recently been reported that tetrahydrofuran (THF) is photocatalytically activated by TBADT, and the alkylation of unsaturated nitriles is obtained in good yield [15]. As an alternative, the C—Br bond in various glycosyl bromides has been homolytically cleaved, and the resulting radical trapped by acrylonitrile to form the corresponding C-glycosides. The halogen abstraction step is initiated by a photolabile iron-based dimeric metal complex [16].

1,3-Dioxolane or 2-substituted-1,3-dioxolanes are useful hydrogen donors that can be used for the introduction of a masked aldehyde or ketone functionality [17, 18]. Accordingly, this photochemical procedure produces masked dicarbonyls selectively in the alkylation of both α,β-unsaturated ketones (monoprotected 1,4-diketones were formed) and of α,β-unsaturated aldehydes (forming monoprotected γ-ketoaldehydes) [13], where the protection involves the less-electrophilic ketone function [7]. This procedure has been recently employed in a total synthesis, as shown in Scheme 3.8. One of the key step in the synthesis of a high-affinity nonpeptidal ligand for the human immunodeficiency virus (HIV) protease inhibitor UIC-94017 is based on the stereoselective photochemical conjugate addition of 1,3-dioxolane (a formyl equivalent) to a protected 5(S)-hydroxymethyl-2(5H)-furanone derivative (9, Scheme 3.8a) [19]. Besides, diketene 10 was smoothly photoalkylated upon irradiation in an acetophenone/2-undecyl-1,3-dioxolane mixture. The end product 11 found application in the synthesis of (−)-tetrahydrolipstatin (a potent pancreatic lipase inhibitor; Scheme 3.8b) [20].

3.2.1.3 Addition of Amines (Hydroaminomethylation) or Amides

Hydroaminomethylation reactions can be accomplished by addition of nucleophilic α-amino alkyl radicals (generated from amines) onto electron-deficient alkenes [7, 21]. Amines have a low oxidation potential, and are easily oxidized by excited aromatic ketones bearing electron-donating substituents (e.g., Michler's ketone) or

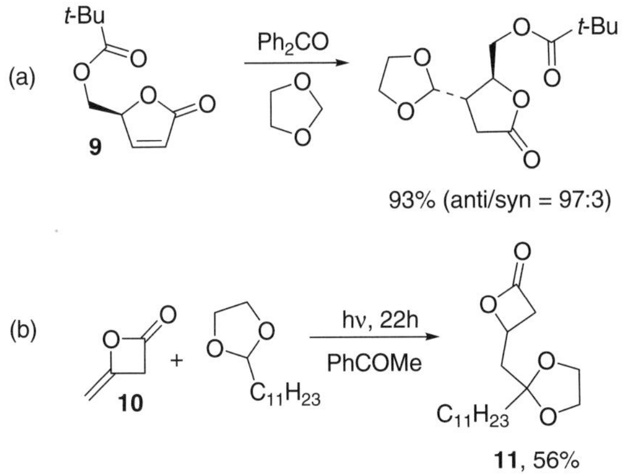

3.2 Addition to C–C Double Bonds | 73

Scheme 3.8

93% (anti/syn = 97:3)

11, 56%

titanium dioxide by a PET reaction. Deprotonation of the radical cations of amines generates the α-amino alkyl radicals. Cyclic tertiary amines such as N-methylpyrrolidine (**13**) are well-suited for the reaction. Scheme 3.9 illustrates two examples where both furan-2(5H)-one (**12**) and (5R)-5-menthyloxy-2-(5H)-furanone (**14**) were functionalized by the photoaddition of **13** under heterogeneous and homogeneous conditions, respectively. As for the last reaction, the process occurred with a high yield (94%) and with complete facial selectivity. This reaction was used for the enantioselective synthesis of some necine bases, namely, (+)-laburnine and (−)-isoretronecanol (Scheme 3.9b) [22]. When thiocarbonyl compounds (e.g., thiocarbamates or xanthates)

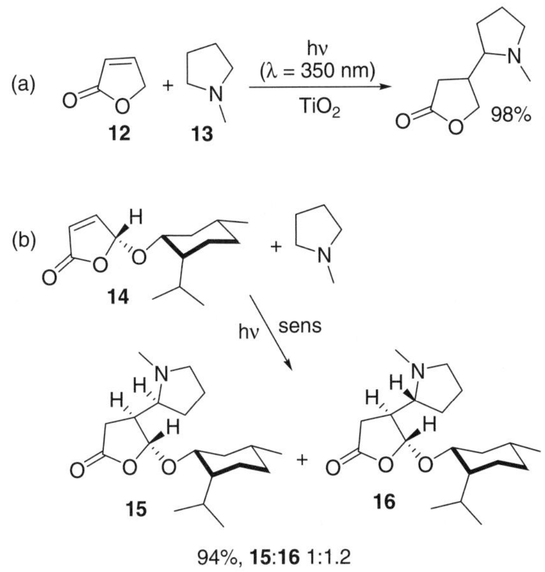

94%, **15:16** 1:1.2

Scheme 3.9

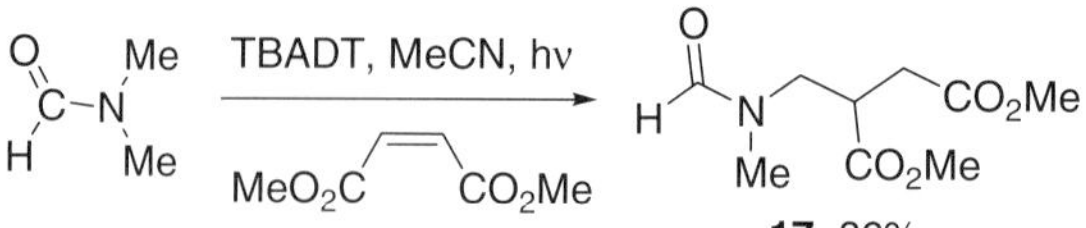

Scheme 3.10

were added to the reaction mixture the reaction was feasible, even with open-chain tertiary amines [21].

Another approach involves the use of an electron-poor olefin acting both as an absorbing electron acceptor and as a radical trap. In this case, a PET reaction between a cyclohexenone derivative and a silylated amine led to a radical ion pair. Desilylation of the silyl amine radical cation intermediate in polar protic solvent (e.g., MeOH) and subsequent aminoalkyl radical attack onto the enone radical anion yielded the alkylated cyclohexanones [23].

In rare instances, hydroaminomethylation can occur by using electron-rich alkenes such as styrene derivatives. Thus, 1-phenylpropene added Me_3N regioselectively to form *N,N*-dimethyl-2-phenylbutan-1-amine upon excitation via a styrene–amine exciplex [24].

The photoamidation of electron-poor olefins has also been reported [7]. A recent example involves the highly selective activation of C−H bonds in amides by excited TBADT (Scheme 3.10) [25]. Thus, this photocatalyst was able to activate selectively the N−Me hydrogens rather than the formyl hydrogen in dimethylformamide. As an example, dimethylmaleate was alkylated under these conditions to form the γ-amino acid derivative **17** in 86% yield. Moreover, the reaction was extended to carbamates [25].

3.2.1.4 Hydrofluoromethylation

Electrophilic rather than nucleophilic radicals can be used in the functionalization of vinyl ethers. Difluoromethyl radicals are readily available via photoinitiated cleavage of the $PhS-CF_2X$ bond in α,α-difluorosulfides, such as 4-[difluoro(phenylthio)methyl]-1,3-dioxolanone (**18**) and 5-[difluoro(phenylthio)methyl]-3-methyl-1,3-oxazolidinone (**19**) [26]. Of note, the difluoromethylene group attacked the terminal position of the vinyl ether (e.g., a dihydrofuran) exclusively (Scheme 3.11). Difluoromethyl radicals bearing ester, phosphonate, nitrile, cyclic carbonate, and carbamate groups were suitable for the syntheses of other CF_2-containing building blocks [27].

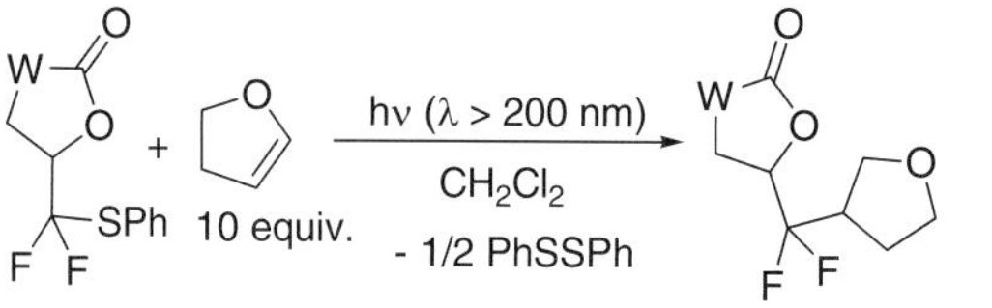

Scheme 3.11

3.2.1.5 Addition of Nitriles or Ketones

The photochemical alkylation of olefins by nitriles and ketones is not straightforward, due mainly to the inefficient abstraction of hydrogen from an electron-withdrawing-substituted carbon by an electrophile such as the photocatalyst excited state. Nevertheless, various methyl ketones have been synthesized by the irradiation of a ketone/olefin mixture dissolved in aqueous acetone. The mechanism of the reaction remains to be clarified, but a water-assisted C—C coupling between an acetonyl radical and the olefin has been postulated (Scheme 3.12). The reaction has several advantages, as it is cheap (an acetone/water mixture is used as the solvent) and occurs under mild metal-free conditions with no need for a photocatalyst [28].

The anti-Markovnikov photochemical addition of malononitrile onto styrenes in the presence of lithium carbonate and a cyanoarene has been reported. In this case, excited 9-cyanophenanthrene (9-CP) oxidizes an olefin by PET, and the resultant radical cation adds to the malononitrile anion. The resulting radical is then reduced by 9-CP$^{\bullet-}$ and protonated to afford the end products [29]. This represents one of the rare examples involving the photoaddition of a carbanion to an olefin, and is explained by the PET-induced Umpolung of the latter, which thus becomes electrophilic.

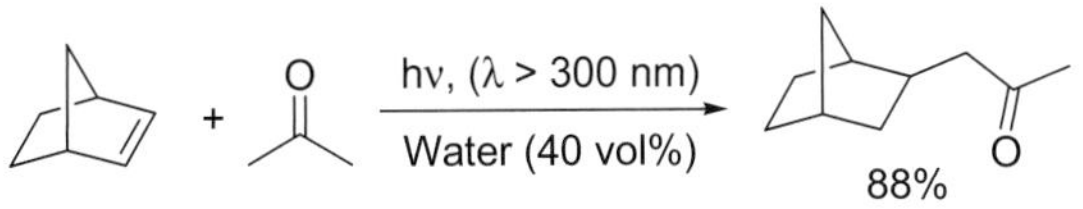

Scheme 3.12

3.2.1.6 Hydroacylation and Hydrocarboxylation Reactions

An efficient acylation (via photogenerated acyl radicals) of the captodative olefin **22** has been obtained through a photochemical reaction of aliphatic (or aromatic) aldehydes by using benzophenone as the photomediator. This is an interesting case of 1,2-asymmetric induction in radical reactions, since the acyl radical intermediate added to **22** with high *syn* selectivity (Scheme 3.13a) [30]. Recently, TBADT was likewise used for the generation of acyl radicals in the photoacylation of electron-poor

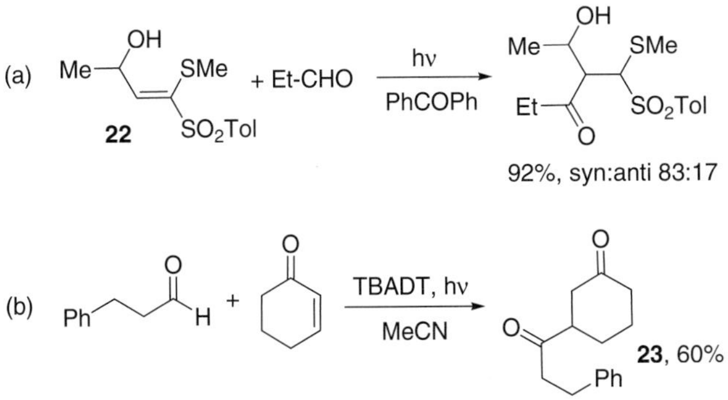

Scheme 3.13

olefins. In the latter case, the method has the advantage of using equimolar amounts of the aldehyde and olefin, and the diketone **23** is easily isolated by bulb-to-bulb distillation (Scheme 3.13b) [31].

A similar reaction has been adopted for the preparation of amides from alkenes by the photoaddition of •CONH$_2$ radicals generated from formamide by benzophenone [7], or via a TBADT [25] photocatalyzed hydrogen abstraction. Solar light has been used in the first case. Furthermore, the introduction of an ester function has been accomplished by generating an alkoxycarbonyl radical by photolysis of [bis(alkoxyoxalyloxy)iodo]benzene at 0–5 °C and ensuing addition to vinyl sulfones in the presence of 1,4-cyclohexadiene [9].

3.2.1.7 Hydroarylation (Photo-EOCAS)

The hydroarylation of olefins is relatively uncommon in photochemistry, despite a high interest in this process which allows the formation of an aryl–carbon bond via the direct activation of an aromatic, with no need for leaving groups in both components of the reaction. The process follows a photo-EOCAS (Electrophile-Olefin Combination Aromatic Substitution) mechanism [32], and is initiated by a PET reaction between an electron-rich aromatic and an electron-poor olefin, as illustrated in Scheme 3.14.

For further reactions involving the formation of aryl–C bond via addition of an aromatic to an olefin, see Chapter 10.

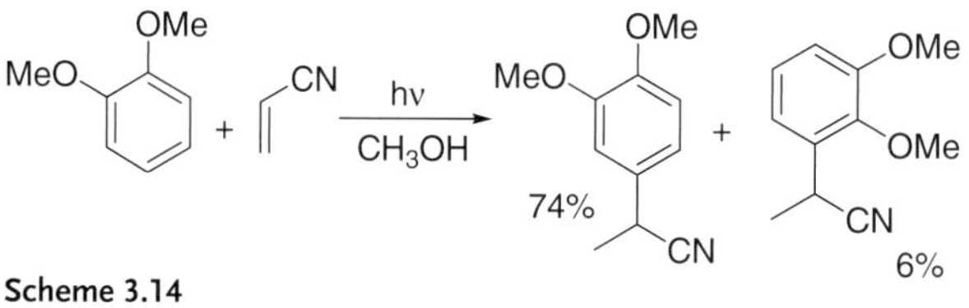

Scheme 3.14

3.2.2
H–N Addition (Hydroamination)

Hydroamination – that is, the direct formation of a new C–N bond by the addition of an amine to an olefin (or an alkyne) – is of particular significance, since this represents a simple route for the synthesis of more complex amines, a class of compounds that is ubiquitous among natural products as well as among artificial drugs [33]. This scheme has been applied to the addition to unsubstituted or electron-donating substituted alkenes, a reaction that in thermal terms would encounter a high activation energy barrier as it involves the combination of two electron-rich reactants. Catalysis by a Brønsted acid or by lanthanoids and transition metals is usually adopted for carrying out this atom-economical reaction. However, whilst under acidic conditions the alkene is activated by protonation, the N-nucleophile is deactivated by formation of the conjugate acid. Thus, the hydroamination of nonactivated alkenes and the anti-Markovnikov addition of H–NR$_2$ remain a challenge.

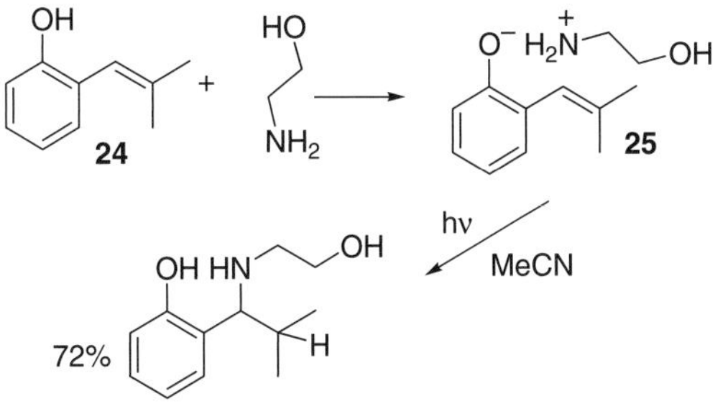

3.2 Addition to C—C Double Bonds

Scheme 3.15

One simple way for arriving at a photoinduced Markovnikov-type amination of olefins is illustrated in Scheme 3.15, for the case of o-alkenylphenols (e.g., **24**). Here, an ammonium salt of the o-alkenylphenolate anion (**25**) is prepared and irradiated; this causes proton transfer within the ion pair (from the amine to the olefin), and the resulting zwitterion undergoes nucleophilic addition of the amine at the benzylic cation center. The addition is selective for N- rather than O-nucleophiles, as shown in the case of ethanolamine [34].

An elegant approach for fostering amine addition is via Umpolung of the olefin by a PET oxidation [35]. The reaction is especially successful when using stilbene and styrene derivatives as the olefin. As an example, the irradiation of stilbene in the presence of a primary amine or ammonia gave no reaction; however, the addition of an electron-acceptor (p-dicyanobenzene, DCNB) allowed an efficient addition of the amine. The mechanism is depicted in Scheme 3.16a, where the excited DCNB selectively oxidizes the olefin and the resulting stilbene radical cation undergoes amine addition, followed by reduction by the DCNB$^{•-}$. The yields of the desired alkylamino derivatives are excellent, exceeding 90% and with all of the amine used [36].

A cosensitizer such as 1,3,5-triphenylbenzene or m-terphenyl can improve the reaction yields, as demonstrated in the photoamination of trans-1-arylpropenes and

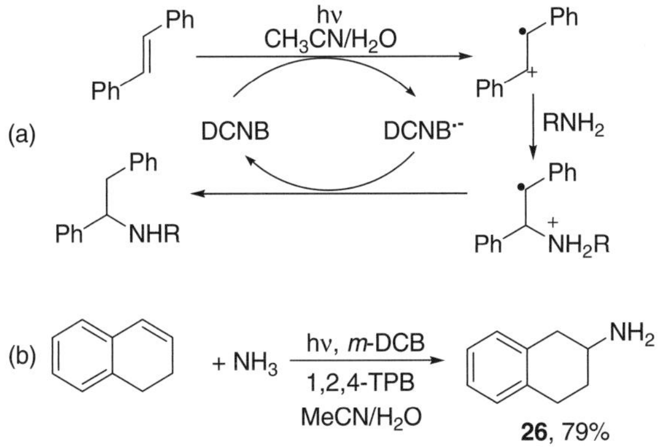

Scheme 3.16

7-methoxy-1,2-dihydronaphthalenes with ammonia. This reaction gave 1-aryl-2-propylamines and 2-amino-6-methoxy-1,2,3,4-tetrahydronaphthalenes, respectively when using 1,3-dicyanobenzene as the photocatalyst [37]. Under similar conditions, the photocatalyzed nucleophilic addition of ammonia (or alkylamines) to methoxy-substituted styrene derivatives took place [38].

Further studies have shown that, to obtain an efficient amination, it is necessary to take into account both the difference in oxidation potential between the photocatalysts and the substrates and the positive charge distribution in the cation radicals of the olefin. The synthetic utility of the method was proven by the successful preparation of an aminotetraline, itself an intermediate for the synthesis of a compound with biological activities such as 2-aminoindan (**26**, Scheme 3.16b) [39]. As with the last synthesis, the redox-photosensitized amination occurred with no need for acids or bases, as usually required when using general protocols.

A PET reaction between excess phthalimide (in equilibrium with its conjugate base) and an alkene led to a clean phthalimidation of nonactivated double bonds. Here, the singlet excited state of phthalimide acts as the oxidant and a radical ion pair is formed. The olefin cation radical is trapped by the phthalimide anion, and back electron transfer, followed by protonation, affords the photoaddition products [40]. Protected phenethylamines are readily accessible in this way. This reaction has been carried out by using NaOH as the base; it has been shown that the amounts (usually equimolar with the alkene) must be carefully chosen in order to avoid the undesired competition with [2 + 2] photocycloaddition.

The reaction can be also applied to azoles (e.g., imidazoles, pyrazoles, triazoles, tetrazole) as the N-nucleophile for the hydroaminations of cyclic olefins. In this case, the presence of methyl benzoate sensitizes the photoisomerization of the double bond to form a highly strained (*E*)-cycloalkene. Protonation of this intermediate by triflic acid (TfOH, 20 mol%) and addition of the azole nucleophile completes the reaction sequence [41]. As an example, the expeditious synthesis of 1-(1-methylcyclohexyl)-1*H*-imidazole (**27**) in 72% yield is shown in Scheme 3.17 [41].

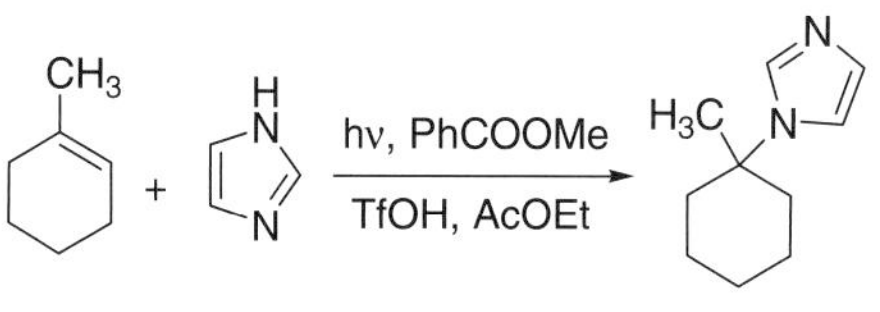

Scheme 3.17

3.2.3
H—P Addition

The photoaddition of secondary phosphines onto alkenes is one of the straightforward approaches for the synthesis of organophosphorus compounds. An anti-Markovnikov phosphorus-centered free radical addition onto the olefin has been

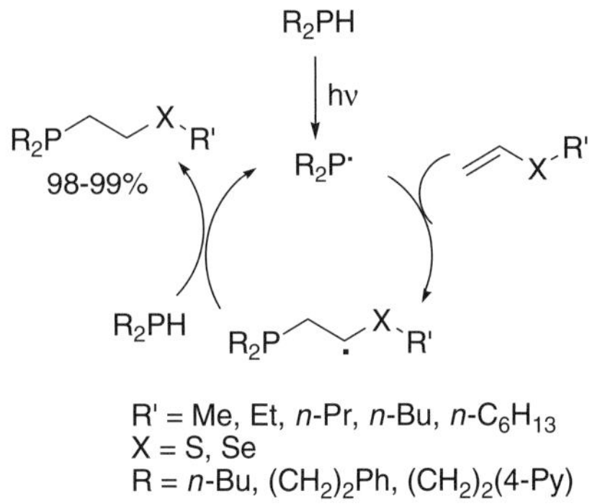

3.2 Addition to C—C Double Bonds

Scheme 3.18

R' = Me, Et, n-Pr, n-Bu, n-C$_6$H$_{13}$
X = S, Se
R = n-Bu, (CH$_2$)$_2$Ph, (CH$_2$)$_2$(4-Py)

envisaged. Various substituted (electron-rich) alkenes have been found suitable for the synthesis of phosphines that are useful ligands. The advantage of the reaction is that, in most cases, it takes place under solvent-free conditions simply upon the irradiation of an equimolar mixture of the two reagents. The hydrophosphination of substituted vinyl sulfides and vinyl selenides occurred in near-quantitative yield (Scheme 3.18) [42]. The compounds formed in the reaction can either serve as reactive building blocks for organic synthesis, or be used as polydentate ligands for designing complex metal catalysts.

The photochemical reaction of bisallyloxynaphthalene with diphenylphosphane gave, smoothly, a bridged bisphosphane (70% yield) that was used to form metallamacrocycles or oligomers with Pt(II) or Mo(0) [43]. Olefins containing a *tris*-ethoxysilyl functionality have been also widely used for photoinduced C—P bond formation. Thus, (*tris*-ethoxysilyl)ethyldiphenylphosphine (or its dimethyl analogue) were prepared starting from vinyltriethoxysilane in more than 90% yield [44]. The bifunctional silylether/phosphines formed in the reaction were also employed as ligands for the preparation of Ir- or Rh-based complexes. Mono- and bidentate phosphine ligands have been likewise prepared from (but-3-enyl)- and (pent-4-enyl)triethoxysilane. New dicarbonylnickel catalysts were accessible in this way, and the ethoxysilane groups of the phosphine were used to immobilize the catalyst on a silica gel surface. This surface-bound catalyst was found to be effective for the cyclotrimerization of phenylacetylene [45]. A (2-phosphinoethyl)silyl chelate ligand has been synthesized in 93% yield by the irradiation in toluene of a vinylsilane derivative and PHPh$_2$ [46]. A photochemical hydrophosphination of 1-alkenes was instead adopted for the synthesis of diphos ligands when reacting with H$_2$P(CH$_2$)$_3$PH$_2$. Water-soluble palladium complexes were then prepared, and proved to be highly active in the carbon monoxide/ethene copolymerization [47]. One interesting variation of the reaction was the hydroselenophosphorylation of alkenes (Scheme 3.19). This process relies on the UV-induced addition of P, Se radicals arising from the cleavage of secondary phosphine selenides **28** onto electron-rich alkenes to give the *anti*-Markovnikov tertiary phosphine selenides **29** regiospecifically, in up to 96% yield [48]. The addition yields depend on the stability of the radical adduct formed; accordingly, non-1-ene was shown to be the least reactive olefin among those studied.

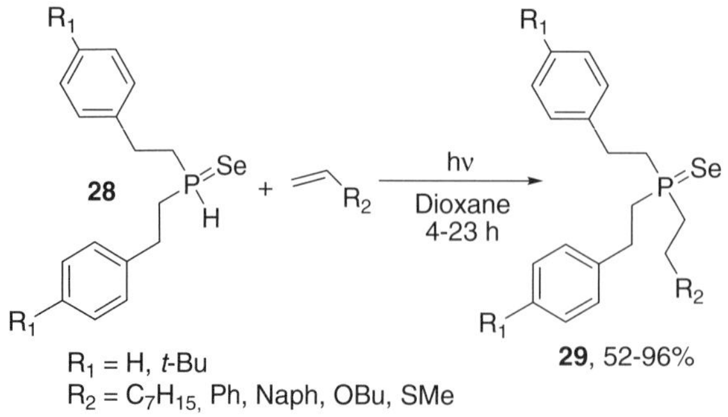

Scheme 3.19

R₁ = H, *t*-Bu
R₂ = C₇H₁₅, Ph, Naph, OBu, SMe

A different approach must be used for the photochemical hydrophosphination of electron-poor olefins, and this involves a PET reaction. Silyl phosphites (e.g., **30**) were used as electron donors, whereas conjugated ketones have the double role of electron acceptors and absorbing species. Thus, the irradiation of a mixture containing 2-cyclohexenone and **30** generated an ion pair. The phosphoniumyl radical cations decomposed to give trimethylsilyl cations (which in turn were trapped by the enone radical anion) and phosphonyl radicals. A radical–radical combination afforded the 4-phosphonylated ketones in yields ranging from 78% to 92% (Scheme 3.20) [49]. This reaction was exploited for the preparation of substituted phosphonates, which serve as key intermediates in the synthesis of a class of biologically active compounds.

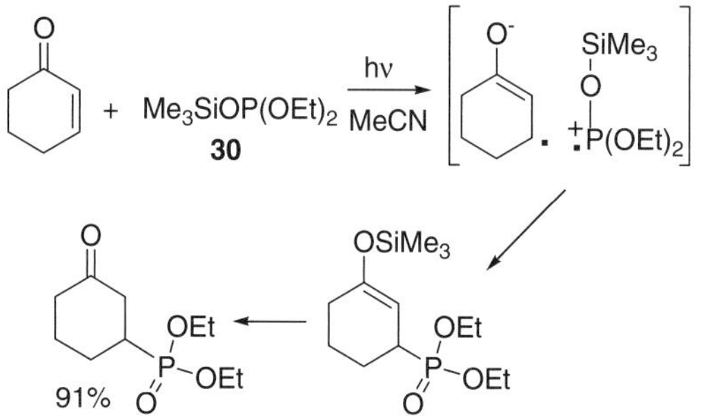

Scheme 3.20

3.2.4
H—O Addition

Both, hydration and hydroalkoxylation of olefins are usually carried out under acid catalysis or by oxymercuration reactions, although reactions mediated by palladium or by other metals have recently begun to emerge. In the latter case, nucleophilic addition of the O—H group is favored by a cationic palladium species that activates

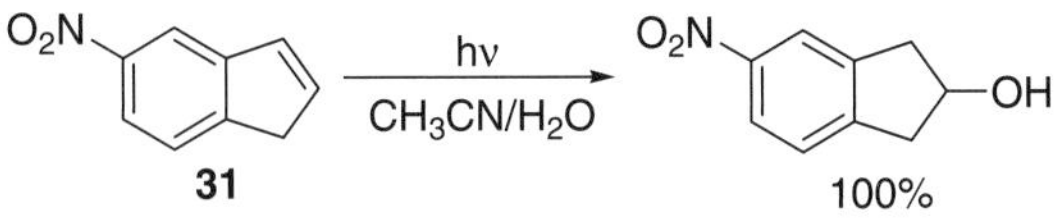

3.2 Addition to C—C Double Bonds

Scheme 3.21

the unsaturated carbon–carbon bonds by coordination. As with the photochemical addition of water or alcohols onto double bonds, there is no unitary mechanism. The direct photoaddition of water to olefins can be achieved in solution only when rapid twisting of the double bond in the alkene excited state is prevented, since otherwise deactivation will take place. Accordingly, some substituted nitrostilbenes do not undergo the photoaddition reaction but only photoisomerization, whereas 5-nitroindene (**31**) smoothly adds water in an anti-Markovnikov fashion upon photolysis in a MeCN/water solution (Scheme 3.21) [50].

In contrast, a Markovnikov addition of water was reported in the irradiation of a variety of o-hydroxystyrenes, again in aqueous acetonitrile, with the formation of 2-(2-hydroxyphenyl)ethanols. In this case, an intramolecular proton transfer from the excited state of the styrene was envisaged as the first step of the reaction [51]. A similar mechanism was postulated in the photohydratation of m-hydroxy-1,1-diaryl alkenes that gave the corresponding 1,1-diarylethanols, although direct protonation of the β-carbon by water competed in some cases [52].

Since alcohols are less effective as hydrogen donors than amines, a PET photoaddition can occur only when the oxidized component of the reaction is the alkene. Furthermore, if the photosensitizer is chiral, the polar addition would occur in an enantiodifferentiating manner to some degree. Thus, the photoaddition of 2-propanol to 1,1-diphenylpropene, when sensitized by chiral naphthalene(di)carboxylates, formed the *anti*-Markovnikov photoadduct with enantiomeric excesses of up to 58% [53]. Unfortunately, the reaction is far from attracting synthetic interest as the yields are still too low.

An ingenious way of adding alcohols to nucleophilic olefins (e.g., 3,4-dihydro-2*H*-pyran) was recently developed which involved the photolysis of 1,5-dichloro-9,10-anthraquinone (DCQ) as a photoacid generator, and which has the advantage of absorbing in the visible spectrum [54]. Accordingly, the acid liberated upon irradiation of DCQ allows the addition of tertiary allyl alcohols (**33**) to the vinyl ether (**32**) to form the acetal (**34**), with no competitive addition onto the double bond of the allyl alcohol (Scheme 3.22). The reaction is relatively insensitive to the presence of oxygen, and is likewise effective by using sunlight.

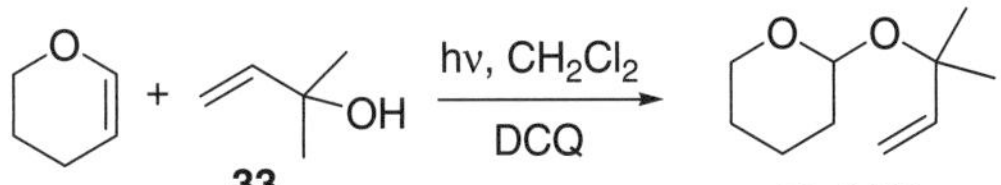

Scheme 3.22

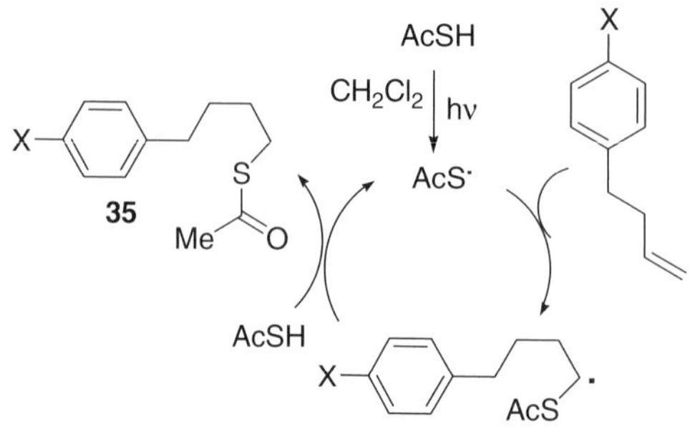

Scheme 3.23

X	Yield (%)
H	100
OMe	97
NMe$_2$	97
CH$_2$NEt$_2$	80

3.2.5
H−S Addition

One of the most useful and widely used applications in the synthesis of natural product derivatives relies on the efficient photoaddition of RS−H onto a double bond (a reaction known as thiol-ene coupling) [55]. The reaction exploits the weakness of the S−H bond that can be cleaved homolytically under irradiation (at ca. 254 nm). The electrophilic sulfur-centered radical attacks a nucleophilic double bond, thus starting a radical chain reaction.

As an example, thioacetic acid was photoadded in high yield to butenyl aromatics in the multigram scale synthesis of compounds **35** (Scheme 3.23) [56]. These products were, in turn, starting materials for the preparation of aryl–alkyl disulfides that were employed as protective agents for gold nanoparticles due to a passivation process.

The introduction of a mercapto group was exploited for the synthesis of novel zinc-chelating group that could be used as inhibitors of matrix metalloproteinases, and recently were tested for the treatment of cancer and rheumatoid arthritis. In this case, the key step in the synthesis of these inhibitors involved the thioacylation of a BOC-protected allylamine (87% yield) upon photolysis in the presence of thioacetic acid [57]. Functionalization by S-bonded groups has been proved useful for stabilizing the otherwise thermally unstable 2,3-diallyltetraazathiapentalene derivatives. The reaction easily took place in acetone by using an excess of thioacetic acid and thiobenzoic acid, and provided end addition products in quantitative yield. Surprisingly, if oxygen was present the process was remarkably faster than that carried out under deaerated conditions [58].

Glycoconjugate polymers were prepared by the copolymerization of different monomers, and immobilized on cellulose membrane for affinity separation. The synthesis of the mannose-based monomer involved the photochemical addition of 2-aminoethanethiol hydrochloride onto an alkenyl α-D-mannoside derivative **36**. Of note, **37** was formed in a quantitative yield upon irradiation in MeOH at room temperature, with no competitive addition of the amino group onto the alkene (Scheme 3.24) [59].

A method for the multigram scale synthesis of *Candida albicans* (1 → 2)-α-D-mannopyranan epitopes has been recently developed which exploits the addition of

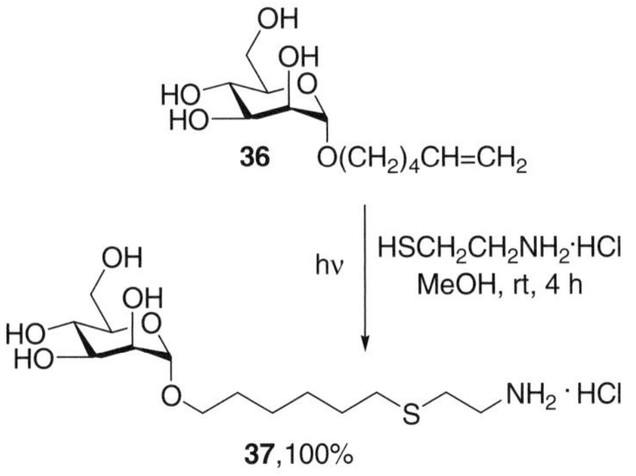

Scheme 3.24

2-aminoethanethiol hydrochloride to an O-allyl mannose polysaccharide. The derivatives thus formed were then tethered to an immunogenic carrier protein, such as tetanus toxoid, and tested for the preparation of an effective vaccine against *C. albicans* infections [60]. Mercaptoacetamides having a long chain were likewise used for the functionalization of disaccharides in the synthesis of moenomycin–bovine serum albumin (BSA) conjugates [61]. This method found an interesting application in the synthesis of substituted D-galactopyranosides by functionalization of an O-allyl chain [62, 63]. Accordingly, cysteamine was added in 94% yield in a 1 : 1 MeOH/water mixture, and the resultant compound was used for the synthesis of tri-sulfated galabioses [62], whereas 3-mercaptopropionic acid was linked (in 83% yield) in neat water and formed a precursor for the synthesis of novel glycodendrimers [63]. Cysteamine was again used to functionalize various N-allyl glycosides that were used for the chemoenzymatic synthesis of neoglycopeptides [64].

When a sulfur derivative having two SH groups (e.g., 1,6-hexanedithiol) is used, it is possible to tether one of these groups to a disaccharide by a photoaddition step onto a pendant O-allyl moiety. The other SH group can then be linked to gold nanoparticles to form water-soluble gold glyconanoparticles, which are important model systems for interaction studies [65]. A thiomethylation reaction can take place in the solid state, as was demonstrated in the photoreaction of a toluene dispersion of crystals of a tetraphenylaminotriazine derivative having eight pendant N-allyl chains in the presence of CH_3SH [66]. The analysis of the resulting thioether showed that 70–85% of the allyl groups had reacted.

3.2.6
Addition of X-Y Reagents to Alkenes

The classification of these reactions is difficult as the panorama is variable. Here, we will present only selected examples that indicate the potential of photochemical methods. Other than some bromination reactions, most attention will be paid to functionalization with the formation of a C—C bond.

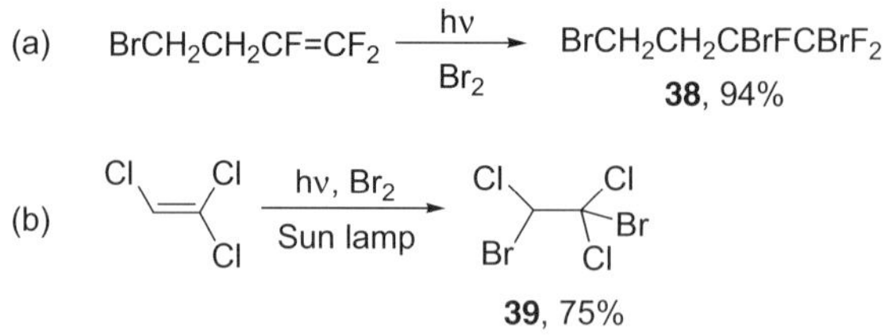

Scheme 3.25

(a) $BrCH_2CH_2CF=CF_2 \xrightarrow[Br_2]{h\nu} BrCH_2CH_2CBrFCBrF_2$
38, 94%

(b) Cl₂C=CHCl $\xrightarrow[\text{Sun lamp}]{h\nu, Br_2}$ CCl₂Br–CHClBr
39, 75%

3.2.6.1 Halogenation

Photoinduced free-radical halogenation is very useful for the functionalization of electron-deficient alkenes such as vinyl halides. Among halides, the addition of bromine is by far the most useful reaction, as shown in Scheme 3.25. In the first example, a polyfluoroalkene was brominated in a high yield by using a 300 W light bulb as the light source. Of note, the synthesis of 1,2,4-tribromo-1,1,2-trifluorobutane (**38**) was carried out under solvent-free conditions in a near-quantitative yield and on a multigram scale (Scheme 3.25a) [67]. 1,2-Dibromo-1,1,2-trichloroethane (**39**) was likewise obtained in about 250 g amounts by the solar lamp irradiation of neat trichloroethylene to which bromine was continuously added (Scheme 3.25b) [68].

3.2.6.2 Addition with the Formation of C−C Bonds

The best-known reactions belonging to this class are based either on the photo-NOCAS process [32] or on the photochemistry of Barton esters [69]. In the first case, a three-component reaction involving a cyanoarene, an olefin, and a nucleophile (usually the solvent) occurs. The reaction is generally initiated by a PET process between the aromatic and the olefin. The examples presented below are chosen from among the most representative and most recent. A typical reaction is illustrated in Scheme 3.26, where an enolized β-dicarbonyl compound acts as an added nucleophile.

Thus, an initial PET reaction between tetracyanobenzene **40** and styrene **41** causes the formation of a radical ion pair. The radical cation of the olefin is trapped by a carbon nucleophile (acetylacetone) and, after deprotonation, the resultant radical reacts with **40**•⁻ to give an anion that in turn undergoes cyanide loss and yields a

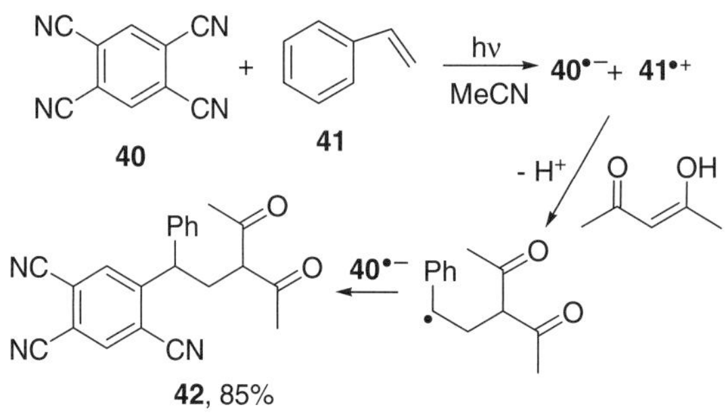

Scheme 3.26

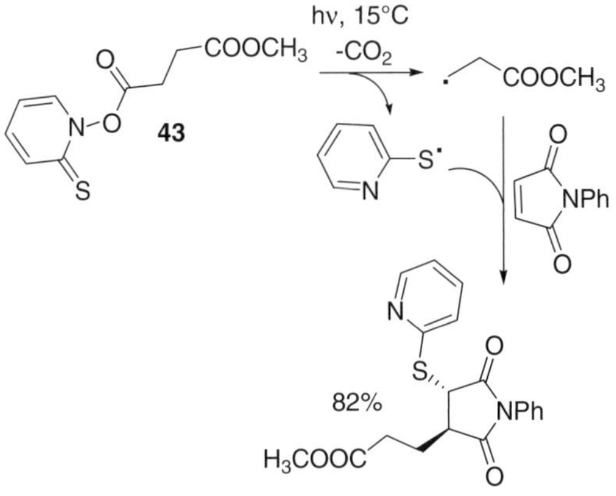

Scheme 3.27

highly functionalized tricyanobenzene **42** in good yield [70]. Alternatively, ammonia and alkylamines can be used as nucleophiles. In this way, a 1,2,4-triphenylbenzene cosensitized reaction between p-dicyanobenzene, 3,4-dihydro-2H-pyran and ammonia gave 4-(3-amino-tetrahydro-2-pyranyl)benzonitrile as a mixture of isomers [71].

The photolysis of Barton esters (N-hydroxy-2-thiopyridone esters, **43**) proved to be an efficient method for generating carbon-centered radicals that are exploited for the regioselective alkylation of electron-deficient olefins; a thiopyridyl unit is likewise incorporated into the end products. In a recent application, Barton esters were found useful in the synthesis of natural and unnatural disubstituted maleimides or maleic anhydrides by way of two consecutive radical addition steps, as described in Scheme 3.27 [72].

The thiopyridyl group can be easily removed or used as the leaving group to restore the maleic amide framework.

The photoaddition of 1,3-diketones to olefins (de Mayo reaction) gives a cyclobutanol that in turn is susceptible to undergo a retro-aldol ring opening leading to 1,5-dicarbonyl compounds. As a result, both positions of the double bond are substituted through the formation of new C−C bonds. Recently, 1,3-diketonatoboron difluorides derivatives have been used as activated 1,3-diketones in the photocycloaddition reaction with an electron-poor olefin. Accordingly, the reaction of dibenzoylmethanatoboron difluoride **44** with isophorone gave a tricarbonyl derivative **45** in 44% yield (Scheme 3.28) [73].

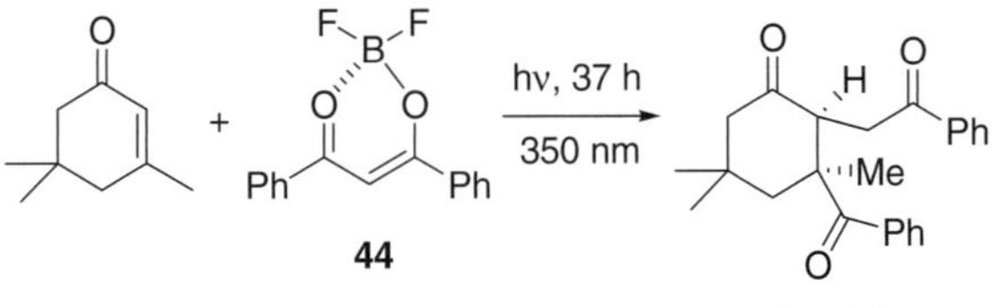

Scheme 3.28

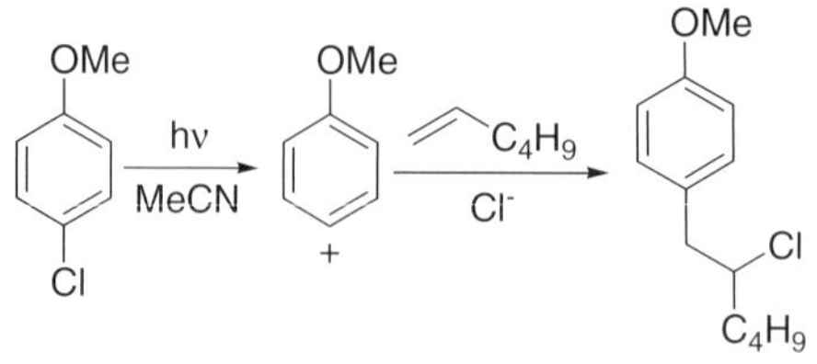

Scheme 3.29

Another useful reaction for the difunctionalization of olefins involves a group transfer carbonylation starting from a α-(phenylseleno)carbonyl (or related derivatives) and a terminal alkene under 80 atm of CO. An alkyl radical is first formed by the photocleavage of a C–SePh bond. The addition of this radical to the olefin, followed by the incorporation of CO and radical coupling with PhSe·, gave substituted selenoesters via a three-component coupling reaction [74]. The intermolecular formation of C–C bonds via phenylseleno group transfer has been likewise adopted in the reaction between ester-substituted O,Se-acetals and an olefin [75].

An unusual process that has attracted much interest recently is the photochemical reaction between chloroaromatic and an alkene, with the aryl ring and chloride adding across the double bond. This offers an alternative approach for the synthesis of alkylaromatics, as illustrated in Scheme 3.29. The irradiation of 4-chloroanisole caused the photoheterolysis of the Ar–Cl bond, thus forming a reactive phenyl cation and Cl⁻. The cation added regiospecifically to the double bond of 1-hexene, and the resulting adduct cation was trapped by the chloride anion to yield 4-(2-chlorohexyl)anisole (**46**) in moderate yield [76].

Aryl cation chemistry allows a double functionalization of vinyl ethers in a three-component reaction, as depicted in Scheme 3.30. In this case, a 3-arylacetal (e.g., **47**) was synthesized after the initial addition of the 4-hydroxyphenyl cation onto ethyl vinyl ether, followed by trapping with the nucleophilic solvent MeOH [77]. For related arylation reactions, see Chapter 10.

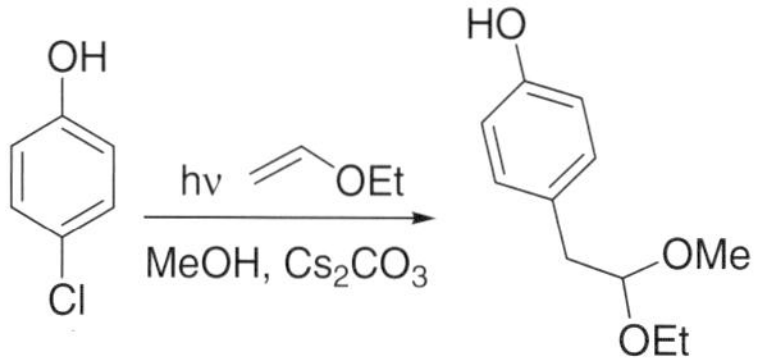

Scheme 3.30

3.3
Addition to C–C Triple Bonds

Photoaddition reactions involving alkynes have been studied much less than those of alkenes. This can be ascribed not only to the lower reactivity of electron-poor alkynes towards radical addition, but also to the fact that even when the functionalization is successful the resultant olefins are usually formed as E/Z mixtures. Usually, the olefins absorb competitively the radiation used for alkynes and undergo geometric isomerization, so that the E/Z ratio often corresponds to the establishment of the photostationary state. X–Y additions to alkynes, however, have been occasionally used, particularly when one of the two groups introduced can be further elaborated easily, as is the case for PhSe or PhTe.

3.3.1
Hydroalkylation Reactions

The photoaddition of alkanes onto electron-poor alkynes (e.g., propiolate or acetilendicarboxylate esters) can be accomplished by a radical conjugate addition reaction [7]. Radicals have been generated either via hydrogen abstraction from cycloalkanes or via electron transfer from 2-alkyl-2-phenyl-1,3-dioxolanes. In the first case, the irradiation was pursued on an alkane solution of an aromatic ketone (used as the photomediator) and the alkyne. Under these conditions, methyl propiolate was alkylated upon irradiation in the presence of 4-trifluoromethylacetophenone to form acrylate **48** in 97% yield (E/Z = 1.3 : 1; Scheme 3.31) [78].

Moreover, this alkylation can be also carried out under heterogeneous conditions (by using silica gel-supported benzophenone) or having recourse to solar light [79]. A vinyl radical is envisaged as the intermediate of the reaction, and since this high-energy intermediate is able to abstract a hydrogen atom from the alkane, a less than stoichiometric amount of the photomediator (20% mol. equiv) was used in some cases.

In the second case, 2-*tert*-butyl dimethylmaleate (as a mixture of isomers) was obtained in 62% yield by a PET reaction between 2-*tert*-butyl-2-phenyl-1,3-dioxolane and dimethylacetilendicarboxylate, catalyzed by 1,2,4,5-tetracyanobenzene (TCB) [80].

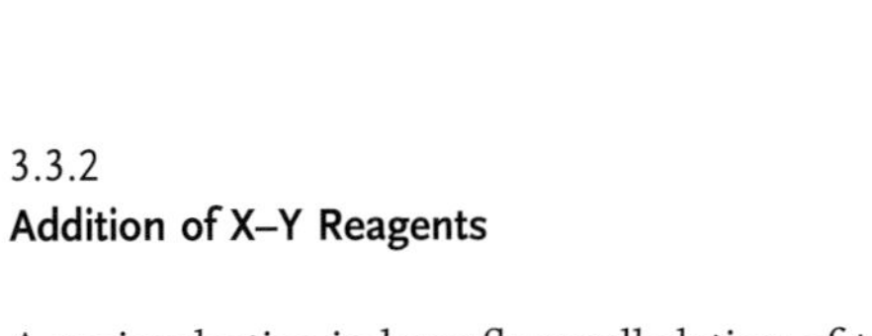

Scheme 3.31

3.3.2
Addition of X–Y Reagents

A regioselective iodoperfluoroalkylation of terminal alkynes (R–C≡CH) has been reported, and is based on photolysis of the C–I bond in perfluoroalkyl iodides (R_F–I). Addition of the thus-formed $R_F^{\bullet}$ radical onto the alkyne afforded a vinyl radical that in turn abstracts an iodine atom from the starting R_F–I to form the end olefin R–C(I)=CH–R_F. A xenon lamp through Pyrex (hv > 300 nm) was used for the reaction, where aliphatic alkynes gave a better alkylation yield with respect to phenylacetylene [81].

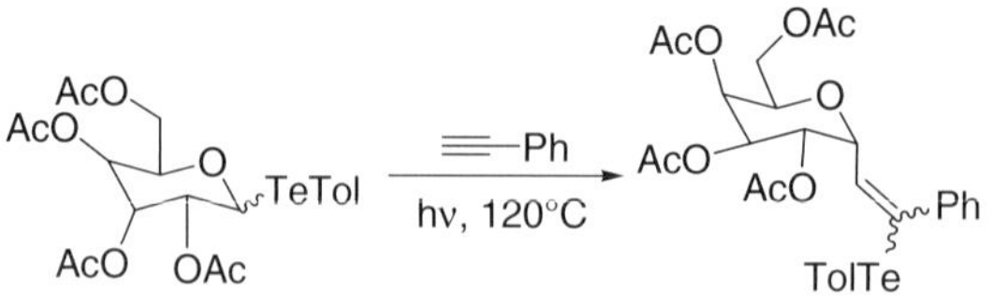

Scheme 3.32

The preparation of a variety of vinylic functionalized C-glycosides involves a photochemical reaction between telluroglycosides with aryl alkynes at 120 °C. Glycosyl radicals were initially formed by photohomolysis of the C–Te bond and readily added to the alkyne; a vinyl telluride (e.g., **49**) was finally obtained by a further attack of RTe· (Scheme 3.32) [82].

An interesting three-component reaction occurred when a solution of diphenyl diselenide (**51**), an electron-deficient alkyne (e.g., ethyl propiolate), and an excess of an isocyanide (**50**) was irradiated through Pyrex with a tungsten lamp (hν > 300 nm). The reaction is based on a selective sequentially radical addition, as illustrated in Scheme 3.33, and the coupling products were formed in 58–85% yield [83]. As for the example reported in Scheme 3.33, the elaboration of compound **52** yielded a precursor for the construction of the carbapenem framework.

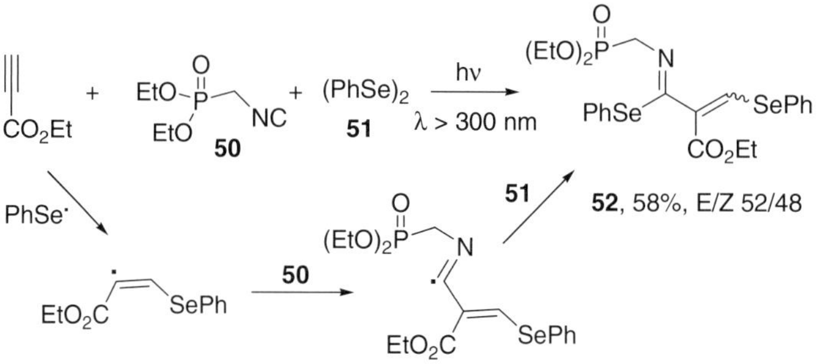

Scheme 3.33

3.4
Concluding Remarks

In this chapter we have highlighted a diverse range of mild and efficient photochemical methods for the intermolecular addition of different molecules onto carbon–carbon multiple bonds. The examples reported demonstrate that, in several cases, C–X bonds were formed through highly selective and efficient photochemical processes. In many cases, the procedures described compared favorably with thermal analogues, allowing a controlled synthetic approach to a wide variety of organic products.

In addition, the above-discussed photochemical methods can be thought of as offering a "clean" (or "green") alternative method [84, 85], which represents an active area of current research.

The obvious practical relevance and potential economic benefits of using light as a cheap, readily available, nontoxic reagent for bond-forming reactions should suffice to drive research towards all aspects of the photochemistry. Accordingly, this chapter may hopefully lead readers to embark on new research projects that further enhance such utility, and to promote the more widespread use of photochemical processes in organic chemistry.

References

1 Semmelhack, M.F.(vol. ed.) (1991) Additions to and substitutions at C—C π-bonds, in *Comprehensive Organic Synthesis, Selectivity, Strategy & Efficiency in Modern Organic Chemistry*, Vol. 4 (eds B.M. Trost and I. Fleming), Pergamon Press, Oxford.

2 Beller, M., Seayad, J., Tillack, A., and Jiao, H. (2004) Catalytic Markovnikov and anti-Markovnikov functionalization of alkenes and alkynes: recent developments and trends. *Angewandte Chemie, International Edition*, **43**, 3368–3398.

3 Chianese, A.R., Lee, S.J., and Gagnè F M. R. (2007) Electrophilic activation of alkenes by Platinum(II): so much more than a slow version of Palladium(II). *Angewandte Chemie, International Edition*, **46**, 4042–4059.

4 Srikanth, G.S.C. and Castle, S.L. (2005) Advances in radical conjugate additions. *Tetrahedron*, **61**, 10377–10441.

5 Horspool, W. and Lenci, F. (eds) (2004) *CRC Handbook of Organic Photochemistry and Photobiology*, 2nd edn, CRC Press, Boca Raton.

6 Mella, M., Fagnoni, M., Freccero, M., Fasani, E., and Albini, A. (1998) New synthetic method via cation fragmentation. *Chemical Society Review*, **27**, 81–89.

7 Fagnoni, M., Dondi, D., Ravelli, D., and Albini, A. (2007) Photocatalysis for the formation of the C—C bond. *Chemical Reviews*, **107**, 2725–2756.

8 Polykarpov, A.Y. and Neckers, D.C. (1995) Tetramethylammonium phenyltrialkylborates in the photoinduced electron transfer reaction with benzophenone. Generation of alkyl radicals and their addition to activated alkenes. *Tetrahedron Letters*, **36**, 5483–5486.

9 Togo, H., Aoki, M., and Yokohama, M. (1993) Reductive addition to electron-deficient olefins with trivalent iodine compounds. *Tetrahedron*, **49**, 8241–8256.

10 Dondi, D., Cardarelli, A.M., Fagnoni, M., and Albini, A. (2006) Photomediated synthesis of β-alkylketones from cycloalkanes. *Tetrahedron*, **62**, 5527–5535.

11 Dondi, D., Fagnoni, M., Molinari, A., Maldotti, A., and Albini, A. (2004) Polyoxotungstate photoinduced alkylation of electrophilic alkenes by cycloalkanes. *Chemistry – A European Journal*, **10**, 142–148.

12 Cermenati, L., Richter, C., and Albini, A. (1998) Solar light induced carbon–carbon bond formation via TiO_2 photocatalysis. *Chemical Communications*, 805–806.

13 Dondi, D., Caprioli, I., Fagnoni, M., Mella, M., and Albini, A. (2003) A convenient route to 1,4-monoprotected dialdehydes, 1,4-ketoaldehydes, γ-lactols and γ-lactones through radical alkylation of α,β-unsaturated aldehydes in organic and, organic-aqueous media. *Tetrahedron*, **59**, 947–957.

14 Církva, V., Böhm, S., and Paleta, O. (2000) Radical addition reactions of fluorinated species. Part 8. Regioselectivity of radical additions to perfluoroalkylethylenes and quantum chemical calculations. Highly selective two-step synthesis of 4-(perfluoroalkyl)butane-1,2-diols. *Journal of Fluorine Chemistry*, **102**, 159–168.

15 Dondi, D., Fagnoni, M., and Albini, A. (2006) Tetrabutylammonium decatungstate-photosensitized alkylation of electrophilic alkenes: convenient functionalization of aliphatic C–H bonds. *Chemistry – A European Journal*, **12**, 4153–4163.

16 Giese, B. and Thoma, G. (1991) Dimeric metal complexes as mediators for radical C–C bond-forming reactions. *Helvetica Chimica Acta*, **74**, 1135–1142.

17 Manfrotto, C., Mella, M., Freccero, M., Fagnoni, M., and Albini, A. (1999) Photochemical synthesis of 4-oxobutanal acetals and of 2-hydroxycyclobutanone ketals. *Journal of Organic Chemistry*, **64**, 5024–5028.

18 Mosca, R., Fagnoni, M., Mella, M., and Albini, A. (2001) Synthesis of monoprotected 1,4-diketones by photo-induced alkylations of enones with 2-substituted-1,3-dioxolanes. *Tetrahedron*, **57**, 10319–10328.

19 Ghosh, A.K., Leshchenko, S., and Noetzel, M. (2004) Stereoselective photochemical 1,3-dioxolane addition to 5-alkoxymethyl-2(5H)-furanone: synthesis of bis-tetrahydrofuranyl ligand for HIV protease inhibitor UIC-94017 (TMC-114). *Journal of Organic Chemistry*, **69**, 7822–7829.

20 Parsons, P.J. and Cowell, J.K. (2000) A rapid synthesis of (−)-tetrahydrolipstatin. *Synlett*, 107–109 and references therein.

21 Hoffmann, N., Bertrand, S., Marinković F S. and Pesch, J. (2006) Efficient radical addition of tertiary amines to alkenes using photochemical electron transfer. *Pure and Applied Chemistry*, **78**, 2227–2246.

22 Bertrand, S., Hoffmann, N., and Pete, J.-P. (2000) Highly efficient and stereoselective radical addition of tertiary amines to electron-deficient alkenes. Application to the enantioselective synthesis of necine bases. *European Journal of Organic Chemistry*, 2227–2238.

23 Hasegawa, E., Xu, W., Mariano, P.S., Yoon, U.-C., and Kim, J.-U. (1988) Electron-transfer-induced photoadditions of the silyl amine Et_2NCH_2TMS to α,β-unsaturated cyclohexenones. Dual reaction pathways based on ion-pair-selective cation-radical chemistry. *Journal of the American Chemical Society*, **110**, 8099–8111.

24 Mai, J.-C., Lin, Y.-C., Hseu, T.-M., and Ho, T.-I. (1993) Styrene-amine exciplexes: substituent effects. *Journal of Photochemistry and Photobiology, A: Chemistry*, **71**, 237–243.

25 Angioni, S., Ravelli, D., Emma, D., Dondi, D., Fagnoni, M., and Albini, A. (2008) Tetrabutylammonium decatungstate (chemo)selective photocatalyzed, radical C–H functionalization in amides. *Advanced Synthesis and Catalysis*, **350**, 2209–2214.

26 Murakami, S. and Fuchigami, T. (2006) Synthesis of precursors of *gem*-difluorodiols and amino alcohols using electrochemical and photochemical reactions. *Synlett*, 1015–1020.

27 Natura, H., Muratami, S., and Fuchigami, T. (2007) Photochemical generation of difluoromethyl radicals having various functional groups and their highly regioselective addition to olefins and aromatic substitution. *Tetrahedron*, **63**, 10237–10245.

28 Shiraishi, Y., Tsukamoto, D., and Hirai, T. (2008) Highly efficient methyl ketone synthesis by acetone. *Organic Letters*, **10**, 3117–3120

29 Ohashi, M., Nakatani, K., Maeda, H., and Mizuno, K. (2008) Photochemical monoalkylation of propanedinitrile by electron-rich alkenes. *Organic Letters*, **10**, 2741–2743.

30 Ogura, K., Aray, T., Kayano, A., and Akazome, M. (1999) Efficient 1,2-asymmetric induction in radical reactions:

addition of acyl radicals to 3-hydroxy-1-(methylthio)-1-(*p*-tolylsulfonyl)-1-alkenes and their acetates. *Tetrahedron Letters*, **40**, 2537–2540.

31 Esposti, S., Dondi, D., Fagnoni, M., and Albini, A. (2007) Acylation of electrophilic olefins through decatungstate-photocatalyzed activation of aldehydes. *Angewandte Chemie, International Edition*, **46**, 2531–2534.

32 Mangion, D. and Arnold, D.R. (2004) The photochemical Nucleophile-Olefin Combination, Aromatic Substitution (Photo-NOCAS) reaction, in *Handbook of Organic Photochemistry and Photobiology*, 2nd edn (eds W. Horspool and F. Lenci), CRC Press, Boca Raton, pp. 40-1–40-17.

33 Müller, T.E., Hultzsch, K.C., Yus, M., Foubelo, F., and Tada, M. (2008) Hydroamination: direct addition of amines to alkenes and alkynes. *Chemical Reviews*, **108**, 3795–3892 and references therein.

34 Yasuda, M., Sone, T., Tanabe, K., and Shima, K. (1994) Amination of *o*-alkenylphenols via photoinduced proton transfer. *Chemistry Letters*, 453–456.

35 Yasuda, M., Shiragami, T., Matsumoto, J., Yamashita, T., and Shima, K. (2006) Photoamination with ammonia and amines, in *Molecular and Supramolecular Photochemistry (Organic Photochemistry and Photophysics)*, Vol. 14 (eds V. Ramamurthy and K.S. Schanze), CRC Press, Boca Raton, pp. 207–253.

36 Yasuda, M., Isami, T., Kubo, J., Mizutani, M., Yamashita, T., and Shima, K. (1992) Regiochemistry on photoamination of stilbene derivatives with ammonia via electron transfer. *Journal of Organic Chemistry*, **57**, 1351–1354.

37 Yamashita, T., Yasuda, M., Isami, T., Nakano, S., Tanabe, K., and Shima, K. (1993) Synthesis of phenethylamine moiety by photoamination of styrene derivatives with ammonia. *Tetrahedron Letters*, **34**, 5131–5134.

38 Yamashita, T., Yasuda, M., Isami, T., Tanabe, K., and Shima, K. (1994) Photoinduced nucleophilic addition of ammonia and alkylamines to methoxy-substituted styrene derivatives. *Tetrahedron*, **50**, 9275–9286.

39 Yasuda, M., Kojima, R., Tsutsui, H., Utsunomiya, D., Ishii, K., Jinnouchi, K., Shiragami, T., and Yamashita, T. (2003) Redox-photosensitized aminations of 1,2-benzo-1,3-cycloalkadienes, arylcyclopropanes, and quadricyclane with ammonia. *Journal of Organic Chemistry*, **68**, 7618–7624.

40 Suau, R., García-Segura, R., Sánchez-Sánchez, C., Pérez-Inestrosa, E., and Pedraza, A.-M. (2003) Photoinduced SET phthalimidation of unactivated double bonds and its application to the synthesis of protected phenethylamines. *Tetrahedron*, **59**, 2913–2919.

41 Moran, J., Cebrowski, P.H., and Beauchemin, A.M. (2008) Intermolecular hydroaminations via strained (*E*)-cycloalkenes. *Journal of Organic Chemistry*, **73**, 1004–1007.

42 Trofimov, B.A., Gusarova, N.K., Malysheva, S.F., Ivanova, N.I., Sukhov, B.G., Belorgolova, N.A., and Kuimov, V.A. (2002) Hydrophosphination of vinyl sulfides and vinyl selenides: first examples. *Synthesis*, 2207–2210.

43 Armstrong, S.K., Cross, R.J., Farrugia, L.J., Nichols, D.A., and Perry, A. (2002) Semi-rigid bis-phosphane ligands for metallamacrocycle formation. *European Journal of Inorganic Chemistry*, 141–152.

44 Kröcher, O., Köppel, R.A., Fröba, M. and Baiker, A. (1998) Silica hybrid gel catalysts containing group(VIII) transition metal complexes: preparation, structural, and catalytic properties in the synthesis of *N*,*N*-dimethylformamide and methyl formate from supercritical carbon dioxide. *Journal of Catalysis*, **178**, 284–298.

45 Reinhard, S., Soba, P., Rominger, F., and Bluemel, J. (2003) New silica-immobilized nickel catalysts for cyclotrimerizations of acetylenes. *Advanced Synthesis and Catalysis*, **345**, 589–602.

46 Okazaki, M., Kawano, Y., Tobita, H., Inomata, S., and Ogino, H. (1995) Light- and heat-induced isomerization of chloro(hydrido)iridium(III) complex containing a (2-phosphinoethyl)silyl chelate ligand. *Chemistry Letters*, 1005–1006.

47 Lindner, E., Schmid, M., Wald, J., Queisser, J.A., Geprägs, M., Wegner, P., and Nachtigal, C. (2000) Catalytic activity of cationic diphospalladium(II) complexes in the alkene/CO copolymerization in organic solvents and water in dependence on the length of the alkyl chain at the phosphine ligands. *Journal of Organometallic Chemistry*, **602**, 173–187.

48 Gusarova, N.K., Malysheva, S.F., Belogorlova, N.A., Sukhov, B.G., and Trofimov, B.A. (2007) Radical addition of secondary phosphine selenides to alkenes. *Synthesis*, 2849–2852.

49 Sekhar, B.B.V.S. and Bentrude, W.G. (1999) Photochemical SET induced 1,4-conjugate additions of silyl phosphites to cyclic enones. *Tetrahedron Letters*, **40**, 1087–1090.

50 Wan, P., Davis, M.J., and Teo, M.A. (1989) Photoaddition of water and alcohols to 3-nitrostyrenes. Structure-reactivity and solvent effects. *Journal of Organic Chemistry*, **54**, 1354–1359.

51 Kalanderopoulos, P. and Yates, K. (1986) Intramolecular proton transfer in photohydration reactions. *Journal of the American Chemical Society*, **108**, 6290–6295.

52 Cole, J.G. and Wan, P. (2002) Mechanistic studies of photohydration of *m*-hydroxy-1,1-diaryl alkenes. *Canadian Journal of Chemistry*, **80**, 46–54.

53 Asaoka, S., Wada, T., and Inoue, Y. (2003) Microenvironmental polarity control of electron-transfer photochirogenesis. Enantiodifferentiating polar addition of 1,1-diphenyl-1-alkenes photosensitized by saccharide naphthalenecarboxylates. *Journal of the American Chemical Society*, **125**, 3008–3027.

54 Oates, R.P. and Jones, P.B. (2008) Photosensitized tetrahydropyran transfer. *Journal of Organic Chemistry*, **73**, 4743–4745.

55 Dondoni, A. (2008) The emergence of thiol–ene coupling as a click process for materials and bioorganic chemistry. *Angewandte Chemie, International Edition*, **47**, 8995–8997.

56 Kanehara, M., Oumi, Y., Sano, T., and Teranishi, T. (2004) Syntheses of the novel acidic and basic ligands and superlattice formation from gold nanoparticles through interparticle acid-base interaction. *Bulletin of the Chemical Society of Japan*, **77**, 1589–1597.

57 Fujisawa, T., Odake, S., Ogawa, Y., Yasuda, J., Morita, Y., and Morikawa, T. (2002) Design and synthesis of sulphur based inhibitors of matrix metalloproteinase-1. *Chemical and Pharmaceutical Bulletin*, **50**, 239–252.

58 Matsumura, N., Mori, O., Takeguchi, T., Okumura, Y., and Mizuno, K. (2004) Photoaddition of thiocarboxylic acids to 2,3-diallyltetraazathiapentalenes. *Journal of Heterocyclic Chemistry*, **41**, 873–876.

59 Miyagawa, A., Kasuya, M.C.Z., and Hatanaka, K. (2006) Immobilization of glycoconjugate polymers on cellulose membrane for affinity separation. *Bulletin of the Chemical Society of Japan*, **79**, 348–356.

60 Wu, X. and Bundle, D.R. (2005) Synthesis of glycoconjugate vaccines for *Candida albicans* using novel linker methodology. *Journal of Organic Chemistry*, **70**, 7381–7388.

61 Buchynskyy, A., Stembera, K., Knoll, D., Vogel, S., Kempin, U., Biallaß, A., Hennig, L., Findeisen, M., Müller, D., and Welzel, P. (2002) Synthesis of tools for raising antibodies against moenomycin epitopes and initial immunological studies. *Tetrahedron*, **58**, 7741–7760.

62 Yoshida, T., Chiba, T., Yokochi, T., Onozaki, K., Sugiyama, T., and Nakashima, I. (2001) Synthesis of a set of di- and tri-sulfated galabioses. *Carbohydrate Research*, **335**, 167–180.

63 Roy, R., Baek, M.-G., and Rittenhouse-Olson, K. (2001) Synthesis of N,N'-bis(acrylamido)acetic acid-based T-antigen glycodendrimers and their mouse monoclonal IgG antibody binding properties. *Journal of the American Chemical Society*, **123**, 1809–1816.

64 Ramos, D., Rollin, P., and Klaffke, P. (2001) Chemoenzymatic synthesis of neoglycopeptides: application to an α-Gal-terminated neoglycopeptide. *Journal of Organic Chemistry*, **66**, 2948–2956.

65 Carvalho de Souza, A., Halkes, K.M., Meeldijk, J.D., Verkleij, A.J., Vliegenthart, J.F.G., and Kamerling, J.P. (2004) Synthesis of gold glyconanoparticles: possible probes for the exploration of carbohydrate-mediated self-recognition of marine sponge cells. *European Journal of Organic Chemistry*, 4323–4339.

66 Brunet, P., Demers, E., Maris, T., Enright, G.D., and Wuest, J.D. (2003) Designing permeable molecular crystals that react with external agents to give crystalline products. *Angewandte Chemie, International Edition*, **42**, 5303–5306.

67 Shellhamer, D.F., Allen, J.L., Allen, R.D., Gleason, D.C., Schlosser, C., Powers, B.J., Probst, J.W., Rhodes, M.C., Ryan, A.J., Titterington, P.K., Vaughan, G.G., and Heasley, V.L. (2003) Ionic reaction of halogens with terminal alkenes: the effect of electron-withdrawing fluorine substituents on the bonding of halonium ions. *Journal of Organic Chemistry*, **68**, 3932–3937.

68 Beit-Yannai, M., Rappoport, Z., Shainyan, B.A., and Danilevich, Y.S. (1997) Intramolecular geminal and vicinal element effects in substitution of simple bromo(chloro)alkenes by methoxide and thiolate ions. An example of a single step substitution? *Journal of Organic Chemistry*, **62**, 8049–8057.

69 Dalko, P.I. (2004) The photochemistry of Barton esters, in *Handbook of Organic Photochemistry and Photobiology*, 2nd edn (eds W. Horspool and F. Lenci), CRC Press, Boca Raton, pp. 67-1–67-23.

70 Zhi-Feng, L., Yong-Miao, S., Jia-Jun, Y., Hong-Wen, H., and Jian-Hua, X. (2008) Photoinduced three-component reactions of tetracyanobenzene with alkenes in the presence of 1,3-dicarbonyl compounds as nucleophiles. *Journal of Organic Chemistry*, **73**, 8010–8015.

71 Yamashita, T., Itagawa, J., Sakamoto, D., Nakagawa, Y., Matsumoto, J., Shiragami, T., and Yasuda, M. (2007) Redox-photosensitized amination of alkenes and alkadienes with ammonia and alkylamines. *Tetrahedron*, **63**, 374–380.

72 Brault, L., Denance, M., Banaszak, E., Maadidi, S.E., Battaglia, E., Bagrel, D., and Samadi, M. (2007) Synthesis and biological evaluation of dialkylsubstituted maleic anhydrides as novel inhibitors of Cdc25 dual specificity phosphatases. *European Journal of Medicinal Chemistry*, **42**, 243–247.

73 Chow, Y.L., Cheng, X., Wang, S.S., and Wu, S.P. (1997) Further examples of photocycloadditions of 1,3-diketonatoboron difluorides to olefins: a description of the reaction pattern and mechanism. *Canadian Journal of Chemistry*, **75**, 720–726.

74 Ryu, I., Muraoka, H., Kambe, N., Komatsu, M., and Sonoda, N. (1996) Group transfer carbonylations: photoinduced alkylative carbonylation of alkenes accompanied by phenylselenenyl transfer. *Journal of Organic Chemistry*, **61**, 6396–6403.

75 Renaud, P. and Abazi, S. (1996) Use of O,Se-acetals for radical-mediated phenylseleno group transfer reactions. *Synthesis*, 253–258.

76 Manet, I., Monti, S., Grabner, G., Protti, S., Dondi, D., Dichiarante, V., Fagnoni, M., and Albini, A. (2008) Revealing phenylium, phenonium, vinylenephenonium, and benzenium ions in solution. *Chemistry – A European Journal*, **14**, 1029–1039.

77 Lazzaroni, S., Protti, S., Fagnoni, M., and Albini, A. (2009) Photoinduced three-component reaction: A convenient access

to 3-arylacetals or 3-arylketals. *Organic Letters*, **11**, 349–352.
78 Doohan, R.A., Hannan, J.J., and Geraghty, N.W.A. (2006) The photomediated reaction of alkynes with cycloalkanes. *Organic and Biomolecular Chemistry*, **4**, 942–952.
79 Doohan, R.A. and Geraghty, N.W.A. (2005) A comparative analysis of the functionalisation of unactivated cycloalkanes using alkynes and either sunlight or a photochemical reactor. *Green Chemistry*, **7**, 91–96.
80 Mella, M., Fagnoni, M., and Albini, A. (1994) Radicals through photoinduced electron transfer. Addition to olefin and addition to olefin-aromatic substitution reactions. *Journal of Organic Chemistry*, **59**, 5614–5622.
81 Tsuchii, K., Imura, M., Kamada, M., Irao, T., and Ogawa, A. (2004) An efficient photoinduced iodoperfluoroalkylation of carbon-carbon unsaturated compounds with perfluoroalkyl iodides. *Journal of Organic Chemistry*, **69**, 6658–6665.
82 Yamago, S., Miyazoe, H., and Yoshida, J. (1999) Synthesis of vinylic C-glycosides from telluroglycosides. Addition of photochemically and thermally, generated glycosyl radicals to alkynes. *Tetrahedron Letters*, **40**, 2343–2346.
83 Ogawa, A., Doi, M., Tsuchii, K. and Hirao, T. (2001) Selective sequential addition of diphenyl diselenide to ethyl propiolate and isocyanides upon irradiation with near UV-light. *Tetrahedron Letters*, **42**, 2317–2319.
84 Albini, A. and Fagnoni, M. (2008) Photochemistry as a green synthetic method, in *New Methodologies and Techniques for a Sustainable Organic Chemistry* (eds A. Mordini and F. Faigl), Springer Science & Business Media B.V., pp. 279–293.
85 Albini, A., Fagnoni, M., and Mella, M. (2000) Environment-friendly organic synthesis. The photochemical approach. *Pure and Applied Chemistry*, **72**, 1321–1326.

4
Formation of a Three-Membered Ring
Takashi Tsuno

4.1
Introduction

Cyclopropanes have been used as prominent synthetic precursors in organic synthesis, since through ring-opening reactions they form more functionalized cycloalkanes and acyclic compounds. In addition, the cyclopropane motif is often found as a key structural element in a wide range of naturally occurring and biologically active compounds. The thematic issue of "cyclopropanes and related ring" in *Chemical Reviews* from the American Chemical Society in 2003, which was devoted to the cyclopropane ring systems, contains a wealth of worthy synthetic targets [1].

The synthesis of cyclopropanes by transition metal-mediated carbene transfer from diazo compounds to C=C bonds represents a major synthetic method. However, because such transition metals are usually unstable under air- and/or moisture-sensitive conditions, their handling requires special techniques. Moreover, the toxicity of transition metals requires special consideration for the synthesis of drugs. In order to circumvent these problems, photochemical reactions represent a good choice for organic synthesis, and would produce small-ring compounds by interaction between a reactant in an excited state and a ground-state molecule. Hence, photochemical methods are widely applied to key steps in the syntheses of pharmaceutical products, natural compounds, agrochemicals, fine chemicals, intermediates, or new chemicals. Since the end of the twentieth century, photochemistry has been especially applied to reaction controls on the molecular level, to the construction of artificial photosynthesis, and to a photochemical process for green and sustainable chemistry. Because a specific position in a molecule can be excited by control at the molecular level, the target molecules are obtained in good yields. Such molecular control has the advantage of reducing the number of reaction steps.

The sections of this chapter are arranged according to the formation of a three-membered ring by photochemical methods; that is, di-π-methane rearrangement, oxa-di-π-methane rearrangement including aza-di-π-methane rearrangement, [2 + 1] cycloaddition of alkenes with carbenes, intramolecular hydrogen abstractions, [3 + 2]

Handbook of Synthetic Photochemistry. Edited by Angelo Albini and Maurizio Fagnoni
Copyright © 2010 WILEY-VCH Verlag GmbH & Co. KGaA, Weinheim
ISBN: 978-3-527-32391-3

cycloaddition of arenes with alkenes, and the photochemical synthesis of three-membered heterocycles. The first examples of the di-π-methane rearrangement, the oxa-di-π-methane rearrangement, and the [3 + 2] cycloaddition were reported in 1966, since which time four decades have elapsed. A number of reviews detailing these reactions, and including their historical backgrounds and mechanisms, have been published in recent years. In this chapter we describe these reactions, with special emphasis on issues of chemoselectivity, regioselectivity, stereoselectivity, and enantioselectivity, as well as on natural product synthesis. The preparation of cycloalkanes – and particularly of cyclopropanes – by the elimination of a group from a cyclic derivative are presented in Chapter 2.

4.2
Di-π-Methane Rearrangement

The first discovery of a di-π-methane (DPM) rearrangement was the photochemical rearrangement of barrelene to semibullvalene (Scheme 4.1) [2]. This DPM rearrangement is also referred to as the Zimmerman rearrangement. The DPM rearrangement has especially been examined with respect to energetics by Zimmerman and colleagues, and several excellent reports are available involving the reaction mechanism, multiplicities, and substituent effects. In 1996, Zimmerman and Armesto reviewed a large number of studies on DPM rearrangement and related rearrangements [3], and other reviews have subsequently been published [4]. The mechanistic details of DPM are described in these reviews [2–4].

4.2.1
Di-π-Methane Rearrangement of Barrelene, Benzobarrelene, Dibenzobarrelene, and Related Derivatives

Barrelene, benzobarrelene and dibenzobarrelene derivatives are readily synthesized by Diels–Alder reactions of arenes with alkenes. For barrelene and related derivatives, the DPM rearrangement proceeds solely from the triplet state, giving semibullvalenes (Scheme 4.1) [3, 4].

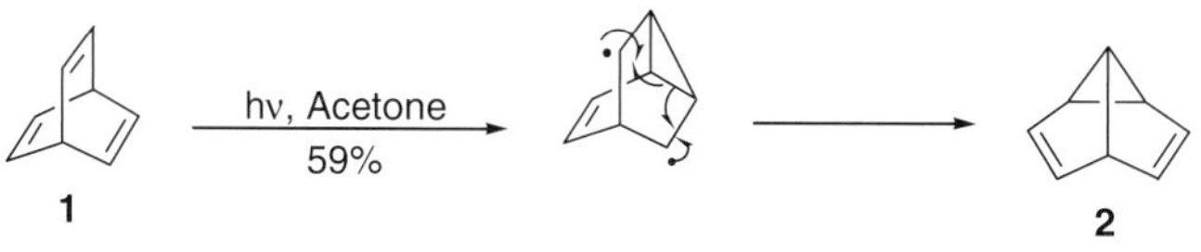

Scheme 4.1

However, homobarrelene derivatives **3** undergo a DPM rearrangement upon direct photolysis with no sensitizers and give photoisomerized products **4** in good yields (Scheme 4.2) [5].

The acetophenone-sensitized irradiation of benzobarrelene **5** gives benzosemibullvalene **6** in a quantitative yield (Scheme 4.3) [6]. However, the direct irradiation

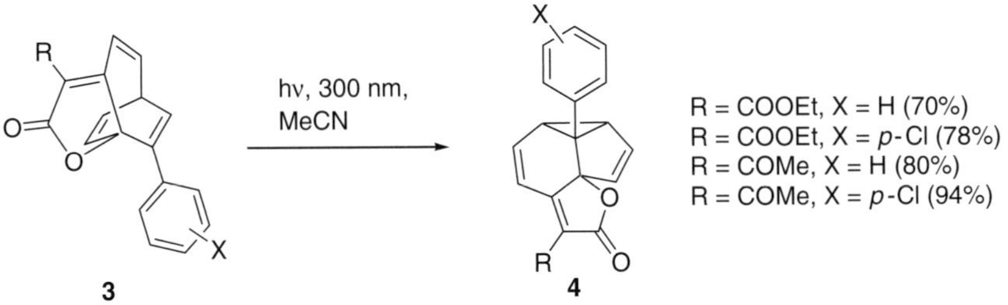
hν, 300 nm,
MeCN
3
4
R = COOEt, X = H (70%)
R = COOEt, X = *p*-Cl (78%)
R = COMe, X = H (80%)
R = COMe, X = *p*-Cl (94%)

Scheme 4.2

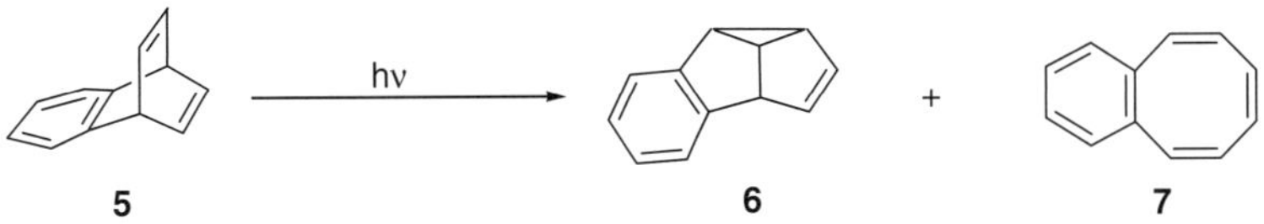
hν
5
6
7

Medium	Yield of **6** (%)	Yield of **7** (%)
Hexane-PhCOMe	100	
Hexane	4	96
TlX-Slurry	92	8
TlX-Solid	95	5

Scheme 4.3

leads to benzocyclooctatetraene **7** in a high yield. When Tl^+-exchange X-zeolite is used as the reaction media, the triplet product **6** is obtained under both TlX-slurry in hexane and TlX-solid conditions, in an excellent yield (Scheme 4.3) [6]. The direct irradiation of **8** (Scheme 4.4, upper part) in TlX also gives **9a**. Although the acetone-sensitized irradiation of substituted dibenzobarrelenes yields two regioisomers **9b** and **9b′**, the direct irradiation in TlX leads to **9b** only (Scheme 4.4, bottom part). Thus, heavy cations such as Tl^+ enhance the intersystem crossing rate of **5** and **8** and also control the regioselectivity for the DPM rearrangement of **8** [6].

The direct irradiation of a crown ether-annulated dibenzobarrelene **10** in the presence of MBF_4 ($M = Na^+$, K^+, Cs^+) in solid state leads preferentially to **11** because of a strong cation effect (Scheme 4.5) [7].

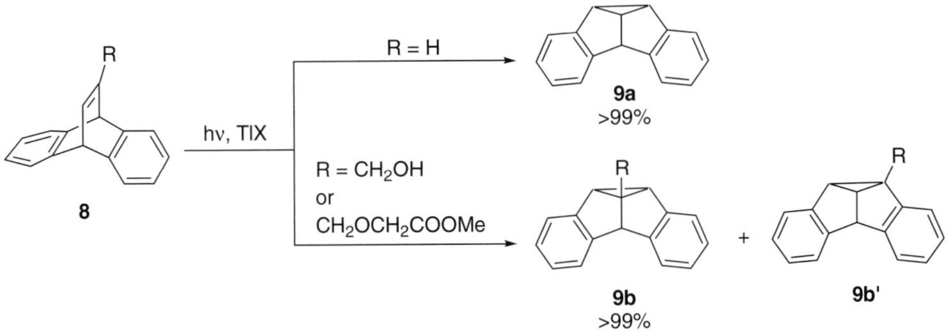
R
8
hν, TlX
R = H
9a
>99%
R = CH_2OH
or
CH_2OCH_2COOMe
9b
>99%
9b'

Scheme 4.4

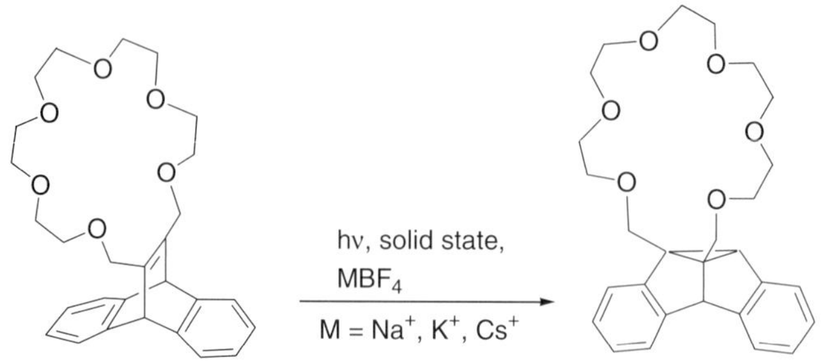

4 Formation of a Three-Membered Ring

Scheme 4.5

MBF$_4$	Conv. (%)	Yield (%)
NaBF$_4$	13	>95
KBF$_4$	47	83
CsBF$_4$	20	>95

Dibenzobarrelene diisopropylester **12** is an achiral molecule, but it forms dimorphic crystals. A chiral crystal, $P2_12_12_1$, can be obtained by crystallization from the molten state. Irradiation of such crystals gives **13** in over 95% enantiomeric excess (ee) (Scheme 4.6) [8].

The 4-benzoylphenyl ester of L-valine acts as a good chiral sensitizer for the DPM rearrangement of **14** in the solid state (Scheme 4.7) [9].

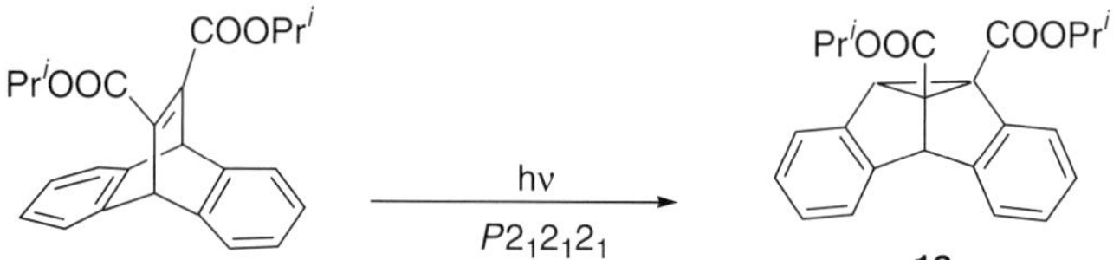

Scheme 4.6

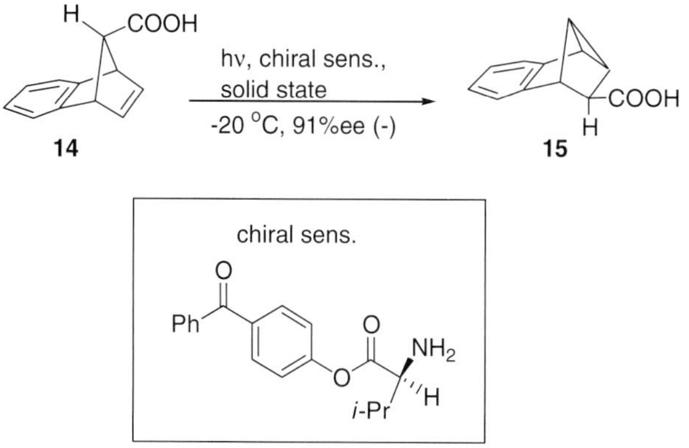

Scheme 4.7

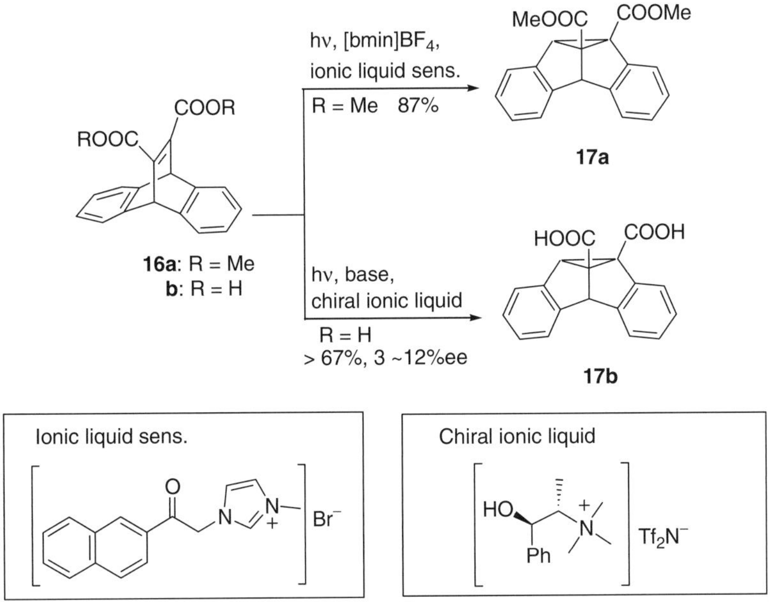

4.2 Di-π-Methane Rearrangement

Scheme 4.8

The DPM rearrangement of **16a** in the presence of a soluble ionic liquid sensitizer in the ionic liquid [bmin]BF$_4$ gives **17a** in 87% yield (upper part in Scheme 4.8) [10]. Chiral ionic liquids have been evaluated as chiral induction solvents for the DPM rearrangement of **16b** (bottom part in Scheme 4.8) [11].

Diazinobarrelene derivatives undergo a chemoselective DPM rearrangement. Liao *et al.* have reported that a direct and acetone-sensitized irradiation of pyrazinobarrelene **18a** affords semibullvalene **19a** in excellent yield, while the direct irradiation of naphthobarrelene **18b** in benzene leads exclusively to **19b** in 97% yield (Scheme 4.9) [12]. The barrelenes **18a,b** undergo predominantly aza-di-π-methane (ADPM) rearrangement (C=C–C–C=N bridging) rather than DPM rearrangement (C=C–C–C=C bridging). The details of ADPM rearrangement are described in Section 4.3.2.

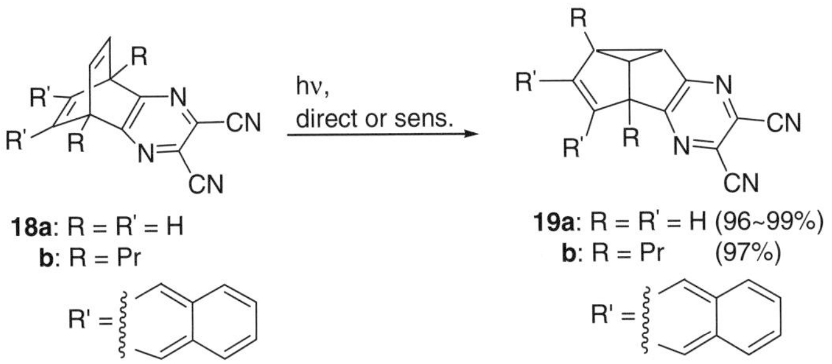

Scheme 4.9

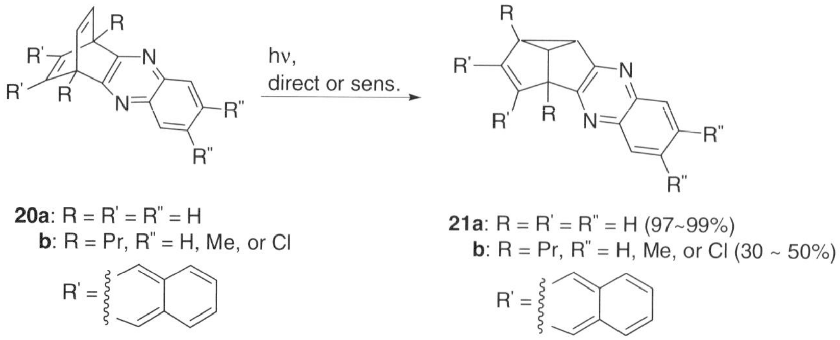

Scheme 4.10

20a: R = R' = R" = H
b: R = Pr, R" = H, Me, or Cl

21a: R = R' = R" = H (97~99%)
b: R = Pr, R" = H, Me, or Cl (30 ~ 50%)

Under similar conditions as with **18a,b**, quinoxalinobarrelenes **20a,b** provide semibullvalenes **21a,b** (Scheme 4.10) [13]. The products **21a** and **21b** are generated via DPM and ADPM rearrangements, respectively. Recently, Liao and colleagues reported the details of substituent effects on the photochemical rearrangements of pyrazino-, quinoxalino-, and benzoquinolinobarrelenes [14].

Substituted sultams **21** undergo DPM rearrangement by acetone sensitization to give **22** in moderate yields (Scheme 4.11) [15].

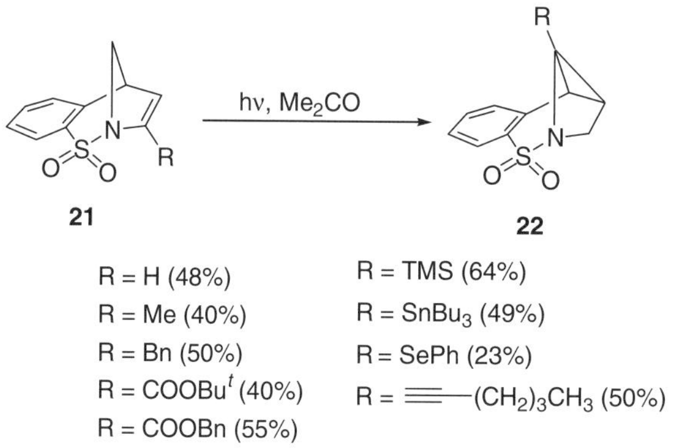

21
R = H (48%)
R = Me (40%)
R = Bn (50%)
R = COOBu^t (40%)
R = COOBn (55%)

22
R = TMS (64%)
R = SnBu$_3$ (49%)
R = SePh (23%)
R = ≡—(CH$_2$)$_3$CH$_3$ (50%)

Scheme 4.11

4.2.2
Di-π-Methane Rearrangement of Acyclic Systems

The DPM rearrangement of acyclic systems, divinyl-methanes, tends to perform better in the singlet excited state by direct irradiation [3]. This is because the triplet excited state of acyclic systems decays preferentially to the ground state via the free rotor effect, and the further geometric isomerization of a vinyl moiety takes place. Regioselective DPM rearrangement upon direct irradiation has been obtained in a number of examples in the photochemistry of acyclic systems. For example, divinylmethanes substituted by either an electron-withdrawing or an electron-

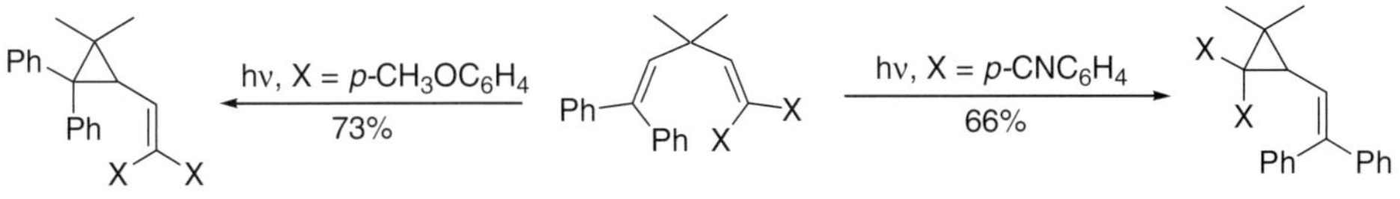

4.2 Di-π-Methane Rearrangement

Scheme 4.12

donating group on one of the vinyl moieties (**23**) undergo a regioselective DPM rearrangement. As shown in Scheme 4.12, the p-cyanophenyl groups are found on the cyclopropane ring of **24** [16], while the p-methoxyphenyl groups remain on the vinyl moiety of **25** [17].

Divinylmethanes, in which a methylene moiety has bulky substituents such as isopropyl groups, undergo DPM rearrangement by the triplet-sensitization, because free rotation of the vinyl moiety is inhibited (Scheme 4.13) [18].

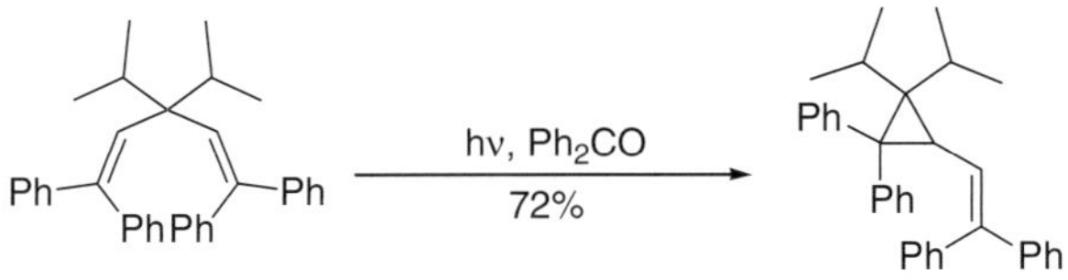

Scheme 4.13

The direct and sensitized photolyses of **26** possessing methoxycarbonyl groups at the geminal position give different regioisomeric vinylcyclopropanes (**27** and **28**) via the regioisomeric DPM rearrangement under multiplicity control (Scheme 4.14) [19].

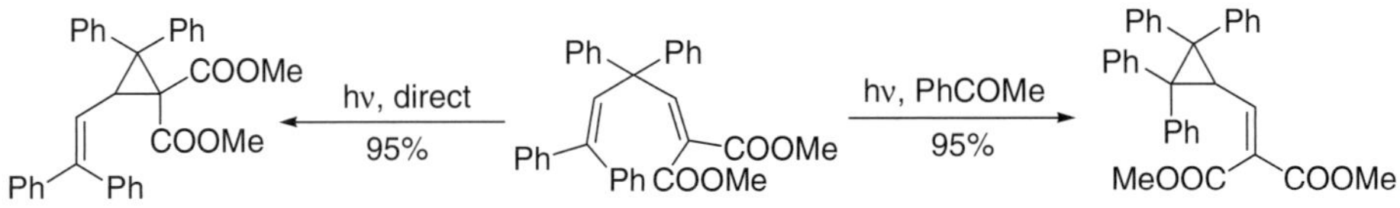

Scheme 4.14

Direct photolysis of 1,1-dicyano-3,3-dimethyl-1,4,5-hexatriene **29** gives an allenylcyclopropane **30** via DPM rearrangement in 57% yield, while acetone-sensitization yields a hausan **31** via an intramolecular [2 + 2] cycloaddition (Scheme 4.15) [20].

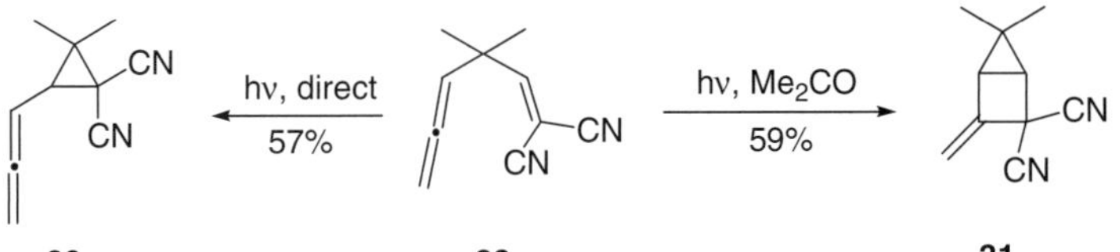

Scheme 4.15

Table 4.1 Photoproducts obtained on direct and acetophenone-sensitized irradiation of **32a,b**.

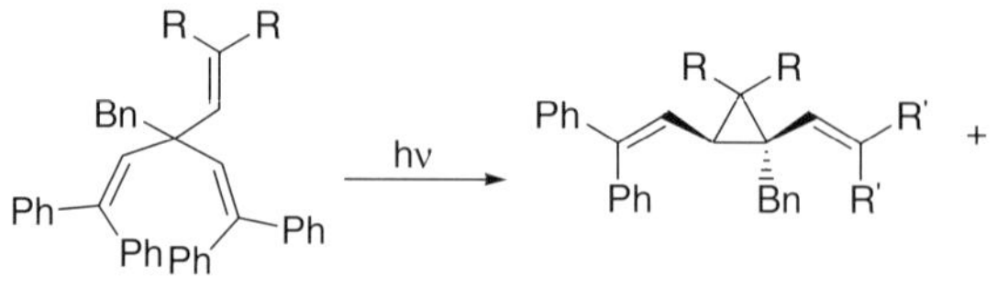

32a: R = Me
b: R = p-C$_6$H$_4$CN

33a: R = Ph, R' = Me
b: R = p-CNC$_6$H$_4$, R' = Ph

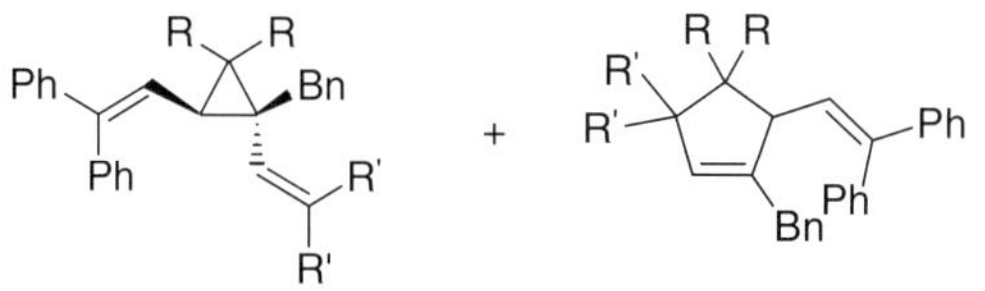

34a: R = Ph, R' = Me
b: R = p-CNC$_6$H$_4$, R' = Ph

35a: R = Ph, R' = Me
b: R = p-CNC$_6$H$_4$, R' = Ph

Reactant	Additive	Conversion (%)	Products (Yield%)		
32a	None	100	33a (35)	34a (23)	35a (42)
	Acetophenone	100	(67)	(28)	(5)
32b	None	100	33b (34)	34b (27)	35b (39)
	Acetophenone	100	(53)	(47)	

When trivinyl-methanes **32** undergo DPM rearrangement, the resultant cyclopropanes **33** and **34** bear two vinyl functional groups on the ring [21]. Although the DPM and tri-π-methane rearrangements compete under the direct irradiation of trivinylmethanes, the resultant divinylcyclopropanes further isomerize via a DPM rearrangement to vinylcyclopentenes by photolysis. However, triplet-sensitization using acetophenone leads preferentially to the divinylcyclopropanes **33** and **34** (Table 4.1) [21a].

Each of the fused alicyclic divinyl-methanes **36** and **37** undergoes DPM rearrangement under direct irradiation to give the same product **38** (Scheme 4.16) [22].

The DPM rearrangement of 4H-thiopyran-1,1-dioxide **39** takes place under both direct and triplet-sensitized irradiation, to yield bicyclic sulfones **40** and **41** (Scheme 4.17). The singlet excited state of **39** leads to only the *syn*-form **41**. Although

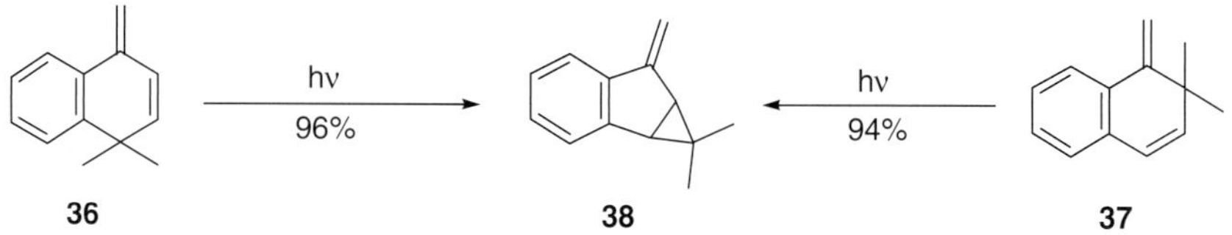

Scheme 4.16

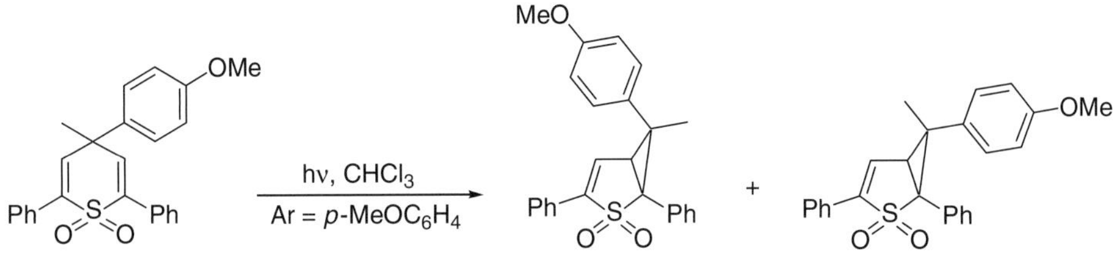

Scheme 4.17

Condition	40	41
254 nm/10 min	35.6%	30.8%
254 nm/anthra-quinone/45 min	44%	2%

the triplet excited state competitively gives **40** and **41**, the *anti*-form **40** is obtained as the main product under anthraquinone-sensitization [23].

4.2.3
Di-π-Methane Rearrangement in Natural Compounds

A cyclopropane unit is often found to be a key structural element in a large range of naturally occurring compounds. It has been suggested that part of the cyclopropanes are actually formed in DPM rearrangements. Photodeoxytridachione **44** is formed via the photorearrangement of 9,10-deoxytridachione **43** derived from the pyrolysis or photolysis of a polyene **42** [24]. The photorearrangement from **43** to **44** involves a triplet-DPM rearrangement mechanism (Scheme 4.18) [25]. Photodeoxytrida-chione **44** arises from the assembling of seven propionate units, with the same building blocks used in the biosynthesis of spectinabilin and the sucrose nonfer-mentable family (SNF) of compounds.

The diterpenoid aquariolide A **45**, having an aquariane skeleton, has been isolated from cultured specimens of *Erythropodium caibaeroum* with the known antimitotic

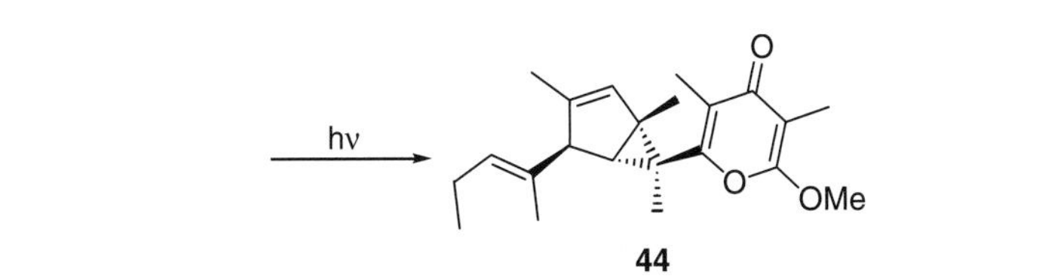

Scheme 4.18

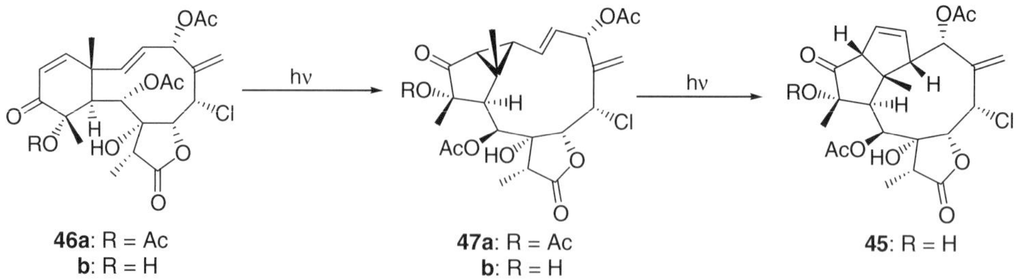

Scheme 4.19

agent eleutherobin and the erythrolides A **46a** and B **47a**. Aquariolide A **45** can be derived from **46b** by sequential DPM rearrangement and vinyl-cyclopropane rearrangement (Scheme 4.19) [26].

It is well known that the photolysis of α-santonin **48** rearranges photolytically to lumisantonin **49**; the latter is further transformed into mazdasantonin **50** (Scheme 4.20) [4a, 27]. Although this rearrangement of 1,5-cyclohexadien-3-one to bicyclo[3.1.0]hex-3-en-2-one differs from a DPM rearrangement with respect to a zwitterionic intermediate occurring and the triplet n-π* state involved, there is a resemblance to DPM rearrangement. This rearrangement is termed the "lumi or lumiketone rearrangement." Pregna-1,4-diene-3,20-diones, which are used as anti-inflammatory drugs, contain a 1,5-cyclohexadien-3-one moiety; these steroids undergo lumi rearrangements by both 254 and 365 nm irradiation (Scheme 4.21) [28]

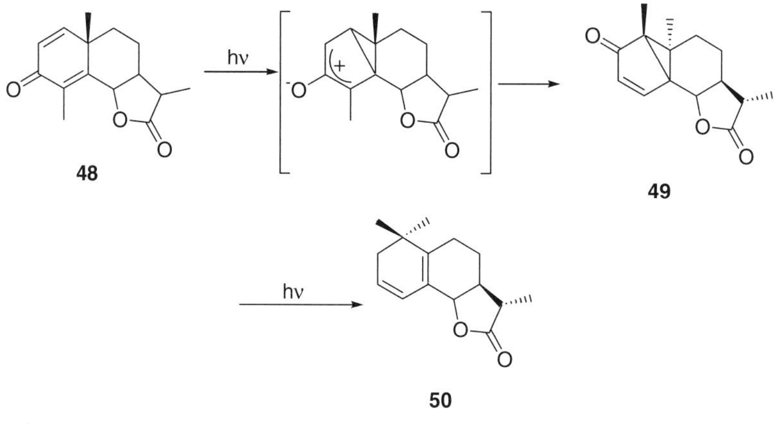

Scheme 4.20

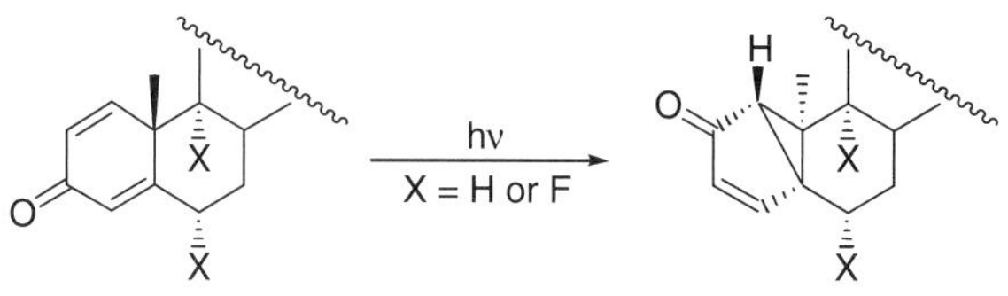

Scheme 4.21

4.3
Oxa-di-π-Methane Rearrangement and Related Rearrangements

A great many studies have been performed on the oxa-di-π-methane (ODPM) rearrangement of β,γ-unsaturated ketones and aldehydes since the first reports of ODPM rearrangement by Tenney et al. [29]. All ODPM rearrangements have been found to produce regioselectively cyclopropyl ketones (Scheme 4.22) [3, 30]. In 1977, the aza-di-π-methane (ADPM) rearrangement was first discovered [30], and subsequently a series of photochemical ADPM rearrangements have been demonstrated by Armesto and colleagues [3, 4b]. Some excellent comprehensive reviews of ODPM and ADPM rearrangements have been prepared by Zimmerman [3], Armesto [3, 4b, 30], and others [31].

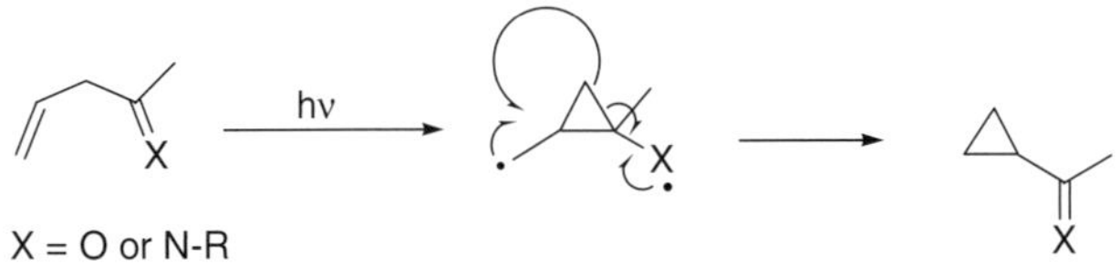

X = O or N-R

Scheme 4.22

4.3.1
Oxa-di-π-Methane Rearrangement of β,γ-Unsaturated Ketones and Aldehydes

It is generally known that the β,γ-unsaturated ketones and aldehydes undergo the ODPM rearrangement in the presence of a triplet sensitizer, while 1,3-acyl migration occurs preferentially on direct irradiation. Therefore, synthetic chemists should consider effective pathways via the triplet as a strategy for synthesizing cyclopropyl ketones from β,γ-unsaturated ketones. Consequently, in many ODPM rearrangements acetone is used as a triplet sensitizer. This is convenient, because β,γ-unsaturated ketones usually have a weak absorption band of the n-π* transition state at ∼300 nm, while the acetone used as solvent is the main light absorber and thus an efficient sensitizer. The irradiation of ketones **51a,b** in acetone furnished the ODPM rearrangement products **52a,b** in good yields (Scheme 4.23) [32].

Acetophenone is also a convenient sensitizer for the ODPM rearrangement. Bicyclo[2.2.2]oct-7-ene-2,5-diones **53a–c** undergo a highly chemoselective ODPM rearrangement under acetophenone-sensitized irradiation (350 nm) to yield **54a,c** and **55b**, in excellent yields (Table 4.2) [33]. However, phenols **56a,b**, which presum-

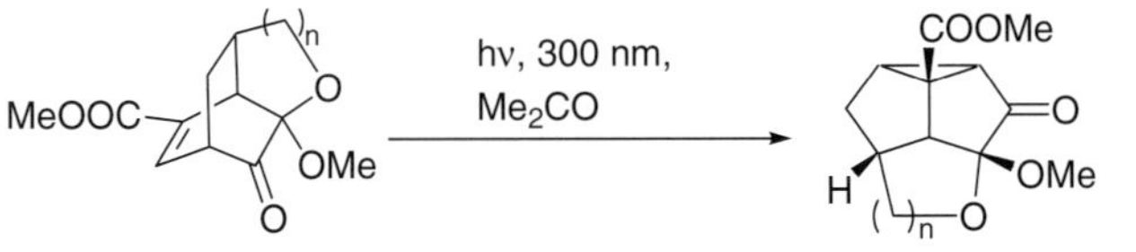

51a: n = 1
b: n = 2

52a: n = 1 (77%)
b: n = 2 (51%)

Scheme 4.23

Table 4.2 Photolyses of bicyclo[2.2.2]oct-7-ene-2,5-diones 53a–d.

53a: R^1 = COOEt, R^2 = H
b: R^1 = R^2 = COOMe
c: R^1 = Ph, R^2 = H
d: R^1 = R^2 = Ph

54a: R^1 = COOEt, R^2 = H
b: R^1 = R^2 = COOMe
c: R^1 = Ph, R^2 = H
d: R^1 = R^2 = Ph

55a: R^1 = COOEt, R^2 = H
b: R^1 = R^2 = COOMe
c: R^1 = Ph, R^2 = H
d: R^1 = R^2 = Ph

56a: R^1 = COOEt, R^2 = H
b: R^1 = R^2 = COOMe
c: R^1 = Ph, R^2 = H
d: R^1 = R^2 = Ph

Reactant	λ (nm)	Solvent	Products (Yield %)		
53a	300	Methanol	54a (20)		56a (50)
		Acetone	(56)		(24)
	350	Acetophenone	(98)		
53b	300	Methanol			56b (76)
		Acetone		55b (40)	(55)
	350	Acetophenone		(90)	
53c	300	Methanol			56c (64)
		Acetone	54c (84)		
	350	Acetophenone	(95)		
53d	300	Methanol			56d (86)
		Acetone			(58)

ably arise via α-cleavage and dimethylketene elimination, were obtained under different conditions, such as irradiation in acetone. On the other hand, a compound such as **53d** is unreactive, as a result of the localization of excitation energy in the triplet on the stilbene moiety, which is too low in energy to allow the rearrangement to occur [33].

Bicyclo[2.2.2]oct-5,7-dien-2-ones **57a–c**, which are endowed with the structural features required for both the ODPM and DPM rearrangements, undergo acetophenone-sensitized DPM rearrangement to afford the compounds **58a–c** in good yields (Scheme 4.24) [34].

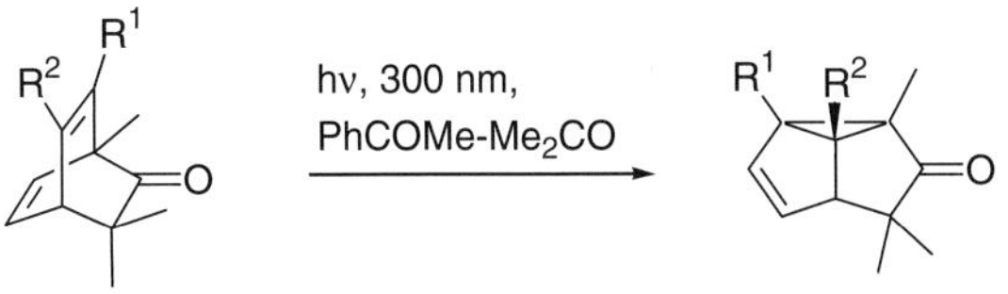

4.3 Oxa-di-π-Methane Rearrangement and Related Rearrangements | 107

Scheme 4.24

57a: $R^1 = H$, $R^2 = COOE$
b: $R^1 = R^2 = COOMe$
c: $R^1 = R^2 = Ph$

58a: $R^1 = H$, $R^2 = COOEt$: 60%
b: $R^1 = R^2 = COOMe$: 62%
c: $R^1 = R^2 = Ph$: 58%

As mentioned above, although acetophenone is a good sensitizer for the ODPM rearrangement, problems sometimes occur when separating the sensitizer and the photoproducts, due to their volatilities. When methyl and heptyl 1-aza-3-carboxybicyclo[2.2.2]oct-2-en-5-one **59** (R = CH$_3$ and C$_7$H$_{15}$) were irradiated in acetophenone, the ODPM rearrangement products **59** were formed (Scheme 4.25). The isolation of methyl ester **59** (R = CH$_3$) by distillation was difficult, but the sensitizer could be readily removed by distillation under reduced pressure of a mixture of acetophenone and the heptyl ester **59** (R = C$_7$H$_{15}$). The cyclopropane moiety of the resultant product **60** could be opened via hydrogenolysis or with Me$_2$CuLi to produce pyrrolizidinones **61** [35].

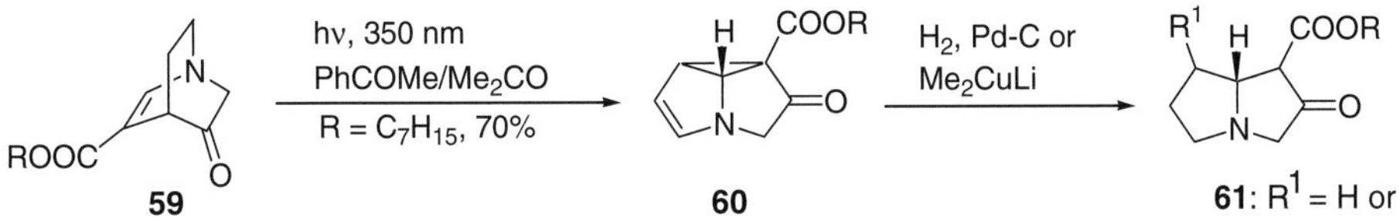

61: R^1 = H or Me

Scheme 4.25

As shown in Scheme 4.3, a heavy cation Tl$^+$ effectively promoted intersystem crossing (ISC) to the triplet excited state within X-type zeolites. While the direct photolysis of bicyclo[2.2.1]hept-5-en-2-one **62** in TlX gave **64** as a main product (89%), in CsY or TlY the ODPM rearrangement mainly proceeded to afford **63** [36]. Furthermore, the ODPM rearrangement of **62** by the triplet-sensitization of 4-methoxyacetophenoene within KX occurred preferentially to afford **63** in 97% yield (Scheme 4.26) [37].

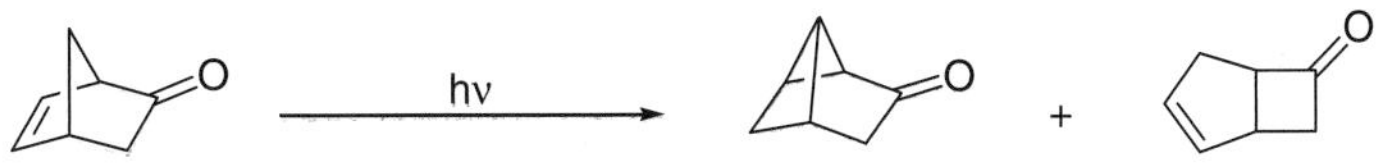

	63	64
TlX:	11%	89%
CsY:	42%	
TlY:	61%	
KX-p-MeOC$_6$H$_4$COMe	97%	3%

Scheme 4.26

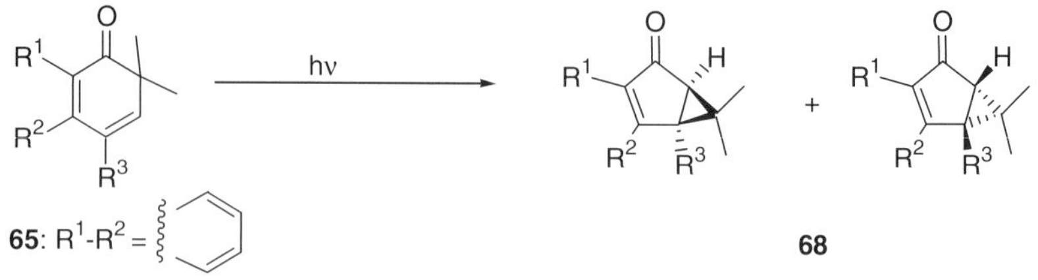

65: R¹-R² = ⟨benzo⟩
R³ = H or COOMe
66: R¹ = H, R² = R³ = Me
67: R¹ = Me, R² = H, R³ = (S)-NHCH(Me)Ph
Scheme 4.27

Ramamurthy et al. have examined enantioselective and diastereoselective ODPM rearrangements of cyclohexadienone and naphthalenone derivatives within MY zeolites, where M is an alkali ion [38, 39]. For example, in Scheme 4.27, in NaY in the presence of several chiral inductors such as ephedrine, pseudoephedrine, and camphorquinone-3-oxime, an enantioselective ODPM rearrangement of **65** and **66** took place to afford **68** in ∼30% ee yield. In the frame of the chiral auxiliary approach, the compound **67** linked to (S)-ephedrine was irradiated within NaY to give **68** in moderate diastereoselective excess (de) (59%). Interestingly, the reaction media of KY, RbY, and CsY reverse the diastereoselectivities. The cation-dependent diastereomeric switch has been discussed with respect to the N- or O-functional group in **67** [39]. Recently, Arumugan reported that the irradiation of naphthalenones **65** linked chiral auxiliaries (R³ = COO(−)-Ment or (S)-NHCH(Me)Ph) in Li⁺ or Na⁺ Nafion resulted in chiral products (∼14% de) [40].

It is known that simple β,γ-unsaturated ketones **69** do not undergo ODPM rearrangement by the acetophenone-sensitization. Instead, under these conditions **69** undergoes alkene E/Z isomerization. However, when thioxanthone is used as a sensitizer, the β,γ-unsaturated ketones **69** react to form ODPM rearrangement photoproducts **70** (Scheme 4.28) [41]. This result shows clearly that the energy of the triplet sensitizer plays an important role in determining the outcome of the triplet ODPM rearrangement of β,γ-unsaturated ketones.

5-Benzoyl-4-methylbicyclo[2.2.2]oct-5-en-2-one **71**, prepared by the Diels–Alder reaction of 2-trimethylsioxy-4-methyl-1,3-cyclohexadiene with benzoylethyne, underwent ODPM rearrangement in the absence of a sensitizer to give **72** in good yield (Scheme 4.29). The irradiation of **72** in 40% triethylamine (TEA)-methanol gave a bicyclo[3.2.1]octanone **73** in 86% yield. On direct irradiation, a one-pot conversion

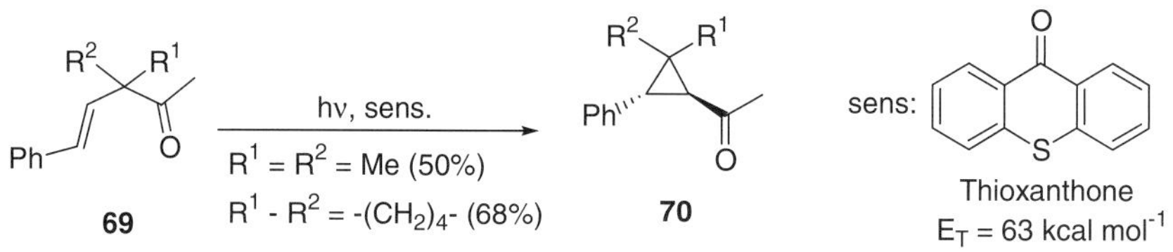

Thioxanthone
E_T = 63 kcal mol⁻¹

Scheme 4.28

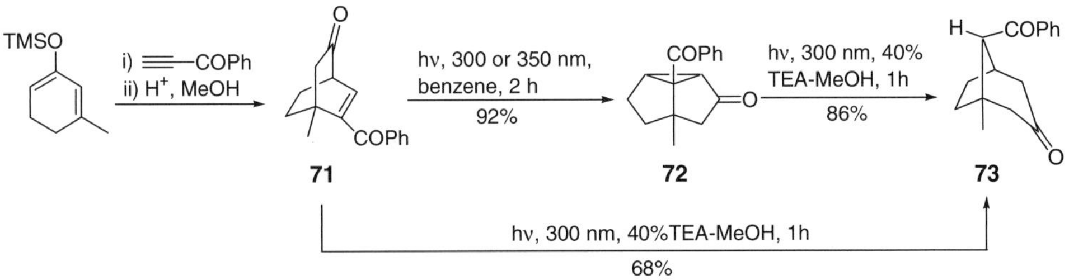

4.3 Oxa-di-π-Methane Rearrangement and Related Rearrangements | 109

Scheme 4.29

of **71** to **73** took place in 40% TEA-methanol [42]. Compound **73** is a very useful intermediate for the construction of different bioactive natural molecules, and has also been considered as a possible phototrigger for liquid crystal-based optical switches [43].

A dione **74** also undergoes ODPM rearrangement under direct irradiation to give compound **75** in 94% yield (Scheme 4.30) [44]. The product **75** was crystallized from ethyl acetate/hexane at room temperature to lead to two polymorphs. Polymorphism is recognized as being a vital branch of solid-state supramolecular chemistry.

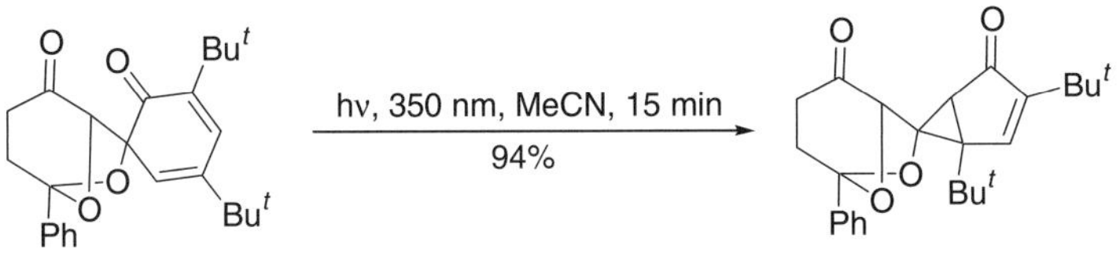

Scheme 4.30

4.3.2
Aza-di-π-Methane Rearrangement

β,γ-Unsaturated imines and their derivatives undergo ADPM rearrangement upon triplet-sensitized irradiation in the presence of acetone and acetophenone (Scheme 4.31) [45–47]. The alkyl-substituted cyclopropylimines readily underwent hydroxylation on SiO_2 to afford the corresponding cyclopropanecarbaldehydes

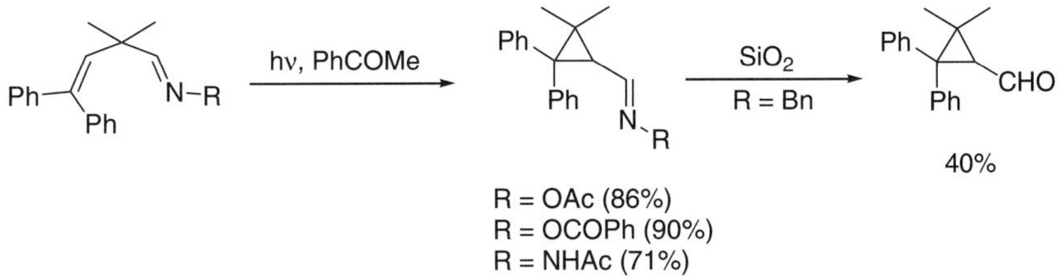

Scheme 4.31

(Scheme 4.31) [48]. The ADPM rearrangement has been reviewed by Armesto *et al.* [3, 4b, 30].

4.3.3
Synthetic Applications of Oxa-di-π-Methane Rearrangement

As shown in Schemes 4.23 and 4.24, the bicyclo[2.2.2]oct-5-en-2-ones gave the tricyclo[3.3.0.0^{2,8}]oct-3-ones via ODPM rearrangements. The tricyclo[3.3.0.0^{2,8}]oct-3-ones are important key precursors for natural compounds such as polyquinanes. Hence, ODPM rearrangement has a high synthetic potential, and many chemists have utilized this as a key reaction in the syntheses of natural and non-natural compounds. Some recent reports of the synthesis of natural compounds are outlined at this point.

Banwell *et al.* have reported that an enantiomerically pure *cis*-1,2-dihydrocatechol was converted through 17 steps into (−)-hirsutene [49]. More recently, the same group succeeded in synthesizing the linear triquinane-type sesquiterpenes (+)-hirsutic acid and (−)-complicatic acid (Scheme 4.32) [50]. The ODPM rearrangements by sensitized irradiation have been used as a key reaction in their total synthesis. Furthermore, Singh *et al.* were successful in obtaining the stereoselective synthesis of (±)-hirsutene from salicyl alcohol (Scheme 4.33) [51], and recently identified stereoselective routes to sterpuranes and polyquinanes [52]. Yen and Liao have also reported an efficient total synthesis of the *Lycopodium* alkaloid, magellanine (Scheme 4.34) [53].

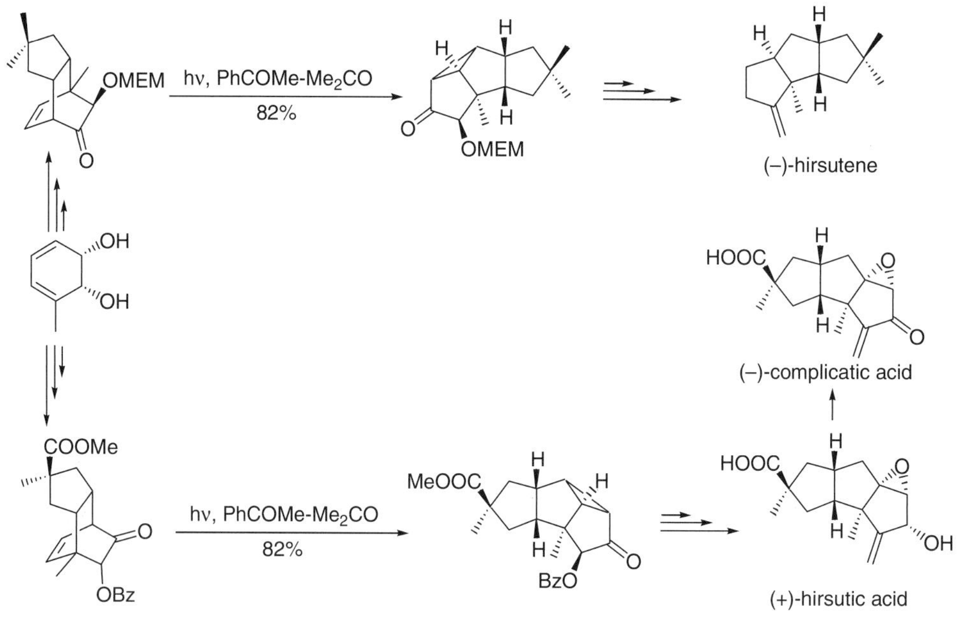

Scheme 4.32

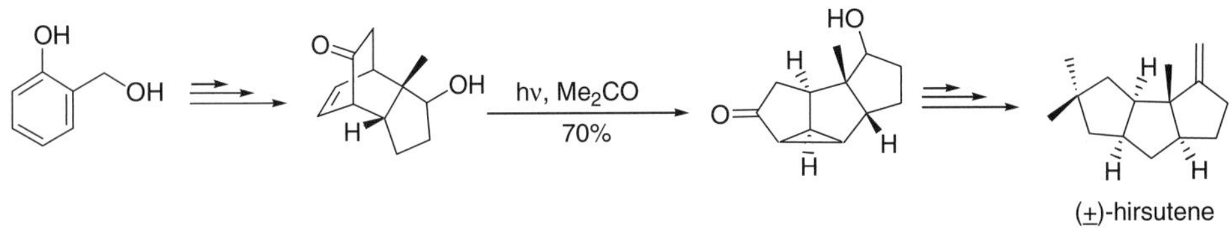

Scheme 4.33

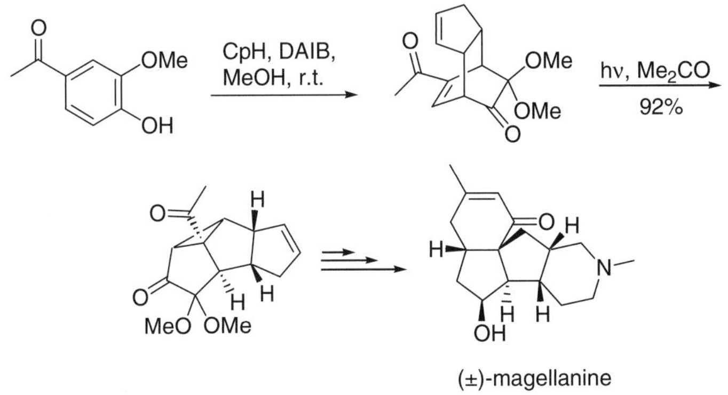

Scheme 4.34

4.4
[2 + 1] Cycloaddition of Alkenes with Carbenes

Carbenes, as synthetic species, are fascinating chemicals, and the synthesis of cyclopropanes by a [2 + 1] cycloaddition of alkenes with carbenes represents an extremely fruitful approach. Recently, highly effective intramolecular [2 + 1] cycloadditions, novel triplet sensitizers, metal-catalyzed cyclopropanations, and novel precursors of carbenes have been developed.

4.4.1
Intramolecular [2 + 1] Cycloaddition

It is known that diazo compounds and diazirines are photochemical precursors to carbenes [54].

The direct photolysis of α-(alkenyloxy)silyl-substituted diazoacetate **76a** leads to a carbene, which undergoes an intramolecular [2 + 1] cycloaddition with the terminal alkene to afford **77a** in 55% yield (Scheme 4.35) [55]. An α-alkenysilyl-substituted diazoacetate **76b** also leads to **77b** in 68% yield (Scheme 4.35) upon direct irradiation [56].

4.4.2
Novel Triplet Sensitizers for the Generation of Carbenes

In the photochemistry of α-diazo carbonyl compounds, singlet carbenes undergo a Wolff rearrangement while triplet carbenes react with alkenes to afford cyclopropanes.

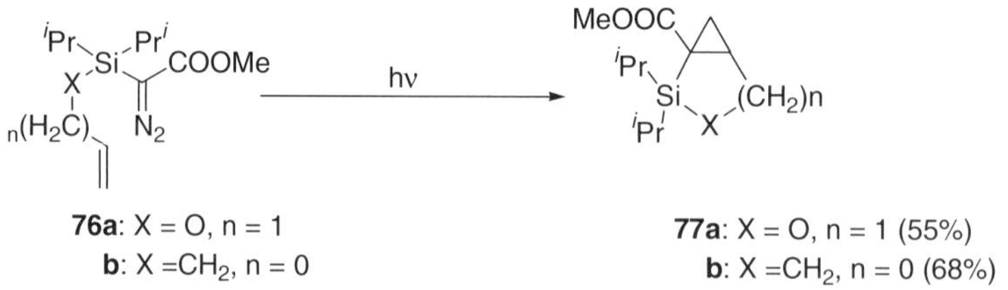

Scheme 4.35

76a: X = O, n = 1
b: X = CH$_2$, n = 0

77a: X = O, n = 1 (55%)
b: X = CH$_2$, n = 0 (68%)

Benzophenone is a convenient triplet sensitizer for the generation of triplet carbenes, but must be used in a large excess in order that it absorbs a large fraction of light. At any rate, because α-diazo carbonyl compounds absorb at longer wavelengths than benzophenone, it is difficult to avoid some direct photolysis. Moreover, benzophenone and some of its derivatives that have an n-π* triplet as a lowest excited state easily abstract hydrogen from the solvent. These shortcomings have been overcome by using 2,2′,4,4′-tetramethoxybenzophenone (TMBP), which Pastor-Pérez and colleagues have shown to be an efficient triplet-sensitizer for the intramolecular [2 + 1] cycloaddition of α-diazo carbonyl compounds **78** in polar solvents such as methanol and acetonitrile (Scheme 4.36) [57]. The same group has also recently reported studies of the photochemistry of **78** in the presence of various hyperbranched polyether polymers, with benzophenone core as triplet sensitizers [58].

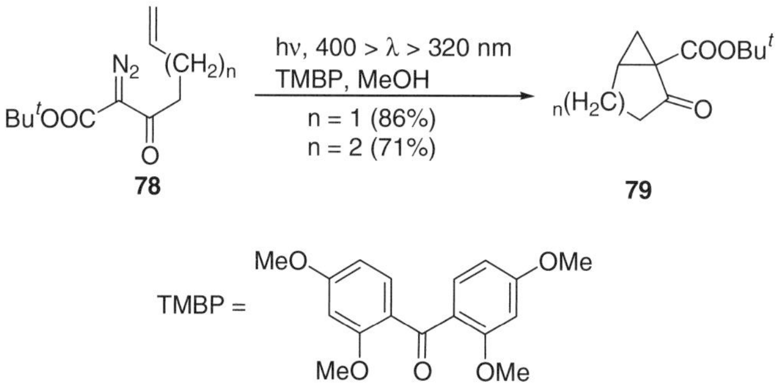

Scheme 4.36

4.4.3
Metal-Catalyzed Cyclopropanation-Supported Photochemistry

Optically active metal complexes have been recognized as excellent catalysts for the enantioselective cyclopropanation of carbenes with alkenes. Normally, diazo compounds react under metal catalysts in the dark to afford carbenoid complexes as key intermediates. Katsuki *et al.* have reported the *cis*-selective and enantioselective cyclopropanation of styrene with α-diazoacetate in the presence of optically active (R,R)-(NO$^+$)(salen)ruthenium complex **80**, supported under illumination (440 nm light or an incandescent bulb) [59]. The irradiation causes dissociation of the apical ligand ON$^+$ in **80**, and thus avoids the splitting of nitrogen from the α-diazoacetate.

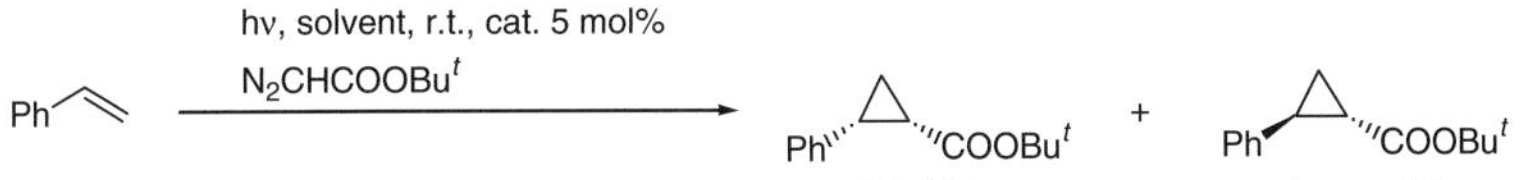

4.4 [2 + 1] Cycloaddition of Alkenes with Carbenes

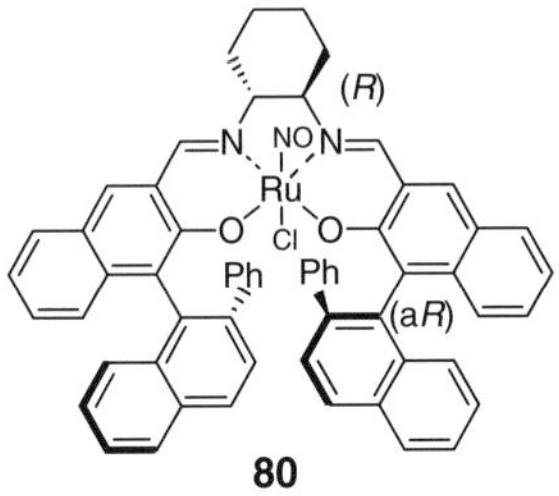

Scheme 4.37

Solvent	Yield/%	cis:trans	%ee (cis)
THF-styrene (10:1)	18	96:4	99 (1S,2R)
THF-styrene (1:1)	45	93:7	97 (1S,2R)
Hexane	10	68:32	83 (1R,2S)

A tetrahydrofuran (THF)-styrene (10:1, v/v) solution of *tert*-butyl diazoacetate in the presence of **80** (5 mol%) under incandescent light gives the (1S,2R)-cyclopropane *cis*-**81** in 99% ee, but the yield is only 18% (Scheme 4.37).

However, the chemical yield increases considerably as the proportion of THF is reduced, although the stereoselectivity is slightly decreased. Interestingly, the sense of enantioselection in the formation of *cis*-**81** in hexane is opposite to that in THF. Katsuki *et al.* applied this reaction to the intramolecular cyclopropanation, when the irradiation of alkenyl diazoketones in the presence of **80** in THF afforded bicyclo [3.1.0]hexan-2-ones in a highly enantioselective cyclopropanation, and in moderate yields (Table 4.3) [60].

Table 4.3 Intramolecular cyclopropanation of alkenyl diazo ketones.

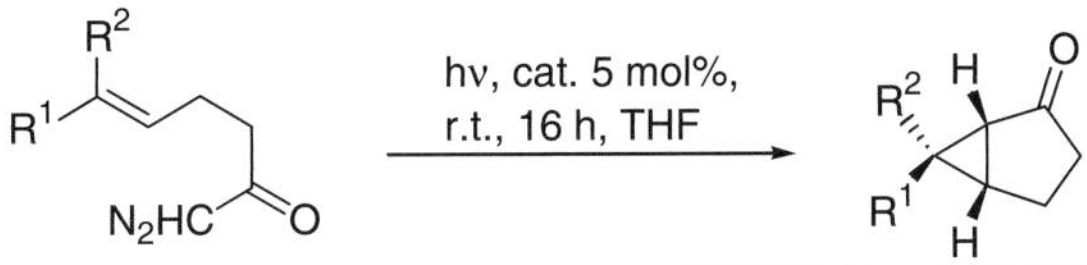

Entry	R¹	R²	Yield (%)	ee (%)
1	Ph	H	78	94
2	i-Pentyl	CH₃	72	93
3	CH₃	CH₃	65	87
4	PhC≡C	H	82	84
5	H	Ph	62	79

4.4.4
Novel Precursors of Carbenes

The influence of substituents on the electronic structure and stability of carbenes has been investigated, and many stable carbenes have been prepared. Bertrand *et al.*

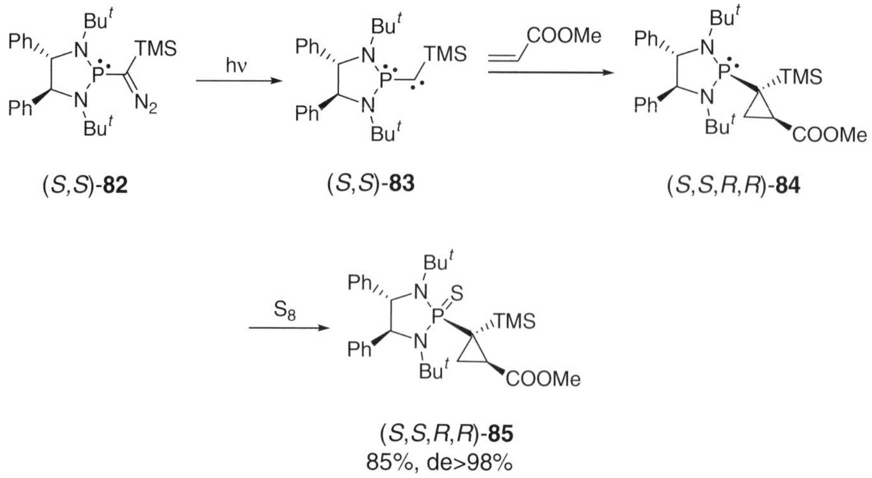

Scheme 4.38

reported the synthesis of stable, optically pure phosphino(silyl)carbenes, and subsequently a highly diastereoselective cyclopropanation [61]. The photolysis of enantiomerically pure phosphino(silyl)diazo compound (S,S)-**82** in toluene gives (S,S)-**83**. The addition of two equivalents of methyl acrylate to the toluene solution yielded (S,S,R,R)-**84**, which was then converted into the thioxo compound (S,S,R,R)-**85** (85% yield, >98% de) by the addition of elemental sulfur (Scheme 4.38). The enantiomer (R,R,S,S)-**85** could be readily obtained from the reaction of (R,R)-**82**.

Recently, new photochemical carbene precursors which differed from diazo compounds and diazirines have been reported [62, 63]. Platz et al. synthesized a photosensitive precursor **86** to phenylsulfanylcarbene **87** (Scheme 4.39) [62]. The photolysis of **86** in the presence of 2,3-dimethyl-2-butene yielded the [2 + 1] cycloadduct **88** in 80% yield. In contrast, Jenks et al. reported that S,C-ylides of malonates were photochemical precursors of carbene **90** [63]. Thus, the S,C-ylide **89** was irradiated at 254 nm in acetonitrile with 10% cyclohexene to give **91** in 44% yield (Scheme 4.40).

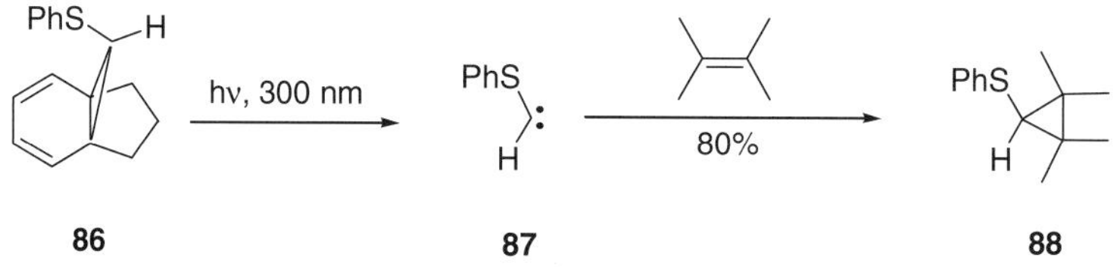

Scheme 4.39

4.5
Formation of a Cyclopropane via Intramolecular Hydrogen Abstraction

In photochemistry, intermolecular and intramolecular hydrogen abstractions by excited carbonyl oxygen are often encountered [64]. If an intramolecular β-hydrogen

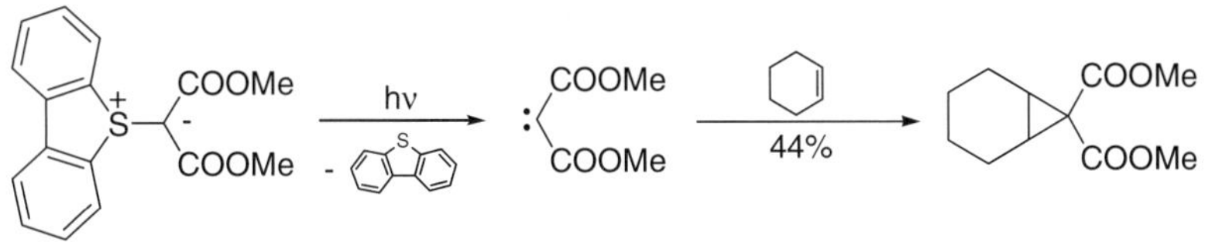

Scheme 4.40

abstraction takes place, then cyclopropanols will be obtained. Although reports of the formation of cyclopropanols by intramolecular β-hydrogen abstraction are fewer in number than those involving γ- or δ-hydrogen abstraction, the synthesis of cyclopropanes via a photochemical intramolecular hydrogen abstraction has the potential to become an extremely useful synthetic tool for molecular design and retrosynthesis.

4.5.1
Formation of Cyclopropanol via Intramolecular β-Hydrogen Abstraction

In 1970, the first example of the formation of cyclopropanols by β-hydrogen abstraction was reported by Roth [65]. In addition, β-morphorinopropiophenones **92** were irradiated to give cyclopropanols **93** stereoselectively in good yields (Scheme 4.41). The initial step was a PET from the β-amino group to the excited triplet-carbonyl in β-morpholinopropiophenones, rather than from a straightforward β-hydrogen abstraction. Likewise, β-aminopropiophenones **94** and **98** and β-(p-amino or methoxy)phenylpropiophenones **93** were shown to undergo an intramolecular PET to give the cyclopropanols **95**, **97**, and **99** (Schemes 4.42 to 4.44) [66–68]. The efficient formation of cyclopropanols should be considered when planning the molecular structure for substituents at the α- and β-positions in β-aminoketones.

Nonetheless, there are some examples of cyclopropanol formation by straightforward β-hydrogen abstraction. The Diels–Alder adduct **100** prepared from o-quino-

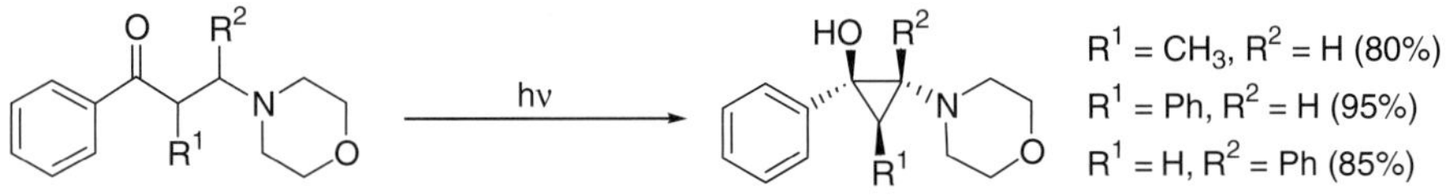

Scheme 4.41

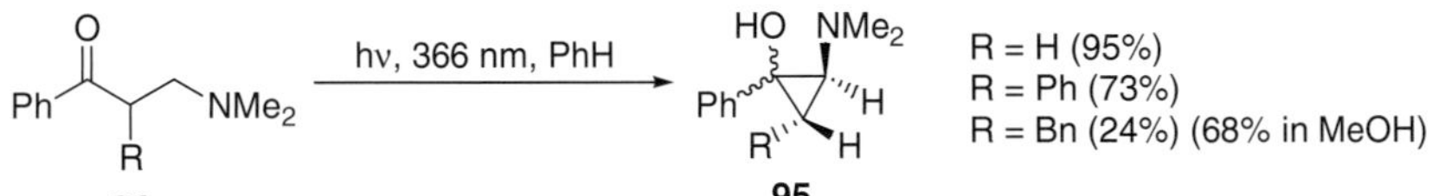

Scheme 4.42

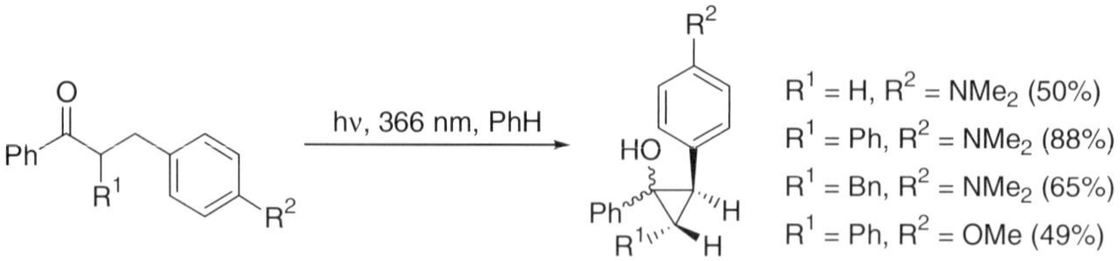

96 **97**

Scheme 4.43

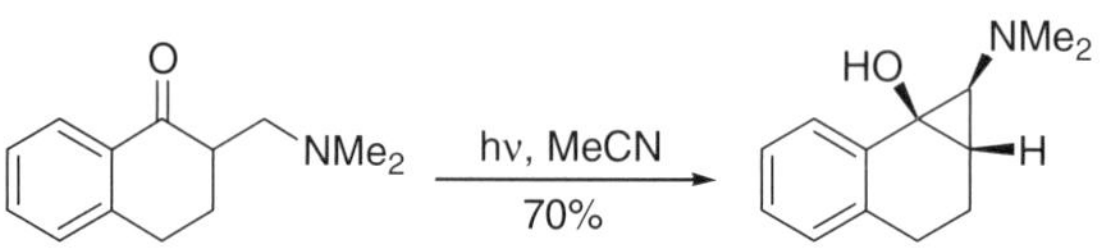

98 **99**

Scheme 4.44

dimethane and 2,3-dimethyl-1,4-naphthoquinone was irradiated in acetonitrile to give a cyclopropanol **101** in 35% yield (see Scheme 4.45) [69]. Also, β-hydroxy or β-alkohoxypropiophenones **102** were shown to give cyclopropanes **103** via a straightforward β-hydrogen abstraction by excited carbonyl oxygen (Scheme 4.46) [70].

Upon direct photolysis, the α-(1-hydroxyalkyl)-substituted α,β-unsaturated ketones **104** gave 1,4-diketones **106** in moderate yields (Scheme 4.47) [71], with the reaction involving an intermolecular β-hydrogen abstraction. The resultant vinylidenecyclopropanols **105** underwent a formal double tautomerization to afford **106**.

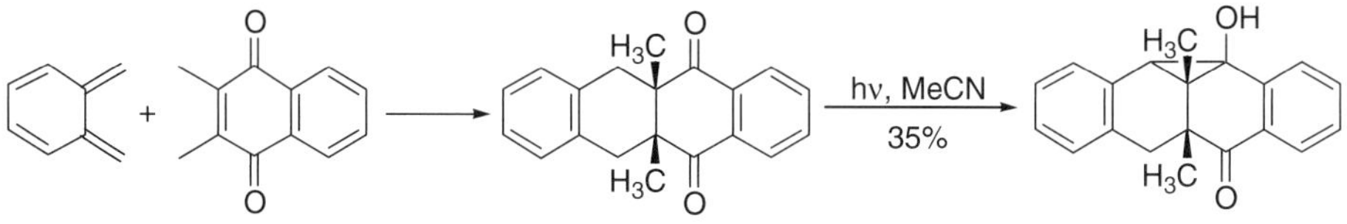

100 **101**

Scheme 4.45

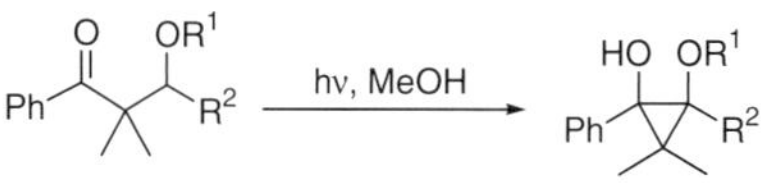

102 **103**

R^1	R^2	Yield / %	*cis:trans*
H	Ph	95	89:6
Me	Me	82	6:4
Me	Ph	92	9:1
Me	OMe	88	
Et	Me	82	5:2
Et	Ph	92	6:1

Scheme 4.46

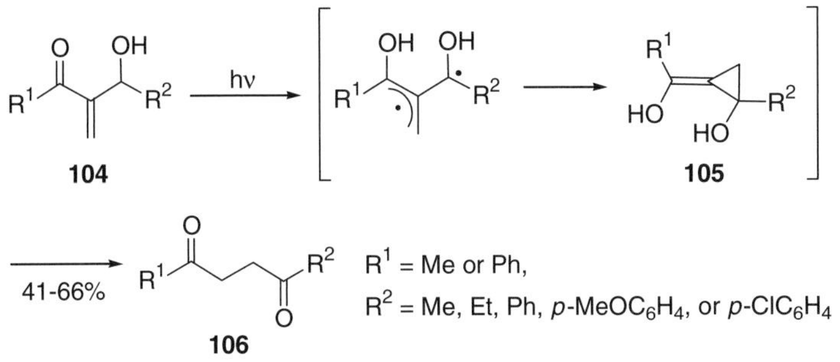

Scheme 4.47

4.5.2
Formation of Cyclopropane Ring via Intramolecular γ-Hydrogen Abstraction

The intramolecular γ-hydrogen abstraction by the excited carbonyl oxygen of ketones is a favorable pathway, and the resultant 1,4-biradical subsequently cyclizes to cyclobutanol [64]. This reaction, which is referred to as the "Yang reaction," will be described in detail in Chapters 5 and 6. Wessig and colleagues have shown that ketones with a variety of leaving groups, such as OMs (OSO_2CH_3), OTs ($OSO_2C_6H_4CH_3$-p), OPO(OEt)$_2$, or ONO$_2$, at the α-position lead to cyclopropyl ketones via intramolecular γ-hydrogen abstraction upon direct irradiation (Scheme 4.48) [72]. These result from acid elimination from the 1,4-biradical intermediates, and for a successful reaction N-methylimidazole as an acid scavenger must be added to the solution [72]. Wessig et al. applied this reaction to the syntheses of bicyclo[n.1.0]alkyl ($n = 4$, 3, or 2) (Scheme 4.49) and heterobicyclo[4.1.0]alkyl ketones [72a]. In addition, the photochemistry of enantiomer-rich samples of the ketones **107** showed a 1,2-chirality transfer (Scheme 4.50) [73]. Although the ketones **107** give the cyclopropyl ketones **108** at room temperature in 28–52% ee yields, the rise in temperature enhances the stereoselectivity. This behavior has been explained by the compensation of entropic and enthalpic influences.

Furthermore, Wessig and Mühling were successful in synthesizing highly strained spiro[2.2]pentanes in 1,1,2-trichloro-1,2,2-trifluoroethane (R113) (Scheme 4.51) [74].

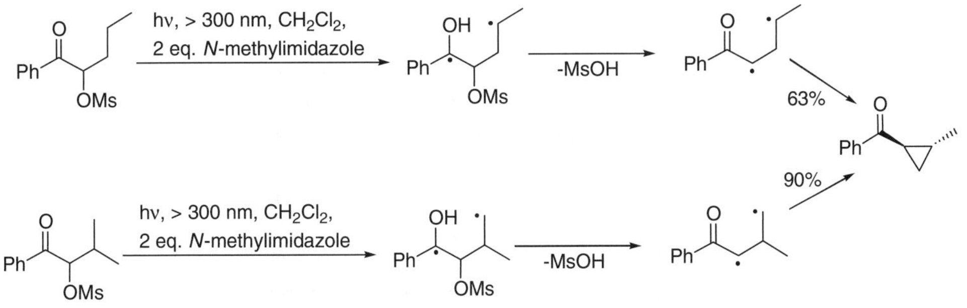

Scheme 4.48

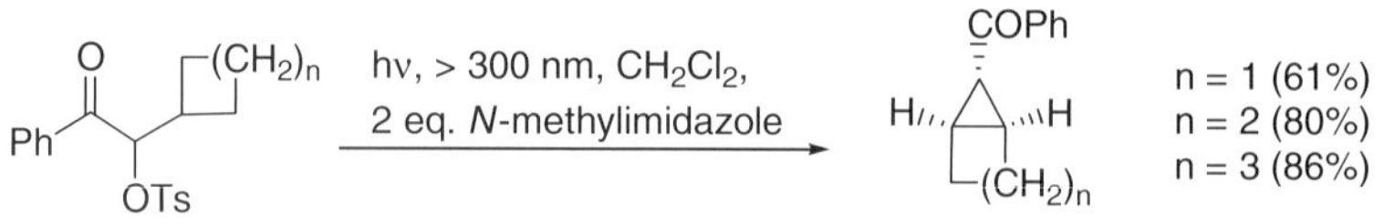

118 | *4 Formation of a Three-Membered Ring*

Scheme 4.49

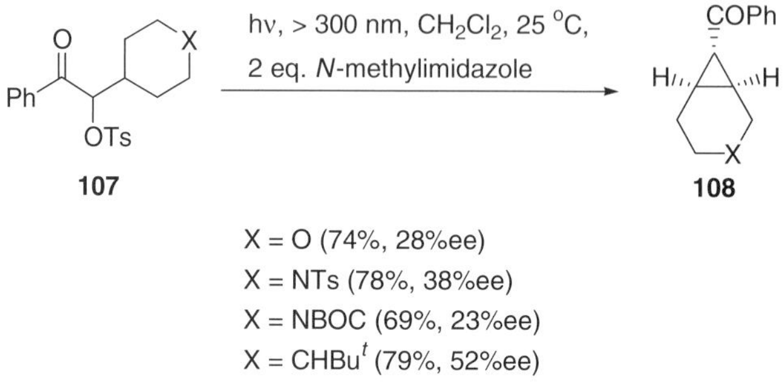

Scheme 4.50

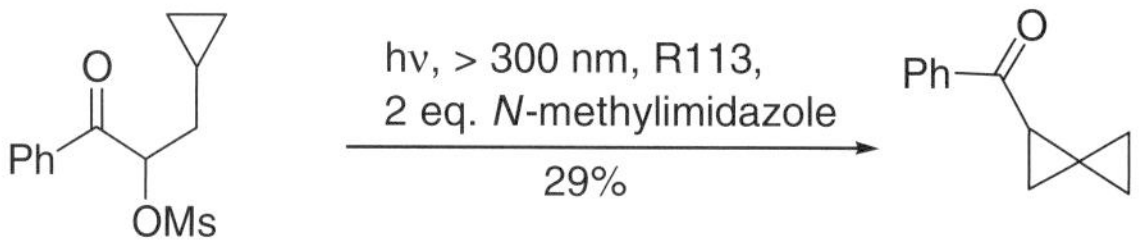

X = O (74%, 28%ee)
X = NTs (78%, 38%ee)
X = NBOC (69%, 23%ee)
X = CHBu^t (79%, 52%ee)

Scheme 4.51

The direct irradiation of 4-propanoyl-6-methylpyrimidine **109** in *tert*-butyl alcohol containing 5% benzene (>340 nm) leads to cyclopropanol **110** in 98% yield (Scheme 4.52) [75]. This reaction proceeds through γ-hydrogen abstraction by N(3), but not by N(1) or by O.

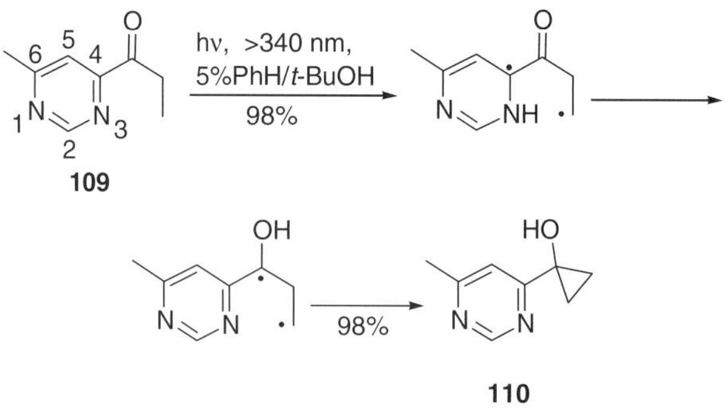

Scheme 4.52

4.6
[3 + 2] Cycloaddition of Arenes with Alkenes

The first photochemical [3 + 2] cycloadditions of arenes with alkenes were reported by two groups in 1966 [76, 77]. This is one of three types of cycloaddition reaction (*ortho*, *meta*, and *para*) of arenes with alkenes, and is referred to as the *meta*-cycloaddition reaction. There are many combinations of arenes and alkenes and, indeed, a large number of reports and reviews have been made on the regiochemistry and stereochemistry [78]. Because this reaction can be used easily to construct a tricyclo[3.3.0.0^{2,8}]oct-3-ene framework, the organic syntheses of many natural compounds using this method have been attempted [79], in which the photochemical [3 + 2] cycloaddition plays a key role.

4.6.1
Intermolecular [3 + 2] Cycloaddition

The β-silyl effect has been found to be operative in excited states, where this group exerts a remarkable effect on the yield and selectivity of intermolecular [3 + 2] cycloaddition (Scheme 4.53) [80]. Although cyclohexene and *cis*-cyclooctene each produce poor yields and poor selectivities with substrates such as toluene and anisole, this reaction is capable of producing intermolecular [3 + 2] cycloadducts with a few cycloalkenes.

The reaction of methyl 1,3-dihydroisoindole-2-carboxylate **111** with cyclopentene affords two *endo*-products **112** and **113** in 90% yield (Scheme 4.54) [81].

4.6.2
Intramolecular [3 + 2] Cycloaddition

The photochemical intramolecular [3 + 2] cycloaddition of arenes with a tethered alkene leads to polycyclic compounds. This reaction is better suited to the synthesis of natural products than to intermolecular [3 + 2] cycloaddition. A carbonyl group in the tether fails to produce any good results, because this chromophore can quench the

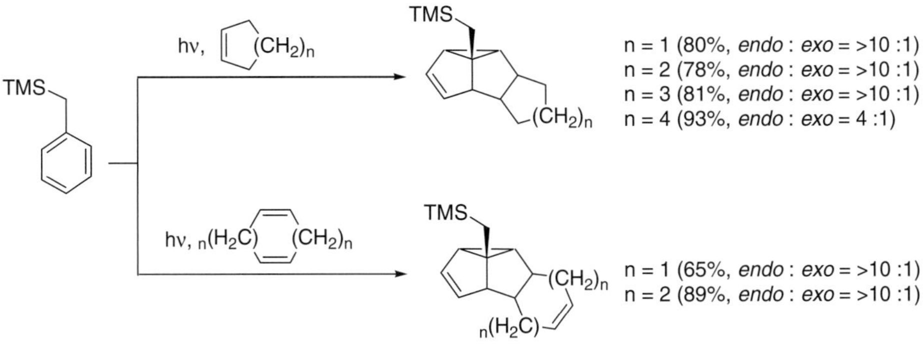

Scheme 4.53

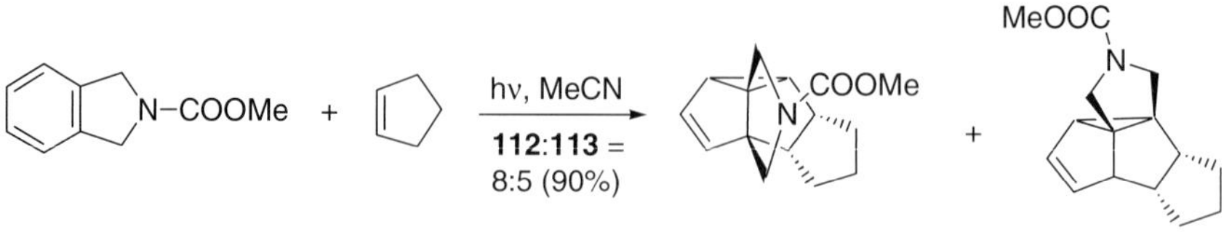

Scheme 4.54

111 + [cyclopentene] → (hv, MeCN; 112:113 = 8:5 (90%)) → 112 + 113

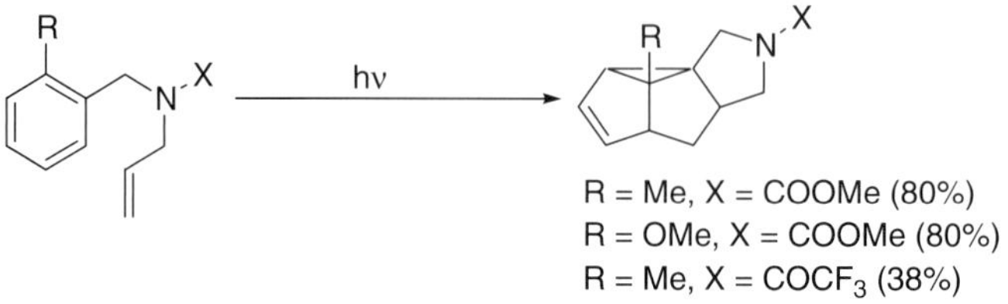

Scheme 4.55

R = Me, X = COOMe (80%)
R = OMe, X = COOMe (80%)
R = Me, X = COCF$_3$ (38%)

exciplex between an arene and an alkene [82]. The length of tether length should be $(-C-)_{3 \text{ or } 4}$ due to the formation of a favorable intramolecular charge-transfer (CT) complex between an arene and an alkene. When a heteroatom such as N [83, 84], O [85], or Si [86] is incorporated in the place of a carbon atom in the tether, the [3 + 2] cycloadducts are formed in good yields with high regioselectivity (Schemes 4.55 to 4.57). When the β-cyclodextrin complex of 5-(3-butenyloxy)-1,2,3,4-tetrahydronaphthalene was irradiated in the solid phase, an effect of the cyclodextrin (CD)-hosting on the regioselectivity and enantioselectivity was found (Scheme 4.57) [87].

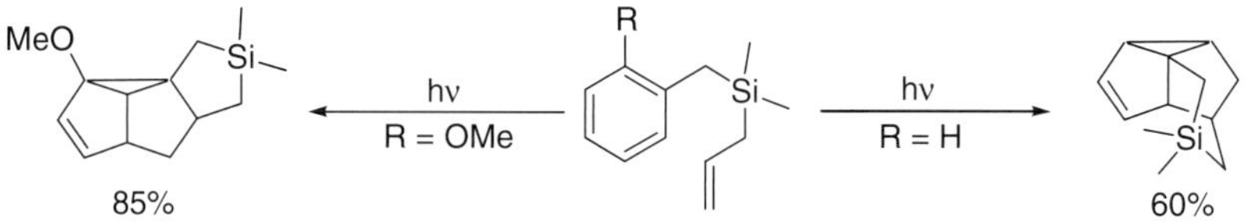

Scheme 4.56

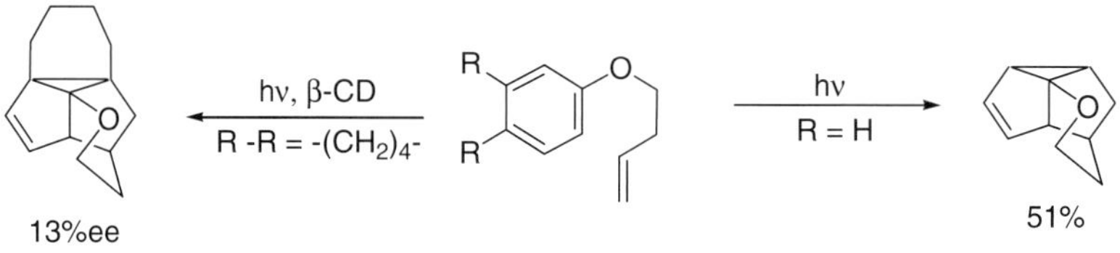

85% (R = OMe) ← hv — [arene-Si-allyl] — hv (R = H) → 60%

Scheme 4.57

13% ee (hv, β-CD; R-R = -(CH$_2$)$_4$-) ← [aryl ether] — hv (R = H) → 51%

Jenkins et al. have reported on diastereocontrol in the intramolecular [3 + 2] cycloaddition. Here, (3R,5S)/(3S,5R)-5-phenyl-1-hepten-3-ol **114** was irradiated to give a diastereopure [3 + 2] cycloadduct **115** (Scheme 4.58) [88]. The other diastereomer (3S,5S)/(3R,5R)-**114** gave an unresolved mixture.

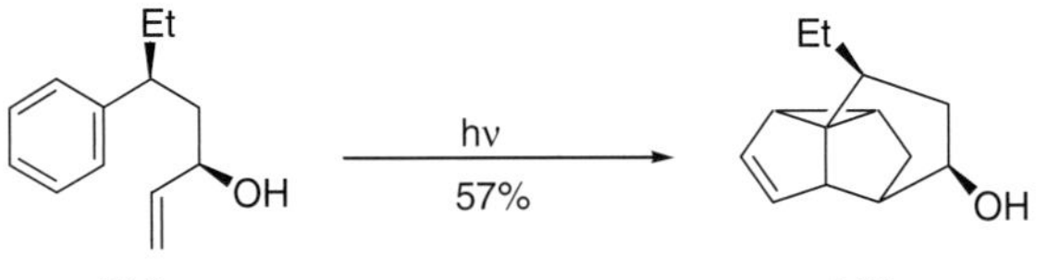

114 **115**

Scheme 4.58

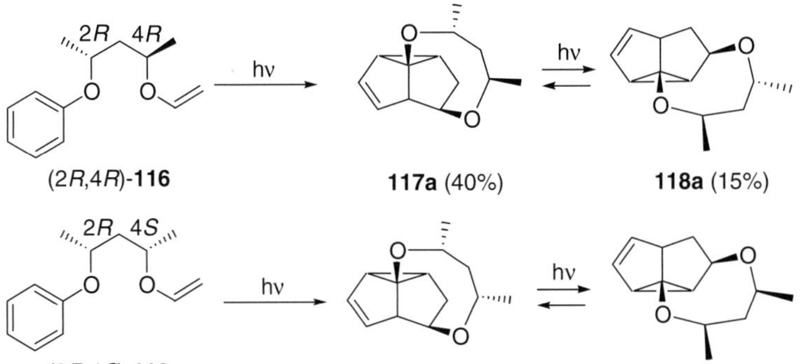

(2R,4R)-**116** **117a** (40%) **118a** (15%)

(2R,4S)-**116** **117b** (< 2%) **118b** (70%)

Scheme 4.59

Diastereomers **116** where the tether is an optically active 2,4-pentanediol gave a pair of regioisomers **117** and **118** (Scheme 4.59). In the photolysis of (2*R*,4*S*)-**116**, regioisomer **118b** was stereoselectively obtained. The regioisomers **118a,b** are not the initial products from **116**, whereas **117a,b** are initial products in the reaction of **116**; hence, compounds **118a,b** are derived from the photoreaction of **117a,b**. Thus, isomerization between **117** and **118** is attributed to the secondary photoreaction [89].

Mizuno *et al.* recently have reported a novel intramolecular [3 + 2] cycloaddition of 1-cyano-2-(4-pentenyl)naphthalene derivatives **119** [90]. The direct photolysis of **119** in acetonitrile gave the [3 + 2] cycloadducts **120** in good yields (Scheme 4.60).

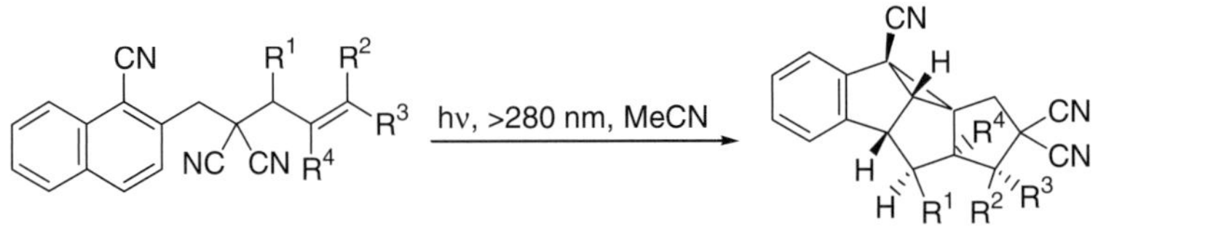

119 **120**

$R^1 = R^2 = R^3 = R^4 = H$ (70%)
$R^1 = R^4 = H, R^2 = R^3 = Me$ (74%)
$R^1 = H, R^2 = R^3 = R^4 = Me$ (64%)
$R^1 - R^2 = -(CH_2)_2-, R^3 = R^4 = H$ (74%)
$R^1 - R^2 = -(CH_2)_3-, R^3 = R^4 = H$ (85%)

Scheme 4.60

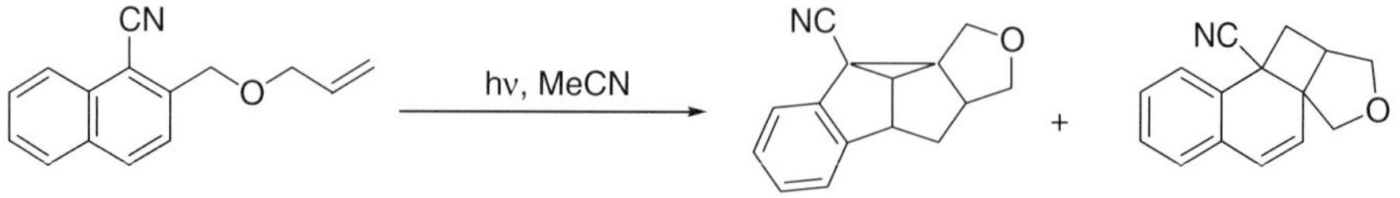

122 | 4 Formation of a Three-Membered Ring

Scheme 4.61

Condition	Irradiation time / min	Yields / %	
Batch	90	97	3
Flow	2.9	90	10

The same group also investigated the photochemistry in a glass microreactor, and reported that the reaction was completed within a short time period (Scheme 4.61) [91].

4.6.3
Application of the Photochemical [3 + 2] Cycloaddition in the Synthesis of Natural Products

The use of photochemical [3 + 2] cycloaddition in natural product synthesis (up to July 2006) has recently been reviewed by Chappell and Russell [79]. Subsequently, one report has been made concerning natural product syntheses up to October 2008. Wang and Chen have used a photochemical [3 + 2] cycloaddition to build the 5,6,7-tricyclic skeleton of lancifodilactone F and buxapentalactone (Scheme 4.62), which were isolated from the Chinese herb *Schisandra* [92].

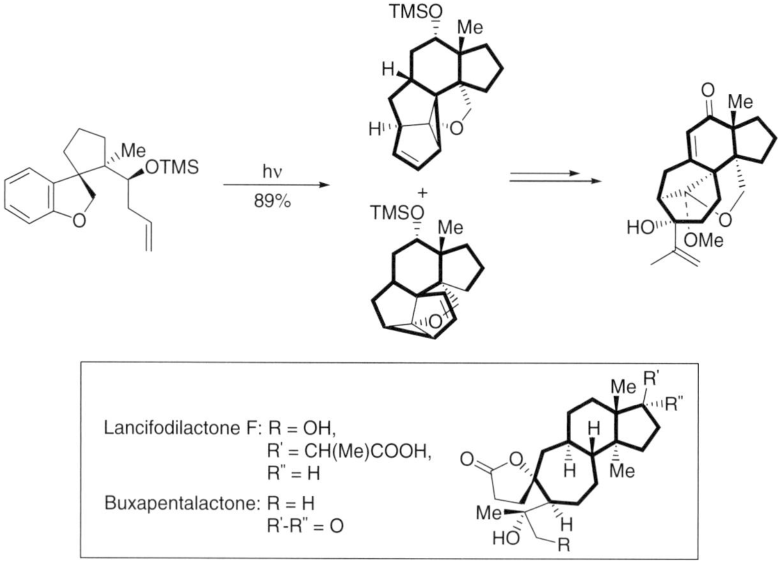

Lancifodilactone F: R = OH,
R' = CH(Me)COOH,
R" = H
Buxapentalactone: R = H
R'-R" = O

Scheme 4.62

4.7
Photochemical Synthesis of Three-Membered Heterocycles

Photochemical preparations of three-membered heterocycles, such as epoxides and aziridines, are fewer in number than those of cyclopropanes. Some recent examples of synthetic interest are described below.

4.7.1
Epoxides

It is well known that singlet oxygen reacts with furan derivatives to yield *endo*-peroxides. The latter are generally unstable, and may rearrange into bisepoxides, epoxylactones, enediones, *cis*-diacyoxiranes, enol esters, butenolides, and others [93]. Astarita *et al.* have recently reported that the dye-sensitized photooxygenation of carbohydrate furans leads to (1S,4R)-*endo*-peroxide in a highly diastereoselective process (Scheme 4.63) [94]. In turn, the *endo*-peroxide rearranges into a *syn*-(1R,2S:3S,4R)-bisepoxide and a keto ester.

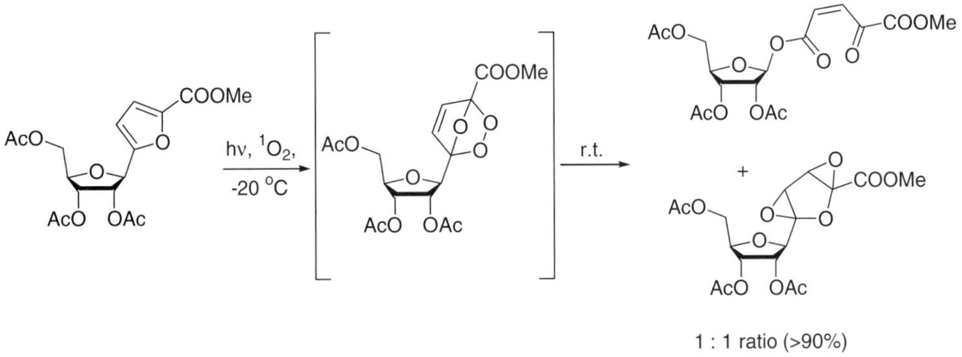

Scheme 4.63

Nakamura *et al.* have reported a one-step multiple addition of amine to [60]fullerene [95]. Here, a mixture of [60]fullerene and *N*-methylpiperazine was irradiated by means of an incandescent lamp in air to give a tetra(amino)fullerene epoxide in excellent yield (Scheme 4.64).

The salen complex, (*R*,*S*)-**80** in Scheme 4.37, which has a (1S,2S)-configuration at the 1,2-diiminocyclohexyl moiety and (*R*) at the binaphthyl moiety, was found to serve as an efficient catalyst for the epoxidation of conjugated olefins in the presence of 2,6-dichloropyridine *N*-oxide upon irradiation with incandescent lamps (Table 4.4) [59b, 96].

4.7.2
Aziridines

It is known that singlet acyl nitrenes derived from the acyl azides upon direct irradiation may undergo stereospecific addition to C=C bonds to yield aziridines [97].

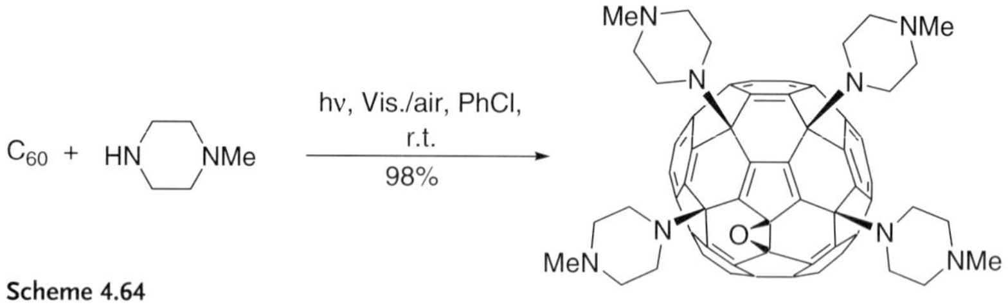

Scheme 4.64

C$_{60}$ + HN(piperazine)NMe →[hv, Vis./air, PhCl, r.t.] [98%] hexakis-adduct with MeN-piperazine-N groups

Table 4.4 Selected asymmetric epoxidation of olefins.

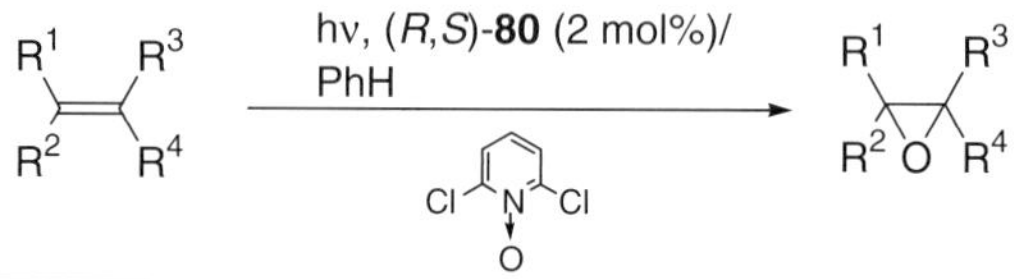

$$\underset{R^2\ \ R^4}{R^1\ \ R^3}\ \xrightarrow[\text{PhH, 2,6-dichloropyridine N-oxide}]{h\nu,\ (R,S)\text{-}80\ (2\ \text{mol\%})}\ \underset{R^2\ O\ R^4}{R^1\ \ R^3}$$

Substrate	Time (h)	Yield (%)	ee (%)	Product (configuration)
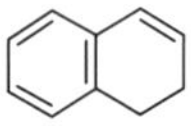 dihydronaphthalene	2	51	87	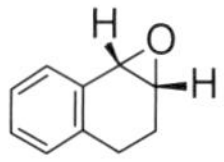 epoxide (1S,2S)
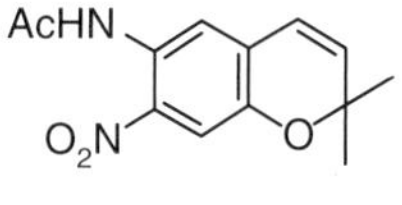 AcHN, O$_2$N-chromene	5	54	98	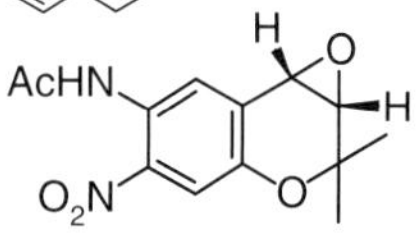 epoxide (3S,4S)
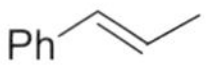 Ph-CH=CH-CH$_3$ (trans)	32	64	75	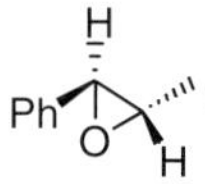 epoxide (1S,2S)
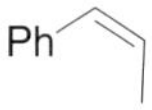 Ph-CH=CH-CH$_3$ (cis)	30	60	89	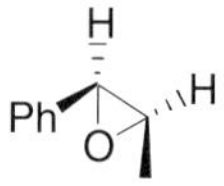 epoxide (1S,2R)
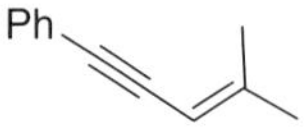 Ph-C≡C-CH=C(CH$_3$)$_2$	18	56	80	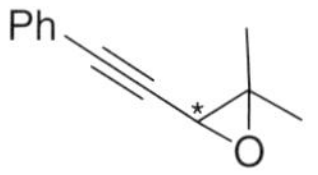 epoxide (not determined)

Loway et al. have prepared an aziridine by the application of photoinduced intramolecular aziridination; the aziridine was subsequently converted to L-daunosamine glycoside (Scheme 4.65) [98].

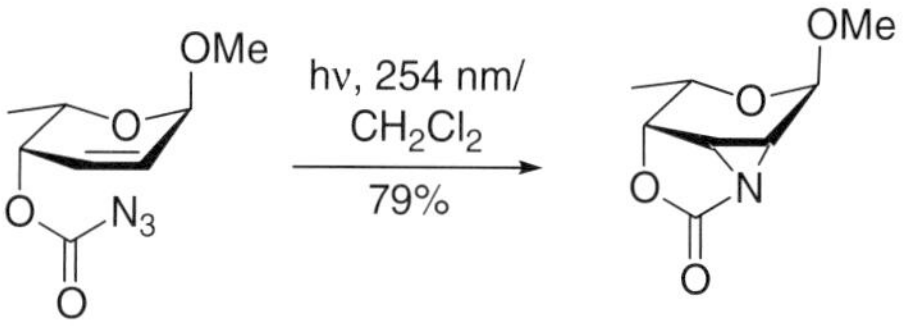

sugar-N$_3$ carbamate →[hv, 254 nm/ CH$_2$Cl$_2$] [79%] bicyclic aziridine-carbamate

Scheme 4.65

In 1972, it was reported that N-methylpyridinium chloride undergoes a photoinduced cyclization to produce the bicyclic aziridine upon irradiation in aqueous base [99]. Recently, Mariano et al. have applied this for the synthesis of the natural (+)-tetrazoamine and the unnatural (−)-enatiomer, via the photoinduced cyclization of pyridinium salts **121** in aqueous NaHCO$_3$ (Scheme 4.66) [100]. The resultant isomeric N-glucosyl-bicyclic-aziridines **122–125** could be separated using silica gel chromatography to give pure **122** (15%) and a mixture of **123–125** (30%). It is possible that (+)-tetrazoamine could be derived from **122**, whereas the (−)-enatiomer was obtained from the reaction of the mixture of **123–125**.

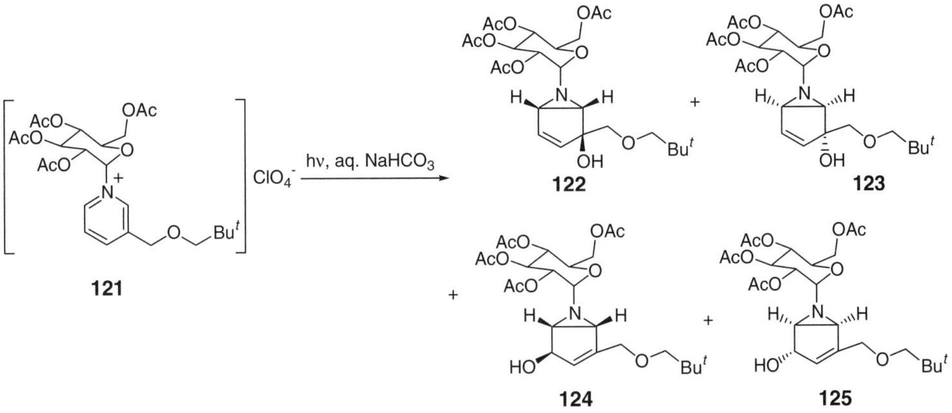

Scheme 4.66

The irradiation of 3-methylisoxazolo[5,4-b]pyridine in the presence of NaBH$_4$ as a nucleophile affords diastereoisomeric spiro aziridines (Scheme 4.67) [101].

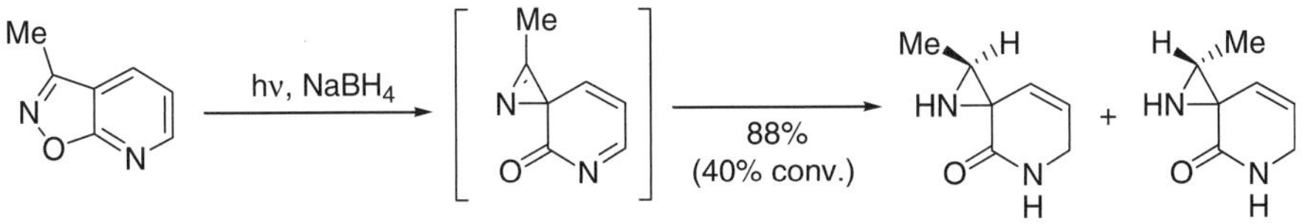

Scheme 4.67

Sakamoto et al. have provided an example of absolute asymmetric synthesis involving β-hydrogen abstraction by thiocarbonyl sulfur (Scheme 4.68) [102]. The achiral benzothioamides **126** crystallize in chiral space group $P2_12_12_1$. When the chiral crystals were irradiated in the solid state at −45 °C, followed by acetylation with AcCl at −78 °C, the aziridines **127** were obtained as a main product in a highly enantioselective cyclization.

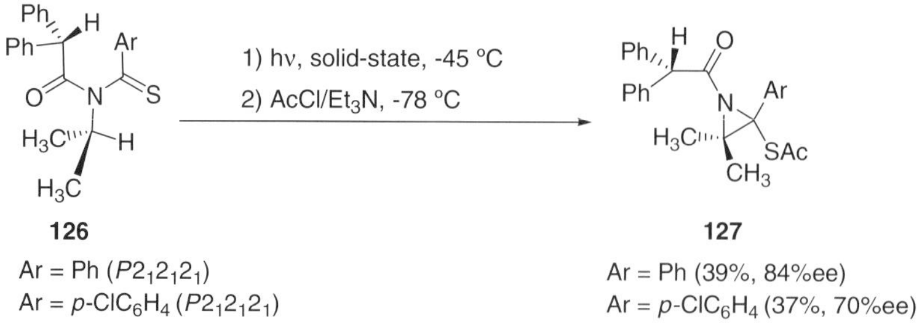

Scheme 4.68

126
Ar = Ph ($P2_12_12_1$)
Ar = p-ClC$_6$H$_4$ ($P2_12_12_1$)

127
Ar = Ph (39%, 84%ee)
Ar = p-ClC$_6$H$_4$ (37%, 70%ee)

References

1 de Meijere, A. (2003) Introduction: cyclopropanes and related rings. *Chemical Reviews*, **103** (4), 931–932.

2 Zimmerman, H.E. and Grunewald, G.L. (1966) Chemistry of barrelene. III. A unique photoisomerization to semibullvalene. *Journal of the American Chemical Society*, **88** (1), 183–184.

3 Zimmerman, H.E. and Armesto, D. (1996) Synthetic aspects of the di-π-methane rearrangement. *Chemical Reviews*, **96** (8), 3065–3112.

4 (a) Zimmerman, H.E. (2006) Five decades of mechanistic and exploratory organic photochemistry. *Pure and Applied Chemistry*, **78** (12), 2193–2203; (b) Armesto, D., Ortiz, M.J., and Agarrabeitia, A.R. (2005) Di-π-methane rearrangement, in *Molecular and Supramolecular Photochemistry: Synthetic Organic Photochemistry*, Vol. 12 (eds A.G. S Griesbeck and J. S Mattay), Marcel Dekker, Inc., New York, pp. 161–187; (c) Ramaiah, D., Sajimon, M.C., Joseph, J., and George, M.V. (2005) Photoisomerisation of dibenzobarrelenes – a facile route to polycyclic synthons. *Chemical Society Reviews*, **34** (1), 48–57.

5 Nair, V., Nandakumar, M.V., Anilkumar, G.N., Maliakal, D., Vairamani, M., Prabhakar, S., and Rath, N.P. (2000) Cycloaddition of 2-oxo-2H-cyclohepta[b]furan derivatives with arylacetylenes and di-π-methane rearrangement of homobarrelene derivatives. *Journal of the Chemical Society, Perkin Transactions 1*, 3795–3798.

6 Pitchumani, K., Warrier, M., Kaanumalle, L.S., Lakshmi, S., and Ramamurthy, V. (2003) Triplet photochemistry within zeolites through heavy atom effect, sensitization and light atom effect. *Tetrahedron*, **59** (30), 5763–5772.

7 Ihmels, H., Schneider, M., and Waidelich, M. (2002) Medium-dependent type selectivity in photoreactions of a crown ether-annelated dibenzobarrelene derivative. *Organic Letters*, **4** (19), 3247–3250.

8 (a) Garcia-Garibay, M., Scheffer, J.R., Trotter, J., and Wireko, F. (1989) Determination of the absolute steric course of a solid-state photorearrangement by anomalous dispersion x-ray crystallography. *Journal of the American Chemical Society*, **111** (13), 4985–4986; (b) Scheffer, J.R., Trotter, J., Garcia-Garibay, M., and Wireko, F. (1988) Studies on the di-π-methane photorearrangement in the solid state. *Molecular Crystals and Liquid Crystals*, **156** (Pt A), 63–84; (c) Evans, S.V., Garcia-Garibay, M., Omkaram, N., Scheffer, J.R., Trotter, J., and Wireko, F. (1986) Use of chiral single crystals to convert achiral reactants to chiral products in high optical yield: application to the di-π-methane and Norrish type II photorearrangements.

Journal of the American Chemical Society, **108** (18), 5648–5650.

9 Janz, K.M. and Scheffer, J.R. (1999) The use of ionic chiral sensitizers in the crystalline state: application to the di-π-methane photorearrangement of a benzonorbornadiene derivative. *Tetrahedron Letters*, **40** (50), 8725–8728.

10 Hubbard, S.C. and Jones, P.B. (2005) Ionic liquid soluble photosensitizers. *Tetrahedron*, **61** (31), 7425–7430.

11 Ding, J., Desikan, V., Han, X., Xiao, T.L., Ding, R., Jenks, W.S., and Armstrong, D.W. (2005) Use of chiral ionic liquids as solvents for the enantioselective photoisomerization of dibenzobicyclo [2.2.2]octatrienes. *Organic Letters*, **7** (2), 335–337.

12 Chen, A.-C., Chuang, G.J., Villarante, N.R., and Liao, C.-C. (2007) Chemoselective photorearrangements of diazinobarrelenes. Deuterium labeling study. *Journal of Organic Chemistry*, **72** (25), 9690–9697.

13 Chou, C.-H., Peddinti, R.K., and Liao, C.-C. (2001) Photochemistry of pyrazino-and quinoxalino-fused naphthobarrelenes. *Heterocycles*, **54** (1), 61–64.

14 Chen, A.-C., Lin, S.-Y., Villarante, N.R., Chuang, G.J., Wu, T.-C., and Liao, C.-C. (2008) Substituent effects on the bridging modes of photochemical rearrangements of pyrazino-, quinoxalino-, and benzo-quinoxalinobarrelenes. *Tetrahedron*, **64** (37), 8907–8921.

15 Dura, R.D. and Paquette, L.A. (2006) Ring contraction of bridgehead sultams by photoinduced di-π-methane rearrangement. *Journal of Organic Chemistry*, **71** (6), 2456–2459.

16 Zimmerman, H.E. and Cotter, B.R. (1974) Substituent effects and the di-π-methane rearrangement. Mechanistic and exploratory organic photochemistry. *Journal of the American Chemical Society*, **96** (24), 7445–7453.

17 Zimmerman, H.E. and Gruenbaum, W.T. (1978) Mechanistic and exploratory photochemistry. 111. Unusual regioselectivity in the di-π-methane rearrangement. Inhibition and control by electron-donating substituents. *Journal of Organic Chemistry*, **43** (10), 1997–2005.

18 Zimmerman, H.E. and Schissel, D.N. (1986) Di-π-methane rearrangements of highly sterically congested molecules: inhibition of free-rotor energy dissipation. Mechanistic and exploratory organic photochemistry. *Journal of Organic Chemistry*, **51** (2), 196–207.

19 Zimmerman, H.E. and Factor, R.E. (1981) Mechanistic and exploratory organic photochemistry. 126. Di-π-methane hypersurfaces and reactivity; multiplicity and regioselectivity; relationship between the di-π-methane and bicycle rearrangements. *Tetrahedron*, **37** (Suppl. 9), 125–141.

20 (a) Tsuno, T., Hoshino, H., Okuda, R., and Sugiyama, K. (2001) Allenyl(vinyl) methane photochemistry. Photochemistry of γ-(3-methyl-1-phenyl-1,2-butadienyl)-substituted α,β-unsaturated ester and nitrile derivatives. *Tetrahedron*, **57** (23), 4831–4840; (b) Tsuno, T. and Sugiyama, K. (2004) The photochemical reactivity of the allenyl-vinyl methane systems, in *CRC Handbook of Organic Photochemistry and Photobiology*, 2nd edn (eds W. M. Horspool and F. Lenci), CRC press, Boca Roton, FL, pp. 30/1–30/15.

21 (a) Zimmerman, H.E. and Novak, T. (2003) Regioselectivity of the tri-π-methane rearrangement: mechanistic and exploratory organic photochemistry. *Journal of Organic Chemistry*, **68** (13), 5056–5066; (b) Zimmerman, H.E. and Církva, V. (2001) Excited- and ground-state versions of the tri-π-methane rearrangement: Mechanistic and exploratory organic photochemistry. *Journal of Organic Chemistry*, **66**, (5), 1839–1851; (c) Zimmerman, H.E. and Církva, V. (2000) The tri-π-methane rearrangement: mechanistic and exploratory organic photochemistry. *Organic Letters*, **2**, (15), 2365–2367.

22 Zimmerman, H.E. and Factor, R.E. (1980) The bicycle rearrangement. 5. Relationship to the di-π-methane rearrangement and control by bifunnel distortion. Mechanistic and exploratory organic photochemistry. *Journal of the American Chemical Society*, **102** (10), 3538–3548.

23 (a) Jafarpour, F., Ramezani, F., and Pirelahi, H. (2008) Effects of electron-withdrawing group on the photoisomerization of tetraaryl-4H-thiopyran-1,1-dioxides. *Heteroatom Chemistry*, **19** (6), 557–561; (b) Jafarpour, F. and Pirelahi, H. (2006) An approach to the novel stereoselectivity in photorearrangement of 4,4-dialkyl-2,6-diphenyl-4H-thiopyran-1,1-dioxides. *Journal of Photochemistry and Photobiology, A: Chemistry*, **181** (2–3), 408–413; (c) Rezanejadebardajee, G., Jafarpour, F., and Pirelahi, H. (2006) A mechanistic approach to photochemical behavior of 4-anisyl-4-methyl-2,6-diphenyl-4H-thiopyran-1,1-dioxide. *Journal of Heterocyclic Chemistry*, **43** (1), 167–170; (d) Pirelahi, H., Jafarpour, F., Rezanejadebardajee, G., Amanishamsabaad, J., and Mouradzadegun, A. (2005) The effects of the electron-donating methoxy group on the photo-isomeriza-tion of 4-methyl-2,4,6-triaryl-4H-thiopyran-1,1-dioxides. *Phosphorus, Sulfur and Silicon and the Related Elements*, **180** (11), 2555–2561.

24 (a) Eade, S.J., Walter, M.W., Byrne, C., Odell, B., Rodriguez, Ra., Baldwin, J.E., Adlington, R.M., and Moses, J.E. (2008) Biomimetic synthesis of pyrone-derived natural products: exploring chemical pathways from a unique polyketide precursor. *Journal of Organic Chemistry*, **73** (13), 4830–4839; (b) Moses, J.E., Baldwin, J.E., Bruckner, S., Eade, S.J., and Adlington, R.M. (2003) Biomimetic studies on polyenes. *Organic & Biomolecular Chemistry*, **1**, (21), 3670–3684; (c) Brückner, S., Baldwin, J.E., Moses, J., Adlington, R.M., and Cowley, A.R. (2003) Mechanistic evidence supporting the biosynthesis of photodeoxytridachione. *Tetrahedron Letters* **44**, (40), 7471–7473.

25 Zuidema, D.R., Miller, A.K., Trauner, D., and Jones, P.B. (2005) Photosensitized conversion of 9,10-deoxytridachione to photodeoxytridachione. *Organic Letters*, **7** (22), 4959–4962.

26 Taglialatela-Scafati, O., Deo-Jangra, U., Campbell, M., Roberge, M., and Andersen, R.J. (2002) Diterpenoids from cultured *Erythropodium caribaeorum*. *Organic Letters*, **4** (23), 4085–4088.

27 (a) Natarajan, A., Tsai, C.K., Khan, S.I., McCarren, P., Houk, K.N., and Garcia-Garibay, M.A. (2007) The photo-arrangement of α-santonin is a single-crystal-to-single-crystal reaction: a long kept secret in solid-state organic chemistry revealed. *Journal of the American Chemical Society*, **129** (32), 9846–9847; (b) Blay, G. (2004) Photochemical rearrangements of 6/6- and 6/5-fused cross-conjugated cyclohexadienones, in *CRC Handbook of Organic Photochemistry and Photobiology*, 2nd edn (eds W. M. Horspool and F. Lenci), CRC press, Boca Roton, FL, pp. 80/1–80/21.

28 (a) Ricci, A., Fasani, E., Mella, M., and Albini, A. (2004) Photochemistry of some steroidal bicyclo[3.1.0]hexenones. *Tetrahedron*, **60** (1), 115–120; (b) Ricci, A., Fasani, E., Mella, M., and Albini, A. (2003) General patterns in the photochemistry of pregna-1,4-dien-3,20-diones. *Journal of Organic Chemistry*, **68**, (11), 4361–4366; (c) Ricci, A., Fasani, E., Mella, M., and Albini, A. (2001) Noncommunicating photoreaction paths in some pregna-1,4-diene-3,20-diones. *Journal of Organic Chemistry*, **66**, (24), 8086–8093; (d) Ibal, J., Gupta, A., and Husain, A. (2006) Photochemistry of clobetasol propionate, a steroidal anti-inflammatory drug. *ARKIVOC* (11), 91–98; (e) Iqbal, J., Husain, A., and Gupta, A. (2006) Photochemistry of desonide, a non-

fluorinated steroidal anti-inflammatory drug. *Chemical & Pharmaceutical Bulletin*, **54** (6), 836–838.

29 Tenney, L.P., Boykin, D.W. Jr, and Lutz, R.E. (1966) Novel photocyclization of a highly phenylated β,γ-unsaturated ketone to a cyclopropyl ketone, involving benzoyl group migration. *Journal of the American Chemical Society*, **88** (8), 1835–1836.

30 Armesto, D., Ortiz, M.J., and Agarrabeitia, A.R. (2004) Novel di-π-methane rearrangements promoted by photoelectron transfer and triplet sensitization, in *CRC Handbook of Organic Photochemistry and Photobiology*, 2nd edn (eds W. M. Horspool and F. Lenci), CRC Press, Boca Roton, FL, pp. 95/1–95/16.

31 (a) Rao, V.J. and Griesbeck, A.G. (2005) Oxa-di-π-methane rearrangements, in *Molecular and Supramolecular Photochemistry: Synthetic Organic Photochemistry*, Vol. 12 (eds A.G. Griesbeck and J. S Mattay), Marcel Dekker, Inc., New York, pp. 189–210; (b) Singh, V. (2004) Photochemical rearrangements in β,γ-unsaturated enones: the oxa-di-π-methane rearrangement, in *CRC Handbook of Organic Photochemistry and Photobiology*, 2nd edn (eds W. M. Horspool and F. Lenci), CRC Press, Boca Roton, FL, pp. 78/1–78/34.

32 Lee, T.-H., Rao, P.D., and Liao, C.-C. (1999) Photochemistry of bicyclo[2.2.2] octenones: an uncommon oxidative decarbonylation. *Chemical Communications*, (9), 801–802.

33 Yang, M.-S., Lu, S.-S., Rao, C.P., Tsai, Y.-F., and Liao, C.-C. (2003) Photochemistry of Bicyclo[2.2.2]oct-7-ene-2,5-diones and the corresponding 5-hydroxyimino and 5-methylene derivatives. *Journal of Organic Chemistry*, **68** (17), 6543–6553.

34 Chang, S.-Y., Huang, S.-L., Villarante, N.R., and Liao, C.-C. (2006) Photo-chemical reactions of 1,3,3-trimethyl-bicyclo[2.2.2]octa-5,7-dien-2-ones. *European Journal of Organic Chemistry*, (20), 4648–4657.

35 McClure, C.K., Kiessling, A.J., and Link, J.S. (2003) Oxa-di-π-methane photochemical rearrangement of quinuclidinones. Synthesis of pyrrolizidinones. *Organic Letters*, **5** (21), 3811–3813.

36 Sadeghpoor, R., Ghandi, M., Najafi, H.M., and Farzaneh, F. (1998) The oxa-di-π-methane rearrangement of β,γ-unsaturated ketones induced by the external heavy atom cation effect within a zeolite. *Chemical Communications*, (3), 329–330.

37 Pitchumani, K., Warrier, M., Kaanumalle, L.S., Lakshmi, S., and Ramamurthy, V. (2003) Triplet photochemistry within zeolites through heavy atom effect, sensitization and light atom effect. *Tetrahedron*, **59** (30), 5763–5772.

38 Uppili, S. and Ramamurthy, V. (2002) Enhanced enantio- and diastereo-selectivities via confinement: photo-rearrangement of 2,4-cyclohexadienones included in zeolites. *Organic Letters*, **4** (1), 87–90.

39 Sivasubramanian, K., Kaanumalle, L.S., Uppili, S., and Ramamurthy, V. (2007) Value of zeolites in asymmetric induction during photocyclization of pyridones, cyclohexadienones and naphthalenones. *Organic & Biomolecular Chemistry*, **5** (10), 1569–1576.

40 Arumugam, S. (2008) Nafion as an efficient reaction medium for diastereoselective photochemical reactions. *Tetrahedron Letters*, **49** (15), 2461–2465.

41 Armesto, D. and Ortiz, M.J. (2005) The effects of triplet sensitizers, energies on the photoreactivity of β,γ-unsaturated methyl ketones. *Angewandte Chemie, International Edition*, **44** (47), 7739–7741.

42 Yadav, S., Banerjee, S., Maji, D., and Lahiri, S. (2007) Synthesis of bicyclo[3.2.1] octanones via ketyl radical promoted rearrangements under reductive PET

conditions. *Tetrahedron*, **63** (45), 10979–10990.

43 Zhang, Y. and Schuster, G.B. (1995) Photoresolution of an axially chiral bicyclo[3.2.1]octan-3-one: phototriggers for a liquid crystal-based optical switch. *Journal of Organic Chemistry*, **60** (22), 7192–7197.

44 Kumar, V.S.S., Sheela, K.C., Nair, V., and Rath, N.P. (2004) Concomitant polymorphism in a spirobicyclic dione. *Crystal Growth & Design*, **4** (6), 1245–1247.

45 (a) Armesto, D., Horspool, W.M., Langa, F., and Ramos, A. (1991) Extension of the aza-di-π-methane reaction to stable derivatives. Photochemical cyclization of β,γ-unsaturated oxime acetates. *Journal of the Chemical Society, Perkin Transactions 1*, 223–228; (b) Armesto, D., Horspool, W.M., and Langa, F. (1987) The aza-di-π-methane rearrangement of O-acetyl 2,2-dimethyl-4,4-diphenylbut-3-enal oxime. *Journal of the Chemical Society, Chemical Communications*, (24), 1874–1875.

46 (a) Armesto, D., Horspool, W.M., Gallego, M.G., and Agarrabeitia, A.R. (1992) Steric and electronic effects on the photochemical reactivity of oxime acetates of β,γ-unsaturated aldehydes. *Journal of the Chemical Society, Perkin Transactions 1*, 163–169; (b) Armesto, D., Horspool, W.M., Mancheno, M.J., and Ortiz, M.J. (1990) The aza-di-π-methane rearrangement of stable derivatives of 2,2-dimethyl-4,4-diphenylbut-3-enal. *Journal of the Chemical Society, Perkin Transactions 1*, 2348–2349.

47 Armesto, D. Horspool, William, M., Mancheno, M.J., and Ortiz, M.J. (1993) A new photochemical synthesis of dihydropyrazoles. Novel mode of photocyclization of some 1-iminobut-3-enes derivatives. *Journal of the Chemical Society, Chemical Communications*, (9), 721–722.

48 (a) Armesto, D., Horspool, W.M., Martin, J.-A.F., and Perez-Ossorio, R. (1986) The synthesis and photochemical reactivity of β,γ-unsaturated imines. An azadi-π-methane rearrangement of 1-azapenta-1,5-dienes. *Journal of Chemical Research, Synopses* (2), 46–47; (b) Armesto, D., Martin, J.-A.F., Perez-Ossorio, R., and Horspool, W.M. (1982) A novel aza-di-π-methane rearrangement. The photoreaction of 4,4-dimethyl-1,6,6-triphenyl-2-azahexa-2,5-diene. *Tetrahedron Letters*, **23** (20), 2149–2152.

49 (a) Banwell, M.G., Edwards, A.J., Harfoot, G.J., and Jolliffe, K.A. (2004) A chemoenzymatic synthesis of the linear triquinane (−)-hirsutene and identification of possible precursors to the naturally occurring (+)-enantiomer. *Tetrahedron*, **60** (3), 535–547; (b) Banwell, M.G., Edwards, A.J., Harfoot, G.J., and Jolliffe, K.A. (2002) A chemoenzymatic synthesis of (−)-hirsutene from toluene. *Journal of the Chemical Society, Perkin Transactions 1*, 2439–2441.

50 (a) Banwell, M.G., Austin, K.A.B., and Willis, A.C. (2007) Chemoenzymatic total syntheses of the linear triquinane-type natural products (+)-hirsutic acid and (−)-complicatic acid from toluene. *Tetrahedron*, **63** (28), 6388–6403; (b) Austin, K.A.B., Banwell, M.G., Harfoot, G.J., and Willis, A.C. (2006) Chemoenzymatic syntheses of the linear triquinane-type sesquiterpenes (+)-hirsutic acid and (−)-complicatic acid. *Tetrahedron Letters*, **47** (41), 7381–7384.

51 (a) Singh, V., Vedantham, P., and Sahu, P.K. (2004) Reactive species from aromatics and oxa-di-π-methane rearrangement: a stereoselective synthesis of (±)-hirsutene from salicyl alcohol. *Tetrahedron*, **60** (37), 8161–8169; (b) Singh, V., Vedantham, P., and Sahu, P.K. (2002) A novel stereoselective total synthesis of (±)-hirsutene from saligenin. *Tetrahedron Letters*, **43** (3), 519–522.

52 Singh, V., Praveena, G.D., Karki, K., and Mobin, S.M. (2007) Cycloaddition of cyclohexa-2,4-dienones, ring-closing metathesis, and photochemical reactions: a common stereoselective approach to duprezianane, polyquinane and

sterpurane frameworks. *Journal of Organic Chemistry*, **72** (6), 2058–2067.
53. Yen, C.-F. and Liao, C.-C. (2002) Concise and efficient total synthesis of Lycopodium alkaloid magellanine. *Angewandte Chemie, International Edition*, **41** (21), 4090–4093.
54. (a) Abdel-Wahab, A.-M.A., Ahmed, S.A., and Dürr, H. (2004) Carbene formation by extrusion of nitrogen, in *CRC Handbook of Organic Photochemistry and Photobiology*, 2nd edn (eds W. M. Horspool and F. Lenci), CRC Press, Boca Roton, FL, pp. 91/1–19/37; (b) Dürr, H. and Abdel-Wahab, A.-M.A. (1994) Carbene formation by extrusion of nitrogen, in *CRC Handbook of Organic Photochemistry and Photobiology* (eds W. M. Horspool and P.-S. Song), CRC Press, Boca Roton, FL, pp. 954–983.
55. Maas, G., Krebs, F., Werle, T., Gettwert, V., and Striegler, R. (1999) Silicon-oxygen heterocycles from thermal, photo-chemical, and transition-metal-catalyzed decomposition of α-(alkoxysilyl and alkenyloxysilyl)- α-diazoacetates. *European Journal of Organic Chemistry*, (8), 1939–1946.
56. Maas, G., Daucher, B., Maier, A., and Gettwert, V. (2004) Synthesis and ring opening reactions of a 2-silabicyclo[2.1.0]pentane. *Chemical Communications*, (2), 238–239.
57. Pastor-Perez, L., Wiebe, C., Perez-Prieto, J., and Stiriba, S.-E. (2007) A tetramethoxybenzophenone as efficient triplet photocatalyst for the transformation of diazo compounds. *Journal of Organic Chemistry*, **72** (4), 1541–1544.
58. (a) Pastor-Perez, L., Barriau, E., Berger-Nicoletti, E., Kilbinger, A.F.M., Perez-Prieto, J., Frey, H. and Stiriba, S.-E. (2008) Incorporation of a photosensitizer core within hyperbranched polyether polyols: effect of the branched shell on the core properties. *Macromolecules*, **41** (4), 1189–1195; (b) Pastor-Perez, L., Barriau, E., Frey, H., Perez-Prieto, J., and Stiriba, S.-E. (2008) Photocatalysis within hyperbranched polyethers with a benzophenone core. *Journal of Organic Chemistry*, **73** (12), 4680–4683.
59. (a) Uchida, T., Irie, R., and Katsuki, T. (2000) cis- and Enantio-selective cyclopropanation with chiral (ON$^+$)Ru-salen complex as a catalyst. *Tetrahedron*, **56** (22), 3501–3509; (b) Katsuki, T. (2000) Asymmetric reactions using optically active (Salen)ruthenium complexes as catalysts. *Nippon Kagaku Kaishi*, (10), 669–676; (c) Uchida, T., Irie, R., and Katsuki, T. (1999) Highly cis- and enantioface-selective cyclopropanation using (R,R)-Ru-salen complex. Solubility dependent enantioface selection. *Synlett*, (11), 1793–1795; (d) Uchida, T., Irie, R., and Katsuki, T. (1999) Chiral (ON) Ru-salen-catalyzed cyclopropanation. High cis- and enantioselectivity. *Synlett*, (7), 1163–1165.
60. (a) Saha, B., Uchida, T., and Katsuki, T. (2003) Asymmetric intramolecular cyclopropanation of diazo compounds with metallosalen complexes as catalyst: structural tuning of salen ligand. *Tetrahedron: Asymmetry*, **14** (7), 823–826; (b) Saha, B., Uchida, T., and Katsuki, T. (2002) Highly enantioselective intramolecular cyclopropanation of alkenyl diazo ketones using Ru(salen) as catalyst. *Chemistry Letters*, (8), 846–847; (c) Saha, B., Uchida, T., and Katsuki, T. (2000) Intramolecular asymmetric cyclopropanation with (nitroso)(salen) ruthenium(II) complexes as catalysts. *Synlett*, (1), 114–116.
61. (a) Krysiak, J., Lyon, C., Baceiredo, A., Gornitzka, H., Mikolajczyk, M., and Bertrand, G. (2004) Stable optically pure phosphino(silyl)carbenes: reagents for highly enantioselective cyclopropanation reactions. *Chemistry – A European Journal*, **10** (8), 1982–1986; (b) Krysiak, J., Kato, T., Gornitzka, H., Baceiredo, A., Mikolajczyk, M., and Bertrand, G. (2001) The first asymmetric cyclopropanation

reactions involving a stable carbene. *Journal of Organic Chemistry*, **66** (24), 8240–8242.

62 Condon, S.E., Buron, C., Tippmann, E.M., Tinner, C., and Platz, M.S. (2004) Generation and characterization of phenylsulfanylcarbene. *Organic Letters*, **6** (5), 815–818.

63 Stoffregen, S.A., Heying, M., and Jenks, W.S. (2007) S,C-Sulfonium ylides from thiophenes: potential carbene precursors. *Journal of the American Chemical Society*, **129** (51), 15746–15747.

64 (a) Wagner, P.J. (2005) Abstraction of γ-hydrogens by excited carbonyls, in *Molecular and Supramolecular Photochemistry: Synthetic Organic Photochemistry* (eds A.G. S Griesbeck and J. S Mattay), Marcel Dekker, Inc., New York, pp. 11–39; (b) Wessig, P. and Mühling, O. (2005) Abstraction of (γ ± n)-hydrogen by excited carbonyls, in *Molecular and Supramolecular Photo-chemistry: Synthetic Organic Photo-chemistry* (eds A.G. S Griesbeck and J. S Mattay), Marcel Dekker, Inc., New York, pp. 41–87; (c) Wagner, P.J. (2004) Yang photocyclization: coupling of biradicals formed by intramolecular hydrogen abstraction of ketones, in *CRC Handbook of Organic Photochemistry and Photo-biology*, 2nd edn (eds W. M. Horspool and F. Lenci), CRC Press, Boca Roton, FL, pp. 58/1–58/70.

65 (a) Roth, H.J. and George, H. (1970) Photoreactions of anilino-benzyl-alkylaryl ketones. II. Photochemical reactions of anilino-benzyl ketones. *Archiv der Pharmazie und Berichte der Deutschen, Pharmazeutischen Gesellschaft*, **303** (9), 725–732; (b) Roth, H.J. and El Raie, M.H. (1970) Photocyclisierung von 3-aminoketonen zu 2-aminocyclopropanolen. *Tetrahedron Letters*, **11**, 2445–2446.

66 Weigel, W., Schiller, S., and Henning, H.-G. (1997) Stereoselective photocyclization to 2-aminocyclopropanols by photolysis of β-aminoketones and oxidative ring opening to enaminones. *Tetrahedron*, **53** (23), 7855–7866.

67 (a) Weigel, W. and Wagner, P.J. (1996) Photocyclization to cyclopropanols and exciplex emission of β-Arylpropiophenones. *Journal of the American Chemical Society*, **118** (50), 12858–12859; (b) Weigel, W. and Wagner, P.J. (1996) Hydrogen abstraction and charge transfer in β-anilinoketones. *Journal of Information Recording*, **23**, 31–34.

68 Kraus, G.A. and Chen, L. (1991) Intramolecular photocyclizations of amino ketones to form fused and bridged bicyclic amines. *Tetrahedron Letters*, **32** (49), 7151–7154.

69 Ariel, S., Askari, S.H., Scheffer, J.R., and Trotter, J. (1986) Cyclopropanol formation via β-hydrogen atom abstraction: the five membered transition state analog of the Norrish type II reaction. *Tetrahedron Letters*, **27** (7), 783–786.

70 (a) Yoshioka, M., Miyazoe, S., and Hasegawa, T. (1993) Photochemical reaction of 3-hydroxy-1-(o-methylaryl)alkan-1-ones: formation of cyclopropane-1,2-diols and benzocyclobutenols through β- and γ-hydrogen abstractions. *Journal of the Chemical Society, Perkin Transactions 1*, 2781–2786; (b) Yoshioka, M. (2004) Intramolecular hydrogen abstraction by excited carbonyl oxygen, particularly on β-hydrogen abstraction. *MaLS Forum*, **2**, 3–10.

71 Matsumoto, S., Okubo, Y., and Mikami, K. (1998) New type of photochemical carbon skeletal rearrangement: transformation of α, β-unsaturated carbonyl to 1,4-dicarbonyl compounds. *Journal of the American Chemical Society*, **120** (16), 4015–4016.

72 (a) Wessig, P. and Mühling, O. (2003) Photochemical synthesis of highly functionalized cyclopropyl ketones. *Helvetica Chimica Acta*, **86** (3), 865–893; (b) Wessig, P. and Mühling, O. (2001) A new photochemical route to cyclopropanes. *Angewandte Chemie, International Edition*, **40** (6), 1064–1065.

73 Wessig, P. and Mühling, O. (2005) 1,2-chirality transfer in the synthesis of cyclopropanes. *Angewandte Chemie, International Edition*, **44** (41), 6778–6781.

74 Mühling, O. and Wessig, P. (2006) Photochemical synthesis of benzoylspiro[2.2]pentanes. *Photochemical & Photobiological Sciences*, **5** (11), 1000–1005.

75 Brumfield, M.A. and Agosta, W.C. (1988) Two triplets mediating intramolecular photochemical abstraction of hydrogen by nitrogen in 4-acyl-6-alkylpyrimidines. *Journal of the American Chemical Society*, **110** (20), 6790–6794.

76 Wilzbach, K.E. and Kaplan, L. (1966) Photochemical 1,3-cycloaddition of olefins to benzene. *Journal of the American Chemical Society*, **88** (9), 2066–2067.

77 Bryce-Smith, D., Gilbert, A., and Orger, B.H. (1966) Photochemical 1,3-cycloaddition of olefins to aromatic compounds. *Chemical Communications*, (15), 512–513.

78 (a) Mattay, J. (2007) Photochemistry of arenes – reloaded. *Angewandte Chemie, International Edition*, **46** (5), 663–665; (b) Hoffmann, N. (2005) Ortho-, meta-, and para-photocycloaddition of arenes, in *Molecular and Supramolecular Photochemistry 12 (Synthetic Organic Photochemistry)* (eds A.G. S Griesbeck and J. S Mattay), Marcel Dekker, Inc., New York, pp. 529–552; (c) Hoffmann, N. (2004) Photochemical cycloaddition between benzene derivatives and alkenes. *Synthesis*, (4), 481–495; (d) Gilbert, A. (2004) Intra- and intermolecular cycloadditions of benzene derivatives, in *CRC Handbook of Organic Photochemistry and Photobiology*, 2nd edn (eds W. S Horspool and F. Lenci), CRC Press, Boca Roton, FL, pp. 41/1–41/11; (e) Wender, P. (1995) Intra- and intermolecular cycloadditions of benzene derivatives, in *CRC Handbook of Organic Photochemistry and Photobiology* (eds W. M. Horspool and P.-S. Song), CRC Press, Boca Roton, FL, pp. 280–290; (f) Wender, P.A., Ternansky, R., DeLong, M., Singh, S., Olivero, A., and Rice, K. (1990) Arene-alkene cycloadditions and organic synthesis. *Pure and Applied Chemistry*, **62** (8), 1597–1602; (g) Cornelisse, J. (1993) The meta photocycloaddition of arenes to alkenes. *Chemical Reviews*, **93** (2), 615–669; (h) De Keukeleire, D. and He, S.L. (1993) Photochemical strategies for the construction of polycyclic molecules. *Chemical Reviews*, **93** (1), 359–380.

79 Chappell, D. and Russell, A.T. (2006) From α-cedrene to crinipellin B and onward: 25 years of the alkene-arene *meta*-photocycloaddition reaction in natural product synthesis. *Organic & Biomolecular Chemistry*, **4** (24), 4409–4430.

80 de Long, M.A., Mathews, J.A., Cohen, S.L., and Gudmundsdottir, A. (2007) β-Silyl arenes in the arene-olefin photocyclization reaction. *Synthesis*, (15), 2343–2350.

81 Piet, D., de Bruijn, S., Sum, J., and Lodder, G. (2005) New molecular entities via intermolecular meta photocycloaddition. *Journal of Heterocyclic Chemistry*, **42** (2), 227–231.

82 Penkett, C.S., Byrne, P.W., Teobald, B.J., Rola, B., Ozanne, A., and Hitchcock, P.B. (2004) The use of temporary tethers in the meta photocycloaddition reaction. *Tetrahedron*, **60** (12), 2771–2784.

83 Blakemore, D.C. and Gilbert, A. (1994) Intramolecular meta photocycloaddition of 3-benzylazaprop-1-enes. *Tetrahedron Letters*, **35** (29), 5267–5270.

84 Guo, X.-C. and Chen, Q.-Y. (1999) Photoinduced intramolecular arene-olefin meta-cycloaddition of 5-phenyl-fluorinated-pent-1-enes. *Journal of Fluorine Chemistry*, **97** (1–2), 149–156.

85 Blakemore, D.C. and Gilbert, A. (1995) Heteroatom control of isomer formation in the intramolecular meta photocycloaddition of arene-ethene bichromophores. *Tetrahedron Letters*, **36** (13), 2307–2310.

86 De Keukeleire, D., He, S.-L., Blakemore, D., and Gilbert, A. (1994) Intramolecular photocycloaddition reactions of 4-phenoxybut-1-enes. *Journal of Photo-*

chemistry and Photobiology, A: Chemistry, 80 (1–3), 233–240.

87 Vizvardi, K., Desmet, K., Luyten, I., Sandra, P., Hoornaert, G., and Van der Eycken, E. (2001) Asymmetric induction in intramolecular meta photocycloaddition: cyclodextrin-mediated solid-phase photochemistry of various phenoxyalkenes. Organic Letters, 3 (8), 1173–1175.

88 Morales, R.C., Lopez-Mosquera, A., Roper, N., Jenkins, P.R., Fawcett, J., and Garcia, M.D. (2006) Diastereocontrol in the intramolecular meta-photocycloaddition of arenes and olefins. Photochemical & Photobiological Sciences, 5 (7), 649–652.

89 (a) Sugimura, T., Yamasaki, A., and Okuyama, T. (2005) Stereocontrolled intramolecular meta-arene-alkene photocycloaddition reactions using chiral tethers: efficiency of the tether derived from 2,4-pentanediol. Tetrahedron: Asymmetry, 16 (3), 675–683; (b) Hagiya, K., Yamasaki, A., Okuyama, T., and Sugimura, T. (2004) Asymmetric meta-arene-alkene photocycloaddition controlled by a 2,4-pentanediol tether. Tetrahedron: Asymmetry, 15 (9), 1409–1417.

90 Mukae, H., Maeda, H., and Mizuno, K. (2006) One-step synthesis of benzotetra- and benzopentacyclic compounds through intramolecular [2 + 3] photocycloaddition of alkenes to naphthalene. Angewandte Chemie, International Edition, 45 (39), 6558–6560.

91 Mukae, H., Maeda, H., Nashihara, S., and Mizuno, K. (2007) Intramolecular photocycloaddition of 2-(2-alkenyloxymethyl)-naphthalene-1-carbonitriles using glass-made microreactors. Bulletin of the Chemical Society of Japan, 80 (6), 1157–1161.

92 Wang, Q. and Chen, C. (2008) An approach to the core skeleton of lancifodilactone F. Organic Letters, 10 (6), 1223–1226.

93 Adam, W., Bosio, S., Bartoschek, A., and Griesbeck, A.G. (2004) Photooxygenation of 1,3-dienes, in CRC Handbook of Organic Photochemistry and Photobiology, 2nd edn (eds W. M. Horspool and F. Lenci), CRC Press, Boca Roton, FL, pp. 25/1–25/19.

94 Astarita, A., Cermola, F., Iesce, M.R., and Previtera, L. (2008) Dye-sensitized photooxygenation of sugar furans: novel bis-epoxide and spirocyclic C-nucleosides. Tetrahedron, 64 (28), 6744–6748.

95 Isobe, H., Tomita, N., and Nakamura, E. (2000) One-step multiple addition of amine to [60]fullerene. Synthesis of tetra(amino)fullerene epoxide under photochemical aerobic conditions. Organic Letters, 2 (23), 2663–3665.

96 (a) Nakata, K., Takeda, T., Mihara, J., Hamada, T., Irie, R., and Katsuki, T. (2001) Asymmetric epoxidation with a photoactivated [Ru(salen)] complex. Chemistry - A European Journal, 7 (17), 3776–3782; (b) Takeda, T., Irie, R., Shinoda, Y., and Katsuki, T. (1999) Ru-salen catalyzed asymmetric epoxidation. Photoactivation of catalytic activity. Synlett, (7), 1157–1159.

97 Bucher, G. (2004) Photochemical reactivity of azides, in CRC Handbook of Organic Photochemistry and Photobiology, 2nd edn (eds W. M. Horspool and F. Lenci), CRC Press, Boca Roton, FL, pp. 44/1–44/31.

98 Mendlik, M.T., Tao, P., Hadad, C.M., Coleman, R.S., and Lowary, T.L. (2006) Synthesis of L-daunosamine and L-ristosamine glycosides via photoinduced aziridination. Conversion to thioglycosides for use in glycosylation reactions. Journal of Organic Chemistry, 71 (21), 8059–8070.

99 (a) Mariano, P.S. (2004) A new look at pyridinium salt photochemistry, in CRC Handbook of Organic Photochemistry and Photobiology, 2nd edn (eds W. M. Horspool and F. Lenci), CRC Press, Boca Roton, FL, pp. 100/1–100/10; (b) Kaplan, L., Pavlik, J.W., and Wilzbach, K.E. (1972) Photohydration of pyridinium ions. Journal of the American Chemical Society, 94 (9), 3283–3284.

100 Feng, X., Duesler, E.N., and Mariano, P.S. (2005) Pyridinium salt photochemistry in a concise route for synthesis of the trehazolin aminocyclitol, trehazolamine. *Journal of Organic Chemistry*, **70** (14), 5618–5623.

101 Donati, D., Fusi, S., and Ponticelli, F. (2002) Trapping of photochemical intermediates as a tool in organic synthesis. Preparation of spiroaziridinopyridones, a new heterocyclic system. *Tetrahedron Letters*, **43** (52), 9527–9530.

102 (a) Sakamoto, M., Takahashi, M., Shimizu, M., Fujita, T., Nishio, T., Iida, I., Yamaguchi, K., and Watanabe, S. (1995) "Absolute" asymmetric synthesis using the chiral crystal environment: photochemical hydrogen abstraction from achiral acyclic monothioimides in the solid state. *Journal of Organic Chemistry*, **60** (22), 7088–7089; (b) Sakamoto, M. (2006) Spontaneous chiral crystallization of achiral materials and absolute asymmetric photochemical transformation using the chiral crystalline environment. *Journal of Photochemistry and Photobiology, C: Photochemistry Reviews*, **7** (4), 183–196.

5
Formation of a Four-Membered Ring
Norbert Hoffmann

5.1
Introduction

Cyclobutanes and cyclobutenes such as other small ring systems are versatile molecular building blocks. Due to inherent ring strain, they are also interesting intermediates in multistep syntheses. Many ring-opening or enlarging reactions enable their application to the synthesis of complex structures [1–3], while ring contraction to cyclopropane is also possible. A variety of synthetic methods for the preparation of four-membered rings have been reported in the literature [4, 5]. When compared to their ground state, photochemically excited compounds possess a significantly different chemical reactivity [6]. Photochemical reactions are particular attractive as these transformations easily enable the preparation of otherwise hardly accessible compounds [7, 8]. For these reasons, four-membered ring systems – and in particular cyclobutanes and cyclobutenes, which belong to very different product families and possess a large substitution spectrum – can be synthesized more easily via photochemical than via ground-state reactions. Two prominent reactions: (i) the addition of alkenes to α,β-unsaturated carbonyl, carboxyl and related compounds; and (ii) the synthesis of oxetanes (Paternò–Büchi reaction), are reported in Chapters 6 and 7, respectively. In this chapter, a variety of other photochemical methods for the synthesis of four-membered rings are discussed.

5.2
[2 + 2]-Photocycloaddition of Nonconjugated Alkenes

According to the Woodward–Hoffmann rules, the concerted $[2_s + 2_s]$ cycloaddition with two alkenes is photochemically symmetry-allowed, but is symmetry-forbidden at the ground state [9]. Photochemical [2 + 2] cycloaddition, in which one of two alkene partners is electronically excited, has been applied to the synthesis of cage hydrocarbons [10]. In such transformations, the intramolecular version of the reaction is particular efficient. The transformation of compound **1**, in which two

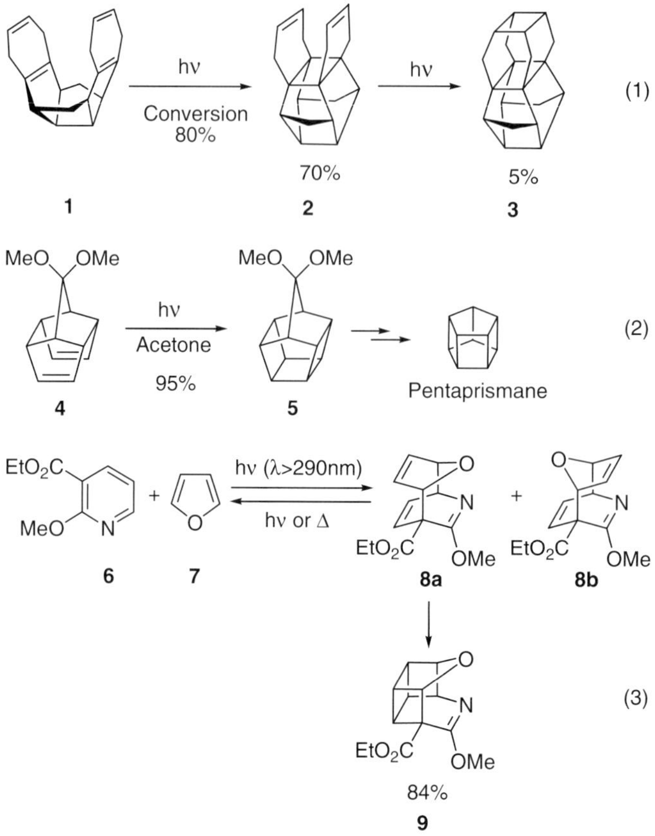

Scheme 5.1 Intramolecular [2 + 2] photocycloaddition.

alkene functions are arranged close to each other, produced in good yields the cyclobutane derivative **2** (Scheme 5.1, reaction 1) [11]. The reaction was carried out with an irradiation at λ = 254 nm in cyclohexane as solvent. These conditions indicate that the reaction took place at the first excited singlet state (S_1), and the cage compound **3** was obtained as a byproduct. The same type of [2 + 2] cycloaddition can also be carried out under photosensitization; thus, compound **5** was obtained in high yields from transformation of the diene **4** (Scheme 5.1, reaction 2) [12], the reaction being carried out in acetone as solvent. Under these conditions, acetone acts as triplet sensitizer. The transformation was also performed as part of the tentative synthesis of hexaprismane and similar compounds [13]. The [2 + 2] photocycloaddition is frequently observed as a consecutive reaction; for instance, in the transformation of the pyridine derivative **6** with furan **7**, a [4 + 4] cycloaddition first takes place leading to the adducts **8a,b** (Scheme 5.1, reaction 3) [14]. This step is reversible and the primary products are not stable enough to be isolated in pure form. However, two alkene functions are well orientated to react readily in a [2 + 2] photocycloaddition, leading to a high overall yield of the final cage hydrocarbon **9**. These reactions are also frequently observed with condensed aromatic compounds, such as naphthalene derivatives [15].

A [2 + 2] photocycloaddition with two alkenes can also be induced by photochemical electron transfer [16, 17]. In such cases, sensitizers are frequently used and the reactions therefore occur under photocatalysis [18]. Under photochemical electron transfer (PET) conditions, the diene **10** yielded in an intramolecular reaction the cyclobutane **11** (Scheme 5.2) [19], such that in this reaction a 12-membered cyclic polyether is built up. The reaction starts with excitation of the sensitizer 1,4-dicyanonaphthalene (DCN); only 0.1 equivalents of the sensitizer are added to the reaction mixture. Electron transfer occurs from the substrate **10** to the excited sensitizer, leading to the radical cation **I**. This intermediate then undergoes cyclization to the radical cation of the cyclobutane (**II**). Electron transfer from the radical anion of the sensitizer to the intermediate **II** leads to the final product **11**, and regenerates the sensitizer. In some cases, for example the cyclodimerization of N-vinylcarbazole, the efficiency is particularly high because a chain mechanism is involved [20].

Copper-catalyzed [2 + 2] photocycloadditions are related to the latter reactions. These transformations have been extensively studied, frequently in the context of application to organic synthesis [21]. When irradiated in the presence of copper(I) triflate, norbornene **12** was efficiently transformed into its dimer **13** (Scheme 5.3, reaction 4) [22]. Although complexes such as **III** are involved in the reaction mechanism [22, 23], it is unclear whether MLCT (metal to ligand charge transfer) or LMCT (ligand to metal charge transfer) excitation induces the transformation.

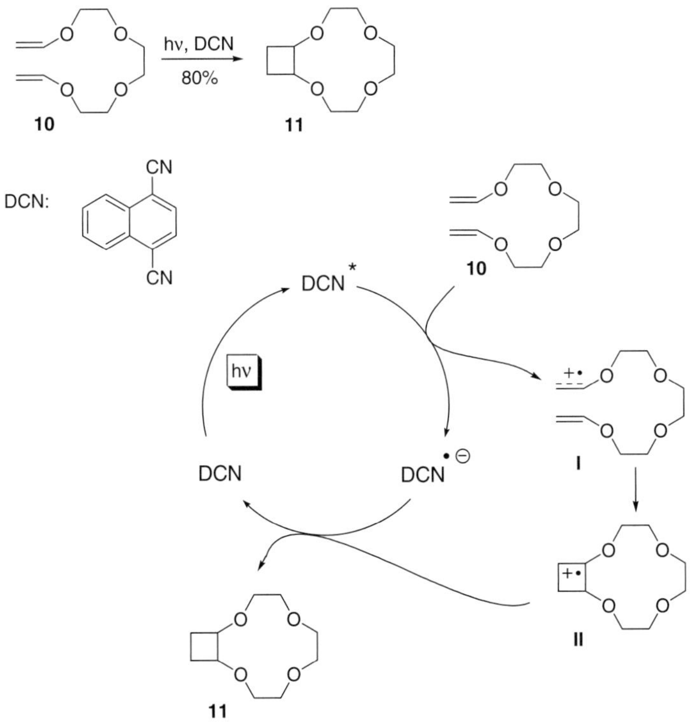

Scheme 5.2 [2 + 2] Photocycloaddition induced by photochemical electron transfer.

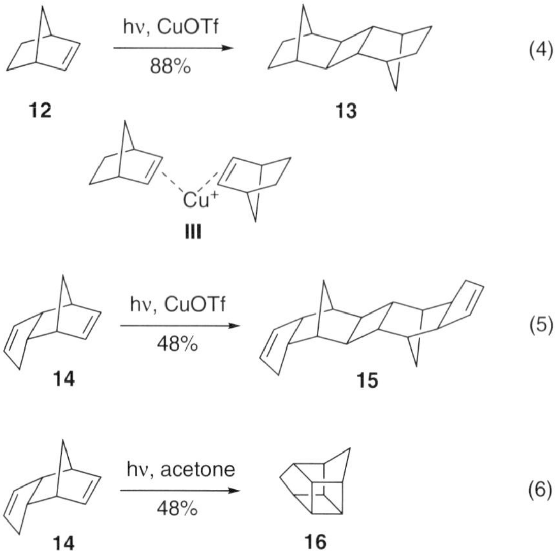

Scheme 5.3 Intermolecular [2 + 2] photocycloadditions catalyzed by CuOTf.

Interestingly, when using copper(I)triflate, the cyclopentadiene dimer **14** reacts in an intermolecular way, leading to the cyclobutane **15** (reaction 5) [22]. When the same substrate is transformed in the presence of the triplet sensitizer acetone, an intramolecular [2 + 2] cycloaddition takes place and the cage hydrocarbon compound **16** is formed. Obviously, the formation of a copper complex intermediate involving both alkene double bonds of the substrate is unfavorable in this case.

Many intramolecular reactions of this type have been described, mainly in the context of applications to organic synthesis or to the synthesis of natural products. The irradiation of compound **17** in the presence of CuOTf leads to the stereoisomers **18a,b** (Scheme 5.4, reaction 7) [24]. In this case, the *exo* isomer **18a** is formed in slight excess. This stereochemistry was explained by an equilibrium between the two copper complexes **IV** and **V**. In **IV**, the copper atom is orientated in an *exo* position; due to steric hindrance, this structure is generally discussed in such reactions. Nevertheless, in the present case, structure **V** with the copper atom in an *endo* position is also formed in considerable amounts, which can be explained by a complexation of the hydroxyl function. In the corresponding transformation of the enantiomerically pure allylalcohol **19**, the *endo* isomer **20b** is formed in excess; this is explained by the increased stability of intermediate **VI** (reaction 8) [25]. Compound **20b** was transformed into (−)-grandisol, which is part of the insect pheromone of the boll weevil (*Anthonomus grandis*). The same strategy was used previously for a further asymmetric synthesis of grandisol [26].

More complex ring systems have been built up using the CuOTf-catalyzed [2 + 2] photocycloaddition. For instance, transformation of the cyclopentene derivative **21** leads in high yields to the tricyclic compounds **22a,b** (Scheme 5.5, reaction 9) [27], with the *endo* isomer **22a** being obtained in excess. The reaction was applied to

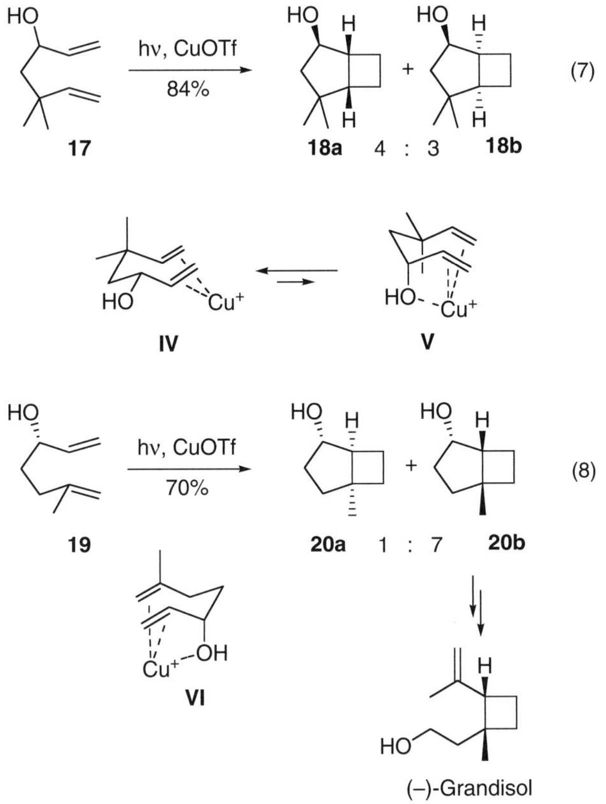

5.2 [2 + 2]-Photocycloaddition of Nonconjugated Alkenes | 141

Scheme 5.4 Intramolecular [2 + 2] photocycloaddition of 3-hydroxy-1,6-heptadiene derivatives catalyzed by CuOTf.

the synthesis of analogues of angular triquinane sesquiterpenes, where various functional groups are tolerated. In the case of compound **23**, a vinyl substituent is added to one of the olefinic double bonds (reaction 10), and this compound was then selectively transformed into the cyclobutane derivative **24** [28]. No Diels–Alder reaction was observed which may occur under PET or Lewis acid catalysis. Enolethers such as **25** react in the same way (reaction 11), the resultant product **26** being transformed into β-necrodol [29], an insect repellent of the defense spray of the red-carrion beetle (*Necrodes surinaminsis*), and into the sesquiterpene herbertene [30]. A similar transformation of an enolether was used as key step in the synthesis of the sequiterpene α-cedrene [31]. Esters [32], carbamates [33], or carbohydrate derivatives [34] possessing two alkene double bonds have all been successfully transformed, and the resulting products applied to organic synthesis.

Due to its versatile applicability, the CuOTf-catalyzed [2 + 2] photocycloaddition was used successfully to study the topology of the intermolecular and intramolecular dimerization of norbornene derivatives. When a racemic mixture of compound **27** is transformed in the presence of CuOTf, a 1 : 1 mixture of two stereoisomers (**28a,b**) is

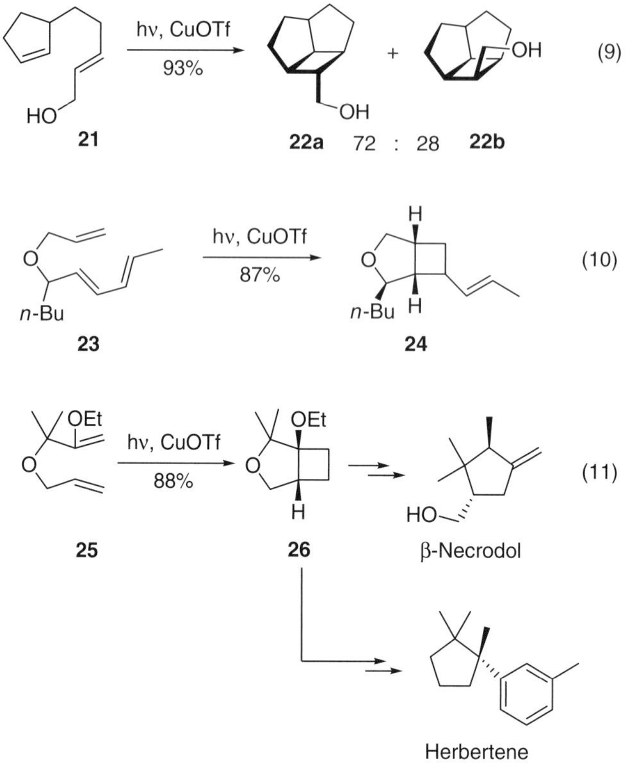

142 | *5 Formation of a Four-Membered Ring*

Scheme 5.5 Intramolecular CuOTf-catalyzed [2 + 2] photocycloaddition, and its application to organic synthesis.

obtained (Scheme 5.6, reaction 12) [35]. In both isomers, the polycyclic system possess an *exo-trans-exo* configuration. Compound **28a** results from a dimerization of molecules of the same absolute configuration (one enantiomer), while **28b** is formed from two molecules possessing the opposite absolute configuration (two enantiomers). When the molecules are linked together by an adamantane tether, two substrate molecules are obtained (reaction 13). In **29a**, norbornene moieties have the same absolute configuration, while in **29b** they possess opposite absolute configurations. The transformation of a 1 : 1 mixture of **29a,b** yielded a 1 : 1 mixture of compounds **30a,b**. The polycyclic molecule **30a** possesses an *exo-trans-exo* configuration (compare **28a,b**), and the norbornene moieties have the same absolute configuration; therefore, **30a** results from the transformation of **29a**. The *meso* isomer **29b** is transformed into an achiral *exo-cis-exo* polycyclic product **30b**. In the case of compound **31**, which possesses a larger tether derived from adamantane, one diastereoisomer (the racemic product) was also obtained. By using the CuOTf-catalyzed [2 + 2] photocycloaddition, this compound can be selectively transformed into the polycyclic *exo-trans-exo* product **32** (reaction 14). It should also be mentioned here that the irradiation time is significantly shorter in the case of the intramolecular reactions. In these cases, under identical reaction conditions, an irradiation of only

5.2 [2 + 2]-Photocycloaddition of Nonconjugated Alkenes

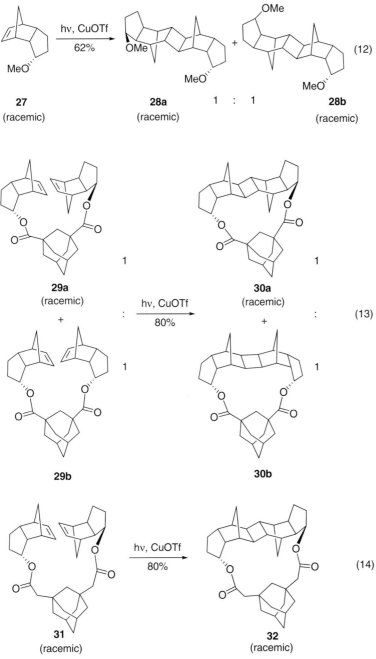

Scheme 5.6 CuOTf-catalyzed [2 + 2] photocycloaddition used for a topology study of inter and intramolecular dimerization of a norbornene derivatives.

3 h was required (see reactions 13 and 14), while the intermolecular reaction was complete after 56 h (reaction 12).

The inter- and intramolecular CuOTf-catalyzed [2 + 2] photocycloaddition of two alkenes was also performed in ionic liquids [36]. The best results were obtained when trimethyl(butyl)ammonium bis(trifluoromethylsulfonyl)imide was used as the reaction medium. This ionic liquid does not absorb light to any significant degree around 254 nm. Imidazolium salts, which are frequently used for ground-state reactions, are not appropriate for this transformation because they absorb light in the range between 200 and 350 nm, which in turn induces a significant decomposition of the ionic liquid.

A large number of [2 + 2] photocycloadditions of vinylarene compounds have been reported in the literature [37], and these are particular suitable for the synthesis of cyclophanes. Based on the fact that conjugation between the olefinic and aromatic π-systems is rather low, these reactions can be compared to corresponding reactions with nonconjugated alkenes.

5.3
[2 + 2]-Photocycloaddition of Aromatic Compounds

A variety of four-membered ring compounds can be obtained with photochemical reactions of aromatic compounds, mainly with the [2 + 2] (*ortho*) photocycloaddition of alkenes. In the case of aromatic compounds of the benzene type, this reaction is often in competition with the [3 + 2] (*meta*) cycloaddition, and less frequently with the [4 + 2] (*para*) cycloaddition (Scheme 5.7) [38–40]. When the aromatic reaction partner is electronically excited, both reactions can occur at the $\pi\pi^*$ singlet state, but only the [2 + 2] addition can also proceed at the $\pi\pi^*$ triplet state. Such competition was also discussed in the context of redox potentials of the reaction partners [17]. Most frequently, it is the electron-active substituents on the aromatic partner and the alkene which direct the reactivity. The [2 + 2] photocycloaddition is strongly favored when electron-withdrawing substituents are present in the substrates. In such a reaction, crotononitrile **34** was added to anisole **33** (Scheme 5.8, reaction 15) [41], and only one regioisomer (**35**) was obtained in good yield. In this transformation, the

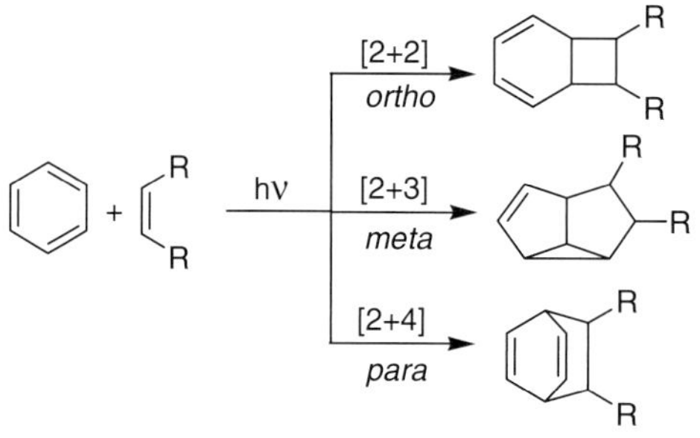

Scheme 5.7 Photocycloadditions of alkenes with benzene.

5.3 [2+2]-Photocycloaddition of Aromatic Compounds | 145

aromatic reaction partner was excited and the addition occurred at the ππ* singlet state. When an additional methoxy group was present on the benzene ring, PET took place and products possessing no cyclobutane ring were formed.

The [2 + 2] cycloaddition of benzene derivatives with alkenes was also carried out using photosensitization. Maleimide **36** was added to benzene in high yields (Scheme 5.8, reaction 16) [42]. In this case, the sensitizer acetophenone **37** transfers its triplet energy to **36**, after which the cycloadduct **VII** reacts immediately with an

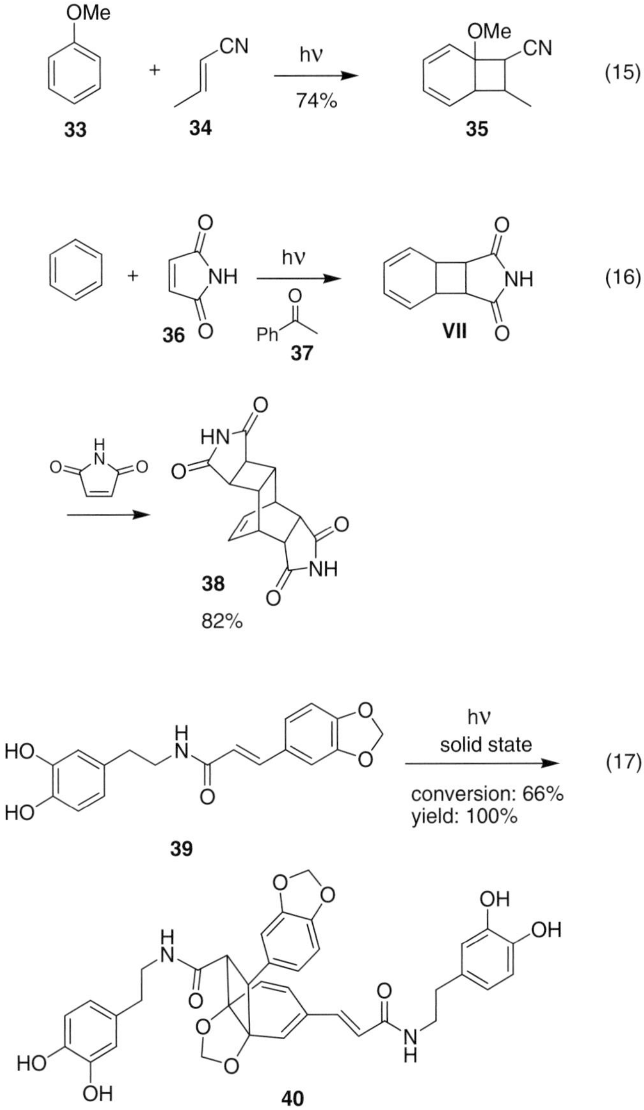

Scheme 5.8 Intermolecular [2 + 2] photocycloaddition of benzene derivatives.

additional maleimide molecule to generate the stable Diels–Alder adduct **38**. A variety of benzene derivatives have been transformed in the same way by using triplet sensitization [43]. These reactions were also performed without sensitization, but with direct excitation of the aromatic reaction partner [38, 44]. A [2 + 2] photocycloaddition may also be favored with respect to other competing reactions, when the reaction partners are well orientated one to another; such an orientation can be established within a crystal structure. The dimerization of compound **39** occurs via [2 + 2] photocycloaddition, leading to **40** (reaction 17) [45]; however, when irradiated in solution, **39** yields a 1 : 1 mixture of E/Z isomers of the starting compound.

Many examples of the intramolecular [2 + 2] photocycloaddition of alkenes to benzene derivatives have been reported. The acetophenone derivative **41** undergoes an efficient [2 + 2] photocycloaddition, leading to the cyclobutane derivative **42** (Scheme 5.9, reaction 18) [46, 47]. It was shown that, in this case, a $\pi\pi^*$ triplet state is involved. The presence of a nitrile group in compound **43** induces a [2 + 2] cycloaddition at position 1,2 of the aromatic moiety, leading to intermediate **VIII** (reaction 19) [48]. Following tautomerization, the final product **44** is formed.

Due to their reduced aromatic character, naphthalene derivatives more easily undergo [2 + 2] photocycloaddition to give access to cyclobutane derivatives [49]. In some cases, the formal [4 + 2] addition is competitive. In contrast to the ground-state Diels–Alder reaction, the [4 + 2] products from these photochemical reactions result from a multistep reaction. Upon irradiation, the 1-cyanonaphthalene derivative **45**

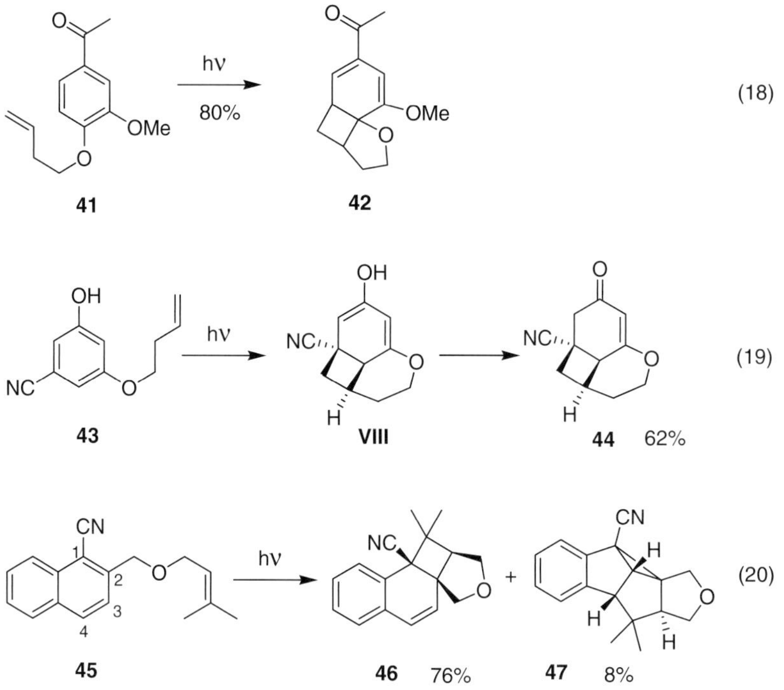

Scheme 5.9 Intramolecular [2 + 2] photocycloaddition of benzene derivatives.

undergo [2 + 2] photocycloaddition leading to the cyclobutane derivatives **46** (Scheme 5.9, reaction 20) [50]. As in the case of **43**, the presence of a nitrile group in compound **45** mainly directs the cycloaddition into the 1,2-position with respect to this substituent. The minor product **47** results from a [3 + 2] photocycloaddition; the structure of this was determined later using X-ray crystallography analysis. The resultant main product **46** undergoes photochemical decomposition, leading to the substrate **45**, while **47** is photostable. Currently, microreactor technology is particularly developed with the aim of more easily controlling exothermic reactions, of reducing the formation of byproducts, and of increasing safety [51]. These reactors have also been optimized for photochemical reactions; hence, the transformation of **45** (Scheme 5.9, reaction 20), for example, was performed in such a reactor [52].

Depending on the substitution pattern in naphthalene derivatives, the solvent polarity, and particularly also the irradiation time, the formation of the *meta* adducts can be increased [50].

Frequently, [2 + 2] photocycloaddition products of aromatic compounds possess low stability. In an *in situ* reaction which may be acid-catalyzed, such intermediates are transformed into stable final products; for example, resorcinol derivatives **48** have been transformed into benzocyclobutene derivatives **49a,b** (Scheme 5.10, reaction 21) [53]. After photochemical excitation of the benzene moiety, a [2 + 2] cycloaddition takes place, although under the irradiation conditions ($\lambda = 254$ nm) this step is reversible. However, in the presence of a strong acid the adducts are protonated (**IX** and **IX'**). After deprotonation at the six-membered ring, stable benzocyclobutene isomers **49a,b** are generated, but in the absence of acid no significant transformation was detected due to a reversibility of the photocycloaddition step. In some cases, an acid-catalyzed rearrangement at the bicyclic intermediate takes place such that monocyclic benzene derivatives (e.g., **50**) are formed via the spirocyclic intermediate **X**. Benzocyclobutenes represent interesting intermediates for the synthesis of polycyclic molecules [3]; compounds such as **49a** have been transformed into nitrogen-containing heterocyclic compounds (**51**) that possess an affinity towards dopamine receptors [54].

In a similar way, the dihydroxybenzoic acid derivative **52** was transformed into the bicyclic compounds **53a,b** (Scheme 5.10, reaction 22) [48]. As with the corresponding nitrile derivative **43** (Scheme 5.9, reaction 19), the electron-withdrawing substituent directs the [2 + 2] photocycloaddition into the 1,2-position to the ester function, leading to the tricyclic intermediate **XI**. The α,β-unsaturated cyclohexanone **XII** derivative was formed in a first acid-catalyzed step. Yet, when the reaction is carried out in methanol as solvent, one solvent molecule adds to the intermediate **XII** such that the bicyclic 1,3-cyclohexandione derivatives **53a,b** are obtained. When compared to nitrile derivatives (**43**, Scheme 5.9), the benzoic ester compounds (**52**, Scheme 5.10) are significantly less reactive, and have only be transformed in an acidic reaction medium. The latter products represent interesting intermediates for the synthesis of compounds derived from 1,3-cyclohexandione that possess herbicidal activities. Similar reactions have been carried out with alkenyl-4-chromanone derivatives [55].

Frequently, the intramolecular [2 + 2] photocycloaddition of an alkene to a benzene ring is followed by further pericyclic reactions. Such transformations yield

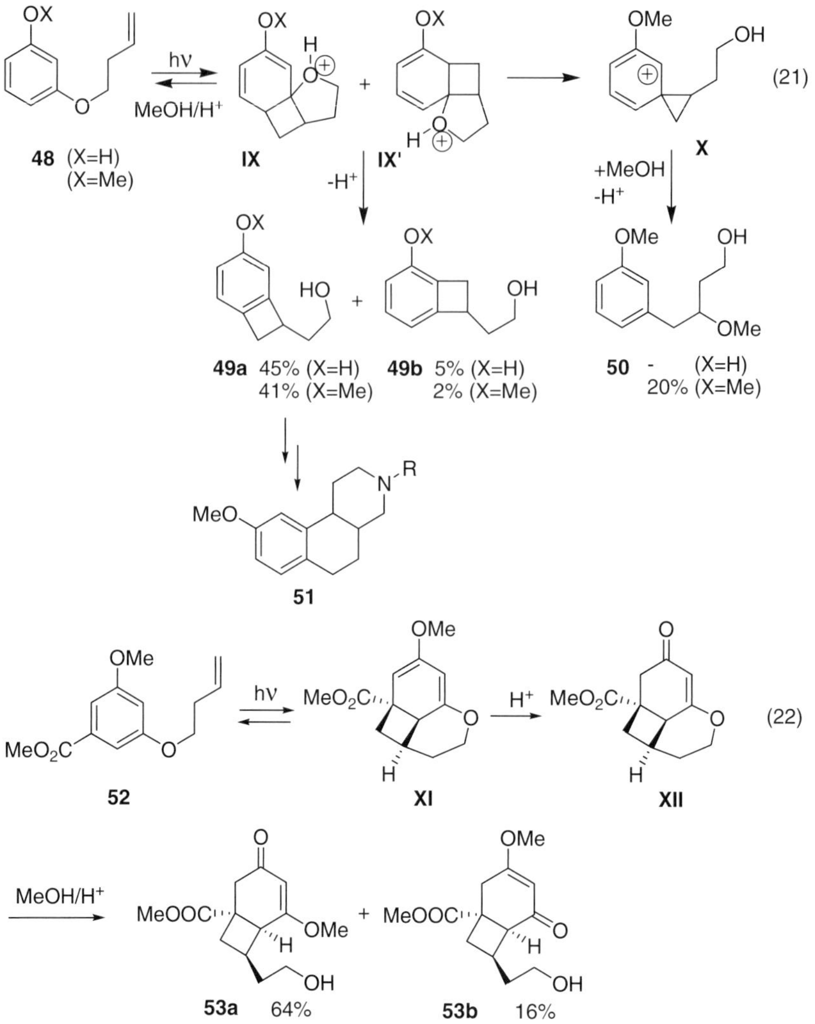

Scheme 5.10 Unstable [2 + 2] photocycloadducts are transformed into stable products by acid-catalyzed reactions.

polycyclic compounds that possess a cyclobutene moiety, and occur at the singlet and triplet state, either by direct excitation or by sensitization [47]. These reactions are frequently observed when electron-withdrawing substituents are attached to the benzene ring. The reaction sequence starts with a [2 + 2] photocycloaddition, as illustrated for the *para*-substituted benzene derivative **54**. The resultant intermediate **XIII** then undergoes a reversible electrocylization, leading to the cyclooctatriene derivative **XIV** (Scheme 5.11, reaction 23). A further, generally photochemical, electrocyclic ring contraction leads to the angular tricyclic cyclobutene derivative **55**, or to the linear analogue **56**. The product ratio here depends on the substitution pattern, and for many products – in particular acetophenone derivatives and angular isomers such as **55** – this last step is thermically reversible. Ring enlargement and contraction are each facilitated by the presence of an electron-donating and -accepting group. For example, in the case of the transformation of **55** into **XIV**, a C—C single bond of the cyclobutene moiety possesses this substitution, which favors heterolytic fragmentation, such that the isolation of tricylic products is hampered. In some cases, the cyclooctatriene product is therefore easily available. Due to ring strain, a ground-state conrotatory ring enlargement according to the Woodward–Hoffmann rules for concerted reactions is not possible [9]. As mentioned above, such equilibria may be shifted towards the desired products in the presence of acid; this has been demonstrated for the transformation of O-alkenyl salicylic acid derivatives into tricyclic cyclobutene derivatives [56].

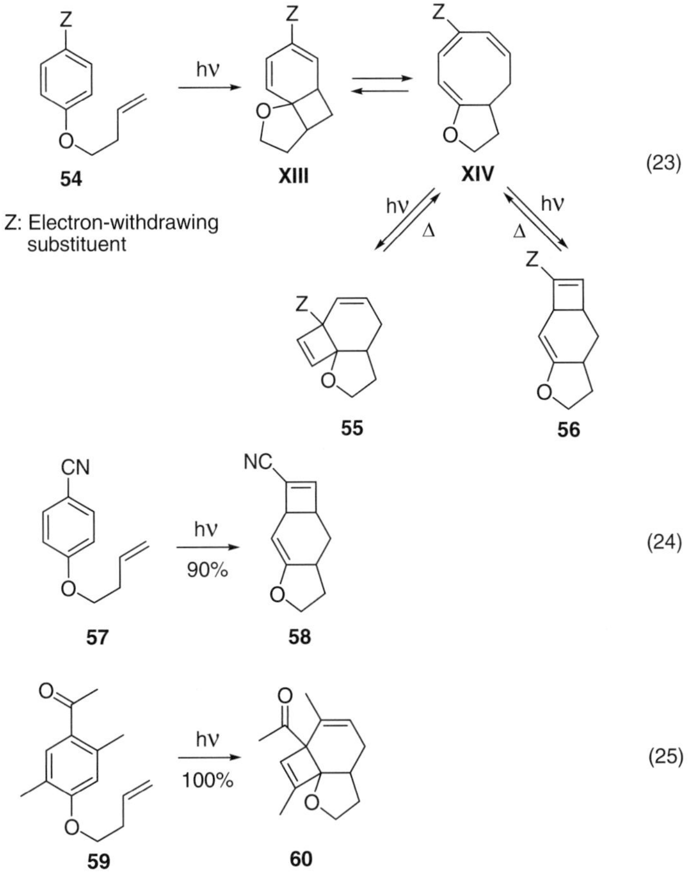

(23)

(24)

(25)

Scheme 5.11 Synthesis of tricyclic cyclobutene derivatives with intramolecular [2 + 2] photocycloadditions of aromatic compounds. The addition is followed by rearrangements leading to the final products.

The *p*-cyanophenol derivative **57** was selectively transformed into the linear tricyclic cyclobutene derivative **58** (Scheme 5.11, reaction 24) [57]. The hydroxyacetophenone derivative **59**, carrying two additional methyl groups at the benzene ring, yields selectively the angular tricyclic cyclobutene derivative **60** (reaction 25) [58]. It should be stressed here that the back-reaction to the corresponding cyclooctatriene compound is inhibited by the methyl substituent at the cyclobutene moiety of **60**. Chiral induction was performed with derivatives carrying a chiral ester or amide function on the hydroxyacetophenone ring [59].

5.4
Photochemical Electrocyclic Reactions

As indicated in Section 5.3, four-membered rings can be obtained from the electrocyclic reactions of polyenes, with cyclic polyenes most frequently being transformed via a ring-contraction process [60]. Such transformations have often been reported for a wide variety of 2-pyridone derivatives (Scheme 5.12, reaction 26) [61]. The resultant products of the photochemical disrotatory electrocyclization contain a cyclobutene and a β-lactam moiety. In general, the reactions must be performed at low concentration in order to prevent dimerization, for instance, via [4 + 4] cycloaddition, which is the major competitive process. Hence, compound **61** (reaction 27) was efficiently transformed into the bicyclic derivative **62** [62]. The same transformation was applied to the synthesis of a large variety of β-lactams, such as **63** [62, 63]. The same electrocyclic reaction was conducted with the ribonucleoside analogue **64** [64]. In this case, the photostationary equilibrium was investigated. Here, when the reaction solution was irradiated at $\lambda > 300$ nm, the equilibrium was almost completely shifted to the products, and the bicyclic product **65** was obtained in high yields as a mixture of two diastereoisomers. Various aspects of asymmetric synthesis have been investigated – and are currently being investigated – by using this reaction. For example, when the menthyloxy substituent was used as the chiral auxiliary the transformation was performed with moderate diastereoselectivity [65]. The major stereoisomers have been obtained in pure form after recrystallization. Photochemical reactions can be carried out very easily in supramolecular structures, and in many such cases chiral induction plays an important role. In contrast to many ground-state reaction conditions, these structures are not destroyed by thermal energy or aggressive reagents when particular photochemical conditions are applied [7, 66–68]. Numerous photochemical reactions have been performed in crystals [68, 69]. For example, when cocrystallized with TADDOL or chiral hexadiine derivatives, pyridone derivatives undergo electrocyclization to yield the bycyclic lactams with ee-values up to 99.5% [70]. On seldom occasions, achiral compounds may crystallize in chiral space groups; for example, the pyridone derivative **68** generates crystals with the chiral space group $P2_12_12_1$ (Scheme 5.12, reaction 29) [71]. In such crystals, only one enantiomeric conformation of the molecules exists, and the enantiomeric excess of the products obtained from the reactions is therefore high. In the case of formation of compound **69**, the ee was observed to diminish as the conversion progressed, but this finding

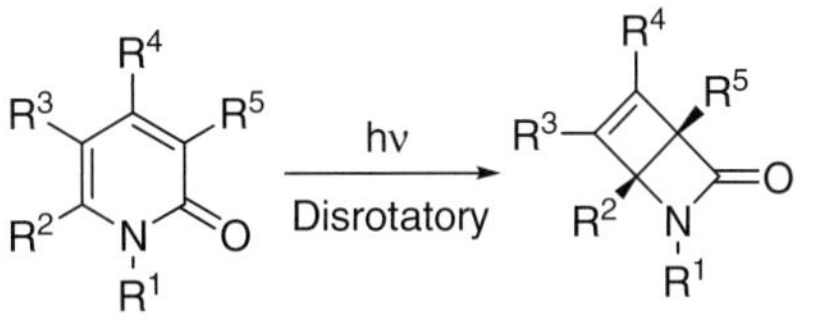

Scheme 5.12 Photochemical electrocyclization of pyridone derivatives.

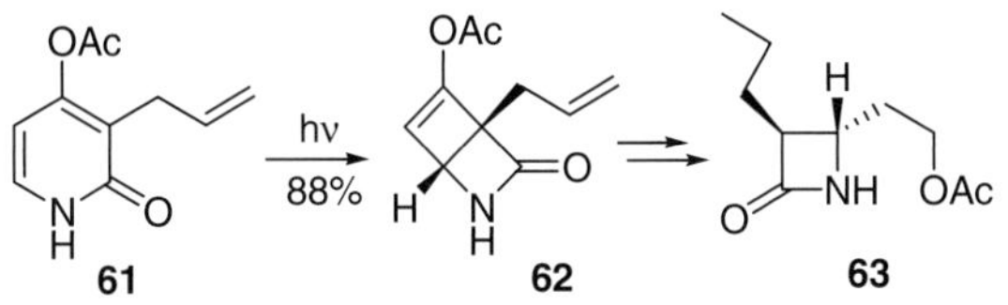

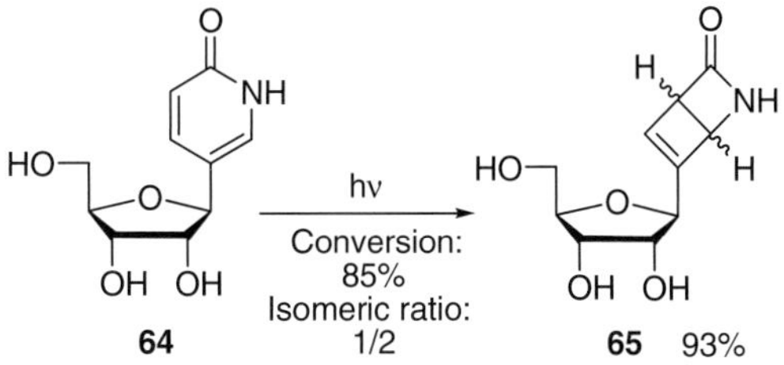

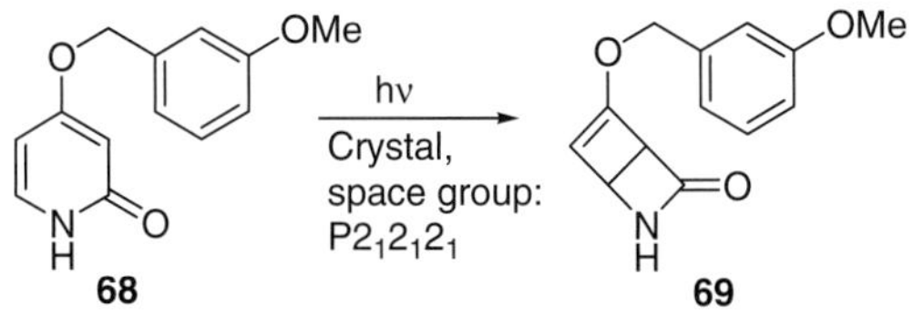

Irradiation time (min)	Conversion (%)	Enantiomeric excess (%)
15	5.6	~100
30	10.3	91
60	20.5	90
90	27.3	90

was explained by the fact that, during the transformation, the crystal structure progressively decomposed and the substrate lost its chiral environment. At this point it should be stressed that *meta*-disubstituted benzene derivatives more likely crystallize in noncentrosymmetric space groups than are their *ortho* or *para* isomers [72]. As all chiral space groups are noncentrosymmetric, those achiral molecules that possess a *meta*-disubstituted benzene moiety should more likely crystallize in the chiral space groups. Based on this consideration, the method for asymmetric synthesis illustrated by the example in reaction 29 (Scheme 5.12) may perhaps be generalized. This finding might also contribute to a better understanding of the crystallization of organic molecules such that, within limits, a prediction of the crystal structure based on the molecular structure would become possible [73].

Many photochemical disrotatory 4π-electrocyclic reactions have been reported with tropolone derivatives such as **70** (Scheme 5.13, reaction 30) [74], with bicyclic cyclobutene derivatives such as **71** being obtained from the electrocyclic reaction. During prolonged irradiation, these compounds undergo further transformation via the intermediates **XV** and **XVI**, which leads in turn to the rearranged enolether derivative **72**. Upon irradiation, colchicine and several of its derivatives undergo 4π-electrocyclization [75, 76]. For example, the colchicine derivative demecolcine **73** was efficiently transformed into lumicolcine **74** (Scheme 5.13, reaction 31) [76]; here, only one stereoisomer was isolated, whereas in other cases – and after prolonged irradiation – two isomers were obtained [75]. Since lumicolechecine derivatives have also been detected in plant extracts [77], this transformation was considered to be either biosynthetic or biomimetic [60]. In fact, the formation of lumicolchicine in plants was discussed as being sensitized by a chemically (enzymatically) produced excited molecule ("photobiochemistry without light") [78]. Many photochemical

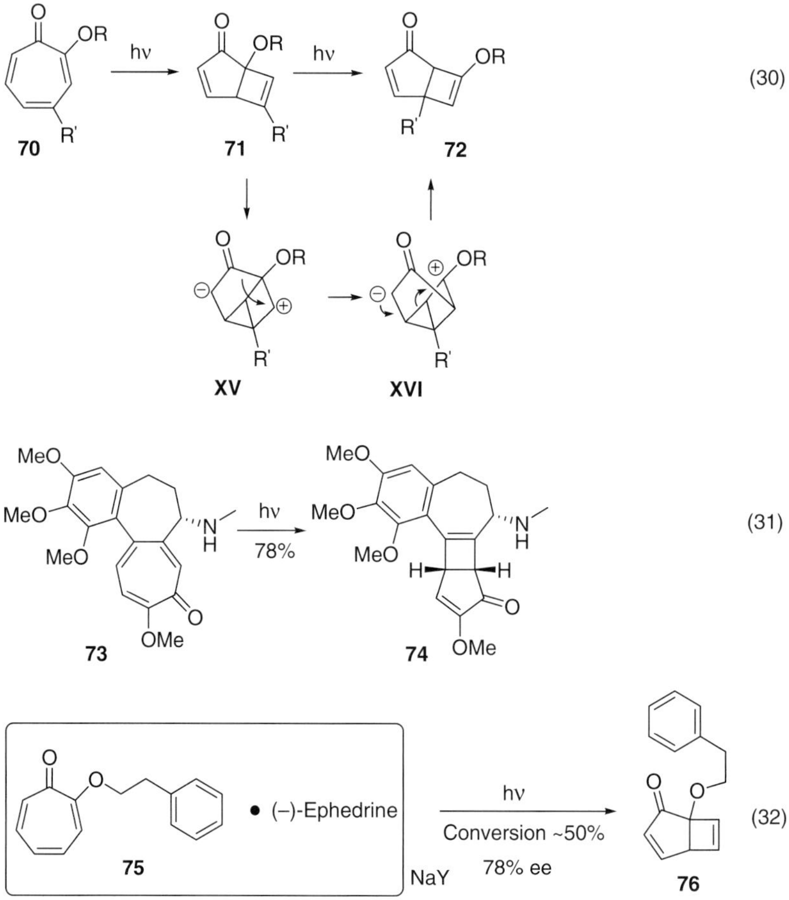

Scheme 5.13 Photochemical 4π-electrocyclization of tropolone derivatives.

reactions have been carried out with the compounds included in zeolite cavities [79]. Under these conditions, the conformational flexibility of the molecules is significantly reduced, and the conformational restriction is assisted by interactions between the cations in the zeolite structure and the substrate. The consequences of this effect for chiral induction have also been investigated for the 4π electrocyclization of tropolone derivatives. In solution, tropolones carrying a chiral O-alkyl substituent react with low diastereoisomeric excesses (generally <10%), which is explained by the high conformational flexibility. However, when the same compounds are included in zeolites the diastereoselectivities are increased by up to 88% [80]. The chiral auxiliary does not need to be attached to the tropolone substrate by a covalent bond; rather, the tropolone derivative **75** was coadsorbed with (−)-ephedrine into the NaY zeolite (reaction 32) [80, 81]. Subsequently, after irradiation, the bicyclic product **76** was obtained with ee-values up to 78%.

5.5
Intramolecular γ-Hydrogen Abstraction (Yang Reaction)

The Yang reaction provides a broad access to four-membered cyclic alcohols. The transformation is linked to the Norrish type II reaction. Following the photochemical excitation of a corresponding ketone, γ-hydrogen abstraction occurs either at the singlet (**XVII**) or the triplet state (**XVIII**) (Scheme 5.14) [82]. The resultant 1,4-biradicals **XIX** and **XX** may now split, leading to an enol and an alkene fragment (Norrish type II reaction). The triplet radical **XIX** can undergo radical combination, leading to cyclobutanol (this is Yang cyclization), with the competition between the two procedures depending on the conformation equilibrium, as shown in Scheme 5.15. Upon irradiation and after intersystem crossing (ISC), the $n\pi^*$ triplet state (**XXI**) of the isoleucine derivative **77a** is generated (reaction 33) [83], while the 1,4-biradical intermediate **XXII** is obtained after γ-hydrogen abstraction. Two conformations (**XXIIIa,b**) of this intermediate are stabilized by a hydrogen bond

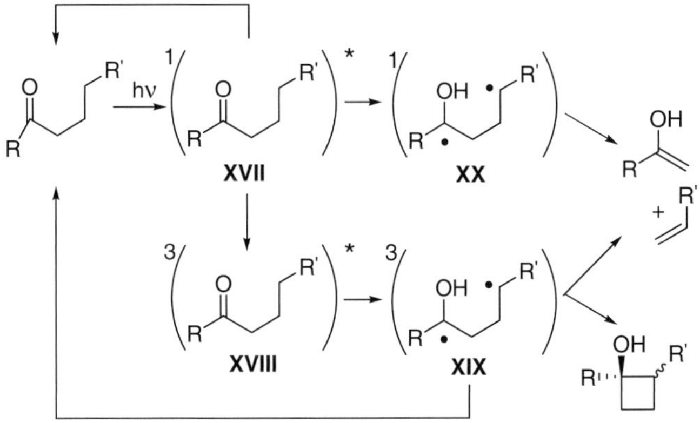

Scheme 5.14 General mechanistic scheme for the Norrish–Yang reaction.

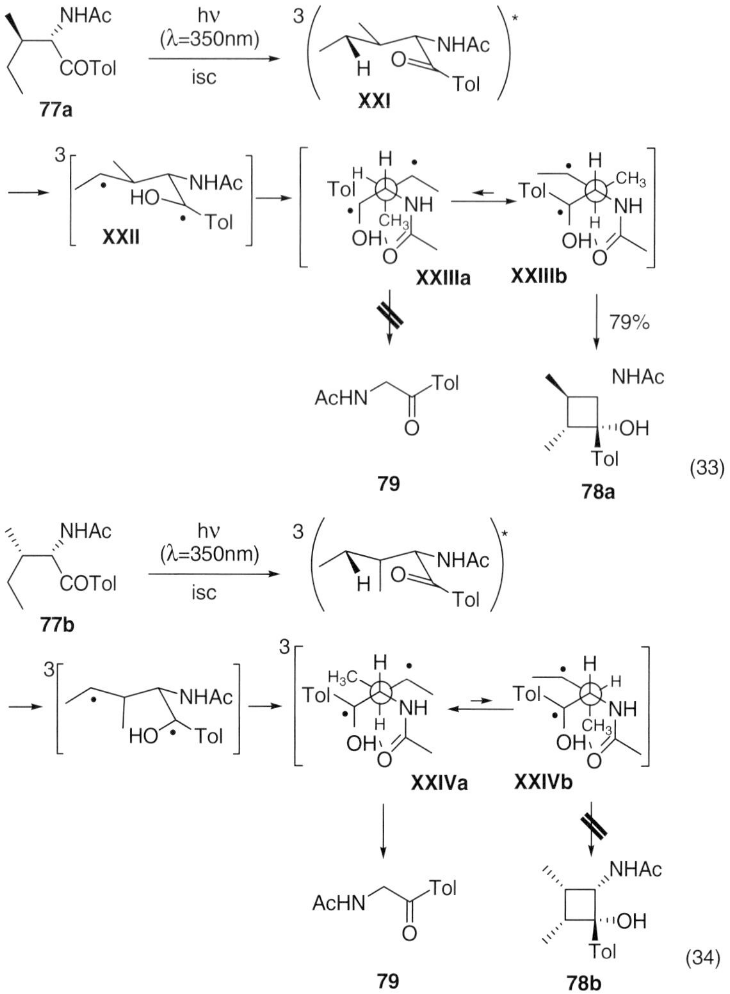

Scheme 5.15 Norrish type II fragmentation and Yang cyclization of isoleucine derivatives, illustrating the influence of conformational equilibria.

between the hydroxyl group and the acetamide function. **XXIIIb** is additionally stabilized with respect to **XXIIIa** by steric hindrance of the methyl group in β position in the latter conformation. In **XXIIIb**, the two radical centers are close to each other, which favors cyclization and leads to the final cyclobutanol derivative **78a**. Under the same reaction conditions, the epimeric product **77b** yields the intermediates **XXIVa, b**. Due to the same steric interactions, the conformer **XXIVa** is predominant in this case (reaction 34). The radical centers are now opposed to each other, which favors fragmentation (Norrish type II reaction). It is not only the competition between cyclization and fragmentation that is determined by geometric requirements, but

5.5 Intramolecular γ-Hydrogen Abstraction (Yang Reaction)

also the hydrogen abstraction in the first step. Recently, the structural parameters in favor of this transformation were identified by a series of transformations carried out in crystals [84].

When using UV or sunlight irradiation, the glucose derivative **80** possessing a α-diketone function was transformed into the spirocyclic hydroxyketone **81** (Scheme 5.16, reaction 35) [85]. After electronic excitation, hydrogen abstraction at the carbohydrate moiety takes place, while the corresponding hydrogen abstraction in the 5-position is facilitated by a pseudoanomeric effect of the ring oxygen and the BnO substituent in the 6-position. Cyclization of the biradical intermediate **XXV** is directed by hydrogen bond between the hydroxy group and the BnO substituent in the 6-position, and the final product was obtained in quantitative yield. This

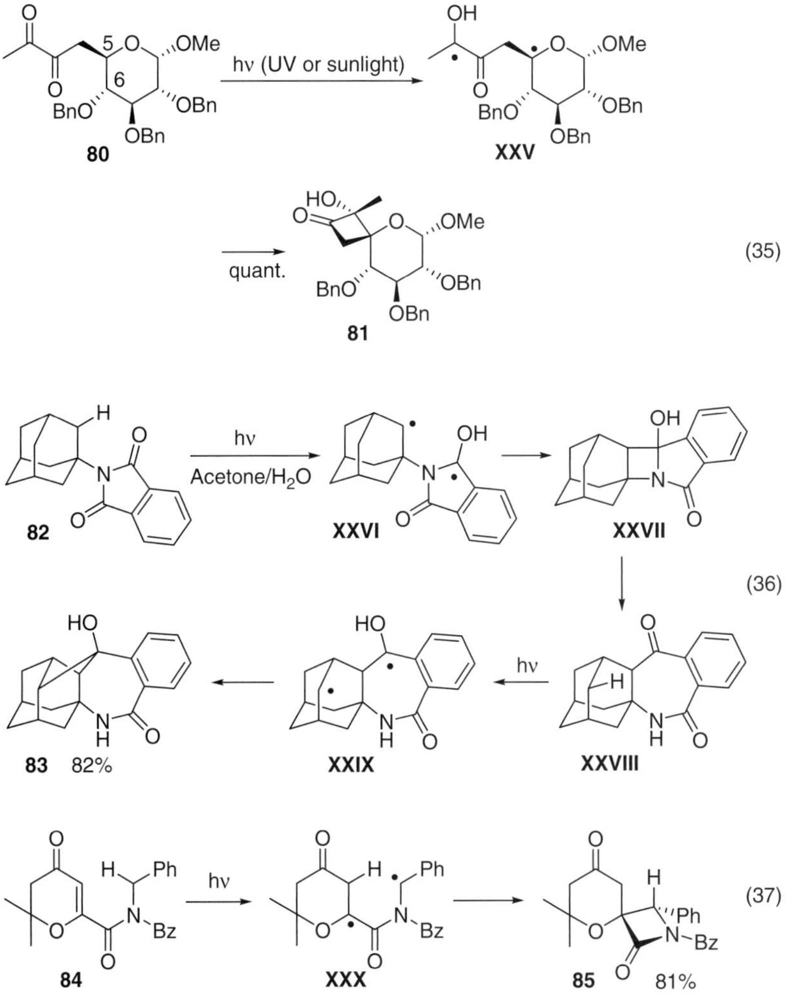

(35)

(36)

(37)

Scheme 5.16 Yang cyclization leading to cyclobutanol derivatives and β-lactams.

transformation was carried out with a variety of similar carbohydrate derivatives, such as hexose derivatives carrying the buta-2,3-dione substituent in the anomeric position. Irradiation at $\lambda = 300$ nm of the phthalimidyl-substituted adamantane derivative **82**. In the presence of acetone yielded the cyclobutanol derivative **83** (reaction 36) [86]. In this transformation, two Yang cyclizations are involved and, after excitation of the substrate **82** the biradical intermediate **XXVI** is formed (this step is sensitized by acetone). Radical combination then leads to the hydroxyazetidine **XXVII**. Cleavage of the half-aminal structure in this intermediate leads to the benzoazepine structure **XXVIII**. In a second photochemical reaction, hydrogen abstraction occurs via the ketone function, leading to the biradical intermediate **XXIX**; a second Yang cyclization then furnishes the final cyclobutanol derivative **83**.

The latter example (reaction 36) already indicates that the Yang cyclization can also be used to synthesize four-membered heterocycles. After light absorption, the α,β-unsaturated carbonyl compound **84** undergoes intramolecular hydrogen abstraction at the α-position of the carbonyl moiety (reaction 37), leading to the 1,4-biradical intermediate **XXX** [87]. A radical combination then efficiently yields the spirocyclic β-lactam derivative **85**, and only one stereoisomer is formed in this case. In this transformation, the α,β-unsaturated carbonyl function can be considered as being vinylogous to a simple ketone.

5.6
Metal-Catalyzed Reactions

Many metal-catalyzed reactions are accelerated by light irradiation [88]. In the case of the Vollhardt reaction [89], which is performed with conveniently available Co catalysts, the irradiation with visible light is included in the "standard" conditions. In a triple [2 + 2 + 2] cycloaddition, the nonaalkyne derivative **86** is transformed into the [7]Phenylene **87** (Scheme 5.17) [90]. Compound **87**, containing six benzocyclobutene moieties, is a partial structure of the archimedene C_{120} **88**. In this way a variety of similar benzocyclobutene structures such as helical phenylenes [91] can be built up.

Four-membered rings possessing a large variety of substituents have also been obtained from the photochemical [2 + 2] photocycloaddition of Fischer-type carbene complexes with alkenes [92]. Most frequently, chromium derivatives have been transformed. When irradiated, the chromium carbene complexes such as **89** (Scheme 5.18, reaction 38) decompose in order to generate the ketene intermediate **XXXI** [93]. The formation of a chromium cyclopropanone intermediate (**XXXII**) has also been discussed, where the β-lactam **91** is formed in a cycloaddition of the imidazoline **90** onto **89**. This lactam was a key intermediate in the synthesis of lanthanide ligands such as **92** and **93**. The corresponding Gd^{3+} complexes have been used as magnetic imaging contrast agents. The reaction can be performed on a stereoselective basis; typically, reaction of the carbene complex **94** with the chiral enamine derivative **95** yielded the corresponding cyclobutanone **96** in high diaster-

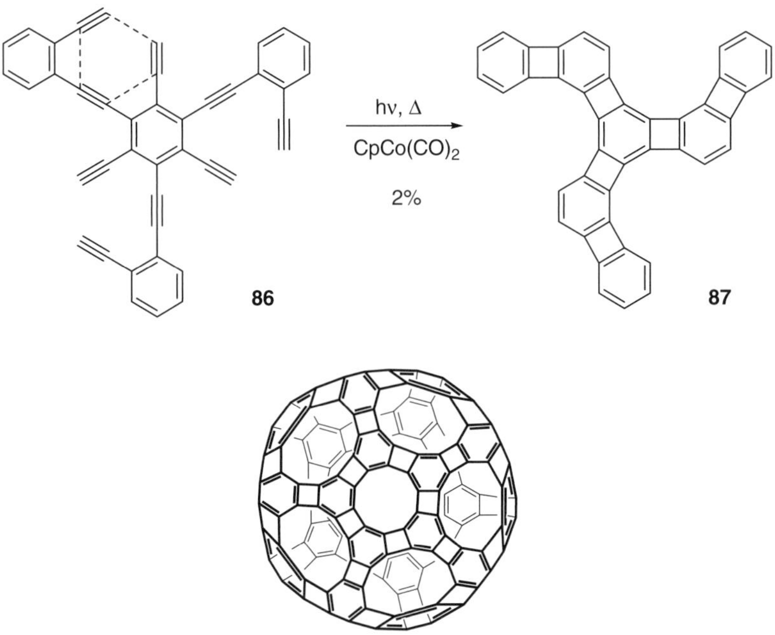

Scheme 5.17 Light-supported triple [2 + 2 + 2] cycloaddition. Synthesis of a [7]Phenylene **87** as partial structure of the archimedene C$_{120}$ **88**.

eoselectivity (reaction 39) [94], with no other stereoisomer being detected. Under these reaction conditions (CO pressure), chromium can be recovered in up to 90% yield as chromium hexacarbonyl when the reaction solution is triturated with methanol (in which chromium hexacarbonyl is insoluble). A number of cyclization reactions have also been performed; upon irradiation of the aldehyde **97** an intramolecular addition of the carbonyl function to the intermediately generated ketene takes place, such that the bicyclic lactone **98** was isolated in good yield (reaction 40) [95]. A seven-membered ring was formed during this transformation.

5.7
Other Methods

Four-membered rings may be generated in a less systematic manner by a variety of photochemical transformations, many of which are multistep reactions. In the field of radical chemistry, the photochemical generation of radical intermediates is frequently applied such that, when the stereoelectronic requirements are fulfilled, the cyclization steps lead to four-membered rings. The oxime derivative **99** undergoes photochemical sensitized cleavage at the fragile N—O bond (Scheme 5.19, reaction

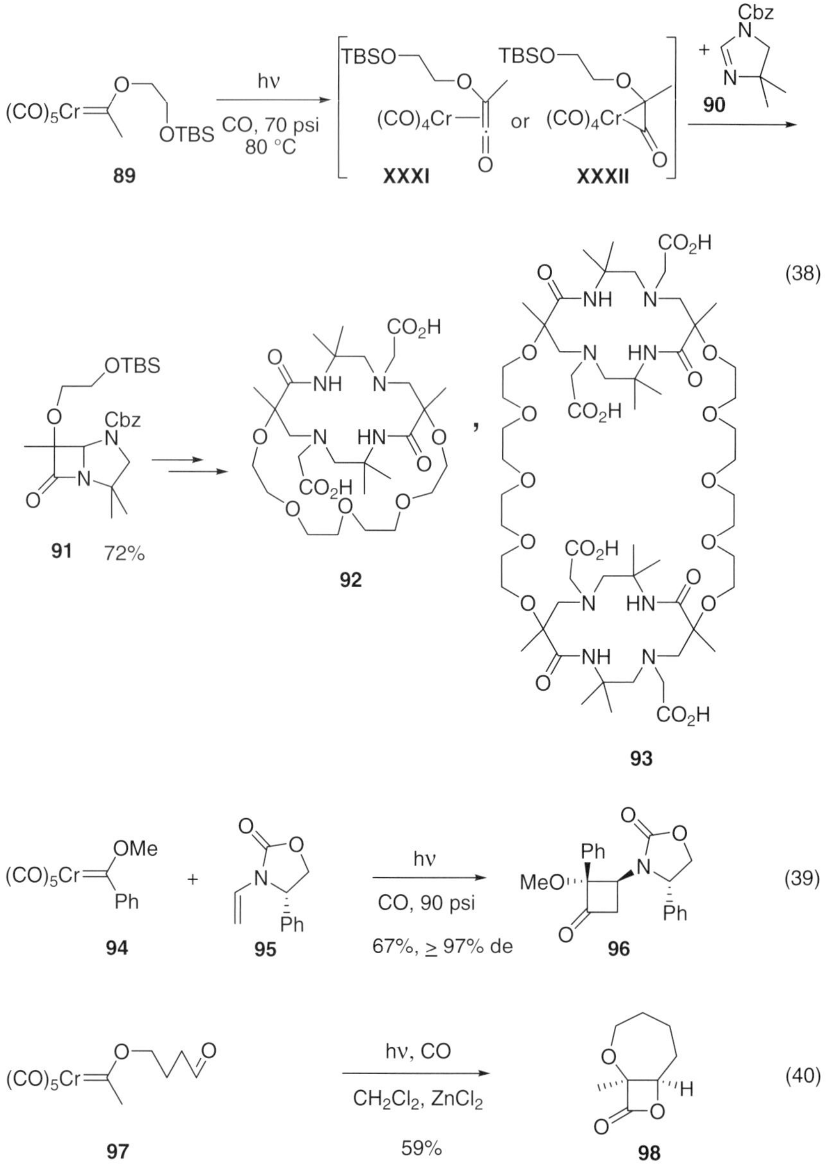

Scheme 5.18 Photochemical transformations of Fischer-type chromium carbene complexes.

41) [96], and the resultant acyl oxyl radical **XXXIII** rapidly loses CO_2, leading to the carbamoyl radical **XXXIV**. The latter intermediate undergoes cyclization to yield the bicyclic radical **XXXV**. Subsequent reaction with oxygen and toluene (used as a solvent) finally leads to the bicyclic β-lactame **100**. The two diastereoisomers were isolated in a ratio of 5:1. Upon UV irradiation, the bridged tricyclic ketone **101**

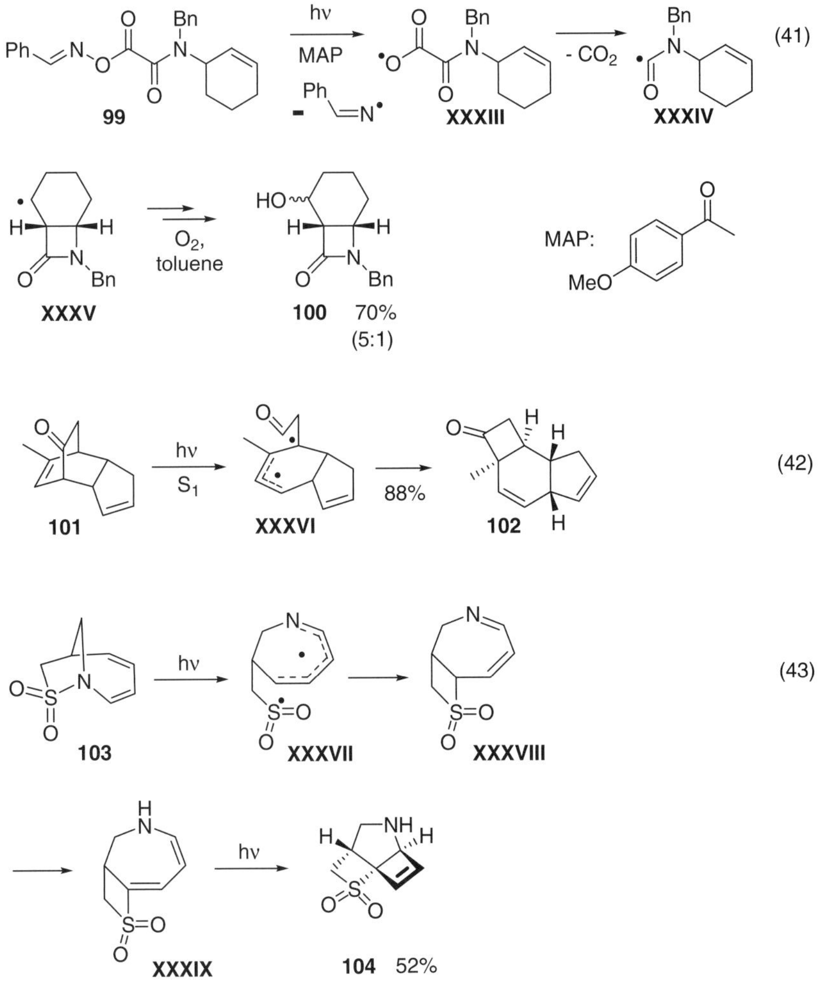

Scheme 5.19 Four-membered ring systems are also generated in less systematic ways by various multistep transformations.

undergoes a Norrish type I cleavage, leading to the biradical intermediate **XXXVI** (reaction 42) [97]. A 1,3-shift then takes place such that the cyclobutanone derivative **102** is formed in high yield by radical combination. It should be pointed out here that this transformation occurs at the singlet state. Triplet excitation of these compounds leads to an oxa-di-π-methane rearrangement. The tricyclic system of **102** serves as the core structure of the sesquiterpene family of protoilludanoids. The S−N bond of the bridgehead sultam **103** was transformed into the tricyclic fenestrane-like compound **104** (Compound 104 is not a fenestrane but resembles it. Frequently, such tricyclic compounds are intermediates in the synthesis of fenestranes.) (reaction 43) [98]. For this, the following mechanism has been discussed: the reaction starts with photochemical cleavage of the S−N bond, followed by a radical combination to form the biradical intermediate **XXXVII** that in turn yields the bicyclic sulfone **XXXVIII**. A 1,5-hydrogen shift (tautomerization) leads to the dienamine derivative **XXXIX**, while the final product **104** is generated by a photochemical disrotatory cyclization.

5.8
Concluding Remarks

Photochemical reactions provide a classical access to four-membered ring compounds that generate major interest in organic synthesis, notably as intermediates in multistep syntheses. The [2 + 2] photocycloaddition of α,β-unsaturated carbonyl and carboxyl compounds with alkenes and [2 + 2] photocycloaddition of ketones with alkenes (the Paternò–Büchi reaction) are discussed in Chapters 6 and 7, respectively. Yet, aside from these transformations, a variety of further reactions provides a systematic access to four-membered rings that possess a wide structural variation. Four-membered ring compounds may also be created via less-systematic photochemical transformations, many of which can be carried out without additional chemical activation. As a consequence, such transformations are rendered not only very convenient but also extremely interesting within the context of "green chemistry."

References

1 Namyslo, J.C. and Kaufmann, D.E. (2003) The application of cyclobutane derivatives in organic synthesis. *Chemical Reviews*, **103**, 1485–1535.

2 (a) Lee-Ruff, E. and Mladenova, G. (2003) Enantiomerically pure cyclobutane derivatives and their use in organic synthesis. *Chemical Reviews*, **103**, 1449–1483; (b) Fleming, S.A. (2005) Photocycloaddition of alkenes to excited alkenes, in *Synthetic Organic Photochemistry* (eds A.G. Griesbeck and J. Mattay), Molecular and Supramolecular Photochemistry, Vol. 12 (eds V. Ramamurthy and K.S. Schanze, Series), Marcel Dekker, New York, pp. 141–160; (c) Griesbeck, A.G. and Fiege, M. (2000) Stereoselectivity of photocycloadditions and photocyclizations, in *Organic, Physical, and Materials Photochemistry* (eds V. Ramamurthy and K.S. Schanze), Molecular and Supramolecular Photochemistry, Vol. 6, Marcel Dekker, New York, pp. 33–100; (d) Bach, T. (1998) Stereoselective intermolecular [2 + 2] photocycloaddition reactions and their application in synthesis. *Synthesis*, 683–703; (e) Mattay, J., Conrads, R., and Hoffmann, R. (1996) [2 + 2] Photocycloaddtions of α,ß-unsaturated carbonyl compounds, in *Houben-Weyl, Methods of Organic Chemistry, Vol. E 21: Stereoselective Synthesis* (eds G. Helmchen, R.W. Hoffmann, J. Mulzer and E. Schaumann), Thieme Verlag, Stuttgart, pp. 3085–3178.

3 Sadana, A.K., Saini, R.K., and Billups, W.E. (2003) Cyclobutarenes and related compounds. *Chemical Reviews*, **103**, 1539–1602.

4 de Meijere, A. (ed.) (1997) *Houben-Weyl, Methods of Organic Chemistry, Vol. E 17e: Carbocyclic Four-Membered Ring Compounds*, Thieme Verlag, Stuttgart.

5 de Meijere, A. (ed.) (1997) *Houben-Weyl, Methods of Organic Chemistry, Vol. E 17 f:*

Carbocyclic Four-Membered Ring Compounds, Thieme Verlag, Stuttgart.

6 Turro, N.J. (1986) Geometric and topological theory in organic chemistry. *Angewandte Chemie, International Edition in English*, **25**, 882–901.

7 Hoffmann, N. (2008) Photochemical reactions as key steps in organic synthesis. *Chemical Reviews*, **108**, 1052–1103.

8 (a) Griesbeck, A.G. and Mattay, J. (eds) (2005) *Molecular and Supramolecular Photochemistry*, Vol. 12 (eds V. Ramamurthy and K.S. Schanze), Marcel Dekker, New York; (b) Mattay, J. and Griesbeck, A. (1994) *Photochemical Key Steps in Organic Synthesis*, VCH, Weinheim.

9 Woodward, R.B. and Hoffmann, R. (1969) The conservation of orbital symmetry. *Angewandte Chemie, International Edition in English*, **8**, 781–852.

10 (a) Hopf, H. (2000) *Classics in Hydrocarbon Chemistry*, Wiley-VCH, Weinheim; (b) Gleiter, R. and Werz, D.B. (2005) Reactions of metal-complexed carbocyclic 4π systems. *Organometallics*, **24**, 4316–4329.

11 Wollenweber, M., Etzkorn, M., Reinbold, J., Wahl, F., Voss, T., Melder, J.-P., Grund, C., Pinkos, R., Hunkler, D., Keller, M., Wörth, J., Knothe, L., and Prinzbach, H. (2000) Photochemical transformations, 85 [2.2.2.2]/[2.1.1.1]Pagodanes and [1.1.1.1]/[2.2.1.1]/[2.2.2.2]Isopagodanes: syntheses, structures, reactivities – benzo/ene- and benzo/benzo-photocycloadditions. *European Journal of Organic Chemistry*, 3855–3886.

12 (a) Eaton, P.E., Or, Y.S., and Branca, S.J. (1981) Pentaprismane. *Journal of the American Chemical Society*, **103**, 2134–2136; (b) Eaton, P.E., Or, Y.S., Branca, S.J., and Shankar, B.K.R. (1986) The synthesis of pentaprismane. *Tetrahedron*, **42**, 1621–1631.

13 (a) Mehta, G. and Padma, S. (1991) Synthetic studies towards prismanes: Seco-[6]-prismane. *Tetrahedron*, **47**, 7783–7806; (b) Mehta, G. and Padma, S. (1987) D2h-bishomohexaprismane ("garudane"). Design of the face-to-face 2 + 2 dimer of norbornadiene. *Journal of the American Chemical Society*, **109**, 7230–7232; (c) For a review on photochemical reactions to the synthesis of prismanes, see: Shinmyozu, T., Nogita, R., Akita, M., and Lim, C., (2004) Photochemical Approaches to the Synthesis of [n]Prismanes, in *CRC Handbook of Organic Photochemistry and Photobiology*, 2nd edn (eds W. Horspool and F. Lenci), CRC Press, Boca Raton, pp. 23/1–23/11.

14 Sakamoto, M., Yagi, T., Fujita, S., Ando, M., Mino, T., Yamaguchi, K., and Fujita, T. (2002) Diastereoselective formation of cage-type adducts via a novel photoreaction of nicotinic acid with furan. *Tetrahedron Letters*, **43**, 6103–6105.

15 For a recent example see: Wu, X.-L., Lei, L., Wu, L.-Z., Liao, G.-H., Luo, L., Shan, X.-F., Zhang, L.P., and Tung, C.-H., (2007) Synthesis, structure, and chirality of hydroxyl- and carboxyl-functionalized cubane-like photodimers of 2-naphthalene. *Tetrahedron*, **63**, 3133–3137.

16 (a) Balzani, V. (ed.) (2001) *Electron Transfer Chemistry*, Vols. 1–5, Wiley-VCH, Weinheim; (b) Fox, M.A. and Chanon, M. (1988) *Photoinduced Electron Transfer*, Elsevier, Amsterdam; (c) Mattay, J. (1987) Charge transfer and radical ions in photochemistry. *Angewandte Chemie, International Edition in English*, **26**, 825–845; (d) Julliard, M. and Chanon, M. (1983) Photoelectron-transfer catalysis: its connections with thermal and electrochemical analogs. *Chemical Reviews*, **83**, 425–508.

17 Müller, F. and Mattay, J. (1993) Photocyclo-additions: control by energy and electron transfer. *Chemical Reviews*, **93**, 99–117.

18 (a) Hoffmann, N. (2008) Efficient photochemical electron transfer sensitisation of homogeneous organic reactions. *Journal of Photochemistry and Photobiology C: Photochemical Reviews*, **9**, 43–60; (b) Fagnoni, M., Dondi, D., Ravelli,

D., and Albini, A. (2007) Photocatalysis for the formation of the C−C bond. *Chemical Reviews*, **107**, 2725–2756.

19 Mizuno, K., Hashizume, T., and Otsuji, Y. (1983) Synthesis of cyclobutanocrown ethers by intramolecular photocycloaddition. *Journal of the Chemical Society, Chemical Communications*, 977–978.

20 Ledwith, A. (1972) Cation radicals in electron transfer reactions. *Accounts of Chemical Research*, **5**, 133–139.

21 (a) Ghosh, S. (2004) Copper(I)-Catalyzed Inter- and Intramolecular [2 + 2]-Photocycloaddition Reactions of Alkenes, in *CRC Handbook of Organic Photochemistry and Photobiology*, 2nd edn (eds W. Horspool and F. Lenci), CRC Press, Boca Raton, pp. 18/1–18/15; (b) Langer, K. and Mattay, J. (1995) Copper(I)-catalyzed inter- and intramolecular [2 + 2]-photocycloaddition reactions of alkenes, in *CRC Handbook of Organic Photochemistry and Photobiology* (eds W.M. Horspool and P-.S. Song), CRC Press, Boca Raton, pp. 84–104; (c) Henning, H., Rehorek, D., and Archer, R.D. (1985) Photocatalytic systems with light-sensitive coordination compounds and possibilities of their spectroscopic sensitization – an overview. *Coordination Chemistry Reviews*, **61**, 1–53; (d) Kutal, C. (1985) Photochemistry of transition metal-organic systems. *Coordination Chemistry Reviews*, **64**, 191–206; (e) Salomon, R.G. (1983) Homogeneous metal catalysis in organic photochemistry. *Tetrahedron*, **39**, 485–575.

22 Salomon, R.G. and Kochi, J.K. (1974) Copper(I) catalysis in photocycloadditions. I. Norbornene. *Journal of the American Chemical Society*, **96**, 1137–1144.

23 (a) van den Hende, J.H. and Baird, W.C. Jr. (1963) The structure of Cuprous Chloride-Cyclooctadiene-1,5 Complex. *Journal of the American Chemical Society*, **85**, 1009–1010; (b) Trecker, D.J., Henry, J.P., and McKeon, J.E. (1965) Photodimerization of metal-complexed olefins. *Journal of the American Chemical Society*, **87**, 3261–3263; (c) Salomon, R.G. and Kochi, J.K. (1973) Cationic olefin complexes of copper(I). Structure and bonding in group Ib metal-olefin complexes. *Journal of the American Chemical Society*, **95**, 1889–1897; (d) Trecker, D.J., Foote, R.S., Henry, J.P., and McKeon, J.E. (1966) Photochemical reactions of metal-complexed olefins. II. Dimerization of norbornene and derivatives. *Journal of the American Chemical Society*, **88**, 3021–3026.

24 Salomon, R.G., Coughlin, D.J., Ghosh, S., and Zagorski, M.G. (1982) Copper(I) catalysis of olefin photoreactions. 9. Photobicyclization of α-, β-, and γ-alkenylallyl alcohols. *Journal of the American Chemical Society*, **104**, 998–1007.

25 Langer, K. and Mattay, J. (1995) Stereoselective intramolecular Copper(I)-catalyzed [2 + 2]- photocycloadditions. enantioselective synthesis of (+)- and (−)-grandisol. *Journal of Organic Chemistry*, **60**, 7256–7266.

26 (a) Rosini, G., Carloni, P., Iapalucci, M.C., and Marotta, E. (1990) Resolution, specific rotation and absolute configuration of 2,6,6-trimethylbicyclo[3.2.0]heptan-*endo*-2-ol and of 2,5-dimethylbicyclo[3.2.0]heptan-*endo*-2-ol, key intermediate in the synthesis of grandisol. *Tetrahedron Asymmetry*, **1**, 751–758; (b) Rosini, G., Marotta, E., Raimondi, A., and Righi, P. (1991) Resolution and EPC synthesis of both enantiomers of 2,5-dimethylbicyclo[3.2.0] heptan-*endo*-2-ol, key intermediate in the synthesis of grandisol. *Tetrahedron Asymmetry*, **2**, 123–138; (c) Panda, J., Ghosh, S., and Ghosh, S. (2001) Synthesis of cyclobutane fused γ-butyrolactones through intramolecular [2 + 2] photocycloaddition. Application in a formal synthesis of grandisol. *ARKIVOC*, **viii**, 146–153; (d) Sarkar, N., Nayek, A., and Ghosh, S. (2004) Copper(I)-catalyzed intramolecular asymmetric [2 + 2] photocycloaddition. Synthesis of both enantiomers of cyclobutane derivatives. *Organic Letters*, **6**, 1903–1905; (e) For an overview on asymmetric syntheses of grandisol, see:

Hoffmann, N. and Scharf, H.-D. (1991) Efficient and diastereoselective synthesis of (+)- and (−)-grandisol and 2-([1R, 2S]-2-isopropenylcyclobutane)ethanol (Demethylgrandisol) in high purity. *Liebigs Annalen der Chemie*, 1273–1277; (f) Marotta, E., Righi, P., and Rosini, G. (1999) The bicyclo[3.2.0]heptan-endo-2-ol and bicyclo [3.2.0]hept-3-en-6-one approaches in the synthesis of grandisol: the evolution of an idea and efforts to improve versatility and practicality. *Organic Process Research & Development*, **3**, 206–219.

27 Salomon, R.G., Ghosh, S., Zagorski, M.G., and Reitz, M. (1982) Copper(I) catalysis of olefin photoreactions. 10. Synthesis of multicyclic carbon networks by photobicyclization. *Journal of Organic Chemistry*, **47**, 829–836.

28 Ghosh, S., Raychaudhuri, S.R., and Salomon, R.G. (1987) Copper(I) catalysis of olefin photoreactions. 15. Synthesis of cyclobutanated butyrolactones via copper(I)-catalyzed intermolecular photocycloadditions of homoallyl vinyl or diallyl ethers. *Journal of Organic Chemistry*, **52**, 83–90.

29 Samajdar, S., Ghatak, A., and Ghosh, S. (1999) Stereocontrolled total synthesis of (±)-β-necrodol. *Tetrahedron Letters*, **40**, 4401–4402.

30 Nayek, A. and Ghosh, S. (2002) Enantiospecific synthesis of (+)-Herbertene. *Tetrahedron Letters*, **43**, 1313–1315.

31 Patra, D. and Ghosh, S. (1995) Photocycloaddition-cyclobutane rearrangement to spiro cyclopentanones: application in a formal synthesis of (±)-α-cedrene. *Journal of the Chemical Society, Perkin Transactions 1*, 2635–2641.

32 Bach, T. (2002) Stereoselective total synthesis of the tricyclic sesquiterpene (±)-kelsoene by an intramolecular Cu(I)-catalyzed [2 + 2]-photocycloaddition reaction. *Synlett*, 1305–1307.

33 Bach, T., Krüger, C., and Harms, K. (2000) The stereoselective synthesis of 2-substituted 3-azabicyclo[3.2.0]heptanes by intramolecular [2 + 2]-photocycloaddition reactions. *Synthesis*, 305–320.

34 Holt, D.J., Barker, W.D., Jenkins, P.R., Ghosh, S., Russell, D.R., and Fawcett, J. (1999) The copper(I) catalysed [2 + 2] intramolecular photoannulation of carbohydrate derivatives. *Synlett*, 1003–1005.

35 Galoppini, E., Chebolu, B., Gilardi, R., and Zhang, W. (2001) Copper(I)-catalyzed [2 + 2] photocycloadditions with tethered linkers: synthesis of syn-photodimers of dicyclopentadienes. *Journal of Organic Chemistry*, **66**, 162–168.

36 Malik, C.K., Vaultier, M., and Ghosh, S. (2007) Copper(I)-catalyzed [2 + 2] photocycloaddition of nonconjugated alkenes in room-temperature ionic liquids. *Synthesis*, 1247–1250.

37 (a) Nishimura, J., Nakamura, Y., Yamazaki, T., and Inokuma, S. (2004) Photochemical synthesis of cyclophanes, in *CRC Handbook of Organic Photochemistry and Photobiology*, 2nd edn (eds W. Horspool and F. Lenci), CRC Press, Boca Raton, pp. 19/1–19/15; (b) de Meijere, A. and König, B. (1997) What is new in [2.2] Paracyclophane chemistry? *Synlett*, 1221–1232.

38 Cornelisse, J. and de Haan, R. (2001) Ortho photocycloaddition of alkenes and alkynes to the benzene ring, in *Understanding & Manipulating Excited State Processes, Molecular and Supramolecular Photochemistry*, Vol. 8 (eds V. Ramamurthy and K.S. Schanze). Marcel Dekker, New York, pp. 1–126.

39 (a) Hoffmann, N. (2004) Photocycloaddition between benzene derivatives and alkenes. *Synthesis*, 481–495; (b) Mattay, J. (2007) Photochemistry of arenes–reloaded. *Angewandte Chemie, International Edition*, **46**, 663–665.

40 (a) De Keukeleire, D. and He, S.-L. (1993) Photochemical strategies for the construction of polycyclic molecules. *Chemical Reviews*, **93**, 359–380; (b) Vízvárdi, K., Toppet, S., Hoonaert, G.J.,

De Keukeleire, D., Bakó, P., and Van der Eycken, E. (2000) Intramolecular ortho and meta photocycloadditions of 4-phenoxybut-1-enes substituted in the arene residue with carbomethoxy, carbomethoxymethyl, and 2-carbomethoxyethyl groups. *Journal of Photochemistry and Photobiology A: Chemistry*, **133**, 135–146; (c) Wender, P.A. and Dore, T.M. (1995) Intra and intermolecular cycloadditions of benzene derivatives, in *CRC Handbook of Organic Photochemistry and Photobiology* (eds W.M. Horspool and P-.S. Song), CRC Press, Boca Raton, pp. 280–290.

41 Ohashi, M., Tanaka, Y., and Yamada, S. (1977) The [2 + 2]cycloaddition *vs* substitution in photochemical reactions of methoxybenzene-acrylonitrile systems. *Tetrahedron Letters*, **18**, 3629–3632.

42 (a) Bradshaw, J.S. (1966) The photosensitized addition of maleimide to benzenes in the absence of a charge-transfer complex. *Tetrahedron Letters*, **7**, 2039–2042; (b) See also: Schenck, G.O. and Steinmetz, R., (1960) Neuartige durch Benzophenon photosensibilisierte Additionen von Maleinsäureanhydrid an Benzol und andere Aromaten. *Tetrahedron Letters*, **1** (42), 1–8.

43 (a) Shaikhrazieva, V.Sh., Tal'vinskii, E.V., and Tolstikov, G.A. (1978) Photochemical transformations of organic compounds. X. Photochemical cycloaddition of maleimide to certain benzene derivatives. *Zhournal Organicheskoi Khimii*, **14**, 1522–1529; (b) Zhubanov, B.A., Almabekov, O.A., and Ismailova, Zh.M. (1981) Photoinitiated addition of maleic anhydride to monoalkylbenzenes. *Zhournal Organicheskoi Khimii*, **17**, 996–999.

44 See for example: Angus, H.J.F. and Bryce-Smith, D. (1960) Liquid phase photolysis. IV. A stable adduct of benzene and maleic anhydride. *Journal of the Chemical Society*, 4791–4795.

45 Itoh, Y., Horie, S., and Shindo, Y. (2001) A novel [2 + 2] photodimerization of N-[(E)-3,4-Methylenedioxycinnamoyl]dopamine in the solid state. *Organic Letters*, **3**, 2411–2413.

46 Wagner, P.J., Sakamoto, M., and Madkour, A.E. (1992) Regioselectivity in intramolecular cycloaddition of double bonds to triplet benzenes. *Journal of the American Chemical Society*, **114**, 7298–7299.

47 Wagner, P.J. (2001) Photoinduced ortho [2 + 2] cycloaddition of double bonds to triplet benzenes. *Accounts of Chemical Research*, **34**, 1–8.

48 Hoffmann, N. and Pete, J.-P. (2001) Intramolecular [2 + 2] photocycloaddition of bichromophoric derivatives of 3,5-dihydroxybenzoic acid and 3,5-dihydroxybenzonitrile. *Synthesis*, 1236–1242.

49 (a) McCullough, J.J. (1987) Photoadditions of aromatic compounds. *Chemical Reviews*, **87**, 811–860; (b) Döpp, D. (2000) Photocycloadditions with captodative alkenes, in *Organic, Physical, and Materials Photochemistry. Molecular and Supramolecular Photochemistry*, Vol. 6 (eds V. Ramamurthy and K.S. Schanze), Marcel Dekker, New York, pp. 101–148.

50 (a) Yoshimi, Y., Konishi, S., Maeda, H., and Mizuno, K. (2001) Site-selective intramolecular photocycloaddition of 2-alkenyl-substituted 1-cyanonaphthalenes depending on additives, solvents, and substituents. *Synthesis*, 1197–1202; (b) Mukae, H., Maeda, H., and Mizuno, K. (2006) One-step synthesis of benzotetra- and benzopentacyclic compounds through intramolecular [2 + 3] photocycloaddition of alkenes to naphthalene. *Angewandte Chemie, International Edition*, **45**, 6558–6560.

51 (a) Jähnisch, K., Hessel, V., Löwe, H., and Baerns, M. (2004) Chemistry in microstructured reactors. *Angewandte Chemie, International Edition*, **43**, 406–446; (b) Geyer, K., Codée, J.D.C., and Seeberger, P.H. (2006) Microreactors as tools for synthetic chemists – the chemists' round-bottomed flask of the 21st century? *Chemistry – A European Journal*, **12**,

8434–8442; (c) Mason, B.P., Price, K.E., Steinbacher, J.L., Bogdan, A.R., and McQuade, D.T. (2007) Greener approaches to organic synthesis using microreactor technology. *Chemical Reviews*, **107**, 2300–2318; (d) Fukuyama, T., Rahman, M.T., Sato, M., and Ryu, I. (2008) Adventures in inner space: microflow systems for practical organic synthesis. *Synlett*, 151–163; (e) Coyle, E.E. and Oelgemöller, M. (2008) Microphotochemistry: photochemistry in microstructured reactors. The new photochemistry of the future? *Photochemical & Photobiological Sciences*, **7**, 1313–1322.

52 Mukae, H., Maeda, H., Nashihara, S., and Mizuno, K. (2007) Intramolecular photocycloaddition of 2-(2-alkenyl-oxymethyl)-naphthalene-1-carbonitriles using glass-made microreactors. *Bulletin of the Chemical Society of Japan*, **80**, 1157–1161.

53 (a) Hoffmann, N. and Pete, J.-P. (1996) Acid catalyzed intramolecular photochemical reactions of 3-alkenyl-oxyphenols. *Tetrahedron Letters*, **37**, 2027–2030; (b) Hoffmann, N. and Pete, J.-P. (1997) Intramolecular photochemical reactions of bichromophoric 3-(alkenyloxy)phenols and 1-(alkenyloxy)-3-(alkyloxy)benzene derivatives. Acid-catalyzed transformations of the primary cycloadducts. *Journal of Organic Chemistry*, **62**, 6952–6960.

54 (a) Verrat, C., Hoffmann, N., and Pete, J.-P. (2000) An easy access to Benzo[f] isoquinoline derivatives using benzocyclobutenes derived from resorcinol. *Synlett*, 1166–1168; (b) Verrat, C. (2000) Photocycloadditions [2 + 2] intramoléculaires d'éthers de polyphénols: accès au squelette de produits naturels hétérocycliques. PhD Thesis, Université de Reims Champagne-Ardenne.

55 Kalena, G.P., Pradhan, P., and Banerji, A. (1999) Stereo- and regioselectivity of intramolecular 1,2-arene-alkene photocycloaddition in 2-alkenyl-4-chromanones. *Tetrahedron*, **55**, 3209–3218.

56 Hoffmann, N. and Pete, J.-P. (1995) Intramolecular photochemical reactions of 3-Alkenyloxyphenols. *Tetrahedron Letters*, **36**, 2623–2626.

57 (a) Al-Qaradawi, S.Y., Cosstick, K.B., and Gilbert, A. (1992) Intramolecular photocycloaddition of 4-phenoxybut-1-enes: a convenient access to the 4-oxatricyclo[7.2.0.0^{3,7}]undeca-2,10-diene skeleton. *Journal of the Chemical Society, Perkin Transactions 1*, 1145–1148; (b) Cosstick, K.B. and Gilbert, A. (1994) 11-Cyano-4-oxatricyclo[7.2.0.0^{3,7}]undeca-2,10-diene, in *Photochemical Key Steps in Organic Synthesis* (eds J. Mattay and A. Griesbeck), VCH, Weinheim, pp. 175–176.

58 Smart, R.P. and Wagner, P.J. (1995) Regioselectivity in intramolecular cycloaddition of double bonds to triplet acylbenzenes II. Effects of substituents meta to the tether. *Tetrahedron Letters*, **36**, 5131–5134.

59 Wagner, P.J. and McMahon, K. (1994) Chiral auxiliaries promote both diastereoselective cycloaddition and kinetic resolution of products in the ortho photocycloaddition of double bonds to benzene rings. *Journal of the American Chemical Society*, **116**, 10827–10828.

60 Beaudry, C.M., Malerich, J.P., and Trauner, D. (2005) Biosynthetic and biomimetic electrocyclizations. *Chemical Reviews*, **105**, 4757–4778.

61 Sieburth, S.McN. (2004) Photochemical reactivity of pyridones, in *CRC Handbook of Organic Photochemistry and Photobiology*, 2nd edn (eds W. Horspool and F. Lenci), CRC Press, Boca Raton, pp. 103/1–103/18.

62 Kaneko, C., Katagiri, N., Sato, M., Muto, M., Sakamoto, T., Saikawa, S., Naito, T., and Saito, A. (1986) Cycloadditions in syntheses. Part 27. rel-(1R,4R,5S)-5-Hydroxy-2-azabicyclo[2.2.0]hexan-3-one and its derivatives: synthesis and transformation to azetidin-2-ones. *Journal*

of the Chemical Society, Perkin Transactions 1, 1283–1288.

63 Katagiri, N., Sato, M., Yoneda, N., Saikawa, S., Sakamoto, T., Muto, M., and Kaneko, C. (1986) Cycloadditions in syntheses. Part 28. 2-Azabicyclo[2.2.0]hexane-3,5-dione and its derivatives: synthesis and transformation to azetidin-2-ones. *Journal of the Chemical Society, Perkin Transactions 1*, 1289–1296.

64 Wenska, G., Skalski, B., Gdaniec, Z., Adamiak, R.W., Matulic-Adamic, J., and Beigelman, L. (2000) Photophysical and photochemical properties of C-linked ribosides of pyridin-2-one. *Journal of Photochemistry and Photobiology A: Chemistry*, **133**, 169–176.

65 Sato, M., Katagiri, N., Muto, M., Haneda, T., and Kaneko, C. (1986) Cycloadditions in synthesis. Part 33. Practicable synthesis of (1R,4R)-5-(l-menthoxy)-2-azabicyclo[2.2.0]hex-5-en-3-one and its derivatives. New building blocks for carbapenem nuclei. *Tetrahedron Letters*, **27**, 6091–6094.

66 For chiral induction in photochemical reactions see: (a) Inoue, Y. and Ramapurthy, V. (eds) (2004) Chiral Photochemistry, in *Molecular and Supramolecular Photochemistry*, Vol. 11 (eds V. Ramamurthy and K.S. Schanze, Series), Marcel Dekker, New York; (b) Rau, H. (1983) Asymmetric photochemistry in solution. *Chemical Reviews*, **83**, 535–547; (c) Inoue, Y. (1992) Asymmetric photochemical reactions in solution. *Chemical Reviews*, **92**, 741–770; (d) Griesbeck, A.G. and Meierhenrich, U.J. (2002) Asymmetric photochemistry and photochirogenesis. *Angewandte Chemie, International Edition*, **41**, 3147–3154.

67 Svoboda, J. and König, B. (2006) Templated photochemistry: toward catalysts enhancing the efficiency and selectivity of photoreactions in homogeneous solutions. *Chemical Reviews*, **106**, 5413–5430.

68 Ramamurthy, V. and Venkatesan, K. (1987) Photochemical reactions of organic crystals. *Chemical Reviews*, **87**, 433–481.

69 (a) Green, B.S., Lahav, M., and Rabinovich, D. (1979) Asymmetric synthesis via reactions in chiral crystals. *Accounts of Chemical Research*, **12**, 191–197; (b) Toda, F. (1995) Solid state organic chemistry: efficient reactions, remarkable yields, and stereoselectivity. *Accounts of Chemical Research*, **28**, 480–486; (c) Sakamoto, M. (2006) Spontaneous chiral crystallization of achiral materials and absolute asymmetric photochemical transformation using the chiral crystalline environment. *Journal of Photochemistry and Photobiology C: Photochemical Reviews*, **7**, 183–196.

70 (a) Tanaka, K., Fujiwara, T., and Urbanczyk-Lipkowska, Z. (2002) Highly enantioselective photocyclization of 1-alkyl-2-pyridones to β-lactams in inclusion crystals with optically active host compounds. *Organic Letters*, **4**, 3255–3257; (b) Fujiwara, T., Tanaka, N., Tanaka, K., and Toda, F. (1986) X-ray structural study of a 1:1 complex of 4-methoxy-1-methylpyridone and (R,R)-(−)-1,6-bis(o-chlorophenyl)-1,6-diphenylhexa-2,4-diyne-1,6-diol the irradiation of which gives optically pure β-lactam. *Journal of the Chemical Society, Perkin Transactions 1*, 663–664; (c) Toda, F. and Tanaka, K. (1988) Enantioselective photoconversion of pyridones into β-lactam derivatives in inclusion complexes with optically active host compounds. *Tetrahedron Letters*, **29**, 4299–4302.

71 Wu, L.-C., Cheer, C.J., Olovsson, G., Scheffer, J.R., Trotter, J., Wang, S.-L., and Liao, F.-L. (1997) Crystal engineering for absolute asymmetric synthesis through the use of *meta*-substituted aryl groups. *Tetrahedron Letters*, **38**, 3135–3138.

72 Curtin, D.Y. and Paul, I.C. (1981) Chemical consequences of the polar axis in organic solid-state chemistry. *Chemical Reviews*, **81**, 525–541.

73 See for example: (a) Jacques, J., Collet, A., and Wilen, S.H., (1994) *Enantiomers*,

Racemates, and Resolution, Krieger Publishing, Melbourne FL; (b) Dunitz, J.D. (2003) Are crystal structures predictable? *Chemical Communications*, 545–548.

74 Dauben, W.G., Koch, K., Smith, S.L., and Chapman, O.L. (1963) Photoisomerizations in the α-Tropolone Series: The mechanistic path of the α-tropolone methyl ether to methyl 4-Oxo-2-cyclopentenylactetate conversion. *Journal of the American Chemical Society*, **85**, 2616–2621.

75 (a) Grewe, R. and Wulf, W. (1951) Die Umwandlung des Colchicins durch Sonnenlicht. *Chemische Berichte*, **84**, 621–625; (b) Forbes, E.J. (1955) Colchicine and related compounds. part XIV. Structure of β- and γ-lumicolchicine. *Journal of the Chemical Society*, 3864–3870; (c) Bellet, P. and Gérard, D. (1961) Les lumicolchicosides β et γ. *Annales Pharmaceutiques Françaises*, **19**, 587–592; (d) Chapman, O.L., Smith, H.G., and King, R.W. (1963) The structure of β-lumicolchicine. *Journal of the American Chemical Society*, **85**, 803–806; (e) Dauben, W.G. and Cox, D.A. (1963) Photochemical transformations. XIV. Isocolchicine. *Journal of the American Chemical Society*, **85**, 2130–2134; (f) Chapman, O.L., Smith, H.G., and Barks, P.A. (1963) Photoisomerization of isocholchicine. *Journal of the American Chemical Society*, **85**, 3171–3173; (g) Bussotti, L., Cacelli, I., D'Auria, M., Foggi, P., and Lesma, G. (2003) Photochemical isomerization of colchicine and thiocolchicine. *Journal of Physical Chemistry A*, **107**, 9079–9085.

76 Neumüller, O.-A., Kuhn, H.J., Schenck, G.O., and Šantavý, F. (1964) Photoisomerisierung von 1-Demecocin und 1-Demecolcinacetat. *Liebigs Annalen der Chemie*, **674**, 122–129.

77 (a) Ellington, E., Bastida, J., Viladomat, F., Šimánek, V., and Codina, C. (2003) Occurrence of colchicines derivatives in plants of the genus *Androcymbium*. *Biochemical Systematics and Ecology*, **31**, 715–722; (b) Alali, F.Q., El-Elimat, F., Li, C., Qandil, A., Alkofahi, A., Tawaha, K., Burgess, J.P., Nakanishi, Y., Kroll, D.J., Navarro, H.A., Falkinham, J.O. III, Wani, M.C., and Oberlies, N.H. (2005) New cholchicinoids from a native Jordanian meadow saffron, *Colchicum brachyphyllum*: isolation of the first naturally occurring dextrorotatory colchicinoid. *Journal of Natural Products*, **68**, 173–178.

78 Brunetti, I.L., Bechara, E.J.H., Cilento, G., and White, E.H. (1982) Possible *in vivo* formation of lumicholchicines from colchicine by endogenously generated triplet species. *Photochemistry and Photobiology*, **36**, 245–249.

79 Joy, A. and Ramamurthy, V. (2000) Chiral photochemistry within zeolites. *Chemistry – A European Journal*, **6**, 1287–1293.

80 Joy, A., Kaanumalle, L.S. and Ramamurthy, V. (2005) Role of cations and confinement in asymmetric photochemistry: enantio- and diastereoselective photocyclization of tropolone derivatives within zeolites. *Organic & Biomolecular Chemistry*, **3**, 3045–3053.

81 Joy, A., Scheffer, J.R. and Ramamurthy, V. (2000) Chirally modified zeolites as reaction media: photochemistry of an achiral tropolone ether. *Organic Letters*, **2**, 119–121.

82 (a) Wagner, P.J. (2005) Abstraction of γ-hydrogens by excited carbonyls, in *Synthetic Organic Photochemistry* (eds A.G. Griesbeck and J. Mattay), Molecular and Supramolecular Photochemistry, Vol. 12 (Series eds V. Ramamurthy and K.S. Schanze), Marcel Dekker, New York, pp. 11–39; (b) Wessig, P. (2004) Regioselective photochemical synthesis of carbo- and heterocyclic compounds: the Norrish/Yang reaction, in *CRC Handbook of Organic Photochemistry and Photobiology*, 2nd edn (eds W. Horspool and F. Lenci), CRC Press, Boca Raton, pp. 57/1–57/20; (c) Weiss, R.G. (1995) Norrish type II processes of ketones: influence of environment, in *CRC Handbook of Organic Photochemistry and Photobiology* (eds W.M. Horspool and

P.-S. Song), CRC Press, Boca Raton, pp. 471–483.

83 Griesbeck, A.G. and Heckroth, H. (2002) Stereoselective synthesis of 2-amino-cyclobutanols via photocyclization of α-amidoalkyl aryl ketones: mechanistic implications for the Norrish/Yang reaction. *Journal of the American Chemical Society*, **124**, 396–403.

84 Leibovitch, M., Olovsson, G., Scheffer, J.R., and Trotter, J. (1998) An investigation of the yang photocyclization reaction in the solid state: asymmetric induction studies and crystal structure-reactivity relationships. *Journal of the American Chemical Society*, **120**, 12755–12769.

85 Herrera, A.J., Rondón, M., and Suárez, E. (2008) Stereocontrolled photocyclization of 1,2-diketones: application of a 1,3-acetyl group transfer methodology to carbohydrates. *Journal of Organic Chemistry*, **73**, 3384–3391.

86 Basarić, N., Horvat, M., Mlinarić-Majerski, K., Zimmermann, E., Neudörfl, J., and Griesbeck, A.G. (2008) Novel 2,4-methanoadamantane-benzazepine by domino photochemistry of *N*-(1-adamantyl)phthalimide. *Organic Letters*, **10**, 3965–3968.

87 Le Blanc, S., Pete, J.-P., and Piva, O. (1992) New access to spiranic β-lactams. *Tetrahedron Letters*, **33**, 1993–1996.

88 Hennig, H. (1999) Homogeneous photo catalysis by transition metal complexes. *Coordination Chemistry Reviews*, **182**, 101–123.

89 (a) Chopade, P.R. and Louie, J. (2006) [2 + 2 + 2] Cycloaddition catalyzed by transition metal complexes. *Advances in Synthesis & Catalysis*, **348**, 2307–2327; (b) Varela, J.A. and Saa, C. (2008) Recent advances in the synthesis of pyridines by transition-metal-catalyzed [2 + 2 + 2] cycloaddition. *Synthesis*, 2571–2578.

90 Bruns, D., Miura, H., Vollhardt, K.P.C., and Stanger, A. (2003) En route to archimedene: total synthesis of c_{3h}-Symmetric [7]Phenylene. *Organic Letters*, **5**, 549–552.

91 Han, S., Bond, A.D., Disch, R.L., Holms, D., Schulman, J.M., Teat, S.J., Vollhardt, K.P.C., and Whitener, G.D. (2002) Total synthesis of structures of angular [6]- and [7]Phenylene: The first helical phenylenes (Heliphenes). *Angewandte Chemie, International Edition*, **41**, 3223–3230 and references cited therein.

92 (a) Barluenga, J., Santamaría, J., and Tomás, M. (2004) Synthesis of heterocycles via group VI Fischer carbene complexes. *Chemical Reviews*, **104**, 2259–2284; (b) Kiehl, O. and Schmalz, H.-G. (2000) B. Transition metal organometallic methods – photolysis of Fischer carbene complexes, in *Organic Synthesis Highlights IV* (ed. H.-G. Schmalz), Wiley-VCH, Weinheim, pp. 71–76; (c) Hegedus, L.S. (1997) Chromium carbene complex photochemistry in organic synthesis. *Tetrahedron*, **53**, 4105–4128.

93 Brugel, T.A. and Hegedus, L.S. (2003) N-functionalization of poly(ethylene glycol)-linked mono- and bis-dioxocyclams as potential ligands for Gd^{3+}. *Journal of Organic Chemistry*, **68**, 8409–8415.

94 Hegedus, L.S., Bates, R.W., and Söderberg, B.C. (1991) Synthesis of optically active cyclobutanones by photolysis of chromium-alkoxycarbene complexes in the presence of optically active ene-carbamates. *Journal of the American Chemical Society*, **113**, 923–927.

95 Colson, P.-J. and Hegedus, L.S. (1994) Synthesis of β-lactones by the photochemical reactions of chromium alkoxycarbene complexes with aldehydes. *Journal of Organic Chemistry*, **59**, 4972–4976.

96 Scanlan, E.M., Slawin, A.M.Z., and Walton, J.C. (2004) Preparation of β- and γ-lactams from cabamoyl radicals derived form oxime oxalate amides. *Organic & Biomolecular Chemistry*, **2**, 716–724.

97 Singh, V. and Porinchu, M. (1996) Sigmatropic 1,2- and 1,3-acyl shifts in excited states: a novel, general protocol for the synthesis of tricyclopentanoids and protoilludanes. *Tetrahedron*, **52**, 7087–7126.

98 Paquette, L.A., Barton, W.R.S., and Gallucci, J.C. (2004) Synthesis of 1-aza-8-thiabicyclo[4.2.1]nona-2,4-diene 8,8-dioxide and its conversion to a strained spirocycle via photoinduced SO_2-N bond cleavage. *Organic Letters*, **6**, 1313–1315.

6
Formation of a Four-Membered Ring: From a Carbonyl-Conjugated Alkene
Jörg P. Hehn, Christiane Müller, and Thorsten Bach

6.1
Introduction

The photochemical cycloaddition of two different alkenes leads to multiply substituted cyclobutanes, and allows for a general access to this class of compounds. More specifically, ever since Ciamician observed the light-induced isomerization of carvone ($\mathbf{1} \rightarrow \mathbf{2}$) (Scheme 6.1) in 1908 [1], the inter- and intramolecular reaction between an α,β-unsaturated carbonyl compound and an alkene has become the most intensively studied and most widely used class of [2 + 2]-photocycloaddition reactions [2–9].

Indeed, many α,β-unsaturated carbonyl, carboxyl, and related heterocyclic compounds of general structure **A** (Scheme 6.2) can be excited with light of relatively long wavelength ($\lambda \geq 250$ nm). Cycloaddition occurs in most cases – after intersystem crossing (ISC) – from the lowest-lying triplet state (T_1), which has often a $\pi\pi^*$-character. In an extremely simplified fashion, using a valence bond description, the excited state can be written in biradical form (**B**), which explains the intermediacy of 1,4-biradicals **C** or **C**' *en route* to the product cyclobutane **D**. The mechanism of the reaction has been extensively studied, and it would be neither scientifically correct nor appropriate to the length of this chapter to attempt a detailed explanation. Instead, the reader is referred to more comprehensive treatises on this issue [10, 11]. It can be said, however, that certain predictions are possible based on the simple model of a radical pathway depicted in Scheme 6.2. In 1,6-dienes, for example, intramolecular five-membered ring formation is expected to be fast, and this hypothesis explains the regioselectivity of many reactions so well that it has been called the "rule of five" [12, 13]. In intermolecular reactions, the substituent pattern of the alkene influences the regioselectivity to some extent, with the general notion being that the T_1-state **B** shows reverted polarity (*Umpolung*) compared to the ground state [14]. As a consequence, a donor substituent in the alkene should be found in a 1,3-relationship to the carbonyl group in the product cyclobutane [head-to-tail (HT) product], an acceptor in a 1,2-relationship [head-to-head (HH) product]. The intermediacy of an

Handbook of Synthetic Photochemistry. Edited by Angelo Albini and Maurizio Fagnoni
Copyright © 2010 WILEY-VCH Verlag GmbH & Co. KGaA, Weinheim
ISBN: 978-3-527-32391-3

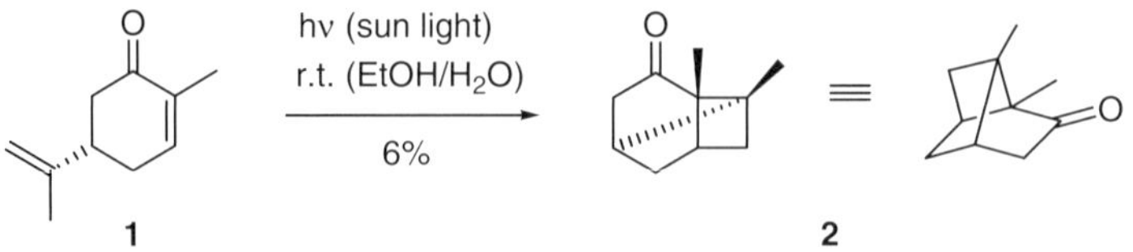

6 Formation of a Four-Membered Ring

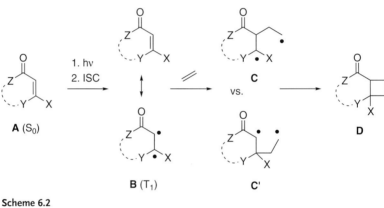

Scheme 6.1

exciplex prior to biradical formation has been proposed and is supported experimentally for specific substrate couples [15].

The representation of the starting material **A** in Scheme 6.2 already alludes to the fact that most α,β-unsaturated carbonyl, carboxyl, and related heterocyclic compounds used in [2 + 2]-photocycloaddition chemistry are cyclic five- and six-membered ring compounds. This choice avoids the rotation around the C–C single bond in intermediate **B**, which would act as a rapid decay pathway to the ground state. In this chapter we will focus exclusively on five- and six-membered compounds as substrates. Moreover, the general structure **A** was divided into eight substructures **A1**–**A8**, according to which the material was organized (Figure 6.1). Here, R

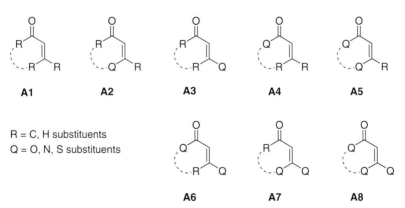

Scheme 6.2

R = C, H substituents
Q = O, N, S substituents

Figure 6.1 General substructures **A1**–**A8**, as used to organize the chapter.

represents the carbon and hydrogen substituents including carbon chains, while Q represents the oxygen, nitrogen and sulfur substituents, which can also be part of a five- or six-membered ring. The last substrate types **A7** and **A8** have not been intensively studied [16], and will therefore not be treated in an individual section.

In the figures and schemes, depicted in the individual sections of this chapter, emphasis was placed on representative examples, which explain clearly the key message of the respective paragraph. If two or more equally useful examples were found in the literature, the more recent work was included preferably. For the representation of absolute versus relative configuration, the Maehr convention was followed. Wedged lines indicate the absolute configuration in enantiomers, and bold lines (hashed or solid) the relative configuration in racemic molecules [17]. A dot (•) next to a carbon center represents an unpaired electron (see Scheme 6.2). A dot at a carbon center represents an upward hydrogen atom in achiral and racemic compounds. In allenes, the dot marks the internal carbon atom. Photodimerization reactions, that is, the [2 + 2]-photocycloaddition of two identical substrates, are beyond the scope of this chapter.

6.2
[2 + 2]-Photocycloaddition of Enones (Substrate Type A1)

Substrates for enone photocycloaddition are represented by the general structure **A1** (Figure 6.1). An extensive array of precedence exists for both cyclopentenone and cyclohexenone photocycloaddition reactions. Typically, the reactions follow the pathway depicted in Scheme 6.2. Electron-rich alkenes are the preferred reaction partners in intermolecular photocycloaddition reactions avoiding triplet energy transfer to the alkene. Based on the photophysical properties of cyclopentenone ($\lambda_{max} = 308$ nm in CH_2Cl_2) [18] and cyclohexenone ($\lambda_{max} = 319$ nm in EtOH) [19], which both exhibit a weak absorption maximum at long wavelength, irradiation experiments are normally conducted with a medium- or high-pressure mercury lamp (Hg lamp, e.g., Hanovia 450 W, Original Hanau TQ 150, Philips HPK-125, Toshiba 400 P) in glassware, which transmits only long wavelength light ($\lambda > 290$ nm, e.g., Pyrex), or with light sources, which emit light at a specific wavelength (e.g., doped lamps, light emitting diodes, or lasers). Commonly used solvents should be transparent above 280 nm, and should contain only relatively stable bonds. Aliphatic and aromatic hydrocarbons generally represent a good choice, provided that the substrates are soluble in a nonpolar solvent. In the individual reaction schemes, specific irradiation conditions have been provided, if available. Further key features are discussed in the individual sections.

6.2.1
Cyclopentenones

Cyclopentenones undergo a clean [2 + 2]-photocycloaddition to a variety of alkenes, resulting in products with a bicyclo[3.2.0]heptanone skeleton. The two annelated

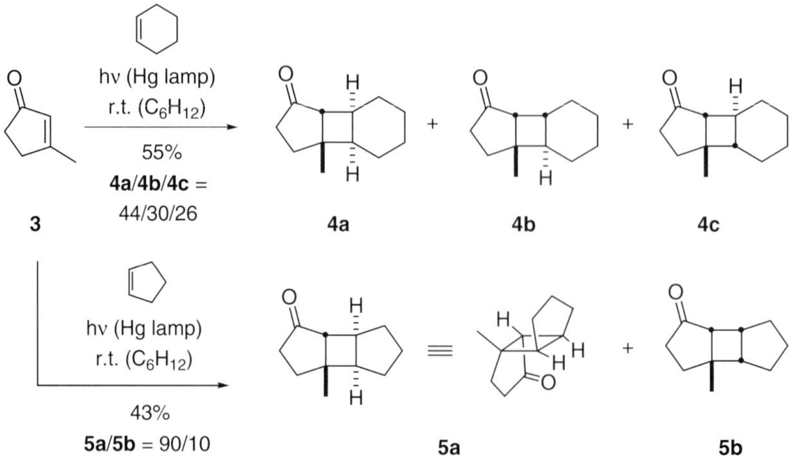

Scheme 6.3

rings are exclusively *cis*-fused; that is, the reaction proceeds stereospecifically relative to the cyclopentenone double bond without isomerization. Alkene double bond isomerization is possible depending on the lifetime of intermediate **C** or **C'**. Four- and five-membered cyclic alkenes as reaction partners react stereospecifically for steric reasons, whereas larger cyclic alkenes can produce diastereomeric mixtures. As an example, the reaction of 3-methylcyclopenten-2-enone (**3**) with cyclohexene resulted not only in the expected *exo*-cycloaddition product **4a** but also in the diastereomeric products **4b** and **4c** (Scheme 6.3) [20]. Reaction with cyclopentene gave the single *exo*-product **5a**, albeit in a relatively low yield. The *endo*-product **5b** was formed as byproduct. The preferential formation of *exo*-products is frequently observed, if a cyclic enone reacts intermolecularly with a cyclic alkene.

The formation of *exo*-products is also the predominant pathway if acyclic alkenes are employed as reaction partners. The reaction of cyclopentenone and ethyl vinyl ether serves as an instructive example (Scheme 6.4) for two reasons. First, it exemplifies the regiochemical outcome (r.r. = regioisomeric ratio) of the [2 + 2]-photocycloaddition with HT-products **7** being predominantly formed (versus HH-products **8**). Second, it illustrates the *exo*-preference with compound **7a** prevailing over **7b** (d.r. = diastereomeric ratio) [21]. However, it is also clear from this

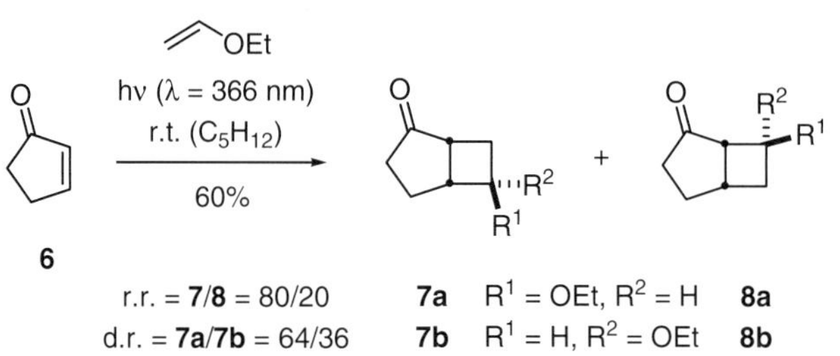

Scheme 6.4

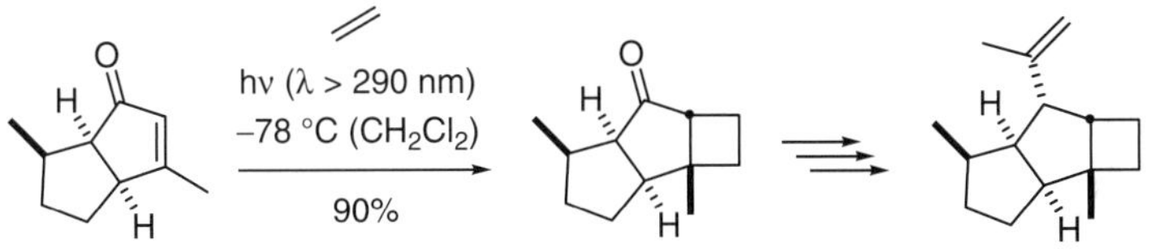

6.2 [2 + 2]-Photocycloaddition of Enones

Scheme 6.5

example, that perfect regioselectivity in intermolecular [2 + 2]-photocycloaddition reactions is difficult to achieve.

With regard to the simple diastereoselectivity, one can generalize that large substituents normally adopt opposite (*trans*) positions in the cyclobutane product whenever feasible.

The facial diastereoselectivity of intermolecular cyclopentenone [2 + 2]-photocycloaddition reactions is predictable if the cyclopentenone or a cyclic alkene reaction partner is chiral. Addition occurs from the more accessible side, and good stereocontrol can be expected if the stereogenic center is located at the α-position to the double bond. In their total synthesis of (±)-kelsoene (**11**), Piers *et al.* [22] utilized cyclopentenone **9** in the [2 + 2]-photocycloaddition to ethylene (Scheme 6.5). The cyclobutane **10** was obtained as a single diastereoisomer. In a similar fashion, Mehta *et al.* have frequently employed the fact that an approach to diquinane-type *cis*-bicyclo[3.3.0]octenones occurs from the more accessible convex face. Applications can be found in the syntheses of (+)-kelsoene [23], (−)-sulcatine G [24], and (±)-merrilactone A [25].

The single stereogenic center of γ-silylsubstituted cyclopentenone **12** allowed for an excellent stereocontrol in the [2 + 2]-photocycloaddition to the strained cyclobutene **13**, which in turn had been obtained by a [2 + 2]-photocycloaddition–ring contraction sequence (Scheme 6.6) [26]. Alkene **13** dictates the *exo*-approach of the cyclopentenone, but as a *meso*-compound is of course not capable of controlling the absolute configuration. Further elaboration of product **14** led to (+)-pentacycloanammoxic acid (**15**).

In a pioneering study, Zandomeneghi and Cavazza showed that the [2 + 2]-photocycloaddition of racemic 4-acetyloxy-2-cyclopentenone can lead to a small, but detectable, enantiomeric excess (ee) of the product if circularly polarized light (CPL) is used as an irradiation source. By preferential excitation of the (−)-antipode of the cyclopentenone with *l*-CPL (351–363 nm laser) in the presence of acetylene, an

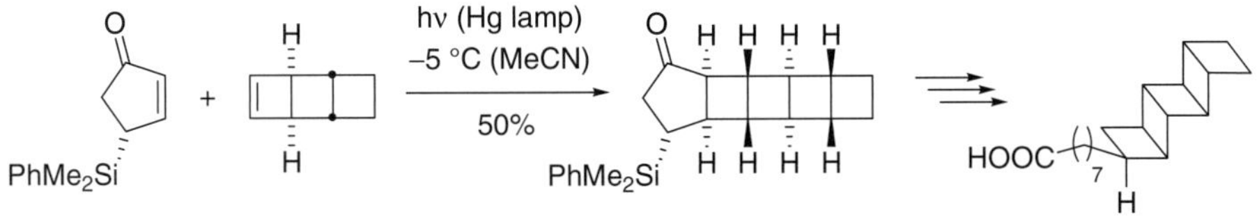

Scheme 6.6

enantiomerically enriched product (1% ee) was obtained. The unreacted starting material was optically enriched in the (+)-enantiomer [27].

The regioselectivity of intramolecular [2 + 2]-photocycloaddition reactions is predictable if five-membered ring formation is possible in the formation of biradicals of type **C** or **C′** (rule of five, *vide supra*). If five-membered ring formation is not feasible, then six-membered rings are most readily formed. The facial diastereoselectivity is efficiently controlled by a stereogenic center in the cyclopentenone if the intramolecular alkene is attached via a tether to this stereogenic center. The key step **16 → 17** in the stereoselective synthesis of (−)-incarvilline (**18**) illustrates the point (Scheme 6.7) [28]. The side chain attached to C-4 in the cyclopentenone **16** carries the terminal alkene, which reacts intramolecularly with perfect regio- and diastereoselectivity to cyclobutane **17**.

If a stereogenic center is present in the chain, which connects the two olefin partners, then conformational aspects must be taken into account. In the typical scenario of a five-membered ring formation, the larger substituent at the stereogenic center resides frequently in the pseudoequatorial position of a chairlike conformation. Enone **19** underwent intramolecular [2 + 2]-photocycloaddition to a furan ring via conformation **19′** in which both groups Et_3SiO and CMe_3 at the stereogenic center can adopt a pseudoequatorial position (Scheme 6.8) [29]. The other chairlike

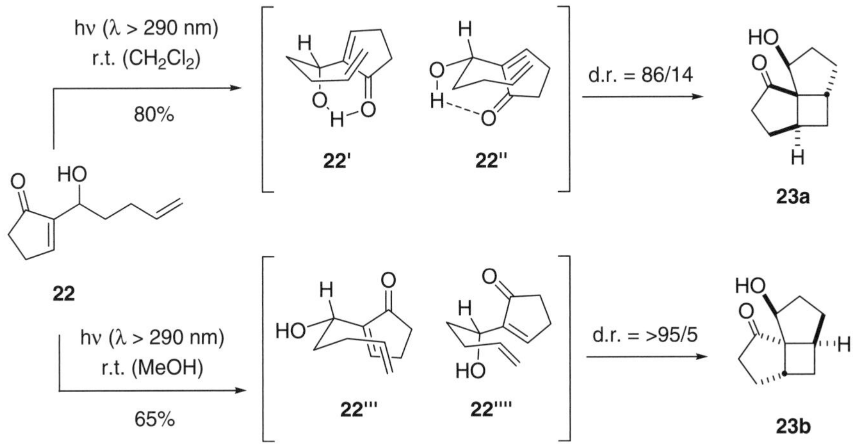

Scheme 6.9

conformation **19″** with two pseudoaxial substituents is not populated in the transition state of the photocycloaddition, and neither are boatlike conformations. Product **20** was formed exclusively and further converted into the structurally complex natural product (±)-ginkgolide B (**21**).

While the steric bulk of a given substituent is normally invariable, hydrogen-bonding interactions can alter the size of a polar group, for example, of a hydroxyl group [30]. A study by Snapper et al. showed that intramolecular [2 + 2]-photocycloaddition of hydroxyenone **22** occurred in nonpolar solvents (e.g., CH_2Cl_2) with a preference for product **23a**, whereas product **23b** was formed as a single diastereoisomer in protic solvents (e.g., MeOH) (Scheme 6.9) [31]. An explanation for this preference is based on the intramolecular hydrogen bonding favoring the boatlike conformation **22′** (and/or chairlike conformation **22″**) in nonpolar solvents, and on the intermolecular hydrogen bonding favoring boatlike conformation **22‴** (and/or chairlike conformation **22⁗** with a pseudoaxial hydroxyl group) [31].

In general, the stereoelectronic influence of substituents in [2 + 2]-photocycloaddition reactions is minor, and the preferred ground-state conformation often accounts for the formation of the major diastereoisomer. Inspection of molecular models and force field calculations provide a good picture of possible transition states leading via 1,4-biradicals to cyclobutane products. The total synthesis of (+)-guanacastepene represents another recent example for the use of stereoselective intramolecular cyclopentenone–olefin photocycloadditions in natural products synthesis [32].

6.2.2
Cyclohexenones

Cyclohexenones require essentially identical irradiation conditions as cyclopentenones (*vide supra*). The outcome of the intermolecular [2 + 2]-photocycloaddition to alkenes is somewhat more complex as compared to cyclopentenones, because the

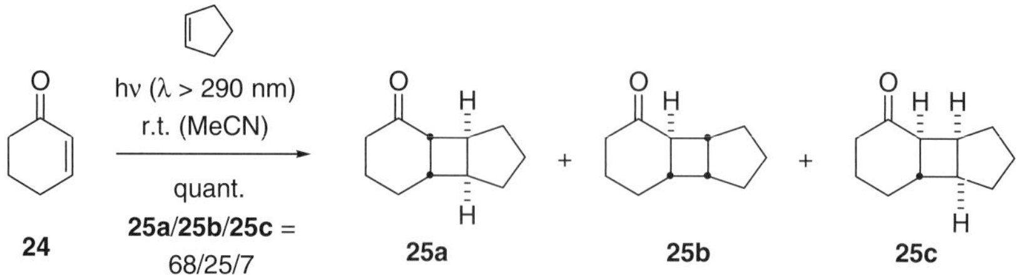

Scheme 6.10

[Scheme 6.10: Cyclohexenone **24** + cyclopentene, hv (λ > 290 nm), r.t. (MeCN), quant., **25a/25b/25c** = 68/25/7, giving products **25a**, **25b**, **25c**.]

twisted nature of the intermediate T_1 state leads in the cyclobutane product not only to a *cis*-configuration along the former cyclohexenone α,β-bond but also to a *trans*-configuration. This behavior is illustrated by the reaction of cyclohexenone (**24**) to cyclopentene, which was one of the first reactions of this type ever studied [33] and which has been reinvestigated in recent years (Scheme 6.10) [34, 35]. The *exo*-product **25a** with a *cis-anti-cis* arrangement of hydrogen atoms was the preferred product, but the *trans*-products **25b** (*trans-anti-cis*) and **25c** (*trans-syn-cis*) were formed in significant amounts. The *cis-syn-cis* product was detected in low (<1%) quantities. Isomerization into the thermodynamically more stable *cis*-products was possible upon base treatment generating two diastereoisomers, *cis-anti-cis* product **25a** (from **25a** and **25c**) and the *cis-syn-cis* product (from **25b**) in a ratio of 75 : 25. Cyclopentene reacted stereospecifically, that is, the former C–C double bond retained its *cis* configuration.

The formation of *trans*-products is observed to a lesser extent in the reaction of 3-alkoxycarbonyl-substituted cyclohexenones, in the reaction with electron-deficient alkenes and in the reaction with olefinic reaction partners, such as alkynes and allenes, in which the four-membered ring is highly strained (Scheme 6.11). The ester **26** reacted with cyclopentene upon irradiation in toluene to only two diastereomeric products **27** [36]. The *exo*-product **27a** (*cis-anti-cis*) prevailed over the *endo*-product **27b** (*cis-syn-cis*); the formation of *trans*-products was not observed. The well-known [2 + 2]-photocycloaddition of cyclohexenone (**24**) to acrylonitrile was recently reinvestigated in connection with a comprehensive study [37]. The product distribution, with the two major products **28a** and **28b** being isolated in 90% purity, nicely illustrates the preferential formation of HH (head-to-head) cyclobutanes with electron-acceptor substituted olefins. The low simple diastereoselectivity can be interpreted by the fact that the cyano group is relatively small and does not exhibit a significant preference for being positioned in an *exo*-fashion.

The last example depicted in Scheme 6.11 illustrates the exclusive *cis*-product formation observed with cyclohexenones and allenes in conjunction with a high HH-preference [38]. In addition, substrate **29** is chiral, and perfect facial diastereoselectivity was observed due to cyclic stereocontrol. Product **30** served as intermediate in a formal total synthesis of the triquinane (±)-pentalenene (**31**).

Chiral cyclohexenones have been frequently employed in intermolecular [2 + 2]-photocycloaddition reactions directed towards natural product synthesis. A further case in point is the reaction of cyclohexenone **32** with *trans*-1,2-dichloroethylene

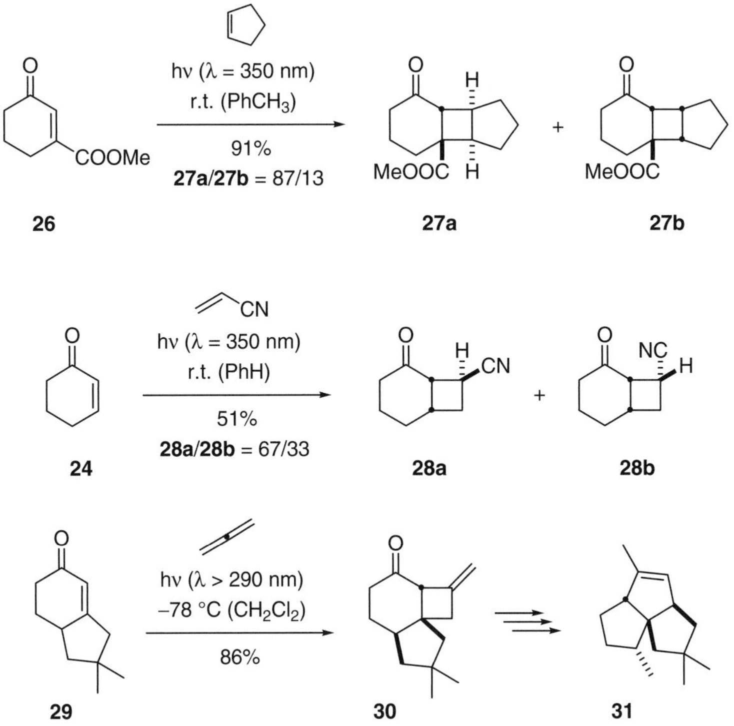

Scheme 6.11

(Scheme 6.12) [39]. Cyclobutane **33** was formed with perfect stereocontrol over the two stereogenic centers in α- and β-position to the carbonyl group. The concave shape of the substrate forces a highly selective approach of the olefin. The relative and absolute configuration at the chlorine-bearing carbon atoms was not relevant, as chlorine was subsequently eliminated under reductive conditions. Compound **33** was further elaborated into (±)-sterpurene (**34**).

Intramolecular reactions of cyclohexenones follow pathways similar to those of cyclopentenones, both with regard to regio- and stereocontrol. The initially mentioned intramolecular [2 + 2]-photocycloaddition of carvone (**1**) is a typical example for five-membered ring formation with high diastereofacial control (Scheme 6.1). In this case, the rule of five requires the terminal carbon atom of the intermolecular

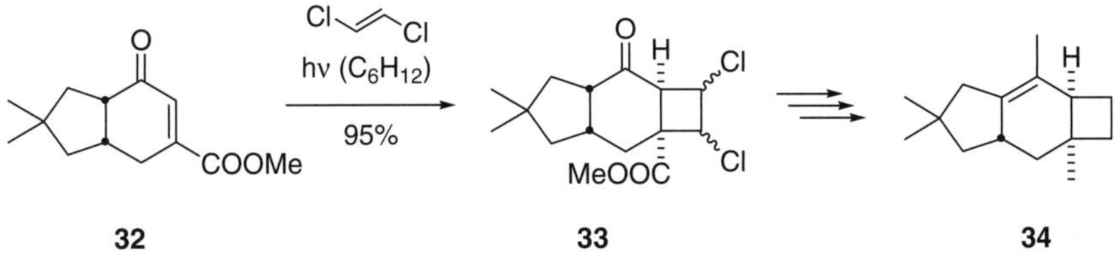

Scheme 6.12

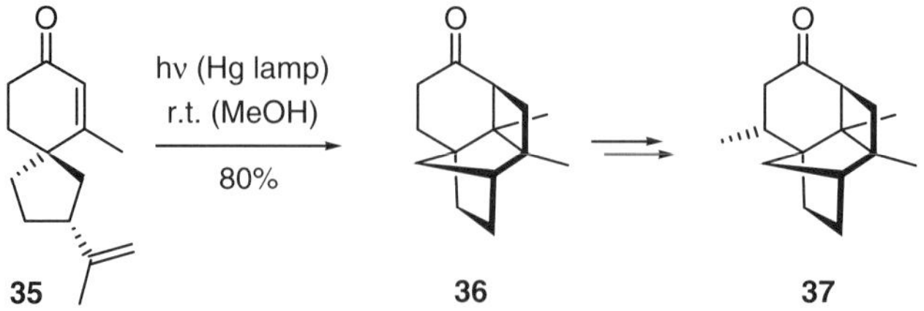

Scheme 6.13

olefin to be attached to the β-carbon atom of the enone (crossed photocycloaddition). The more common addition mode is the straight photocycloaddition (e.g., Schemes 6.7 to 6.9), as is also employed in the biomimetic [2 + 2]-photocycloaddition of enone **35** to cyclobutane **36** (Scheme 6.13) [40]. Intermediate **36** was further transformed into (+)-solanascone (**37**), a natural product, which is presumably formed in nature by intramolecular [2 + 2]-photocycloaddition from the corresponding methyl-bearing precursor solavetivone [41].

Control over the absolute configuration in cyclohexenone photocycloadditions has been achieved by auxiliary-induced diastereoselectivity. In particular, esters related to compound **26**, which are derived from a chiral alcohol but not from methanol, lend themselves as potential precursors, from which the chiral auxiliary can be effectively cleaved [42, 43]. In a recent study, the use of additives was advertised to increase the diastereomeric excess in these reactions [44]. An intriguing auxiliary-induced approach was presented by Piva et al., who employed chiral β-hydroxycarboxylic acids as tethers to control both the regioselectivity and the diastereoselectivity of intramolecular [2 + 2]-photocycloaddition reactions [45]. In Scheme 6.14 the reaction of the (S)-mandelic acid derived substrate **38** is depicted, which led with very good stereocontrol almost exclusively to product **39a**, with the other diastereoisomer **39b** being formed only in minor quantities (**39a**/**39b** = 96/4). Other acids, such as (S)-lactic acid, performed equally well. The chiral tether could be cleaved under basic conditions to afford enantiomerically pure cyclobutane lactones in good yields.

Attempts to achieve absolute stereocontrol by means of chiral sensitizers or chiral complexing agents [46] have seen little success with cyclohexenones and other unfunctionalized enone substrates. Ester **26**, for example, underwent an

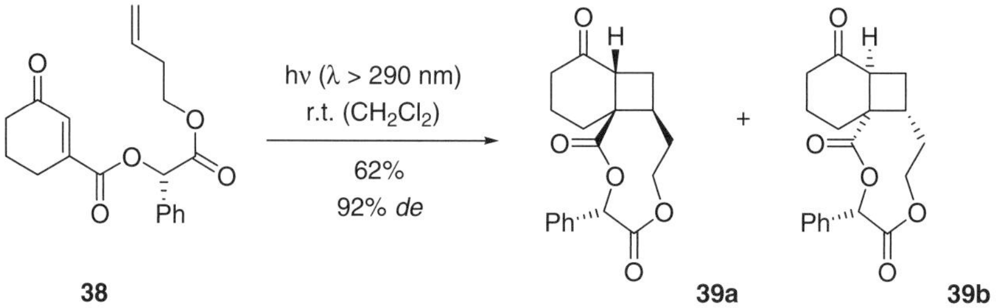

Scheme 6.14

intermolecular [2 + 2]-photocycloaddition to ethylene in the presence of γ-cyclodextrin (γ-CD) with low but detectable enantiomeric excess (2.2%) [47].

6.2.3
para-Quinones and Related Substrates

Depending on the nature of the lowest-lying triplet state ($n\pi^*$ versus $\pi\pi^*$), *para*-quinones undergo either typical carbonyl photochemistry (Paternò–Büchi reaction, hydrogen abstraction) or typical alkene photochemistry ([2 + 2]-photocycloaddition) [48]. The parent 1,4-benzoquinone ($n\pi^*$ triplet ca. 76 kJ mol^{-1} below $\pi\pi^*$ triplet) undergoes mainly spirooxetane formation, while the parent 1,4-naphthoquinone delivers products of both carbonyl and alkene photochemistry depending on the nature of the olefin. Olefins, which are donor-substituted, favor oxetane formation. Electron-donating substituents at the quinone destabilize the $n\pi^*$ triplet, leading to an increased preference for [2 + 2]-photocycloaddition. The relatively long wavelength absorption of *para*-quinones, many of which are colored, enables photochemistry even with visible light. The parent compound, 1,4-benzoquinone, exhibits a longest wavelength absorption at $\lambda_{max} = 458$ nm in hexane [49]. The synthetically undesired intermolecular [2 + 2]-photocycloaddition of substrate **40**, which was meant to – and eventually did – undergo an intramolecular Diels–Alder reaction, illustrates the ease with which *para*-quinone photocycloadditions can occur (Scheme 6.15) [50]. Cyclobutane **41** was formed as a single diastereoisomer in 80% yield upon heating substrate **40** in toluene (120 °C) at ordinary room light. The photochemical reaction course was proven upon irradiation with a visible light source at room temperature.

An intramolecular naphthoquinone [2 + 2]-photocycloaddition (Scheme 6.16) led directly to the formation of the natural product (−)-elecanacin (**43a**) [51]. Here,

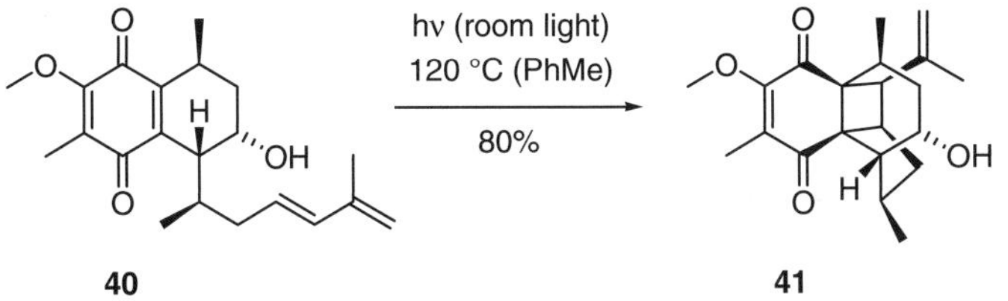

Scheme 6.15

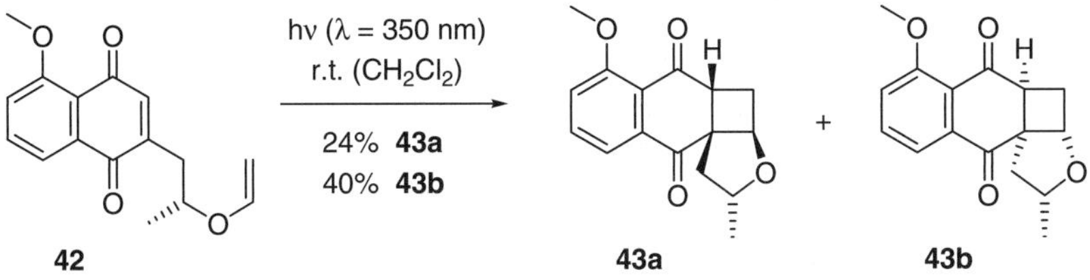

Scheme 6.16

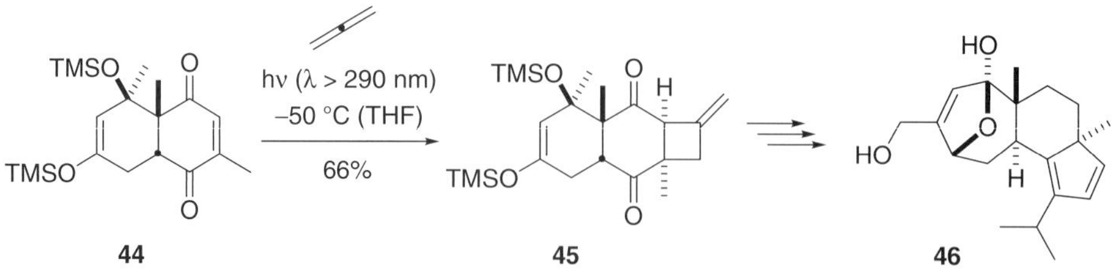

Scheme 6.17

enantiomerically pure substrate **42** was irradiated in CH_2Cl_2 solution yielding the desired natural product **43a** as the minor diastereoisomer, and its isomer **43b** as the major isomer. The excess for the latter product is presumably due to a slight preference of the methyl group to reside in the pseudoequatorial position of one of the chairlike transition states *en route* to the respective products.

The intermolecular [2 + 2]-photocycloaddition of *para*-tetrahydronaphthoquinones has been applied by Ward *et al.* to the synthesis of cyathin diterpenes [52]. An example is represented by the total synthesis of (±)-allocyathin B$_3$ (**46**), during the course of which the diastereoselective [2 + 2]-photocycloaddition of allene to substrate **44** served as one of the pivotal steps (Scheme 6.17) [53]. The addition delivered a mixture of regioisomers (r.r. = 80/20), from which compound **45** was separated. The facial diastereoselectivity was perfect due to the concave shape of the quinone.

Similarly, homobenzoquinones, in which one benzoquinone double bond is replaced by a cyclopropane, react with very high facial diastereoselectivity. An attack to the double bond occurred exclusively from the face opposite to the cyclopropane [54].

6.3
[2 + 2]-Photocycloaddition of Vinylogous Amides and Esters (Substrate Classes A2 and A3)

Heteroatoms in β-position lead to a slight bathochromic shift of the longest wavelength absorption (λ_{max}) as compared to cycloalkenones. This absorption band, to which an nπ* character is assigned, is often difficult to detect by UV/visible spectroscopy due to its low extinction coefficient ($\varepsilon \leq 100$). In addition, the intense ππ* absorption is shifted significantly to higher wavelengths for substrates **A2** and **A3** overlapping for strong donor heteroatoms with the weak nπ* band. The bathochromic shift generally does not require a change in the irradiation set-up as compared to enones. An adjustment to longer wavelengths can be made if light sources, which emit at a specific wavelength, are used.

From a preparative point of view the heteroatom in β-position has an influence because important latent cyclobutane cleavage pathways exist in the product. The prototypical reaction of this type is the [2 + 2]-photocycloaddition/retro-aldol reaction sequence (de Mayo reaction) [55–57], the course of which is illustrated for substrate **A3** (Q = O, PG = protecting group) in the reaction with ethylene as a generic olefin.

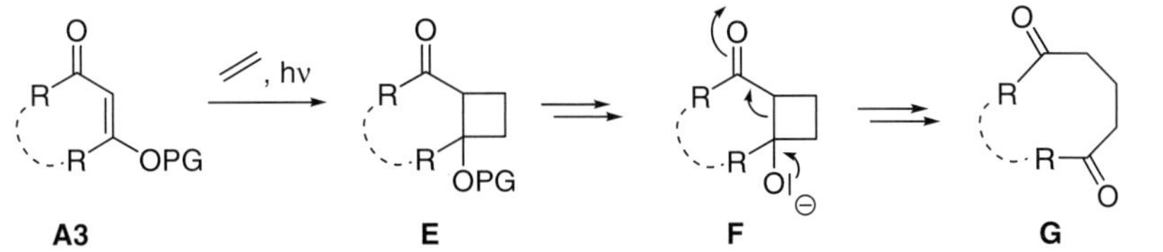

6.3 [2+2]-Photocycloaddition of Vinylogous Amides and Esters

Scheme 6.18

Upon [2+2]-photocycloaddition to product **E** and protecting group removal, the retro-aldol fragmentation can be initiated by base or acid treatment (Scheme 6.18). Under basic conditions, alkoxide **F** generates an enolate which is subsequently protonated to the 1,5-diketone **G**.

With Q = N, a similar fragmentation reaction is possible (retro-Mannich reaction), which leads to an imine or an iminium ion [57]. Fragmentations of this type have been frequently used for substrates of type **A3** but can also be found for substrate class **A2**, **A5**, and **A6** (*vide infra*).

6.3.1
Endocyclic Heteroatom Q in β-Position (Substrate Class A2)

Substrates of this compound class are classified as 4-oxa- (Q = O), 4-aza- (Q = N), and 4-thia-2-cycloalkenones (Q = S). They undergo both intra- and intermolecular [2+2]-photocycloaddition reactions smoothly. The regioselectivity of an intermolecular reaction is in favor of the HT product if an alkene is employed which is substituted by a donor substituent. The regioselectivity is often higher than the regioselectivity achieved in the reaction of the same alkene and a cycloalkenone. 4-Aza-2-cycloalkenones can only be used in [2+2]-photocycloaddition reactions if the nitrogen atom is appropriately substituted by an electron-withdrawing substituent. Otherwise, a single-electron transfer reaction precludes the photocycloaddition pathway [58].

6.3.1.1 4-Hetero-2-Cyclopentenones

3(2H)-Furanones (4-oxa-2-cyclopentenones) have been extensively explored as [2+2]-photocycloaddition substrates [59, 60]. A recent intermolecular example (Scheme 6.19) illustrates the application of their photocycloaddition chemistry to

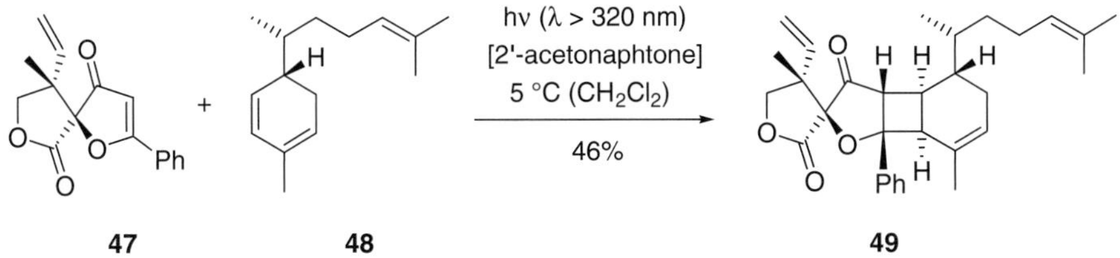

Scheme 6.19

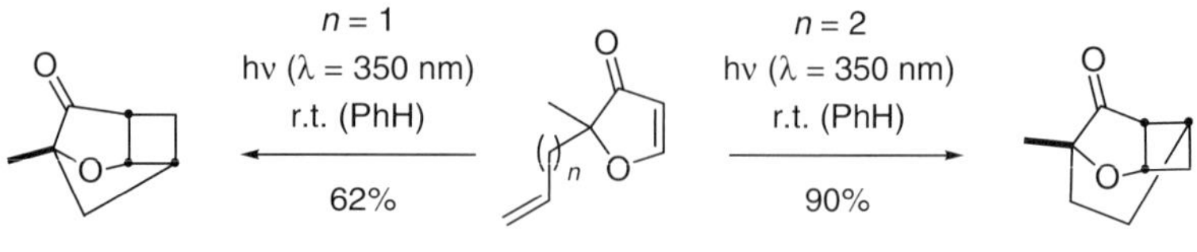

Scheme 6.20

natural product synthesis. The natural product biyouyanagin A (**49**) was obtained from enone **47** and terpene fragment **48** in a stereoselective fashion [61]. The spatial approach of the reaction partners is in line with a cyclic stereocontrol exerted by the ring substituents, leading to the depicted *cis-anti-cis* product. The regioselectivity and position selectivity may be understood based on the stability of the intermediate 1,4-biradical, but they are still remarkable given that four double bonds are available for attack with two different modes of addition each.

The 3(2*H*)-furanones are – as the other 4-hetero-2-cyclopentenones – normally 2,2-disubstituted to avoid enolization to the respective 3-hydroxyfuran. If one of the substituents is an alkenyl side chain, then intramolecular [2 + 2]-photocycloaddition reactions are possible with the regioselectivity being dependent on the chain length (Scheme 6.20). The allyl-substituted substrate **50** ($n=1$) gave predominantly the formal straight product **51** [62], while the butenyl-substituted substrate **50** ($n=2$) resulted in formation of the crossed product **52** [63].

The 4-aza- and 4-thia-2-cyclopentenone exhibit similar reactivity as their oxanalogues. As an example, the reaction with 2,3-dimethyl-2-butene is compared in Scheme 6.21. All three compounds **53a–53c** react in equally high yields, delivering the products **54** as single diastereoisomers [58, 64, 65]. By comparison of the reaction rates under identical conditions it was found that the azacompound **53b** reacts most rapidly, followed by the thiacompound **53c**. The observed relative rate constants in the reaction with 2,3-dimethyl-2-butene were 0.06: 0.11: 1.0 for **53a:53c:53b** [65].

The utility of dioxopyrrolines in [2 + 2]-photocycloaddition reactions was comprehensively demonstrated by Sano *et al.* [66]. Substrate **55**, for instance, underwent a clean reaction with 2-trimethylsilyloxy-butadiene to provide bicyclic HT product **56** as a single diastereoisomer (Scheme 6.22). In line with previous observations [67], the vinyl group was positioned *exo* relative to the five-membered ring, and the silyloxy

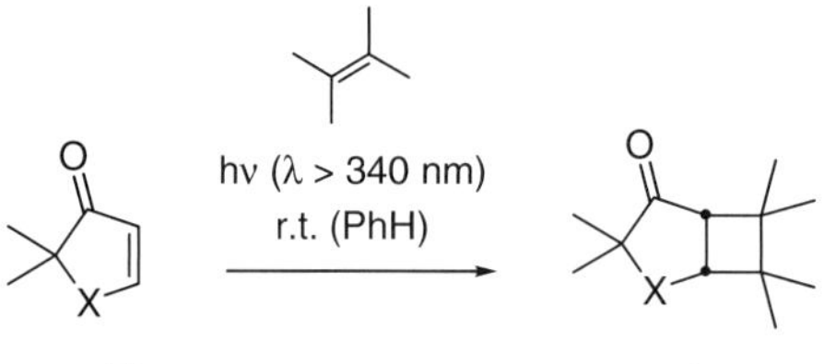

Scheme 6.21

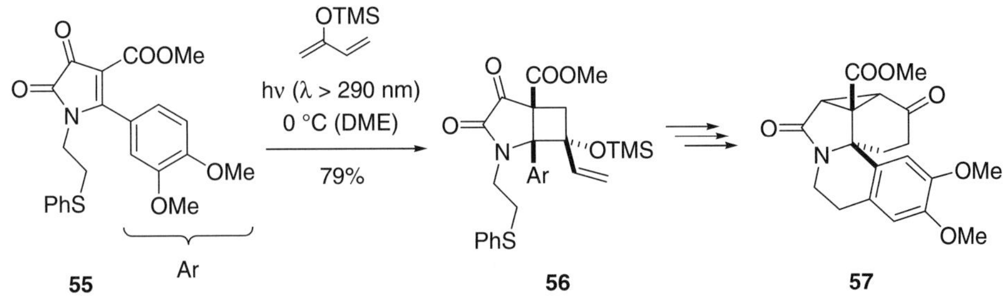

Scheme 6.22

group *endo*. A 1,3-anionic rearrangement was used for ring enlargement of the four- to a six-membered ring. Further elaboration resulted in the cycloerythrinan **57**, which is a versatile precursor for many erythrinan alkaloids, for example, (±)-erysotrine.

6.3.1.2 4-Hetero-2-Cyclohexenones

The photochemistry of 4-oxa-2-cyclohexenones is not significantly different from the corresponding oxacyclopentenones. The reaction outcome of intermolecular reactions is complicated by the fact that – as with the corresponding carbocyclic cyclohexenones – a *trans*-fusion of the cyclobutane to the six-membered ring is partially observed, depending on the nature of the alkene and the enone [59, 68]. The reaction of enone **58** with cyclopentene yielded exclusively the *cis-anti-cis* products **59** with low facial diastereocontrol exerted by the existing stereogenic center (Scheme 6.23). In an interesting consecutive reaction, both diastereoisomers **59** were converted into a single product **60** by an acid-catalyzed retro-Michael/retro-Aldol/Michael reaction sequence [69].

Benzannelated 4-oxa-2-cyclohexenones (chromones) react in a similar fashion to deliver the respective cyclobutanes in a *syn*-selective addition. The reaction has been recently applied to the formal synthesis of (±)-heliannuol D [70].

The utility of the [2 + 2]-photocycloaddition to 4-aza-2-cyclohexenones has been explored by Comins *et al.* in synthetic approaches to different alkaloids [71, 72]. As originally reported by Neier *et al.* the corresponding acyl- and alkoxycarbonyl-substituted 2,3-dihydropyridin-4(1*H*)-ones are particularly useful substrates [73]. In a recent study, the intramolecular reaction of dihydropyridone **61** was found to lead to the tricyclic product **62**, which was further converted into the quinolizidine **63**

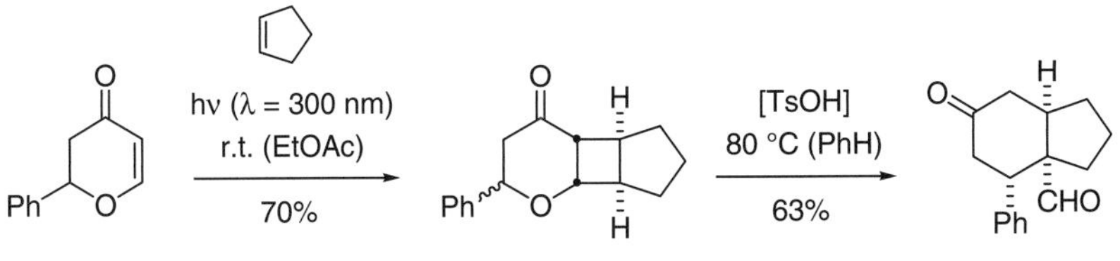

Scheme 6.23

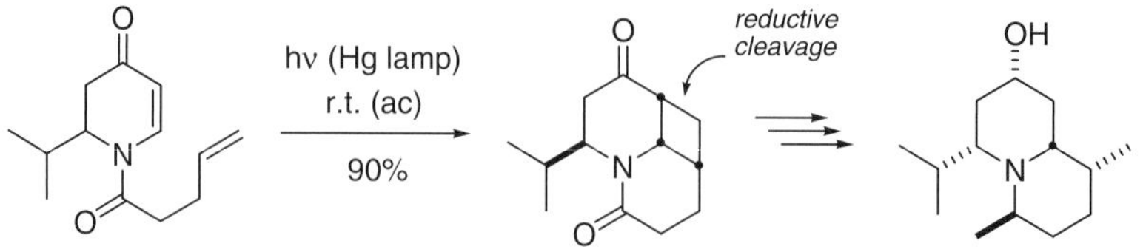

Scheme 6.24

(Scheme 6.24) [74]. The facial diastereoselectivity was controlled by the stereogenic center in the starting material. An adjustment of the relative configuration was performed by an oxidation/reduction sequence, and ring opening at the indicated bond was achieved by a SmI$_2$-induced reduction. It turned out that product **63** was not identical to the lupin alkaloid (+)-plumerinine, to which structure **63** was originally assigned.

The low triplet energy of 4-thia-2-cyclohexenones allows for [2 + 2]-photocycloaddition reactions of these substrates, even with electron-deficient enenitriles and with conjugated dienes, which serve commonly as triplet quenchers [75].

6.3.2
Exocyclic Heteroatom Q in β-Position (Substrate Class A3)

Substrates of class **A3** are formally derived from cyclic 1,3-diketones or the respective enols. The most commonly used substrates in [2 + 2]-photocycloaddition chemistry are the corresponding oxygen derivatives (Q = O), which are vinylogous esters or anhydrides. In general, intermolecular reactions of substrates **A3** are – as compared to the reactions of substrates **A2** – less efficient and lower yielding. The required excitation wavelength is similar to **A2**, and the reactions are commonly initiated by direct excitation but not by sensitization. As depicted in Scheme 6.18, the products of a [2 + 2]-photocycloaddition invite fragmentation reactions, which have been used extensively in synthesis. In a recent approach to the tricyclic skeleton of fusoxysporone (**66**), substrate **64** was converted into the tetracyclic product **65** (Scheme 6.25) which, upon saponification, underwent the desired retro-aldol reaction [76]. The facial diastereoselectivity of the reaction as induced by the stereogenic center in the chain can be understood based on 1,3-allylic strain arguments [77]. The relative configuration in the ring is dictated by the *cis*-fusion of the rings annelated to the central cyclobutane.

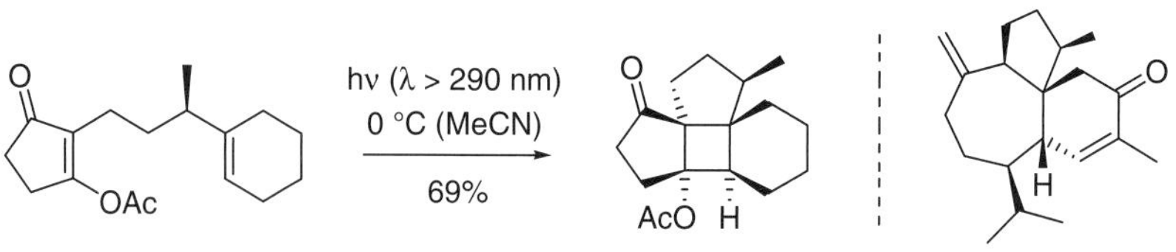

Scheme 6.25

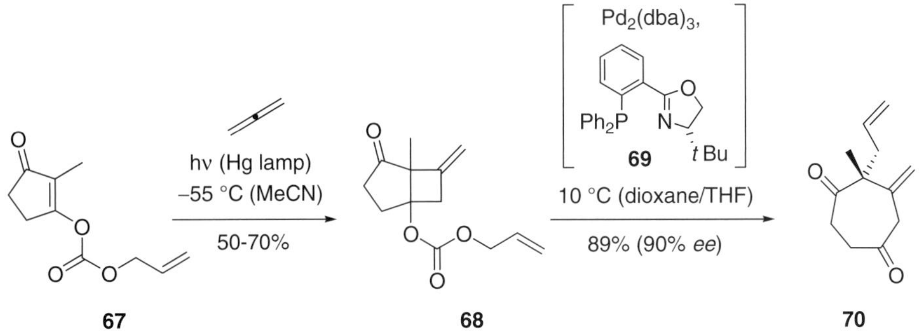

Scheme 6.26

A very useful extension of the de Mayo reaction has been recently introduced by Blechert et al. (Scheme 6.26) [78]. The retro-aldol fragmentation was combined with an intramolecular enantioselective allylation (asymmetric ring-expanding allylation) catalyzed by a chiral Pd complex. Bicycloheptane **68**, for example, was accessible by intermolecular [2 + 2]-photocycloaddition of cyclopentenone **67** with allene. Further transformation in the presence of $Pd_2(dba)_3$ (dba = dibenzylideneacetone) and the chiral oxazoline ligand **69** (tBu-phox) resulted in the enantioselective formation of cycloheptadione **70**.

Substrates **A3** (Q=O) have been employed not only as starting materials for fragmentation reactions but also to probe novel stereoselectivity concepts. The photochemical transformation of axial chirality into central chirality was achieved by Carreira et al., who employed chiral, enantiomerically pure allenes in intramolecular [2 + 2]-photocycloaddition reactions (Scheme 6.27) [79]. The reaction of enantiomerically pure (99% ee) cyclohexenone **71**, for example, yielded the two diastereomeric products **72a** and **72b**, which differed only in the double bond configuration. Apparently, the chiral control element directs the attack at the allene to its re face. The double bond isomerization is due to the known configurational liability of the vinyl radical formed as intermediate after the first C—C bond formation step (see Scheme 6.2, intermediate **C**).

In another approach to control the absolute configuration of enone photocycloaddition products, an intermediate iminium ion with a chiral secondary amine was employed by Mariano et al. (Scheme 6.28) [80]. Irradiation of substrate **73** at relatively short wavelength (direct $\pi\pi^*$ excitation) led, via intermediate **74**, to the chiral

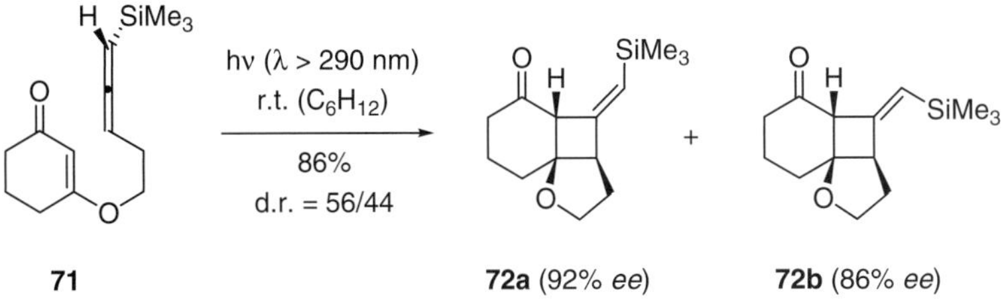

Scheme 6.27

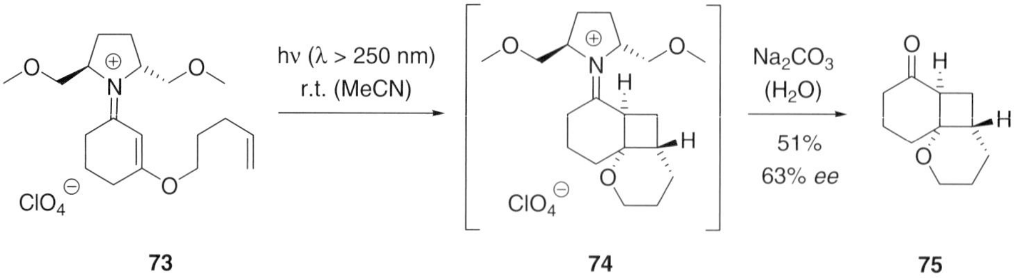

Scheme 6.28

cyclobutane **75**. The enantioselectivity was dependent on the conversion and on the reaction temperature. At 4 °C, 78% ee was achieved at a conversion of 60%, while at room temperature 82% ee was achieved after 40% conversion. The example depicted in Scheme 6.28 relates to a reaction, which was run until conversion was almost complete (90%).

In analogy to the vinylogous esters the corresponding amides of general structure **A3** (Q = N) can be employed in intramolecular [2 + 2]-photocycloaddition reactions. A recent study was concerned with the reactions of cyclic N-alkenoyl-β-enaminones, which underwent a regio- and diastereoselective intramolecular [2 + 2]-photocycloaddition [81]. The total synthesis of the alkaloid (−)-perhydrohistrionicotoxin (**78**) by Winkler et al. includes a classic application of enaminone photochemistry (Scheme 6.29) [82]. The vinylogous amide **76** was converted in a highly diastereoselective fashion into the tetracyclic product **77**. Since the reaction was performed by direct irradiation in a Pyrex vessel, it is likely that the enaminone chromophore is responsible for the reaction but not the dioxenone, which requires short-wavelength excitation or sensitization (*vide infra*). The facial diastereoselectivity originates from a pseudoequatorial orientation of the pentyl group in a chairlike transition state, leading to **77**. Cyclobutane ring opening was achieved by a retro-aldol reaction after transesterification of the dioxenone. Further elaboration eventually led to the target **78**.

Vinylogous thioesters (**A3**, Q = S) have been used less frequently in [2 + 2]-photocycloaddition as compared to their oxygen and nitrogen analogues [83]. More recent applications can be found in the above-mentioned study with chiral allenes [79].

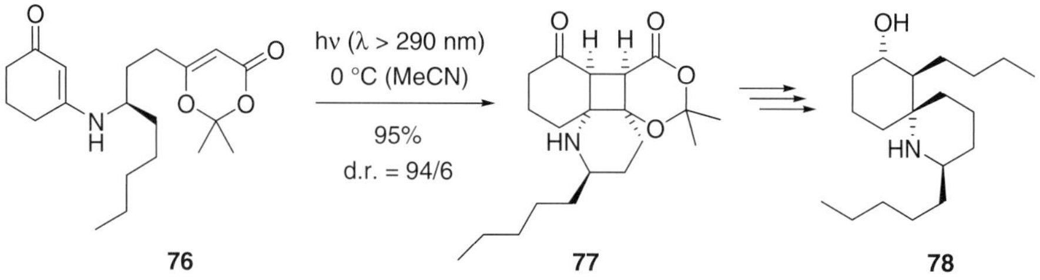

Scheme 6.29

6.4
[2 + 2]-Photocycloadditon of α,β-Unsaturated Carboxylic Acid Derivatives
(Substrate Classes A4, A5, and A6)

6.4.1
No Further Heteroatom Q in β-Position (Substrate Class A4)

The introduction of a heteroatom in α-position to the carbonyl group leads to a hypsochromic shift of the $n\pi^*$ transition. Esters show only a weak absorption at $\lambda = 200$ nm. Conjugation in α,β-unsaturated carboxylic acid derivatives leads to stronger absorption bands at $\lambda = 200$–250 nm, ascribed to the intense $\pi\pi^*$ and the weak $n\pi^*$ transition. In α,β-unsaturated lactams, absorptions in the near ultraviolet (200–220 nm) occur with high intensity ($\varepsilon \leq 10\,000$). An additional weaker band can be detected at $\lambda_{max} \cong 250$ nm. Most substrates belonging to the classes **A4–A6** are transparent at $\lambda \geq 280$ nm. As a consequence, the irradiation set-up needs some adjustment in comparison to enones, and for this there are two options. First, direct excitation can be achieved by the use of mercury lamps employing quartz vessels, which are transparent above 200 nm. Second, one can rely on the irradiation equipment used for enone photocycloaddition attempting a sensitized excitation. Indeed, most α,β-unsaturated carboxylic acid derivatives exhibit a relatively low triplet energy ($E_T \leq 330$ kJ mol^{-1}), so that appropriate ketones can be used for sensitization. Most conveniently, acetone ($E_T = 330$ kJ mol^{-1}) is used as a solvent. It is also possible, however, to use a more sophisticated sensitizer, for example, an acetophenone or benzophenone, in a transparent solvent. Sensitization results in a direct promotion of the substrate into the $\pi\pi^*$-triplet state, from which photocycloaddition chemistry can occur according to Scheme 6.2.

6.4.1.1 α,β-Unsaturated Lactones

The [2 + 2]-photocycloaddition chemistry of α,β-unsaturated lactones has been widely explored. The factors governing regio- and simple diastereoselectivity are similar to what has been discussed in enone photochemistry (substrate class **A1**, Section 6.2). The HT product is the predominant product in the reaction with electron-rich alkenes [84]. A stereogenic center in the γ-position of α,β-unsaturated γ-lactones (butenolides) can serve as a valuable control element to achieve facial diastereoselectivity [85, 86]. The selectivity is most pronounced if the lactone is substituted in the α- and/or β-position. The readily available chiral 2(5H)-furanones **79** and **82** have been successfully employed in natural product total syntheses (Scheme 6.30). In both cases, the intermediate photocycloaddition product with 1,2-dichloroethylene was reductively converted into a cyclobutene. In the first reaction sequence, the two-step procedure resulted diastereoselectively (d.r. = 88/12) in product **80**, which was separated from the minor diastereoisomer (9%). Direct excitation (Hg lamp, quartz) in acetonitrile solution was superior to sensitized irradiation (Hg lamp, Pyrex) in acetone, the former providing the photocycloaddition products in 89% yield, the latter in only 45%. Cyclobutene **80** was further converted into the monoterpenoid pheromone (+)-lineatin (**81**) [87]. In the second reaction

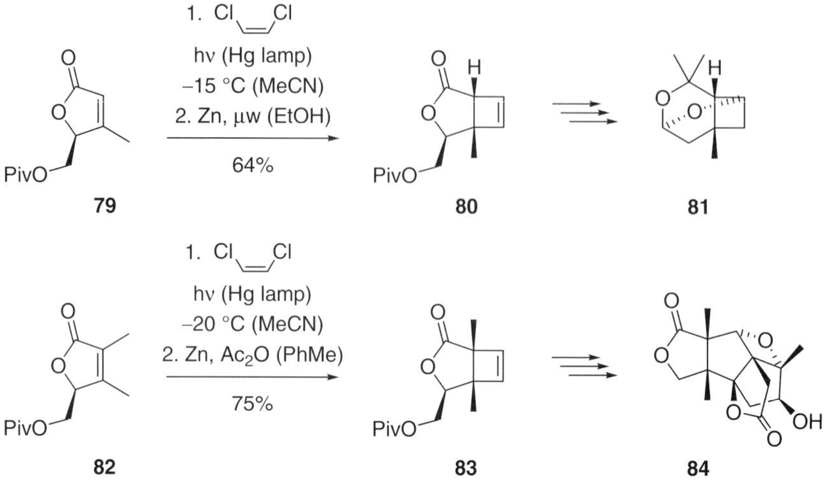

Scheme 6.30

sequence, the photocycloaddition proved under similar conditions to be equally diastereoselective (d.r. = 91/9), yielding after reduction the major diastereoisomer **83**. The two quaternary stereogenic centers established in the photochemical reactions were retained in the further synthetic route, which eventually led to the sesquiterpene (−)-merrilactone A (**84**) [88].

The crossed intramolecular [2 + 2]-photocycloaddition of allenes to α,β-unsaturated γ-lactones has been extensively studied by Hiemstra *et al.* in an approach to racemic solanoeclepin A (**87**). The sensitized irradiation of butenolide **85** in a 9 : 1 mixture of benzene and acetone, for example, led selectively to the strained photocycloadduct **86** (Scheme 6.31) [89]. The facial diastereoselectivity is determined by the stereogenic center, to which the allene is attached. The carbon atom in α-position to the carbonyl carbon atom is attacked from its *re* face, forming a bond to the tertiary allene carbon atom, while the β-carbon atom is being connected to the internal allene carbon atom by a *si* face attack. The method allows facial diastereocontrol over three contiguous stereogenic centers in the bicyclo[2.1.1]heptane part of the natural product.

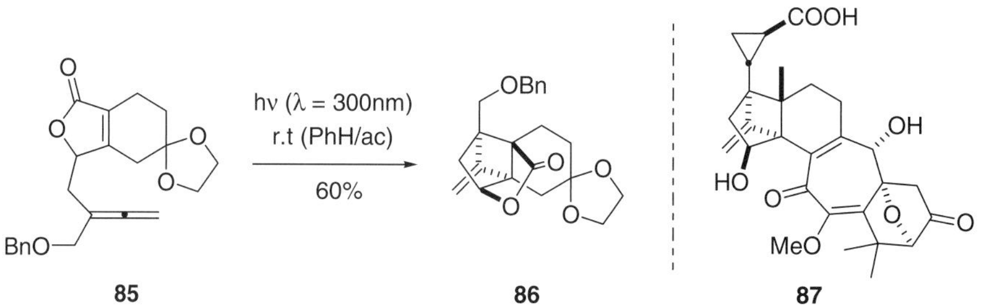

Scheme 6.31

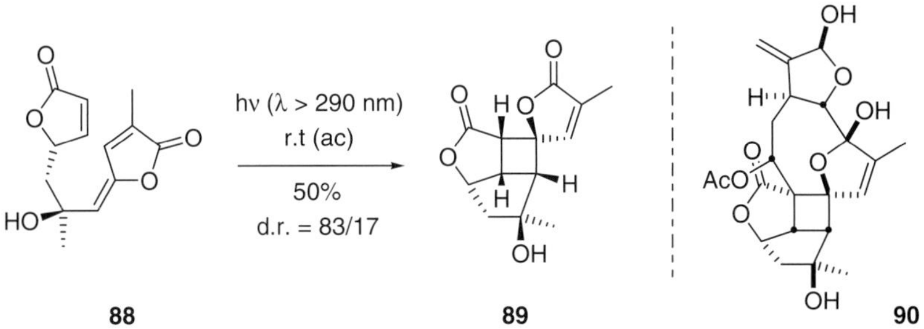

6.4 [2 + 2]-Photocycloadditon of α,β-Unsaturated Carboxylic Acid Derivatives

Scheme 6.32

Another natural product, which contains a cyclobutane ring, is the hexacyclic diterpene bielschowskysin (**90**), the absolute configuration of which has not yet been determined. The retrosynthetic disconnection into a [2 + 2]-photocycloaddition precursor is easily recognized in its structure and appears to be possible in a forward direction (Scheme 6.32) [90]. At least the conversion of butenolide **88** into tetracyclic product **89** was possible. The configuration of the exocyclic double bond was not retained in the [2 + 2]-photocycloaddition because an isomerization was observed during irradiation. In addition, since the reaction is likely to proceed via 1,4-diradical intermediates, it is not expected to be stereospecific. Remarkably, diastereoisomer **89** was the predominant isomer with the other diastereomer, which was formed in minor amounts, being an epimer of **89** with opposite configuration at the spiro center. In close analogy to previously mentioned examples (Schemes 6.7, 6.20 and 6.31) the facial diastereoselectivity is controlled by the endocyclic stereogenic center adjacent to the double bond.

The photocycloaddition chemistry of α,β-unsaturated δ-lactones is similar to the chemistry of γ-lactones. Complications arise as with cyclohexenones because *anti*-addition to the α,β-unsaturated double bond can occur, particularly in the intermolecular addition mode. Even if one product prevails, intermolecular [2 + 2]-photocycloaddition reactions are often sluggish. Despite the fact that alkene **92**, for example, was employed in a twofold excess relative to dihydropyranone **91**, the reaction delivered only 32% of the desired product **93** (43% based on recovered starting material; Scheme 6.33). The relative product configuration, which was established by X-ray crystallography, came as a surprise because the lactone apparently

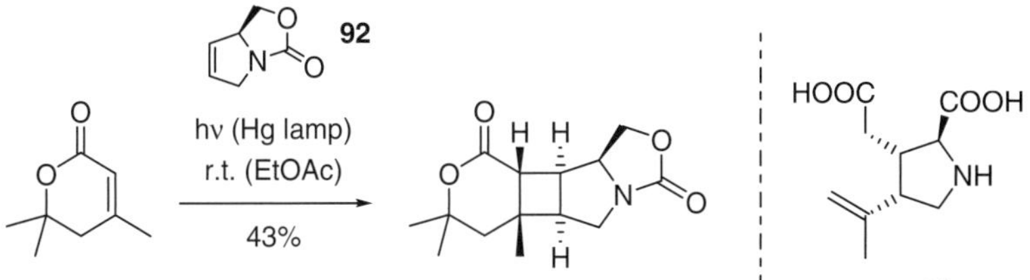

Scheme 6.33

adds onto the sterically hindered concave face of the proline-derived dihydropyrrole **92**. The relative configuration within the doubly anellated cyclobutane was expectedly *cis-anti-cis*. The experiments were conducted in connection with a study towards a total synthesis of (−)-kainic acid (**94**) [91].

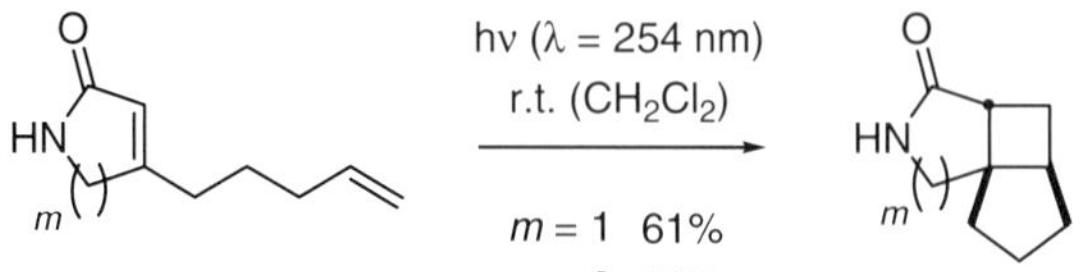

Scheme 6.34

6.4.1.2 α,β-Unsaturated Lactams

Lactams of type **A4** have been successfully employed in intramolecular [2 + 2]-photocycloaddition reactions. Direct excitation has been found most useful to achieve sufficient conversion in these reactions. The reaction of β-substituted α,β-unsaturated lactams **95** in CH_2Cl_2 gave the expected products **96** of five-membered ring closure in decent yields (Scheme 6.34) [92].

A stereogenic center in γ-position provides good diastereofacial control in γ-lactams, with the regioselectivity of the photocycloaddition being dependent on the chain length of the tether. The *N-tert*-butoxycarbonyl(Boc)-protected lactams **97** were subjected to an intramolecular [2 + 2]-photocycloaddition in acetonitrile as the solvent (Scheme 6.35). The substrate with the allyl side chain ($n = 1$) formed predominantly the crossed photocycloaddition product **98** (52%), but minor quantities of the straight photocycloaddition product **99** were also observed. In line with the rule of five, the situation was reversed in the butenyl-substituted substrate ($n = 2$), with **99** prevailing over **98**. The pentenyl-substituted substrate **97** ($n = 3$) yielded a single product **99**, presumably via an initial six-membered ring formation, followed by subsequent ring closure of the resulting 1,4-biradical to the cyclobutane [93].

Pyridones can react photochemically along several reaction channels [94]. Besides [4 + 4]-photodimerization and [4π]-ring closure, [2 + 2]-photocycloaddition reactions are possible in an α,β- or in a γ,δ-mode relative to the carbonyl carbon atom. With regard to the former reaction pathway, the [2 + 2]-photocycloaddition of olefins to 4-alkoxypyridones appears to be synthetically most useful (*vide infra*).

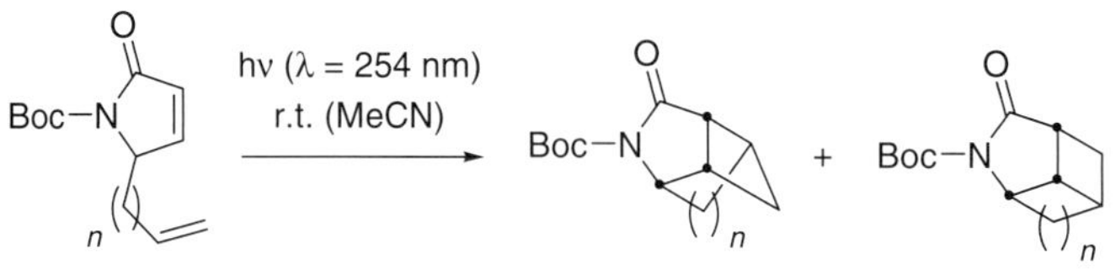

	98	99
$n = 1$	52%	10%
$n = 2$	8%	62%
$n = 3$	–	61%

Scheme 6.35

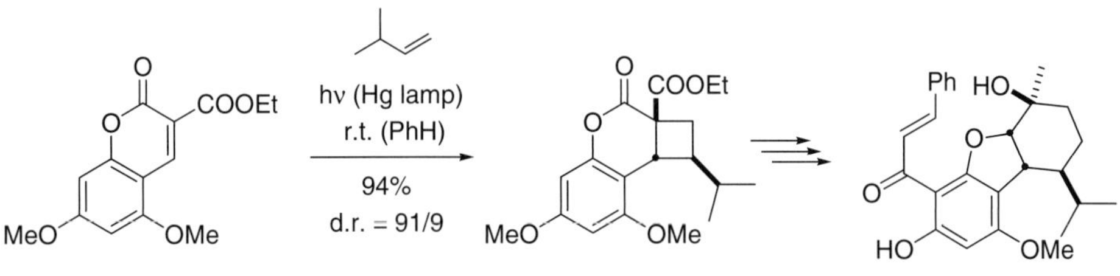

Scheme 6.36

6.4.1.3 Coumarins

The direct irradiation of the parent coumarin in the presence of alkenes results only in an inefficient photodimerization and [2 + 2]-photocycloaddition. Lewis acid coordination appears to increase the singlet state lifetime, and leads to improved yields in the stereospecific [2 + 2]-photocycloaddition [95]. Alternatively, triplet sensitization can be employed to facilitate a [2 + 2]-photocycloaddition. Yields of intramolecular [2 + 2]-photocycloadditions remain, however, even with electron-rich alkenes in the medium range at best. The preference for HT addition and for formation of the *exo*-product is in line with mechanistic considerations discussed earlier for other triplet [2 + 2]-photocycloadditions [96, 97]. Substituted coumarins were found to react more efficiently than the parent compound, even under conditions of direct irradiation. 3-Substituted coumarins, for example, 3-methoxycarbonylcoumarin [98], are most useful and have been exploited extensively. The reaction of 3-ethoxycarbonylcoumarin (**100**) with 3-methyl-1-butene yielded cleanly the cyclobutane **101** (Scheme 6.36) with a pronounced preference for the *exo*-product (d.r. = 91/9). Product **101** underwent a ring-opening/ring-closure sequence upon treatment with dimethylsulfoxonium methylide to generate a tetrahydrodibenzofuran, which was further converted into the natural product (±)-linderol A (**102**) [99].

Coumarin photochemistry has been recently employed to demonstrate that a frozen axial chirality can be used to induce the absolute configuration of stereogenic centers. Coumarin **103** was obtained as a single atropisomer by spontaneous crystallization (Scheme 6.37). Upon warming powdered crystals of **103** in MeOH to −20 °C, sensitized [2 + 2]-photocycloaddition to ethyl vinyl ether gave the almost enantiomerically pure products **104**. The approach to the coumarin double bond occurred preferentially from the less-shielded face to which the amide carbonyl group

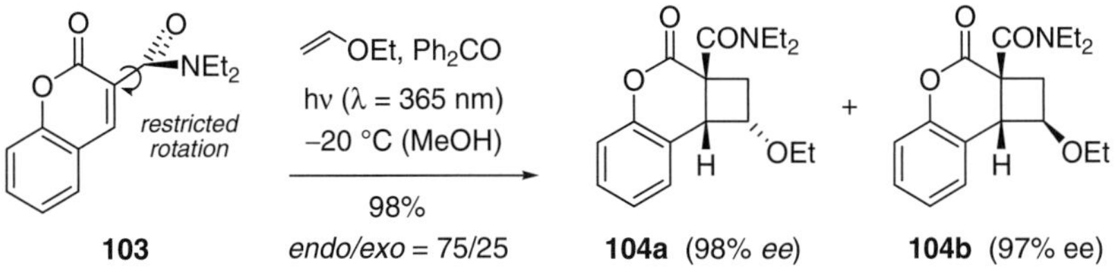

Scheme 6.37

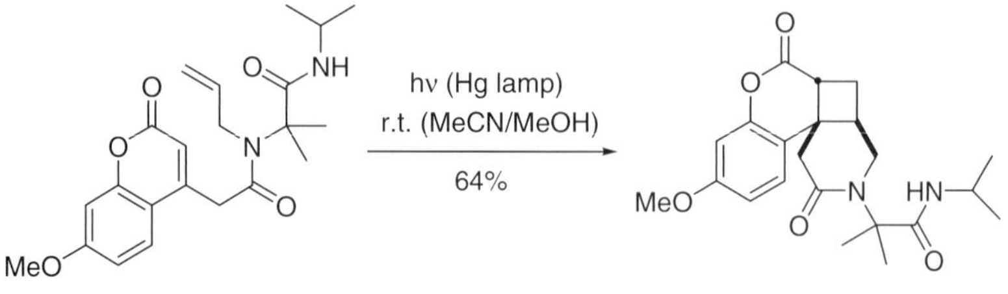

Scheme 6.38

points. Desired homochiral crystals for this process could be selectively prepared by the addition of a corresponding seed crystal during the crystallization process [100].

Intramolecular [2 + 2]-photocycloadditions of coumarins proceed with regard to regio- and stereoselectivity along the guidelines previously discussed for other α,β-unsaturated carbonyl compounds. In a recent example, the products of an Ugi four-component reaction, such as amide **105**, were converted photochemically into uniquely shaped 3-azabicyclo[4.2.0]octan-4-ones, such as cyclobutane **106** (Scheme 6.38) [101].

6.4.1.4 Quinolones

Due to a very efficient singlet- to triplet-state intersystem crossing, the [2 + 2]-photocycloaddition chemistry of 2-quinolones can be initiated easily by direct excitation at 300–350 nm [102]. The addition of a sensitizer is not required. The parent compound has been first employed in a [2 + 2]-photocycloaddition as early as 1968 [103]. With regard to regio- and stereoselectivity, 2-quinolone (**107**) behaves as expected, exhibiting a preference for HT product formation with electron-rich olefins, such as 1,1-dimethoxyethene (Scheme 6.39, DMA = N,N-dimethylacetamide). The highly efficient reaction delivers product **108** quantitatively [104]. The preference for

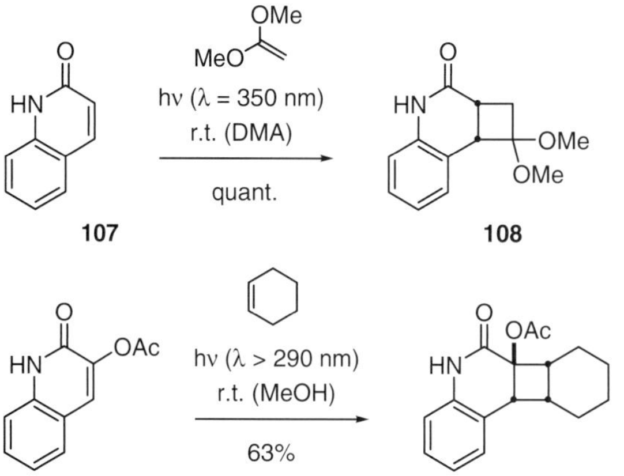

Scheme 6.39

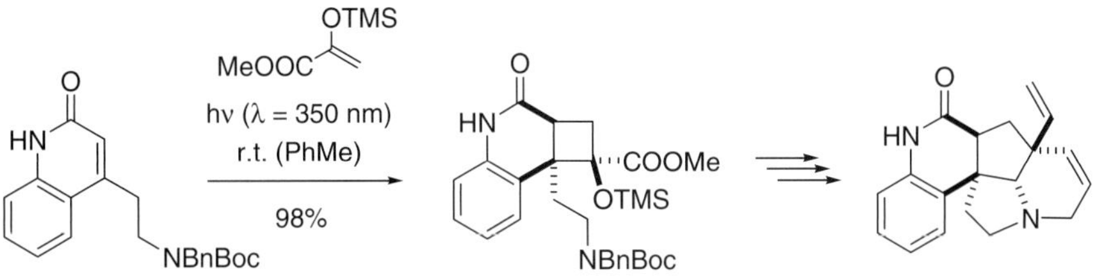

6.4 [2 + 2]-Photocycloadditon of α,β-Unsaturated Carboxylic Acid Derivatives

Scheme 6.40

exo-product formation, as observed for example with a d.r. of 75/24 for the reaction of **107** and cyclopentene, and as typically observed in enone photochemistry, is reversed for 3-acyloxyquinolones such as 3-acetoxyquinolone (**109**). Reaction with cyclopentene resulted in a product mixture, in which the *endo*-product (*cis-syn-cis*) prevailed (d.r. = 70/30), and addition to cyclohexene resulted almost exclusively in the *cis-syn-cis* product **110** [105]. Possibly, hydrogen bonds to the protic solvent methanol enlarge the size of the acetoxy group disfavoring the formation of the *exo*-product.

4-Alkylquinolones react with acrylates regioselectively to deliver the formal HT products [106]. According to the 1,4-biradical model (Scheme 6.2), radical stabilization in the benzylic position overrides other electronic effects. A recent application of quinolone photocycloaddition chemistry to natural product synthesis is depicted in Scheme 6.40. The silylenolether derived from methyl pyruvate underwent a clean addition reaction to the N,N-diprotected 4-(2'-aminoethyl)-substituted quinolone **111**. The stereoselective and high-yielding reaction delivered a single product with the larger substituents (aryl/COOMe) *trans*-positioned. Upon treatment with K_2CO_3 in MeOH, product **112** underwent a ring enlargement to a five-membered ring (retro-benzilic acid rearrangement) under retention of configuration at the stereogenic centers in α- and β-position. Subsequent key steps *en route* to the formation of the pentacyclic monoterpenoid indole alkaloid (±)-meloscine (**113**) were a reductive amination, a Claisen rearrangement, and a ring-closing metathesis [107].

Recent interest in the use of N-unsubstituted 2-quinolones stems from the fact, that they coordinate effectively to chiral lactam-based templates via two hydrogen bonds. The prototypical template to be used in photochemical reactions is compound **115**, which can be readily prepared from Kemp's triacid [108]. The template is transparent at a wavelength $\lambda \geq 290$ nm, and can be nicely used in stoichiometric amounts for enantioselective photochemical and radical reactions [109]. Conditions which favor hydrogen bonding (nonpolar solvent, low temperature) are required to achieve an efficient association of a given substrate. The intramolecular [2 + 2]-photocycloaddition of 4-alkylquinolone **114** proceeded in the presence of **115** with excellent enantioselectivity, and delivered product **116** as the exclusive stereoisomer (Scheme 6.41) [110]. Application of the enantiomer *ent*-**115** of complexing agent **115** to the reaction **111** → **112** depicted in Scheme 6.40 enabled enantioselective access to (+)-meloscine [111].

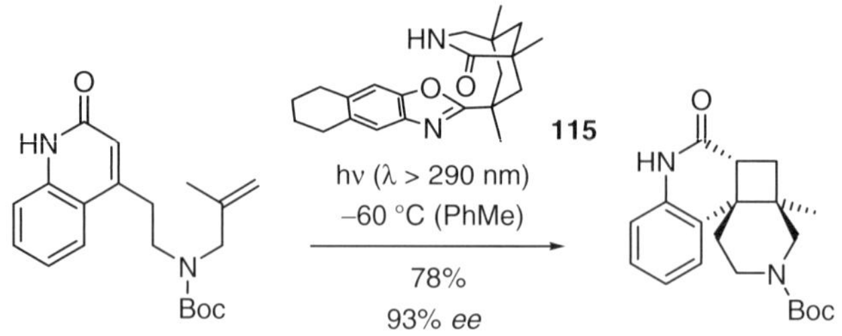

Scheme 6.41

6.4.1.5 Maleic Anhydride and Derivatives

Maleic anhydride (**117**) belongs to the first compounds, the [2 + 2]-photocycloaddition reactions of which were extensively explored [112]. It is preferably converted to the corresponding cyclobutanes by irradiation in the presence of a sensitizer, for example, benzophenone, allowing the addition of a plethora of alkenes (Scheme 6.42). In a recent application the photocycloaddition product **118** of maleic anhydride and 1,4-dichloro-2-butene was converted into the marine alkaloid (±)-sceptrin (**119**) [113].

Substituted maleic anhydrides have been directly excited, but sensitization may also be used. In some cases the first method is better, and in some cases the second. In an approach to merrilactone A, which is closely related to the earlier-mentioned synthesis (Scheme 6.30), 2,3-dimethylmaleic anhydride was employed as a starting material in a sensitized [2 + 2]-photocycloaddition to 1,2-dichlorethene [114]. The reaction of tetrahydrophthalic anhydride (**120**) with alkenols and alkynols was conducted by direct irradiation in a Pyrex vessel. As an example, the reaction with allyl alcohol is depicted. The *exo*-product **121** was the preferred product with the *endo*-product cyclizing spontaneously to lactone **122** (Scheme 6.43) [115]. Other alkenols reacted similarly.

In the same study, maleimides were irradiated in the presence of allyl alcohol and allyl ethyl ether, yielding the respective cyclobutanes with significant *exo*-preference [115]. Diastereofacial stereocontrol was achieved in the [2 + 2]-photocycloaddition of tetrahydrophthalimide by a chiral tether. The valinol-derived sub-

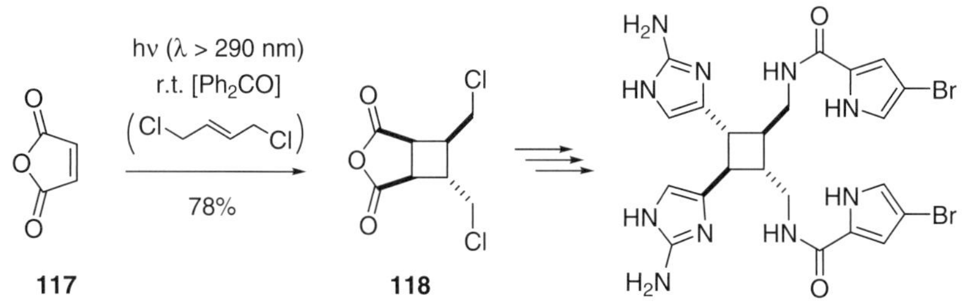

Scheme 6.42

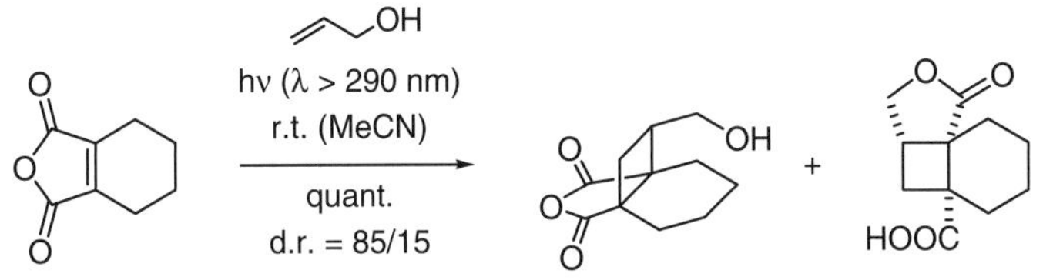

6.4 [2 + 2]-Photocycloadditon of α,β-Unsaturated Carboxylic Acid Derivatives

Scheme 6.43

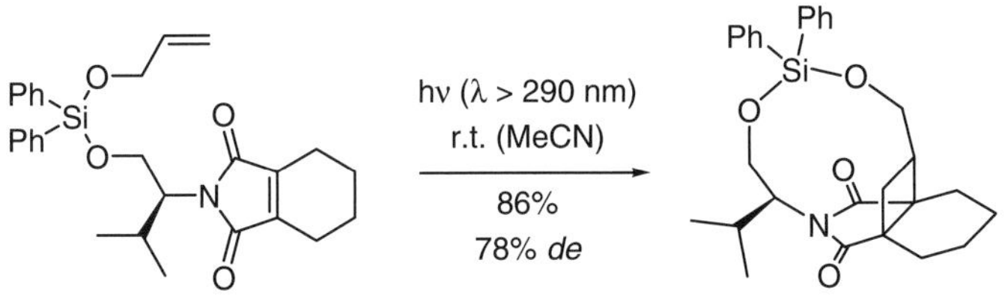

Scheme 6.44

strate **123** was transformed diastereoselectively into the product **124** of an intramolecular [2 + 2]-photocycloaddition (Scheme 6.44) [116].

Enantioselectivity control in a [2 + 2]-photocycloaddition reaction was achieved in a chiral, self-assembled host. Fluoranthenes and N-cyclohexylmaleimide underwent an intramolecular reaction in a cage made of M_6L_4, with the metal M being palladium (II) coordinated to a chiral diamine, and the ligand L being 2,4,6-tris(4′-pyridyl)-1,3,5-triazine. Up to 50% ee was observed [117].

6.4.1.6 Sulfur Compounds

Sulfur analogues of the compounds discussed previously in Sections 4.1.1–4.1.5 have been employed with some success in [2 + 2]-photocycloaddition chemistry. The analogous nonaromatic lactones, such as **125** [118] and **126** [119], have found little use due to the fact that the yields achieved in their cycloaddition reactions remained low (Figure 6.2). Thiophen-2(5H)-one **126** and its 5-substituted derivatives delivered with 2,3-dimethyl-2-butene under direct irradiation conditions (λ = 350 nm in cyclohexane) the corresponding cycloaddition products in yields of only 10–15%.

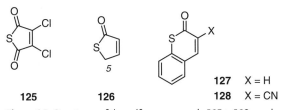

Figure 6.2 Structures of the sulfur compounds **125** to **128** employed in [2 + 2]-photocycloaddition reactions.

Apparently, other photochemical reactions, which occur in the singlet manifold, are faster than intersystem crossing and compete effectively with the [2 + 2]-photocycloaddition [120].

1-Thiocoumarin (**127**) underwent [2 + 2]-photocycloaddition reactions in better yields than **126**. In contrast to coumarin, *cis*- and *trans*-fused products are being found, however, for example, in the reaction with 2,3-dimethyl-2-butene, possibly because the thiopyran ring is more flexible than the pyran ring due to the longer C—S bonds. HT product is favored with electron donor-substituted olefins [121]. Electron acceptor substitution in 3-position, as in 3-cyano-1-thiocoumarin (**128**), leads to an improved performance in [2 + 2]-photocycloaddition reactions [122].

6.4.2
Endocyclic Heteroatom Q in β-Position (Substrate Class A5)

There are two major substrate classes that fall into category **A5**, that is, 1,3-dioxin-4-ones (dioxenones) and 4-pyrimidinones, which will be treated in individual subsections. The latter substrate class includes also the nucleobases uracil and thymine, which are 2,4-pyrimidindiones. Sensitization by carbonyl compounds, for example, acetone, is generally employed to promote the substrates into the excited triplet state, which has $\pi\pi^*$ character and which is responsible for [2 + 2]-photocycloaddition chemistry. Alternatively, direct excitation at relatively short wavelength ($\lambda \leq 300$ nm) may be applicable.

6.4.2.1 1,3-Dioxin-4-ones

1,3-Dioxin-4-ones represent ideal surrogates for β-ketocarboxylic acids, and have been used extensively in photochemistry and in conventional synthetic organic chemistry. Their acetal-type carbon atom at C-2 offers not only the option of easy hydrolysis after a [2 + 2]-photocycloaddition, but also lends itself as a stereogenic center, with the aid of which diastereofacial control can be achieved. It must be noted, however, that the face selectivity achieved in the photocycloaddition of 1,3-dioxin-4-ones is opposite to the selectivity observed in the ground state [123]. An addition occurs from the apparently more shielded face (Scheme 6.45). As an example, the menthone-derived dioxinone **129** underwent [2 + 2]-photocycloaddition to cyclopentene, yielding with significant preference product **130**, which results from

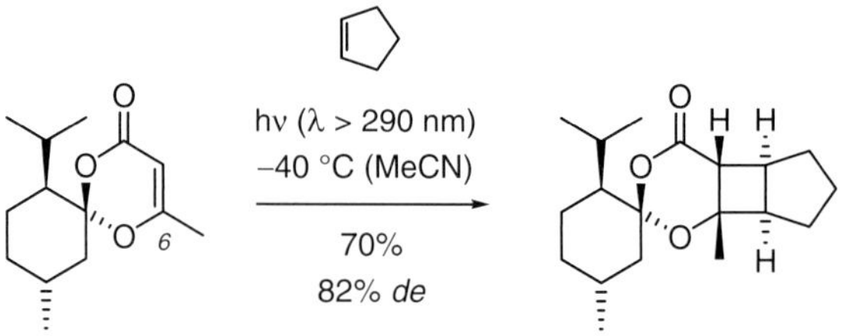

Scheme 6.45

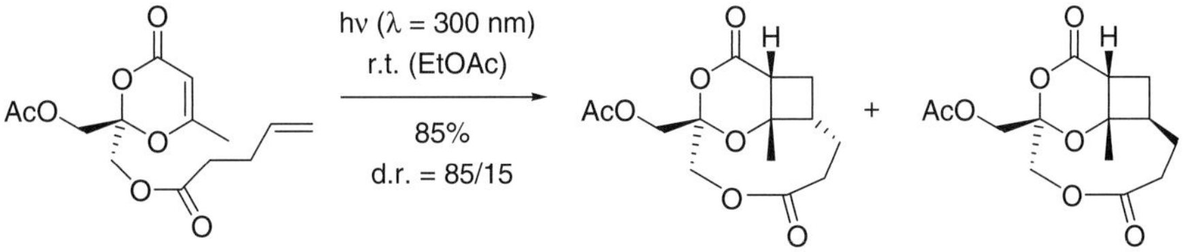

6.4 [2 + 2]-Photocycloadditon of α,β-Unsaturated Carboxylic Acid Derivatives | 199

Scheme 6.46

an approach of the cyclopentene to the face, which seems to be shielded by the *iso*-propyl-substituted part of the spiro-cyclohexane [124].

Explanations for the outcome of the photocycloaddition reactions have been proposed. Seebach *et al.* suggested that carbon atom C-6 is pyramidalized in the excited triplet state in the opposite direction as compared to the ground state [125]. Sato proposed that different dioxinone conformations are responsible which vary depending on the reaction type [126].

If the stereogenic center carries a tethered alkene, the diastereoface selectivity is dictated by the direction of the tether. Diene **131** underwent a diastereoselective [2 + 2]-photocycloaddition to the products **132** of a bottom approach relative to the six-membered dioxenone ring (Scheme 6.46). The simple diastereoselectivity was not perfect, however. The *endo*-product **132a** was preferred over the *exo*-product **132b** [127].

The deMayo-type photochemistry of 1,3-dioxin-4-ones has been beautifully applied by Winkler *et al.* to the synthesis of complex natural products. Substrate **133** gave under sensitized irradiation (with acetone as cosolvent) product **134** as single diastereoisomer (Scheme 6.47). The diastereoselectivity results from cyclic stereo-control exerted by the two stereogenic centers in the spiro-bis-lactone part of the starting material. After installation of the furan, saponification and bond scission in a retro-aldol fashion generated a keto carboxylic acid, which produced the natural product (±)-saudin (**135**) by simultaneous formation of two acetal groups [128].

The intramolecular [2 + 2]-photocycloaddition of the tricyclic substrate **136**, which was employed as an epimeric mixture relative to the chlorine-bearing stereogenic center, resulted in the cyclobutane **137**, which is derived from diastereoisomer **136a** (Scheme 6.48). The reaction established the intrabridgehead configuration at the stereogenic centers C-8 and C-10 in the target compound (±)-ingenol (**139**) [129].

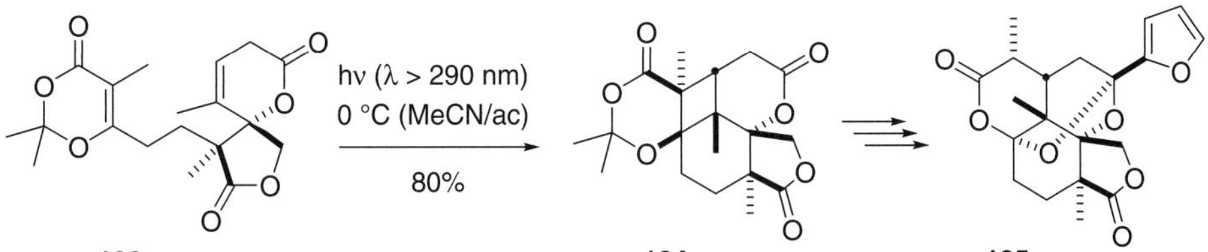

Scheme 6.47

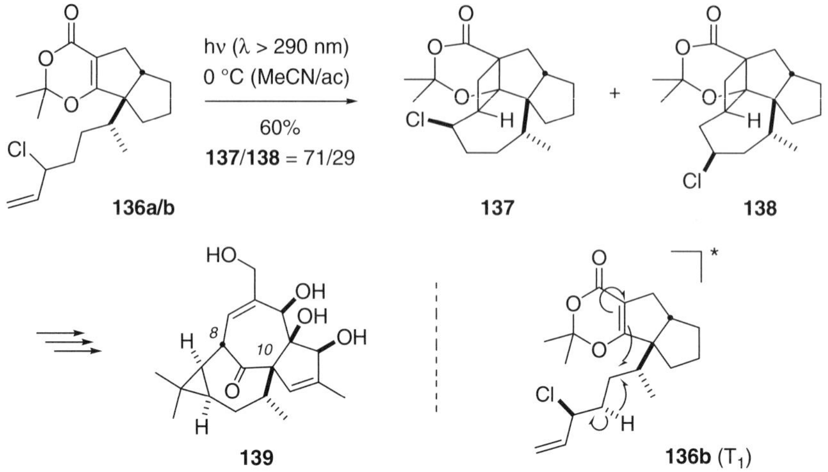

Scheme 6.48

The facial diastereoselectivity can be explained by the rigid conformation of the starting material, which enables attack of the alkene to the chromophore only from one direction. The chain length of the alkenyl group is responsible for the desired *endo*-addition mode. The unexpected formation of product **138** was explained by a hydrogen abstraction in the excited state of diastereoisomer **136b**, which is for steric reasons not suitable for a direct photocycloaddition. The radical adjacent to the C–Cl bond is stabilized by chlorine in a bridged chloronium intermediate which can, after hydrogen transfer to the allylic position, undergo an intramolecular [2 + 2]-photocycloaddition [130].

In other recent examples, 1,3-dioxin-4-ones have been employed in synthetic efforts to prepare solanoeclepin A and kainic acid, modifications of which were previously mentioned (Schemes 6.31 and 6.33) [89, 91].

6.4.2.2 4-Pyrimidinones

The intramolecular photodimerization and [2 + 2]-photocycloaddition in DNA involves thymine or cytosine as the chromophore. This chemistry has been intensively investigated with regards to DNA damage and repair [131]. Despite the fact that the area is of continuous interest [132], the synthetic applications are limited and are not covered here in detail. However, some preparative aspects of 4-pyrimidinone photocycloaddition chemistry will be addressed. Aitken *et al.* have prepared a plethora of constrained cyclobutane β-amino acids by intra- or intermolecular [2 + 2]-photocycloaddition to uracil and its derivatives [133, 134]. In a chiral adaptation of this method, the uracil-derived enone **140** was employed to prepare the diastereomeric cyclobutanes **141** in very good yield (Scheme 6.49). The compounds are easily separated and were – despite the relatively low auxiliary-induced diastereoselectivity – well suited to prepare the *cis*-2-aminocyclobutanecarboxylic acids **142** in enantiomerically pure form. Enantioselective access to the corresponding *trans*-products was feasible by epimerization in α-position to the carboxyl group [135].

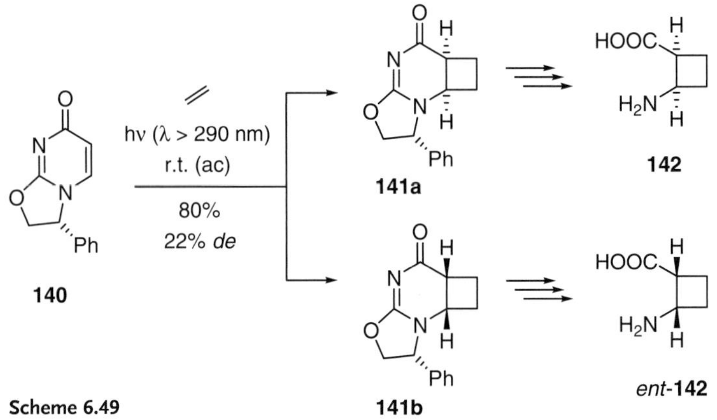

6.4 [2 + 2]-Photocycloadditon of α,β-Unsaturated Carboxylic Acid Derivatives

Scheme 6.49

In an approach to a dihydrooritidine analogue, the intermolecular [2 + 2]-photocycloaddition of 2′,3′-O-isopropylideneuridine to chiral and achiral acrylates was found to be unsatisfactory both with respect to regio- and diastereoselectivity. The intramolecular approach was more successful, and uridine **143** produced selectively the single diastereomerically pure product **144**. Due to concurrent photodimerization and polymerization reactions, however, the yield was only moderate (Scheme 6.50) [136].

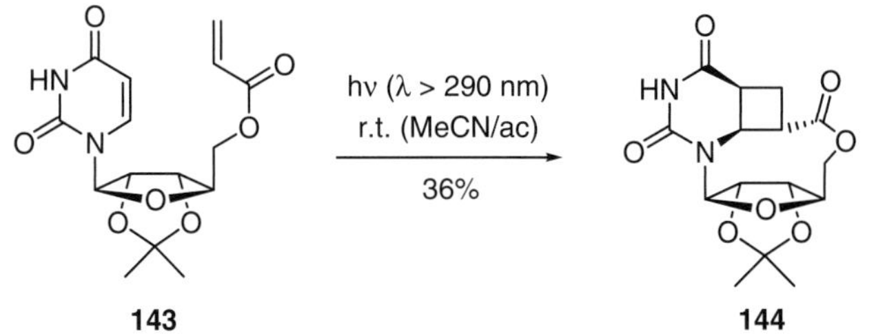

Scheme 6.50

6.4.3
Exocyclic Heteroatom Q in β-Position (Substrate Class A6)

The photocycloaddition chemistry of compounds A6 occurs under similar conditions as the parent compounds A4. Either direct irradiation at low wavelength ($\lambda \cong 254$ nm) or sensitizing conditions are used. Although the additional heteroatom facilitates ring-opening reactions, it can however also be used simply as further starting point for heteroatom–carbon bond formation. In order to avoid a lengthy discussion of the aspects previously addressed in Section 4.1. the present section is divided into two subsections only – one dealing with β-substituted α,β-unsaturated lactones (including pyrones and coumarins), and the other with α,β-unsaturated lactams (including pyridones and quinolones).

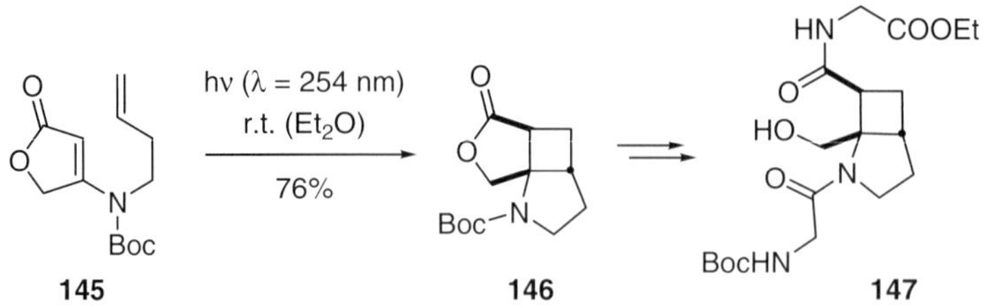

Scheme 6.51

6.4.3.1 Lactones

The starting material for the synthesis of five-membered ring compounds of type **A6** is tetronic acid. The photochemistry of tetronic acid esters and amides has been investigated intensively in recent years, and photocycloaddition reactions have been shown to proceed best under conditions of direct irradiation [137, 138]. Although intermolecular reactions are possible, better yields were obtained if the reactions were conducted intramolecularly, irrespective of whether the alkenyl chain was attached to the α-, β-, or γ-position of the tetronate [139]. In Scheme 6.51, the synthesis of the conformationally restricted tripeptide **147** is depicted. The key step here was the intramolecular photocycloaddition of tetronic acid amide **145**, which delivered the tricyclic product **146** as a single diastereoisomer. N-acylation could be performed after Boc-deprotection, and subsequent lactone ring opening was achieved by simple treatment with a primary amine in refluxing ethanol [138].

Tetronates derived from 1,3-divinyl-2-cyclopentanol were employed to study the possibility of a differentiation of enantiotopic or diastereotopic double bonds in their [2 + 2]-photocycloaddition [140]. It was found that tetronate **148** underwent a selective [2 + 2]-photocycloaddition (r.r. = 75/25) at one of the two possible double bonds to deliver product **149** in 67% yield (Scheme 6.52). The reaction was analyzed regarding the preferred conformations of the cyclopentanol, with the notion that the tetronate resides in a pseudoequatorial position, and the vinyl group in a pseudoaxial position of the envelope conformation. Intermediate **149** served as starting material for the first total synthesis of the tetracyclic sesquiterpene (±)-punctaporonin C (**150**) [141].

Ascorbic acid derivatives such as **151** were shown to undergo a highly efficient transannular [2 + 2]-photocycloaddition (Scheme 6.53). The reaction proceeded with

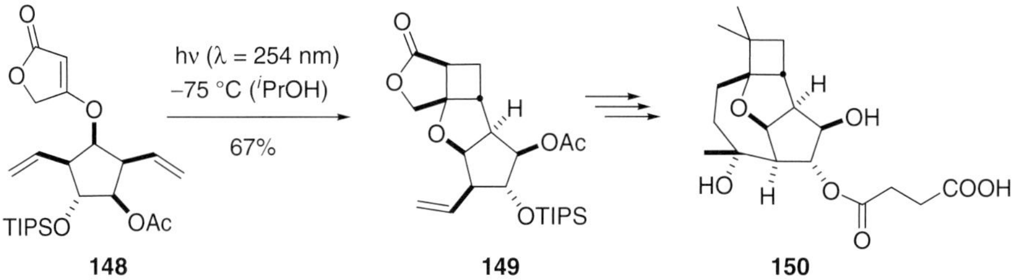

Scheme 6.52

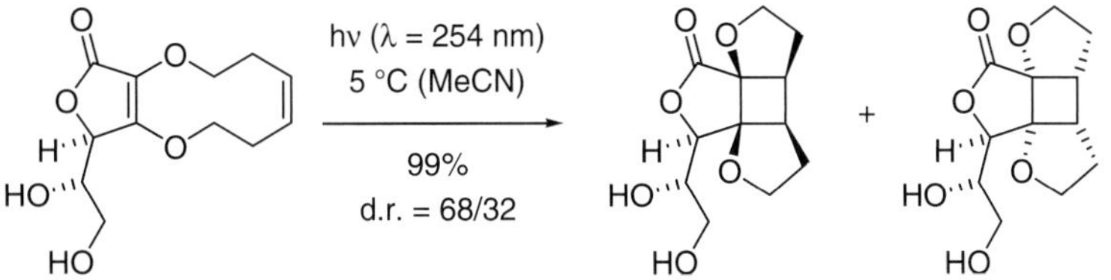

Scheme 6.53

modest stereocontrol by the existing stereogenic center in γ-position. The diastereoisomers **152a** and **152b** were formed in a ratio of 68/32 [142].

4-Hydroxycoumarin and its alkenyl ethers undergo intermolecular or intramolecular [2 + 2]-photocycloaddition upon direct excitation (Pyrex filter) [143, 144]. 4-Butenyloxy- and 4-pentenyloxycoumarin deliver the straight photocycloaddition products, whereas the 4-allyloxycoumarin does not react. Five-membered ring formation is also the preferred mode of addition in the intramolecular reaction of 4-alkenyloxy-2-pyrones (Scheme 6.54). Substrate **153** gave, under sensitizing irradiation conditions, exclusively the straight [2 + 2]-photocycloaddition product **154** in 70% yield. If the chain was longer, the yields decreased to 48% and 32% [145].

Recently, pyrone [2 + 2]-photocycloaddition reactions were used to construct macrocyclic compounds. A dipyrone was irradiated in the presences of α,ω-diolefins yielding 18- to 25-membered rings by a sequential intermolecular and intramolecular cycloaddition [146].

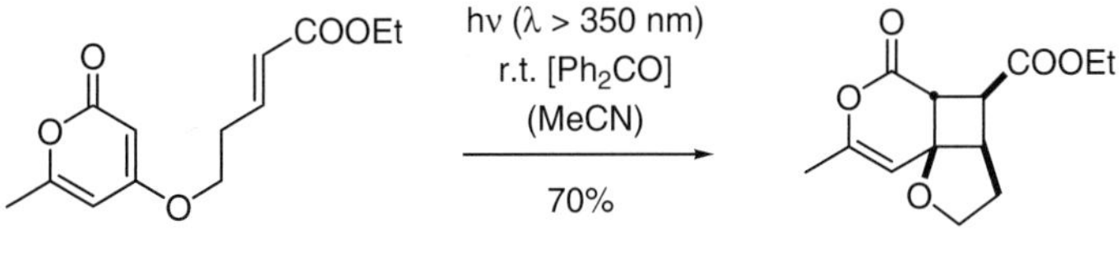

Scheme 6.54

6.4.3.2 Lactams

Tetramic acids and piperidin-2,4-diones are suitable starting materials for the synthesis of the corresponding five- and six-membered β-substituted α,β-unsaturated lactams [92]. The intramolecular [2 + 2]-photocycloaddition of several substrates prepared this way proceeded in reasonable yields (58–79%) upon direct irradiation at $\lambda = 254$ nm in CH_2Cl_2 as the solvent. 5,6-Dihydro-1H-pyridin-2-one **155**, for example, gave the tricyclic product **156** in 60% yield upon irradiation at room temperature for 30 min (Scheme 6.55), although at a lower temperature the reaction slowed down significantly; at −60 °C the conversion in CH_2Cl_2 after 3 h was only 5%. This fact – together with its instability at short wavelength irradiation – hampered the use of chiral template **115** (Scheme 6.41) in this and related reactions. Initially, a 75% ee was recorded in toluene as the solvent at −60 °C (3 h, 6% conversion), which dropped to 59% ee after 29 h (45% conversion) [92].

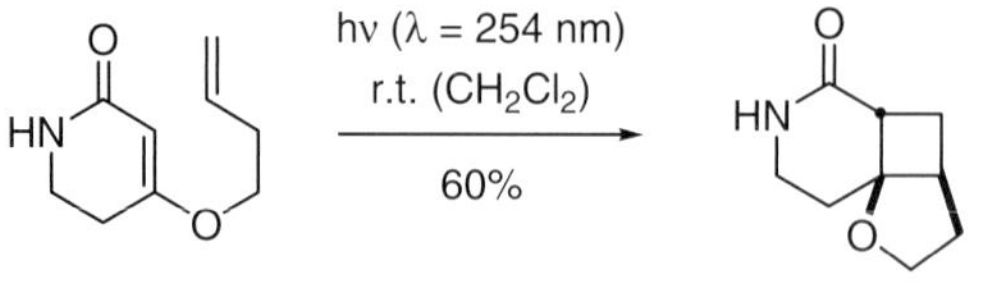

Scheme 6.55

155 → 156

hν (λ = 254 nm), r.t. (CH$_2$Cl$_2$), 60%

Scheme 6.56

157 → 158

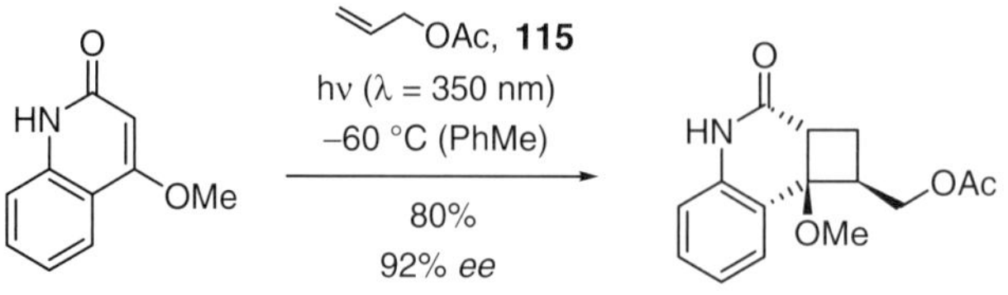

OAc, **115**; hν (λ = 350 nm), −60 °C (PhMe), 80%, 92% ee

In enantioselective photocycloaddition reactions, 4-alkoxyquinolones perform in superior fashion to 1,5-dihydropyrrol-2-ones and 5,6-dihydro-1H-pyridin-2-ones. Both, intermolecular and intramolecular reactions were performed with excellent enantioselectivity in the presence of the chiral template **115**, or of its enantiomer *ent*-**115** [147, 148]. The well-established photocycloaddition reactions [149, 150] enabled access to a variety of chiral dihydroquinolones. 4-Methoxyquinolone (**157**) produced, upon direct irradiation in the presence of allyl acetate, the formal HT product **158** in 80% yield and with 92% ee (Scheme 6.56) [151].

As mentioned in Section 4.1.2 there are also scattered examples of [2 + 2]-photocycloaddition reactions of pyridones. 4-Methoxypyridone (**159**) was reported to form a 1:1-photoadduct **160** with allene (Scheme 6.57), with exclusive HT product formation being observed. Analogously, the reaction with diketene was reported in the same study to be regioselective, but resulted in a mixture of two epimeric products [152].

Among the many consecutive reaction pathways open to the photocycloaddition products of 4-hydroxyquinolones, the formation of furoquinolones is depicted in Scheme 6.58. The photocycloaddition product **161**, which was available by reaction of 4-acetyloxyquinolone and 1-methoxycyclopentene (70%), was converted into the N-methylated hydroxycompound **162**. Upon treatment of this intermediate with I$_2$/HgO the desired ring expansion was observed, providing diastereomerically pure furoquinolone **163** in 62% yield [153].

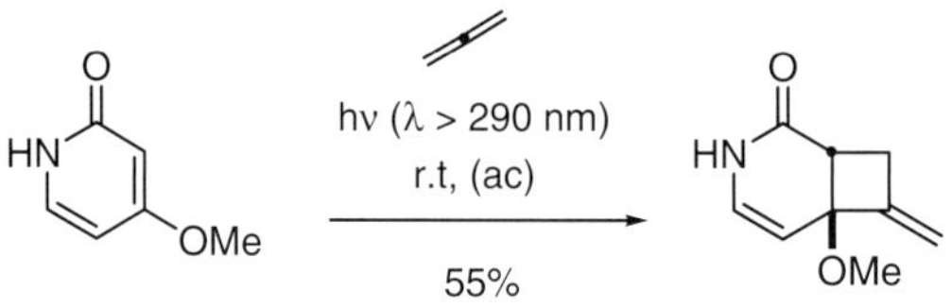

Scheme 6.57

159 → 160

hν (λ > 290 nm), r.t, (ac), 55%

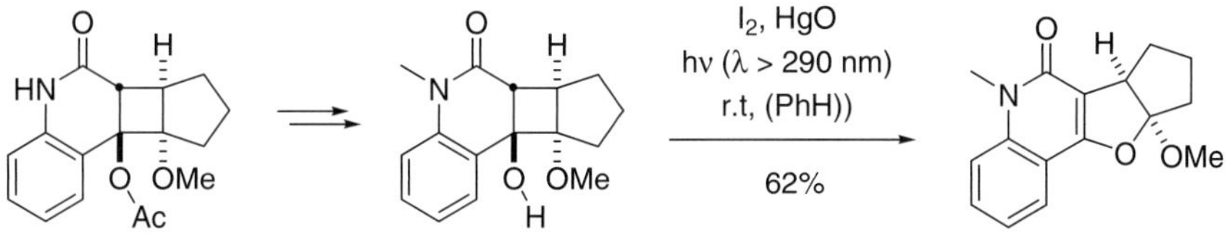

161 → **162** → **163**

Scheme 6.58

6.5
Concluding Remarks

Despite the fact that the [2 + 2]-photocycloaddition reaction of enones has a history of more than 100 years, it has remained a vital and attractive reaction. The continuing interest and many applications to increasingly more complex targets not only bear testimony to its utility but also contradict the myth that photochemical reactions are nonselective and unpredictable. It would be desirable if this most useful chemistry could also be appreciated in the life sciences industry. The first blockbuster drug to be synthesized via a [2 + 2]-photocycloaddition is yet to be developed. Apart from the conventional evolution of the reaction, which involves an increase in scope and an improvement in its practical execution [154], it is expected that sensitization – as a means of catalyzing photochemical reactions in general [155–157] – will become a dominant factor in the development of catalytic enantioselective [2 + 2]-photocycloaddition variants.

References

1 Ciamician, G. and Silber, P. (1908) Chemische Lichtwirkungen. *Berichte der Deutschen Chemischen Gesellschaft*, **41**, 1928–1935.

2 Hoffmann, N. (2008) Photochemical reactions as key steps in organic synthesis. *Chemical Reviews*, **108**, 1052–1103.

3 Iriondo-Alberdi, J. and Greaney, M.F. (2007) Photocycloaddition in natural product synthesis. *European Journal of Organic Chemistry*, 4801–4815.

4 Margaretha, P. (2005) Photocycloaddition of Cycloalk-2-enones to Alkenes, in *Organic Photochemistry (Molecular and Supramolecular Photochemistry)*, Vol. 12 (eds A.G. Griesbeck and J. Mattay), Marcel Dekker, New York, pp. 211–237.

5 Bach, T. (1998) Stereoselective intermolecular [2 + 2]-photocycloaddition reactions and their application in synthesis. *Synthesis*, 683–703.

6 Fleming, S.A., Bradford, C.L., and Gao, J.J. (1997) Regioselective and Stereoselective [2 + 2]-photocycloadditions, in *Organic Photochemistry (Molecular and Supramolecular Photochemistry)*, Vol. 1 (eds V. Ramamurthy and K.S. Schanze), Marcel Dekker, New York, pp. 187–243.

7 Pete, J.-P. (1996) Asymmetric photoreactions of conjugated enones and esters, in *Advances in Photochemistry*, Vol. 21 (eds D.C. Neckers, D.H. Volman and Gv. Bünau), John Wiley & Sons, Inc., New York, pp. 135–216.

8. Mattay, J., Conrads, R., and Hoffmann, R. (1995) Formation of C−C bonds by light-induced [2 + 2] cycloadditions, in *Methods of Organic Chemistry (Houben-Weyl)*, Vol. E21c (eds G. Helmchen, R.W. Hoffmann, J. Mulzer and E. Schaumann), Thieme, Stuttgart, pp. 3085–3132.

9. Crimmins, M.T. and Reinhold, T.L. (1993) Enone-olefin [2 + 2] photochemical cycloadditions, in *Organic Reactions*, Vol. 44, John Wiley & Sons Inc., New York, pp. 297–588.

10. Schuster, D.I. (2004) Mechanistic issues in [2 + 2] photocycloadditions of cyclic enones to alkenes, in *CRC Handbook of Organic Photochemistry and Photobiology*, 2nd edn (eds F. Lenci and W.M. Horspool), CRC Press, New York, pp. 72.1–72.24.

11. Schuster, D.I., Lem, G.N., and Kaprinidis, A. (1993) New insights into an old mechanism: [2 + 2] photocycloaddition of enones to alkenes. *Chemical Reviews*, **93**, 3–22.

12. Srinivasan, R. and Hill Carlough, K. (1967) Mercury ($^3P_1$) photosensitized internal cycloaddition reactions in 1,4-,1,5-, and 1,6-dienes. *Journal of the American Chemical Society*, **89**, 4932–4936.

13. Liu, R.S.H. and Hammond, G.S. (1967) Photosensitized internal addition of dienes to olefins. *Journal of the American Chemical Society*, **89**, 4936–4944.

14. Broeker, J.L., Eksterowicz, J.E., Belk, A.J., and Houk, K.N. (1995) On the regioselectivity of photocycloadditions of triplet cyclohexenones to alkenes. *Journal of the American Chemical Society*, **117**, 1847–1848.

15. Caldwell, R.A., Hrncir, D.C., Munoz, T. Jr, and Unett, D.J. (1996) Photocycloaddition of 4,4-dimethylcyclohexenone to 1,1-diphenylethylene. Evidence for a triplet exciplex intermediate. *Journal of the American Chemical Society*, **118**, 8741–8742.

16. Kaupp, G. and Ringer, E. (1990) Selektivitäten bei Photoadditionen an multifunktionelle Coffein-Derivate. *Chemische Berichte*, **124**, 339–345.

17. Maehr, H. (1985) A proposed new convention for graphic presentation of molecular geometry and topography. *Journal of Chemical Education*, **62**, 114–120.

18. Fox, M.A., Cardona, R., and Ranade, A.C. (1985) Effect of Cu(I) on cyclic enone photodimerisation. *Journal of Organic Chemistry*, **50**, 5016–5018.

19. Kosower, E.M., Wu, G.-S., and Sorensen, T.S. (1961) Effect of solvent on spectra. VI. Detection of the solvent effect on molecular conformation or shape through Z-values. *Journal of the American Chemical Society*, **83**, 3147–3154.

20. Grota, J., Domke, I., Stoll, I., Schröder, T., Mattay, J., Schmidtmann, M., Bögge, H., and Müller, A. (2005) Synthesis, fragmentation, and rearrangement reactions of annelated cyclobutylcarbinols. *Synthesis*, 2321–2326.

21. Termont, D., De Keukeleire, D., and Vandewalle, M. (1977) Regio- and stereoselectivity in [$_\pi 2 + _\pi 2$] photocycloaddition reactions between cyclopent-2-enone and electron-rich alkenes. *Journal of the Chemical Society, Perkin Transactions 1*, 2349–2353.

22. Piers, E. and Orellana, A. (2001) Total synthesis of (±)-kelsoene. *Synthesis*, 2138–2142.

23. Mehta, G. and Srinivas, K. (2001) Enantioselective total syntheses of the novel tricyclic sesquiterpene hydrocarbons (+)- and (−)-kelsoene. Absolute configuration of the natural product. *Tetrahedron Letters*, **42**, 2855–2857.

24. Mehta, G. and Sreenivas, K. (2002) Enantioselective total synthesis of the novel tricyclic sesquiterpene (−)-sulcatine G. Absolute configuration of the natural product. *Tetrahedron Letters*, **43**, 3319–3321.

25 Mehta, G. and Singh, S.R. (2006) Total synthesis of (±)-merrilactone A. *Angewandte Chemie, International Edition*, **45**, 953–955.

26 Mascitti, V. and Corey, E.J. (2006) Enantioselective synthesis of pentacycloanammoxic acid. *Journal of the American Chemical Society*, **128**, 3118–3119.

27 Zandomeneghi, M., Cavazza, M., and Pietra, F. (1984) Enantiomeric enrichment of cis-bicyclo[3.2.0]hept-3,6-dien-2-one with circularly polarized light via the photointerconversion of the enantiomers. *Journal of the American Chemical Society*, **106**, 7261–7262.

28 Ichikawa, M., Aoyagi, S., and Kibayashi, C. (2005) Total synthesis of (−)-incarvilline. *Tetrahedron Letters*, **46**, 2327–2329.

29 Crimmins, M.T., Pace, J.M., Nantermet, P.G., Kim-Meade, A.S., Thomas, J.B., Watterson, S.H., and Wagman, A.S. (2000) The total synthesis of (±)-ginkgolide B. *Journal of the American Chemical Society*, **122**, 8453–8463.

30 Lindemann, U., Reck, G., Wulff-Molder, D., and Wessig, P. (1998) Photocyclization of 4-oxo-4-phenyl-butanoyl amines to δ-lactams. *Tetrahedron*, **54**, 2529–2544.

31 Ng, S.M., Bader, S.J., and Snapper, M.L. (2006) Solvent-controlled intramolecular [2 + 2] photocycloadditions of α-substituted enones. *Journal of the American Chemical Society*, **128**, 7315–7319.

32 Shipe, W.D. and Sorensen, E.J. (2006) Convergent, enantioselective syntheses of guanacastepenes A and E featuring a selective cyclobutane fragmentation. *Journal of the American Chemical Society*, **128**, 7025–7035.

33 Corey, E.J., Bass, J.D., LeMahieu, R., and Mitra, R.B. (1964) A Study of the photochemical reactions of 2-cyclohexenones with substituted olefins. *Journal of the American Chemical Society*, **86**, 5570–5583.

34 Schuster, D.I., Kaprinidis, N., Wink, D.J., and Dewan, J.C. (1991) Stereochemistry of [2 + 2] photocycloaddition of cyclic enones to alkenes: structural and mechanistic considerations in formation of trans-fused cycloadducts. *Journal of Organic Chemistry*, **56**, 561–567.

35 Maradyn, D.J. and Weedon, A.C. (1994) The photochemical cycloaddition reaction of 2-cyclohexenone with alkenes: trapping of triplet 1,4-biradical intermediates with hydrogen selenide. *Tetrahedron Letters*, **35**, 8107–8110.

36 Lange, G.L., Decicco, C.P., Willson, J., and Strickland, L.A. (1989) Ring expansions of [2 + 2] photoadducts. Potential applications in the synthesis of triquinane and taxane skeletons. *Journal of Organic Chemistry*, **54**, 1805–1810.

37 Meyer, L., Alouane, N., Schmidt, K., and Margaretha, P. (2003) Photocycloaddition of cyclohex-2-enones to acrylonitrile. *Canadian Journal of Chemistry*, **81**, 417–422.

38 Morimoto, T., Horiguchi, T., Yamada, K., Tsutsumi, K., Kurosawa, H., and Kakiuchi, K. (2004) Acid-catalyzed rearrangement of an allene-cyclohexenone photoadduct and its application in the synthesis of (±)-Pentalenene. *Synthesis*, 753–756.

39 Mehta, G. and Sreenivas, K. (2002) A new synthesis of tricyclic sesquiterpene (±)-sterpurene. *Tetrahedron Letters*, **43**, 703–706.

40 Srikrishna, A. and Ramasastry, S.S.V. (2005) Enantiospecific total synthesis of phytoalexins, (+)-solanascone, (+)-dehydrosolanascone, and (+)-anhydro-β-rotunol. *Tetrahedron Letters*, **46**, 7373–7376.

41 Fujimori, T., Kasuga, R., Kaneko, H., Sakamura, S., and Noguchi, M. (1978) Solanascone: a novel sesquiterpene ketone from *Nicotiana tabacum*. X-Ray structure determination of the corresponding oxime. *Journal of the Chemical Society, Chemical Communications*, 563–564.

42 Lange, G.L., Decicco, C., Tan, S.L., and Chamberlain, G. (1985) Asymmetric induction in simple [2 + 2] photo-additions. *Tetrahedron Letters*, **26**, 4707–4710.

43 Herzog, H., Koch, H., Scharf, H.-D., and Runsik, J. (1986) Chiral induction in photochemical reactions V. Regio- and diastereoselectivity in the photochemical [2 + 2] cycloaddition of chiral cyclenone-3-carboxylates with 1,1′-diethoxyethene. *Tetrahedron*, **42**, 3547–3558.

44 Tsutsumi, K., Nakano, H., Furutani, A., Endou, K., Merpuge, A., Shintani, T., Morimoto, T., and Kakiuchi, K. (2004) Novel enhancement of diastereo-selectivity of [2 + 2] photocycloaddition of chiral cyclohexenones to ethylene by adding naphthalenes. *Journal of Organic Chemistry*, **69**, 785–789.

45 Faure, S., Piva-Le-Blanc, S., Bertrand, C., Pete, J.-P., Faure, R., and Piva, O. (2002) Asymmetric intramolecular [2 + 2] photocycloadditions: α- and β-Hydroxy acids as chiral tether groups. *Journal of Organic Chemistry*, **67**, 1061–1070.

46 Müller, C. and Bach, T. (2008) Chirality control in photochemical reactions: enantioselective formation of complex photoproducts in solution. *Australian Journal of Chemistry*, **61**, 557–564.

47 Furutani, A., Katayama, K., Uesima, Y., Ogura, M., Tobe, Y., Kurosawa, H., Tsutsumi, K., Morimoto, T., and Kakiuchi, K. (2006) Asymmetric [2 + 2] photocycloaddition of cycloalkenone-cyclodextrin complexes to ethylene. *Chirality*, **18**, 217–221.

48 Gilbert, A. (2004) 1,4-Quinone Cycloaddition Reactions with Alkenes, Alkynes, and Related Compounds, in *CRC Handbook of Organic Photochemistry and Photobiology*, 2nd edn (eds F. Lenci and W.M. Horspool), CRC Press, New York, pp. 87-1–87-12.

49 Wilson, R.M. and Musser, A.K. (1980) Photocyclizations involving quinone-olefin charge-transfer exciplexes. *Journal of the American Chemical Society*, **102**, 1720–1722.

50 Nicolaou, K.C., Vassilikogiannakis, G., Mägerlein, W., and Kranich, R. (2001) Total synthesis of colombiasin A and determination of its absolute configuration. *Chemistry – A European Journal*, **7**, 5359–5371.

51 Nielsen, L.B. and Wege, D. (2006) The enantioselective synthesis of elecanacin through an intramolecular naphtho-quinone-vinyl ether photochemical cycloaddition. *Organic & Biomolecular Chemistry*, **4**, 868–876.

52 Ward, D.E. and Shen, J. (2007) Enantioselective total synthesis of cyathin A_3. *Organic Letters*, **9**, 2843–2846.

53 Ward, D.E., Gai, Y., Qiao, Q., and Shen, J. (2004) Synthetic studies on cyathin diterpenes – Total synthesis of (±)-allocyathin B_3. *Canadian Journal of Chemistry*, **82**, 254–267.

54 Kokubo, K., Yamaguchi, H., Kawamoto, T., and Oshima, T. (2002) Substituent effects on the stereochemistry in the [2 + 2] photocycloaddition reaction of homobenzoquinone derivative with variously substituted alkenes and alkynes. *Journal of the American Chemical Society*, **124**, 8912–8921.

55 deMayo, P. (1971) Enone photo-annelation. *Accounts of Chemical Research*, **4**, 41–47.

56 Oppolzer, W. (1982) Intramolecular [2 + 2] photoaddition/cyclobutane-fragmentation sequence in organic synthesis. *Accounts of Chemical Research*, **15**, 135–141.

57 Winkler, J.D., Bowen, C.M., and Liotta, F. (1995) [2 + 2] Photoaddition/Fragmentation strategies for the synthesis of natural and unnatural products. *Chemical Reviews*, **95**, 2003–2020.

58 Anklam, E., Ghaffari-Tabrizi, R., Hombrecher, H., Lau, S., and Margaretha, P. (1984) Synthesis and photochemistry of 2,2-dimethyl-3(2H)-thiophenone, a ketonic tautomer of 3-hydroxythiophene. *Helvetica Chimica Acta*, **67**, 1402–1405.

59 Margaretha, P. (1975) Zyklische Oxaenone: Untersuchungen mechanischer und stereochemischer Aspekte der [2 + 2]-Photocycloaddition von Enonen an Olefine. *Chimia*, **29**, 203–209.

60 Baldwin, S.W. and Mazzuckelli, T.J. (1986) Face selectivity in the [2 + 2]-photoannelation of chiral 3(2H)-furanones with alkenes. *Tetrahedron Letters*, **27**, 5975–5978.

61 Nicolaou, K.C., Wu, T.R., Sarlah, D., Shaw, D.M., Rowcliffe, E., and Burton, D.R. (2008) Total synthesis, revised structure, and biological evaluation of biyouyanagin a and analogues thereof. *Journal of the American Chemical Society*, **130**, 11114–11121.

62 Gebel, R.-C. and Margaretha, P. (1992) Photochemical synthesis and some reactions of 7-oxa- and 7-thiatricyclo [3.2.1.0^{3,6}]octan-2-ones. *Helvetica Chimica Acta*, **75**, 1633–1638.

63 Bach, T., Kemmler, M., and Herdtweck, E. (2003) Complete control of regioselectivity in the intramolecular [2 + 2] photocycloaddition of 2-alkenyl-3 (2H)-furanones by the length of the side chain. *Journal of Organic Chemistry*, **68**, 1994–1997.

64 Baldwin, S.W. and Wilkinson, J.M. (1979) Cyclohexenones by the photoannelation of alkenes with 2,2-dimethyl-3(2H)-furanone. *Tetrahedron Letters*, **29**, 2657–2660.

65 Patjens, J. and Margaretha, P. (1989) Photochemistry of ethyl 2,3-dihydro-2,2-dimethyl-3-oxo-1H-pyrrole-1-carboxylate. *Helvetica Chimica Acta*, **72**, 1817–1824.

66 Toda, J., Niimura, Y., Takeda, K., Sano, T., and Tsuda, Y. (1998) General method for synthesis of erythrinan and homo-erythrinan alkaloids (1): synthesis of a cycloerythrinan, as a key intermediate to erythrina alkaloids, by Pummerer-type reaction. *Chemical & Pharmaceutical Bulletin*, **46**, 906–912, and literature cited therein.

67 Sano, T., Toda, J., Tsuda, Y., Yamaguchi, K., and Sakai, S.-I. (1984) Dioxopyrrolines. XXXII. X-Ray determination of the molecular structure of a photoadduct of 2-trimethylsilyloxybutadiene to 3-ethoxycarbonyl-2-phenyl-Δ^2-pyrroline-4,5-dione. *Chemical & Pharmaceutical Bulletin*, **32**, 3255–3258.

68 Bahaji, M. and Margaretha, P. (2007) Photocycloaddition of six-membered cyclic enones to propen-2-yl isocyanate. *Helvetica Chimica Acta*, **90**, 1455–1460.

69 Sato, M., Sunami, S., Kogawa, T., and Kaneko, C. (1994) An Efficient synthesis of cis-hydroindan-5-ones by novel modified de mayo reactions using 2,3-dihydro-4-pyrones as the enone chromophore. *Chemistry Letters*, 2191–2194.

70 Sabui, S.K. and Venkateswaran, R.V. (2004) Synthesis of heliannuol D, an allelochemical from *Helianthus annus*. *Tetrahedron Letters*, **45**, 983–985.

71 Comins, D.L., Zhang, Y., and Zheng, X. (1998) Photochemical reactions of the chiral 2,3-dihydro-4(1H)-pyridones: asymmetric synthesis of (−)-perhydrohistrionicotoxin. *Chemical Communications*, 2509–2510.

72 Comins, D.L. and Williams, A.L. (2001) Model studies toward the total synthesis of the lycopodium alkaloid spirolucidine. *Organic Letters*, **3**, 3217–3220.

73 Guerry, P., Blanco, P., Brodbeck, H., Pasteris, O., and Neier, R. (1991) 1-Methoxycarbonyl-substituiertes 2,3-Dihydropyridin-4(1H)-on (= Methyl-1,2,3,4-tetrahydro-4-oxopyridin-1-carboxylat) als Chromophor für die photochemische [2 + 2]-Cycloaddition. *Helvetica Chimica Acta*, **74**, 163–178.

74 Comins, D.L., Zheng, X., and Goehring, R.R. (2002) Total synthesis of the putative structure of the lupin alkaloid plumerinine. *Organic Letters*, **4**, 1611–1613.

75 Margaretha, P., Schmidt, K., Kopf, J., and Sinnwell, V. (2007) Photocycloaddition of 2,3-dihydro-2,2-dimethyl-4H-thiopyran-4-one (a 4-thiacyclohex-2-enone) to bona fide triplet quenchers. A contradiction? *Synthesis*, 1426–1433.

76 Hong, B.-C., Chen, S.-H., Kumar, E.S., Lee, G.-H., and Lin, K.-J. (2003) Intramolecular [2 + 2] photocycloaddition-fragmentation: facile entry to a novel tricyclic 5-6-7 ring system. *Journal of the Chinese Chemical Society*, **50**, 917–926.

77 Hoffmann, R.W. (1989) Allylic 1,3-strain as a controlling factor in stereoselective transformations. *Chemical Reviews*, **89**, 1841–1860.

78 Schulz, S.R. and Blechert, S. (2007) Palladium-catalyzed synthesis of substituted cycloheptane-1,4-diones by an asymmetric ring expanding allylation (AREA). *Angewandte Chemie, International Edition*, **46**, 3966–3970.

79 Shepard, M.S. and Carreira, E.M. (1997) Asymmetric photocycloadditions with an optically active allenylsilane: trimethylsilyl as a removable stereocontrolling group for the enantioselective synthesis of exo-methylenecyclobutanes. *Journal of the American Chemical Society*, **119**, 2597–2605.

80 Chen, C., Chang, V., Cai, X., Duesler, E., and Mariano, P.S. (2001) A general strategy for absolute stereochemical control in enone-olefin [2 + 2] photocycloaddition reactions. *Journal of the American Chemical Society*, **123**, 6433–6434.

81 Amougay, A., Pete, J.-P., and Piva, O. (1996) Réactions de photocycloadditions [2 + 2] intramoléculaire de N-alcénoyl-β-énaminones. *Bulletin de la Société Chimique de France*, **133**, 625–635.

82 Winkler, J.D. and Hershberger, P.M. (1989) A stereoselective synthesis of (−)-perhydrohistrinonicotoxin. *Journal of the American Chemical Society*, **111**, 4852–4856.

83 Berenjian, N., de Mayo, P., Sturgeon, M.-E., Sydnes, L.K., and Weedon, A.C. (1982) Biphasic photochemistry: micelle solutions as media for photochemical cycloadditions of enones. *Canadian Journal of Chemistry*, **60**, 425–436.

84 Kosugi, H., Sekiguchi, S., Sekita, R., and Uda, H. (1976) Photochemical cycloaddition reactions of α,β-unsaturated lactones with olefins, and application to synthesis of natural products. *Bulletin of the Chemical Society of Japan*, **49**, 520–528.

85 Alibés, R., Bourdelande, J.L., Font, J., Gregori, A., and Parella, T. (1996) [2 + 2] Photocycloaddition of homochiral 2(5H)-furanones to alkenes. First step for an efficient and diastereoselective synthesis of (+)- and (−)-grandisol. *Tetrahedron*, **52**, 1267–1278.

86 Bertrand, S., Hoffmann, N., and Pete, J.-P. (1998) Photochemical [2 + 2] cycloaddition of cyclic enones to (5R)-5-menthyloxy-2[5H]-furanone. *Tetrahedron*, **52**, 4873–4888.

87 Racamonde, M., Alibés, R., Figueredo, M., Font, J., and de March, P. (2008) Photochemical cycloaddition of mono-1,1-, and 1,2-disubstituted olefins to a chiral 2(5H)-furanone. Diastereoselective synthesis of (+)-lineatin. *Journal of Organic Chemistry*, **73**, 5944–5952.

88 Inoue, M., Sato, T., and Hirama, M. (2006) Asymmetric total synthesis of (−)-merrilactone A: Use of a bulky protecting group as long-range stereocontrolling element. *Angewandte Chemie, International Edition*, **45**, 4843–4848.

89 Buu Hue, B.T., Dijkink, J., Kuiper, S.v., Schaik, S.v., Maarseveen, J.H., and Hiemstra, H. (2006) Synthesis of the tricyclic core of solanoeclepin a through intramolecular [2 + 2] photocycloaddition of an allene butenolide. *European Journal of Organic Chemistry*, 127–137.

90 Doroh, B. and Sulikowski, G.A. (2006) Progress toward the total synthesis of bielschowskys: a stereoselective [2 + 2] photocycloaddition. *Organic Letters*, **8**, 903–906.

91 Greenwood, E.S., Hitchcock, P.B., and Parsons, P.J. (2003) Studies towards a total synthesis of kainic acid. *Tetrahedron*, **59**, 3307–3314.

92 Albrecht, D., Basler, B., and Bach, T. (2008) Preparation and intramolecular [2 + 2]-

photocycloaddition of 1,5-dihydropyrrol-2-ones and 5,6-dihydro-1H-pyridin-2-ones with C-, N-, and O-linked alkenyl side chains at the 4-position. *Journal of Organic Chemistry*, **73**, 2345–2356.
93 Wrobel, M.N. and Margaretha, P. (2003) Photocycloisomerization of Boc-protected 5-alkenyl-2,5-dihydro-1H-pyrrol-2-ones. *Helvetica Chimica Acta*, **86**, 515–521.
94 Sieburth, S.M. (2004) Photochemical reactivity of pyridones, in *CRC Handbook of Organic Photochemistry and Photobiology*, 2nd edn (eds F. Lenci and W.M. Horspool), CRC Press, New York, pp. 103.1–103.18.
95 Lewis, F.D. and Barancyk, S.V. (1989) Lewis Acid catalysis of photochemical reactions. 8. Photodimerization and cross-cycloaddition of coumarin. *Journal of the American Chemical Society*, **111**, 8653–8661.
96 Hanifin, J.W. and Cohen, E. (1966) Some photochemical cycloaddition reactions of coumarin. *Tetrahedron Letters*, **7**, 1419–1424.
97 Suginome, H. and Kobayashi, K. (1988) A reinvestigation of a sensitized photochemical cycloaddition of coumarin with cyclopentene. *Bulletin of the Chemical Society of Japan*, **61**, 3782–3784.
98 Pfoertner, K.-H. (1976) 84. Photoreaktionen von 3-substituierten Cumarinen. *Helvetica Chimica Acta*, **59**, 83–84.
99 Yamashita, M., Inaba, T., Nagahama, M., Shimizu, T., Kosaka, S., Kawasaki, I., and Ohta, S. (2005) Novel stereoconvergent transformation of 1,2a-disubstituted 1,2,2a,8b-tetrahydro-3H-benzo[b]cyclobuta[d]pyran-3-ones to 1,3-disubstituted 1,2,4a,9b-tetrahydro-dibenzofuran-4-ols and its application to the second-generation synthesis of (±)-linderol A. *Organic & Biomolecular Chemistry*, **3**, 2296–2304.
100 Sakamoto, M., Kato, M., Aida, Y., Fujita, K., Mino, T., and Fujita, T. (2008) Photo-sensitized 2 + 2 cycloaddition reaction using homochirality generated by spontaneous crystallization. *Journal of the American Chemical Society*, **130**, 1132–1133.
101 Akritopoulou-Zanze, I., Whitehead, A., Waters, J.E., Henry, R.F., and Djuric, S.W. (2007) Synthesis of novel and uniquely shaped 3-azabicyclo[4.2.0]octan-4-one derivatives by sequential Ugi/[2 + 2] ene-enone photocycloadditions. *Organic Letters*, **9**, 1299–1302.
102 Lewis, F.D., Reddy, G.D., Elbert, J.E., Tillberg, B.E., Meltzer, J.A., and Kojima, M. (1991) Spectroscopy and photochemistry of 2-quinolones and their Lewis acid complexes. *Journal of Organic Chemistry*, **56**, 5311–5318.
103 Evanega, G.R. and Fabiny, D.L. (1968) The photocycloaddition of carbostyril to olefins. *Tetrahedron Letters*, **9**, 2241–2246.
104 Evanega, G.R. and Fabiny, D.L. (1970) The photocycloaddition of carbostyril to olefins. The stereochemistry of the adducts. *Journal of Organic Chemistry*, **35**, 1757–1761.
105 Kobayashi, K., Suzuki, M., and Suginome, H. (1992) Photoinduced molecular transformations. 128. Regioselective [2 + 2] photocycloaddition of 3-acetoxyquinolin-2(1H)-one with alkenes and formation of furo[2,3-c]quinolin-4(5H)-ones, 1-benzazocine-2,3-diones, and cyclopropa[d]benz[1]azepine-2,3-diones via a β-scission of cyclobutanoxyl radicals generated from the resulting [2 + 2] photoadducts. *Journal of Organic Chemistry*, **57**, 599–606.
106 Selig, P. and Bach, T. (2006) Photochemistry of 4-(2′-aminoethyl) quinolones: enantioselective synthesis of tetracyclic tetrahydro-1aH-pyrido [4′,3′:2,3]-cyclobuta[1,2-c] quinoline-2,11 (3H, 8H)-diones by intra- and inter-molecular [2 + 2] photocycloaddition reactions in solution. *Journal of Organic Chemistry*, **71**, 5662–5673.
107 Selig, P. and Bach, T. (2009) Total synthesis of meloscine by a [2 + 2]-Photocycloaddition/Ring expansion

route. *Chemistry – A European Journal*, **15**, 15, 3509–3525.

108 Bach, T., Bergmann, H., Grosch, B., Harms, K., and Herdtweck, E. (2001) Synthesis of enantiomerically pure 1,5,7-trimethyl-3-azabicyclo[3.3.1]nonan-2-ones as chiral host compounds for enantioselective photochemical reactions in solution. *Synthesis*, 1395–1405.

109 Breitenlechner, S., Selig, P., and Bach, T. (2007) Chiral organocatalysts for enantioselective photochemical reactions, in *Ernst Schering Foundation Symposium Proceedings*, vol. **2** (eds. M.T. Reetz, B. List, S. Jaroch and H. Weinmann), Springer-Verlag, Berlin, pp. 255–279.

110 Brandes, S., Selig, P., and Bach, T. (2004) Stereoselective intra- and intermolecular [2 + 2] photocycloaddition reactions of 4-(2′-aminoethyl)quinolones. *Synlett*, 2588–2590.

111 Selig, P. and Bach, T. (2008) Enantioselective total synthesis of the melodinus alkaloid (+)-meloscine. *Angewandte Chemie, International Edition*, **47**, 5082–5084.

112 Kaupp, G. (1975) Cycloadditionen zwischen ungleichen Olefinen, in *Photochemie I (Houben-Weyl)*, Vol. IV/5 a (ed. E. Müller), Thieme, Stuttgart, pp. 360–412.

113 Birman, V.B. and Jiang, X.-T. (2004) Synthesis of sceptrin alkaloids. *Organic Letters*, **6**, 2369–2371.

114 Inoue, M., Lee, N., Kasuya, S., Sato, T., Hirama, M., Moriyama, M., and Fukuyama, Y. (2007) Total synthesis and bioactivity of an unnatural enantiomer of merrilactone a: development of an enantioselective desymmetrization strategy. *Journal of Organic Chemistry*, **72**, 3065–3075.

115 Booker-Milburn, K.I., Cowell, J.K., Jiménez, F.D., Sharpe, A., and White, A.J. (1999) Stereoselective intermolecular [2 + 2] photocycloaddition reactions of tetrahydrophthalic anhydride and derivatives with alkenols and alkynols. *Tetrahedron*, **55**, 5875–5888.

116 Gülten, S., Sharpe, A., Baker, J.R., and Booker-Milburn, K.I. (2007) Use of temporary tethers in the intramolecular [2 + 2] photocycloaddition reactions of tetrahydrophthalimide derivatives: a new approach to complex tricyclic lactones. *Tetrahedron*, **63**, 3659–3671.

117 Nishioka, Y., Yamaguchi, T., Kawano, M., and Fujita, M. (2008) Asymmetric [2 + 2] olefin cross photoaddition in a self-assembled host with remote chiral auxiliaries. *Journal of the American Chemical Society*, **130**, 8160–8161.

118 Scherer, O. and Kluge, F. (1966) Synthese von Dichlormaleinsäure-thioanhydrid und dessen Einsatz für die Synthese der Thiophentetracarbonsäure. *Chemische Berichte*, **99**, 1973–1983.

119 Kiesewetter, R. and Margaretha, P. (1989) Photochemistry of thiophen-2(5H)-ones. *Helvetica Chimica Acta*, **72**, 83–92.

120 Anklam, E. and Margaretha, P. (1984) Synthese von Tetrahydro-3-thienylessig-säureestern durch konsekutive lichtinduzierte Reaktionen. *Angewandte Chemie*, **96**, 360–361.

121 Karbe, C. and Margaretha, P. (1991) Photocycloadditions to 1-thiocoumarin. *Journal of Photochemistry and Photobiology A: Chemistry*, **57**, 231–233.

122 Schwebel, D. and Margaretha, P. (2000) Photocycloaddition of 2H-1-benzopyran-3-carbonitriles and 2H-1-benzothiopyran-3-carbonitriles to alkenes and alkenynes. *Helvetica Chimica Acta*, **83**, 1168–1174.

123 Organ, M.G., Froese, R.D.J., Goddard, J.D., Taylor, N.J., and Lange, G.L. (1994) Photoadditions and dialkylcuprate additions to 2-*tert*-butyl-2,6-dimethyl-1,3-dioxin-4-one and related heterocycles. Experimental, ab initio theoretical, and X-ray structural studies of facial selectivity and enone pyramidalization. *Journal of the American Chemical Society*, **116**, 3312–3323.

124 Demuth, M., Palomer, A., Sluma, H.-D., Dey, A.K., Krüger, C., and Tsay, Y.-H. (1986) Asymmetric photocycloadditions with optically pure, spirocyclic enones.

Simple synthesis of (+)- and (−)-Grandisol. *Angewandte Chemie, International Edition*, **25**, 1117–1119.

125 Seebach, D., Zimmermann, J., Gysel, U., Ziegler, R., and Ha, T.K. (1988) Totally stereoselective additions to 2,6-disubstituted 1,3-dioxin-4-ones (chiral acetoacetic acid derivatives). Synthetic and mechanistic aspects of remote stereoselectivity. *Journal of the American Chemical Society*, **110**, 4763–4772.

126 Sato, M., Abe, Y., and Kaneko, C. (1990) Cycloadditions in syntheses. Part 47. 2-Monosubstituted 1,3-dioxin-4-ones: diastereofacial selectivity in pericyclic reactions and its explanation. *Journal of the Chemical Society, Perkin Transactions 1*, 1779–1783.

127 Murakami, M., Kamaya, H., Kaneko, C., and Sato, M. (2003) Synthesis of optically active 1,3-dioxin-4-one derivatives having a hydroxymethyl group at the 2-position and their use for regio-, diastereo-, and enantioselective synthesis of substituted cyclobutanols. *Tetrahedron: Asymmetry*, **14**, 201–215.

128 Winkler, J.D. and Doherty, E.M. (1999) The first total synthesis of (±)-saudin. *Journal of the American Chemical Society*, **121**, 7425–7426.

129 Winkler, J.D., Rouse, M.B., Greaney, M.F., Harrison, S.J., and Jeon, Y.T. (2002) The first total synthesis of (±)-ingenol. *Journal of the American Chemical Society*, **124**, 9726–9728.

130 Winkler, J.D., Harrison, S.J., Greaney, M.F., and Rouse, M.B. (2002) Mechanistic observations on the unusual reactivity of dioxenone photosubstrates in the synthesis of ingenol. *Synthesis*, 2150–2154.

131 Friedel, M.G., Cichon, M.K., and Carell, T. (2004) DNA damage and repair: photochemistry, in *CRC Handbook of Organic Photochemistry and Photobiology*, 2nd edn (eds F. Lenci and W.M. Horspool), CRC Press, New York, pp. 141.1–141.22.

132 Moriou, C., Thomas, M., Adeline, M.-T., Martin, M.-T., Chiaroni, A., Pochet, S., Fourrey, J.-L., Favre, A., and Clivio, P. (2007) Crystal structure and photochemical behavior in solution of the 3′-N-sulfamate analogue of thymidylyl(3′-5′)thymidine. *Journal of Organic Chemistry*, **72**, 43–50.

133 Mondière, A., Peng, R., Remuson, R., and Aitken, D.J. (2008) Efficient synthesis of 3-hydroxymethylated cis- and trans-cyclobutane β-amino acids using an intramolecular photocycloaddition strategy. *Tetrahedron*, **64**, 1088–1093.

134 Gauzy, C., Saby, B., Pereira, S., Faure, S., and Aitken, D.J. (2006) The [2 + 2] photocycloaddition of uracil derivatives with ethylene as a general route to cis-cyclobutane β-amino acids. *Synlett*, 1394–1398.

135 Fernandes, C., Gauzy, C., Yang, Y., Roy, O., Pereira, E., Faure, S., and Aitken, D.J. (2007) [2 + 2] photocycloadditions with chiral uracil derivatives: access to all four stereoisomers of 2-aminocyclobutanecarboxylic acid. *Synthesis*, 2222–2232.

136 Agócs, A., Batta, G., Jekö, J., and Herczegh, P. (2004) First synthesis of a dihydroorotidine analogue via a diastereoselective [2 + 2] photocycloaddition. *Tetrahedron: Asymmetry*, **15**, 283–287.

137 Kemmler, M. and Bach, T. (2003) [2 + 2] photocycloaddition of tetronates. *Angewandte Chemie, International Edition*, **42**, 4824–4826.

138 Basler, B., Schuster, O., and Bach, T. (2005) Conformationally constrained β-amino acid derivatives by intramolecular [2 + 2]-photocycloaddition of a tetronic acid amide and subsequent lactone ring opening. *Journal of Organic Chemistry*, **70**, 9798–9808.

139 Kemmler, M., Herdtweck, E., and Bach, T. (2004) Inter- and intramolecular [2 + 2]-photocycloaddition of tetronates – stereoselectivity, mechanism, scope and synthetic applications. *European Journal of Organic Chemistry*, 4582–4595.

140 Fleck, M., Yang, C., Wada, T., Inoue, Y., and Bach, T. (2007) Regioselective [2 + 2]-

photocycloaddition reactions of chiral tetronates – influence of temperature, pressure, and reaction medium. *Chemical Communications*, 822–824.

141 Fleck, M. and Bach, T. (2008) Total synthesis of the tetracyclic sesquiterpene (±)-Punctaporonin C. *Angewandte Chemie, International Edition*, **42**, 6189–6191.

142 Redon, S. and Piva, O. (2006) Diastereoselective transannular [2 + 2] photocycloadditions of ascorbic acid derivatives. *Tetrahedron Letters*, **47**, 733–736.

143 Haywood, D.J., Hunt, R.G., Potter, C.J., and Reid, S.T. (1977) Photochemical transformations. Part 10. Photocycloaddition reactions of 4-hydroxycoumarin with cycloalkenes. *Journal of the Chemical Society, Perkin Transactions 1*, 2458–2461.

144 Haywood, D.J. and Reid, S.T. (1979) Intramolecular photocycloaddition reactions of 4-alkenyloxycoumarins. *Tetrahedron Letters*, **20**, 2637–2638.

145 Shimo, T., Tajima, J., Suishu, T., and Somekawa, K. (1991) Intramolecular photochemical reactions of 4-(ω-alkenyloxy)-6-methyl-2-pyrones having an alkoxycarbonyl group at the olefinic carbon chain. *Journal of Organic Chemistry*, **56**, 7150–7154.

146 Miyauchi, H., Ikematsu, C., Shimazaki, T., Kato, S., Shinmyozu, T., Shimo, T., and Somekawa, K. (2008) One-pot synthesis of macrocyclic compounds possessing two cyclobutane rings by sequential inter- and intramolecular [2 + 2] photocycloaddition reactions. *Tetrahedron*, **64**, 4108–4116.

147 Bach, T., Bergmann, H., and Harms, K. (2000) Enantioselective intramolecular [2 + 2]-photocycloaddition reactions in solution. *Angewandte Chemie, International Edition*, **39**, 2302–2304.

148 Bach, T. and Bergmann, H. (2000) Enantioselective intermolecular [2 + 2]-photocycloaddition reactions of alkenes and a 2-quinolone in solution. *Journal of the American Chemical Society*, **122**, 11525–11526.

149 Kaneko, C., Naito, T., and Somei, M. (1979) Synthesis of Cyclobuta[c]quinolin-3-ones. Intra- and inter-molecular photocycloadditions of 4-alkoxy-2-quinolone systems with olefins. *Journal of the Chemical Society, Chemical Communications*, 804–805.

150 Kaneko, C., Suzuki, T., Sato, M., and Naito, T. (1987) Cycloadditions in syntheses. XXXII. intramolecular photocycloaddition of 4-(ω-alkenyloxy) quinolin-2(1H)-one: synthesis of 2-substituted cyclobuta[c]quinolin-3(4H)-ones. *Chemical & Pharmaceutical Bulletin*, **35**, 112–123.

151 Bach, T., Bergmann, H., Grosch, B., and Harms, K. (2002) Highly enantioselective intra- and intermolecular [2 + 2] photocycloaddition reactions of 2-quinolones mediated by a chiral lactam host: host-guest interactions, product configuration, and the origin of the stereoselectivity in solution. *Journal of the American Chemical Society*, **124**, 7982–7990.

152 Chiba, T., Kato, T., Yoshida, A., Moroi, R., Shimomura, N., Momose, Y., Naito, T., and Kaneko, C. (1984) Photocycloaddition of 2-quinolone and 2-pyridone derivatives to diketene: on the regioselectivity of the photoaddition. *Chemical & Pharmaceutical Bulletin*, **32**, 4707–4720.

153 Suginome, H., Kobayashi, K., Itoh, M., and Seko, S. (1990) Photoinduced molecular transformations. 110. Formation of furoquinolinones via β-scission of cyclobutanoxyl radicals generated from [2 + 2] photoadducts of 4-hydroxy-2-quinolone and acyclic and cyclic alkenes. X-ray crystal structure of (6aα,6bβ,10aβ,10bα)-(±)-10b-acetoxy-6a,6b,7,8,9,10,10a,10b-octahydro-5-methylbenzo[3,4]cyclobuta[1,2-c] quinolin-6(5H)-one. *Journal of Organic Chemistry*, **55**, 4933–4943.

154 Hook, B.D.A., Dohle, W., Hirst, P.R., Pickworth, M., Berry, M.B., and Booker-Milburn, K.I. (2005) A practical flow reactor for continuous organic photochemistry. *Journal of Organic Chemistry*, **70**, 7558–7564.

155 Inoue, Y. (2004) Enantiodifferentiating photosensitized reactions, in *Chiral Photochemistry (Molecular and Supramolecular Photochemistry)*, Vol. 11 (eds Y. Inoue and V. Ramamurthy), Marcel Dekker, New York, pp. 129–177.

156 Inoue, Y. (2005) Light on chirality. *Nature*, **436**, 1099–1100.

157 Bauer, A., Westkämper, F., Grimme, S., and Bach, T. (2005) Catalytic enantioselective reactions driven by photoinduced electron transfer. *Nature*, **436**, 1139–1140.

7
Formation of a Four-Membered Ring: Oxetanes
Manabu Abe

7.1
Introduction

Oxetanes, four-membered cyclic ethers, have a ring strain energy (SE) of approximately 110 kJ mol^{-1} [1] and polar properties of the CO bonds (Figure 7.1). Thus, similar to the synthetic utility of oxiranes (epoxides, SE = 114 kJ mol^{-1}), the ring-opening reaction of oxetanes, accompanying bond-formation reactions, would be very useful for synthetic purposes [2]. Since the oxetane ring is an important structural component of biologically active compounds, such as merrilactone A [3], thromboxane A2 [4], oxetanocin [5], oxetin [6], taxane alkaloids [7], and laureacetal-B [8], efficient and selective methods to synthesize the strained structure are currently active areas of research. Moreover, oxetane-ring-containing compounds are important industrial curing agents [9]. As a consequence, there are today over 2900 patents which include the term "oxetane" as a key word.

All of these findings clearly indicate that the demand for synthetic oxetanes is high. There are basically three methods for preparing oxetanes:

1. The intramolecular nucleophilic substitution reaction.
2. The ring-expansion reaction of epoxides.
3. The thermal and photochemical [2 + 2] cycloaddition reaction of alkenes with carbonyls.

The intramolecular nucleophilic substitution reaction – for example, the Williamson-type reaction – represents one of the important methods for preparing oxetane ring structures, and have been widely applied to the synthesis of oxetanes (Scheme 7.1) [10]. Unfortunately, side reactions – which include fragmentation from the intermediary alkoxide anion or elimination from the intermediary carbocation – often decrease the chemical yields of oxetane formation.

The ring-expansion reaction of epoxides was first reported by Okuma and coworkers, to produce less-substituted oxetanes (Scheme 7.2) [11]. The nucleophilic attack by dimethyloxosulfonium methylide is proposed to react with the less-

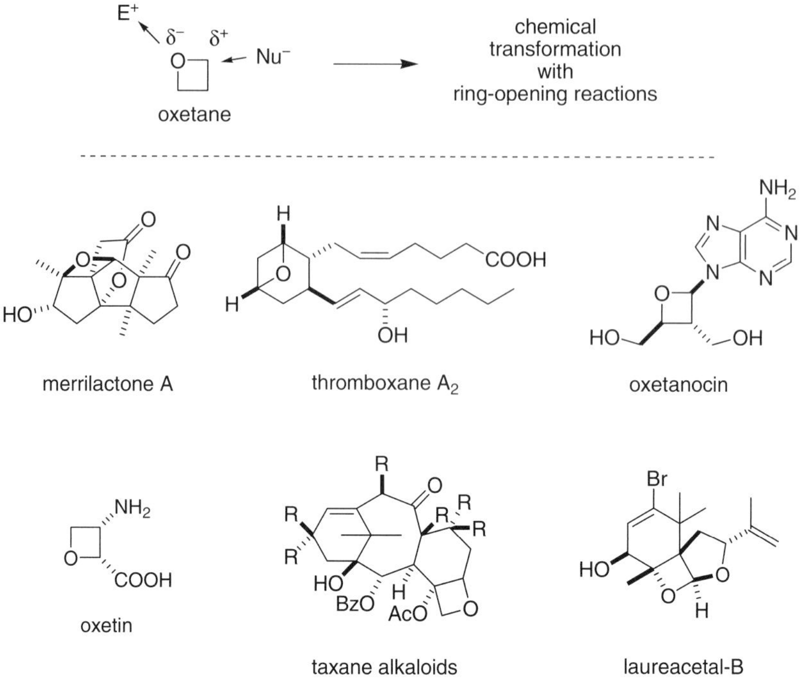

Figure 7.1 The synthetic utility of oxetane-ring and biologically important oxetane derivatives.

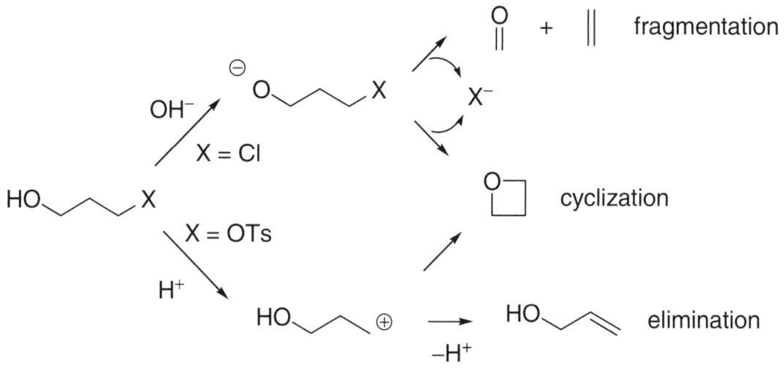

Scheme 7.1 Formation of oxetanes by intramolecular cyclization reactions.

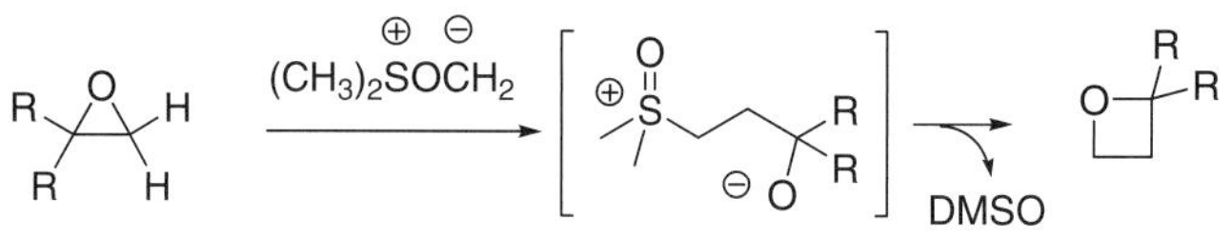

Scheme 7.2 The formation of oxetanes by ring-expansion reactions.

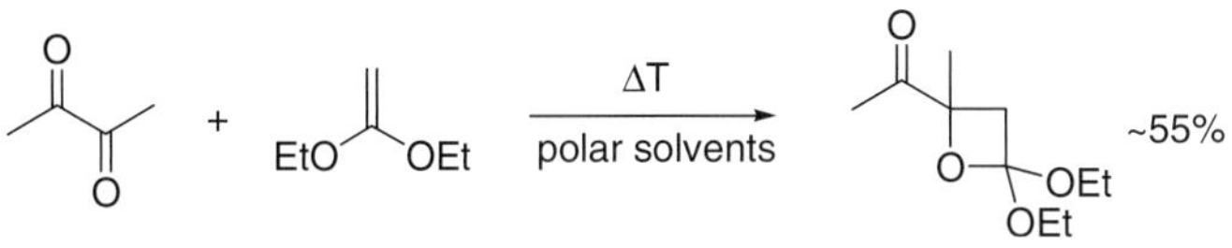

Scheme 7.3 Formation of oxetanes in the thermal [2 + 2] cycloaddition reaction.

hindered carbon of the epoxide ring to give the zwitterionic intermediates; this is followed by cyclization to afford the strained structures.

The [2 + 2] cycloaddition reaction of alkenes with carbonyl compounds represents one of the most promising methods for the synthesis of oxetanes, as it can be applied to a wide variety of alkenes and carbonyl compounds. According to the Woodward–Hoffmann rule, the thermal [2 + 2] cycloaddition reaction would be a "symmetry-forbidden" process. Mattay and coworkers, however, identified the stepwise formation of oxetanes in the thermal reaction of the highly electron-rich alkenes with highly electron-accepting carbonyl compound (Scheme 7.3) [12]. These authors proposed the stepwise mechanism – that is, addition and cyclization – for the thermal [2 + 2] oxetane formation. The thermal reaction is only observed in the reaction of strongly electron-donating and accepting groups. The regio-isomeric oxetane was identified under photochemical conditions in nonpolar solvents.

The photochemical reaction of carbonyl compounds and alkenes, which is referred to as the Paternò–Büchi (PB) reaction, was developed in 1909 [13], and is currently one of the most widely used methods for oxetane synthesis (Scheme 7.4). As exemplified in the PB reaction of benzophenone with 2-methylpropene [14], a selective formation of the oxetane is possible even when the photochemical reaction involves highly unstable molecules; that is, the excited state of carbonyls. Due to its synthetic importance and mechanistic interest, the PB reaction is the most extensively studied synthetic method for oxetanes. Thus, several extensive reviews describing the PB reaction have been published since 1968, and the reader is directed towards these for further information [15]. In this chapter, methods that allow for the control of the regioselective and stereoselective formation of synthetically important oxetanes will be described.

Before describing the regioselective, site-selective and stereo-selective preparation of oxetanes via the PB reaction, the mechanism of the photochemical reaction will be briefly summarized. The reason for this is that an understanding of the reaction

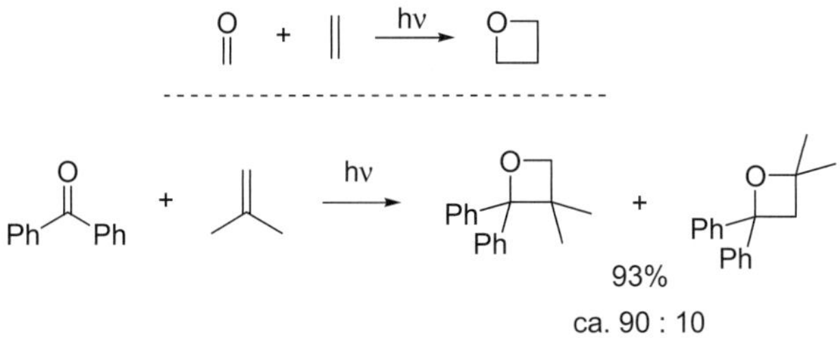

Scheme 7.4 The Paternò–Büchi (PB) reaction, and an example of the highly selective formation of oxetane.

mechanism will enable synthetic chemists to design reactions that allow for selective oxetane synthesis.

7.2
The Generally Accepted Mechanism of the Paternò–Büchi Reaction

The currently accepted mechanism of the PB reaction is summarized in Scheme 7.5 [15a]. First, the carbonyl group absorbs light (hv) to generate the excited singlet state of carbonyls (C=^{1}O*). As the electronic excitation is from the electron of the lone-pair to the π^* orbital of the CO double-bond, the oxygen atom in the excited state has an electrophilic character. The structure of the excited state of carbonyls (C=O*) in Scheme 7.5 is helpful for defining the excited carbonyl as an umpolung reagent [16], and the ground state of the carbonyls reverse electronics with respect to the functional group. In general, the intersystem crossing (ISC) process from the singlet to the triplet state quite rapidly produces the triplet excited state of carbonyls (C=^{3}O*). The singlet excited state of aliphatic carbonyls can react with alkenes, but the intermolecular reaction would be inefficient for the case of aromatic carbonyls, as the rate constant ($k_{ISC} = 10^{11}$ s^{-1}) of the ISC to the triplet beyond the diffusion-controlled rate constant. Thus, normally, the long-lived triplet excited state of the aromatic carbonyls, for example, benzaldehyde and benzophenone derivatives, has the chance to react intermolecularly with alkenes to produce the intermediary triplet 1,4-diradicals (**T-BR**) or radical ion (**RI**) pairs. The regioisomeric biradical, **BR'**, is proposed to be involved in reactions of electron-poor alkenes [17]. For the photochemical [2 + 2] cycloaddition reactions with electron-rich alkenes, the preferred mechanism is largely dependent on the redox potentials between the excited carbonyl compounds and the alkenes used in the photochemical reactions [18]. When the photoinduced electron transfer (PET) reaction is an energetically favorable process, which can be determined using the Rehm–Weller equation: $\Delta G_{et} = E_{ox} - E_{red} < 0$ [19], the **RI** pairs

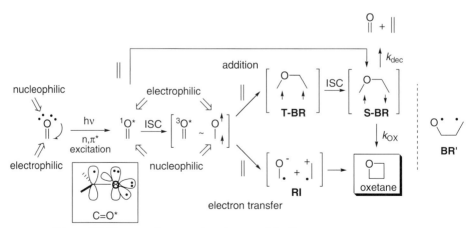

Scheme 7.5 The generally accepted mechanism of the PB reaction.

are generated. When the PET reactions are energetically unfavorable, 1,4-biradicals (**BR**) become important for the oxetane formation. Singlet biradicals (**S-BR**) can form directly from the excited singlet carbonyl compounds, when the lifetime of the excited singlet carbonyl compounds, such as aliphatic carbonyl compounds and naphthaldehyde, is long enough to interact with alkenes.

When unsymmetrical alkenes and/or carbonyl compounds are used in the photochemical reaction, then regioselectivity, site-selectivity, and stereoselectivity will arise in the formation of oxetanes. For reactions that involve a **RI** pair, the oxetane regioselectivity is determined by the spin and charge distribution of the radical cations and anions. The singlet 1,4-biradical intermediate, **S-BR**, has two possible paths: (i) bond formation to give the oxetane, k_{OX}; or (ii) bond cleavage within the 1,4-biradical to give the starting materials, k_{dec}. Thus, the ratio (k_{OX}/k_{dec}) plays an important role in the determination of regioselectivity, site-selectivity and stereoselectivity during oxetane formation [20]. The geometry of the conical intersection is also important for selectivity in the excited singlet reaction [21]. For excited triplet reactions, the geometry of the triplet biradical (**T-BR**) plays an important role in controlling stereoselectivity, as the rate constant for ISC to produce the **S-BR** is largely dependent on the orbital orientation of the two spin centers [15m].

The following sections describe the regioselective, site-selective, and stereoselective synthesis of oxetanes.

7.3
Regioselective and Site-Selective Syntheses of Oxetanes

The energies of carbonyl compounds in the n,π* excited states are greater than the ground-state energies, by about 70–80 kcal mol^{-1}. Thus, the electrophilic oxygen of these highly reactive molecules is supposed to randomly attack both alkene carbons to give regioisomeric 1,4-biradicals, which produce regioisomers of oxetanes (Scheme 7.6). However, as mentioned in Section 7.1, the regioselective formation of 2,2-diphenyl-3,3-dimethyloxetane was found in the PB reaction of 1,1-dimethylethylene with benzophenone (Scheme 7.4). The regioselectivity can be explained by the "biradical-stability rule" (Scheme 7.7). Thus, the intermediary biradical **BR1** is supposed to be more stable than the regioisomeric biradical **BR2**.

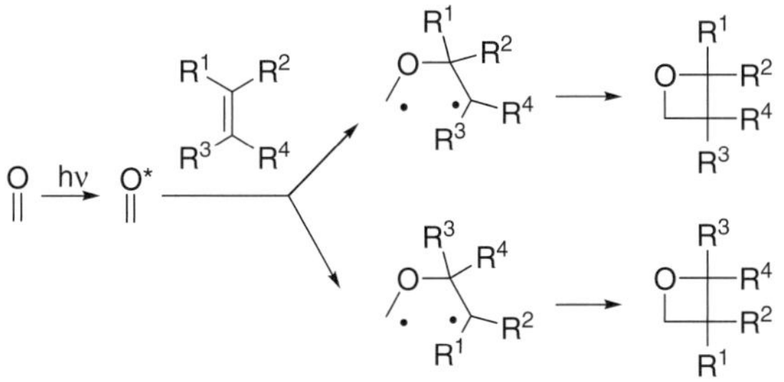

Scheme 7.6 Regioselectivity of the PB reaction of unsymmetrically substituted alkenes.

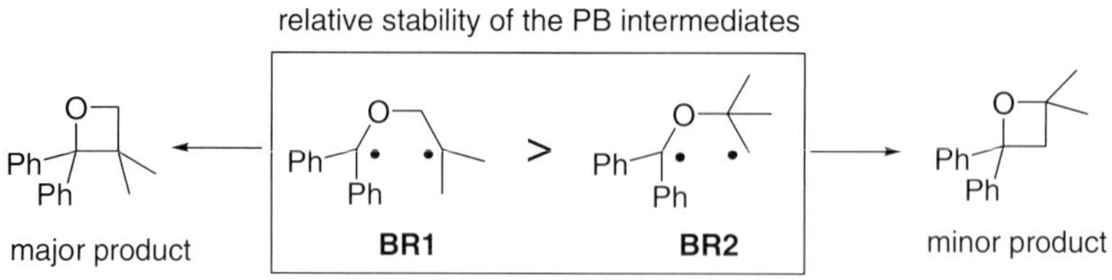

Scheme 7.7 Regioselectivity; the radical stability rule.

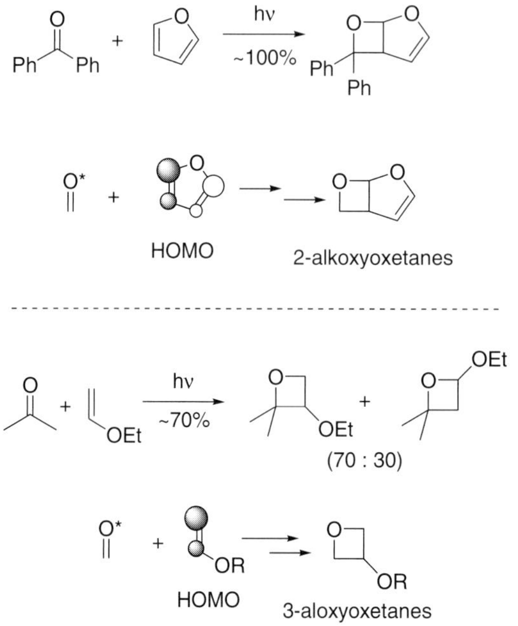

Scheme 7.8 Regioselective formation of oxetanes derived from furans or vinyl ethers; the nucleophilicity rule.

When the nucleophilicities of the two carbons in alkenes differ significantly, the regioselective formation of 1,4-biradicals results. In fact, the regioselective oxetane formation was reported for the PB reaction of both furans [22] and vinyl ethers [23] (Scheme 7.8). Thus, 2-alkoxyoxetanes are formed exclusively during the PB reaction of furan derivatives. In contrast, 3-alkoxyoxetanes were selectively prepared in the PB reaction of vinyl ethers. This dramatic change in regioselectivity can be explained by the difference in the HOMO coefficient. Thus, in a furan ring, the C-2 carbon is known to be more nucleophilic, whereas the β-carbon is the nucleophilic site in vinyl ethers.

The PB reaction of silyl enol ethers also regioselectively produced the 3-siloxyoxetane in high yield (Scheme 7.9), a reaction first identified in 1983 [24]. The synthetic utility of the thermal ring-opening reaction was also reported. The regioselective formation of oxetane was also reported in the PB reaction of allylic silanes (Scheme 7.9) [25], where the reaction yield was moderate but the regioselectivity

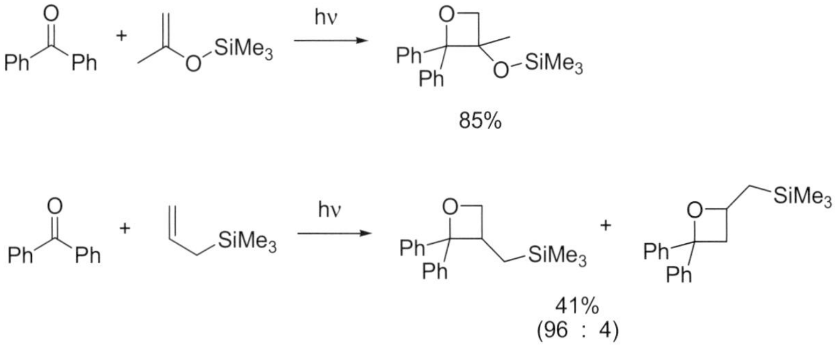

Scheme 7.9 Regioselective formation of oxetanes derived from silyl enol ethers and allylic silanes.

was found to be excellent. The regioselectivity was also explained by the relative nucleophilicity of the carbons in the allylic silane; thus, the β-carbon is much nucleophilic than the α-carbon.

The position of the substituent (R1 and R2) effect was found largely to affect the regioselectivity in the PB reaction of uracils with benzophenone derivatives (Scheme 7.10) [26].

As mentioned in Section 7.2, when the electron transfer reaction between electron-rich alkenes and excited carbonyl compounds is energetically favorable, the **RI** pair becomes an important intermediate in photochemical [2 + 2] cycloaddition reactions (Scheme 7.5). The regioselectivity of these reactions may differ from that observed for the PB reaction involving 1,4-triplet biradical intermediates. Typical examples of PB reactions with very electron-rich alkenes, ketene silyl acetals ($E_{ox} = 0.9$ V vs SCE), have been reported (Scheme 7.11) [27]. Thus, 2-alkoxyoxetanes were selectively formed as a result of the PB reaction with benzaldehyde or benzophenone derivatives, whereas a selective formation of 3-alkoxyoxetanes was observed in less electron-rich alkenes (see Scheme 7.9). When *p*-methoxybenzaldehyde was used in the photochemical reaction, the regioselectivity was less than that observed in the case of benzaldehyde. This dramatic decrease in regioselectivity provided evidence that the selective formation of 2-alkoxyoxetanes occurred via **RI** pair intermediates. It should be noted that the stereoselectivity is also completely different from that associated with triplet 1,4-biradicals (*vide infra*).

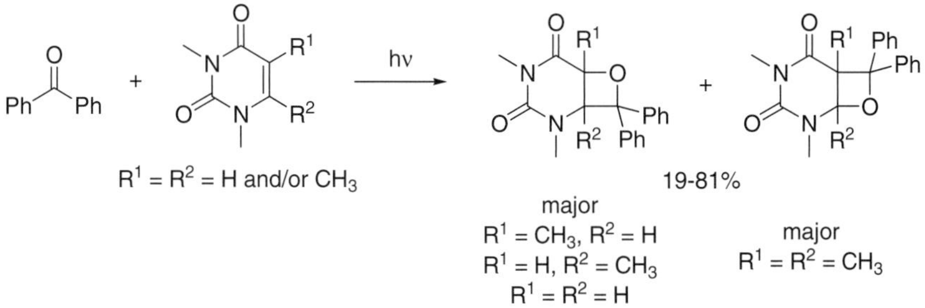

Scheme 7.10 The PB reaction of uracil derivatives with benzophenone.

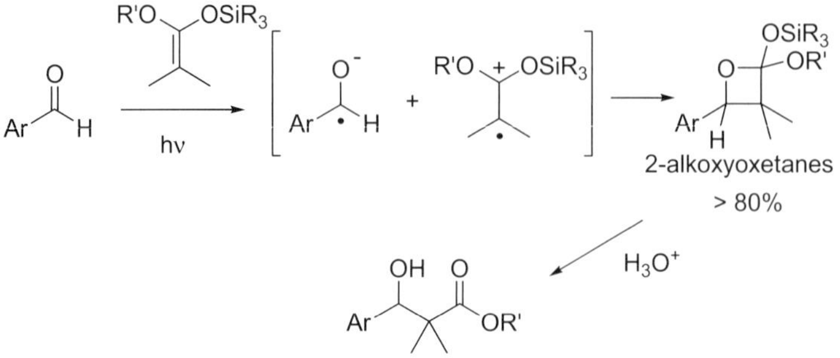

Scheme 7.11 Regioselective formation of oxetanes via radical ion pairs.

The site-selectivity (i.e., chemoselectivity or double-bond selection) of the PB reaction has been widely investigated in the photochemical reaction of unsymmetrically substituted furans with benzophenone (Scheme 7.12) [28]. The oxetane ring was found always to be formed at the double bond bearing alkyl substituents in the furan ring, under irradiation conditions ($-10\,°C$).

Thus, the bicyclic oxetane **OX2** was found to be selectively formed in the PB reaction of 2-methyl-, 2,3-dimethyl-, 2-hydroxymethyl-, and 3-methylfuran. However, in the PB reaction of 2,4-dimethylfuran ($R^1 = H$, $R^2 = H$, $R^3 = CH_3$, $R^4 = H$), a 1:1 mixture of the oxetanes **OX1** and **OX2** was observed under the similar photochemical reaction conditions.

In the PB reaction of unsymmetrically substituted furans with aldehydes, the site-selectivity was reported as quite difficult to control (Scheme 7.13). Thus, a 1.3:1 mixture of oxetanes was formed in the PB reaction of 2-methylfuran with benzaldehyde. Schreiber and coworkers found that the site-selectivity could be controlled by using bulky substituents at the furan ring [29a], and consequently the less-substituted oxetanes were selectively prepared in the PB reaction (Scheme 7.13). On the other hand, a highly site-selective formation of the more-substituted oxetanes was reported in the PB reaction of acetylfurans with aromatic aldehydes (Scheme 7.13) [29b]. A high *exo*-selectivity was also observed in the PB reaction with aldehydes (*vide infra*).

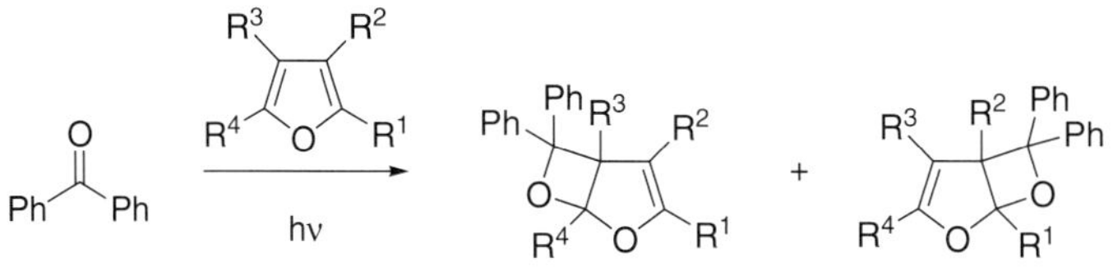

	OX1	OX2
$R^1 = CH_3$, R^2-$R^4 = H$;		98%
$R^1 = R^2 = CH_3$, R^3-$R^4 = H$;		~100%
$R^1 = CH_2OH$, R^2-$R^4 = H$;		80%
$R^1 = H$, $R^2 = CH_3$, R^3-$R^4 = H$;		98%
$R^1 = CH_3$, $R^2 = H$, $R^3 = CH_3$, $R^4 = H$;	50%	50%

Scheme 7.12 Site selectivity (chemoselectivity) in the PB reaction of unsymmetrically substituted furans with benzophenone.

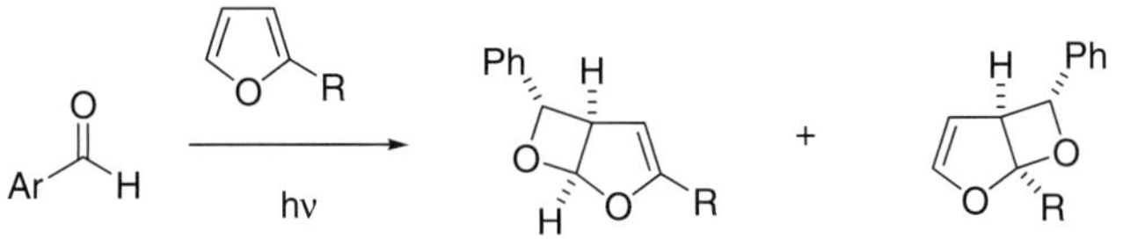

7.3 Regioselective and Site-Selective Syntheses of Oxetanes

Ar	R	ratios (yields)
Ph	CH_3	1 : 1.3 (89%)
Ph	CH_2OH	1 : 1.5 (80%)
Ph	$SiMe_3$	2.5 : 1 (40%)
Ph	Si^iPr_3	> 20 : 1 (56%)
4-CNC_6H_4	CH_3	1 : 2 (65%)
4-CNC_6H_4	$COCH_3$	1 : > 20 (70%)

Scheme 7.13 The site selectivity in the PB reaction of unsymmetrically substituted furans with aromatic aldehydes.

Recently, a notable temperature-related effect was reported for site-selectivity (double-bond selectivity or chemoselectivity) in the PB reaction of unsymmetrically substituted furans (Scheme 7.14) [30]. For example, the selective formation of the more substituted oxetane, **OX1**, was observed during the PB reaction of 2-methylfuran with benzophenone at a high temperature (61 °C). However, a 58 : 42 mixture of the oxetanes, **OX1** and **OX2**, was reported at low temperature (−77 °C). This notable effect of temperature could be explained by the relative population of conformers of the intermediary triplet 1,4-biradicals, **T-BR1** and **T-BR2**. The excited benzophenone was considered to attack the double bonds equally so as to produce a mixture of the conformers of **T-BR1** and **T-BR2**; however, at low temperature the conformational change was suppressed. Thus, the site-random formation of oxetanes **OX1** and **OX2** was observed after the ISC process. Nonetheless, at high

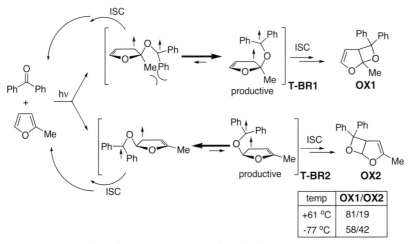

temp	OX1/OX2
+61 °C	81/19
−77 °C	58/42

Scheme 7.14 The effect of temperature on site selectivity. ISC, intersystem crossing.

temperatures a double-bond selection – that is, **OX1** versus **OX2** – was determined by the relative population of the conformers of **T-BR1** and **T-BR2**, as conformational change would be faster than ISC. In fact, the quantum yield for oxetane formation at low temperature was greater than that at high temperature.

7.4
Stereoselective Syntheses of Oxetanes

As mentioned in Section 7.2, stereoselectivity can arise from the bond-formation and bond-breaking step via the intermediary biradicals or **RI** pairs. The bond formation step should essentially be downhill, because bond formation represents the coupling between two radicals, or between the cation and anion parts. Regardless of such barrier-less processes, there are two contrasting examples in which almost perfect stereoselectivity was observed in the PB reaction of benzaldehyde (Scheme 7.15). Thus, in the PB reaction of furan with benzaldehyde, the highly *exo*-selective formation of bicyclic oxetane was found to occur in high yield [31]. But in sharp contrast, Griesbeck and coworkers achieved a breakthrough in the regioselective and stereoselective preparation of oxetanes via the PB reaction of vinyl ethers. Thus, a highly *endo*-selective formation of bicyclic oxetanes was observed as a result of the PB reaction between dihydrofuran and benzaldehyde (Scheme 7.15) [32].

Griesbeck and colleagues proposed a reliable model that would predict the stereoselectivity in the PB reaction of the dihydrofuran derivatives (Scheme 7.16). Thus, the "Griesbeck Model" [33] explains the stereoselectivity of oxetanes formed in the PB reactions of cyclic alkenes.

However, this model can also be generalized to other cyclicalkenes. The ISC reactive conformation of the intermediary triplet biradicals is important for stereoselectivity, as the rate constant for ISC, which is controlled by a spin-orbit-coupling (SOC) mechanism, is heavily dependent on the orientation of the two spin centers. The importance of the ISC process was reasonably proved by the low *endo*-selectivity in PB reactions with naphthaldehydes or aliphatic aldehydes, in which the excited singlet states may react with alkenes [34]. The stereoselectivity observed in the PB

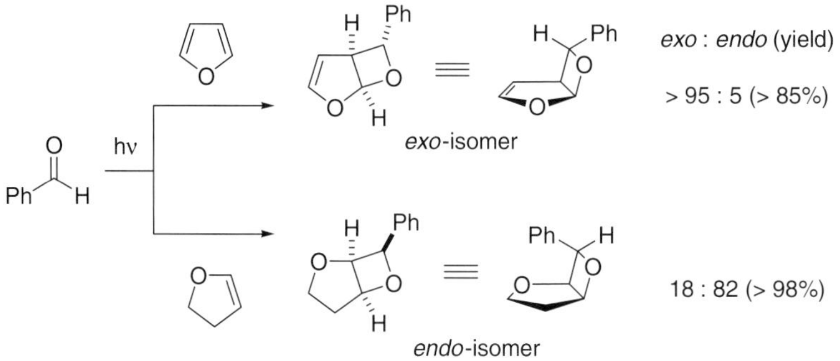

Scheme 7.15 Stereoselectivity of the PB reaction of cyclic ethers with benzaldehyde.

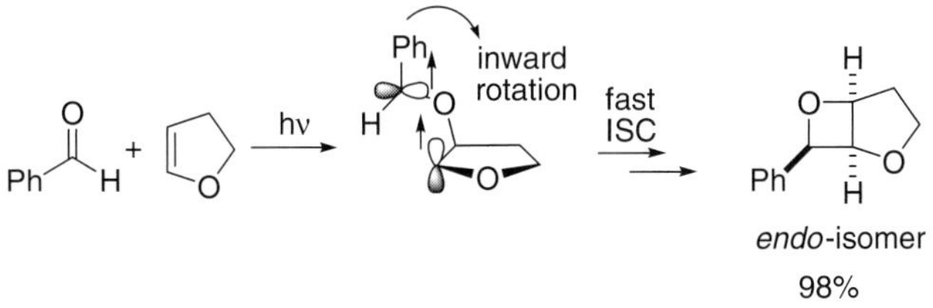

Scheme 7.16 Endo-selective formation of oxetanes in the PB reaction of dihydrofuran with benzaldehyde (the Griesbeck Model).

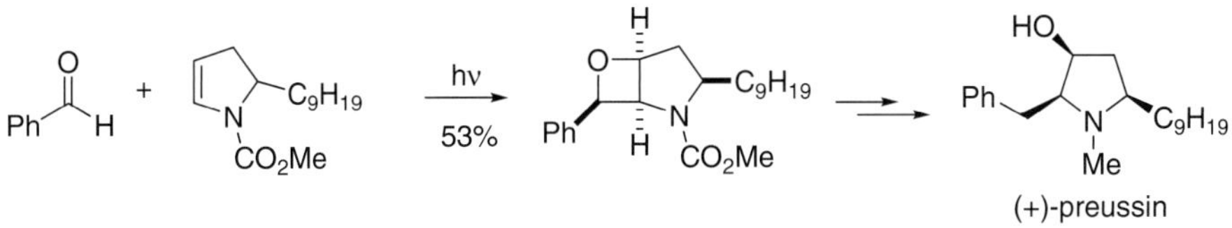

Scheme 7.17 Cis-selective formation of oxetanes in the PB reactions of enamines.

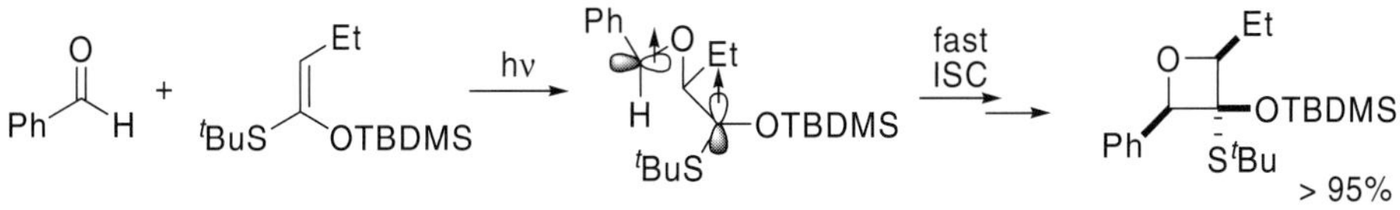

Scheme 7.18 Stereoselective formation of oxetane in the PB reaction of O,S-ketene acetal.

reactions with enamines can also be explained by the Griesbeck Model (Scheme 7.16). Thus, a cis-selective formation of oxetanes was reported by Bach and coworkers, despite the isomer being (in theory) thermodynamically less stable than the trans-isomer (Scheme 7.17) [35]. The selective formation of the endo-iosomer could be converted to the antifungal pyrrolidinol alkaloid (+)-preussin.

In the PB reaction using O,S-ketene silyl ether, oxetane formation was highly stereoselective (Scheme 7.18) [36]. Such stereoselectivity could be explained by the preferred conformation of the intermediary triplet 1,4-biradical for the ISC process.

The exo-selective formation of bicyclic oxetanes of furan derivatives is reported as being successful when applied to the synthesis of natural products (Scheme 7.19) [15e,37]. For example, the total synthesis of asteltoxin and avenaciolide was achieved from 3,4-dimethylfuran and furan, with the PB reaction being used as an initial step of the synthesis.

In the PB reaction of furan derivatives with benzaldehyde (as shown in Scheme 7.13), a site-random – but highly stereoselective – formation of exo-oxetanes was reported (Scheme 7.13) [15q,30]. The highly exo-selective formation of oxetanes was explained by the conformational stability of the intermediary triplet biradicals. The anomeric effect was proposed to play an important role in stabilization of the exo-isomer precursor (Scheme 7.20). Thus, the inside conformer which should

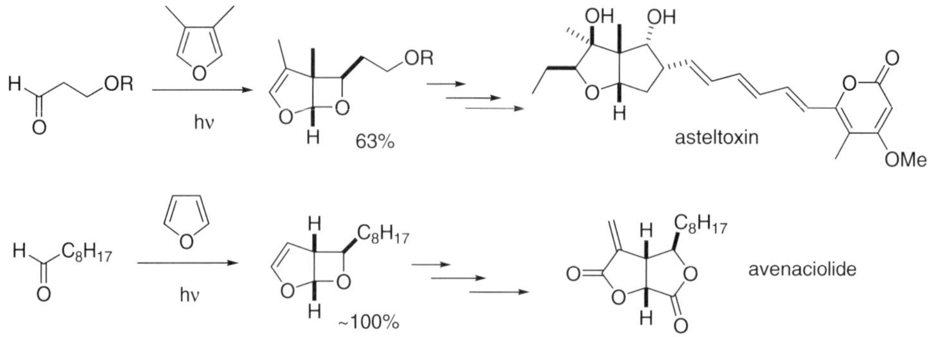

Scheme 7.19 The total synthesis of asteltoxin and avenaciolide.

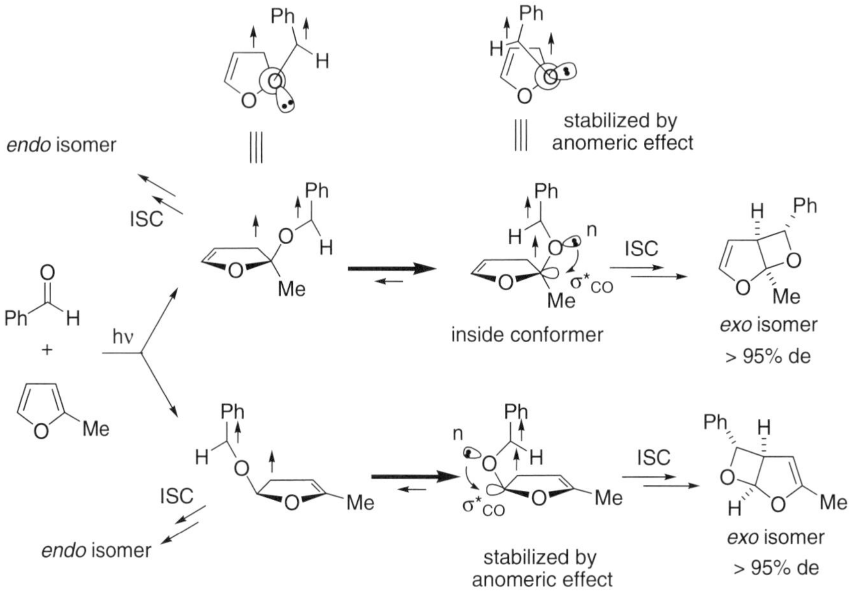

Scheme 7.20 The *exo*-selective formation of oxetanes.

serve as the precursor of the *exo*-bicyclic oxetane was stabilized by the electron delocalization of the lone-pair electrons to the low-lying C−O σ* orbital.

A highly *exo*-selective formation of bicyclic oxetanes was also observed during the PB reaction with oxazole derivatives [38] and vinylene carbonate [39] (Scheme 7.21).

The stereoselective formation of bicyclic oxetanes was also reported in the PB reactions of diketones (Scheme 7.22). Here, Mattay and Griesbeck reported the *endo*-selective formation of oxetanes in the PB reaction of 1,3-dioxol derivatives with some methyl pyruvates [40]. The PB reaction of arylglyoxylates with furan was also found to produce stereoselectively the bicyclic oxetanes (Scheme 7.23) [41]. Neckers and coworker demonstrated the highly efficient formation of oxetanes in the intramolecular PB reaction (Scheme 7.24) [42].

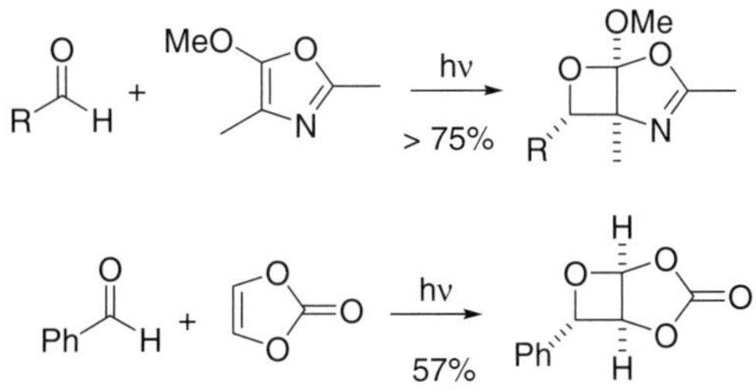

Scheme 7.21 The *exo*-selective formation of bicyclic oxetanes during the PB reactions of oxazoles and vinylene carbonate.

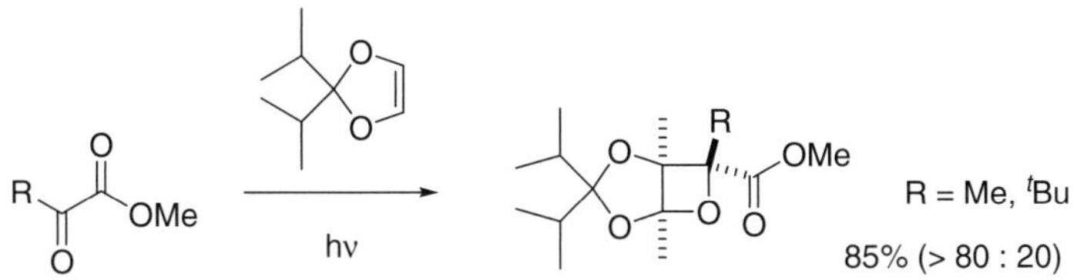

Scheme 7.22 Stereoselective formation of bicyclic oxetanes in the PB reaction of 1,3-dioxol derivatives.

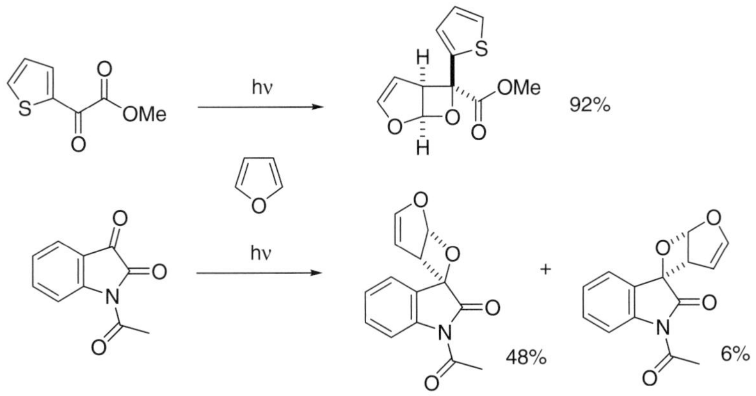

Scheme 7.23 Stereoselective synthesis of bicyclic oxetanes in the PB reaction of arylglyoxylates.

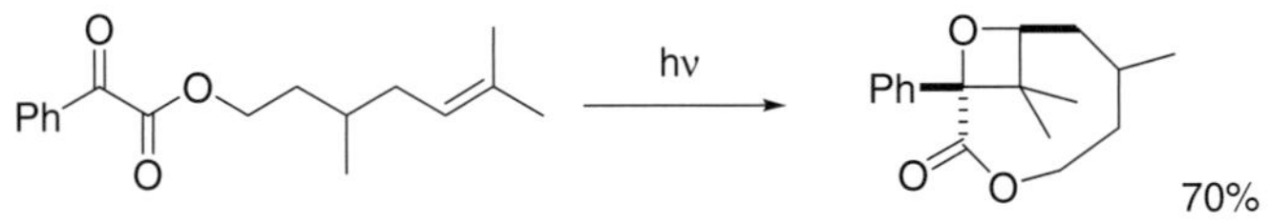

Scheme 7.24 Stereoselective synthesis of oxetanes in the intramolecular PB reaction of arylglyoxylate derivative.

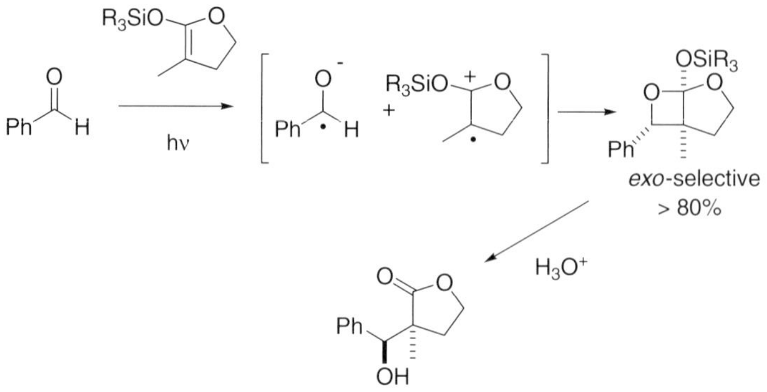

Scheme 7.25 Regioselective and stereoselective formation of oxetanes via radical ion pairs.

It should be noted that the stereoselectivity is also completely different from that associated with triplet 1,4-biradicals. Thus, a highly *exo*-selective formation of bicyclic oxetanes was observed during PET-promoted PB reactions, whereas a highly *endo*-selective formation of bicyclic oxetanes was reported for PB reactions that proceeded via triplet 1,4-biradicals (see Scheme 7.25). The competitive reaction pathway for electron-rich alkenes explained a notable solvent effect on the regioselectivity and stereoselectivity of the PB reaction of dihydrofuran (Scheme 7.15). Thus, an *endo*-selective formation of 3-alkoxyoxetane was observed when using benzene, whereas the *exo*-isomer of 2-alkoxyoxetane was detected as a product of the PB reaction in acetonitrile (Scheme 7.15).

Bach and coworkers observed both regioselective and stereoselective oxetane formation during the PB reaction of acyclic vinyl ethers (Scheme 7.26) [15n]. The stereoselectivity observed for such photochemical reactions cannot be explained using the Griesbeck Model, even though triplet, 14-biradicals were proposed as intermediates. Thus, the stereoselectivity was proposed to be largely dependent on product stability.

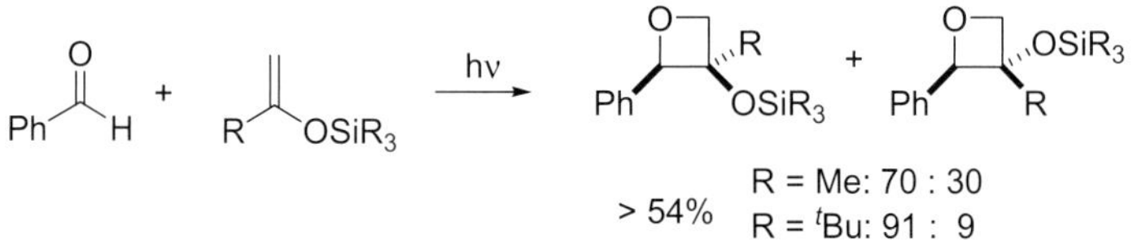

Scheme 7.26 Regioselective and stereoselective formation of oxetanes in the PB reactions of silyl enol ethers.

Adam and coworkers reported the regioselective and diastereoselective formation of oxetanes during the PB reaction of allylic alcohols (Scheme 7.27) [43, 44]. This group proposed that hydrogen-bond interactions in the exciplex played an important role in controlling the selectivity. D'Auria and coworkers also observed a site-selective and diastereoselective formation of oxetanes in the PB reaction of 2-furylmethanol derivatives (Scheme 7.27) [45].

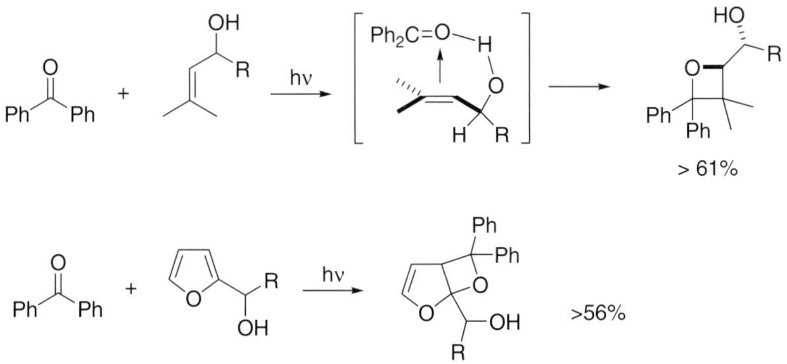

Scheme 7.27 Regioselective and diastereoselective formation of oxetanes in the PB reaction of allylic alcohols.

However, hydroxy-directed diastereoselectivity was not generalized to face-selectivity in the PB reaction of hydroxy-substituted dihydrofuran and furan derivatives (Scheme 7.28) [46].

In an example of solid-state photochemistry, an unexpected *exo*-selective formation of bicyclic oxetane was reported by Kang and Scheffer (Scheme 7.29) [47]. When a solid-state ketone was irradiated using a medium-pressure Hg lamp, via a Pyrex filter (>290 nm), the *exo*-selective oxetane formation of oxetane was predominant (yield 91%). In acetonitrile-solution photochemistry, the radical coupling product (43%) was the only isolable product.

In 2005, Greaney and coworkers applied the PB reaction to the synthesis of merrilactone A (Scheme 7.30) [48]. Very recently, Hammaecher and Portella developed a clean formation of the intramolecular PB reaction of acylsilanes (Scheme 7.31) [49].

Zamojski, Scharf, and Bach and colleagues have shown the PB reaction to be applicable for the asymmetric synthesis of oxetanes. In 1982, Zamojski and

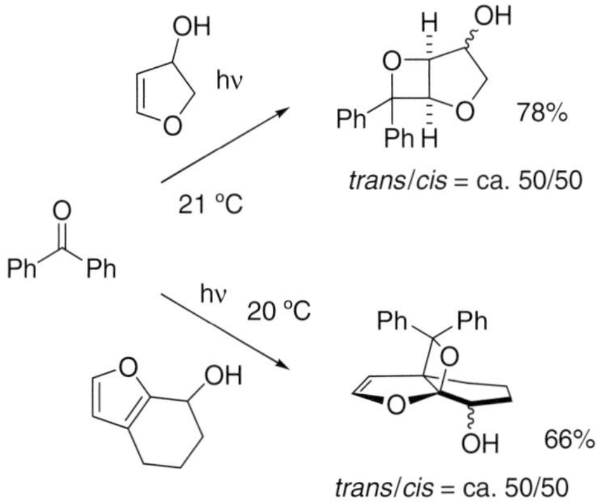

Scheme 7.28 Face selectivity in the PB reactions of hydroxy-substituted dihydrofurans and furan derivatives.

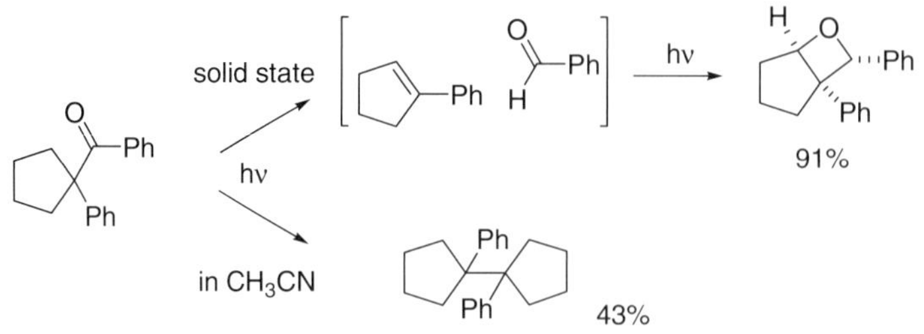

Scheme 7.29 The *exo*-selective formation of oxetane in a solid-state photochemical reaction.

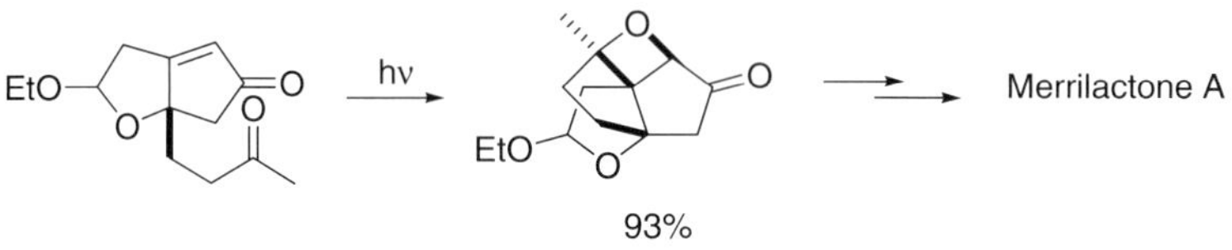

Scheme 7.30 Synthesis of merrilactone A using the intramolecular PB reaction.

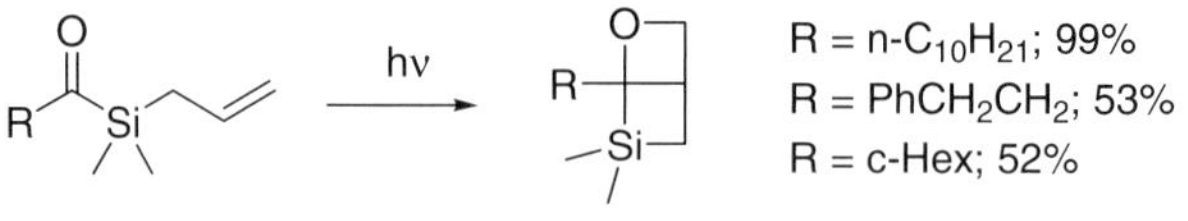

Scheme 7.31 The PB reaction of acylsilanes.

coworkers reported the PB reaction of furan with phenylglyoxylic acid esters [50], while Scharf and colleagues undertook a thorough investigation of chiral induction in the PB oxetane formation reactions of tetramethylethylene, or of diethylketene-acetal, with phenylglyoxylic acid esters (Scheme 7.32) [51]. Subsequently, a moderate to high diastereoselectivity was observed in the oxetane formation reactions, although the chiral auxiliary, such as R* = 8-phenylmenthyl, is separated from the reaction center.

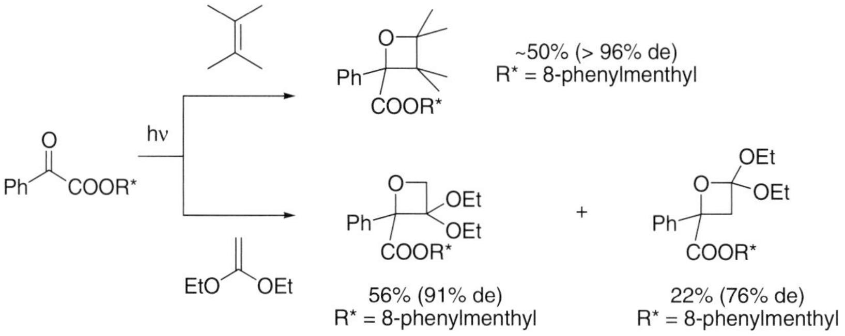

Scheme 7.32 Enantioselective synthesis of oxetanes in the PB reaction of glyoxylate derivatives.

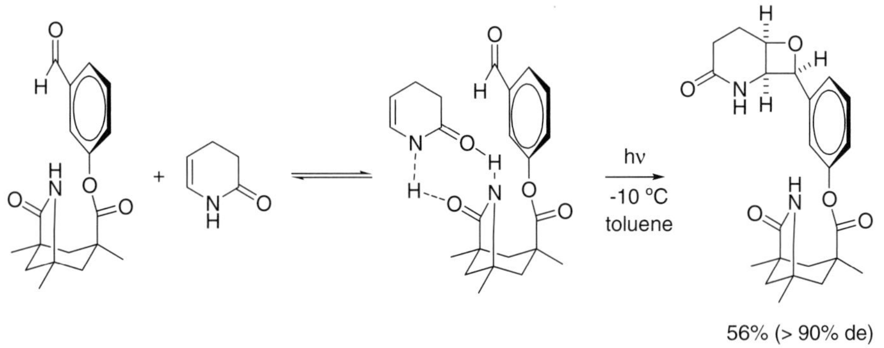

Scheme 7.33 Enantioselective synthesis of oxetanes in the hydrogen-bonding network.

A highly diastereoselective oxetane formation was identified in the PB reaction of dihydropyridone with a *m*-hydroxybenzaldehyde derivative (Scheme 7.33). The chiral auxiliary, when bound to the aldehyde, offered a binding site to which the reaction partner could attach by two hydrogen bonds. In the hydrogen-bonded complex that was produced, the two enantiotopic faces of the alkene could be differentiated [52]

7.5
Concluding Remarks

In this chapter, recent developments in the regioselective, site-selective, and stereoselective preparation of oxetanes have been summarized. The relative nucleophilicity of the alkene carbons was seen to be important for regioselectivity, in addition to the well-known "radical stability rule." Likewise, the three-dimensional structures of the triplet 1,4-biradicals were seen to play an important role in stereoselectivity. For photochemical reactions that proceed via radical ion pairs, the spin and charge distributions are crucial determinants of regioselectivity. It follows that the concepts used in selective oxetane synthesis should stimulate future investigations into the mechanistically and synthetically fascinating Paternò-Büchi-type reactions.

References

1 Ringner, B., Sunner, S., and Watanabe, H. (1971) Enthalpies of combustion and formation of some 3,3-disubstituted oxetanes. *Acta Chemica Scandinavica*, **25**, 141–146.

2 Bertolini, F., Crotti, S., Bussolo, V.D., Macchia, F., and Pineschi, M. (2008) Regio- and stereoselective ring opening of enantiomerically enriched 2-aryl oxetanes and 2-aryl azetidines with aryl borates. *Journal of Organic Chemistry*, **73**, 8998–9007.

3 Huang, J.-M., Yokoyama, R., Yang, C.-S., and Fukuyama, Y. (2000) Merilactone A, a novel neurotrophic sesquiterpene dilactone from *Illicium merrillianum*. *Tetrahedron Letters*, **41**, 6111–6114.

4 Bhagwat, S.S., Hamann, P.R., and Still, W.C. (1985) Synthesis of thromboxane A2. *Journal of the American Chemical Society*, **107**, 6372–6376.

5 Norbeck, D.W. and Kramer, J.B. (1988) Synthesis of (−)-oxetanocin. *Journal of the American Chemical Society*, **110**, 7217–7218.

6 Kawahata, Y., Takatsuto, S., Ikekawa, N., Murata, M., and Omura, S. (1986) Synthesis of a new amino acid-antibiotic, oxetin and its three stereoisomers. *Chemical & Pharmaceutical Bulletin*, **34**, 3102–3210.

7 Miller, R.W.R., Powell, R.G., and Smith, C.R. Jr. (1981) Antileukemic alkaloids from *Taxus wallichiana* Zucc. *Journal of Organic Chemistry*, **46**, 1469–1474.

8 Suzuki, T. and Kurosawa, E. (1979) New bromo acetal from the marine alga, *Laurencia nipponica* Yamada. 1. *Chemistry Letters*, 301–304.

9 (a) Sasaki, H. (2007) Curing properties of cycloaliphatic epoxy derivatives. *Progress in Organic Coatings*, **58**, 227–230; (b) Nishikubo, T., Kameyama, A., and Kudo, H. (2006) Synthesis of Polymers with well-defined structures by novel ring-opening reactions of oxetanes. *Journal of Synthetic Organic Chemistry (Yuki Gosei, Kagaku Kyokaishi)*, **64**, 934–946.

10 (a) Derick, C.G. and Bissel, D.W. (1916) Studies of trimethylene oxide. 1. Preparation and characterization. *Journal of the American Chemical Society*, **38**, 2478–2486; (b) Allen, J.S. and Hibbert, H. (1934) Studies on reacting relating to carbohydrates and polysaccharides. The oxygen valence angle and the structure of glucose and related compounds. *Journal of the American Chemical Society*, **56**, 1398–1403; (c) Searles, S. Jr. (1951) The reaction of trimethylene oxide with Grignard reagents and organolithium compounds. *Journal of the American Chemical Society*, **73**, 124–125; (d) Biggs, J. (1975) Three new convenient preparation of oxetane. *Tetrahedron Letters*, **16**, 4285–4286; (e) Still, W.C. (1976) Allyloxycarbanions, cyclizations to vinyl oxetanes. *Tetrahedron Letters*, **17**, 2115; (f) Fischer, W. and Grob, C.A. (1978) Fragmentation reactions. 28. Competing fragmentation, substitution and elimination in the solvolysis of alkylated 3-chloropropanols and their ethers. *Helvetica Chimica Acta*, **61**, 2336–2350; (g) Koell, P. and Schultz, J. (1978) 1,6-Anhydrofuranosen. Selective Tosylierung der 1,6-Anhydro-β-D-glucofuranose. Darstellung der 1,6;3,5-dianhydro-α-l-idofuranose. *Tetrahedron Letters*, **19**, 49–50; (h) Denmark, S.E. (1981) Facile oxetane formation in a rigid bicycle[2.2.2]octane system. *Journal Organic Chemistry*, **46**, 3144–3147; (i) Bats, J.-P. and Moulines, J. (1976) Transposition D'oxiranes-ethanols par L'intermediatre D'alcoxy-etains. *Tetrahedron Letters*, **17**, 2249–2250.

11 Okuma, K., Tanaka, Y., Kaji, S., and Ohta, H. (1983) Reaction of Dimethyl-oxosulfonium methylide with epoxides. Preparation of oxetanes. *Journal of Organic Chemistry*, **48**, 5133–5134.

12 (a) Mattay, J., Gersdorf, J., and Freudenberg, U. (1983) Thermal and photochemical reactions of biacetyl with 1,1-diethoxyethene. "Umpolung" of the reactivity of biacetyl by photochemical induced electron transfer. *Tetrahedron Letters*, **25**, 817–820; (b) Mattay, J. and Buchkremer, K. (1988) Thermal and photochemical oxetane formation. A contribution to the synthesis of branched-chain aldonolactones. *Helvetica Chimica Acta*, **71**, 981–987; (c) Mattay, J., Gersdorf, J., and Buchkremer, K. (1987) Photoreaction of biacetyls with electron-rich olefins. An excited mechanism. *Chemische Berichte*, **120**, 307–318.

13 (a) Paternó, E. and Chieffi, G. (1909) Sintesi in chimica organic per mezzo della luce. Nota II. Composti degli idrocarburi non saturi con aldeidi e chetoni. *Gazzetta Chimica Italiana*, **39**, 341–361; (b) Büchi, G., Ihman, C.G., and Lipinsky, E.S. (1954) Light-catalyzed organic reactions. I. The reaction of carbonyl compounds with

2-methyl-2-butene in the presence of ultraviolet light. *Journal of the American Chemical Society*, **76** (17), 4327; (c) Yang, N.C., Nussim, M., Jorgenson, M.J., and Murov, S. (1964) Photochemical reactions of carbonyl compounds in solution. The Paternò-Büchi reaction. *Tetrahedron Letters*, **5** (48), 3657–3664.

14 Arnold, D.R., Hinman, R.L., and Glick, A.H. (1964) Chemical properties of the carbonyl n,π* state. The photochemical preparation of oxetanes. *Tetrahedron Letters*, **5** (22), 1425–1430.

15 (a) Arnold, D.R. (1968) The photocycloaddition of carbonyl compounds to unsaturated systems: The syntheses of oxetanes in *Advances in Photochemistry*, Vol. 6, John Wiley & Sons, pp. 301–349; (b) Jones, G. II (1981) Synthetic application of the Paternò-Büchi reaction, in *Organic Photochemistry* (ed. A. Padwa), Marcel Dekker, Inc., pp. 1–122; (c) Searles, S. (1984) Oxetanes and oxetenes, in *Comprehensive Heterocyclic Chemistry*, Vol. 7 (ed. W. Lwowski), pp. 363–402; (d) Carless, H.A.J. (1984) Photochemical synthesis of oxetanes, in *Synthetic Organic Photochemistry* (ed. W.M. Horspool), Plenum Press, pp. 425–487; (e) Porco, J.A. and Schreiber, S.L. (1991) The Paternò-Büchi reaction, in *Comprehensive Organic Synthesis*, Vol. 5 (ed. B.M. Trost), Pergamon Press, Oxford, pp. 151–192; (f) Griesbeck, A.G. (1995) Oxetane formation: intermolecular additions, in *CRC Handbook of Organic Photochemistry and Photobiology* (eds W.M. Horspool and P.-S. Song), CRC Press Inc., pp. 522–535; (g) Rivas, C. (1995) Oxetane formation: addition to heterocycles, in *CRC Handbook of Organic Photochemistry and Photobiology* (eds W.M. Horspool and P.-.S. Song), CRC Press Inc., pp. 536–549; (h) Griesbeck, A.G. (1995) Oxetane formation: stereocontrol, in *CRC Handbook of Organic Photochemistry and Photobiology* (eds W.M. Horspool and P.-.S. Song), CRC Press Inc., pp. 550–559; (i) Carless, H.A.J. (1995) Oxetane formation: intramolecular addition, in *CRC Handbook of Organic Photochemistry and Photobiology* (eds W.M. Horspool and P.-.S. Song), CRC Press Inc., pp. 560–569; (j) Griesbeck, A.G. and Bondock, S. (2004) Oxetane formation: stereocontol, in *CRC Handbook of Organic Photochemistry and Photobiology*, 2nd edn (eds W.M. Horspool and F. Lenci), CRC Press Inc., pp. 59.1–59.19; (k) Griesbeck, A.G. and Bondock, S. (2004) Oxetane formation: intermolecular additions, in *CRC Handbook of Organic Photochemistry and Photobiology*, 2nd edn (eds W.M. Horspool and F. Lenci), CRC Press Inc., pp. 60.1–60.20; (l) Abe, M. (2004) Photochemical oxetane formation: addition to heterocycles, in *CRC Handbook of Organic Photochemistry and Photobiology*, 2nd edn (eds W.M. Horspool and F. Lenci), CRC Press Inc., pp. 1627; (m) Griesbeck, A.G. (2005) Photocycloadditions of alkenes to excited carbonyls, in *Synthetic Organic Photochemistry* (eds A.G. Griesbeck and J. Mattay), Marcel Dekker, pp. 89–135; (n) Griesbeck, A.G. (1994) Intersystem crossing in triplet 1,4-biradicals: Conformational memory effects on the stereoselectivity of photocycloaddition reactions. *Accounts of Chemical Research*, **27**, 70–75; (o) Bach, T. (1997) The Paternò-Büchi reaction of 3-heteroatom-substituted alkenes as a stereoselective entry to polyfunctional cyclic and acyclic molecules. *Liebigs Annalen/Recueil*, 1627–1634; (p) Bach, T. (1998) Stereoselective intermolecular [2 + 2] photocycloaddition reactions and their application in synthesis. *Synthesis* (5), 683–703; (q) Bach, T. (2000) The Paternò-Büchi reaction of N-acyl enamines and aldehydes-the development of a new synthetic method and its application to total synthesis and molecular recognition studies. *Synlett*, (12), 1699–1707; (r) Griesbeck, A.G., Abe, M. and Bondock, S. (2004) Selectivity control in electron spin inversion processes: Regio- and stereochemistry of Paternò-Büchi photocycloadditions as a powerful tool for mapping intersystem crossing. *Accounts of Chemical Research*, **37** (12), 919–928;

(s) Abe, M. (2008) Recent progress regarding regio-, site-, and stereoselective formation of oxetanes in Paternò-Büchi reactions. *Journal of the Chinese Chemical Society*, **55**, 479–486.

16 Wager, P. and Park, B.-S. (1991) Photoinduced hydrogen atom abstraction by carbonyl compounds, in *Organic Photochemistry*, Vol. 11 (ed. A. Padwa), Marcel Dekker, Inc., pp. 227–366.

17 (a) Turro, N.J. and Farrington, G.L. (1980) of the fluorescence of 2-norboranone and derivatives by electron-rich and electron-poor ethylenes. *Journal of the American Chemical Society*, **102**, 6051–6052; (b) Turro, N.J. and Farrington, G.L. (1980) Photoinduced oxetane formation between 2-norbornanone with electron-poor ethylenes. *Journal of the American Chemical Society*, **102**, 6056–6065.

18 (a) Eriksen, J. and Plith, P.E. (1982) *Tetrahedron Letters*, **23**, 481; (b) Mattay, J., Gersdorf, J., and Buchkremer, K. (1987) *Chemische Berichte*, **120**, 307; (c) Gersdorf, J., Mattay, J., and Gorner, H. (1987) *Journal of the American Chemical Society*, **109**, 1203; (d) Eckert, G. and Goez, M. (1994) *Journal of the American Chemical Society*, **116**, 11999; (e) Bosch, E., Hubig, S.M., and Kochi, J.K. (1998) *Journal of the American Chemical Society*, **120**, 386.

19 Rehm, D. and Weller, A. (1970) Kinetics of fluorescence quenching by electron and hydrogen-atom transfer. *Israel Journal of Chemistry*, **8** (2), 259–271.

20 Buschmann, H., Scharf, H.-D., Hoffmann, N., and Esser, P. (1991) The isoinversion principle. A general selection model in chemistry. *Angewandte Chemie, International Edition*, **30** (5), 477–515.

21 Palmer, I.J., Ragazos, I.N., Bernardi, F., Olivucci, M., and Robb, M.A. (1994) An MC-SCF study of the photochemical Paternò-Büchi reaction. *Journal of the American Chemical Society*, **116**, 2121–2132.

22 (a) Schenck, G.O., Hartymann, W., and Steinmetz, R. (1963) Vierringstnthesen durch photosensibilisierte cycloaddition von dimethylmaleinsaureanhydrid an olefin. *Chemische Berichte*, **96**, 498–508; (b) Hammond, G.S. and Turro, N.J. (1963) Organic photochemistry. The study of photochemical reactions provides new information on the excited states of molecules. *Science*, **142**, 1541–1553; (c) Toki, S., Shima, K., and Sakurai, H. (1965) Organic photochemical reactions. I. The synthesis of substituted oxetanes by the photoaddition of aldehydes to furans. *Bulletin Chemical Society of Japan*, **38** (5), 760–762.

23 Schroeter, S.H. and Orlando, C.M. Jr (1969) Photocycloaddition of various ketones and aldehydes to vinyl ethers and ketene diethyl acetal. *Journal of Organic Chemistry*, **34** (5), 1181–1187.

24 Shimizu, N., Yamaoka, S., and Tsuno, Y. (1983) Synthesis of 2,2-diphenyl-3-oxetanol derivatives and their thermal or acid-catalyzed decomposition. *Bulletin of the Chemical Society of Japan*, **56**, 3853–3854.

25 Takuwa, A., Fujii, N., Tagawa, H., and Iwamoto, H. (1989) Photochemical reaction of benzophenones with allylic silanes. *Bulletin of the Chemical Society of Japan*, **62**, 336–338.

26 Kong, F.F., Zhai, B.C., and Song, Q.-H. (2008) Substituent effects on the regioselective formation of the Paternò-Büchi reaction of 5- or/and-6-methyl substituted uracils with 4,4′-disubstituted benzophenones. *Photochemical & Photobiological Sciences*, **7**, 1332–1336.

27 (a) Abe, M., Ikeda, M., Shirodai, Y., and Nojima, M. (1996) Regio- and stereoselective formation of 2-siloxy-2-alkoxyoxetanes in the photoreaction of cyclic ketene silyl acetals with 2-naphthaldehyde and their transformation to aldol-type adducts. *Tetrahedron Letters*, **37** (33), 5901–5904; (b) Abe, M., Shirodai, Y., and Nojima, M. (1998) Regioselective formation of 2-alkoxyoxetanes in the photoreaction of aromatic carbonyl compounds with β,β-dimethyl keten silyl acetals: notable solvent

and silyl group effects. *Journal of the Chemical Society, Perkin Transactions 1*, 3253–3260; (c) Abe, M., Ikeda, M., and Nojima, M. (1998) A stereoselective, tandem [2 + 2] photocycloaddition-hydrolysis route to aldol-type adducts. *Journal of the Chemical Society, Perkin Transactions 1*, 3261–3266.

28 (a) Rivas, C. and Payo, E. (1967) Synthesis of oxetanes by photoaddition of benzophenone to furans. *Journal of Organic Chemistry*, 32, 2918–2920; (b) Nakano, T., Rivas, C., and Perez, C. (1973) Configuration and stereochemistry of photoproducts by application of the nuclear Overhauser effect. Adduct of benzophenone with methyl-substituted furans and 2,5-dimethylthiophenone, and methyl-substituted maleic anhydrides with thiophen and its methyl derivatives, and benzothiophen. *Journal of the Chemical Society, Perkin Transactions 1*, 2322–2327.

29 (a) Schreiber, S.L., Desmaele, D., and Porco, J.A. Jr (1988) On the use of unsymmetrically substituted furans in the furan-carbonyl photocycloaddition reaction: Synthesis of a kaddurenone-ginkgolide hybrid. *Tetrahedron Letters*, 29, 6689–6692; (b) Carless, H.A. and Halfhide, A.F. Highly regioselective [2 + 2] photocycloaddition of aromatic aldehydes to acetylfurans. *Journal of the Chemical Society, Perkin Transactions 1*, 1081–1082.

30 Abe, M., Kawakami, T., Ohata, S., Nozaki, K., and Nojima, M. (2004) Mechanism of Stereo- and regioselectivity in the Paternò-Büchi reaction of furan derivatives with aromatic carbonyl compounds: Importance of the conformational distribution in the intermediary triplet 1,4-diradicals. *Journal of the American Chemical Society*, 126 (9), 2838–2846.

31 (a) Toki, S., Shima, K., and Sakurai, H. (1965) Organic photochemical reactions. I. The synthesis of substituted oxetanes by the photoaddition of aldehydes to furans. *Bulletin of the Chemical Society of Japan*, 38, 760–762; (b) Whipple, E.B. and Evanega, G.R. (1968) The assignment of configuration to the photoaddition products of unsymmetrical carbonyls to furan using pseudocontact shifts. *Tetrahedron*, 24, 1299–1310; (c) Schreiber, S.L., Hoveyda, A.H., and Wu, H.-J. (1983) A Photochemical route to the formation of threo aldols. *Journal of the American Chemical Society*, 105, 660–661; (d) Griesneck, A.G. and Stadtmuller, S. (1990) Regio- und stereoselektive photocycloadditionen aromatischer aldehyde an furan und 2,3-dihydrofuran. *Chemische Berichte*, 123, 357–362; (e) Hambalek, R. and Just, G. (1990) Trisubstituted oxetanes from 2,7-dioxabicyclo-[3.2.0]-hept-3-enes. *Tetrahedron Letters*, 31, 4693–4696.

32 Griesbeck, A.G. and Stadtmuller, S. (1990) Photocycloaddition of benzaldehyde to cyclic olefins: electronic control of endo stereoselectivity. *Journal of the American Chemical Society*, 112, 1281–1283.

33 (a) Abe, M., Torii, E., and Nojima, M. (2000) Paternò-Büchi photocyclo-addition of 2-siloxyfurans and carbonyl compounds. Notable substituent and carbonyl (aldehyde vs. ketone and singlet- vs. triplet-excited state) effects on the regioselectivity (double-bond selection) in the formation of bicyclic exo-oxetanes. *Journal of the Organic Chemistry*, 65, 3426–3431; (b) Kutateladze, A.G. (2001) Conformational analysis of singlet-triplet state mixing in Paternò-Büchi diradicals. *Journal of the American Chemical Society*, 123 (38), 9279–9282.

34 Griesbeck, A.G., Fiege, M., Bondock, S., and Gudipati, M.S. (2000) Spin-directed stereoselectivity of carbonyl-alkene photocycloadditions. *Organic Letters*, 2 (23), 3623–2625.

35 Bach, T. (1996) N-acyl enamines in the Paternò-Büchi reaction: stereoselective preparation of 1,2-amino alcohols by C–C bond formation. *Angewandte Chemie, International Edition in English*, 35 (8), 884–886.

36 Abe, M., Fujimoto, K., and Nojima, M. (2000) Notable sulfur atom effects on the

regio- and stereoselective formation of oxetanes on Paternò-Büchi photocycloaddition of aromatic aldehydes with silyl O,S-ketene acetals. *Journal of the American Chemical Society*, **122** (17), 4005–4010.

37 (a) Schreiber, S.L. and Satake, K. (1983) Application of the furan-carbonyl photocycloaddition reaction to the synthesis of the bis(tetrahydrofuran) moiety of asteltoxin. *Journal of the American Chemical Society*, **105**, 6723–6724; (b) Schreiber, S.L. and Hoveyda, A.H. (1984) Synthetic studies of the furan-carbonyl photocycloaddition reaction. A total synthesis of avenaciolide. *Journal of the American Chemical Society*, **106**, 7200–7202.

38 Griesbeck, A.G., Bonduck, S., and Lex, J. (2003) Synthesis of erythro-α-amino β-hydroxy carboxylic acid esters by diastereoselective photocycloaddition of 5-methoxyoxazoles with aldehydes. *Journal of Organic Chemistry*, **68** (26), 9899–9906.

39 Abe, M., Taniguchi, K., and Hayashi, T. (2007) Exo-selective formation of bicyclic oxetanes in the photocycloaddition reaction of carbonyl compounds with vinylene carbonate: the important role of intermediary triplet diradicals in the stereoselectivity. *ARKIVOC*, viii, 58–65.

40 Buhr, S., Griesbeck, A.G., Lex, J., Mattay, J., and Schroer, J. (1996) Stereoselectivity in the Paternò-Büchi reaction of 2,2-diisopropyl-1,3-dioxol with methyl trimethylpyruvate. *Tetrahedron Letters*, **37**, 1195–1196.

41 (a) Hu, S. and Neckers, D.C. (1999) Photocycloaddition and ortho-hydrogen abstraction reactions of methyl arylglyoxylates: structure dependent reactivities. *Journal of the Chemical Society, Perkin Transactions 2*, 1771–1778; (b) Zhang, Y., Xue, J., Gao, Y., Fun, H.-K., and Xu, J.-H. (2002) Photoinduced [2 + 2] cycloadditions (the Paternò-Büchi reaction) of 1-acetylisatin with enol ethers-regioselectivity, diastereoselectivity and acid catalysed transformations of the spirooxetane products. *Journal of the Chemical Society, Perkin Transactions* **1**, 345–353.

42 Hu, S. and Neckers, D.C. (1997) Rapid regio and diastereoselective Paternò-Büchi reaction of alkyl phenylglyoxylates. *Journal of Organic Chemistry*, **62**, 564–567.

43 Adam, W., Peters, K., Peters, E.M., and Stegmann, V.R. (2000) Hydroxy-group directed regio- and diastereoselective [2 + 2] photocycloaddition of benzophenone to chiral allylic alcohols. *Journal of the American Chemical Society*, **122** (12), 2958–2959.

44 Griesbeck, A.G. and Bondock, S. (2001) Paternò-Büchi reactions of allylic alcohols and acetates with aldehydes: hydrogen-bond interaction in the excited singlet and triplet states? *Journal of the American Chemical Society*, **123** (25), 6191–6192.

45 D'Auria, M., Racioppi, R., and Romaniello, G. (2000) The Paternò-Büchi reaction of 2-furylmethanols. *European Journal of Organic Chemistry*, (19), 3265–3272.

46 (a) Abe, M., Terazawa, M., Nozaki, K., Masuyama, A., and Hayashi, T. (2006) Notable temperature effect on the stereoselectivity in the photochemical [2 + 2] cycloaddition reaction (Paternò-Büchi reaction) of 2,3-dihydrofuran-3-ol derivatives with benzophenone. *Tetrahedron Letters*, **47**, 2527–2530; (b) Yabuno, Y., Hiraga, Y., and Abe, M. (2008) Sitte- and stereoselectivity in the photochemical oxetane formation reaction (Paternò-Büchi reaction) of tetrahydrobenzofuranols with benzophenone: hydroxyl-directed diastereoselectivity? *Chemistry Letters*, **37** (8), 822–823.

47 Kang, T., and Scheffer, J.R. (2001) An unexpected Paternò-Büchi reaction in the crystalline state. *Organic Letters*, **3** (21), 3361–3364.

48 Iriondo, J., Peras-Buceta, J.E., and Greaney, M.F. (2005) A Paternò-Büchi approach to the synthesis of merrilactone A. *Organic Letters*, **7**, 3969–3971.

49 Hammaecher, C. and Portella, C. (2008) New 6-oxa-2-silabicyclo[2.2.0]hexanes by photochemical conversion of acyl(allyl)(dimethyl)silanes. *Chemical Communications*, 5833–5835.

50 Jarosz, S. and Zamojski, A. (1982) Asymmetric photocycloaddition between furan and chiral alkyl glyoxylates. *Tetrahedron*, **38**, 1447–1451.

51 (a) Koch, H., Runsink, J., and Scharf, H.-D. (1983) Investigation of chiral induction in photochemical oxetane formation. *Tetrahedron Letters*, **24**, 3217–3220; (b) Buschmann, H., Scharf, H.-D., Hoffmann, N., Plath, M.W., and Runsink, J. (1989) Chiral Induction in photochemical reactions 10. The principle of isoinversion: A model of stereoselection developed from the diastereoselectivity of the Paternò-Büchi reaction. *Journal of the American Chemical Society*, **111**, 5367–5373.

52 Bach, T., Bergmann, H., and Harms, K. (1999) High facial selectivity in the photocycloaddition of chiral aromatic aldehyde and enamide induced by intermolecular hydrogen bonding. *Journal of the American Chemical Society*, **121**, 10650–10651.

8
Formation of a Five-Membered Ring
Ganesh Pandey and Smita R. Gadre

8.1
Introduction

Five-membered rings are one of the most common core structures found in natural products. The carbocycles of this particular ring size are considered to be important, as they are found in the D ring of steroids, in prostaglandins, and in carbanucleosides. Five-membered oxa- and aza-ring systems are also widely distributed in nature, and have attracted the interest of the scientific community, particularly in conjunction with the total synthesis of polyether natural products, or pyrrolidines with glycosidase-inhibiting activity. Developing methods to synthesize both five-membered carbocycles [1] as well as heterocycles [2, 3] have been intensively pursued by the synthetic groups, and remarkable progress has been made in their syntheses. These important molecular fragments have been synthetically accessed mainly through the following categories of reaction: (i) [3 + 2] cycloadditions between simpler molecules; (ii) radical and electrocyclizations of linear molecules; (iii) ring expansions of smaller carbo- as well as heterocycles such as cyclopropanes, oxiranes and aziridines, and so on. In this chapter we will describe the photochemical aspects of the above reactions in detail, as photochemistry serves as the powerful tool for the construction of variety of five-membered rings.

8.2
Formation of Five-Membered Rings: Intramolecular δ-H Abstraction

The lowest excited state of most ketones has the (n, π*) electronic structure, which gives the carbonyl (C=O) double bond a 1,2-biradical character. Therefore, the electron-deficient oxygen atom of this moiety, obtained upon excitation, acquires a radical reactivity, similar to the alkoxy radical. This excited state property of ketones leads to intramolecular H-abstraction to form 1-hydroxy-1, x-biradicals. Depending upon the structure and reaction conditions of the carbonyl compound, two common competing reactions may follow:

Handbook of Synthetic Photochemistry. Edited by Angelo Albini and Maurizio Fagnoni
Copyright © 2010 WILEY-VCH Verlag GmbH & Co. KGaA, Weinheim
ISBN: 978-3-527-32391-3

1. Coupling to produce cyclic alcohols.
2. Disproportionation back to ketone or to various enols.

The best-known example of such types of excited state chemistry of the ketones are the γ-hydrogen abstraction producing 1,4-biradicals. However, in the absence of a reactive γ-hydrogen, where the definition of "reactive" refers to both C−H bond strength and proximity to the carbonyl group, δ-hydrogen abstraction takes place by 1,5-biradicals which, on coupling, yield five-membered ring compounds. In this section, the reactions pertaining to the type of five-membered ring formation are categorized.

8.2.1
Formation of Cyclopentanol Ring System

It was shown very early by Wagner et al. [4] that several phenyl alkyl ketones containing either intrinsically more reactive δ C−H bonds compared to γ C−H or containing no γ C−H, undergo photocyclizations involving 1,5-biradicals to produce the corresponding cyclopentanol derivatives. For example, δ-methoxyvalerophenone (**1**) which has comparable reactivity at both γ C−H as well as δ C−H towards the phenone triplet, provides roughly equal amounts of the products from both the Type II reaction as well as 1,5-biradical cyclizations (Scheme 8.1). The cyclization of biradical intermediate **3** is reported to be slow in a hydrocarbon solvent, but increases

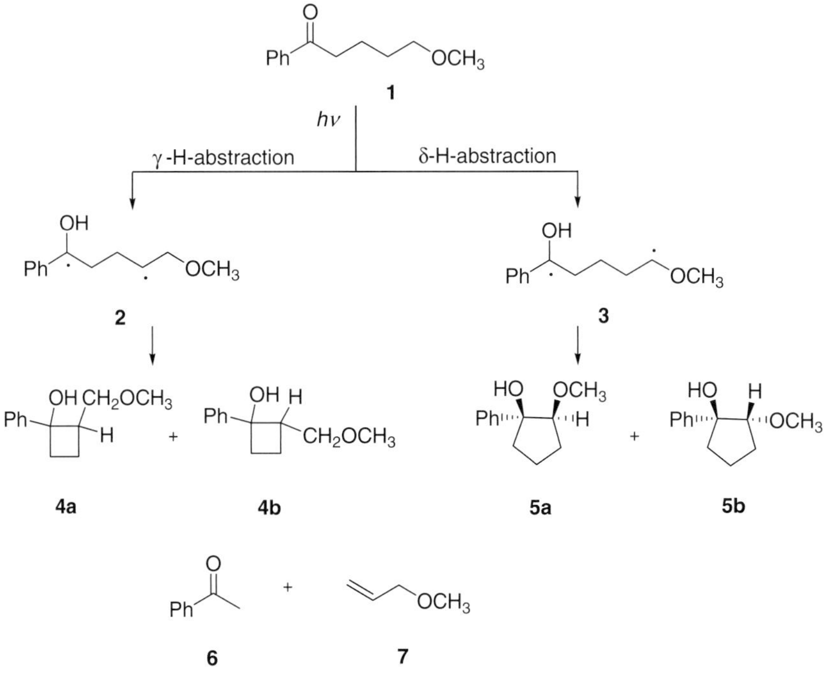

Scheme 8.1

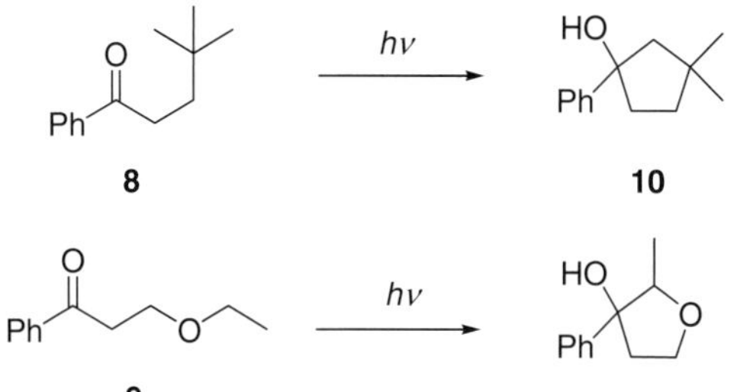

Scheme 8.2

slightly in t-BuOH, favoring the formation of cyclopentanols **5** with the −OH and methoxymethyl groups *trans* to each other. Formation of the *cis*-isomer is not favored in polar solvents in comparison to benzene.

Compounds such as γ-dimethylvalerophenones (**8**) and β-ethoxypropiophenone (**9**), which lack γ−H, produce exclusively cyclopentanol **10** and tetrahydrofuranol **11** derivatives, respectively, albeit in low quantum yields (Scheme 8.2). With benzene as solvent, the diastereomeric ratio for **11** was 5 : 1 (*trans*/*cis*), but was 1 : 1 in acetonitrile. The easy 1,5-H-transfer in these acyclic ketones was rationalized as involving a torsion-free, chair-like six-membered transition state.

Paquette [5] applied this chemistry very elegantly to the synthesis of 1,16-dimethyl-dodecahedrane (**15**), as shown in Scheme 8.3.

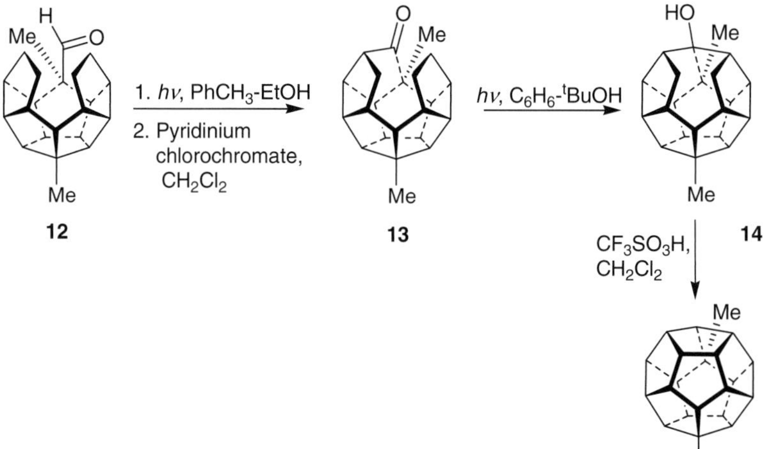

Scheme 8.3

8.2.1.1 Synthesis of Indanols

The irradiation of both 2,4-di-*tert*-butylbenzophenone (**16a**) and 2,4-di-*tert*-butylacetophenone (**16b**) are reported to undergo δ-H-abstraction from their triplet excited state to produce the corresponding 1,5-biradical **17** (Scheme 8.4) [6].

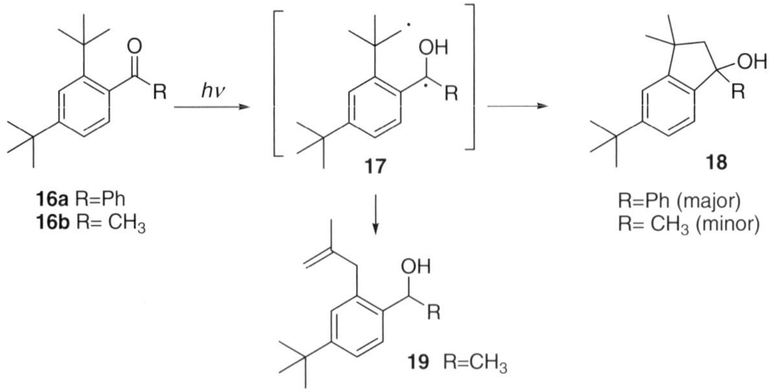

8 Formation of a Five-Membered Ring

Scheme 8.4

The biradical corresponding to **16a** produces indanol **18** quantitatively, whereas the biradical from **16b** undergoes disproportionation to form mainly **19**. The quantum efficiency for the formation of **18** was 0.03 in hydrocarbon solvent, and 1.0 in methanol. In contrast, the total quantum yield for **16b** was rather low (0.02–0.05) in both hydrocarbon and methanol solvents. These differences were ascribed to a considerably smaller dihedral angle between the carbonyl and *t*-butyl benzene in the triplet excited state of **16a** in comparison to **16b**, and a differing rotational freedom in the biradical **17** intermediate species.

The same group [7] has further reported the formation of two diastereomeric sets of indanols (**23Z, 23E**) and (**24Z, 24E**) from the irradiation of *o-tert*-amylbenzophenone (**20**), depending upon the regioselectivity of H-abstraction. (Scheme 8.5).

The product ratio (**23/24** = 1/1.6) in solution indicated the preference for secondary versus primary H-abstraction, while this was opposite ($\leq 7/1$) in solid. A detailed study has shown that there are no indications of any significant stereoelectronic

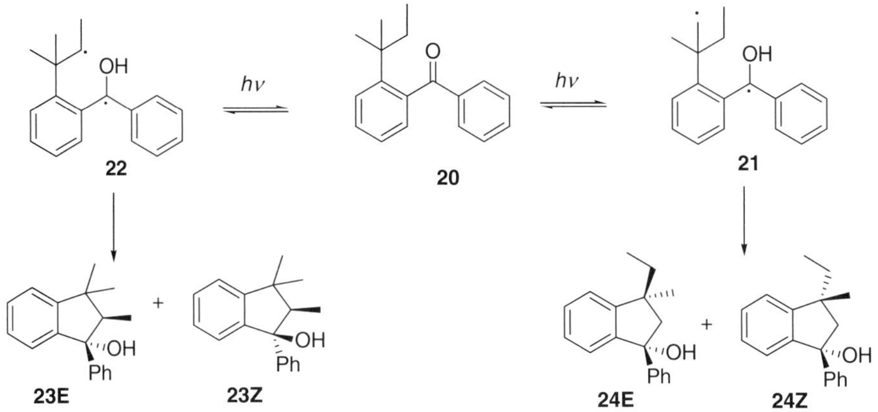

Scheme 8.5

requirements for H-abstraction and for cyclization of the 1,5-biradicals from their initial geometries, which produced indanols with very little Z/E diastereoselectivity in solution.

Wagner et al. [8] have further reported that the irradiation of several ring-substituted α-(o-tolyl) acetophenones (**25**) leads to quantitative photocyclization yielding 2-phenyl-2-indanols (**26**) in high quantum efficiency (0.42–1.0). Based on the large quantum yield and large triplet decay rates ($k_q\tau = 3.4$–$31\,M^{-1}$), it was suggested that the reactive *syn* conformer is readily accessible and the excited conformer attains equilibrium before the δ-H-abstraction takes place (Scheme 8.6).

Furthermore, it was shown that *alpha*- as well as *ortho*-substituents in such ketones retard the photocyclization due to other competing reactions. However, photocyclization showed interesting stereochemical trends, which were also strongly affected by the solvent polarity and the phase (Scheme 8.7). For example, the photocyclization of α-(o-ethyl phenyl acetophenone (**27**) either in benzene or as a solid favored the isomer with the methyl and phenyl group *trans* to R, due to the conformational preferences in the 1,5-biradical intermediate [9].

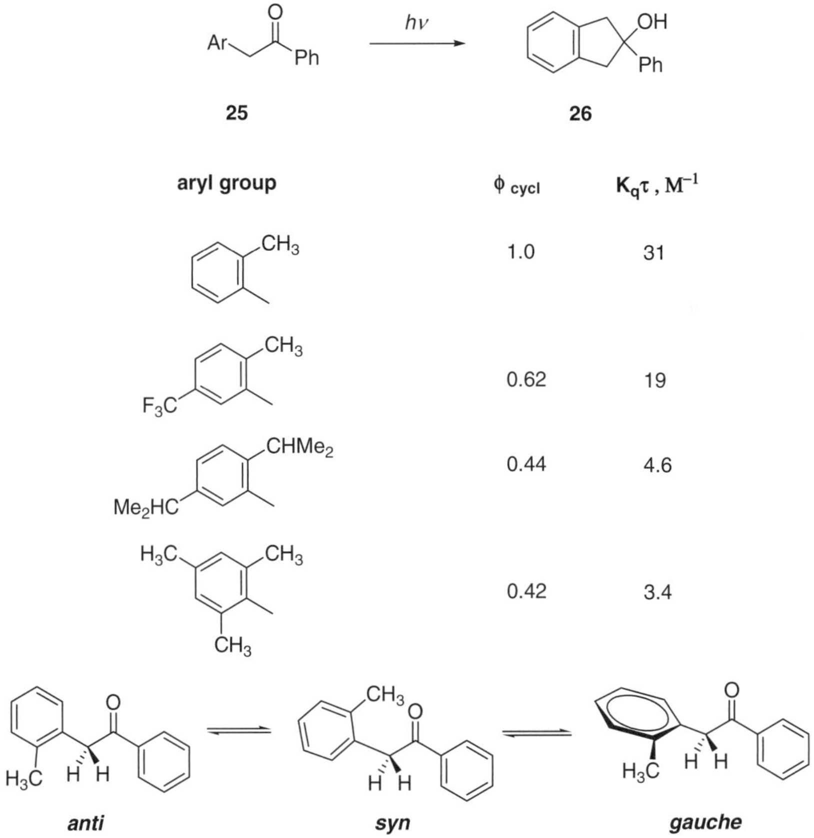

aryl group	Φ cycl	$K_q\tau$, M^{-1}
o-CH₃	1.0	31
F₃C, CH₃	0.62	19
Me₂HC, CHMe₂	0.44	4.6
H₃C, CH₃, CH₃	0.42	3.4

Scheme 8.6

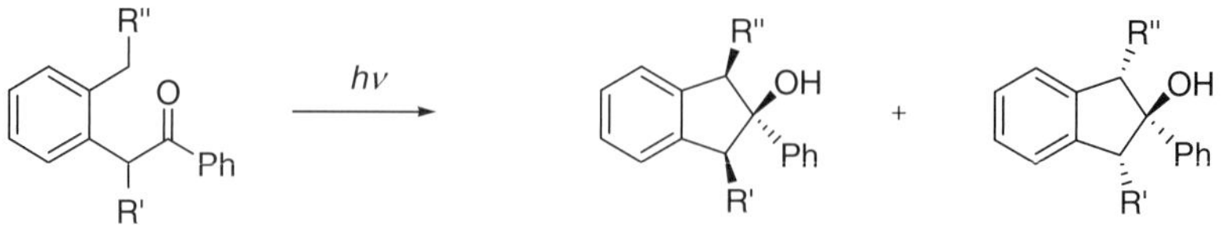

Scheme 8.7

		28a	28b
a R' = H, R" = CH$_3$	Benzene MeOH Crystal	95% 67% 100%	5% 33% ---
b R' = CH$_3$, R" = CH$_3$	Benzene MeOH	97% 70%	3% 30%
c R' = H, R" = Ph	Benzene MeOH	55% 54%	45% 46%

A decrease in the formation of the Z-isomer in MeOH as solvent is caused by a lowering of the energy of the preferred conformer due to H-bonding. The lack of diastereoselectivity in the case of α-(o-benzylphenyl)methyl phenyl ketone can be explained (Figure 8.1) by considering the difficulty for two benzene rings in the 1,5-biradical to conjugate fully, due to the half-occupied p-orbital.

Park et al. [10] have recently reported an efficient photocyclization of 1-benzoyl-1-(o-ethylphenyl) cyclopropane (**29**) to the corresponding indanol **31**. In this ketone, the cyclopropane ring bulk appears to be equivalent to that of two methyl groups, and forces the geometry of the 1,5-biradical intermediate suitable for cyclization (Scheme 8.8).

The Z/E ratio was reported to be opposite to that observed in earlier cases [9], ranging from 1:19 to 1:2.5 at between −72 °C and 90 °C in toluene, and reaching a value as low as 1 : 99 in methanol at 25 °C. The formation of E-indanol is explained by considering a larger rotational barrier of the 1,5-biradical for Z-forming geometry.

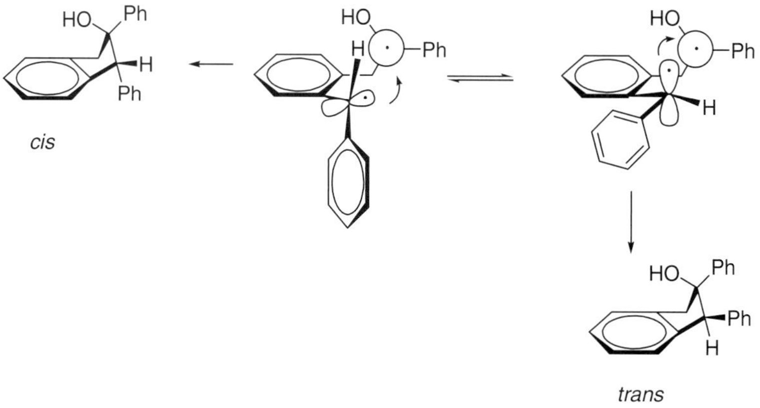

Figure 8.1 Transition state model for 1,5 biradical cyclization.

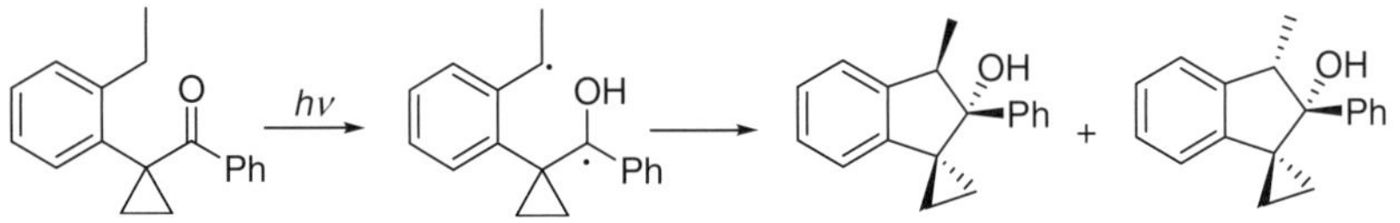

8.2 Formation of Five-Membered Rings: Intramolecular δ-H Abstraction

Scheme 8.8

Despite having the lowest triplet of π-π* character, 8-benzyl-1-benzoylnaphthalene (**32**) is reported to undergo δ-H-abstraction, followed by cyclization of the resultant 1,5-biradical, to produce **33** in high yield (Scheme 8.9) [11]. This is one of the rare examples where a π-π* triplet is involved in H-abstraction.

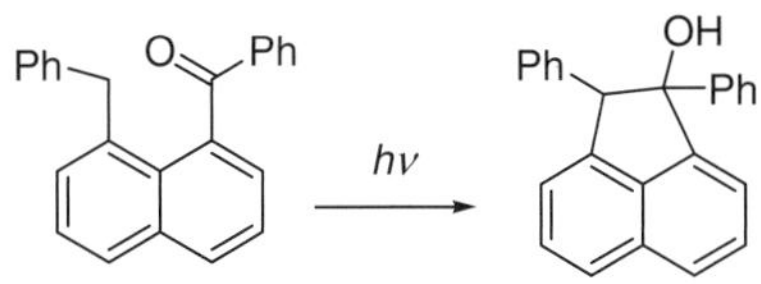

Scheme 8.9

8.2.2
Synthesis of Tetrahydrofuranols

One of the first examples of δ-hydrogen abstraction in acyclic ketones was the photocyclization of β-alkoxy ketones, in particular of β-ethoxypropiophenone (**34**), to the corresponding furanol derivatives **37** (Scheme 8.10). It was revealed that formation of enol **36** as well as a reversion to the starting ketone occur by 1,4-hydrogen transfer from the 1,5-biradical **35** causing the lower quantum efficiency for cyclization [12].

The quantum efficiency of this photocyclization was lower in alcoholic solvents than in hydrocarbons, which was in sharp contrast to the solvent effect observed on the formation quantum efficiencies of products resulting from γ-H abstraction. It was also observed that the addition of an alcoholic solvent lowered the overall

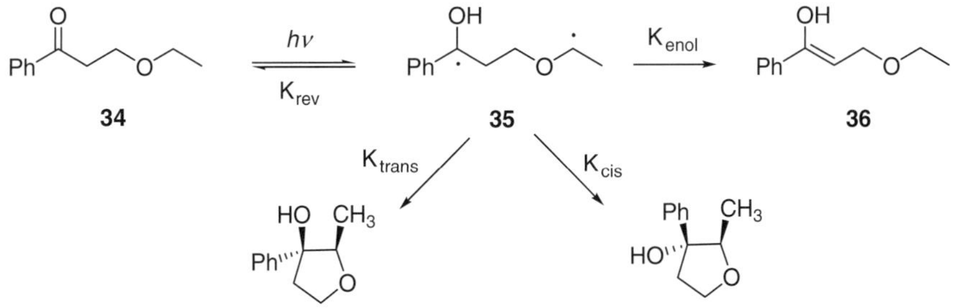

Scheme 8.10

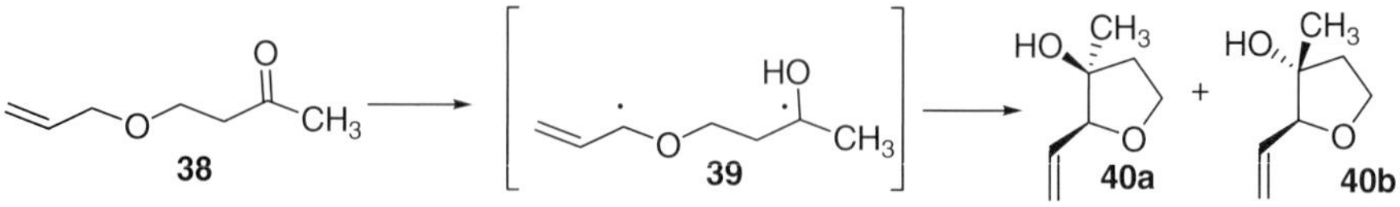

8 Formation of a Five-Membered Ring

Scheme 8.11

quantum efficiency of the reaction and reversed the *trans/cis* ratio of cyclized product due to the solvation of biradicals.

Subsequently, Carless et al. [13] also showed that the irradiation of β-alkoxy carbonyl compound **38** in benzene gives two stereoisomeric tetrahydrofuran-3-ols **40** in high yield, having biradical **39** as an intermediate (Scheme 8.11).

The same group has further shown that the alkenyl double bond geometry of geraniol- as well as nerol-derived ketones **41** and **43** remains unaffected during the photocyclizations. This suggests that the rate of cyclization is much higher than the rotation about the double bond in the allylic radical center of the biradical (Scheme 8.12).

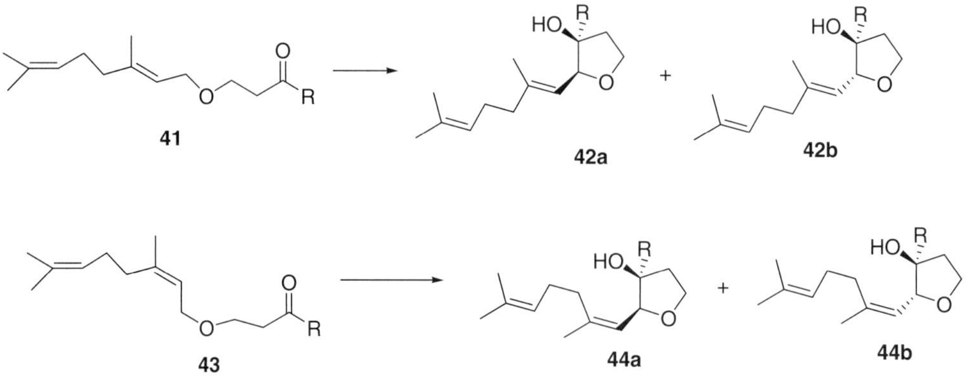

Scheme 8.12

The Descostes group [14] has used this strategy of cyclization for the synthesis of the analogues of the ionophore antibiotic with anomeric spiro-structure **46** by irradiating the 4-hydroxyl-2-butanone glucoside **45** in benzene (Scheme 8.13).

This cyclization was found to be more efficient with the β-anomer, and the new C–C bond was formed at the anomeric center by preferential axial anomeric hydrogen abstraction, followed by carbocyclization. The same group has further

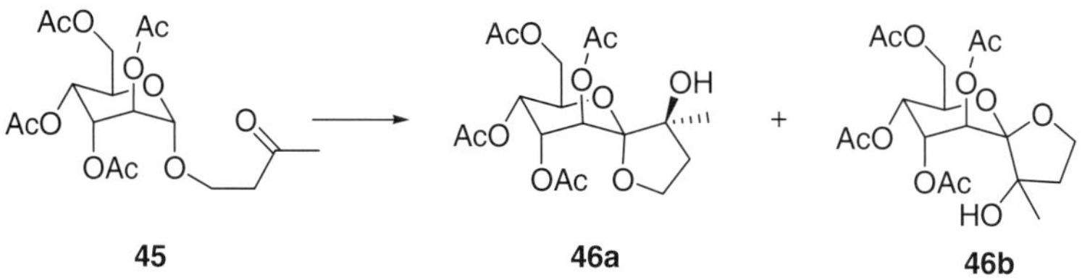

Scheme 8.13

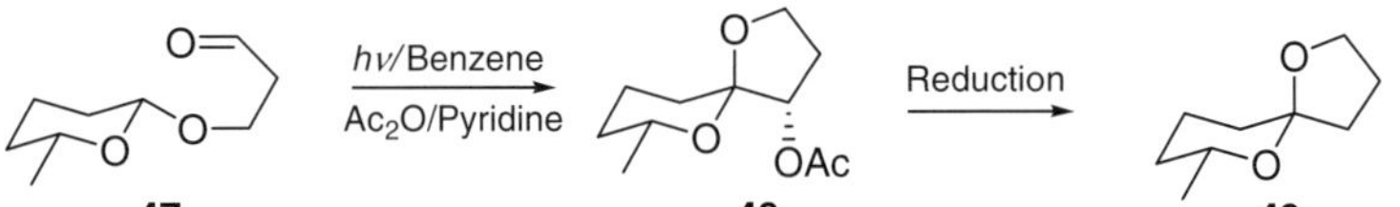

8.2 Formation of Five-Membered Rings: Intramolecular δ-H Abstraction

Scheme 8.14

extended this photocyclization strategy to synthesize dioxa-1,6-spiro[4.5]decane (**49**) as well as pheromones of *Paravespula vulgeris* L, by the irradiation of **47**, as shown in Scheme 8.14 [15].

The furo[3,4-*c*]furan ring system of lignan, (±)-paulownin (**51**) was constructed stereoselectively by Kraus et al. [16] by the photocyclization of **50**, as shown in Scheme 8.15

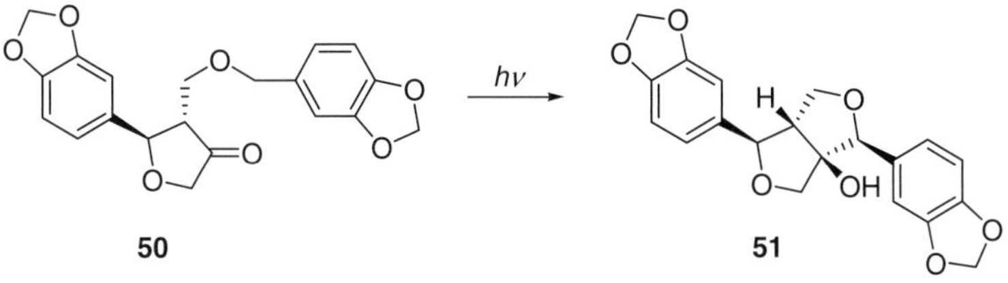

Scheme 8.15

The high stereoselectivity observed in this cyclization reaction was attributed to the location of the carbonyl moiety in a rigid cyclic ring framework, allowing H-abstraction from the same face of the initially formed five-membered ring. The same group has further utilized the constraint imparted by the benzylidene acetal moiety on the excited state photoreactivity of the carbonyl moiety of the D-xylose derivative **52**. Thus, δ-H-abstraction results in the formation of a diastereomeric mixture of cyclized product **53**, which is further transformed into D-glucopyranose **54** and D-idopyranose **55** derivatives, as shown in Scheme 8.16 [17].

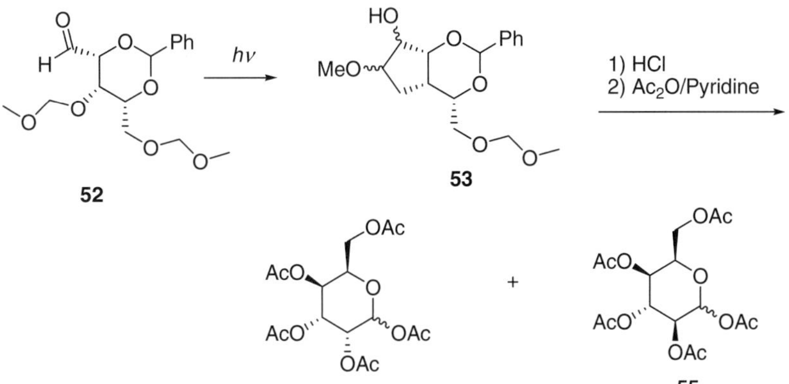

Scheme 8.16

8 Formation of a Five-Membered Ring

Table 8.1 Photocyclization quantum yields of o-alkoxyphenyl ketones in benzene.

S. No	R^1	R^2	R^3	Quantum yields
a	Ph	Me	H	0.30
b	Ph	Et	H	0.62
c	Ph	CH_2Ph	H	0.94
d	CH_3	CH_2Ph	H	0.023
e	Ph	Me	PhCO	0.77
f	Ph	CH_2Ph	PhCO	1.0
g	CH_3	CH_2Ph	CH_3CO	0.17

8.2.2.1 Formation of Benzofuranols

The photocyclization of o-alkoxy phenyl ketones to yield benzofuranols (**57** and **58**) represents one of the earliest example of δ-H-abstraction from the lowest n, π* triplet. Wagner et al. [18] have provided detailed photokinetic data studying the photocyclization of a variety of o-alkoxyphenyl ketones **56**, and have revealed that quantum efficiency for cyclization for **56d** was the lowest (0.023) and that for **56f** the highest (1.00). The diastereoselectivity for cyclization of **56** was found to be higher in benzene and lower in polar solvents. From the estimated k_H values (0.6–25 × 10^6 s^{-1}), it was inferred that the low rate constant for **56e** (8 × 10^6 s^{-1}) compared to that for **56g** (25 × 10^6 s^{-1}) is due to the alkyl chain in the alkoxy groups that points away from the o-carbonyl moiety in the most populated equilibrium conformations (Table 8.1).

Horaguchi and coworkers [19, 20] have studied extensively the effect of solvents and the groups attached to the δ-carbon, and have concluded that polar solvents lower both photocyclization yield and diastereoselectivity. Studies with different substrates have revealed that the development of diastereoselectivity is the result of conformational preferences of the biradical rather than of the steric interactions between the two radicals. The same research group has also studied the photocyclization of ethyl α-(o-benzoylphenoxy)carboxylates (**59**), and reported that the product yields and Z/E ratios (**60a**: **60b**) in benzene and acetonitrile are, interestingly, much the same as for other ketones in which R is not an electron-withdrawing group [20] (Table 8.2).

8.2.3
Synthesis of Pyrrolidine Derivatives

Similar to the synthesis of tetrahydrofuranols by the irradiation of β-alkoxy ketones, Henning's group was the first to show that the photolysis of N-(β-benzoylethyl)-N-

8.2 Formation of Five-Membered Rings: Intramolecular δ-H Abstraction

Table 8.2 Photocyclization of ethyl α-(o-benzoylphenoxy)carboxylates.

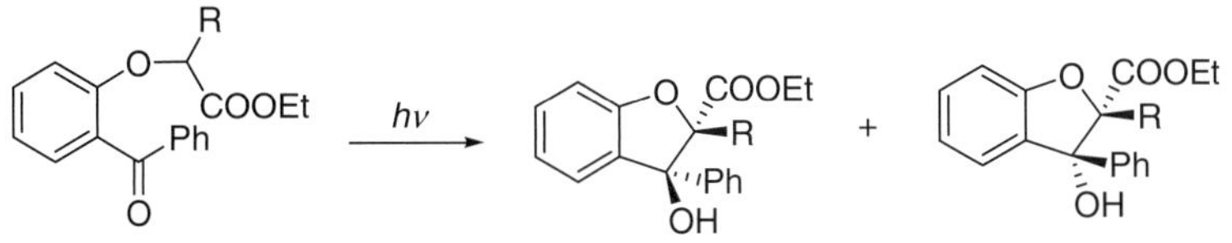

R	Benzene	Acetonitrile	MeOH
H	74 (15/1)	75 (1.5/1)	0
Me	93 (30/1)	72 (4/1)	68 (1/1.4)
Et	84 (50/1)	74 (5/1)	61 (1/5)
i-Pr	70 (1)	75 (6/1)	48 (1/4)
Ph	74 (1/0)	75 (18/1)	72 (18/1)

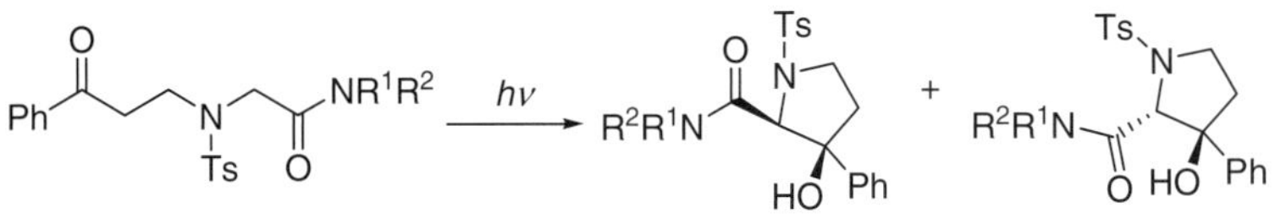

Scheme 8.17

tosyl-glycinamides (**61a**, $R^1 = R^2 = H$ and $R^1 = H$, $R^2 = CH_3$) in benzene leads to photocyclization producing diastereomerically pure *cis*-pyrrolidine derivative **62a** [21]. However, the irradiation of **61b** (R^1, R^2 = alkyl) in acetone under similar reaction conditions gave *trans*-isomer **62b** (Scheme 8.17).

The difference in the diastereoselectivity of **61a** versus **61b** was rationalized by considering the intramolecular dipolar interaction and intramolecular H-bonding to the sulfonamide carbonyl in the hydroxyl biradical intermediate, as shown in Scheme 8.18.

The formation of five-membered γ-lactam **64** in high yield was also reported [22] subsequently by the photolysis of N,N-dibenzyl-2-benzoylacetamide (**63**) in benzene. The presence of acyl moiety in **63** suppressed the basicity of the nitrogen lone pair and prevented its involvement in the quenching of the carbonyl excited state by an electron-transfer processes (Scheme 8.19).

The rate of photocyclization of **63** was greatly enhanced when irradiation was carried out by absorbing it on silica gel surface, possibly due to a higher content of the keto tautomer.

The obvious extension of the above concept was made by Wessig *et al.* [23] to obtain optically active *cis*-3-hydroxyproline derivatives **66** in moderate to good yields by irradiating chiral N-(2-benzoylethyl)-N-tosyl glycinamide (**65**) in cyclohexane/benzene (4:1) mixture. The diastereomeric excess (de) of the photocyclization was

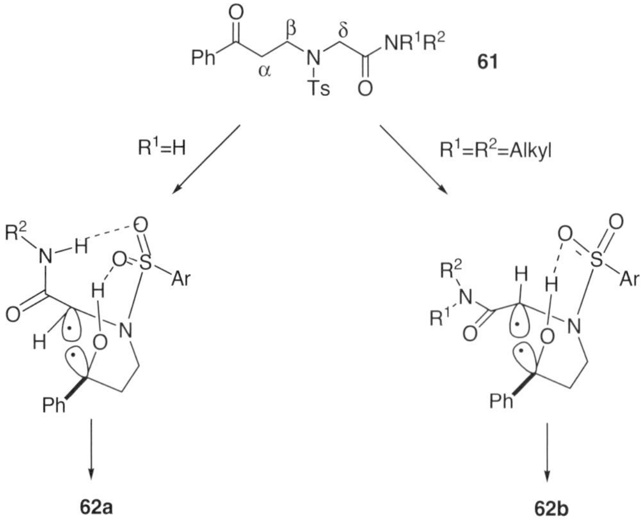

252 | *8 Formation of a Five-Membered Ring*

Scheme 8.18

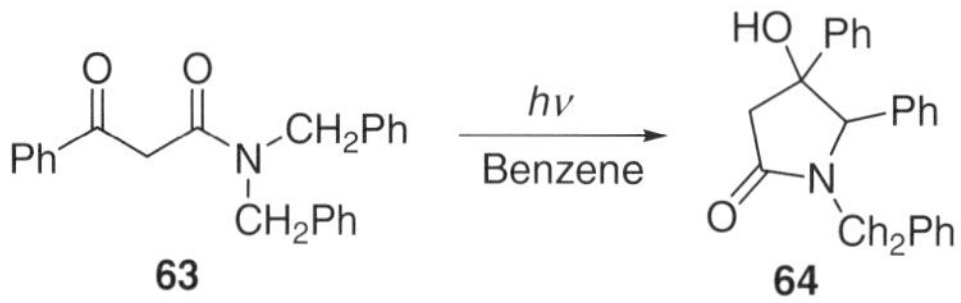

Scheme 8.19

found to depend on the size of the substituents at the C-2 and C-5 positions of the C_2-symmetric pyrrolidine chiral auxiliaries (Scheme 8.20).

Irradiation at a lower temperature increased the de of the products, but the reaction rate was drastically decreased, thus restricting its practical utility. The same group has further explored the diastereoselectivity and regioselectivity of the photocyclizations of chiral N-tosyl glycine esters of type **67** to give the corresponding 3-hydroxy proline derivatives **68** (Scheme 8.21).

It has been noted that the photolysis of **67a** and **67b** gave cis-3-hydroxyproline esters **68** exclusively in moderate to good yields, whereas **67c–67d** gave the isomeric compound **68′**. The selectivity for the formation of **68′** was explained by considering the preferred conformation **B** and **C** due to intramolecular H-bonding as compared to **A** [24], as illustrated in Scheme 8.22.

Giese et al. [25] have compared the steric course of the reaction of T_1 (triplet) and S_1 (singlet) by studying the photocyclization reaction of alanine derivatives **69**, which led to the formation of proline derivatives **71a** and **71b** (Table 8.3)

It was observed that photoreaction from T_1 was almost nonselective as it gave all of the possible isomers, whereas S_1 underwent stereoselective cyclization. This difference in selectivity was attributed to the faster rate of cyclization of the singlet diradical compared to the triplet.

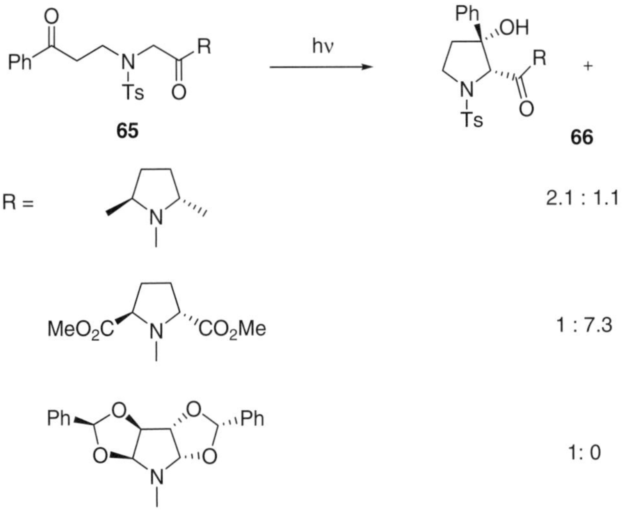

8.2 Formation of Five-Membered Rings: Intramolecular δ-H Abstraction

Scheme 8.20

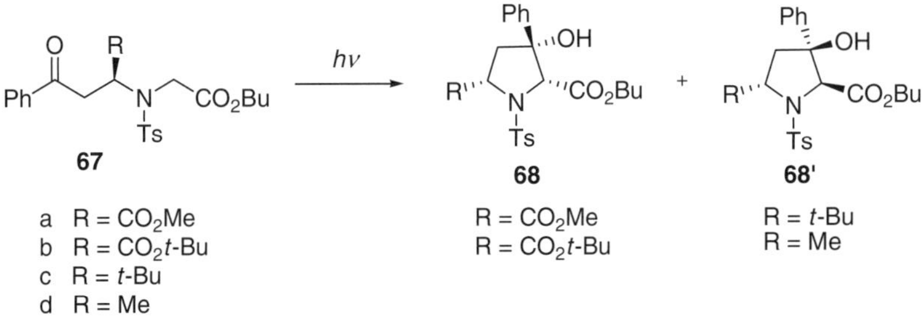

Scheme 8.21

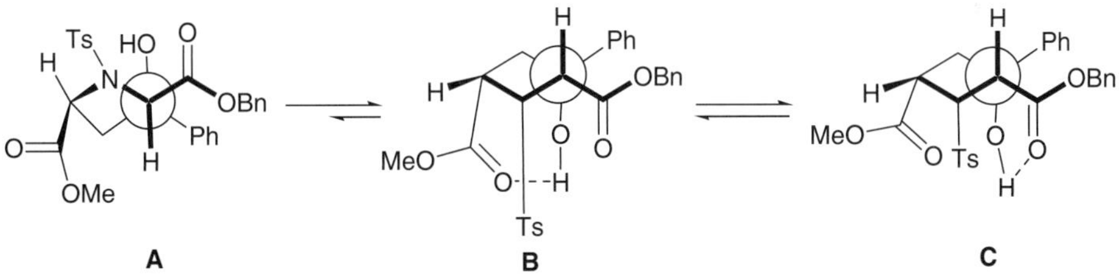

Scheme 8.22

N-Tosyl piperidine derivative **72** lacking abstractable γ-hydrogen on photolysis produced the two different bicyclic compounds **73** and **74**, as shown in Scheme 8.23 [26].

The competition between the formation of **73** (δ-H-abstraction from the Me) and **74** (ring H-abstraction) has been rationalized on the basis of the preferred conformation of the piperazine ring.

Table 8.3 Selectivities of the photocyclizations of alanine derivative **69** leading to *cis* product **71a** and *trans* product **71b** in benzene at 20 °C.

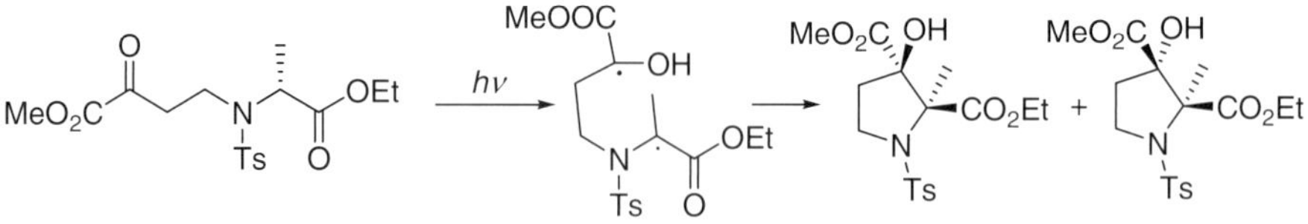

Conditions	71a:enant-71a	71b:enant-71b	cis:trans	Overall yield (%)
hν/naphthalene (1 M)	24	16	5.7	47
hν/naphthalene (0.5 M)	18	13	5.3	50
hν/isoprene (0.5 M)	9.4	3.0	2.9	47
hν/O$_2$	9.6	3.6	2.6	48
hν/Ar	2.4	1.6	0.9	35
hν/benzophenone (1 M)/Ar	1.4	1.4	0.8	10

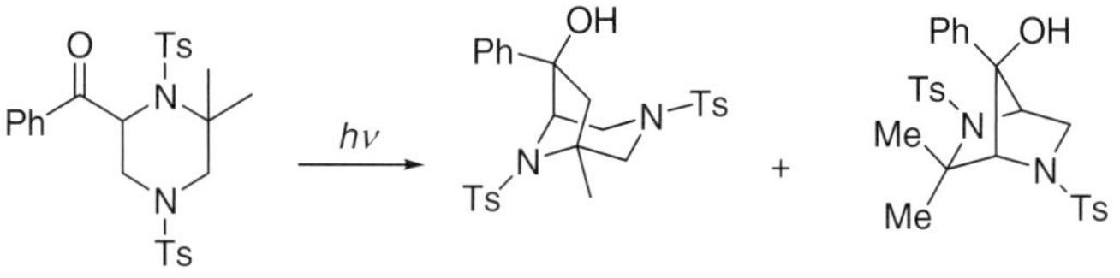

Scheme 8.23

8.3
Formation of Five-Membered Rings via [3 + 2]-Cycloadditions

The ring strain associated with the three-membered ring facilitates the generation of reactive species such as biradicals (cyclopropanes) and ylides (oxiranes and aziridines) upon photoexcitation. These intermediates can undergo cycloaddition reaction with olefins to give five-membered rings. Selected examples from each reactive intermediate are presented in this section.

8.3.1
Photofragmentation of Oxiranes to Carbonyl Ylides: Synthesis of Tetrahydrofurans

Singlet as well as triplet excited states of oxiranes undergo C–C bond scission to produce carbonyl ylides which, upon cycloaddition with dipolarophiles, give tetrahydrofuran (THF) derivatives. For example, *trans*- or *cis*-stilbene oxide on direct photolysis using 254 nm light in the presence of methyl acrylate gave diastereomeric

8.3 Formation of Five-Membered Rings via [3 + 2]-Cycloadditions

Table 8.4 Cycloadduct yields from photolysis of stilbene oxides in the presence of methyl acrylate.

Solvent	Stilbene oxide	Cycloadduct ratios	
		77a : 77b	77c : 77d
CH_3CN	75a	62 : 27	8 : 3
CH_3CN	75b	9 : 2	66 : 24
C_6H_{12}	75a	61 : 26	6 : 7
C_6H_{12}	75b	17 : 5	36 : 42

mixtures of THF derivatives (Table 8.4) [27]. The THF derivatives were formed by the cycloaddition of the carbonyl ylides (**76a** and **76b**) by the allowed 4πn + 2πn cycloaddition rules [28]. Based on a simultaneous irradiation of **75a** and **75b**, it was suggested that **75a** photoisomerizes much more readily to **75b** compared to its reverse processes. The cycloaddition reaches quantitative yield (≈99%) under triplet-sensitized reaction condition. The stereochemistries [29] of the dipolarophiles are preserved in the cycloadducts consistent with a concerted addition process; however, solvent effects and steric hindrance play key roles in determining the product distribution. Similar results have also been reported for the irradiation of 3,3-dicyanostilbene oxide [30] and α-cyanostilbene oxides [31].

In contrast to above observation, Whiting et al. [32] reported that direct or triplet-sensitized irradiation of 2,3-bis (p-methoxy phenyl) oxiranes (**78**) does undergo C—O bond fission, but does not produce carbonyl ylide; rather, carbonyl compounds **79** and **80**, respectively, are produced (Scheme 8.24). However, photolysis under

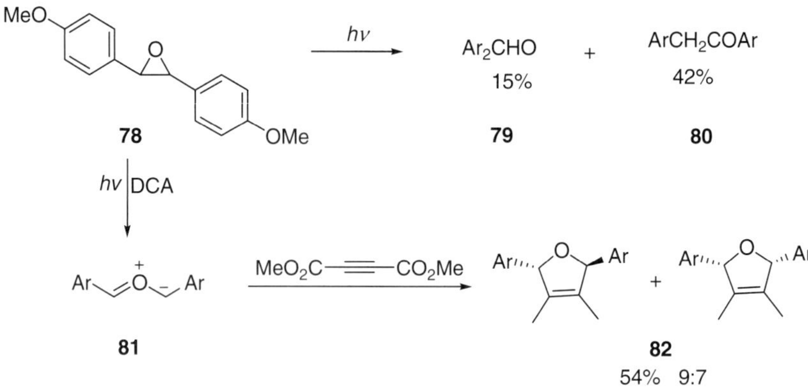

Scheme 8.24

photoinduced electron transfer (PET) reaction conditions employing 9,10-dicyanoanthracene (DCA) as the excited state electron acceptor produced carbonyl ylide **81** which, on cycloaddition with dimethyl acetylendicarboxylate (DMAD), produces a diastereomeric mixture of cycloadducts **82**. However, due to sluggishness and nonstereospecificity of the cycloaddition reaction, the synthetic utility of this reaction for lignan synthesis remained elusive. Continuing studies performed by the same group [33] have led to a generation of push-pull ylides of type **84** by the irradiation of **83**. The ensuing cycloaddition with maleic anhydride produced compound **85** in 66% yield from the addition of both *exo-* and *endo-*ylide **84** to the dipolarophile in *endo-*fashion, as depicted in Scheme 8.25.

Linear, as well as angular, benzannulated perhydroazulenes **88** and **89**, which are related to some important classes of diterpenes, have been synthesized (Scheme 8.26) [34] by the intramolecular cycloaddition of the carbonyl ylide generated by the irradiation of α,β-epoxyketone **86** and **87**, respectively.

Eberbach et al. [35] have also prepared annulated THF derivatives **92a** and **92b** by directly irradiating **91a** and **91b**, respectively. However, a better yield of **92a** was obtained when irradiation was conducted in acetone, which acts as a sensitizer (Scheme 8.27).

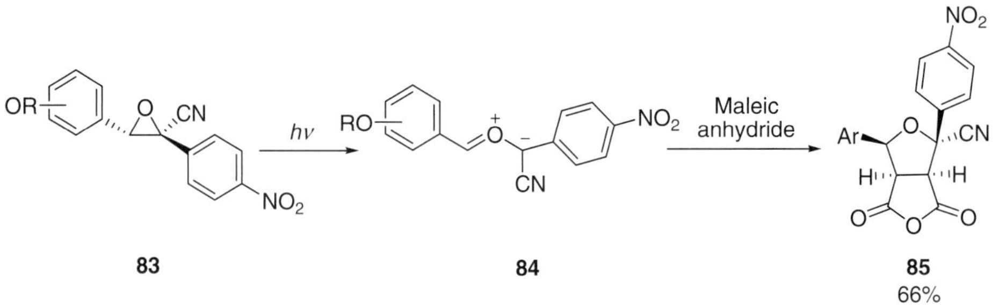

Scheme 8.25

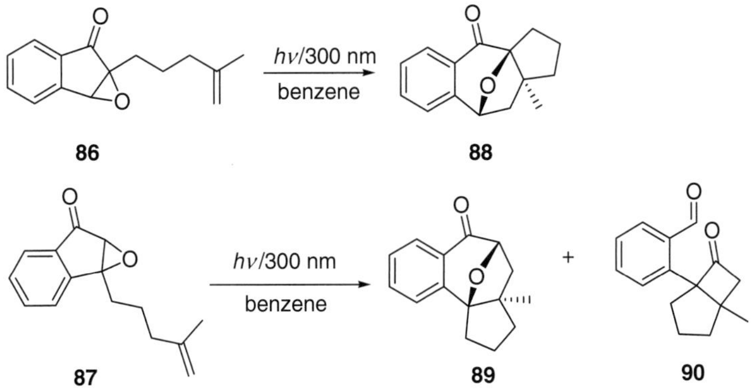

8.3 Formation of Five-Membered Rings via [3 + 2]-Cycloadditions

Scheme 8.26

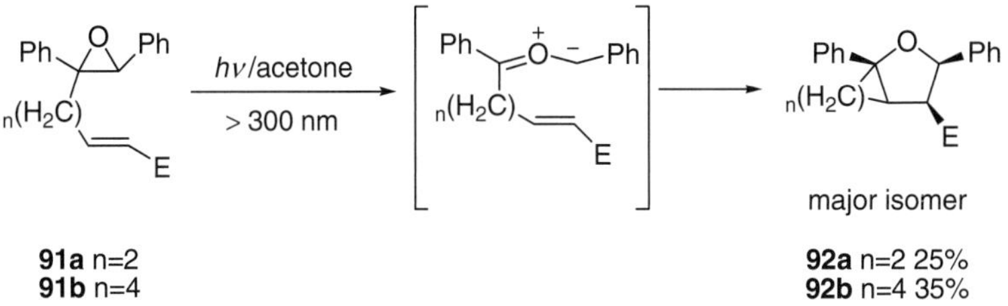

Scheme 8.27

Carbonyl ylides of type **94** were generated, as illustrated in Scheme 8.28, by irradiating α,β-unsaturated-γ,δ-epoxynitrile **93** in acetonitrile at 254 nm; upon trapping with ethyl vinyl ether, these produced THF system **95** with high regioselectivity.

The failure of ylide **93a** to undergo cycloaddition is explained on the basis of the LUMO energy level and the C (γ) and C (δ) distance in ylide [36].

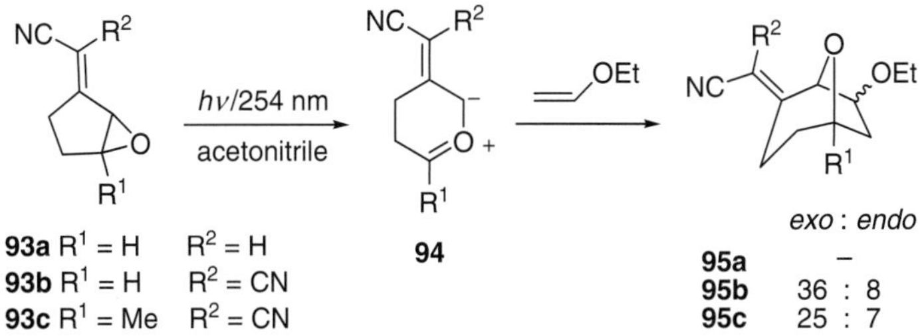

93a R^1 = H, R^2 = H
93b R^1 = H, R^2 = CN
93c R^1 = Me, R^2 = CN

exo : endo
95a —
95b 36 : 8
95c 25 : 7

Scheme 8.28

8.3.2
Generation of Azomethine Ylides by the Photolysis of Aziridines: Synthesis of the Pyrrolidine Framework

As early as 1967, Huisgen and coworkers [37] had shown that, upon photolysis, certain aziridines of type **96** undergo C–C bond fragmentation stereospecifically to produce octet-stabilized azomethine ylides which, on cycloaddition with electron-deficient dipolarophiles, produce pyrrolidine ring systems (Scheme 8.29).

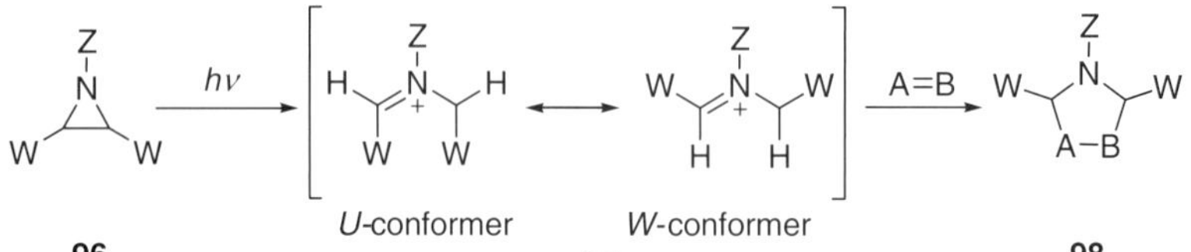

Scheme 8.29

The most important aspect of such cycloadditions for synthetic purposes was the *endo-* addition which followed the $4\pi n + 2\pi n$ rules. The chemical behavior of aziridines under photolytic conditions is dependent on the number and type of substituents present in the aziridine ring, and especially on the nitrogen atom. The *W* conformer of the ylide undergoes cycloaddition more readily than the *U* conformer, due to the lower van der Waals strain (Scheme 8.29). The reaction of azomethine ylide with alkenes is a powerful method for the synthesis of pyrrolidines, an important building block [38–41] in the synthesis of natural products and pharmaceuticals. As many reviews [42–45] and monographs [46–48] have been prepared on this subject, only very recent investigations are highlighted in this section.

Garner *et al.* [49] have synthesized the 6-*exo*-substituted 3,8-diazabicyclo [3.2.1] octane core of DNA reactive quinocarcin alkaloid **99** by cycloaddition of the azomethine ylide **101**, generated by irradiation ($\lambda = 254$ nm) of the corresponding aziridine **100**, with Oppolzer's chiral acryloyl sultam (**102**). This produced 6-*exo*-substituted cycloadducts **103**:**104** in 25:1 diastereoselectivity (Scheme 8.30).

In order to improve the diastereoselectivity further, the cycloaddition of enantiomerically pure azomethine ylide **106**, generated by photolysis (254 nm) of the corresponding aziridine **105**, with dipolarophile **102**, is also reported to give the cycloadduct **107** as a single diastereomer in 61% yield (Scheme 8.31). In order to control the diastereoselectivity, the same group has also explored the intramolecular version of this cycloaddition reaction [50].

In general, aziridines bearing an adjacent electron-withdrawing group (EWG) or phenyl group only undergo a photochemical ring-opening reaction to produce azomethine ylides. However, Ishii's group [51] recently demonstrated the generation of azomethine ylides through the photoreaction of *N*-benzyl β-aziridinylacrylonitrile (**108**). On cycloaddition with acrylate, this produced the head-to-head cycloadducts **109** with moderate diastereoselectivity, as depicted in Scheme 8.32. This cycloaddition strategy was extended to the formal synthesis of the indolizidine fragment of stellettamides [52].

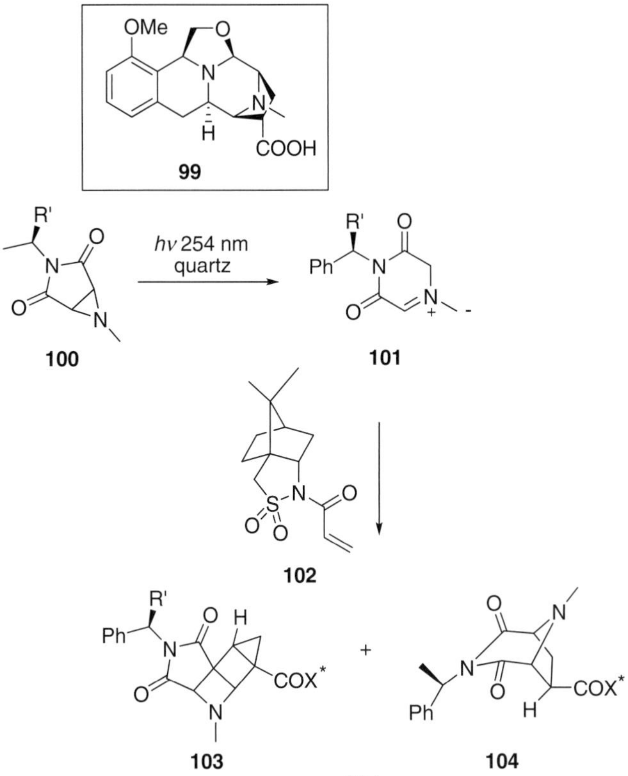

8.3 *Formation of Five-Membered Rings via [3 + 2]-Cycloadditions* | 259

Scheme 8.30

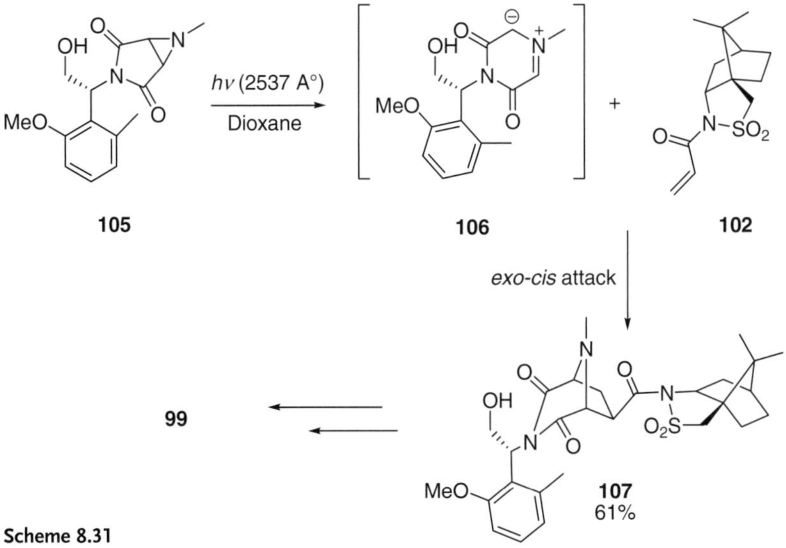

Scheme 8.31

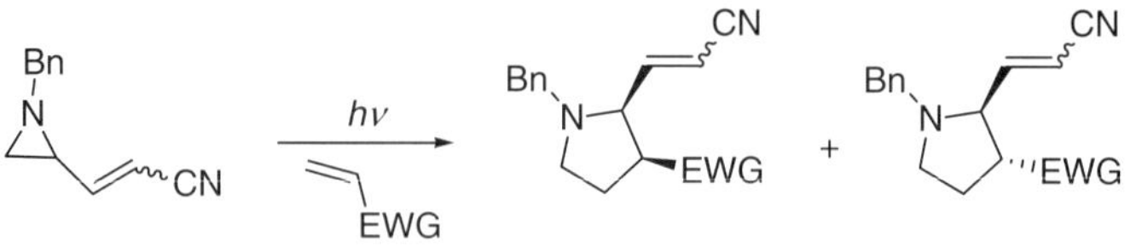

260 | 8 Formation of a Five-Membered Ring

Scheme 8.32

8.3.3
Vinyl Cyclopropane to Cyclopentene Rearrangement

The photochemical rearrangement of vinyl cyclopropanes to cyclopentene derivatives has a limited synthetic scope due to other accompanying reactions, such as E/Z isomerizations of the cyclopropane ring and the vinyl unit and ring-opening reaction leading to conjugate dienes [53–55]. The simple vinyl cyclopropane derivative isopropenylcyclopropane **110**, upon direct irradiation in hexane, has been reported [56] to give methyl cyclopentene (**112**, Scheme 8.33) in 55% yield.

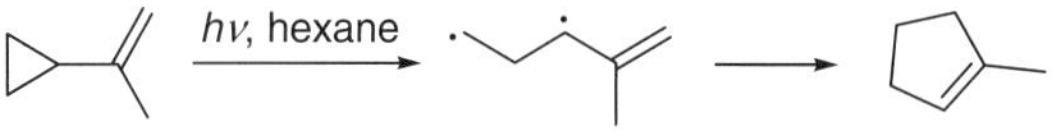

Scheme 8.33

On the basis of triplet sensitization experiments, this rearrangement was shown to occur from the excited singlet state and to involve the biradical intermediate **111**. Interestingly, vinyl cyclopropane esters **113–114** underwent geometric isomerization at a much faster rate than rearrangement on direct excitation (Scheme 8.34) [57, 58]. However, the triplet-sensitized reaction of these esters give both cyclopentene and isomerization products.

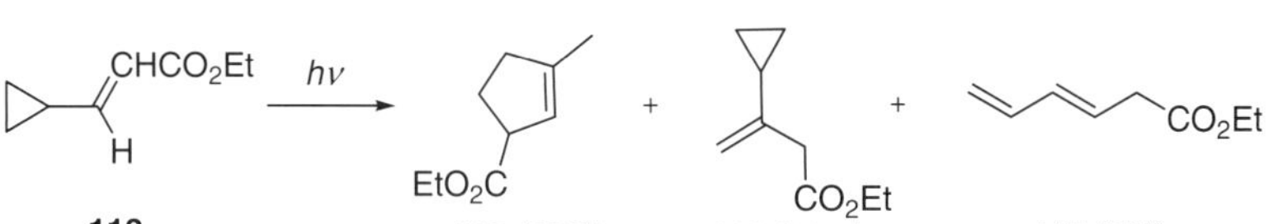

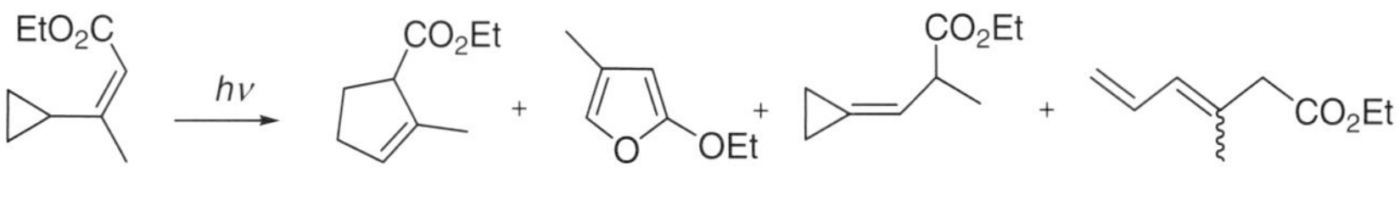

Scheme 8.34

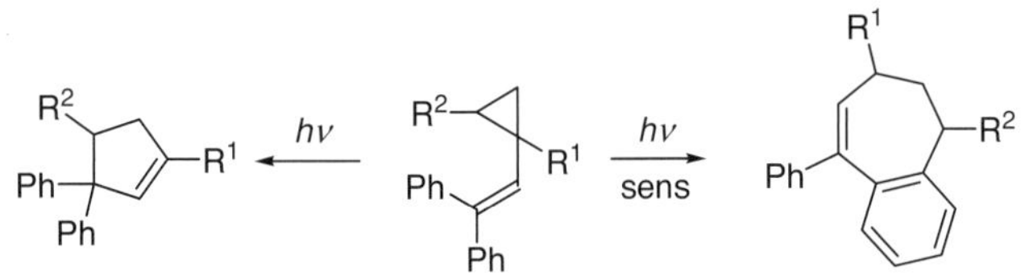

Scheme 8.35

124: $R^1 = CH_2Me$, Me; $R^2 = H$ or Ph

123: $R^1 = COOR$, C=NOAc; $R^2 = H$, Me or Ph

Armesto et al. [59] have reported a novel observation (Scheme 8.35) on the photorearrangement of **122** which is dependent on the electronic characteristics of the C-1 substituents. For example, C-1 having an electron-withdrawing group undergoes a novel rearrangement to produce benzocycloheptene **123** upon triplet-sensitized irradiation, whereas C-1 devoid of an electron-withdrawing group undergoes a conventional ring-opening reaction to produce cyclopentene **124**. The formation of **123** is explained by the involvement of an intramolecular single electron transfer (SET) from the diphenylvinyl unit to the electron acceptor group.

The photochemical rearrangement of azavinylcyclopropane **125** also occurs [60] readily to produce 1-pyrroline derivatives **126** in high yield (50–85%) upon irradiation using Pyrex-filtered light, as shown in Scheme 8.36.

This rearrangement was found to be independent of the nature of the substituents present in **125**.

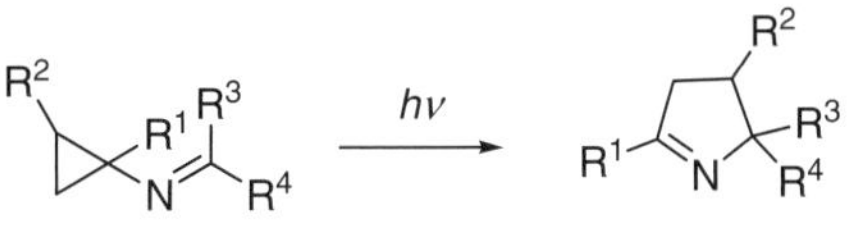

Scheme 8.36

126: 50–85%

8.4
Photochemical Electrocyclization Reactions: Synthesis of Fused, Five-Membered Ring Compounds

Pentadienyl anion-like [61] photocyclizations occur from molecules having a heteroatom connected to two C=C bonds (olefinic and/or aromatic). This results in a reactive zwitterionic intermediate **128** which, upon oxidative dehydrogenation, produces derivatives of pyrroles, furans, thiophenes, and selenophenes **129** (Scheme 8.37) [62].

Mechanistically, in these reactions the lone pair of electrons on the heteroatom contributes to the 6π electron system, which undergoes conrotatory electrocyclizations with a conservation of orbital symmetry. The dehydrogenation of zwitterionic

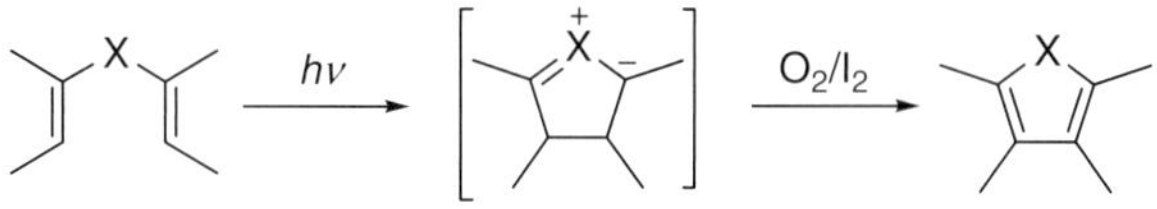

262 | *8 Formation of a Five-Membered Ring*

Scheme 8.37

X= N,O,S,Se

intermediate **128** (re-aromatization being the driving force), in the presence of oxygen or iodine as oxidants, leads to the formation of cyclized product **129**.

The irradiation of diphenyl amine **130** (X = NHPh) [63–65], diphenyl ether **130** (X = O) [66, 67] or diphenyl thioether **130** (X = S) [66] derivatives is reported to cause efficient cyclization reactions and to produce the corresponding carbazoles, dibenzofurans, and dibenzothiophenes, respectively (Scheme 8.38).

There are also known examples [68] where a corresponding zwitterionic intermediate **134** (Scheme 8.39), obtained from the irradiation of aryl amine **133** in acetonitrile containing aqueous HCl, instead of becoming aromatic through oxidation, undergoes protonation to produce dihydrocarbazole intermediate **135** which by

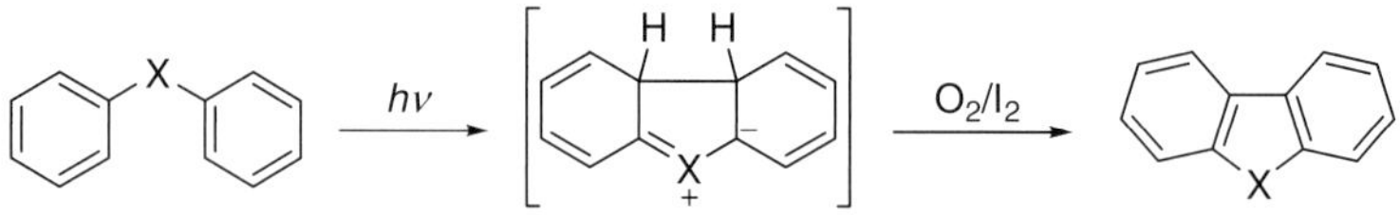

X= NHPh,O,S

Scheme 8.38

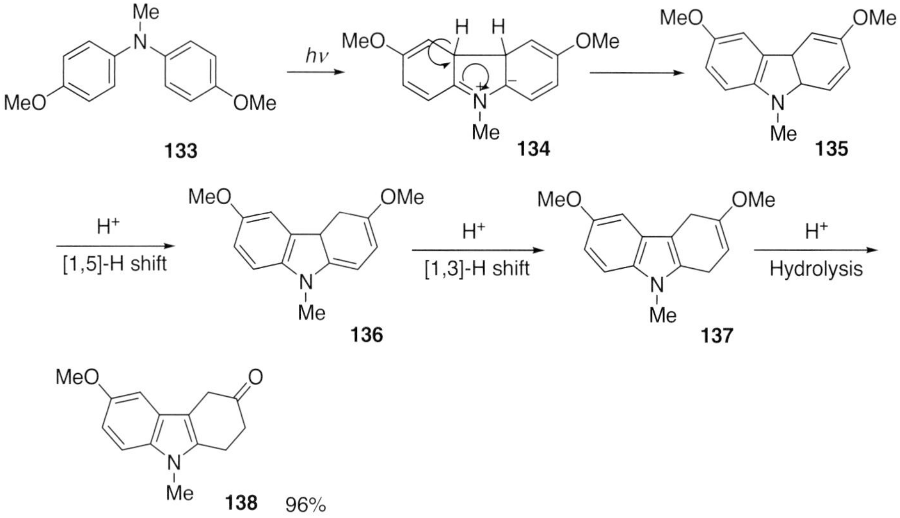

Scheme 8.39

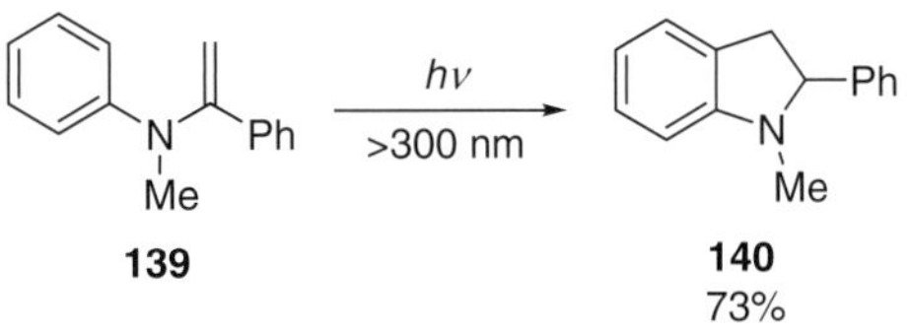

Scheme 8.40

sequential [1,5]-hydrogen and [1,3]-hydrogen shift, followed by proton-assisted hydrolysis, gives 1,2,4-trihydro(4a*H*)-carbazole-3-ones (**138**) in high yield.

Several examples pertaining to the photochemical electrocyclization of compounds possessing an aromatic moiety and an olefinic substituent are also known [69, 70]. For example, one of the earliest reports in this series was concerned with the photocyclization-rearrangement of α-(*N*-methylanilino) styrene (**139**) which, on irradiation in the absence of oxygen, produced 1-methyl-2-phenyl-2,3-dihydroindole (**140**) in 73% yield, as depicted in Scheme 8.40.

The stereochemistry of such photocyclizations was also explored [71] by irradiating enamines of type **141** which gave *trans*-indolines **142** as the major reaction product, along with minor *cis*-**143** (Scheme 8.41).

Interestingly, the enamine of cyclopentanone **144**, upon irradiation [72], gave only the *cis*-indoline **145** in 52% yield; this indicated the effect of ring-strain on the stereochemistry of the indolines (Scheme 8.42). Mechanistically, several pathways

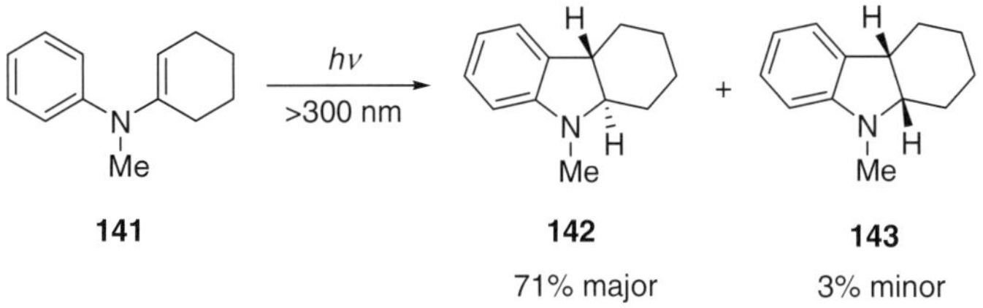

Scheme 8.41

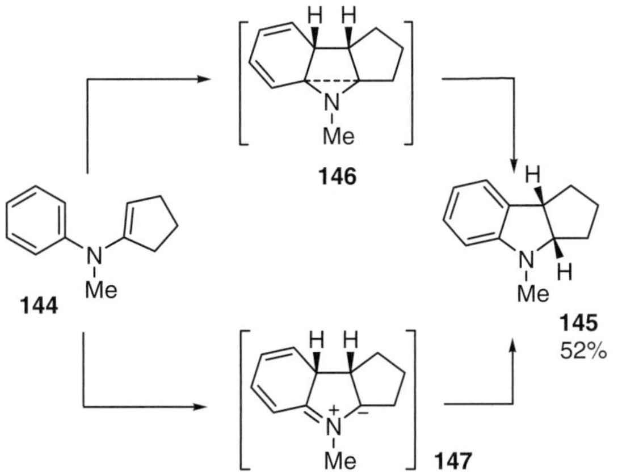

Scheme 8.42

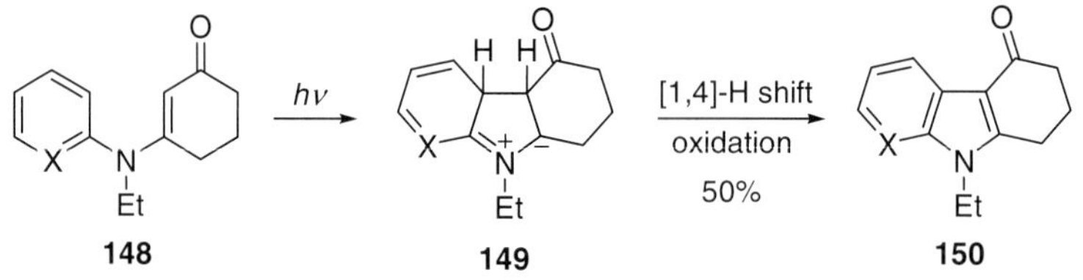

Scheme 8.43

X = -CH-, N

may be considered for such cyclization reactions. For example, an intramolecular [2 + 2]-photocycloaddition of **144** may give *cis*-dihydrocyclobutane **146**, which might change to **145** under photoreaction conditions. Alternatively, this cyclization can also involve a vibrationally excited ground state via an allowed disrotatory process to give the *cis*-dihydro intermediate **147** which, on H-shift, produces **145**. Another mechanism which should also be considered involves ylide protonation from the photoreaction medium.

Enaminones of type **148**, on irradiation in a mixture of benzene/methanol solvent, undergo a smooth cyclization (Scheme 8.43) to produce the corresponding indolones **150** in high yield [73–75].

This enaminone cyclization route to indolones has been applied [76–79] for the synthesis of a variety of aspidospermidine class of alkaloids **153**–**155** from compound **152**, which was obtained in very high yield from the photolysis of **151**, as illustrated in Scheme 8.44.

This type of photocyclization strategy has also been utilized [80] to construct indoline substructures **157** and **159** by irradiating enamine derivatives bearing an electron-withdrawing substituent on the olefinic double bond, such as **156**. This was produced as a mixture of *cis* and *trans* diastereomeric indolines, as shown in Scheme 8.45. The formation of diastereomers (**157** and **159**) is explained by

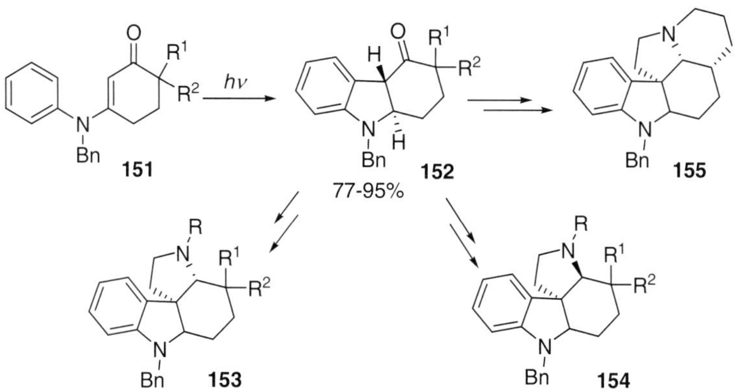

Scheme 8.44

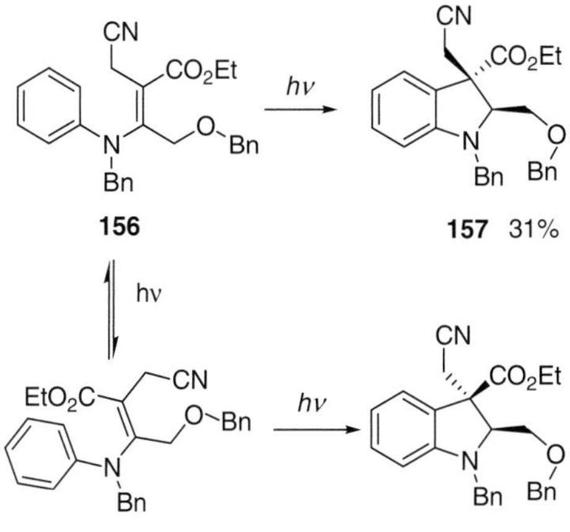

Scheme 8.45

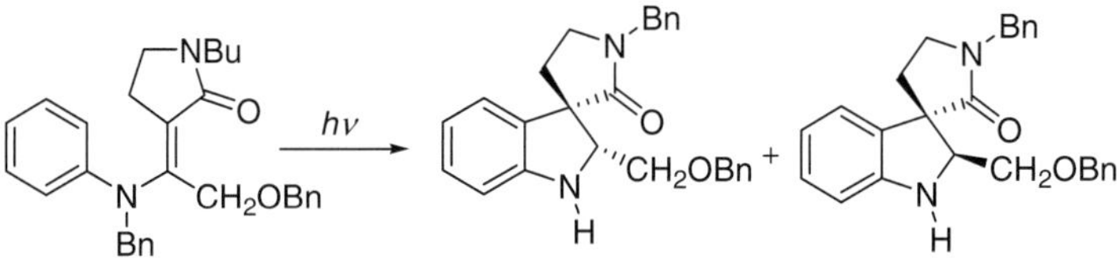

Scheme 8.46

considering the photochemical E/Z isomerization of the double bond before the photocyclization step.

The same group has extended this photocyclization approach to obtain the corresponding debenzylated spiroindolines **161** and **162** in 68% yield and 2.4:1 ratio from the photocyclization of **160** (Scheme 8.46).

The role of the zwitterion intermediate **164** and its rearrangement in the photocyclization of aromatic thioethers **163** to arene dihydro thiophene derivatives **165** is supported [81] by its independent trapping (via 1,3-dipolar cycloaddition) with N-phenyl maleimide to obtain **166** in high yields (Scheme 8.47).

The intramolecular trapping of ylide **164** is also reported [82] in the photoreaction of **167**, which gave unexpected **169** as a major product, as depicted in Scheme 8.48.

The expected intramolecular 1,3-dipolar cycloaddition product **171** was only a minor product (3%). The formation of major product **169** was explained through an intramolecular Michael reaction of the enolate ion.

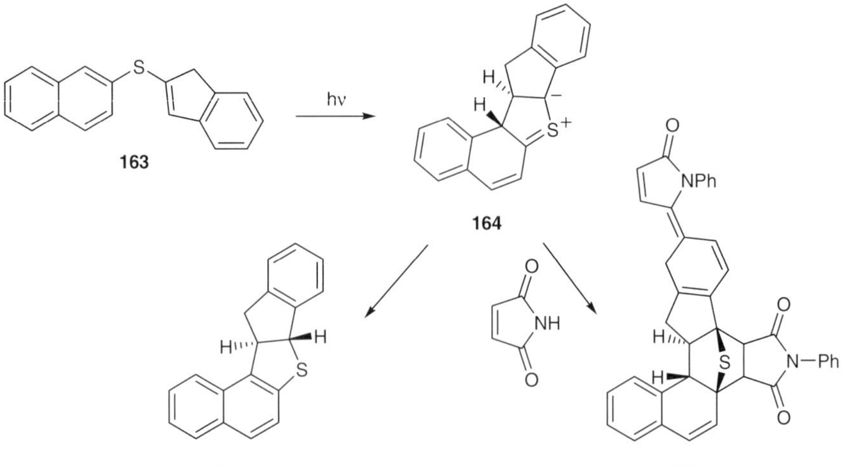

163 164 165 166

Scheme 8.47

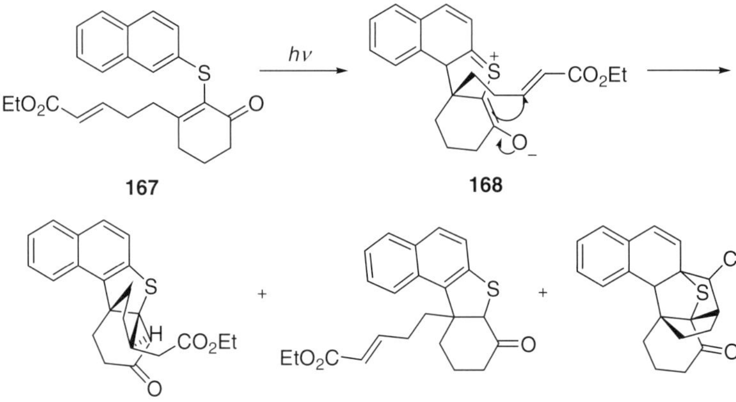

167 168

169 84% 170 5% 171 3%

Scheme 8.48

8.5
Photoinduced Electron Transfer-Mediated Cyclizations: Synthesis of Five-Membered Carbocyclic and Heterocyclic Ring Systems

8.5.1
Radical Cation-Mediated Carbon–Carbon Bond Formation

The reaction of organic radical cations have been the focus of much interest, and their synthetic reactions – including addition to alkenes and nucleophilic capture by alcohols resulting in carbon–carbon and carbon–oxygen bond formation, respec-

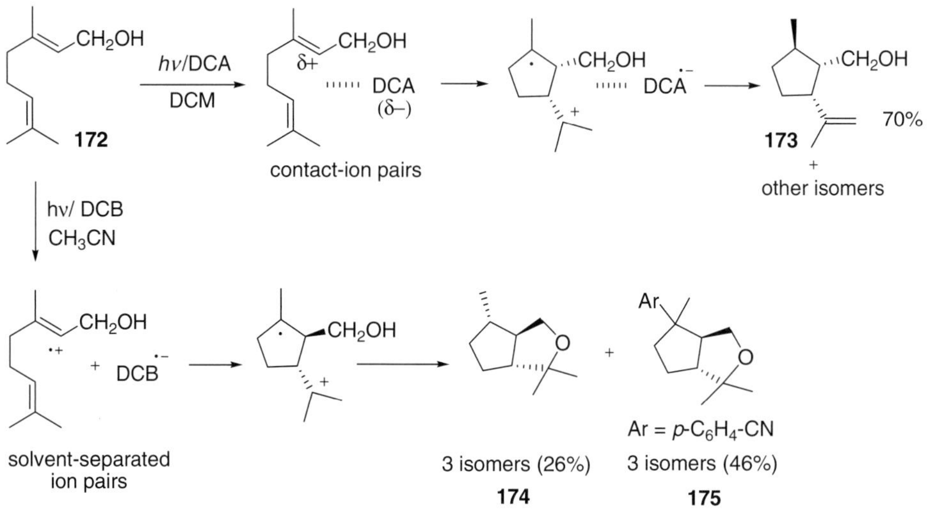

Scheme 8.49

tively – has been investigated in detail. For example, Roth's group [83] studied the electron-transfer photochemistry of geraniol and farnesol (Scheme 8.49), where radical cations were shown to add to olefin and generate five-membered ring systems. The electron-transfer photoreaction between DCA and (*E*)-3,7-dimethylocta-2,6-dien-1-ol (geraniol); (*E*)-**172** in dichloromethane produces mainly *cis*-2-(2-propenyl)-*trans*-5-methylcyclopentanemethanol (**173**, 70%), whereas the photoreaction using 1,4-dicyanobenzene (DCB) in acetonitrile produced all of the possible cyclic ethers (**174**, **175**).

The difference in product distribution was explained by considering the involvement of solvent-separated ion pairs of **172**-DCB and contact-ion pairs **172**-DCA. Similar observations were also made with farnesol.

Demuth et al. [84] have studied, systematically, the PET cyclization of isoprenoid polyalkenes using sterically hindered electron acceptors such as 1,4-dicyano-2,3,5,6-tetramethyl benzene (TMDCB) in combination with biphenyl (BP) as a cosensitizer (Scheme 8.50). The cyclization mode (6-*endo-trig* versus 5-*exo-trig*) yielding five- and six-membered rings was found to depend on the substitution pattern of the polyalkenes. For example, the PET reaction of isoprenoid polyalkene **176a** in acetonitrile/water (4:1) gave **177** (15–30%), while **176b** produced **178** (∼50%). The synthetic scope of such transformations has also been explored with polyalkenes **179** and **181**, which produced functionalized bi- and tricyclic (all-*trans*-fused) compounds **180** and **182**, respectively (Scheme 8.50). The reaction is initiated by the regioselective one-electron transfer oxidation of the ω-alkene site of the acyclic polyalkene, giving rise to a radical cation which is trapped by a nucleophile such as water or methanol present in the reaction medium. The resulting neutral radical initiates a cyclization cascade, resulting in a tertiary radical. Termination of the process is achieved either by hydrogen atom transfer to this radical, or through reduction of the final radical.

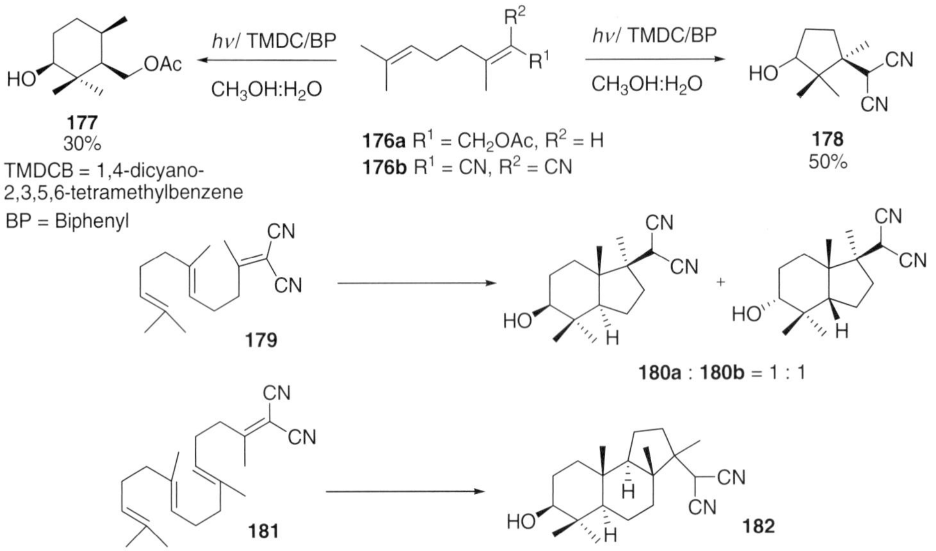

268 | 8 Formation of a Five-Membered Ring

Scheme 8.50

The observed 5-*exo-trig* cyclization in the above reaction is attributed to the formation of more stabilized malononitrile radical intermediate compared to the trialkyl-substituted radical. The stereochemical outcome of these transformations with regards to the malonodinitrile group is consistently β-selective; this suggests least steric interactions between these substituents as compared to the alternative methyl–methyl interaction in the transition state.

Hasegawa *et al.* [85] have explored the electron transfer-promoted regioselective ring opening of cyclopropyl silyl ethers **183** for the generation of cyclopropoxy radical **185** which, upon intramolecular radical cyclization, leads to the formation of spirocyclic compound **184** in moderate yields (Scheme 8.51).

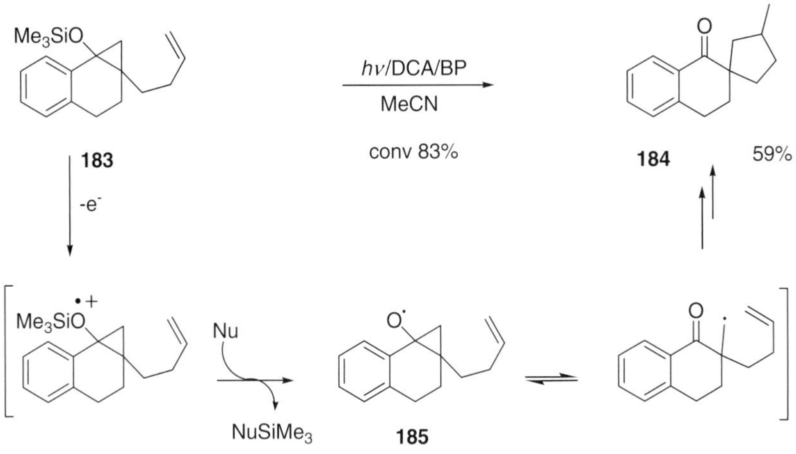

Scheme 8.51

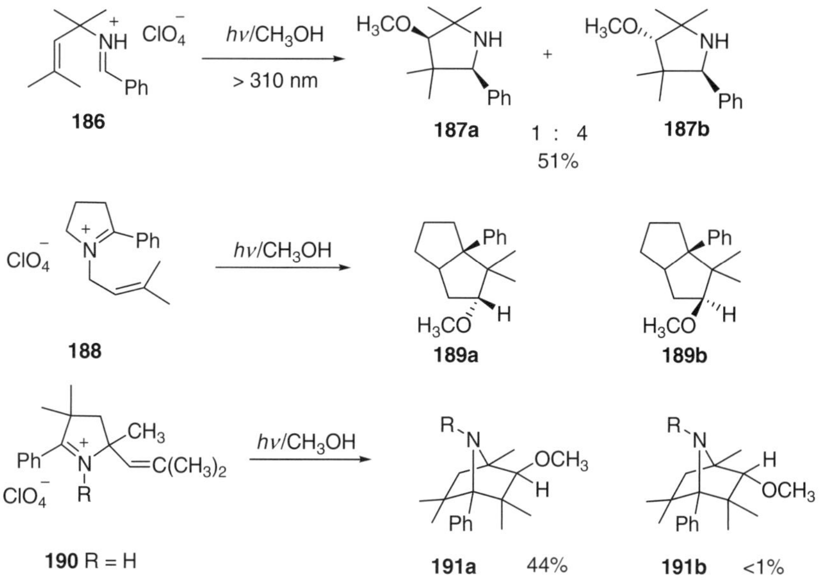

8.5 Photoinduced Electron Transfer-Mediated Cyclizations

Scheme 8.52

A novel class of PET processes in aqueous or methanolic solution leading to C—C bond-forming reactions has been developed by Mariano et al. [86] between electron-rich olefins and iminium salts. This photochemical reaction was used in the construction of pyrrolidine ring system by irradiating (>310 nm) acyclic N-allylimi-nium salt **186** in methanol. This reaction gave stereoisomeric mixtures of **187a** and **187b** in 1:4 ratios in 51% isolated yield (Scheme 8.52). Both fused- as well as bridged-pyrrolidine skeletons (**189** and **191**) were also synthesized by the photolysis of corresponding 2-pheny-5-vinyl-1-pyrrolinium (**188**) and 1-allyl-1-pyrrolinium perchlorates (**190**), respectively, in methanol.

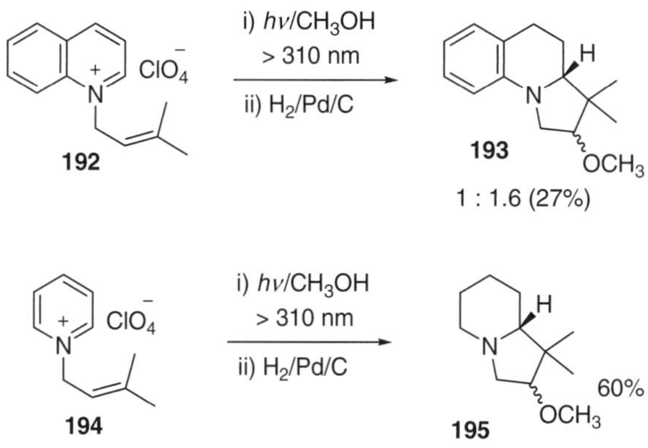

Scheme 8.53

270 | 8 Formation of a Five-Membered Ring

The reaction was rationalized in terms of electron transfer from the olefin π-systems to a singlet excited iminium salt moiety, followed by nucleophilic attack of the solvent (CH_3OH or water) on the cation diradical and subsequent 1,5-radical coupling. The same protocol was extended to the transformation of N-prenylquinolinium perchlorate (**192**) and N-prenylpyridinium perchlorates (**194**) into the corresponding cyclic compounds which, upon hydrogenation, produced benzoindolizidines (1 : 1.6) and perhydroindolizines, respectively (Scheme 8.53) [87].

The spirocyclic CD-ring of **197**, the precursor of the harringtonine family of alkaloid **198**, is constructed efficiently (33% yield) by alyliminium salt **196** photocyclizations [88], as shown in Scheme 8.54.

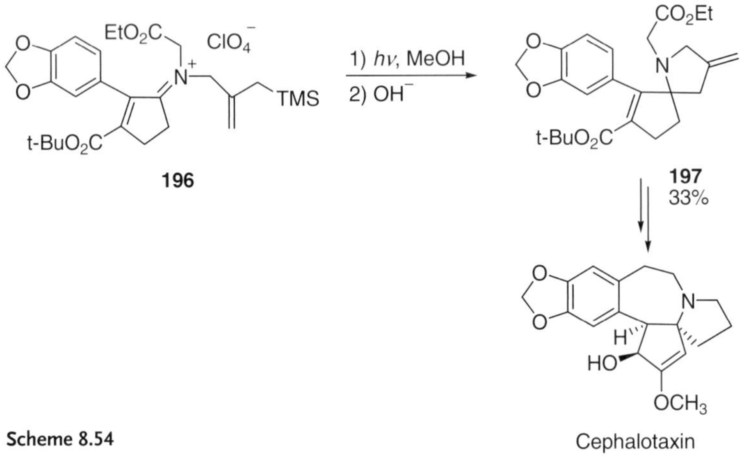

Scheme 8.54

Cephalotaxin

Cha's group [89] has recently developed a convenient method for the ring opening of tertiary aminocyclopropane tethered with an alkenyl moiety **199**. The PET reaction of **199** led to formation of the five-membered bicyclic system **200** through a [3 + 2] annulation reaction in a good yield, as shown in Scheme 8.55.

This approach has been extended to the construction of a medium-sized carbocyclic ring (Scheme 8.56), which is found in an increasing number of natural products [90].

α-Silylmethylamine radical cations **204**, generated by electron transfer from **203** to excited 1,4-dicyanonaphthalene, were reported [91] to undergo efficient cyclization to the tethered olefins producing pyrrolidine **205**, as shown in Scheme 8.57.

Mechanistically, it was suggested [92] that this cyclization does not involve the "free α-amino radical" formed by cleavage of the C–Si bond of the trimethylsilylmethylamine radical cation. Instead, it was pointed out that cleavage of the C–Si σ-bond from the delocalized trimethylsilylmethylamine radical cation, produced by a vertical overlap of the C–Si bond and empty p-orbital of nitrogen, is assisted by the π-orbitals of the olefin. This strategy was applied to the stereoselective synthesis of pyrrolizidine and indolizidine ring systems [93]. The synthetic utility of this reaction is also demonstrated by the synthesis of (±)-iso-retronecanol [94].

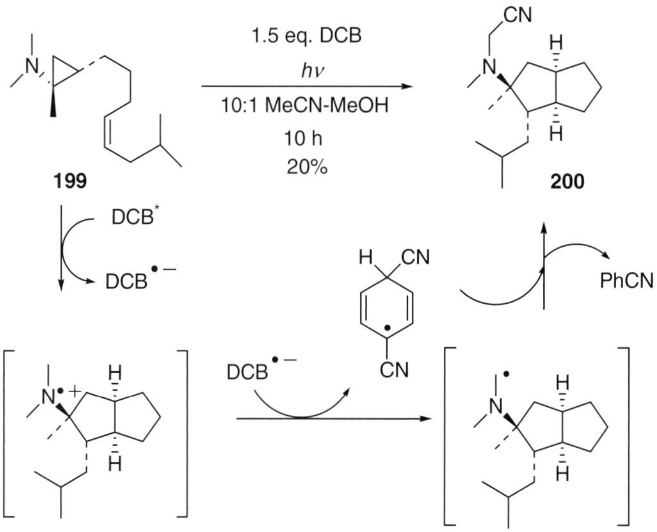

8.5 Photoinduced Electron Transfer-Mediated Cyclizations

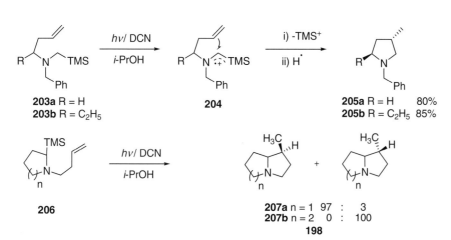

Scheme 8.55

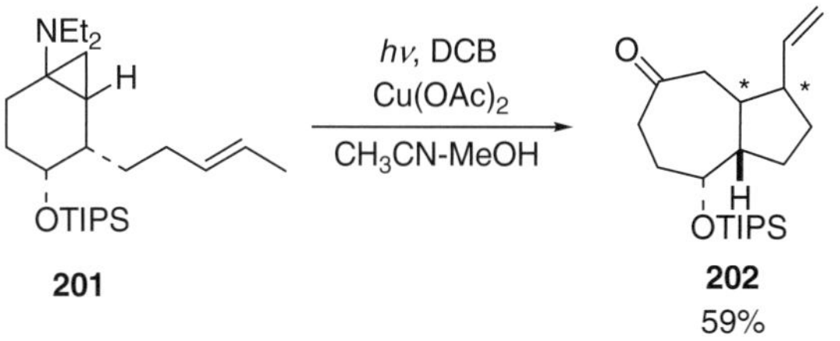

Scheme 8.56

Scheme 8.57

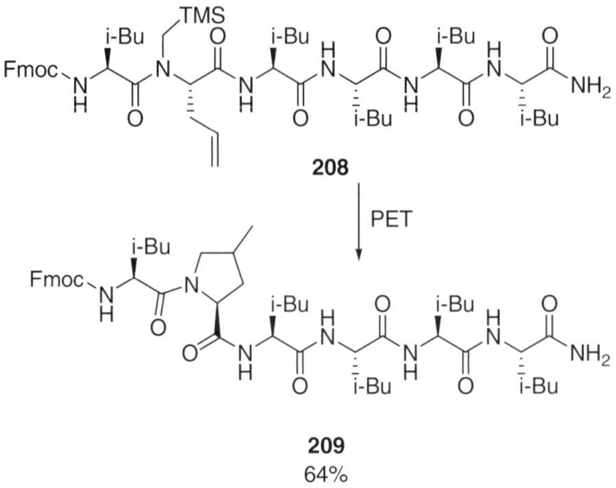

272 | 8 Formation of a Five-Membered Ring

Scheme 8.58

Steckhan's group [95] has also used this cyclization protocol to bring about structural changes in the peptide chemistry by the *in situ* formation of proline residues. For example, the PET reaction of peptide **208** using the DCA/biphenyl photoredox couple produced structurally modified peptide derivative **209** in 64% yield, as depicted in Scheme 8.58.

8.5.2
Radical Anion-Mediated Cyclizations

The chemistry of radical anions, generated by PET processes, is less developed than that of radical cations, possibly due to the nonavailability of suitable photosystems to initiate photosensitized one-electron redox reactions.

Pandey's group [96] has developed a new strategy (Figure 8.2) of carbon–carbon bond-forming reactions through the β-activation of α,β-unsaturated carbonyl compounds **210** tethered to an olefin. This reaction involves electron transfer from the *in situ*-generated DCA$^{•-}$ through a photosystem employing DCA as a light-harvesting electron acceptor and triphenyl phosphine as a sacrificial electron donor. The radical anion **211** adds on to the tethered olefin to produce 1, 2-*trans*-dialkylated cyclopentane derivatives **212** in good yield.

A further improvement in the photosystem has been considered (Figure 8.3) [97] by utilizing 1,4-dimethoxynapthalene (DMN) as the primary electron donor (towards ^{1}DCA*) and ascorbic acid as the sacrificial electron donor.

However, in this system cyclization gives 1,2-*cis*-disubstituted cyclopentane derivative **212a**. An intramolecular H-bonding of ascorbic acid with intermediate **211** is suggested to be the reason for the reversal in stereochemistry. Pandey's group has also developed an electron-transfer reductive activation of the C—SePh group and phenylselenyl group transfer radical cyclization reaction to produce five-membered ring compounds, employing the photosystem shown in Figure 8.4 [97].

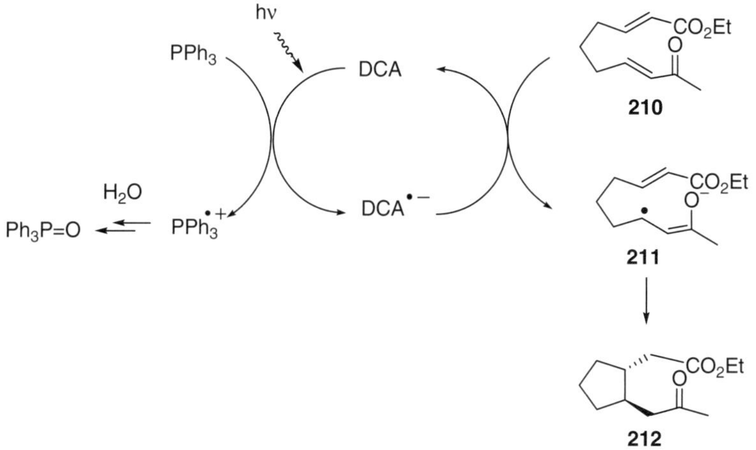

Figure 8.2 Photosystem-I for β-reductive activation of α,β-unsaturated ketones.

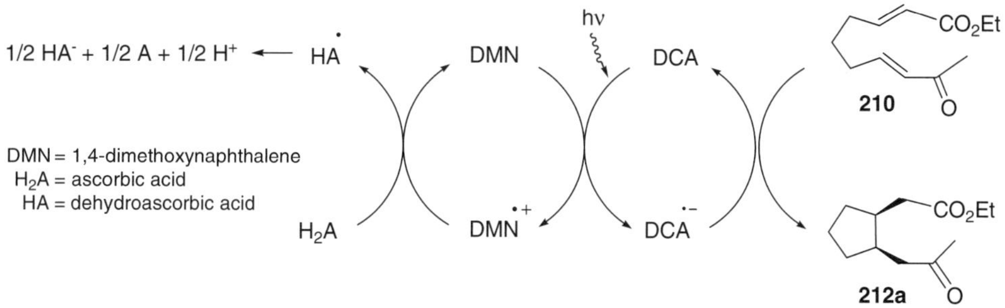

Figure 8.3 Photosystem-II for β-reductive activation of α,β-unsaturated ketones.

Cyclopentanols **214** and **217** are synthesized by intramolecular ketyl radical addition to tethered olefin by a photochemically induced electron-transfer activation of δ,ω-unsaturated ketones **213** and **216**, respectively, using either hexamethylphosphoric triamide (HMPA) or triethylamine in CH_3CN [98] as solvent (see Scheme 8.59).

When several functional groups are present in the starting molecule, electron transfer from the donor occurs selectively to the most easily reducible group. For example, the irradiation of acetylenic ketoester **218** either in HMPA or Et_3N/CH_3CN afforded the allylic bicyclic alcohol **219** as the exclusive product in high yield (Scheme 8.60). Similarly, when allenic compound **220** was irradiated in Et_3N/CH_3CN, it afforded the cyclized products **221** and **222**, in 55% and 20% yield, respectively [99].

The efficiency of this protocol has been adequately demonstrated by applying it to the total synthesis of (±)-Hirsutene (**224**), as depicted in Scheme 8.61 [100].

The synthetic value of the photoinduced electron-transfer reactions of ketones has been expanded (Scheme 8.62) by generating a five-membered carbocyclic framework, as shown in the preparation of compound **226**.

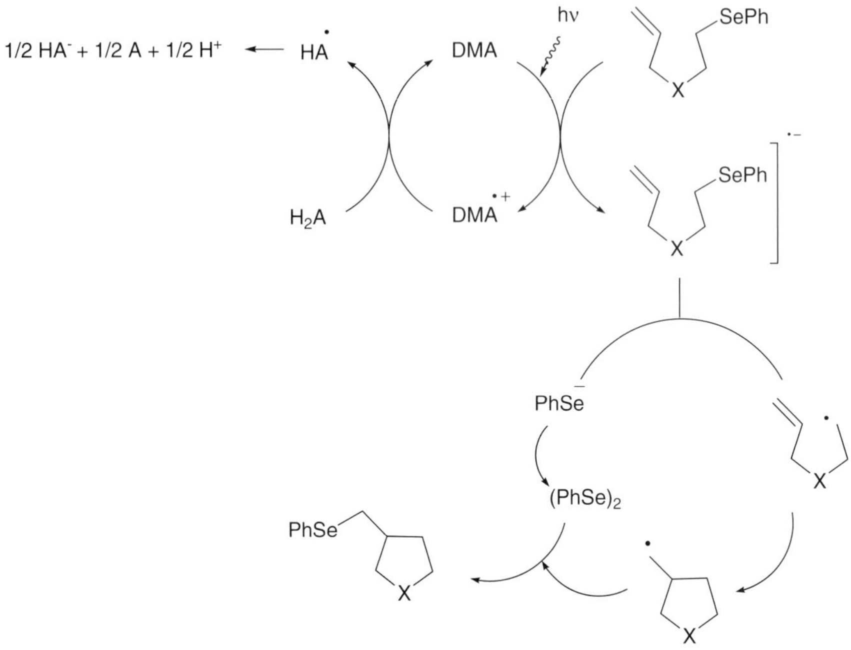

Figure 8.4 PhSe-group transfer radical cyclization reaction.

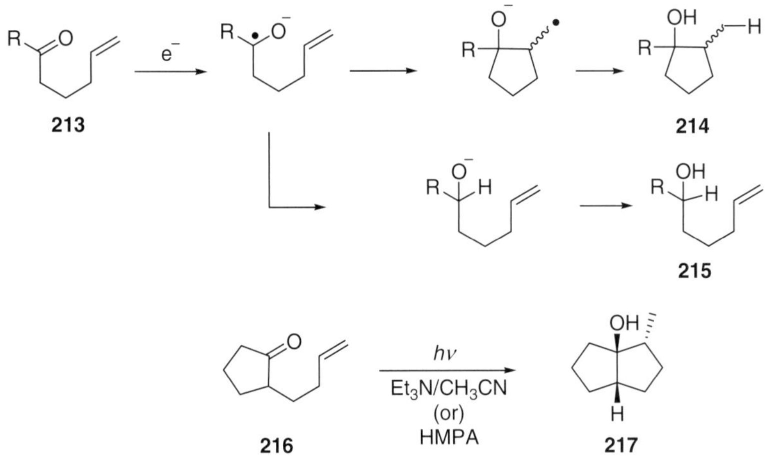

Scheme 8.59

This concept has also been explored for the synthesis of spiro-cyclic compounds **228** (Scheme 8.63) [101].

The generation of ketyl radicals and their intramolecular addition to tethered olefin to obtain cyclopentanol derivatives have also been achieved by Pandey et al. [102] by using such as those shown in Figures 8.2. An interesting application of this strategy is demonstrated by the synthesis of C-furanosides **231** (Scheme 8.64).

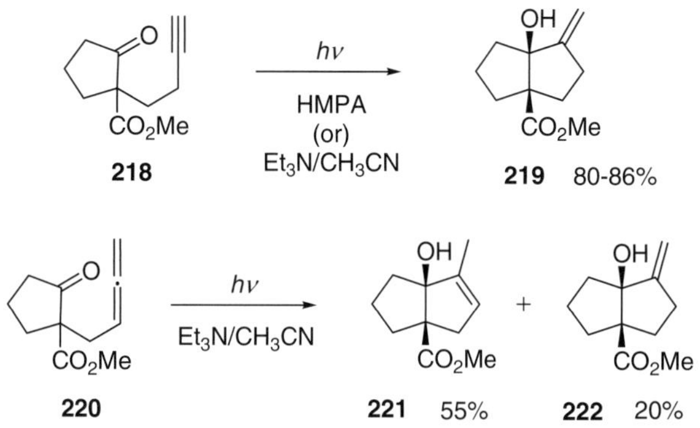

8.5 *Photoinduced Electron Transfer-Mediated Cyclizations* | 275

Scheme 8.60

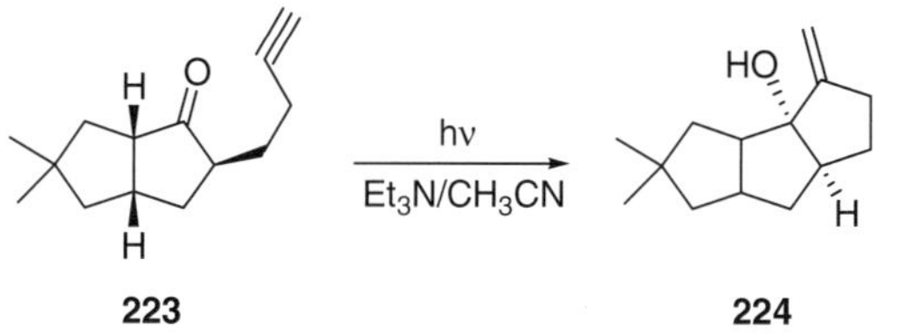

218 → (hv, HMPA (or) Et₃N/CH₃CN) → 219 80–86%

220 → (hv, Et₃N/CH₃CN) → 221 55% + 222 20%

Scheme 8.61

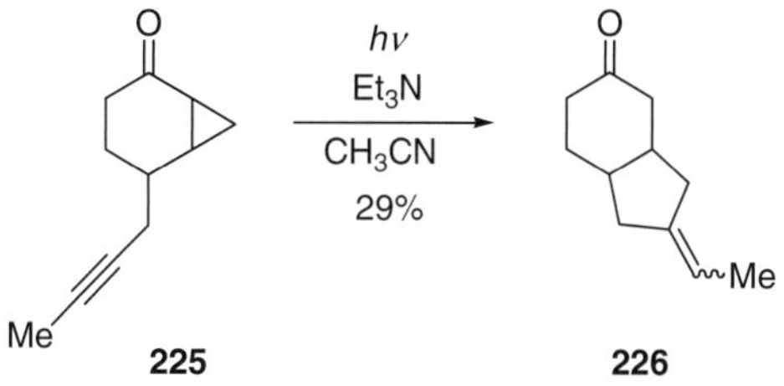

223 → (hv, Et₃N/CH₃CN) → 224

Scheme 8.62

225 → (hv, Et₃N, CH₃CN, 29%) → 226

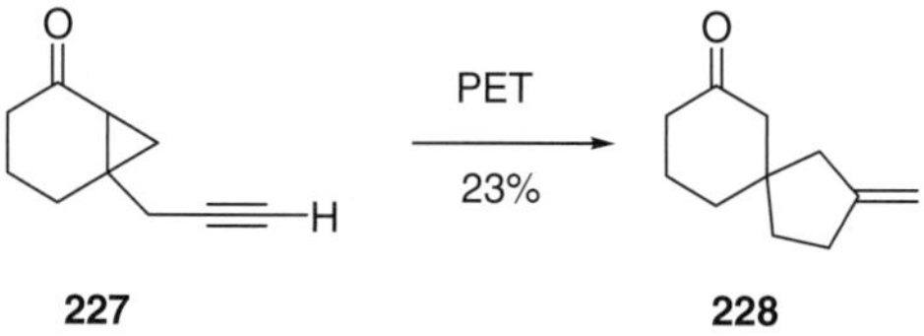

Scheme 8.63

227 → (PET, 23%) → 228

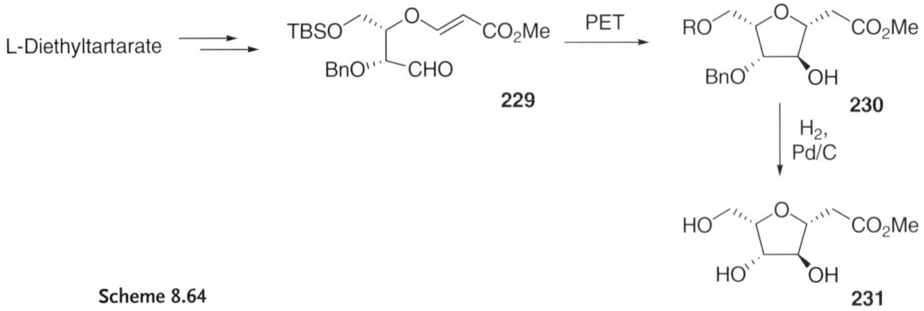

Scheme 8.64

8.5.3
Intramolecular Trapping of Radical Cations by Nucleophiles

Both, strained and unsaturated organic molecules are known to form cation radicals as a result of electron transfer to photoexcited sensitizers (excited-state oxidants). The resulting cation radical–anion radical pairs can undergo a variety of reactions, including back electron transfer, nucleophilic attack on to the cation radical, electrophilic attack on the anion radical, reduction of anion radical, and addition of anion radical to the cation radical. This concept has been nicely demonstrated by Gassman et al. [103, 104], using the photoinduced electron-transfer cyclization of γ,δ-unsaturated carboxylic acid **232** to γ-lactones **233** and **234** as an example (see Scheme 8.65).

Mechanistically, the conversion of acid to lactone is viewed as occurring through the transfer of an electron from the olefinic moiety of **232** to yield the corresponding radical cation, which then undergoes cyclization to give lactone **233**. When 1-cyanonaphthalene (1-CN) is used as the sensitizer, both the quantum and product yields were found to be moderate. However, the use of a sterically hindered sensitizer led to an increase of both yields, as well as of the quantum efficiency.

Homobenzylic ether radical cations **236**, generated photochemically by electron transfer to N-methylquinolinium hexafluorophosphate (NMQPF$_6$), have been trapped by oxygen nucleophiles to give a variety of oxygen heterocycles, namely **237**, **239**, and **241** (Scheme 8.66) [105]. A large number of cyclic acetals have been synthesized in good yield relying on this approach. This cyclization proceeds through an electron transfer-mediated carbon–carbon σ-bond activation pathway, utilizing

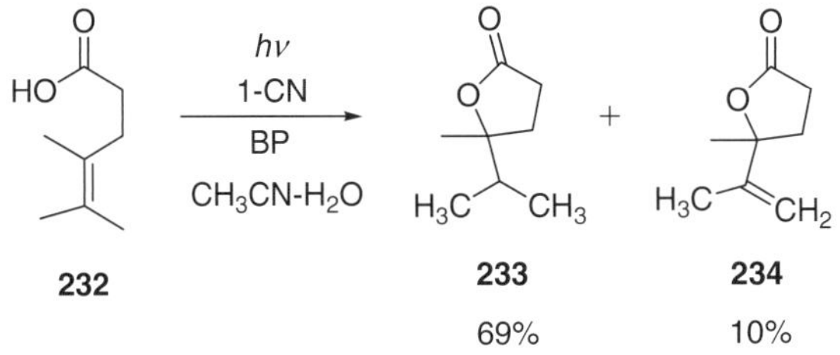

Scheme 8.65

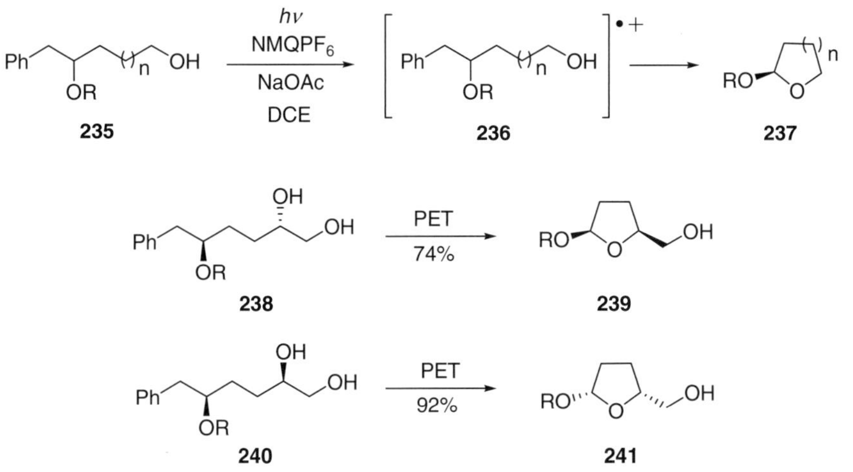

8.5 Photoinduced Electron Transfer-Mediated Cyclizations | 277

Scheme 8.66

the bicyclic oxocarbenium ions as the intermediate. The nucleophilic attack was found to be highly diastereoselective, and in the construction of furanoside and pyranoside system there was full control of the anomeric stereochemistry.

Aryl radical cations generated by electron-transfer processes from methoxy substituted arenes to DCN, tethered by oxygen, nitrogen as well as carbon nucleophile leads to intramolecular cyclizations (Scheme 8.67). The synthetic potentials of

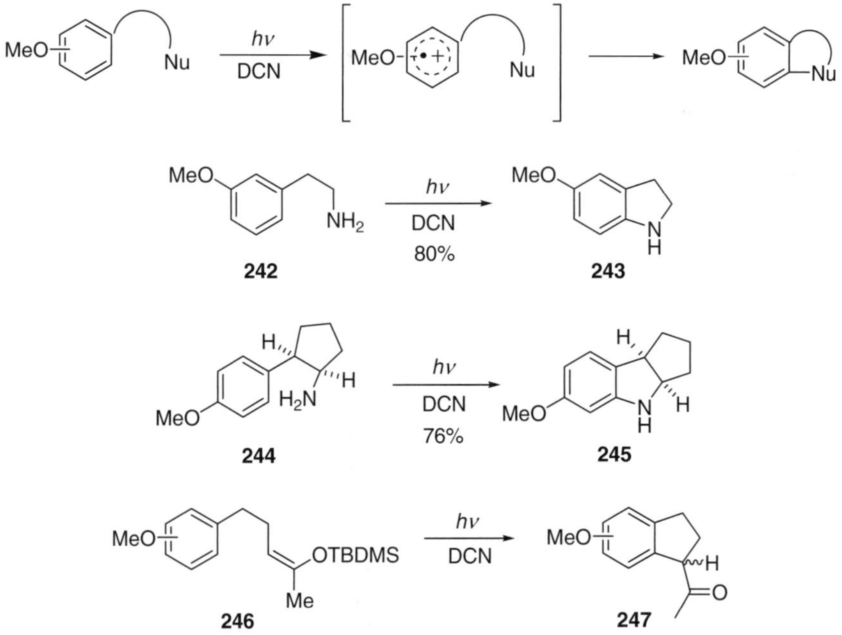

Scheme 8.67

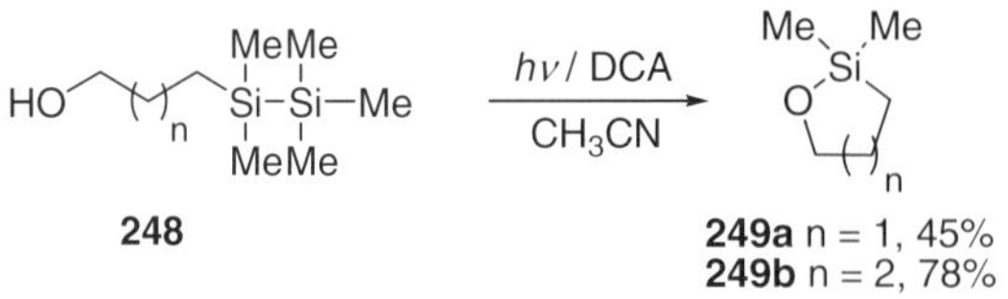

8 Formation of a Five-Membered Ring

Scheme 8.68

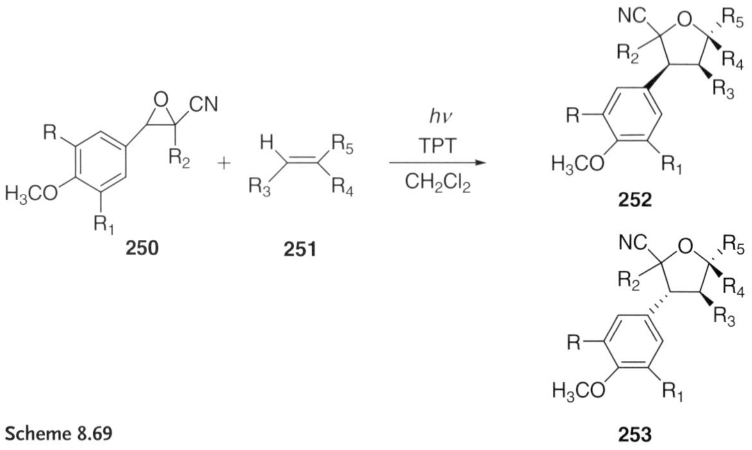

248 → 249a n = 1, 45%; 249b n = 2, 78%

Scheme 8.69

Table 8.5 Cycloadducts from PET-generated nonstabilized azomethine ylide.

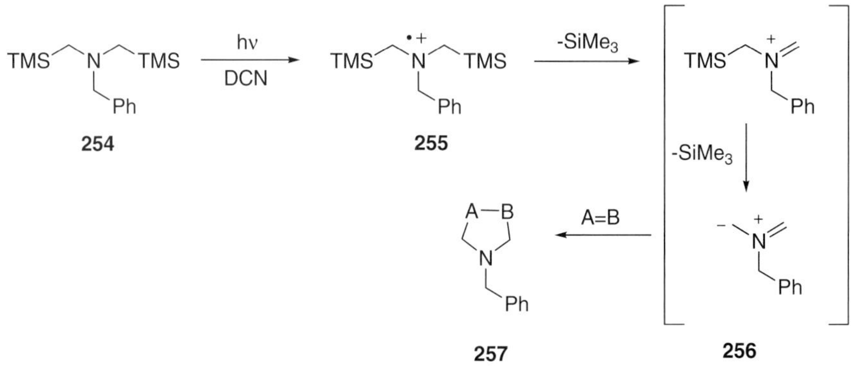

Entry	Dipolarophile	Yield of cycloadduct (%)
1	Benzophenone	80
2	Dimethyl fumarate	60
3	*trans*-Methyl cinnamate	78
4	*N*-Phenylmaleimide	83
5	Ethyl acrylate	58
6	Phenyl vinyl sulfone	75

this strategy have been exploited for the formation of aromatic five-membered heterocyclic derivatives 243 and 245 [106], benzofurans [106], and indanes 247 [107].

The PET activation of Si—Si bond possessing compounds such as 248 using excited DCA as an electron acceptor is reported to lead its dissociation, followed by intramolecular trapping of the resultant cation to produce the corresponding cyclic silyl ethers 249 in 45–78% yield, as depicted in Scheme 8.68 [108].

The 2,4,6-triphenylpyrylium terafluoroborate (TPT)-sensitized electron transfer of aryl-substituted epoxides such as 250 leads to ring opening via selective C—O bond cleavage, while subsequent [3 + 2]-cycloaddition of the resultant carbonyl ylide with electron-rich olefins 251 leads to the synthesis of substituted THF derivatives 252 and 253 (Scheme 8.69) [109].

The generation of nonstabilized azomethine ylide 256 via PET-initiated sequential double desilylation and [3 + 2]-cycloaddition reaction with various dipolarophiles to generate five-membered heterocycles 257, has also been established by Pandey et al., as shown in Table 8.5 [110].

Acknowledgments

The authors gratefully acknowledge the diligent and dedicated collaboration of all their colleagues whose names appear as coauthors in the references. Thanks are also due to Mr K. C. Bharadwaj, Mr Balakrishnan, and Mr K. N. Tiwari for their assistance in the preparation of the manuscript.

References

1 Wilson, J.E. and Fu, G.C. (2006) Synthesis of functionalized cyclopentenes through catalytic asymmetric [3 + 2] cycloadditions of allenes with enones. *Angewandte Chemie, International Edition*, **45**, 1426–1429.

2 Wolfe, J.P. and Hay, M.B. (2007) Recent advances in the stereoselective synthesis of tetrahydrofurans. *Tetrahedron*, **63**, 261–290.

3 Bellina, F. and Rossi, R. (2006) Synthesis and biological activity of pyrrole, pyrroline and pyrrolidine derivatives with two aryl groups on adjacent positions. *Tetrahedron*, **62**, 7213–7256.

4 Wagner, P.J. and Zepp, R.G. (1971) γ- vs δ-hydrogen abstraction in the photochemistry of β-alkoxy ketones. Overlooked reaction of hydroxy biradicals. *Journal of the American Chemical Society*, **93**, 4958–4959.

5 Paquette, L.A. and Balogh, D.W. (1982) An expedient synthesis of 1, 16-dimethyldodecahedrane. *Journal of the American Chemical Society*, **104**, 774–783.

6 Wagner, P.J., Giri, B.P., Pabon, R., and Singh, S.B. (1987) Divergent photochemistry of 2,4-di-*tert*-butylacetophenone and -benzophenone. *Journal of the American Chemical Society*, **109**, 8104–8105.

7 Wagner, P.J., Pabon, R., Park, B.-S., Zand, A.R., and Ward, D.L. (1994) The regioselectivity of internal hydrogen abstraction by triplet o-tert-amylbenzophenone. *Journal of the American Chemical Society*, **116**, 589–596.

8 Meador, M.A. and Wagner, P.J. (1983) 2-Indanol formation from photocyclization of α-arylacetophenones. *Journal of the American Chemical Society*, **105**, 4484–4486.

9 Wagner, P.J., Hasegawa, T., Zhou, B., and Ward, D.L. (1991) Diverse photochemistry of sterically congested α-arylacetophenones: ground-state conformational control of reactivity. *Journal of the American Chemical Society*, **113**, 9640–9654.

10 Chang, D.J., Nahm, K., and Park, B.S. (2002) Reversed diastereoselectivity in the yang photocyclization upon introducing a cyclopropyl group at the α-position to carbonyls. *Tetrahedron Letters*, **43**, 4249–4252.

11 DeBoer, C.D., Herkstroeter, W.G., Marchetti, A.P., Schultz, A.G., and Schlessinger, R.H. (1973) Norrish type II rearrangement from π, π* triplet states. *Journal of the American Chemical Society*, **95**, 3963–3969.

12 Wagner, P.J. and Chiu, C. (1979) Preferential 1,4- vs. 1,6-hydrogen transfer in a 1,5 biradical. Photochemistry of β-ethoxypropiophenone. *Journal of the American Chemical Society*, **101**, 7134–7135.

13 Carless, H.A.J., Swan, D.I., and Haywood, D.J. (1993) Stereoselective synthesis of tetrahydrofuran-3-ols by photochemical δ-hydrogen abstraction of β-allyloxycarbonyl compounds. *Tetrahedron*, **49**, 1665–1674.

14 Descotes, G. (1982) Acetal photoreactivity in the heterocyclic and glycoside series. *Bulletin des Societes Chimique Belges*, **91**, 783–973.

15 Koźluk, T., Cottier, L., and Descotes, G. (1981) Syntheses photochimiques de Dioxa-1,6-Spiro[4.5]decanes pheromones de paravespula vulgaris L. *Tetrahedron*, **37**, 1875–1880.

16 Kraus, G.A. and Chen, L. (1990) A total synthesis of racemic paulownin using a type II photocyclization reaction. *Journal of the American Chemical Society*, **112**, 3464–3466.

17 Kraus, G.A. and Schwinden, M.D. (1991) Type II photocyclizations of carbohydrates. *Journal of Photochemistry and Photobiology, A: Chemistry*, **62**, 241–244.

18 Wagner, P.J., Meador, M.A., and Park, B.-S. (1990) The photocyclization of o-alkoxy phenyl ketones. *Journal of the American Chemical Society*, **112**, 5199–5211.

19 Sharshira, E.M., Okamura, M., Hasegawa, E., and Horaguchi, T. (1997) Photocyclization reactions. Part 6 [1]. Solvent and substituent effects in the synthesis of dihydrobenzofuranols using photocyclization of 2-alkoxybenzophenones and ethyl 2-benzoylphenoxyacetates. *Journal of Heterocyclic Chemistry*, **34**, 861–869.

20 Sharshira, E.M. and Horaguchi, T. (1997) Photocyclization reactions. Part 7 [1]. Solvent and substituent effects in the synthesis of dihydrobenzofuranols using photocyclization of α-(2-acylphenoxy) toluenes and ethyl 2-acylphenoxyacetates. *Journal of Heterocyclic Chemistry*, **34**, 1837–1848.

21 Walther, K., Kranz, U., and Henning, H.-G. (1987) Darstellung and diastereoselektive photocylisierung van N-(β- Benzoulethyl)-N-tosyl-glycinamiden. *Journal für Praktische Chemie*, **329**, 859–870.

22 Hasegawa, T., Moribe, J.-I., and Yoshioka, M. (1988) The photocyclization of N,N-dialkyl β-oxo amides on silica gel. *Bulletin of the Chemical Society of Japan*, **61**, 1437–1439.

23 Wessig, P., Wettstein, P., Giese, B., Neuburger, M., and Zehnder, M. (1994) Asymmetric synthesis of 3-hydroxy-prolines by photocyclization of N-(2-benzoylethyl)glycinamides. *Helvetica Chimica Acta*, **77**, 829–837.

24 Steiner, A., Wessig, P., and Polborn, K. (1996) Asymmetric synthesis of 3-hydroxyprolines by photocyclization of

C(1′)-substituted N-(2-benzoylethyl) glycine esters. *Helvetica Chimica Acta*, **79**, 1843–1862.
25 Giese, B., Wettstein, P., Stähelin, C., Barbosa, F., Neuburger, M., Zehnder, M., and Wessig, P. (1999) Memory of chirality in photochemistry. *Angewandte Chemie, International Edition*, **38**, 2586–2587.
26 Wessig, P., Legart, F., Hoffmann, B., and Henning, H.-G. (1991) Photochemie von Aminoketonen, 14. Syntheses und Transannulare photocyclisierung von 2-benzoyl-1,4-bis(tosyl)piperazinon. *Liebigs Annalen der Chemie*, **10**, 979–982.
27 Lee, G.A. (1976) Photochemistry of cis- and trans-stilbene oxides. *Journal of Organic Chemistry*, **41**, 2656–2658.
28 Woodward, R.B. and Hoffmann, R. (1970) *The Conservation of Orbital Symmetry*, Academic Press, New York, p. 70.
29 Wong, J.P.K., Fahmi, A.A., Griffin, G.W., and Bhacca, N.S. (1981) Photo- and thermoinduced generation of 1,3-diaryl carbonyl ylides from 2,3-diaryloxiranes 1,3-dipolar cycloadditions to dipolarophiles. *Tetrahedron*, **37**, 3345–3355.
30 Lev, I.J., Ishikawa, K., Bhacca, N.S., and Griffin, G.W. (1976) Photogeneration and reactions of acyclic carbonyl ylides. *Journal of Organic Chemistry*, **41**, 2654–2656.
31 Markowski, V. and Huisgen, R. (1976) Disrotatory photoconversion of cis,trans isomeric oxiranes to carbonyl ylides. *Tetrahedron Letters*, **17**, 4643–4646.
32 Clawson, P., Lunn, P.M., and Whiting, D.A. (1990) Synthetic studies on O-heterocycles via cycloadditions. Part 1. Photochemical (electron transfer sensitized) carbon-carbon cleavage of diaryloxiranes. *Journal of the Chemical Society, Perkin Transactions 1*, 153–157.
33 Clawson, P. and Whiting, D.A. (1990) Synthetic studies on O-heterocycles via cycloadditions. Part 3. Regiochemical and mechanistic questions in reactions of polarized diaryl carbonyl ylides. *Journal of the Chemical Society, Perkin Transactions 1*, 1193–1198.
34 Feldman, K.S. (1983) Photochemical activation of α-β-Epoxyketones. Formation of benzannelated perhydroazulenes by intramolecular trapping of a reactive intermediate. *Tetrahedron Letters*, **24**, 5585–5586.
35 Brokatzky-Geiger, J. and Eberbach, W. (1984) Thermal and photochemical studies with stilbene oxides: intramolecular trapping of carbonyl ylide intermediates. *Tetrahedron Letters*, **25**, 1137–1140.
36 (a) Kotera, M., Ishii, K., Tamura, O., and Sakamoto, M. (1998) 1,3-Dipolar cycloadditions of photoinduced carbonyl ylides. Part 2. Photoreactions of α,β-unsaturated γ,δ-epoxy dinitriles and ethyl vinyl ether. *Journal of the Chemical Society, Perkin Transactions 1*, 313–318; (b) Kotera, M., Ishii, K., Tamura, O., and Sakamoto, M. (1994) 1,3-Dipolar cycloadditions of photoinduced carbonyl ylides from α,β-unsaturated, γ,δ-epoxy dinitriles. *Journal of the Chemical Society, Perkin Transactions 1*, 2353–2354.
37 Huisgen, R., Scheer, W., and Huber, H. (1967) Stereospecific conversion of cis-trans isomeric aziridines to open-chain azomethine ylides. *Journal of the American Chemical Society*, **89**, 1753–1755.
38 Daly, J.W., Spande, T.F., and Garraffo, H.M. (2005) Alkaloids from amphibian skin: a tabulation of over eight-hundred compounds. *Journal of Natural Products*, **68**, 1556–1575.
39 O'Hagan, D. (2000) Pyrrole, pyrrolidine, pyridine, piperidine and tropane alkaloids. *Natural Product Reports*, **17**, 435–446.
40 Leclercq, S., Braekman, J.C., Daloze, D., and Pasteels, J.M. (2000) The defensive chemistry of ants. *Progress in the Chemistry of Organic Natural Products*, **79**, 115–229.
41 Daly, J.W., Garraffo, H.M., and Spande, T.F. (1999) Chapter 1 in *Alkaloids: Chemical*

and *Biological Perspectives*, Vol. 13 (ed. S.W. Pelletier), Pergamon, New York.

42 Pandey, G., Banerjee, P., and Gadre, S.R. (2006) Construction of enantiopure pyrrolidine ring system via asymmetric [3 + 2] cycloaddition of azomethine ylides. *Chemical Reviews*, **106**, 4484–4517.

43 Coldham, I. and Hufton, R. (2005) Intramolecular dipolar cycloaddition reactions of azomethine ylides. *Chemical Reviews*, **105**, 2765–2810.

44 Pearson, W.H. and Stoy, P. (2003) Cycloadditions of nonstabilized 2-azaallyllithiums (2-azaallyl anions) and azomethine ylides with alkenes: [3 + 2] cycloaddition approaches to pyrrolidines and application to alkaloid total synthesis. *Synlett*, 903–921.

45 Najera, C. and Sansano, J.M. (2003) Azomethine ylides in organic synthesis. *Current Organic Chemistry*, **7**, 1105–1150.

46 Padwa, A. (1984) *1,3-Dipolar Cycloaddition Chemistry*, Vol. 2 (ed. A. S Padwa), John Wiley & Sons, New York, p 277.

47 Harwood, L.M. and Vickers, R.J. (2002) *Synthetic Applications of 1,3-Dipolar Cycloaddition Chemistry Toward Heterocycles and Natural Products* (eds A. S Padwa and W.H. Pearson), John Wiley & Sons, New York, p. 169.

48 Gothelf, K.V. (2002) Chapter 6 in *Cycloaddition Reactions in Organic Synthesis* (eds S. Kobayashi and K.A. Jørgensen), Wiley-VCH, Weinheim, p. 211.

49 Garner, P., Ho, W.B., and Shin, H. (1993) The asymmetric synthesis of (−)-quinocarcin via a 1,3-dipolar cycloadditive strategy. *Journal of the American Chemical Society*, **115**, 10742–10753.

50 Garner, P.P., Cox, P.B., Klippenstein, S.J., Youngs, W.J., and McConville, D.B. (1994) Tether-mediated stereocontrol in intramolecular azomethine ylide cycloadditions. *Journal of Organic Chemistry*, **59**, 6510–6511.

51 Ishii, K., Sone, T., Shimada, Y., Shigeyama, T., Noji, M., and Sugiyama, S. (2004) Photoreactions of β-aziridinylacrylonitriles and acrylates with alkenes: formation of head-to-head adducts and application to the preparation of pyrrolizidine alkaloid. *Tetrahedron*, **60**, 10887–10898.

52 Ishii, K., Sone, T., Shigeyama, T., Noji, M., and Sugiyama, S. (2006) Photoreactions of β-aziridinylacrylonitriles and acrylates with alkenes: the substituent effects on the formation of [3 + 2] cycloadducts. *Tetrahedron*, **62**, 10865–10878.

53 Hudlicky, T. and Reed, J.W. (1991) *Comprehensive Organic Synthesis*, Vol. 5 (eds B.M. Trost and I. Fleming), Pergamon, Oxford, p. 899.

54 Wong, H.N.C., Hon, M.Y., Tse, C.W., Yip, Y.C., Tanko, J., and Hudlicky, T. (1989) Use of cyclopropanes and their derivatives in organic synthesis. *Chemical Reviews*, **89**, 165–198.

55 Goldschmidt, Z. and Crammer, B. (1988) Vinylcyclopropane rearrangements. *Chemical Society Reviews*, **17**, 229–267.

56 Cooke, R.S. (1970) A photochemical vinylcyclopropane to cyclopentene rearrangement. *Journal of the Chemical Society D: Chemical Communications*, (7), 454–455.

57 Jorgenson, M.J. (1969) Photochemistry of α, β-unsaturated esters. VII. Photolytic behavior of vinylcyclopropanecarboxylates. *Journal of the American Chemical Society*, **91**, 6432–6443.

58 Kristinsson, H. and Hammond, G.S. (1967) Mechanisms of photochemical reactions in solution. XLVIII. rearrangement of phenylvinylcyclo-propanes. *Journal of the American Chemical Society*, **89**, 5970–5971.

59 Armesto, D., Ramos, A., Mayoral, E.P., Ortiz, M.J., and Agarrabeitia, A.R. (2000) A novel photochemical vinylcyclopropane rearrangement yielding 6,7-dihydro-5*H*-benzocycloheptene derivatives. *Organic Letters*, **2**, 183–186.

60 Campos, P.J., Soldevilla, A., Sampedro, D., and Rodríguez, M.A. (2001) N-Cyclopropylimine-1-pyrroline rearrangement. A novel photochemical reaction. *Organic Letters*, **3**, 4087–4089.

61 Schultz, A.G. (1983) Photochemical six-electron heterocyclization reactions. *Accounts of Chemical Research*, **16**, 210–218.

62 Schultz, A.G. and Motkya, L. (1983) *Organic Photochemistry*, Vol. 6 (ed. A. Padwa), Marcel Dekker, New York, p. 1.

63 Grellmann, K.-H., Sherman, G.M., and Linschitz, H. (1963) Photo-conversion of diphenylamines to carbazoles, and accompanying transient species. *Journal of the American Chemical Society*, **85**, 1881–1882.

64 Grellmann, K.-H., Kühnle, W., Weller, H., and Wolff, T. (1981) Photochemical formation of dihydrocarbazoles from diphenylamines and their thermal rearrangement and disproportionation reactions. *Journal of the American Chemical Society*, **103**, 6889–6893.

65 Fox, M.A., Dulay, M.T., and Krosley, K. (1994) Comparison of oxidative and excited state cyclizations of N-benzyl-diphenylamines to N-benzylcarbazoles. *Journal of the American Chemical Society*, **116**, 10992–10999.

66 Zeller, K.-P. and Petersen, H. (1975) Photochemische Herstellung von Dibenzofuranen und Dibenzothiophenen. *Synthesis*, 532–533.

67 Zeller, K.-P. and Berger, S. (1977) Steric hindrance in substituted dibenzofurans. *Journal of the Chemical Society, Perkin Transactions 2*, 54–58.

68 Ho, J.-H. and Ho, T.-I. (2002) Novel photoreaction of N-alkyl(p-methoxyphenyl)arylamines assisted by protic acids. *Chemical Communications*, 270–271.

69 Chapman, O.L. and Eian, G.L. (1968) Photochemical transformations. XXVIII. Photochemical synthesis of 2,3-dihydroindoles from N-aryl enamines. *Journal of the American Chemical Society*, **90**, 5329–5330.

70 Chapman, O.L., Eian, G.L., Bloom, A., and Clardy, J. (1971) Photochemical transformations. XXXVIII. Nonoxidative photocyclization of N-aryl enamines. A facile synthetic entry to trans-hexahydrocarbazoles. *Journal of the American Chemical Society*, **93**, 2918–2928.

71 Wolf, T. (1975) Ph.D. Thesis, Göttingen.

72 Shimizu, J., Murakami, S., Oishi, T., and Ban, Y. (1971) The Fischer indole synthesis with formic acid. II. The synthesis of hexahydrocyclopent(6)indoles. *Chemical & Pharmaceutical Bulletin*, **19**, 2561–2566.

73 Blache, Y., Sinibaldi-Troin, M.-E., Hichour, M., Benezech, V., Chavignon, O., Gramain, J.-C., Teulade, J.-C., and Chapat, J.-P. (1999) Heterocyclic enaminones: photochemical synthesis of 6,7,8,9-tetrahydro-5H-pyrido[2,3-b]indol-9-ones. *Tetrahedron*, **55**, 1959–1970.

74 Arnould, J.C., Cossy, J., and Pete, J.P. (1980) Reactivite photochimique des α-aminoenones: reactions de cyclisation et nouveau type de reaction dans les α-Sulfonamido-cyclohexenones. *Tetrahedron*, **36**, 1585–1592.

75 Tietcheu, C., Garcia, C., Gardette, D., Dugat, D., and Gramain, J.-C. (2002) Efficient photochemical synthesis of tricyclic keto-indoles. *Journal of Heterocyclic Chemistry*, **39**, 965–973.

76 Dugat, D., Gramain, J.-C., and Dauphin, G. (1990) Structure, stereochemistry, and conformation of diastereoisomeric cis- and trans-3-ethyl-1,2,3,4,4a,9a-hexahydrocarbazol-4-ones by means of ^{13}C and two-dimensional ^{1}H nuclear magnetic resonance spectroscopy. An example of diastereoselection in a photocyclisation reaction. *Journal of the Chemical Society, Perkin Transactions 2*, 605–611.

77 Gramain, J.-C., Husson, H.-P., and Troin, Y. (1985) A novel and efficient synthesis of

the aspidosperma alkaloid ring system: N(a)-benzyldeethyl-aspidospermidine. *Journal of Organic Chemistry*, **50**, 5517–5520.

78 Benchekroun-Mounir, N., Dugat, D., Gramain, J.C., and Husson, H.P. (1993) Stereocontrolled formation of octahydro-1H-pyrrolo[2,3-d]carbazoles by reductive cyclization: total synthesis of (±)-N-benzylaspidospermidine. *Journal of Organic Chemistry*, **58**, 6457–6465.

79 Dugat, D., Benchekroun-Mounir, N., Dauphin, G., and Gramain, J.-C. (1998) Reactivity of hexahydrocarbazol-4-ones in Michael reactions: stereocontrolled formation of decahydropyrido[2,3-d] carbazoles. *Journal of the Chemical Society, Perkin Transactions 1*, 2145–2149.

80 Ibrahim-Ouali, M., Sinibaldi, F. M.-È., Troin, Y., Guillaume, D., and Gramain, J.-C. (1997) Diastereoselective synthesis of 3-3′ disubstituted indolines. *Tetrahedron*, **53**, 16083–16096.

81 Schultz, A.G. and DeTar, M.B. (1976) Thiocarbonyl ylides. Photogeneration, rearrangement, and cycloaddition reactions. *Journal of the American Chemical Society*, **98**, 3564–3572.

82 Dittami, J.P., Nie, X.-Y., Buntel, C.J., and Rigatti, S. (1990) Photoinitiated intramolecular ylide-alkene cycloaddition reaction. *Tetrahedron Letters*, **31**, 3821–3824.

83 Weng, H., Scarlata, C., and Roth, H.D. (1996) Electron transfer photochemistry of geraniol and (E, E)-farnesol. A novel "tandem", 1,5-cyclization, intramolecular capture. *Journal of the American Chemical Society*, **118**, 10947–10953.

84 Warzecha, K.-D., Xing, X. and Demuth, M. (1997) Cyclization of terpenoid polyalkenes via photo-induced electron transfer – versatile single-step syntheses of mono- and polycycles. *Pure & Applied Chemistry*, **69**, 109–112.

85 Hasegawa, E., Yamaguchi, N., Muraoka, H., and Tsuchida, H. (2007) Electron transfer promoted regioselective ring-opening reaction of cyclopropyl silyl ethers. *Organic Letters*, **9**, 2811–2814.

86 Stavinoha, J.L., Mariano, P.S., Leone-Bay, A., Swanson, R., and Bracken, C. (1981) Photocyclizations of N-allyliminium salts leading to the production of substituted pyrrolidines. *Journal of the American Chemical Society*, **103**, 3148–3160.

87 Yoon, U.C., Quillen, S.L., Mariano, P.S., Swanson, R., Stavinoha, J.L., and Bay, E. (1983) Exploratory and mechanistic aspects of the electron-transfer photochemistry of olefin-N-heteroaromatic cation systems. *Journal of the American Chemical Society*, **105**, 1204–1218.

88 Chiu, F.-T., Ullrich, J.W., and Mariano, F P. S. (1984) Model studies examining the application of allyliminium salt photospirocyclization methodologies in synthetic approaches to the harringtonine alkaloids. *Journal of Organic Chemistry*, **49**, 228–236.

89 Ha, J.D., Lee, J., Blackstock, S.C., and Cha, J.K. (1998) Intramolecular [3 + 2] annulation of olefin-tethered cyclopropylamines. *Journal of Organic Chemistry*, **63**, 8510–8514.

90 Lee, H.B., Sung, M.J., Blackstock, S.C., and Cha, J.K. (2001) Radical cation-mediated annulation. stereoselective construction of bicyclo[5.3.0]decan-3-ones by aerobic oxidation of cyclopropylamines. *Journal of the American Chemical Society*, **123**, 11322–11324.

91 Pandey, G., Kumaraswamy, G., and Bhalerao, U.T. (1989) Photoinduced SET generation of α-amine radicals: A practical method for the synthesis of pyrrolidines and piperidines. *Tetrahedron Letters*, **30**, 6059–6062.

92 Pandey, G., Devi Reddy, G., and Kumaraswamy, G. (1994) Photoinduced electron transfer (PET) promoted cyclisations of 1-[N-alkyl-N-(trimethylsilyl)methyl]amines tethered to proximate olefin: mechanistic and synthetic perspectives. *Tetrahedron*, **50**, 8185–8194.

93 Pandey, G. and Devi Reddy, G. (1992) Stereoselectivity in the cyclisation of photoinduced electron transfer (PET) generated cyclic α-amino radicals: first general stereoselective entry to 1-azabicyclo (m:n:o) alkane systems. *Tetrahedron Letters*, **33**, 6533–6536.

94 Pandey, G., Devi Reddy, G., and Chakrabarti, D. (1996) Stereoselectivity in the photoinduced electron transfer (PET) promoted intramolecular cyclisations of 1-alkenyl-2-silyl-piperidines and -pyrrolidines: rapid construction of 1-azabicyclo [m.n.o] alkanes and stereoselective synthesis of (±)-isoretronecanol and (±)-epilupinine. *Journal of the Chemical Society, Perkin Transactions* 1, 219–224.

95 Jonas, M., Blechert, S., and Steckhan, E. (2001) Photochemically induced electron transfer (PET) catalyzed radical cyclization: A practical method for inducing structural changes in peptides by formation of cyclic amino acid derivatives. *Journal of Organic Chemistry*, **66**, 6896–6904.

96 Pandey, G., Hajra, S., Ghorai, M.K., and Kumar, K.R. (1997) Designing photosystems for harvesting photons into electrons by sequential electron-transfer processes: reversing the reactivity profiles of α,β-unsaturated ketones as carbon radical precursor by one electron reductive β-activation. *Journal of the American Chemical Society*, **119**, 8777–8787.

97 Pandey, G., Rao, F K.S.S.P., and Rao, K.V.N. (2000) PhSeSiR$_3$-catalyzed group transfer radical reactions. *Journal of Organic Chemistry*, **65**, 4309–4314.

98 Belotti, D., Cossy, J., Pete, J.P., and Portella, C. (1985) Photoreductive cyclization of δ,ε-unsaturated ketones. *Tetrahedron Letters*, **26**, 4591–4594.

99 Belotti, D., Cossy, J., Pete, J.P., and Portella, C. (1986) Synthesis of bicyclic cyclopentanols by photoreductive cyclization of δ,ε-unsaturated ketones. *Journal of Organic Chemistry*, **51**, 4196–4200.

100 Cossy, J., Belotti, D., and Pete, J.P. (1990) Synthese totale du (±) hirsutene. *Tetrahedron*, **46**, 1859–1870.

101 Fagnoni, M., Schmoldt, P., Kirschberg, T., and Mattay, J. (1998) Reductive cyclization of α-cyclopropylketones with alkynyl- and aryl-tethered substituents. *Tetrahedron*, **54**, 6427–6444.

102 Pandey, G., Hajra, S., Ghorai, M.K., and Kumar, K.R. (1997) Visible light initiated photosensitized electron transfer cyclizations of aldehydes and ketones to tethered α,β-unsaturated esters: stereoselective synthesis of optically pure C-furanosides. *Journal of Organic Chemistry*, **62**, 5966–5973.

103 Gassman, P.G. and Bottorff, K.J. (1987) Photoinduced lactonization. a useful but mechanistically complex single electron transfer process. *Journal of the American Chemical Society*, **109**, 7547–7548.

104 Gassman, P.G. and De Silva, S.A. (1991) Use of sterically hindered sensitizers for improved photoinduced electron-transfer reactions. *Journal of the American Chemical Society*, **113**, 9870–9872.

105 Kumar, V.S. and Floreancig, P.E. (2001) Electron transfer initiated cyclizations: cyclic acetal synthesis through carbon–carbon σ-bond activation. *Journal of the American Chemical Society*, **123**, 3842–3843.

106 Pandey, G., Sridhar, M., and Bhalerao, U.T. (1990) Regiospecfic dihydroindoles directly from β-arylethylamines by photoinduced SET reaction: one-pot "wavelength switch" approach to benzopyrrolizidines related to mitomycin. *Tetrahedron Letters*, **31**, 5373–5376.

107 Pandey, G., Karthikeyan, M., and Murugan, A. (1998) New intramolecular α-arylation strategy of ketones by the reaction of silyl enol ethers to photosensitized electron transfer generated arene radical cations: construction of benzannulated and benzospiroannulated compounds. *Journal of Organic Chemistry*, **63**, 2867–2872.

108 Nakadaira, Y., Sekiguchi, A., Funada, Y., and Sakurai, H. (1991) Photo-induced electron transfer reaction of polysilanes. intramolecular trapping of a transient radical cation with a nucleophile. *Chemistry Letters*, 327–330.

109 Pan, J., Zhang, W., Zhang, J., and Lu, S. (2007) Photochemically catalyzed ring opening of oxirane carbonitriles and [3 + 2] cycloaddition with olefins: synthesis of polysubstituted tetrahydrofurans. *Tetrahedron Letters*, 48, 2781–2785.

110 Pandey, G., Lakshmaiah, G., and Kumaraswamy, G. (1992) A new and efficient strategy for non-stabilized azomethine ylide via photoinduced electron transfer (PET) initiated sequential double desilylation. *Journal of the Chemical Society, Chemical Communications*, 1313–1314.

9
Formation of Six-Membered (and Larger) Rings
Julia Pérez-Prieto and Miguel Angel Miranda

9.1
Introduction

The wide variety of reactions presented in this chapter illustrates the power of photochemistry for the construction of six-membered (and larger) rings, as well as polycyclic systems [1]. These structures are present in many products obtained from natural sources or developed for technological applications, which are often difficult to obtain by alternative thermal procedures. The potential of photoinduced electron transfer (PET), photocycloaddition, 6π-electrocyclization, electrocyclic ring opening and ring enlargement, or intramolecular hydrogen abstraction, among others, in organic synthetic photochemistry is demonstrated here by the provision of specific examples.

9.2
Photoelectron Transfer-Initiated Cyclizations

Electronically excited carbonyl chromophores in ketones, aldehydes, amides, imides, or electron-deficient aromatic compounds may act as electron acceptors (A) versus alkenes, amines, carboxylates, carboxamides, and thioethers (D, donors). In addition, PET processes can also occur from aromatic rings with electron-donating groups to chloroacetamides. These reactions can be versatile procedures for the synthesis of nitrogen-containing heterocyclic compounds with six-membered (or larger) rings [2].

9.2.1
Phthalimides as Electron Acceptors

N-Substituted phthalimides are easily synthesized via solvent-free condensation of phthalic anhydrides with amines or via coupling methods [3]. The phthalimide moiety has a reversible reduction potential in dimethylformamide (DMF) which is

Handbook of Synthetic Photochemistry. Edited by Angelo Albini and Maurizio Fagnoni
Copyright © 2010 WILEY-VCH Verlag GmbH & Co. KGaA, Weinheim
ISBN: 978-3-527-32391-3

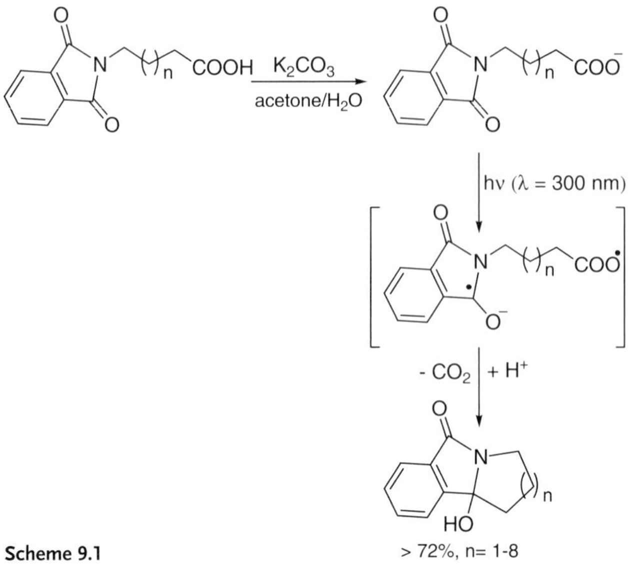

Scheme 9.1

> 72%, n= 1-8

close to −1.4 V (versus SCE) and typically absorbs in the 300 nm range, with molar absorption coefficients of approximately 10^3. Electronically excited phthalimides are strong electron acceptors that can oxidize substrates such as carboxylates, thioethers, and alkenes, in carboxyalkyl-, thioalkyl-, and alkenyl-N-substituted derivatives, respectively. In addition, electron-transfer processes can occur in phthalimides that contain N-linked α-trimethylsilylmethyl-terminated polyether, polysulfonamide, and polypeptide chains.

Photoinduced electron transfer from the carboxylate ion to the excited triplet phthalimide ($E_T = 293–300$ kJ mol^{-1}) appears to be followed by a rapid protonation of the radical anion and cyclization via a biradical recombination reaction (Scheme 9.1). Acetone (which also acts as photosensitizer) containing a small amount of water is the solvent of choice, whereas potassium carbonate is the ideal base to enhance cyclization versus simple decarboxylation and ring opening of the phthalimide [4].

Crown ethers and cyclic oligopeptides are also accessible by this method (Schemes 9.2 and 9.3). Thus, C-unprotected di- and tri-peptides activated by the

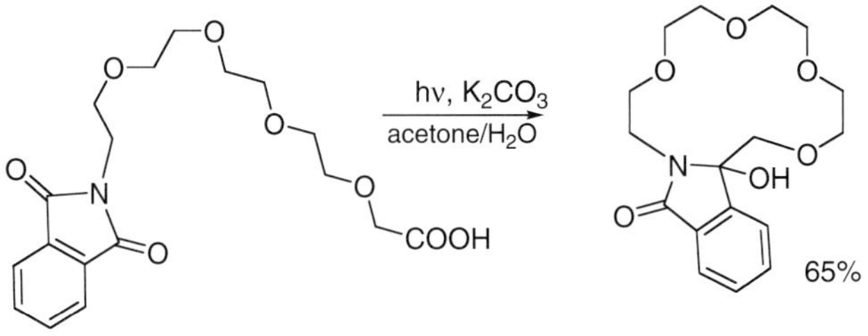

Scheme 9.2

65%

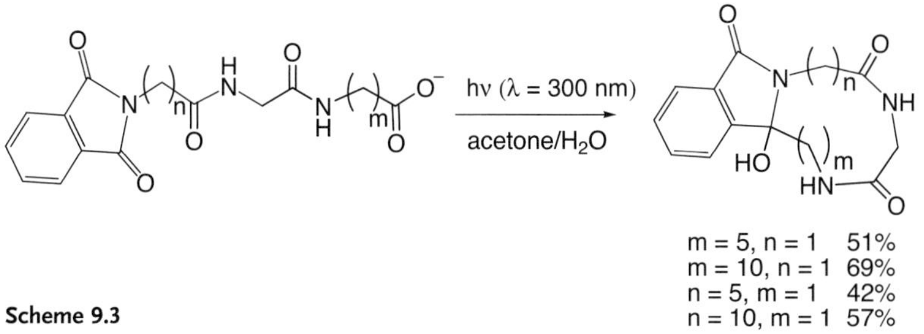

9.2 Photoelectron Transfer-Initiated Cyclizations

Scheme 9.3

m = 5, n = 1   51%
m = 10, n = 1   69%
n = 5, m = 1   42%
n = 10, m = 1   57%

N-substituted phthalimide functionality and containing a flexible N-terminal spacer or a flexible primary amino acid undergo efficient cyclization, as shown in Scheme 9.3 [5].

Interestingly, a high chiral memory has been observed for this photochemical reaction, which probably involves triplet biradicals. For example, triplet-photosensitized electron transfer in the substrate of Scheme 9.4 leads to decarboxylation, followed by cyclization to a pyrrolobenzodiazepine with 86% enantiomeric excess (ee) [6].

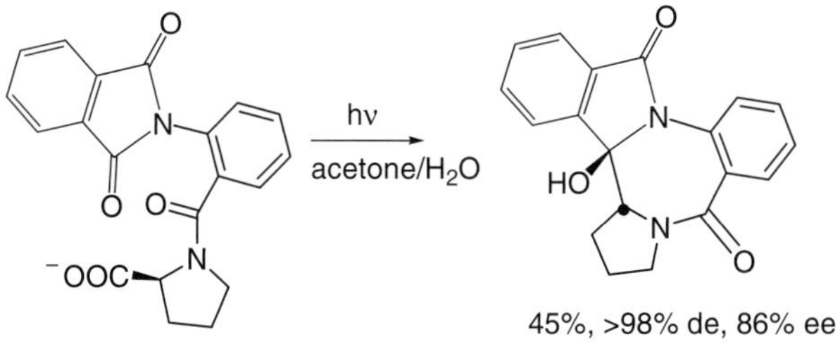

45%, >98% de, 86% ee

Scheme 9.4

Moreover, clean and high-yielding photodecarboxylation reactions have also been performed in water. In this context, irradiation of a suspension of a carboxyalkyl-N-substituted phthalimide (50 mmol) in water (2 l) produces a suspension of the cyclization product which, after partial elimination of the solvent, precipitates and is isolated by filtration in 82% yield (Scheme 9.5) [7].

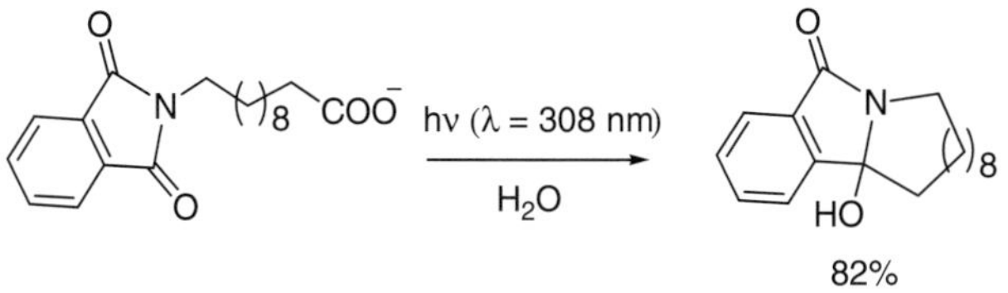

82%

Scheme 9.5

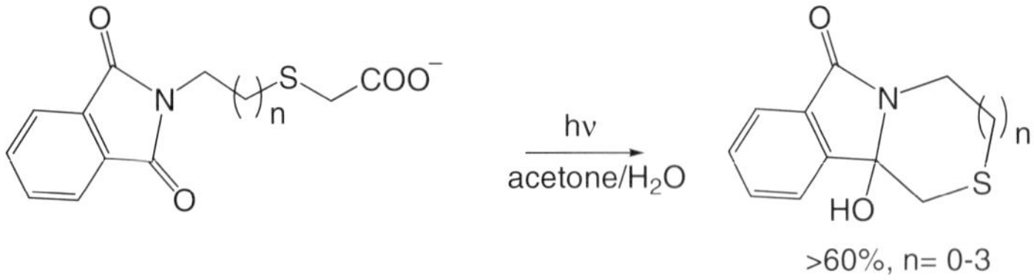

Scheme 9.6

Sulfur can also act as an electron donor towards the phthalimide group competing with the carboxylate ion. Therefore, in bifunctional compounds incorporating the two electron donors (carboxylate and thioether groups), the connecting linker and the solvent is crucial for the result of the reaction (Scheme 9.6) [8].

Under acetone-sensitized conditions, an intermolecular electron transfer from the sulfur to triplet acetone has been postulated in thioalkyl-N-substituted phthalimides. Proton transfer from the terminal α-position of the sulfur, followed by C—C bond formation, efficiently leads to macrocyclization with long and flexible chains as linkers (Scheme 9.7) [9].

Moreover, styrylalkyl-N-substituted phthalimides undergo medium to macro ring photocyclomerization via electron transfer between the arylalkenyl and the imide

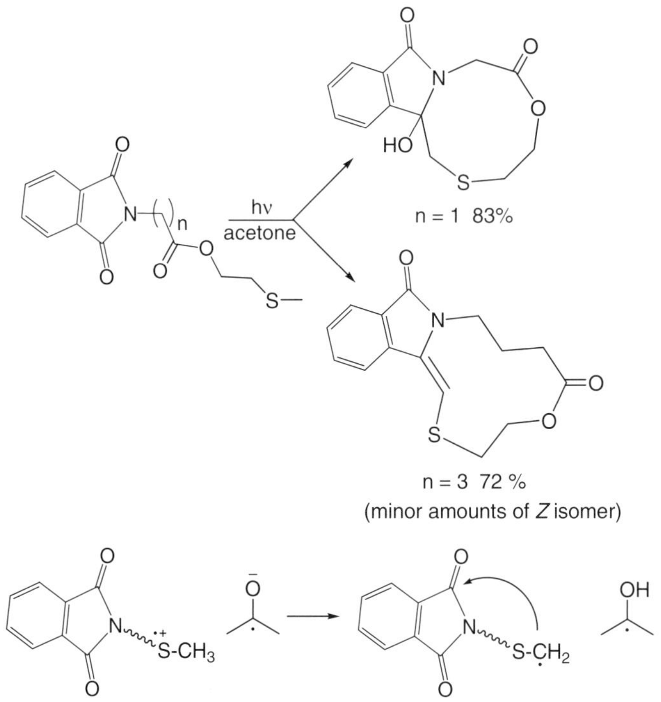

Scheme 9.7

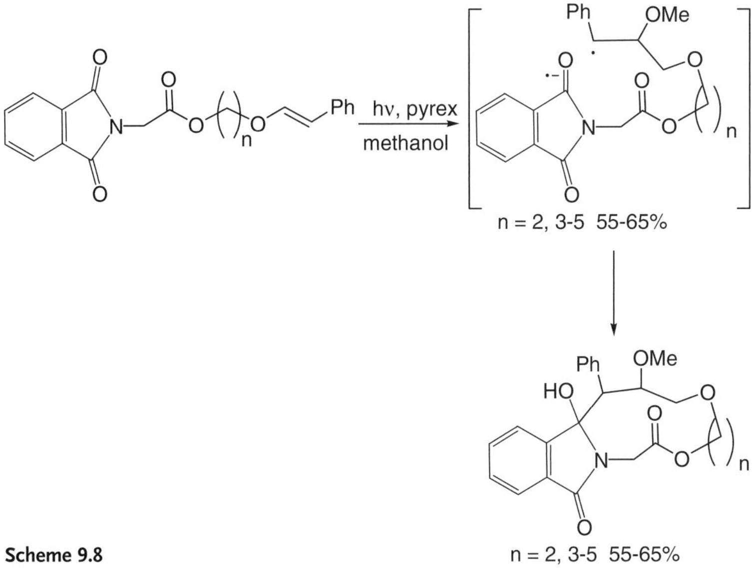

Scheme 9.8

n = 2, 3-5 55-65%

moieties separated by six to 13 bonds. The reaction occurs with solvent (methanol) incorporation, as shown in Scheme 9.8 [10].

N-Phthalimide-linked peptides that contain C-terminal α-amidotrimethylsilyl groups undergo macrocyclization reactions via sequential single-electron transfer processes [11, 12]. Thus, electron transfer from the neighboring amide to the excited phthalimide chromophore leads to an amide radical cation. Hole migration to the α-amidotrimethylsilane center is followed by desilylation to form an 1,ω-biradical intermediate, which finally cyclizes to the macrocyclic peptide (Scheme 9.9).

9.2.2
Aromatic Ketones as Electron Acceptors

Electron transfer from tertiary amines to excited aromatic ketones generates α-aminoalkyl radicals, which add efficiently to electron-deficient alkenes (Scheme 9.10) [13].

Tandem radical addition/cyclization reactions have been performed using unsaturated tertiary amines (Scheme 9.11) [14, 15]. Radical attack is highly stereoselective *anti* with respect to the 5-alkoxy substituent of 2-(5H)-furanones, which act as the electron-deficient alkenes. However, the configuration of the α position of the nitrogen cannot be controlled. Likewise, tandem addition cyclization reactions occur with aromatic tertiary amines (Scheme 9.12); in this case, acetone (mild oxidant) must be added to prevent the partial reduction of the unsaturated ketone [14].

Further information on the radical alkylation of electrophilic alkenes is available in Chapter 3.

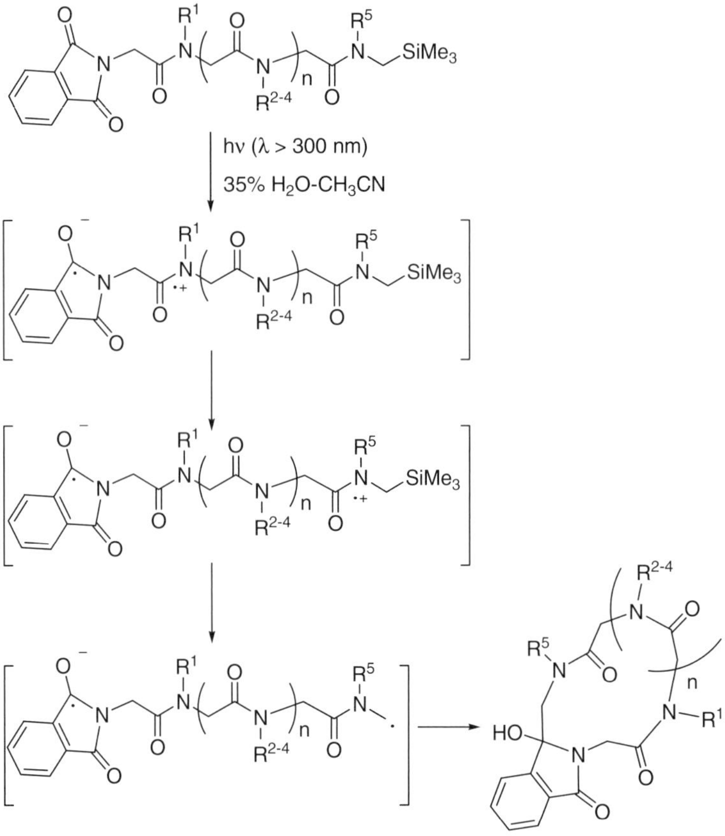

Scheme 9.9

R¹ = CH$_3$, PhCH$_2$
R² = CH$_3$, PhCH$_2$
R³ = R⁴ = R⁵ = PhCH$_2$

15-70%, n = 0-3

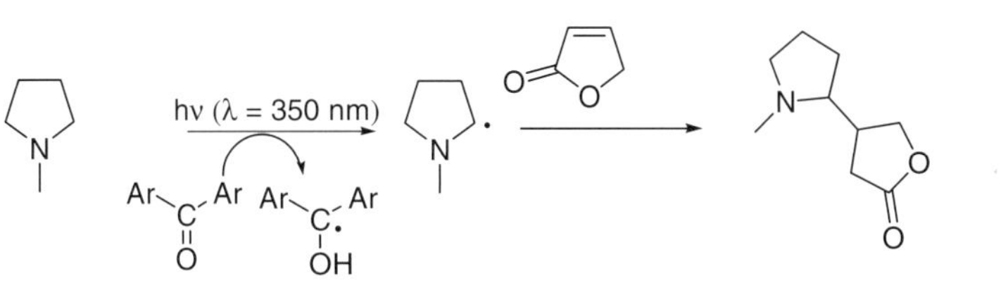

Scheme 9.10

9.2.3
Chloroacetamides as Electron Acceptors

Chloroacetamides can act as electron acceptors versus excited aromatic rings. The key step of this reaction is the formation of a radical (via intramolecular PET from the

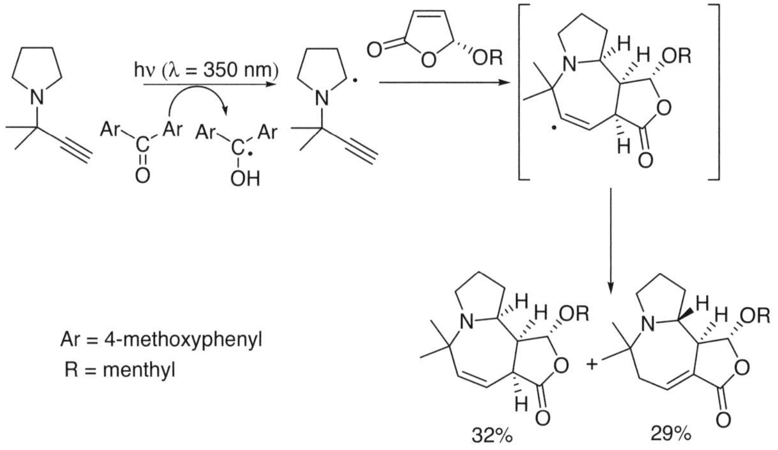

Scheme 9.11

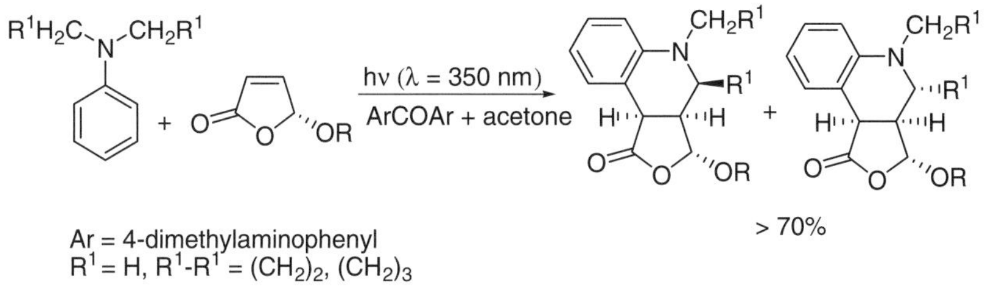

Ar = 4-methoxyphenyl
R = menthyl

Scheme 9.12

Ar = 4-dimethylaminophenyl
R¹ = H, R¹-R¹ = (CH$_2$)$_2$, (CH$_2$)$_3$

aromatic ring to the amide, followed by elimination of a chloride ion) which reacts with the aromatic ring and leads to the aromatic substitution product. The photocyclization of aromatic chloroacetamides has been applied to the synthesis of azepinoindole derivatives, such as those depicted in Scheme 9.13, which are not easily accessible by ground-state radical reactions or under Friedel–Craft catalytic conditions [16–19].

9.2.4
Electron-Deficient Aromatic Compounds as Electron Acceptors

Substrates containing an electron-rich double bond, such as enol ethers and enol acetates, are easily oxidized by means of PET to electron-deficient aromatic compounds, such as dicyanoanthracene (DCA) or dicyanonaphthalene (DCN), which act as photosensitizers. Cyclization reactions of the initially formed silyloxy radical cation in cyclic silyl enol ethers tethered to an olefinic or an electron-rich aromatic ring, can produce bicyclic and tricyclic ketones with definite stereochemistry (Scheme 9.14) [20, 21].

In addition, the photocyclization of haloarenes having a reactive group on an *ortho*-side chain can be used as a versatile strategy for the synthesis of cyclic

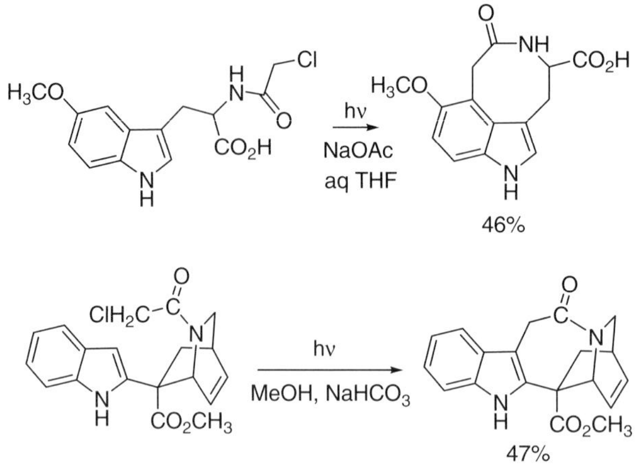

Scheme 9.13

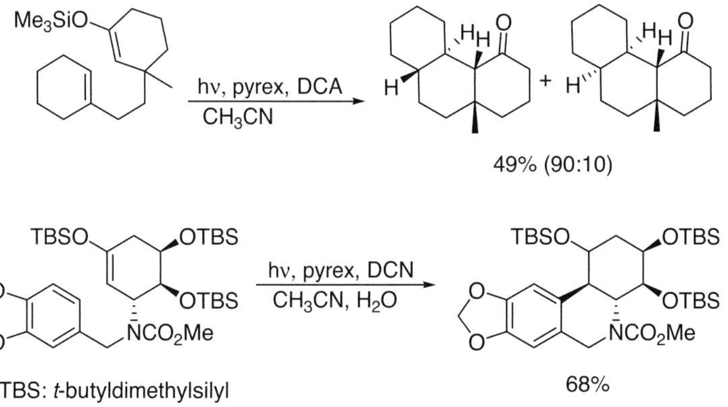

TBS: *t*-butyldimethylsilyl

Scheme 9.14

compounds [22]. PET from a negatively charged center to the haloarene may be followed by a loss of halide ion and subsequent coupling of the two radical termini. Thus, medium-sized (six- to ten-membered) rings can be obtained by the photocyclization of enolate anions (Scheme 9.15) [23].

An alternative mechanism can be $S_{RN}1$, in which an aryl radical arising from intermolecular electron transfer reacts with the anionic center to give a radical anion;

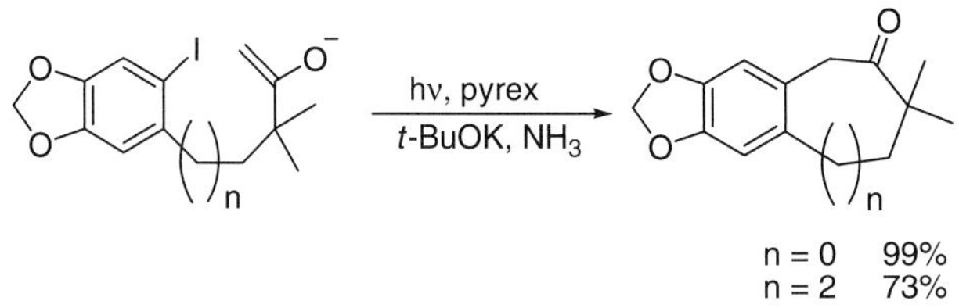

Scheme 9.15

this then transfers the electron to the starting aryl halide to propagate a chain reaction (for further details, see Chapter 10).

9.3
Photoinduced 6π-Electrocyclization

Electrocyclic reactions are a type of pericyclic rearrangement reaction where the net result is a conversion of one π-bond into a σ-bond (Scheme 9.16). These reactions can be either photochemically or thermally induced. Selected examples of the potential of photoinduced 6π-electrocyclization for organic synthesis are shown below.

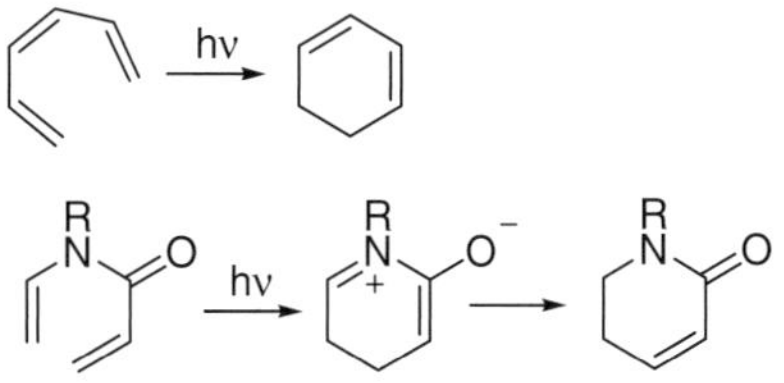

Scheme 9.16

9.3.1
Stilbene-Like Photocyclization

Cis-stilbenes, which are easily obtained from their trans-isomers by trans, cis-photoisomerization, can give rise to dihydrophenanthrene derivatives upon direct irradiation (Scheme 9.17). The latter compounds are easily converted into phenanthrenes under oxidizing conditions (oxygen, I_2, or π-electron acceptors such as tetracyanoethene) [24].

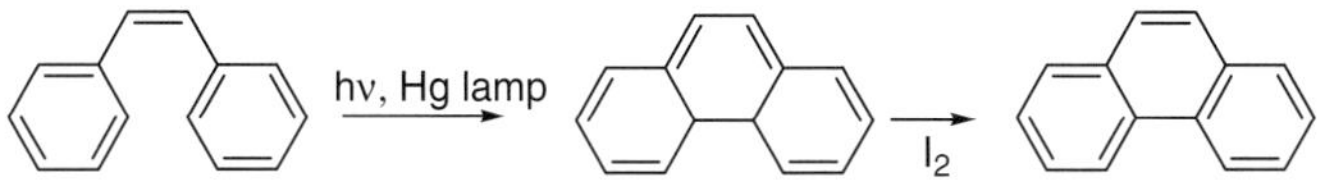

Scheme 9.17

The reaction is considered to arise from the stilbene singlet π,π* excited state and is useful for the synthesis of a great variety of polynuclear aromatic compounds, such as helicenes and large [n]phenacenes ($n = 3$ up to $n = 7$) [25]. The arene positions involved in photocyclization are predictable; the preferred route corresponds to the dihydro intermediate with the higher degree of aromaticity (Scheme 9.18).

This strategy has also been applied to the preparation of heteroaromatic analogues of phenanthrenes [26, 27], and crown-containing heteroaromatic cations, which are difficult to synthesize by other means. The reaction is accelerated in the presence of Ag^+ or Hg^{2+} cations, as well as iodine (Scheme 9.19) [28]. Although 4,5-disubstituted phenanthrenes are difficult to obtain using this methodology, due to steric hindrance associated with cyclization, this problem has been overcome by using rigid cis-stilbene compounds (Scheme 9.20) [29].

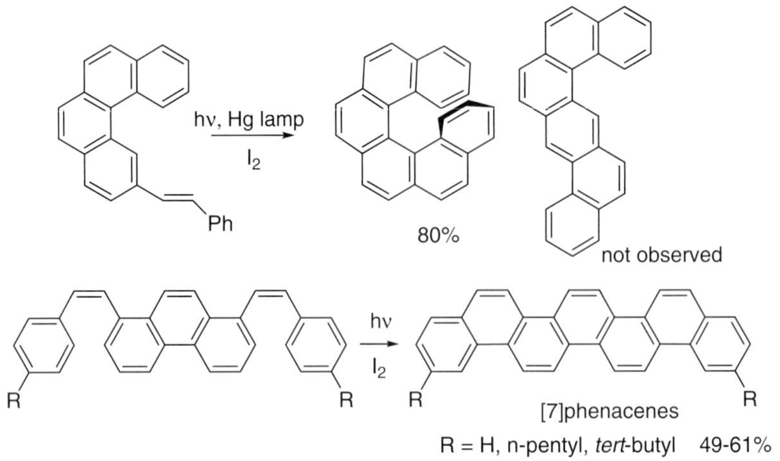

Scheme 9.18

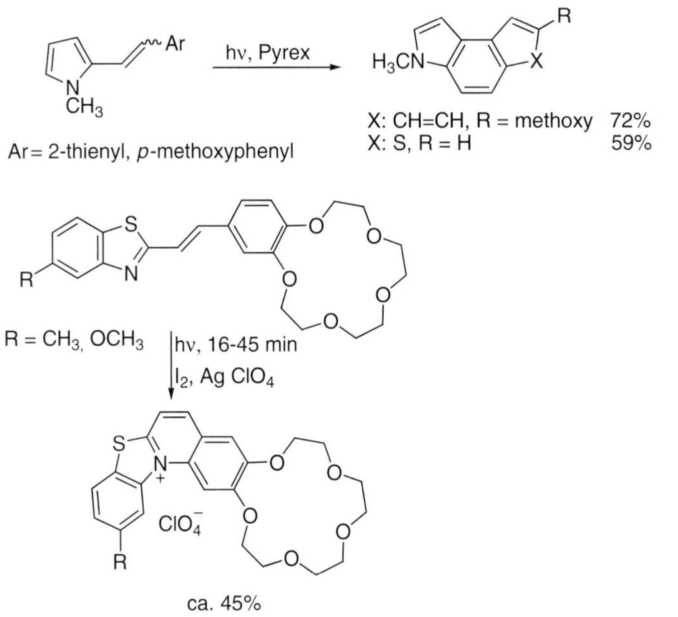

Scheme 9.19

The reaction is prevented in the case of low-lying $n\pi^*$ excited states, which are responsible for the lack of photocyclization of azobenzenes and benzylidene anilines. However, when the electronic configuration switches to a low-lying $\pi\pi^*$ excited state upon complexation with Lewis acids, photocyclization takes place under these conditions [30]. Similarly, the presence of perchloric acid allows that, upon irradiation, (*E*)-4-benzyloxy-1,1,3,4-tetraphenyl-2-azabuta-1,3-diene is transformed in an isoquinoline derivative (Scheme 9.21) [31, 32].

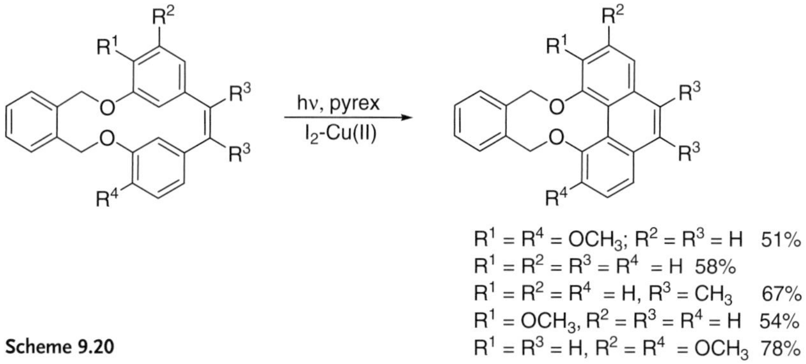
hν, pyrex
I_2-Cu(II)
R1 = R4 = OCH3; R2 = R3 = H 51%
R1 = R2 = R3 = R4 = H 58%
R1 = R2 = R4 = H, R3 = CH3 67%
R1 = OCH3, R2 = R3 = R4 = H 54%
R1 = R3 = H, R2 = R4 = OCH3 78%

Scheme 9.20

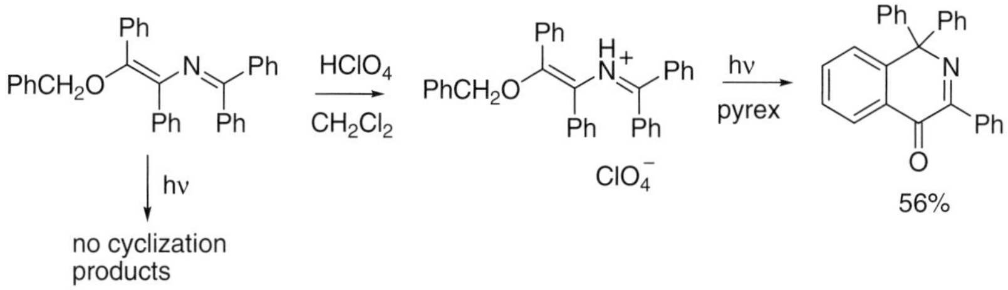
HClO4
CH2Cl2
hν
pyrex
hν
no cyclization
products
56%

Scheme 9.21

In addition, oxidative photocyclization has been applied for the formation of tetra-, penta-, and hexa-cyclic azonia-aromatic compounds via photoinduced quaternization (Scheme 9.22). In order to avoid photodecomposition of the photoproduct, the substrates are irradiated selectively by using Pyrex-filtered light through an aqueous nickel sulfate solution filter [33].

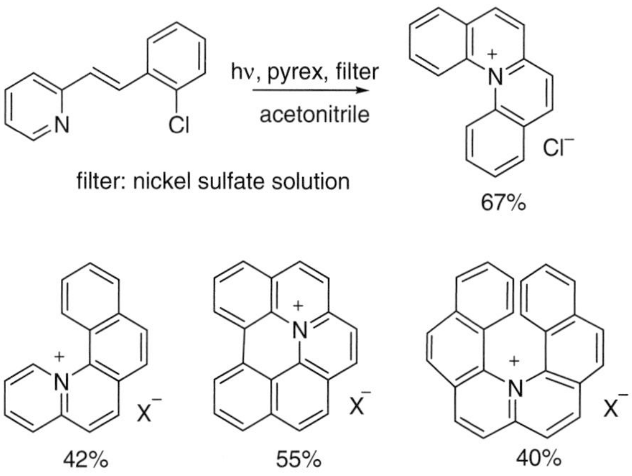
hν, pyrex, filter
acetonitrile
filter: nickel sulfate solution
67%
42%
55%
40%

Scheme 9.22

The photochemical pericyclic reaction leading from 1,3,5-hexatrienes to 1,3-cyclohexadienes has been exploited for the design of photochromic compounds. Suitable diarylethenes are thermally stable photochromic compounds, which have been developed for application as versatile photoresponsive systems [34, 35]. In general, the irradiation of hexatrienes with UV light gives colored cyclohexadienes, whereas irradiation of the latter with visible light leads back to the original hexatriene derivatives (Scheme 9.23). Fused diarylethenes in solution can display different colors by irradiation with light of different wavelengths, and for different periods of time. Thermal irreversibility and fatigue resistance are some of the advantages over other photochromic compounds.

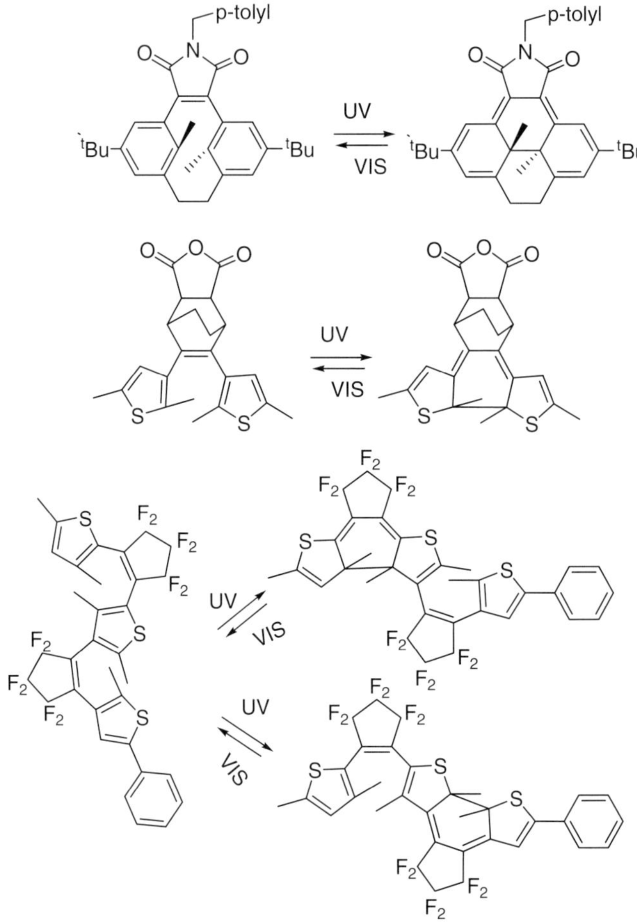

Scheme 9.23

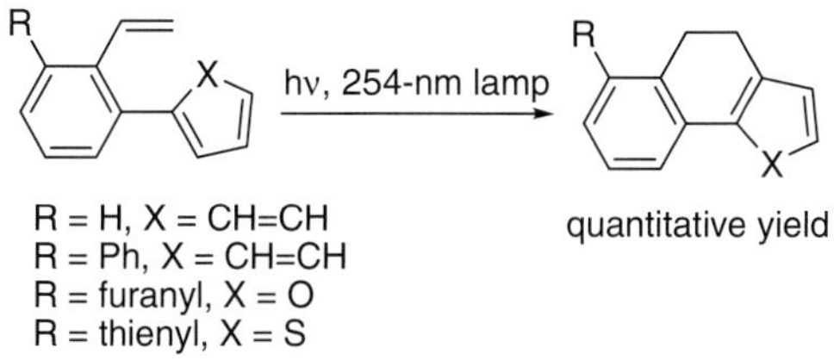
R
X
hν, 254-nm lamp
R
X
R = H, X = CH=CH
R = Ph, X = CH=CH
R = furanyl, X = O
R = thienyl, X = S
quantitative yield

Scheme 9.24

9.3.2
Vinyl-Biphenyls Photocyclization

One process which is related to the photocyclization of stilbene-like substrates is the conversion of 2-vinyl biphenyls to 8,9a-dihydrophenanthrenes, which undergo a facile thermal isomerization to 9,10-dihydrophenanthrenes (Scheme 9.24) [36]. For the synthesis of 5,6-dihydro-1,10-phenanthroline and 1,10-phenanthroline derivatives, metal-chelation of the nitrogen atoms has shown to be productive (Scheme 9.25) [37].

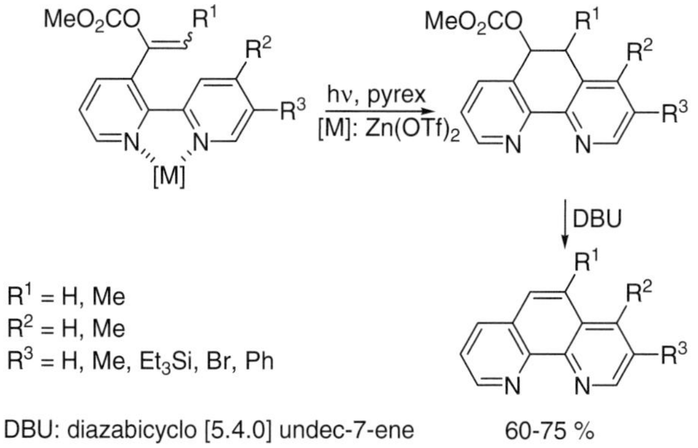
MeO2CO
R1
R2
R3
N
N
[M]
hν, pyrex
[M]: Zn(OTf)2
DBU
R1 = H, Me
R2 = H, Me
R3 = H, Me, Et3Si, Br, Ph
DBU: diazabicyclo [5.4.0] undec-7-ene
60-75 %

Scheme 9.25

9.3.3
Anilides and Enamides Photocyclization

The photochemical electrocyclic reaction of acrylamides represents a versatile strategy for alkaloid synthesis. Thus, (*S*)-pipecoline has been synthesized using the photochemical cyclization of enantiomerically pure acrylamide derivatives in the presence of $NaBH_4$, which causes reduction of the imonium intermediate. The lactam may then easily be transformed into the desired heterocyclic compound (Scheme 9.26) [38].

N-Benzoylenamines also undergo photocyclization; after the photochemical cyclization, the aromaticity is re-established via a thermal 1,5-H shift (Scheme 9.26).

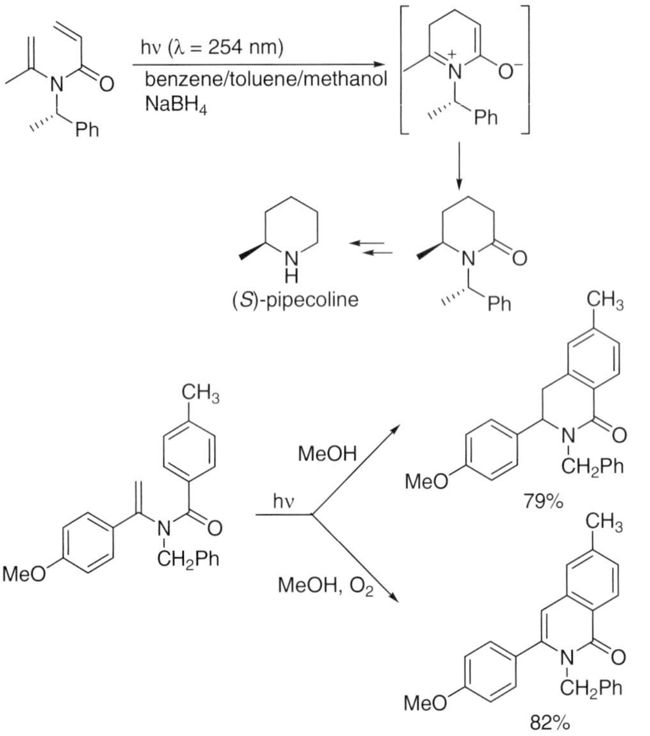

Scheme 9.26

When the reaction is performed in the presence of oxygen or iodine, the double bond of the enamine moiety is reformed in the final product [39].

Acrylanilides also undergo this photoinduced process; the reaction can occur either in the solid state or in aqueous suspension. In this way, optically active photoproducts can be obtained by irradiating prochiral substrates in the presence of optically active partners (1:1 complexes) (see Scheme 9.27). The substrate is highly oriented and immobilized in the chiral environment. Although photocyclization in the solid state takes a long time (close to 150 h), the photoreaction of a powdered complex in a water suspension containing sodium alkylsulfate as a surfactant proceeds more efficiently (less than 50 h) [40].

9.4
Photocycloaddition Reactions

Photocycloaddition represents an important strategy for the synthesis of three- to eight-membered rings, offering tremendous scope for stereocontrol [41]. Consequently, this type of reaction can enable the preparation of cyclic systems that are difficult to access by alternative thermal methods. Examples of such efficient

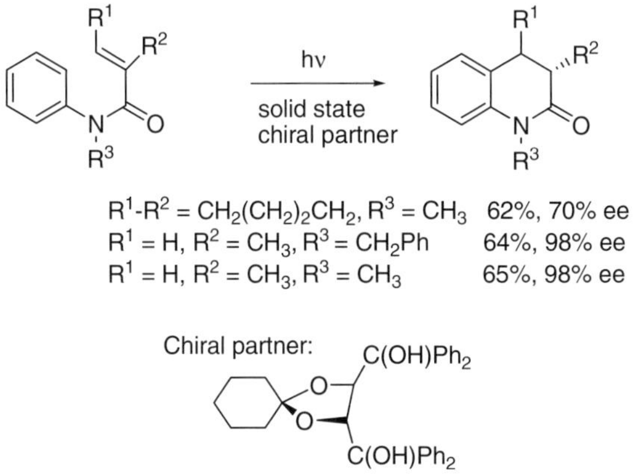
hν
solid state
chiral partner
R1-R2 = CH2(CH2)2CH2, R3 = CH3 62%, 70% ee
R1 = H, R2 = CH3, R3 = CH2Ph 64%, 98% ee
R1 = H, R2 = CH3, R3 = CH3 65%, 98% ee
Chiral partner:
C(OH)Ph2
C(OH)Ph2

Scheme 9.27

application to the formation of six-membered or larger rings are presented here; these are related to [4π + 2π], [5π + 2π], and [4π + 4π] processes.

9.4.1 Photochemical Diels–Alder Reaction

The Diels–Alder reaction is a well-established synthetic method that allows the creation of two new carbon–carbon bonds and leads to the formation of six-membered rings. Eventually, the photochemical reaction can advantageously compete with the thermal process. For instance, anthracene undergoes thermal and photochemical Diels–Alder reactions with alkenes, but the photoinduced addition of maleic anhydride to the homochiral anthracene, as depicted in Scheme 9.28, is faster than the thermal reaction and occurs with excellent diastereoselectivity (only one diastereoisomer) [42].

The involvement of styryl as a diene partner has also been demonstrated in the photosensitized electron transfer [4 + 2] intramolecular cycloaddition of 1,1,8,8-tetraphenyl-1-7-octadiene (Scheme 9.29) [43]. This reaction is promoted by 1,4-dicyanobencene (DCB), and a single electron-transfer mechanism has been suggested.

In addition, the thermal [4 + 2] reaction between dienes and alkenes is generally inefficient when both partners are electron-rich compounds [44]. In such cases, the

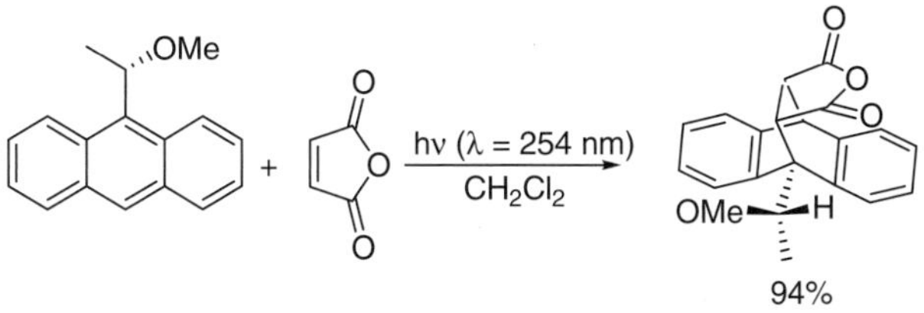
OMe
hν (λ = 254 nm)
CH2Cl2
OMe
H
94%

Scheme 9.28

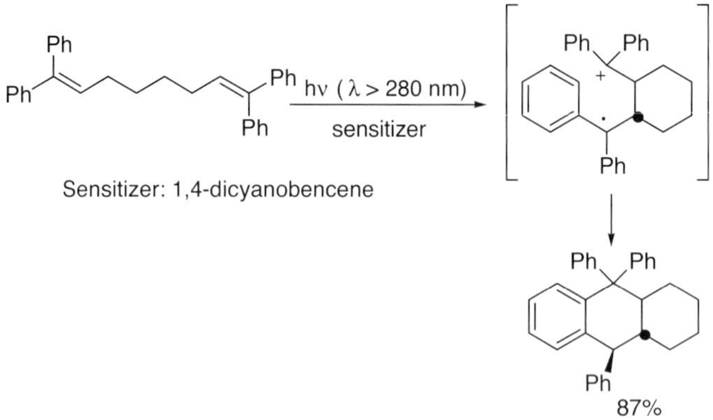

Scheme 9.29

photochemical Diels–Alder reaction is a powerful tool as it uses electron-deficient arenes and triarylpyrylium (TPP) salts as photosensitizers (Scheme 9.30) [45–49]. Recently, it has been reported that ππ* triplet ketones, in particular benzoylthiophenes, can also photosensitize the cycloaddition between indoles and cyclohexadienes [50, 51].

Examples of the versatility of photochemical hetero Diels–Alder photocycloadditions have also been reported. Thus, the 2,4,6-triphenylpyrylium terafluoroborate (TPT) photocatalyzed reaction of arylimines with N-vinyl-2-pyrrolidinone and N-vinylcarbazole represents a convenient approach for the preparation of tetrahydroquinoline derivatives under mild conditions (Scheme 9.31) [52].

Likewise, isothiochroman derivatives can be prepared via a (thia)pyrilium salts-promoted photocycloaddition between thiobenzophenone and arylalkenes (Scheme 9.32) [53].

9.4.2
Photoenolization/Diels–Alder Reaction

The photochemical generation of hydroxy-o-quinodimethanes from o-alkylbenzaldehydes may be followed by a thermal Diels–Alder reaction with alkenes (Scheme 9.33) [54]. This reaction is effective in both intermolecular and intramolecular modes, and has been applied to the synthesis of various natural products [55].

9.4.3
[4 + 4]-Photocycloaddition

The photodimerization of 2-pyridones is an efficient, regiospecific, and stereoselective [4 + 4]-cycloaddition [56] that converts two achiral aromatics into a highly functionalized tricyclic cyclooctadiene with four stereogenic centers (Scheme 9.34). For tethered pyridones, the *trans* isomer is usually the major product when one or both pyridine nitrogens are methylated. By contrast, in the unsubstituted systems,

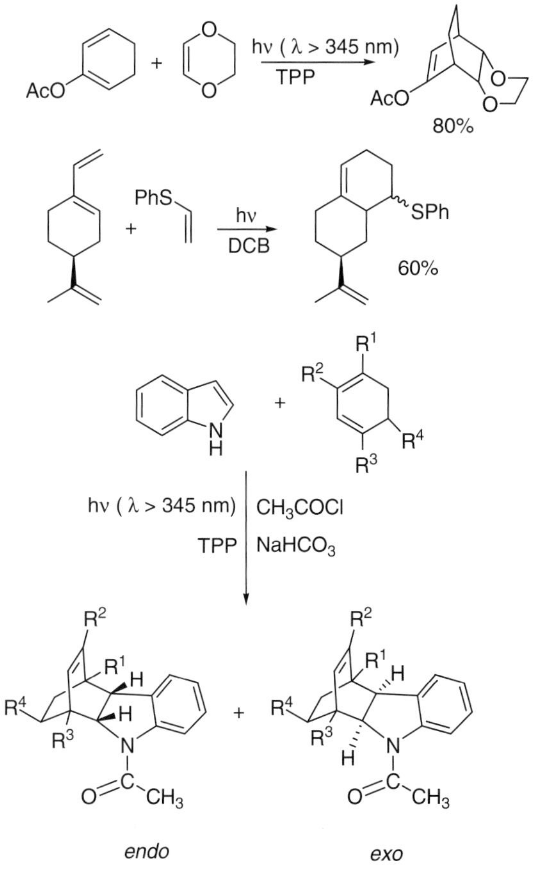

Scheme 9.30

	endo:exo	
$R^1 = R^2 = R^3 = R^4 = H$	70%	3.3:1.0
$R^1 = CH_3, R^2 = H, R^3 = {}^iPr, R^4 = H$	43%	1.7:1.0
$R^1 = H, R^2 = CH_3, R^3 = H, R^4 = {}^iPr$	57%	2.4:1.0
$R^1 = H, R^2 = OAc, R^3 = R^4 = H$	59%	3.0:1.0

solvent-dependent stereoselectivity is found, which provides a level of stereochemical control useful for synthetic applications [57].

Selective cycloadditions of 2-pyridone mixtures, as well as the analogous cross-cycloadditions with furan, naphthalene, and 1,3-dienes, have been intensively investigated.

Likewise, the irradiation of pyran-2-ones bearing pendant furans in aqueous methanol, followed by heating, is a convenient method for the one-pot synthesis of fused bicyclic products containing a cyclooctatriene ring (Scheme 9.35) [58]. This is a tandem photocycloaddition/decarboxylation sequence, the generality of which has been demonstrated in the preparation of several cyclooctanoid natural products.

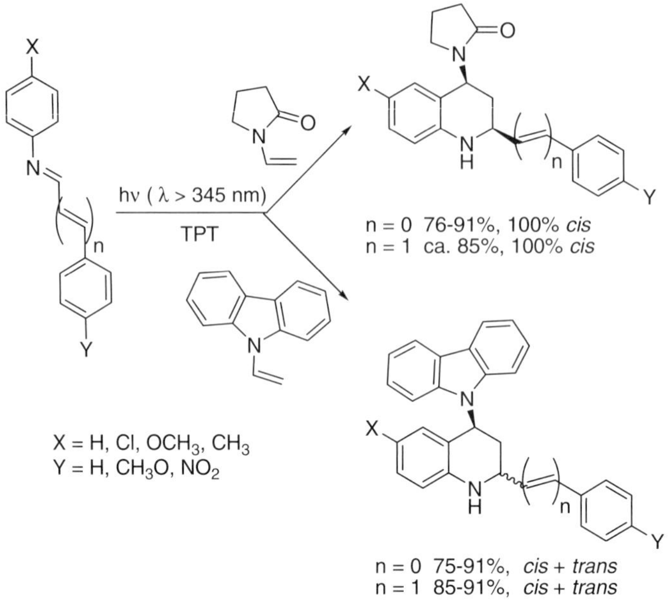

304 | *9 Formation of Six-Membered (and Larger) Rings*

X = H, Cl, OCH$_3$, CH$_3$
Y = H, CH$_3$O, NO$_2$

n = 0 76–91%, 100% *cis*
n = 1 ca. 85%, 100% *cis*

n = 0 75–91%, *cis* + *trans*
n = 1 85–91%, *cis* + *trans*

Scheme 9.31

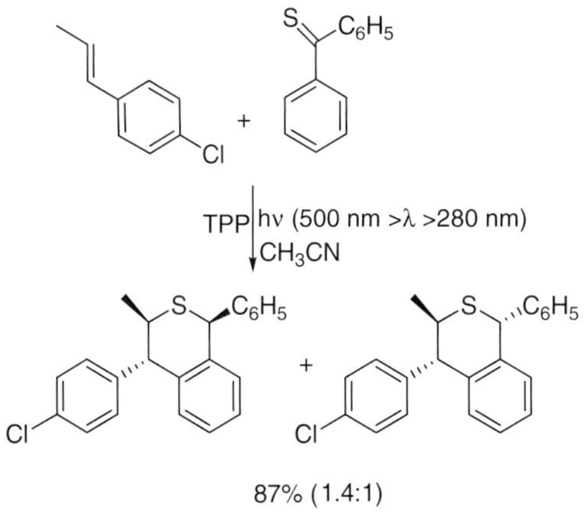

87% (1.4:1)

Scheme 9.32

9.4.4
Transition Metal Template-Controlled Reactions

The photochemical reactivity of metal complexes can also be exploited to efficiently synthesize six-membered or larger rings. Cycloaddition of the type [6π + 2π], [6π + 4π], and [4π + 4π] typically occurs with low yield. However, a procedure has

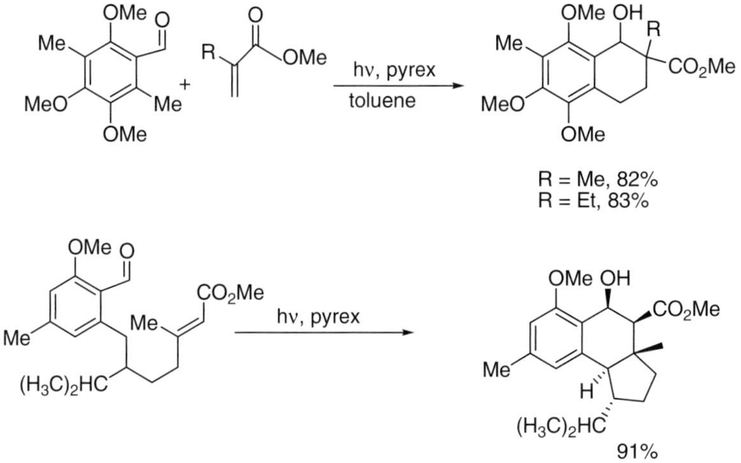

Scheme 9.33

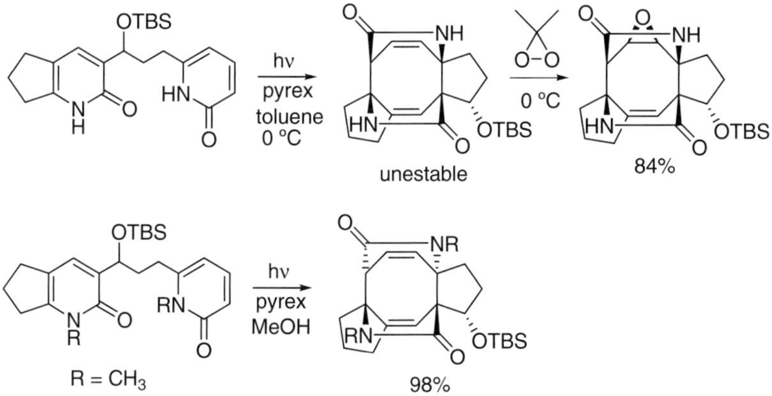

Scheme 9.34

been developed for synthesizing medium-size carbocycles, using an appropriate metal complexed with one of the reaction partners. Thus, cyclic triene chromium complexes can efficiently react with dienes, leading to the [6 + 4] cycloadduct (Scheme 9.36) [59, 60].

This strategy has also been extended to [6 + 2]-cycloaddition [61]. Thus, substituted cyclooctatetraenes can be prepared in two steps employing a Cr-(0)-promoted [6π + 2π] thiepin dioxide-alkyne cycloaddition, followed by photoactivated sulfur dioxide extrusion (Scheme 9.37) [62].

The use of metals as templates in the higher-order cycloadditions is similar to the [2 + 2]-cycloadditions of alkenes using Cu(I) catalysts.

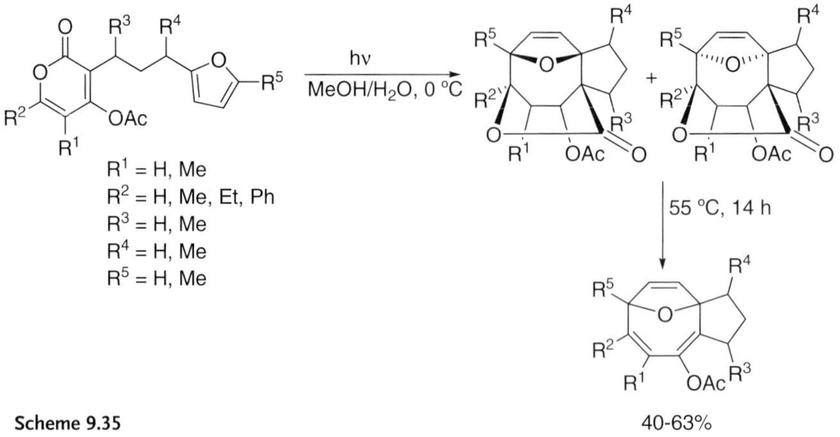

9 Formation of Six-Membered (and Larger) Rings

Scheme 9.35

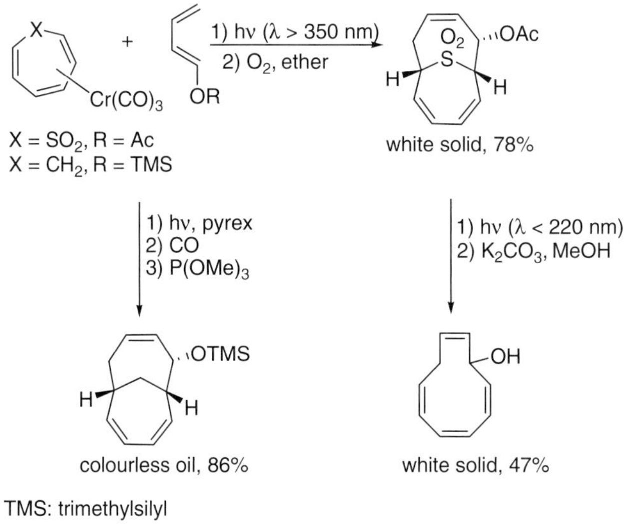

R¹ = H, Me
R² = H, Me, Et, Ph
R³ = H, Me
R⁴ = H, Me
R⁵ = H, Me

40–63%

Scheme 9.36

X = SO₂, R = Ac
X = CH₂, R = TMS

white solid, 78%

colourless oil, 86%

white solid, 47%

TMS: trimethylsilyl

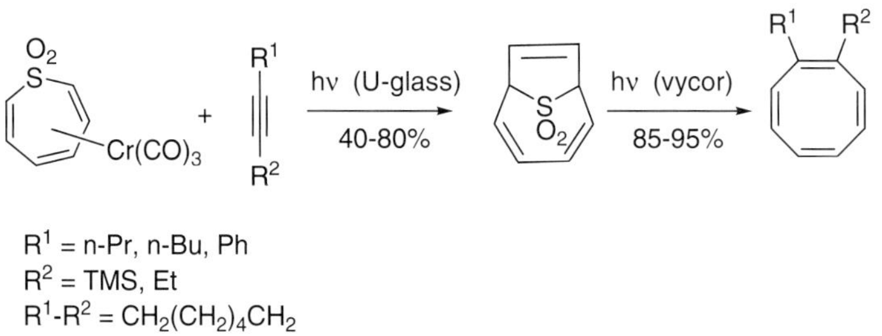

Scheme 9.37

R¹ = n-Pr, n-Bu, Ph
R² = TMS, Et
R¹-R² = CH₂(CH₂)₄CH₂

40–80%   85–95%

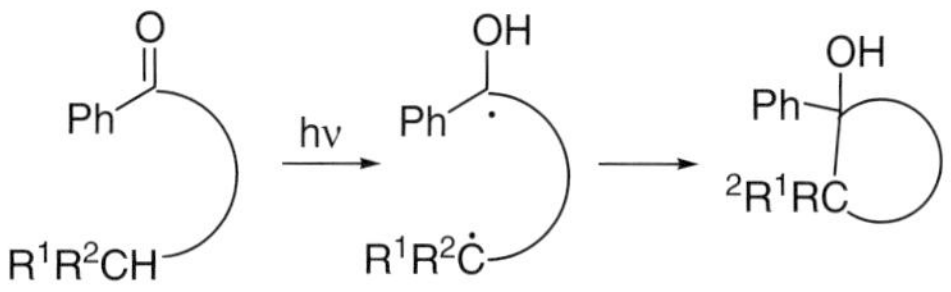

Scheme 9.38

9.5
Remote Intramolecular Hydrogen Abstraction

The Norrish–Yang reaction [63] is a widely used photochemical reaction, which consists of an intramolecular hydrogen abstraction by a photoexcited ketone, followed by carbon–carbon bond formation in the (1,n)-biradical intermediates (Scheme 9.38).

This process is an important tool in synthetic organic photochemistry for the preparation of small rings. The difficulties in the preparation of medium-size and larger cyclic compounds, caused by entropic factors, have been circumvented by the presence of a rigid molecular scaffold (cyclopropanes, aromatic rings, cyclohexenes and cis-alkenes), which prevents hydrogen abstraction from positions close to the carbonyl group (Scheme 9.39) [64–66].

This strategy allows the preparation of macrocyclic carbinols, lactones, and lactams. For example, highly regioselective and stereoselective macrocyclization has been found for ketoprofen–tetrahydrofuran (THF) conjugates, due to exclusive H-abstraction from the less-substituted carbon and a preferred *cisoid* ring junction over the *transoid* junction (Scheme 9.40) [67].

The irradiation of cholesterol-ketoprofen dyads also leads to stereoselective cyclization, giving rise to macrocyclic lactones with moderate yield (Scheme 9.41) [68].

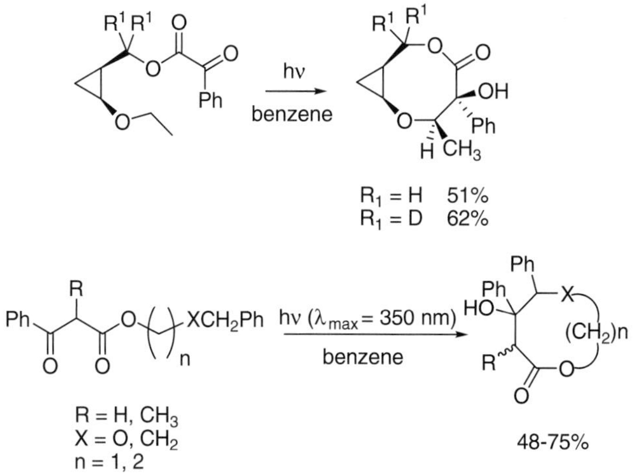

Scheme 9.39

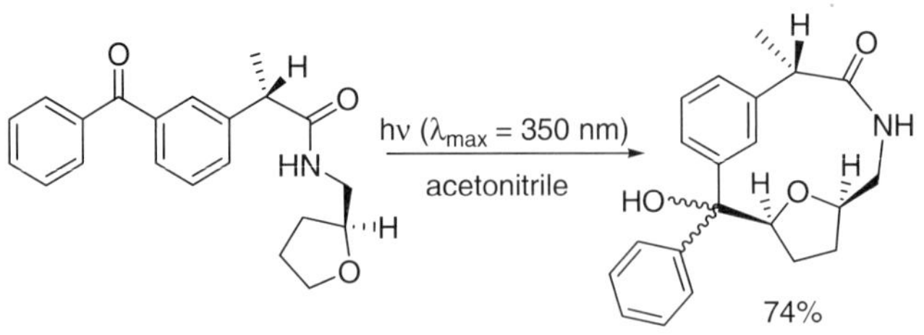

hν (λmax = 350 nm)
acetonitrile
74%

Scheme 9.40

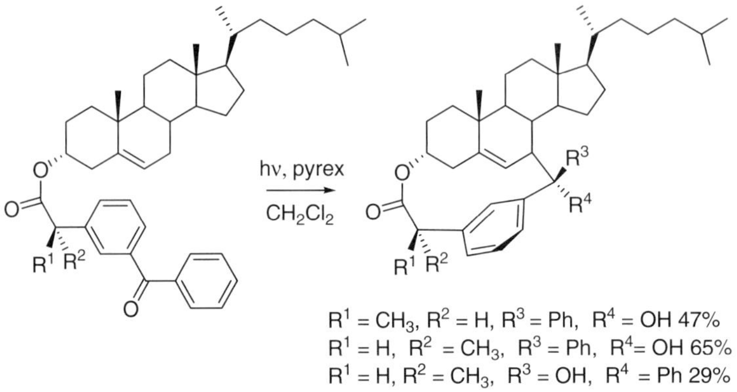

hν, pyrex
CH2Cl2
R1 = CH3, R2 = H, R3 = Ph, R4 = OH 47%
R1 = H, R2 = CH3, R3 = Ph, R4= OH 65%
R1 = H, R2 = CH3, R3 = OH, R4 = Ph 29%

Scheme 9.41

One remarkable process is the photochemical synthesis of 3,4-dihydro-2*H*-1,3-oxazin-4-ones from α-sulfonyloxy-β-keto amides (obtained by coupling of β-keto-carboxylic acids with amines, followed by treatment with an iodanyl mesylate). This allows the regioselective oxidation of less-activated C−H bonds and a C−O bond formation which is unusual for a Norrish–Yang reaction [69]. The formation of a 1,6-O−C biradical has been postulated as an intermediate (Scheme 9.42).

9.6 Ring Contraction and Ring Enlargement

Ring contraction via the extrusion of small molecules (SO_2, CO_2, CO) can represent a relatively efficient strategy for the preparation of medium to large ring systems [70, 71]. One essential requisite when applying this reaction is the stability of the biradical intermediates. It has been reported that 4,4′-(1,*n*-alkanediyl)bisbenzyl biradicals, generated from photocleavage of [3,*n*]paracyclophan-2-ones followed by decarbonylation, produce paracyclophanes as the only reaction. The lifetime of these species increases with their decreasing chain length; indeed, it has been shown that when $n = 2$ the biradical neither thermally nor photochemically decomposes to *p*-xylylene (Scheme 9.43) [72, 73].

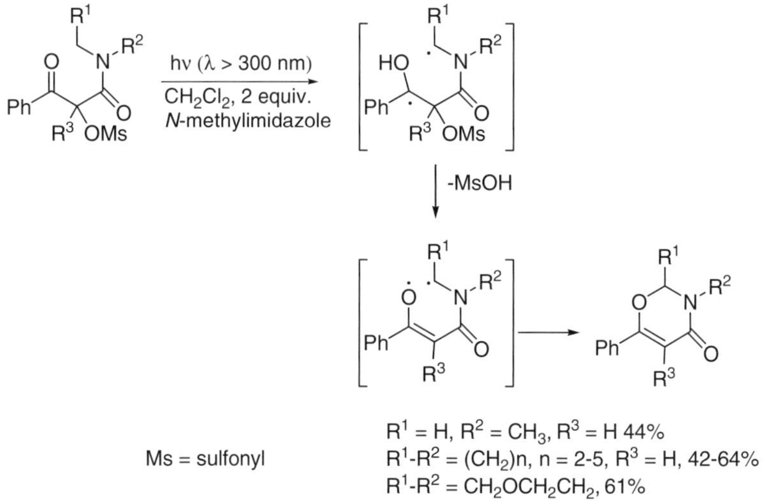

9.6 Ring Contraction and Ring Enlargement

Scheme 9.42

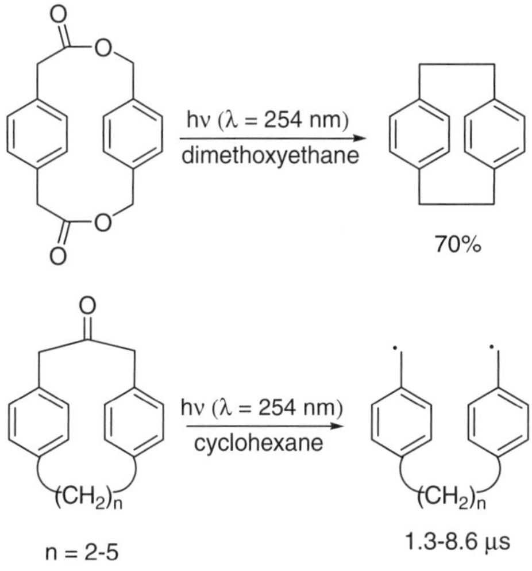

Ms = sulfonyl

R^1 = H, R^2 = CH_3, R^3 = H 44%
R^1-R^2 = $(CH_2)_n$, n = 2-5, R^3 = H, 42-64%
R^1-R^2 = $CH_2OCH_2CH_2$, 61%

Scheme 9.43

n = 2-5

1.3–8.6 μs

Conversely, the photoinduced ring-opening of cyclobutenes allows the construction of medium-sized hetero- or carbocycles, starting from bicyclic frameworks. Alkenes can undergo thermal (metal-assisted) or photochemical [2 + 2]-cycloaddition to a variety of alkynes; the subsequent two-carbon ring-enlargement can serve as a versatile method for the preparation of hydroazulenes and dioxacyclooctadiene derivatives (Scheme 9.44) [74, 75].

On the other hand, N-alk-4-enyl-substituted maleimides, which can be readily synthesized using the maleimide-alkenol Mitsunobu coupling procedure, give rise to

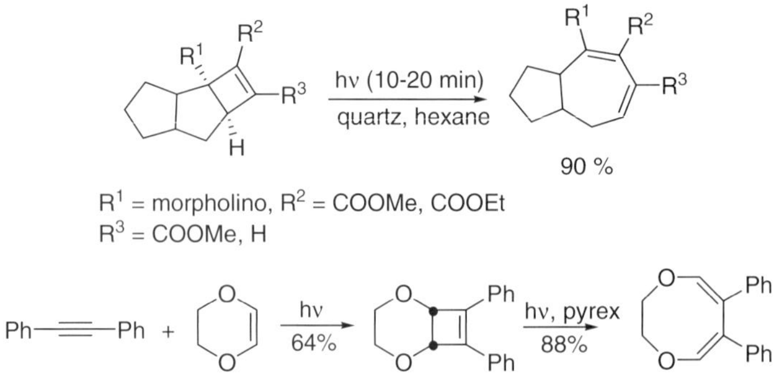
R1
R2
R3
H
hν (10-20 min)
quartz, hexane
90 %
R1 = morpholino, R2 = COOMe, COOEt
R3 = COOMe, H
Ph
Ph +
hν
64%
hν, pyrex
88%
O

Scheme 9.44

formal [5 + 2]-cycloaddition products upon irradiation (Scheme 9.45) [76]. This reaction proceeds with good yields, yielding tri- and tetracyclic products with a high degree of diastereoselectivity when using cycloalkenyl side chains. Some of the synthesized compounds constitute the core skeleton of complex alkaloids [77].

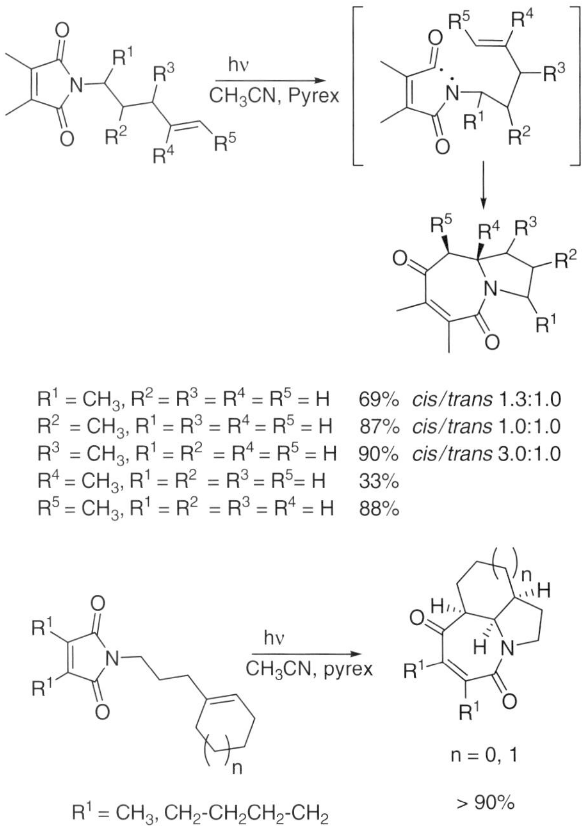
hν
CH3CN, Pyrex
R1
R2
R3
R4
R5
N
O
R1 = CH3, R2 = R3 = R4 = R5 = H 69% cis/trans 1.3:1.0
R2 = CH3, R1 = R3 = R4 = R5 = H 87% cis/trans 1.0:1.0
R3 = CH3, R1 = R2 = R4 = R5 = H 90% cis/trans 3.0:1.0
R4 = CH3, R1 = R2 = R3 = R5= H 33%
R5 = CH3, R1 = R2 = R3 = R4 = H 88%
hν
CH3CN, pyrex
n
H
n = 0, 1
> 90%
R1 = CH3, CH2-CH2CH2-CH2

Scheme 9.45

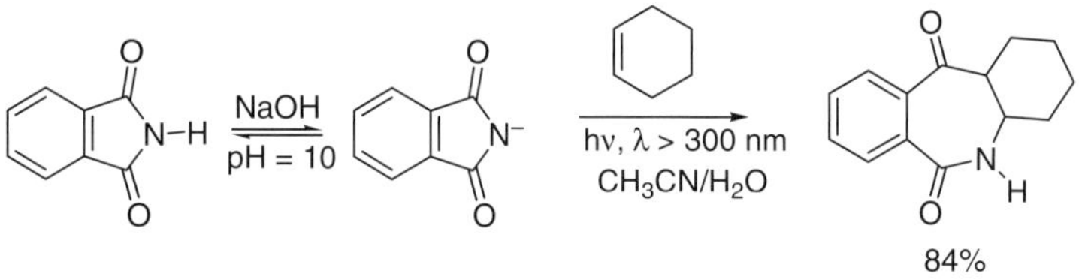

Scheme 9.46

Interestingly, compact flow reactors have been applied to prepare hundred of grams of bicyclic azepines using this photochemical strategy [78].

In addition, the phthalimide anion undergoes photoaddition with a wide variety of alkenes, providing a preparative method for the synthesis of [2]benzazepine-1,5-diones via a [2 + 2] cycloaddition followed by a thermal ring expansion (Scheme 9.46) [79].

9.7
Other Reactions

9.7.1
Intramolecular [2 + 2]-Cycloadditions

The [2 + 2]-photocycloaddition of nonconjugated alkenes was described in Chapter 5. This strategy can be used for synthesizing macrocyclic rings by using a long linker between the two alkene moieties (Scheme 9.47). Thus, bicyclic cyclobutanes can be obtained starting from enol ethers and cinnamates by means of electron-transfer sensitizers, such as DCN or DCA [80], and triplet photosensitizers such as benzophenone (BP) [81].

9.7.2
Photocyclization of Cinnamylanilides

The photolysis of cinnamylanilides produced six-membered rings with a high yield (Scheme 9.48). It has been suggested that this reaction occurs via intramolecular charge-transfer exciplexes [82].

9.7.3
Photocycloaddition of Aromatic Compounds

The most characteristic photochemical reaction of aromatic compounds is their cycloaddition with alkenes. The intramolecular reaction is suitable for the synthesis of complex structures, such as those depicted in Scheme 9.49, where [3 + 2]-photocycloaddition leads to structures which resemble natural products (aphidicoline and stemoclinone). An interaction of the arene singlet excited state with the alkene ground state gives rise to the meta adduct [83, 84].

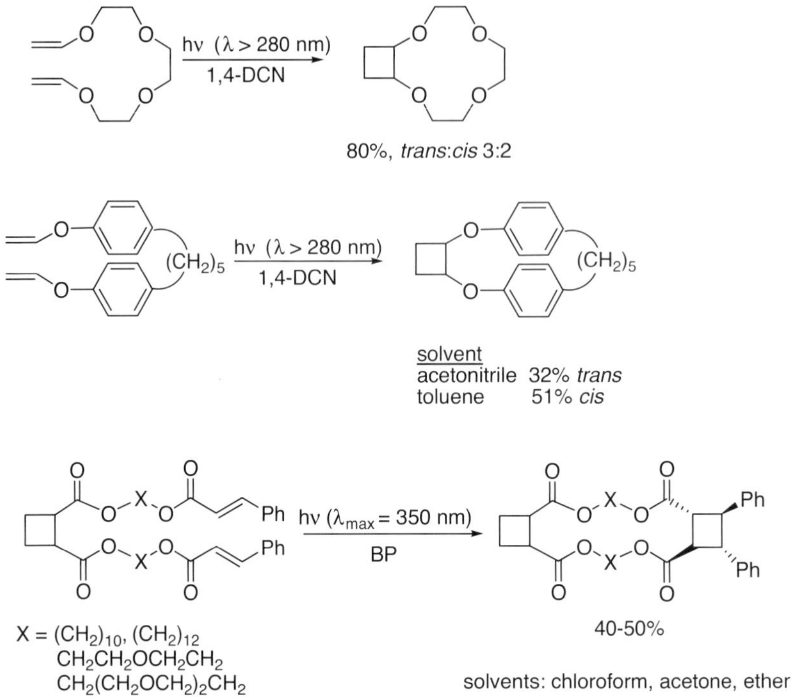

Scheme 9.47

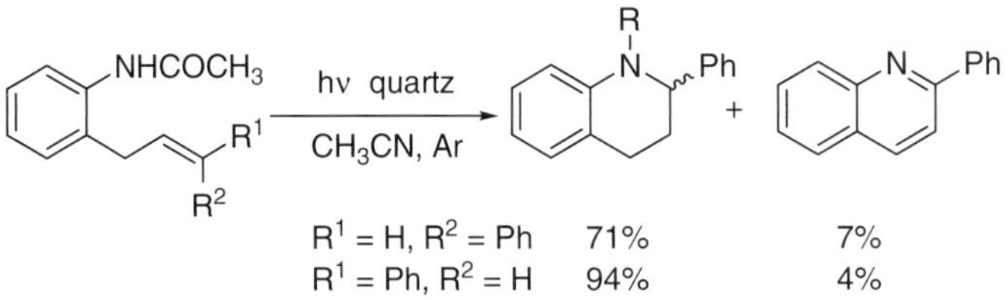

Scheme 9.48

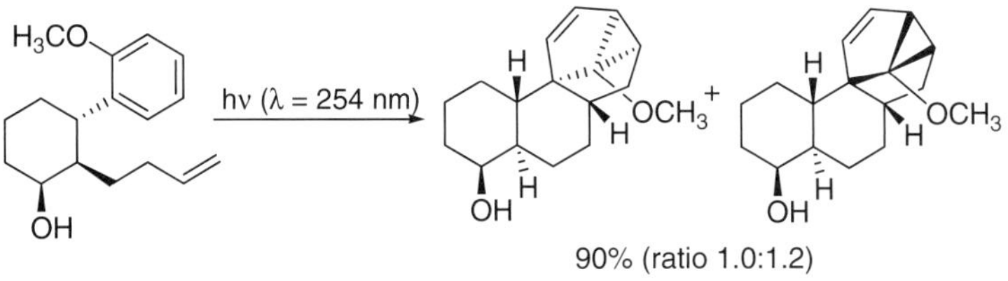

Scheme 9.49

9.8
Concluding Remarks

The synthesis of a variety of organic compounds containing six-membered (and larger) rings can be achieved using photochemical reactions as the key step. In this chapter, an attempt has been made to select typical examples with a marked preparative value. In a number of cases, the photochemical synthesis is the method of choice, and occasionally there is no thermal counterpart. It should be noted that the list of reactions included here is not meant to be exhaustive, and additional examples are provided in several related reviews [1, 2, 41, 85, 86].

References

1 Hoffmann, N. (2008) Photochemical reactions as key steps in organic synthesis. *Chemical Reviews*, **108**, 1052–1103.

2 Griesbeck, A.G., Hoffmann, N., and Warzecha, K.D. (2007) Photoinduced-electron-transfer chemistry: from studies on PET processes to applications in natural products synthesis. *Accounts of Chemical Research*, **40**, 128–140.

3 Casimir, J.R., Guichard, G., and Briand, J.-P. (2002) Methyl-2-(succinimidyl-N-oxycarbonyl)benzoate (MBS). A new, efficient reagent for N-phthaloylation of amino acid and peptide derivatives. *Journal of Organic Chemistry*, **67**, 3764–3768.

4 Griesbeck, A.G., Henz, A., Kramer, W., Lex, J., Nerowski, F., and Oelgemöller, M. (1997) Synthesis of medium- and large-ring compounds initiated by photochemical decarboxylation of ω-phthalimidoalkanoates. *Helvetica Chimica Acta*, **80**, 912–933.

5 Griesbeck, A.G., Heinrich, T., Oelgemöller, M., Lex, J., and Molis, A. (2002) A photochemical route for efficient cyclopeptide formation with a minimum of protection and activation chemistry. *Journal of the American Chemical Society*, **124**, 10972–10973.

6 Griesbeck, A.G., Kramer, W., and Lex, J. (2001) Diastereo- and enantioselective synthesis of pyrrolo(1,4)benzodiazepines through decarboxylative photocyclization. *Angewandte Chemie, International Edition*, **40**, 577–579.

7 Griesbeck, A.G., Kramer, W., and Oelgemöller, M. (1999) Photoinduced decarboxylation reactions. *Green Chemistry*, **180**, 205–207.

8 Griesbeck, A.G., Mauder, H., Müller, I., Peters, K., Peters, E.-M., and von Schnering, H.G. (1993) Photochemistry of N-phthaloyl derivatives of methionine. *Tetrahedron Letters*, **34**, 453–456.

9 Görner, H., Griesbeck, A.G., Heinrich, T., Kramer, W., and Oelgemöller, M. (2001) Time-resolved spectroscopy of sulfur- and carboxy-substituted N-alkylphthalimides. *Chemistry – A European Journal*, **7**, 1530–1538.

10 Maruyama, K. and Kubo, Y. (1978) Solvent-incorporated medium to macrocyclic compounds by the photochemical cyclization of N-alkenylphthalimides. *Journal of the American Chemical Society*, **100**, 7772–7773.

11 Cho, D.W., Choi, J.H., Oh, S.W., Quan, C., Yoon, U.C., Wang, R., Yang, S., and Mariano, P.S. (2008) Single electron transfer-promoted photocyclization reactions of linked acceptor-polydonor systems: effects of chain length and type on the efficiencies of macrocyclic ring-forming photoreactions of tethered α-silyl

ether phthalimide substrates. *Journal of the American Chemical Society*, **130**, 2276–2284.

12 Yoon, U.C., Jin, Y.X., Oh, S.W., Park, C.H., Park, J.H., Campana, C.F., Cai, X., Duesler, E.N., and Mariano, P.S. (2003) A synthetic strategy for the preparation of cyclic peptide mimetics based on SET-promoted photocyclization processes. *Journal of the American Chemical Society*, **125**, 10664–10671.

13 Bertrand, S., Hoffmann, N., and Pete, J.P. (2000) Highly efficient and stereoselective radical addition of tertiary amines to electron-deficient alkenes-application to the enantioselective synthesis of necine bases. *European Journal of Organic Chemistry*, **82**, 2227–2238.

14 Bertrand, S., Hoffmann, N., Humbel, S., and Pete, J.P. (2000) Diastereoselective tandem addition-cyclization reactions of unsaturated tertiary amines initiated by photochemical electron transfer (PET). *Journal of Organic Chemistry*, **65**, 8690–8703.

15 Marinkovic, S., Brule, C., Hoffmann, N., Prost, E., Nuzillard, J.-M., and Bulach, V. (2004) Origin of chiral induction in radical reactions with the diastereoisomers (5R)- and (5S)-5-l-menthyloxyfuran-2[5H]-one. *Journal of Organic Chemistry*, **69**, 1646–1651.

16 Kobayashi, T., Spande, T.F., Aoyagi, H., and Witkop, B. (1969) Tricyclic analogs of melatonin. *Journal of Medicinal Chemistry*, **12**, 636–638.

17 Sundberg, R.J. and Bloom, J.D. (1980) Chloroacetamide photocyclization. Synthesis of 20-deethylcatharanthine. *Journal of Organic Chemistry*, **45**, 3382–3387.

18 Bosch, J., Amat, M., Sanfeliu, E., and Miranda, M.A. (1985) Studies on the synthesis of pentacyclic strychnos indole alkaloids. Photocyclization of N-chloroacetyl-1,2,3,4,5,6-hexahydro-1,5-methanoazocino[4,3-b]indole derivatives. *Tetrahedron*, **41**, 2557–2566.

19 Feldman, K.S. and Ngernmeesri, P. (2005) Dragmacidin E synthesis studies. Preparation of a model cycloheptannelated indole fragment. *Organic Letters*, **7**, 5449–5452.

20 Hintz, S. Frölich, R. and Mattay, J. (1996) PET-oxidative cyclization of unsaturated silyl enol ethers. Regioselective control by solvent effects. *Tetrahedron Letters*, **37**, 7349–7352.

21 Pandey, G., Murugan, A., and Balakrishnan, M. (2002) A new strategy towards the total synthesis of phenanthridone alkaloids: synthesis of (+)-2,7-dideoxypancratistatin as a model study. *Chemical Communications*, 624–625.

22 Kessar, S.V. and Mankotia, A.K.S. (1995) Photocyclization of haloarenes, in *CRC Handbook of Organic Photochemistry and Photobiology*, 2nd edn (ed. W. Horspool), CRC Press, pp. 1218–1228.

23 Semmelhack, M.F. and Bargar, T. (1980) Photostimulated nucleophilic aromatic substitution for halides with carbon nucleophiles. Preparative and mechanistic aspects. *Journal of the American Chemical Society*, **102**, 7765–7774.

24 Gilbert, A. (2004) Cyclization of stilbene and its derivatives, in *CRC Handbook of Organic Photochemistry and Photobiology*, 2nd edn (ed. W. Horspool and F. Lenci), CRC Press, pp. 33/1–33/11.

25 Mallory, F.B., Butler, K.E., and Evans, A.C. (1996) Phenacenes: a family of graphite ribbons. 1. Syntheses of some [7]phenacenes by stilbene-like photocyclization. *Tetrahedron Letters*, **37**, 7173–7176.

26 Rawal, V.H. and Cava, M.P. (1984) Photocyclization of pyrrole analogues of stilbene: an expedient approach to antitumour agent CC-1065. *Journal Chemical Society, Chemical Communications*, 1526–1527.

27 Hendrickson, J.B. and Vries, J.G. (1982) A convergent total synthesis of methoxatin. *Journal of Organic Chemistry*, **47**, 1148–1150.

28 Andryukhina, E.N., Fedorova, O.A., Fedorov, Y.V., Panfilov, M.A., Ihmels, H., Alfimov, M.V., and Gromov, S.P. (2005) Electrocyclic reaction of crown-containing 2-styrylbenzothiazoles. *Russian Chemical Bulletin*, **54**, 1328–1330.

29 Ben, I., Castedo, L., Saa, J.M., Seijas, J.A., Suau, R., and Tojo, G. (1985) 4,5-O-Substituted phenanthrenes from cyclophane. The total synthesis of cannithrene II. *Journal of Organic Chemistry*, **50**, 2236–2240.

30 Thompson, C.M. and Docter, S. (1988) Lewis acid promoted photocyclization of arylimines. Studies directed towards the synthesis of pentacyclic natural products. *Tetrahedron Letters*, **29**, 5213–5216.

31 Armesto, D., Hoorspool, W.M., Ortiz, M.J., and Romano, S. (1992) Reaction of anions from monoimines of benzil with alkylating agents. Photochemical reactivity of some 4-alkoxy-2-aza-1,3-dienes. *Journal of the Chemical Society, Perkin Transactions 1*, 171–175.

32 For an example of photocyclization of 2-azadienes in neutral medium see: Campos, P.J., Caro, M., and Rodriguez, M.A. (2001) Synthesis of isoquinolines by irradiation of 1-methoxy-2-azabuta-1,3-dienes in a neutral medium. *Tetrahedron Letters*, **42**, 3575–3577.

33 Arai, S., Ishikura, M., and Yamagishi, T. (1988) Synthesis of polycyclic azoniaaromatic compounds by photo-induced intramolecular quaternization: azonia derivatives of benzo[c]phenanthrene, [5]helicene and [6]helicene. *Journal of the Chemical Society, Perkin Transactions 1*, 1561–1567.

34 Irie, M. (2000) Diarylethenes for memories and switches. *Chemical Reviews*, **100**, 1685–1716.

35 Matsuda, K. and Irie, M. (2006) Recent developments of 6π-electrocyclic photochromic systems. *Chemistry Letters*, **35**, 1204–1209.

36 Lewis, F.D., Crompton, E.M., Sajimon, M.C., Gevorgyan, V., and Rubin, M. (2006) Ring-closing photoisomerization of some 2,6-diarylstyrenes. *Photochemistry and Photobiology*, **82**, 119–122.

37 Takahashi, A., Hirose, Y., Kusama, H., and Iwasawa, N. (2008) Chelation-assisted electrocyclic reactions of 3-alkenyl-2,2′-bipyridines: an efficient method for the synthesis of 5,6-dihydro-1,10-phenanthroline and 1,10-phenanthroline derivatives. *Chemical Communications*, 609–611.

38 Bois, F., Gardette, D., and Gramain, J.L. (2000) A new asymmetric synthesis of (S)-(+)-pipecoline and (S)-(+)- and (R)-(−)-coniine by reductive photocyclization of dienamides. *Tetrahedron Letters*, **41**, 8769–8772.

39 Couture, A., Grandclaudon, P., and Hooijer, S.O. (1991) Photocyclization in an alcohol containing dissolved base. A new development in enamide photochemistry. *Journal of Organic Chemistry*, **56**, 4977–4980.

40 Tanaka, K., Kakinoki, O., and Toda, F. (1992) Control of the stereochemistry in the photocyclisation of acrylanilides to 3,4-dihydroquinolin-2(1H)-ones. Delicate dependence on the host compound. *Journal of the Chemical Society, Chemical Communications*, 1053–1054.

41 Iriondo-Alberdi, J. and Greaney, M.F. (2007) Photocycloaddition in natural product synthesis. *European Journal of Organic Chemistry*, **78**, 4801–4815.

42 Jones, S. and Atherton, J.C. (2001) Highly diastereoselective photochemical Diels-Alder reactions: towards the development of a photoactivated chiral auxiliary. *Tetrahedron Asymmetry*, **12**, 1117–1119.

43 Mangion, D., Borg, R.M., and Errington, W. (1995) The photosensitised electron transfer [4e + 2e] intramolecular cycloaddition of 1,1,8,8-tetraphenyl-1,7-octadiene; X-ray crystal structure of 4a,9a-*trans*-4a, 10-*trans*-1,2,3,4,4a,9,9a, 10-octahydro-9,9,10-triphenylanthracene. *Journal of the Chemical Society, Chemical Communications*, 1963–1964.

44 Some thermal Diels–Alder reactions between electron-rich components are

metal-catalyzed processes promoted by aminium salts: Hilt, G. and Smolko, K.I. (2002) Cobalt(I)-catalyzed neutral Diels-Alder reactions of 1,3-diynes with acyclic 1,3-dienes. *Synthesis*, **61**, 686–692.

45 Bauld, N. (1989) Cation radical cycloadditions and related sigmatropic reactions. *Tetrahedron*, **45**, 5307–5363.

46 Miranda, M.A. and García, H. (1994) 2,46-Triphenylpyrylium tetrafluoroborate as an electron transfer photosensitizer. *Chemical Reviews*, **94**, 1063–1089.

47 Müller, F. and Mattay, J. (1993) Photocycloadditions: control by energy and electron transfer. *Chemical Reviews*, **93**, 99–117.

48 Harirchian, B. and Bauld, N.L. (1989) Cation radical Diels-Alder cycloadditions in organic synthesis: a formal total synthesis of (−)-β-selinene. *Journal American Chemical Society*, **111**, 1826–1828.

49 Gieseler, A., Steckhan, E., Wiest, O., and Knoch, F. (1991) Photochemically induced radical cation Diels-Alder reaction of indole and electron-rich dienes. *Journal of Organic Chemistry*, **56**, 1405–1411.

50 Pérez-Prieto, J., Stiriba, S.E., González-Béjar, M., Domingo, L.R. and Miranda, M.A. (2004) Mechanism of triplet photosensitized Diels-Alder reaction between indoles and cyclohexadienes: theoretical support for an adiabatic pathway. *Organic Letters*, **6**, 3905–3908.

51 González-Béjar, M., Stiriba, S.E., Domingo, L.R., Pérez-Prieto, J., and Miranda, M.A. (2006) The mechanism of triplet photosensitized Diels-Alder reaction between indoles and cyclohexadienes: theoretical support for an adiabatic pathway. *Journal of Organic Chemistry*, **71**, 6932–6941.

52 Zhang, W., Guo, Y., Liu, Z., Jin, X., Yang, L., and Liu, Z.-L. (2005) Photochemically catalysed Diels-Alder reaction of arylimines with *N*-vinylpyrrolidinone and *N*-vinylcarbazole by 2,4,6-triphenylpyrylium salt: synthesis of 4-heterocyclesubstituted tetrahydroquinoline derivatives. *Tetrahedron*, **61**, 1325–1333.

53 Argüello, J.E., Pérez-Ruiz, R., and Miranda, M.A. (2007) Novel [4 + 2] cycloaddition between thiobenzophenone and aryl-substituted alkenes via photo-induced electron transfer. *Organic Letters*, **9**, 3587–3590.

54 Nicolau, K.C. and Gray, D.L.F. (2004) Total synthesis of hybocarpone and analogues thereof. A facile dimerization of naphthazarins to pentacyclic systems. *Journal of the American Chemical Society*, **126**, 607–612.

55 Nicolau, K.C., Gray, D., and Tae, J. (2001) Total synthesis of hamigerans: Part 1. Development of synthetic technology for the construction of benzannulated polycyclic systems by the intramolecular trapping of photogenerated hydroxy-*o*-quinodimethanes and synthesis of key building blocks. *Angewandte Chemie, International Edition*, **40**, 3675–3678.

56 Sieburth, S.M. (2004) Photochemical reactivity of pyridones, in *CRC Handbook of Organic Photochemistry and Photobiology*, 2nd edn (ed. W. Horspool), CRC Press, pp. 103/1–103/18.

57 McGee, K.F. Jr., Al-Tel, T.H., and Sieburth, S.McN. (2001) Fusicoccin synthesis by intramolecular [4 + 4]photocycloaddition of 2-pyridones: stereocontrol of the cycloaddition and elaboration of the pentacyclic product. *Synthesis*, 1185–1196.

58 Li, L., Chase, C.E., and West, F.G. (2008) Cyclooctatrienes from pyran-2-ones via a tandem [4 + 4] photocycloaddition/decarboxylation process. *Chemical Communications*, 4025–4027.

59 Rigby, J.H., Ateeq, H.S., and Krueger, A.C. (1992) Metal-promoted higher order cycloaddition reactions. A facile entry into eight- and ten-membered carbocycles. *Tetrahedron Letters*, **33**, 5873–5876.

60 Rigby, J.H. and Ateeq, H.S. (1990) Synthetic studies on transition-metal-mediated higher order cycloaddition

reactions: highly stereoselective construction of substituted bicyclo[4.4.1]undecane systems. *Journal of the American Chemical Society*, **112**, 6442–6443.

61 Rigby, J.H. and Henshilwood, J.A. (1991) Transition metal template controlled cycloaddition reactions. An efficient chromium(0)-mediated [6π + 2π] cycloaddition. *Journal of the American Chemical Society*, **113**, 5122–5123.

62 Rigby, J.H. and Warshakoon, N.C. (1997) A convenient synthesis of 1,2-disubstituted cyclooctatetraenes. *Tetrahedron Letters*, **38**, 2049–2052.

63 Yang, N.C. and Yang, D.-D.H. (1958) Photochemical reactions of ketones in solution. *Journal of the American Chemical Society*, **80**, 2913–2914.

64 Kraus, G.A. and Wu, Y. (1992) 1,5- and 1,9-Hydrogen atom abstractions. Photochemical strategies for radical cyclizations. *Journal of the American Chemical Society*, **114**, 8705–8707.

65 Hasegawa, T., Horikoshi, Y., Iwata, S., and Yoshioka, M. (1991) Photocyclization of β-oxoesters: limiting factors for remote hydrogen abstraction through a medium-sized cyclic transition state. *Journal of the Chemical Society, Chemical Communications*, 1617–1618.

66 Hasegawa, T., Miyata, K., Ogawa, T., Yoshihara, N., and Yoshioka, M. (1985) Remote photocyclization of (dibenzylamino)ethyl benzoylacetate. Intramolecular hydrogen abstraction through a ten-membered cyclic transition state. *Journal of the Chemical Society, Chemical Communications*, 363–364.

67 Abad, S., Bosca, F., Domingo, L.R., Gil, S., Pischel, U., and Miranda, M.A. (2007) Triplet reactivity and regio-/stereo-selectivity in the macrocyclization of diastereomeric ketoprofen-quencher conjugates via remote hydrogen abstractions. *Journal of the American Chemical Society*, **129**, 7407–7420.

68 Andreu, I., Bosca, F., Sánchez, L., Morera, I.M., Camps, P., and Miranda, M.A. (2006) Efficient and selective photogeneration of cholesterol-derived radicals by intramolecular hydrogen abstraction in model dyads. *Organic Letters*, **8**, 4597–4600.

69 Wessig, P., Schwarz, J., Linderman, U., and Holthausen, M.C. (2001) Photochemical synthesis of 3,4-dihydro-2*H*-1,3-oxazin-4-ones. *Synthesis*, 1258–1262.

70 Kaplan, M.L. and Truesdale, E.A. (1976) [2.2]Paracyclophane by photoextrusion of carbon dioxide from a cyclic diester. *Tetrahedron Letters*, **17**, 3665–3666.

71 Hilbert, M. and Solladie, G. (1980) Substituent effect during the synthesis of substituted [2.2] paracyclophane by photoextrusion of carbon dioxide from a cyclic diester. *Journal of Organic Chemistry*, **45**, 4496–4498.

72 Miranda, M.A., Font-Sanchis, E., Pérez-Prieto, J., and Scaiano, J.C. (2001) The 4,4′-(1,2-ethanediyl)bisbenzyl biradical: its generation, detection and (photo)chemical behavior in solution. *Journal of Organic Chemistry*, **66**, 2717–2721.

73 Font-Sanchis, E., Miranda, M.A., Pérez-Prieto, J., and Scaiano, J.C. (2002) Laser flash photolysis of [3,*n*]paracyclophan-2-ones. Direct observation and chemical behavior of 4,4′-(1,*n*-alkanediyl)bisbenzyl biradicals. *Journal of Organic Chemistry*, **67**, 6131–6135.

74 Mislin, G.L. and Miesch, M. (2003) Photochemical, thermal and base-induced access to hydroazulene derivatives via two-carbon ring-enlargement reactions of condensed electrophilic cyclobutenes. *Journal of Organic Chemistry*, **68**, 433–441.

75 Kaupp, G. and Stark, M. (1977) 6,7-Diphenyl-2,3-dihydro-1,4-dioxocin. *Angewandte Chemie, International Edition*, **16**, 552.

76 Davies, D.M.E., Murray, C., Berry, M., Orr-Ewing, A.J., and Booker-Milburn, K.I. (2007) Reaction optimisation and mechanism in maleimide photocyclo-addition: a dual approach using tunable UV laser and time-dependent DFT. *Journal of Organic Chemistry*, **72**, 1449–1457.

77 Booker-Milburn, K.I., Anson, C.E., Clissold, C., Costin, N.J., Dainty, R.F.,

Murray, M., Patel, D., and Sharpe, A. (2001) Intramolecular photocycloaddition of *N*-alkenyl substituted maleimides: a potential tool for the rapid construction of perhydroazaazulene alkaloids. *European Journal of Organic Chemistry*, 1473–1482.

78 Hook, B.D.A., Dohle, W., Hirst, P.R., Pickworth, M., Berry, M.B., and Booker-Milburn, K.I. (2005) A practical flow reactor for continuous organic photochemistry. *Journal of Organic Chemistry*, **70**, 7558–7564.

79 Suau, R., Sánchez-Sánchez, C., García-Segura, R., and Pérez-Inestrosa, E. (2002) Photocycloaddition of phthalimide anion to alkenes. A highly efficient, convergent method for [2]benzazepine synthesis. *European Journal of Organic Chemistry*, 1903–1911.

80 Mizuno, K., Kagano, H., and Otsuji, Y. (1983) Regio- and stereoselective intramolecular photocycloaddition: synthesis of macrocyclic 2,ω-dioxabicyclo [n. 2. 0] ring system. *Tetrahedron Letters*, **24**, 3849–3850.

81 Ors, J.A. and Srinivasan, R. (1978) Synthesis of macrocyclic rings by internal photocycloaddition of α,ω-dicinnamates. *Journal of the Chemical Society, Chemical Communications*, 400–401.

82 Benali, O., Miranda, M.A., and Tormos, R. (2002) Electron transfer versus proton transfer in excited states of bichromophoric aniline/olefin systems. *European Journal of Organic Chemistry*, 2317–2322.

83 Boyd, J.W., Greaves, N., Kettle, J., Russell, A.T., and Steed, J.W. (2005) Alkene-Arene *meta* photocycloadditions with a four-carbon-atom tether: efficient approach toward the polycyclic ring systems of aphidicolin and stemodinone. *Angewandte Chemie, International Edition*, **44**, 944–946.

84 Gilbert, A. (2004) Intra- and intermolecular cycloadditions of benzene derivatives, in *CRC Handbook of Organic Photochemistry and Photobiology*, 2nd edn (ed. W. Horspool), CRC Press, pp. 41-1–41-11.

85 Griesbeck, A.G., Henz, A., and Hirt, J. (1996) Photochemical synthesis of macrocycles. *Synthesis*, 1261–1276.

86 Fagnoni, M., Dondi, D., Ravelli, D., and Albini, A. (2007) Photocatalysis for the formation of the C−C bond. *Chemical Reviews*, **107**, 2725–2756.

10
Aromatic and Heteroaromatic Substitution by $S_{RN}1$ and S_N1 Reactions

Alicia B. Peñéñory and Juan E. Argüello

10.1
Introduction

Aromatic nucleophilic substitution can be accomplished by different mechanisms: (i) the addition–elimination or $S_N Ar$ process; (ii) the benzyne reaction; (iii) halogen–metal exchange (HME); (iv) vicarious nucleophilic substitution of hydrogen atoms; (v) the $S_N 1$ reaction with the intermediacy of aryl cations; and (vi) the Unimolecular Radical Nucleophilic Substitution or $S_{RN}1$ mechanism involving electron transfer (ET) steps and radical and radical anions as intermediates. The $S_N Ar$ process is restricted to aromatic substrates activated by electron- withdrawing groups (EWGs) *ortho* and *para* to the leaving group; the benzyne mechanism in most cases involves the use of very strong bases; the HME substitution of halobenzenes generally proceeds with nucleophiles derived from tin, silicon, and germanium; for the vicarious reaction, the presence of an EWG is also required. In view of this, processes (i) to (iv) involve serious limitations in their synthetic scope. On the other hand, $S_{RN}1$ and aryl $S_N 1$ mechanisms complement each other and proceed efficiently under photostimulation, providing good access to a diversity of aromatic compounds.

The $S_{RN}1$ process has proven to be a versatile mechanism for replacing a suitable leaving group by a nucleophile at the *ipso* position. This reaction affords substitution in nonactivated aromatic (ArX) compounds, with an extensive variety of nucleophiles (Nu$^-$) derived from carbon, nitrogen, and oxygen to form new C–C bonds, and from tin, phosphorus, arsenic, antimony, sulfur, selenium, and tellurium to afford new C–heteroatom bonds.

Several examples involve the formation of two bonds leading to (hetero)cyclic products. In addition to halides, other leaving groups have been reported, including: $(EtO)_2P(O)O$, RS (R = Ph, alkyl), PhSO, $PhSO_2$, PhSe, Ph_2S^+, RSN_2 (R = *t*-Bu, Ph), BF_4N_2, and Me_3N^+. This reaction is also compatible with many substituents such as alkyl groups, OR, OAr, SAr, CF_3, CO_2R, NH_2, NHCOR, NHBoc, SO_2R, CN, COAr, NR_2, and F. Even though the reaction is not inhibited by the presence of negatively charged substituents as carboxylate ions, other charged groups such as oxyanions will hinder the process.

Handbook of Synthetic Photochemistry. Edited by Angelo Albini and Maurizio Fagnoni
Copyright © 2010 WILEY-VCH Verlag GmbH & Co. KGaA, Weinheim
ISBN: 978-3-527-32391-3

Aromatic substrates bearing two leaving groups can be substituted by consecutive $S_{RN}1$ reactions. The aryl stannanes obtained by means of this approach are substrates suitable for Pd-catalyzed crosscoupling reactions to afford bi-, tri-, and polyaryl compounds. Moreover, the $S_{RN}1$ mechanism represents an excellent route for the synthesis of carbocycles and heterocycles by ring-closure reactions, such as indoles, isoquinolinones, chromanes, phenanthridines, benzothiazoles, and binaphthyls, and for a significant number of natural products.

The most relevant results regarding the $S_{RN}1$ substitutions have been extensively reviewed [1–5]; consequently, in this chapter we include only the most recent and major representative synthetic examples in the aromatic system, relating to C−C and C−heteroatom bond formation with sulfur and tin nucleophiles, and ring-closure reactions.

During the past ten years, the S_N1 photoarylation reaction and some of its synthetic implications have been extensively explored and recently reviewed [6–8]. This reaction involves the mediation of aryl cations (Ar^+) formed by photoheterolysis of the Ar−X bond which allows the use of a variety of precursors, such as readily available aryl chlorides and aryl esters (i.e., mesylates and phosphates). Interestingly, the reaction applies also to aryl fluorides, although this is difficult by other methods with nonactivated aromatic derivatives.

The reactions of phenyl cations with various carbon nucleophiles, including olefins, alkynes, arenes and heteroarenes, have shown to offer a convenient access to Ar−C bonds. These processes are photochemical alternatives to the metal-catalyzed Heck and crosscoupling reactions, with the advantage of avoiding the use of expensive catalysts and of operating under mild conditions. In addition, aryl cations are able to add to enamines and silyl enol ethers, giving their corresponding α-arylketones in satisfactory yields. This procedure can be extended to ketene silyl acetals which, in a single step, afford α-arylpropionic esters in good yields.

Heterocyclic compounds can also be achieved by intramolecular and intermolecular nucleophilic trapping of the Ar^+ intermediates.

10.2
General Mechanistic Features

10.2.1
$S_{RN}1$ Mechanism

The mechanism of this reaction involves a chain process with radicals and radical anions as intermediates, the initiation step of which consists of an ET to the substrate to give the radical anion. The generally accepted pathways for the propagation cycle for the aromatic $S_{RN}1$ process are outlined in Scheme 10.1. The radical anion of the substrate fragments into the aryl radical and the anion of the leaving group (reaction 1). The aryl radical can react with the Nu^- to yield the radical anion of the substitution product which, by ET to the ArX, forms the intermediates required for continuing the propagation cycle (reactions 2 and 3). The overall process amounts to

$$(ArX)^{-\bullet} \longrightarrow Ar^{\bullet} + X^{-} \quad (1)$$

$$Ar^{\bullet} + Nu^{-} \longrightarrow (ArNu)^{-\bullet} \quad (2)$$

$$(ArNu)^{-\bullet} + ArX \longrightarrow ArNu + (ArX)^{-\bullet} \quad (3)$$

$$ArX + Nu^{-} \longrightarrow ArNu + X^{-} \quad (4)$$

Scheme 10.1

a nucleophilic substitution (reaction 4). The mechanism has termination steps that depend on the nature of ArX and Nu$^-$, and on the experimental conditions.

The most widely used initiation method is photoinduced electron transfer (PET) from a charged nucleophile to the substrate [1]. This involves one of the following pathways:

- Photoexcitation of a charge transfer complex (CTC) formed between the nucleophile and the substrate, followed by ET.
- Homolytic cleavage of the C—X bond.
- Electron transfer from an excited nucleophile to the substrate.
- Electron transfer from a nucleophile to an excited substrate.

For nucleophiles which are unable to initiate the reaction but are quite reactive in the propagation steps, the addition of minute amounts of another nucleophile capable of initiating the reaction increases the generation of intermediates. This allows the less-reactive initiation-nucleophile to start its own propagation. This *entrainment* reaction allows an extension of the $S_{RN}1$ mechanism to nucleophiles that are poor donors, supporting the chain nature of the reaction [9].

The addition of a nucleophile to a radical to form the radical anion of the substitution product constitutes the main feature of a $S_{RN}1$ process, although the chain can be short or even nonexistent (reaction 2). For a photoinduced reaction, a quantum yield higher than 1 can be taken as evidence of a chain reaction, although a global quantum yield below 1 cannot be used as a criterion against a chain reaction [9, 10].

Experimental evidence for the presence of radical intermediates is provided by the identification of expected products from radical rearrangements, by the use of appropriate radical probes and by direct detection by electron spin resonance (ESR). Other mechanistic evidence includes inhibition by radical traps, such as di-*t*-butylnitroxide (DTBN), TEMPO (2,2,6,6-tetramethyl-1-piperidinyloxy), galvinoxyl and oxygen, and by radical anion scavengers such as *p*-dinitrobenzene (*p*-DNB).

Finally, fragmentation of the radical anion of the substitution product (ArNu$^{\bullet-}$) into new reactive species also allows an assessment of the participation of these intermediates in the reaction. For example, the photostimulated reactions of Ph$_2$As$^-$, Ph$_2$Sb$^-$, PhSe$^-$ and PhTe$^-$ ions with some ArI, have shown a scrambling of the aromatic rings, as illustrated in Scheme 10.2 [1].

Fragmentation at the Ph–Se bond of the radical anion formed by the coupling of *p*-anisyl radicals and PhSe$^-$ ions into phenyl radicals and the new nucleophile *p*-AnSe$^-$ ions accounts for the formation of scrambled products.

$$\mathrm{AnI} + \mathrm{PhSe}^- \xrightarrow[\mathrm{NH_3}]{h\nu} \underset{20\%}{\mathrm{AnSePh}} + \underset{25\%}{\mathrm{An_2Se}} + \underset{19\%}{\mathrm{Ph_2Se}}$$

An = *p*-$MeOC_6H_4$

Scheme 10.2

The photostimulated reactions of thiourea, thioacetate, and thiobenzoate anions with ArX are further examples of fragmentation of the $ArNu^{\bullet -}$ into new reactive radicals and anions capable of continuing the propagation cycle of the $S_{RN}1$ [9]. The mechanism of these reactions has been studied in detail, and the propagation steps are described in Scheme 10.3. After coupling of the aryl radical with the sulfur-centered nucleophile, the radical anion intermediate fragments into ArS^- and a new carbon-centered radical. Deprotonation of the latter can afford a new radical anion intermediate capable of transferring the odd electron to the ArX to allow for a chain to be built up [9].

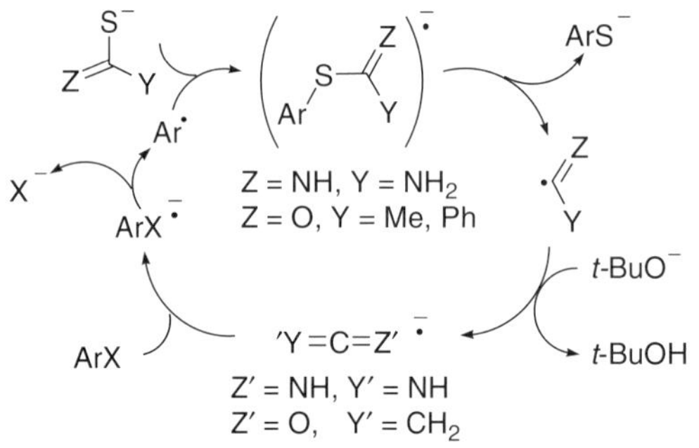
S⁻
Z
Y
Ar•
Ar
S
Z
Y
ArS⁻
X⁻
ArX•⁻
Z = NH, Y = NH2
Z = O, Y = Me, Ph
Z
•
Y
t-BuO⁻
ArX
'Y=C=Z'•⁻
t-BuOH
Z' = NH, Y' = NH
Z' = O, Y' = CH2

Scheme 10.3

10.2.2
S_N1 Mechanism

The aryl S_N1 reaction results from the unimolecular heterolytic photofragmentation of an Ar–X bond, with the generation of an aryl cation (Scheme 10.4, reaction 5). Subsequent addition of this cation to the nucleophile affords the substitution product (Scheme 10.4, reaction 6).

Aryl cations are highly reactive intermediates with two possible electronic configurations, the singlet and triplet states. The former is a localized cation with a vacant σ orbital at the dicoordinated carbon atom, whereas the latter has a diradical character with single occupancy of the σ orbital and the charge being delocalized in the

$$\mathrm{ArX} \longrightarrow \mathrm{Ar^+} + \mathrm{X^-} \quad (5)$$

$$\mathrm{Ar^+} + \mathrm{NuH} \longrightarrow \mathrm{ArNu} + \mathrm{H^+} \quad (6)$$

Scheme 10.4

aromatic ring. Recently, theoretical studies of these intermediates have been conducted considering the effect of the substituent on singlet–triplet gap energy [11].

Singlet phenyl cations are unselective electrophiles, and reaction via such intermediacy mostly results in solvolysis, giving new C−O or C−N bonds. Triplet phenyl cations react selectively with π nucleophiles such as electron-rich olefins, alkynes and aromatics, forming new Ar−C bonds.

The thermal methods available for the formation of aryl cation in solution are limited to the solvolysis of suitable perfluorophenylalkyl esters and aromatic diazonium salts [6]. Furthermore, these methods lead to unselective chemistry due to the singlet nature of the aryl cation formed. Under photostimulation, and according to the substituents, aromatic halides give either the singlet or triplet phenyl cation. The mediation of triplet aryl cations is essential for achieving the chemoselectivity desired in these reactions [12]. In order to ensure the formation of such intermediates, certain conditions should be met:

- The aromatic ring should bear an electron-donor group (EDG) in order to make photoheterolysis of the carbon–halogen bond feasible. In the case of aromatics bearing EWGs, diazonium salts can be employed.
- Intersystem crossing (ISC) should be efficient for fragmentation to occur from the triplet in order to generate the phenyl cation with the same multiplicity; if this condition is not fulfilled, sensitization with high-energy triplet donors can be used instead [13].
- A polar solvent, such as ethyl acetate, acetonitrile (MeCN), methanol (MeOH) or 2,2,2-trifluoroethanol (TFE) should be employed.

In addition to the synthetic capability of the $S_{RN}1$ arylation reactions, their mechanism has been studied in depth, and their intermediates detected using transient spectroscopy [14].

10.3
Carbon–Carbon Bond Formation

10.3.1
Carbanions from Ketones, Esters, Acids, Amides, and Imides as the Nucleophiles

One of the most relevant achievements of the $S_{RN}1$ is the formation of a new C−C bond by the reaction of carbanions with aromatic compounds bearing halides, SPh, SO_2Ph, and N_2SBu-t as leaving groups. Therefore, compounds with acidic hydrogen α to carbonyl, carboxylic acid, ester, amide, imide and cyano groups can be α-arylated and α-heteroarylated in good yields by the $S_{RN}1$ reaction [1].

The enolate ions from acyclic and cyclic aliphatic ketones, and mainly those derived from acetone and pinacolone, are the carbanions most extensively studied so far. In general, the substitution products are obtained in good yield under irradiation in both liquid ammonia and dimethylsulfoxide (DMSO). For example, the anti-inflammatory drug fluorobiprophen **1** can be synthesized by the reaction of

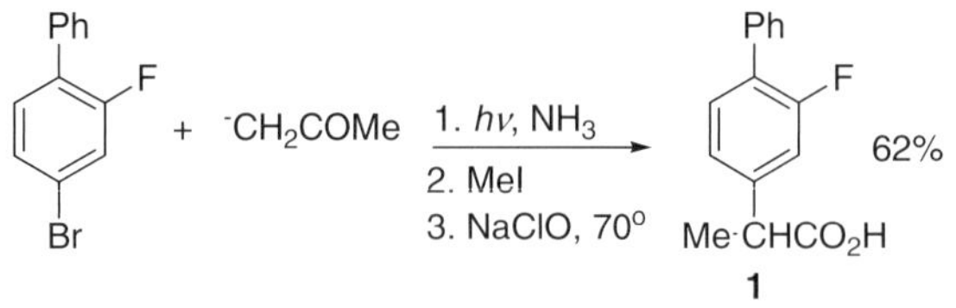

Scheme 10.5

4-bromo-2-fluorobiphenyl with $^-CH_2COMe$ ions, followed by methylation and oxidative demethylation (Scheme 10.5). This is possible because the fluoride ion is not a suitable leaving group in $S_{RN}1$ reactions [15].

α,α-Diarylation is observed in some cases in low yields (<15%). The anions of cyclic ketones can likewise be α-phenylated and heteroarylated. Arylation of the enolate ion of (+)-camphor is a remarkable reaction which affords almost exclusively *endo*-arylation at C_3 in excellent yields (Scheme 10.6) [16].

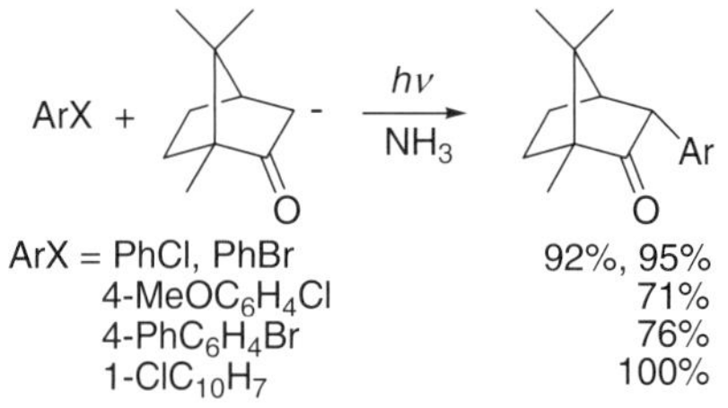

ArX = PhCl, PhBr — 92%, 95%
4-MeOC$_6$H$_4$Cl — 71%
4-PhC$_6$H$_4$Br — 76%
1-ClC$_{10}$H$_7$ — 100%

Scheme 10.6

No substitution product is formed with the enolate ion of cyclohex-2-en-1-one. However, this anion is quite reactive at the initiation step and is often used as an entrainment reagent [17]. Under photostimulation, the best yields of heteroarylation and phenylation of the enolate ions from aromatic ketones are obtained in DMSO [18].

Disubstitution products are obtained when dihalobenzenes (Cl, Br, I) react with aliphatic ketone enolate anions. Conversely, the reactions of *o*-iodohalobenzenes (X = I, Br, Cl) with the enolate anions of aromatic ketones, such as acetophenone, propiophenone and 2-naphthyl methyl ketone in DMSO yield mainly monosubstitution with the retention of one halogen (Scheme 10.7). The extent of dehalogenation is explained in terms of the energetics of the intramolecular ET from the ArCO-π-system to the C—X bond in the monosubstituted radical anions proposed as intermediates [19].

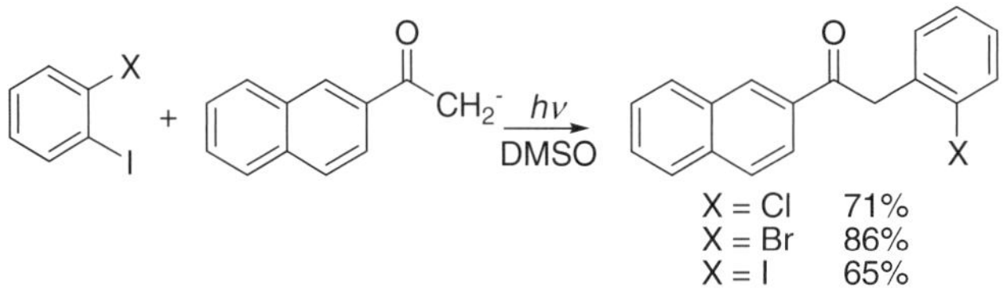

X = Cl 71%
X = Br 86%
X = I   65%

Scheme 10.7

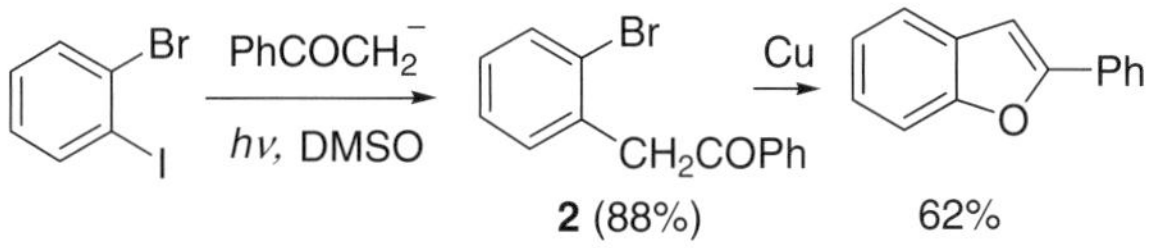

10.3 Carbon–Carbon Bond Formation

Scheme 10.8

The photostimulated reaction of o-bromoiodobenzene with $^-CH_2COPh$ ions gives monosubstitution product 2 in 88% yield, which cyclizes to 2-phenylbenzofuran under catalysis by Cu (activated copper bronze) (Scheme 10.8) [19].

The synthesis of α-aryl esters can be achieved through the reaction of aryl and heteroaryl halides with $^-CH_2CO_2Bu$-t, $^-CHMeCO_2Bu$-t and $^-CHPhCO_2Et$ ions under photoinitiation in 57–89% yields [20]. Moreover, phenylacetic acid dianion reacts with ArX upon irradiation to give aryl substitution products at the p- or α-carbon, depending on the counterion used (Scheme 10.9) [21].

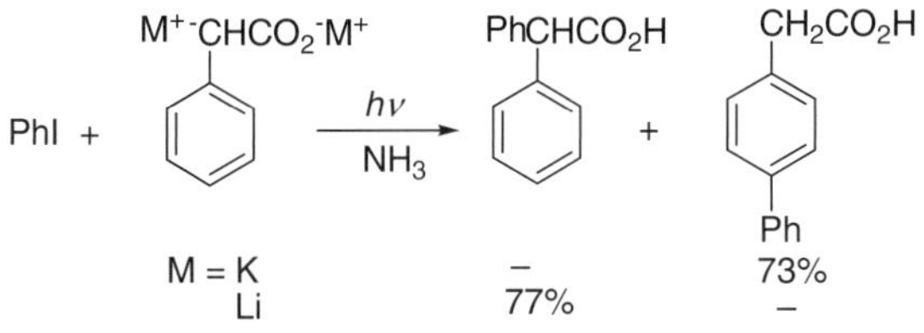

Scheme 10.9

The anions derived from acetamide and N-methyl acetamide are unreactive towards ArX by the $S_{RN}1$ mechanism. On the other hand, carbanions from N,N-disubstituted amides give their corresponding α-arylated compounds in good yields by reaction with ArX in liquid ammonia under irradiation [15, 22]. This procedure is a feasible alternative for the synthesis of N,N-dimethyl-α-aryl and α,α-diarylacetamides, some of which are used as herbicides [15, 22]. A novel approach to the synthesis of α-aryl propionic acids is based on the photoinduced reaction of ArX with anion $^-CH_2CONMe_2$ followed by the addition of MeI. After hydrolysis, the acids are obtained in good overall yields (50–88%) (Scheme 10.10) [15].

The reaction of 1-iodonaphthalene with chiral-assisted imide enolate ion 3 provides an interesting example of the stereoselective coupling of an aromatic radical with a nucleophile. In this reaction, the diastereomeric isomers of the substitution compound are formed (43–64%), while the selectivity observed is highly dependent on the metal counterion used and its chelation properties (Scheme 10.11). All of the ions studied (Li, Na, K, Cs, Ti(IV)) are selective; however, the highest stereoselectivity

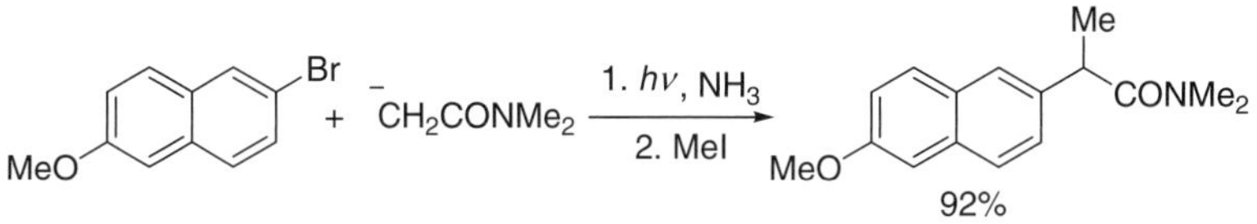

Scheme 10.10

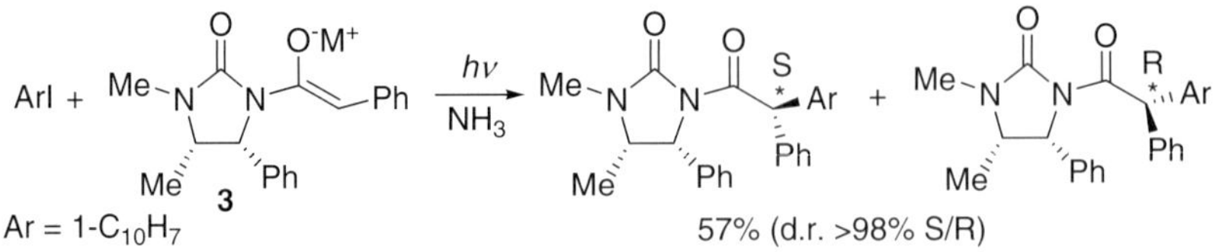

Scheme 10.11

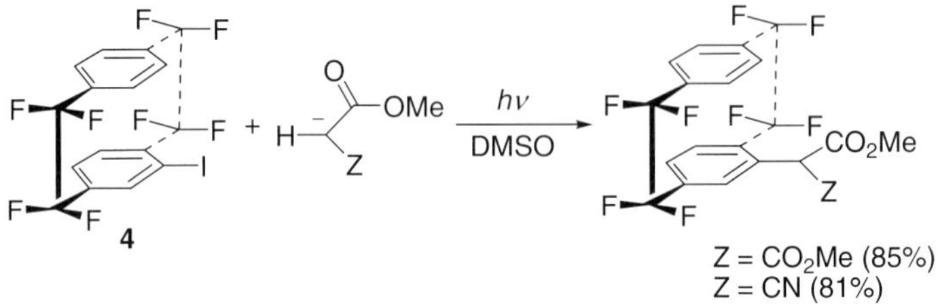

Scheme 10.12

found is with Li at low temperature (−78 °C) and with Ti (IV) [diastereomeric ratio (d.r.) >98%, *S/R*] [23].

Monosubstitution is obtained in the photoinitiated reaction of different carbanions, such as ⁻CH(COMe)$_2$, ⁻CH(CO$_2$Et)$_2$, ⁻CH(COMe)CO$_2$Me, ⁻CMe(COMe)CO$_2$Me and ⁻CEt(CO$_2$Et)$_2$ (60–92%), with electrophilic substrates. Investigations have shown that 4-iodo-1,1,2,2,9,9,10,10-octafluoro[2.2]paracyclophane (**4**) exhibits excellent S$_{RN}$1 reactivity with some of these stabilized enolates (Scheme 10.12) [24].

Other stabilized carbanions suggested to react via the S$_{RN}$1 process include the anions from 2,4,4-trimethyl-2-oxazoline, 2-benzyl-4,4-dimethyl-2-oxazoline, 2,4-dimethylthiazole, 2-benzyl-4,4-dimethylthiazole, and dimethyl methylphosphonate [25]. Some examples are described in Scheme 10.13. Competition with a benzyne process has been determined in some of these reactions.

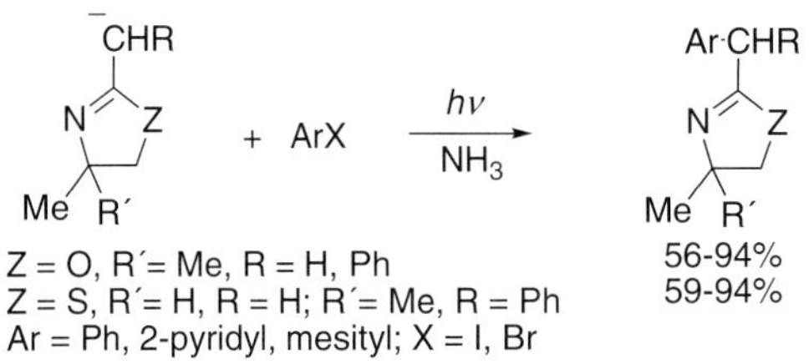

Scheme 10.13

Z = O, R′ = Me, R = H, Ph   56–94%
Z = S, R′ = H, R = H; R′ = Me, R = Ph   59–94%
Ar = Ph, 2-pyridyl, mesityl; X = I, Br

10.3.2
Alkenes, Alkynes, Enols, and Vinyl Amines as the Nucleophiles

Arylation is possible by the addition of triplet aryl cations to neutral nucleophiles such as alkenes, alkynes, enols, and enamines (photo S$_N$1 mechanism). For alkenes, the

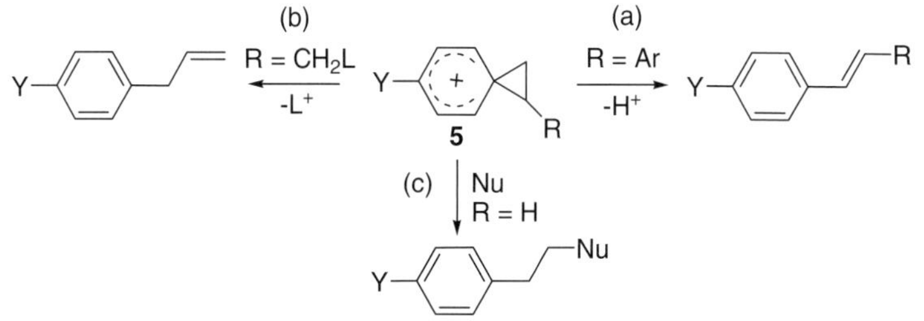

10.3 Carbon–Carbon Bond Formation

Scheme 10.14

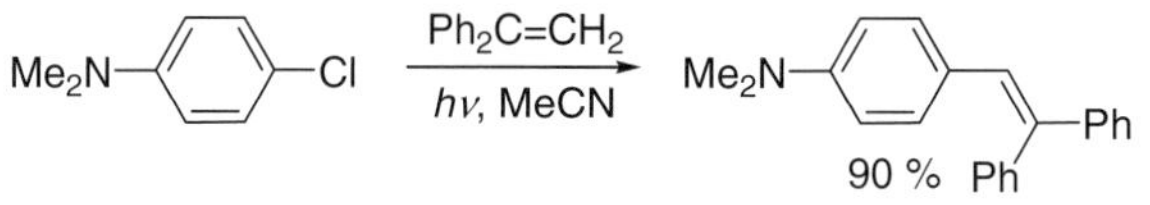

Scheme 10.15

phenonium ion (5) is proposed as the intermediate, and a variety of arylated products can be obtained according to the structure of the olefin and the reaction conditions utilized (Scheme 10.14).

The presence of a good electrofugal group (L^+), such as H^+ or Me_3Si^+, makes elimination the predominant path (Scheme 10.14, pathways a and b). When R is an aromatic ring, pathway (a) is favorable and a benzylic proton is lost to form styrenes; this results in a photochemical process analogous to the Heck reaction (Scheme 10.15) [12].

With allyltrimethylsilane or alkylated alkenes, the leaving group L^+ is lost and allylbenzenes are efficiently formed (Scheme 10.14, pathway b). This reaction has been fully explored, and typical examples are illustrated in Scheme 10.16 for allyltrimethylsilane and 2,3-dimethyl-2-butene [12, 13, 26, 27]. This methodology has been effectively applied to the synthesis of naturally occurring phenols with medicinal activity [28].

Finally, with a poor electrofugal group, the course of the reaction may be directed by choosing suitable conditions:

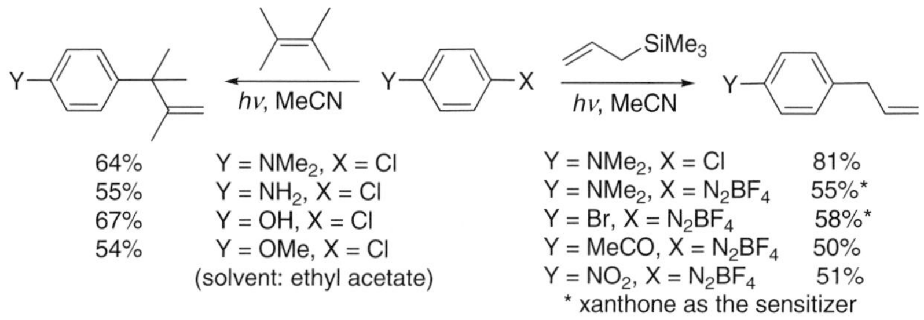

Left side:
64%   Y = NMe₂, X = Cl
55%   Y = NH₂, X = Cl
67%   Y = OH, X = Cl
54%   Y = OMe, X = Cl
(solvent: ethyl acetate)

Right side:
Y = NMe₂, X = Cl   81%
Y = NMe₂, X = N₂BF₄   55%*
Y = Br, X = N₂BF₄   58%*
Y = MeCO, X = N₂BF₄   50%
Y = NO₂, X = N₂BF₄   51%
* xanthone as the sensitizer

Scheme 10.16

- When using a polar, non-nucleophilic medium, the leaving group remains paired to the cation **5** and is incorporated into the adduct (Scheme 10.14, pathway c, Nu = X).
- When employing a more charge-stabilizing solvent (e.g., TFE, H$_2$O-MeCN) that favors reaction via the free ions rather than the ion pair.
- When using a nucleophilic solvent or an added nucleophile that overcomes competition with the counterion generated in the photoheterolysis (Scheme 10.14, pathway c), although this procedure is liable to produce mixtures.

The arylation of alkynes is also successful when starting from aryl derivatives bearing several leaving groups (Cl, F, OMs, OTf and OP(O)(OEt)$_2$) [29]. These reactions offer an alternative to the Sonogashira reaction for arylalkynes synthesis. The addition of an aryl cation to alkynes probably gives a cyclic vinylenebenzenium ion as the intermediate, forming the corresponding arylalkyne after the loss of an electrofugal group, H$^+$ or Me$_3$Si$^+$ (Scheme 10.17). In these cases, triethylamine (TEA) has been added in order to buffer the reaction medium. When the starting aryl derivative absorbs poorly at the wavelength used, the photoreaction can be conveniently carried out by using acetone as solvent/sensitizer, giving the desired triplet aryl cation.

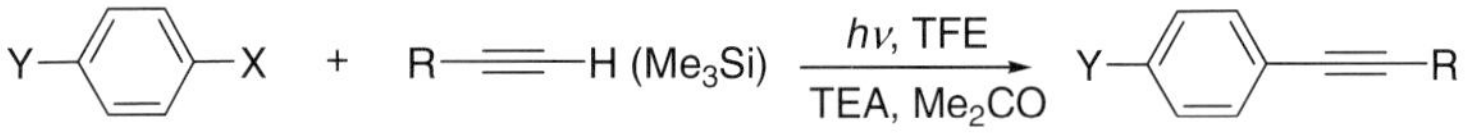

X = Cl, F, OMs, OTf, OP(O)(OEt)$_2$
Y = OMe   R = C$_4$H$_9$, SiMe$_3$, Me   50–90%
Y = SMe   R = C$_4$H$_9$, SiMe$_3$ 55–58%
Y = SiMe$_2$(t-Bu) R = C$_4$H$_9$, SiMe$_3$, Me   51–71%

Scheme 10.17

The reactivity of different dienes towards 4-(N,N-dimethylamino)phenyl cation (**6**) has been studied, and for open-chain dienes, such as 2,5-dimethyl-2,4-hexadiene, only the *trans*-pentadienylaniline is found in a 52% yield [30]. In the case of cyclic dienes, transannular cyclization takes place, leading to an arylnortricyclene from norbornadiene and to 1-arylbicyclo[3.3.0]octanes from 1,5-cyclooctadiene in moderate yields (Scheme 10.18) [30].

The photo-S$_N$1 reaction with particular classes of alkenes – namely enamines, silyl enol ethers and silyl ketene acetals – affords a smooth synthesis of α-aryl ketones,

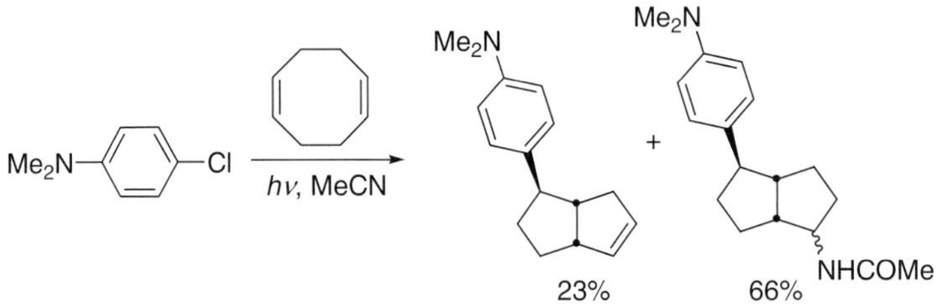

Scheme 10.18

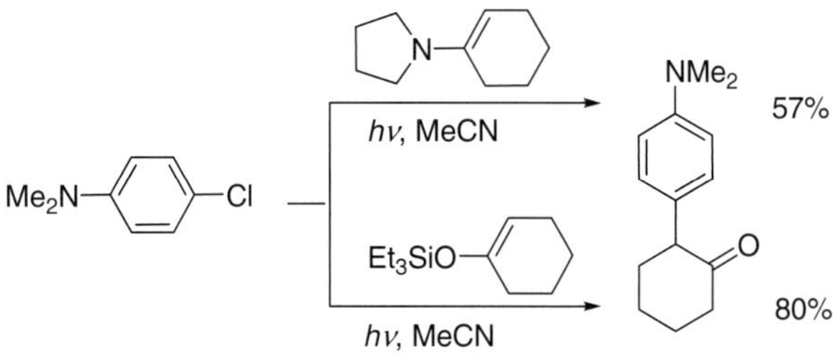

Scheme 10.19

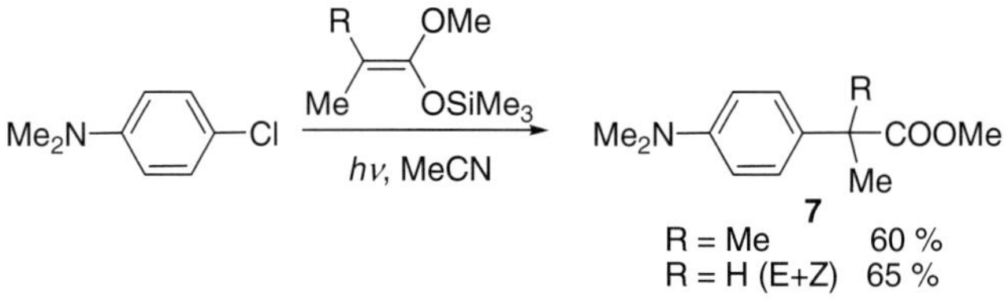

Scheme 10.20

aldehydes and arylacetic esters, respectively [31]. The N,N-dimethylaminophenyl cation (**6**) for instance, adds to an enamine of cyclohexanone and gives the corresponding α-arylketone in 57% yield (Scheme 10.19). The yield of the same ketone is increased when silyl enol ethers rather than enamines are used. The addition to silyl enol ethers of aldehydes occurs with lower yields, exhibiting no particular interest for the synthesis of α-arylaldehydes.

The procedure is successful with ketene silyl acetals prepared from the enolate anions of methyl 2-methylpropanoate and methylpropanoate, and provides good yields of the propionic esters **7** in a single step (Scheme 10.20). The latter are known to be intermediates in the preparation of analgesic compounds. No significant difference in the yields is observed when 4-chloroaniline is used as the starting material in place of its dimethylated analogue [31].

The reaction of the aryl cation **6** with Danishefsky's diene gives α-arylated enone **8** in 73% yield (Scheme 10.21) [31].

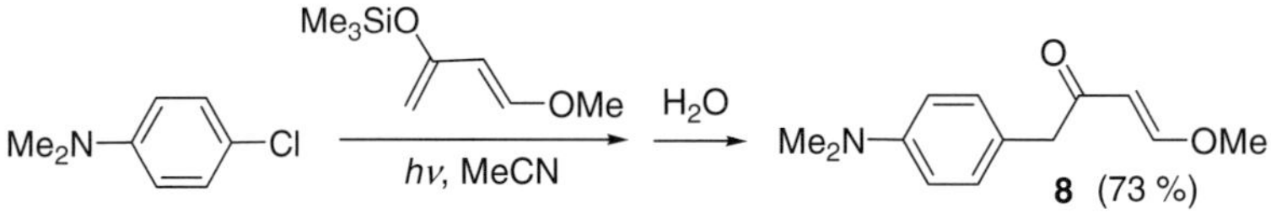

Scheme 10.21

10.3.3
Aryl Alkoxide and Aryl Amide Anions as the Nucleophiles

The photoinduced $S_{RN}1$ reaction of ArX with ArO⁻ ions, mainly 2,6 and 2,4-di-t-butyl phenoxides and 1- and 2-naphthoxide ions, represents another approach to C—C

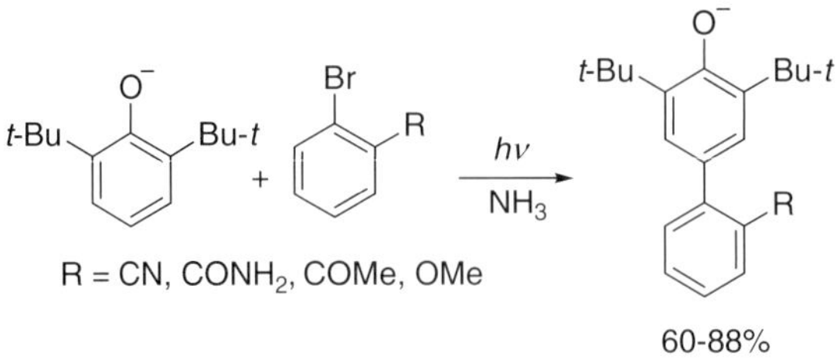

Scheme 10.22

bond formation, as these anions couple with radicals at carbon [1]. As a consequence, this reaction represents a good method for the synthesis of biaryls unsymmetrically substituted by EWGs and EDGs, as well as for cyclic compounds (see Section 10.6.1).

The unsubstituted PhO$^-$ ions afford the *ortho*- ($\approx$40%) and *para*- ($\approx$20%) arylation products. By contrast, the selective synthesis of either the *ortho* or the *para* isomer can be performed when the *t*-Bu groups substitute two of the three possible coupling positions [20]. In the case of the 2,6-disubstituted phenoxide ion, the *para*-substituted compound is obtained (Scheme 10.22), and the *t*-Bu substituents can be easily removed later [20].

In the reaction of 1-naphthoxide ions, a mixture of 2- and 4-aryl-, along with 2,4-diaryl-1-naphthol, is formed. However, substitution occurs only at C$_4$ with the 2-Me-substituted anion (50–70% yields) [1]. On the other hand, 2-naphthoxide ions react with ArX to give substitution only at C$_1$ of the naphthalene ring [32, 33]. The reactivity of the 2-naphthoxide ions allows the synthesis of naphthylpyridines, naphthylquinolines, and naphthylisoquinolines via their coupling reactions with the corresponding halo arenes, in good to excellent yields (50–95%) [33]. The photostimulated reaction between 2-naphthoxide ions and 1-iodo-2-methoxy-naphthalene was explored in liquid ammonia, as a novel approach to the synthesis of [1,1′]binaphthalenyl-2,2′-diol (BINOL) derivatives (Scheme 10.23). This procedure has also been applied to the synthesis of BINOL in moderate yield (40%), which represents the first report of an S$_{RN}$1 reaction in water [34].

The anion of 9-phenanthrol shows a similar behavior in the photostimulated reaction with PhI, and the product from C$_{10}$ arylation is obtained in 53% yield [35].

Aryl amide anions are bidentate nucleophiles allowing C−C bond formation and the synthesis of aminobiaryls. PhNH$^-$ gives low yields of substitution with ArI under irradiation, whereas 2-naphthylamide ions react by a photoinduced S$_{RN}$1 process

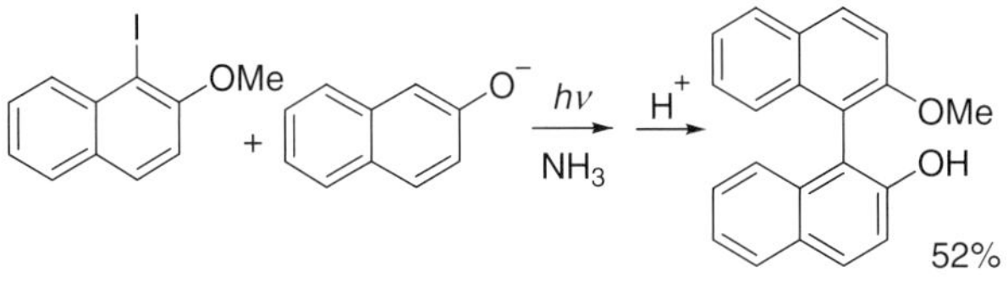

Scheme 10.23

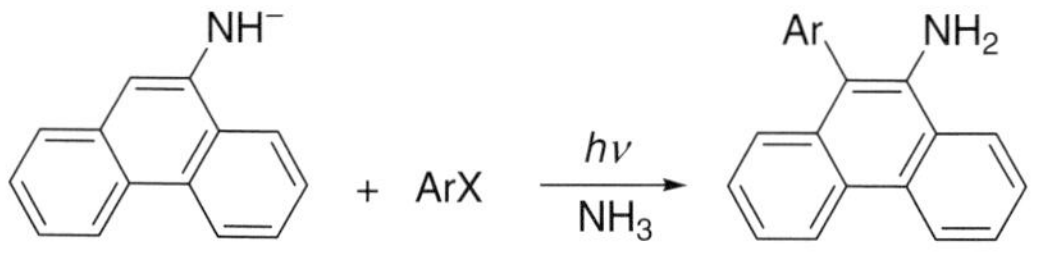

Scheme 10.24

ArX	Yield
PhI	100%
4-MeOC$_6$H$_4$I	81%
4-MeC$_6$H$_4$I	75%
9-Br-phenanthrene	93%

with PhI, 4-MeOC$_6$H$_4$I and 1-iodonaphthalene in liquid ammonia, and 1-aryl-2-napthylamines are formed in 47%, 63%, and 45% yields, respectively [36]. This approach has been used for the synthesis of 10-aryl-9-aminophenanthrene by reaction of the anion of 9-aminophenanthrene with ArX in good yields (Scheme 10.24) [35].

10.3.4
Cyanide Ions as the Nucleophile

An alternative to metal catalysis for the synthesis of aryl cyanides involves the use of light to activate the aromatic-leaving group bond by photoheterolysis, giving aryl cations. Thus, the irradiation of phenyl derivatives bearing O-, N- or S-bonded EDGs with an adequate leaving group such as fluoride, chloride or esters (mesylates, triflates and phosphates) in the presence of KCN gives the corresponding benzonitriles in good to excellent yield (Scheme 10.25) [37]. For these reactions, the mediation of a triplet phenyl cation through an Aryl S_N1 process has been proposed. Direct irradiation is effective for compounds that absorb efficiently (aryl chlorides), whereas sensitization by acetone is required for the aryl esters. Under these conditions up to 70% of the 4-(N,N-dimethylamino)benzonitrile is formed with the dimethylphosphate derivative (Scheme 10.25).

In contrast to the deactivation observed in the metal-catalyzed process, the photochemical reaction tolerates the use of an excess of KCN, a relatively inexpensive cyanating agent, to generally improve the yield.

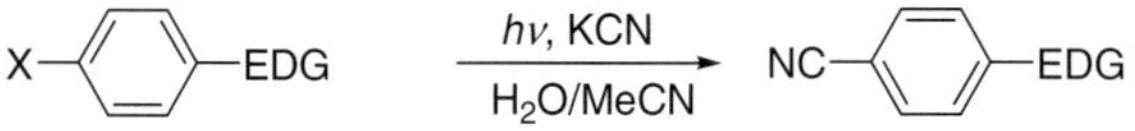

Substrate	Yield
X = F, EDG = OH	69%
X = Cl, EDG = OMe	90%
X = Cl, EDG = NH$_2$	100%
X = Cl, EDG = NMe$_2$	93%
X = OP(O)(OEt)$_2$, EDG = NMe$_2$	70%*
X = Cl, EDG = SMe	57%

*acetone as sensitizer

Scheme 10.25

10.4
Carbon–Heteroatom Bond Formation

10.4.1
Tin Nucleophiles

Me$_3$Sn$^-$ ions are one of the most reactive nucleophiles in the S$_{RN}$1 reaction, affording substitution products in high yield (88–100%) with ArCl in liquid ammonia and under irradiation. Conversely, ArBr and ArI react via an HME pathway [38].

The photostimulated reaction of *o*-, *m*-, and *p*-C$_6$H$_4$Cl$_2$ with Me$_3$Sn$^-$ ions gives disubstitution in 58%, 90%, and 88% yield, respectively [39]; in addition, methyl 2,5-dichlorobenzoate and methyl 3,6-dichloro-2-methoxybenzoate afford bistannylated aromatic compounds in 99% and 64% yields, respectively [40]. Disubstitution products are also obtained by the reaction of 2,6-and 3,5-dichloropyridine under photoinitiation in about 80% yield (Scheme 10.26). Even trisubstitution is possible by the photoinduced reaction of 1,3,5-C$_6$H$_3$Cl$_3$ with Me$_3$Sn$^-$ ions in liquid ammonia, in 71% yield [39].

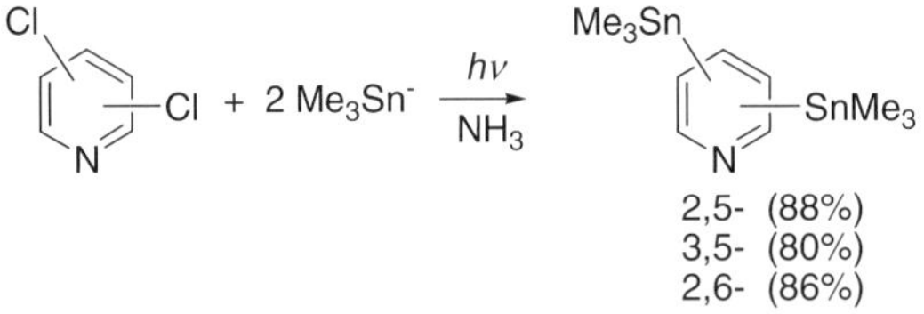

Cl–[pyridine]–Cl + 2 Me$_3$Sn$^-$ $\xrightarrow[\text{NH}_3]{h\nu}$ Me$_3$Sn–[pyridine]–SnMe$_3$

2,5- (88%)
3,5- (80%)
2,6- (86%)

Scheme 10.26

It has also been shown that, upon irradiation, ArCl reacts in diglyme with Me$_3$Sn$^-$ ions according to the S$_{RN}$1 mechanism [41]. This solvent is particularly useful for aryl chlorides that are highly insoluble in liquid ammonia.

Me$_3$SnAr can also be synthesized by the photoinduced reaction of aryldiethylphosphate esters [42] or aryltrimethylammonium salts [43] with Me$_3$Sn$^-$ ions in liquid ammonia, in excellent yields (85–100%). These routes allow for the synthesis of arylstannanes in very good yields from inexpensive and commercially available phenols and anilines, which are easily converted to the corresponding phosphate esters and ammonium salts, respectively.

The synthesis of bistannylated aromatic compounds can be achieved in good yields via a simple two-step sequence from phenols [44]. This methodology enables, for example, the transformation of BINOL into the corresponding aryl diethylphosphate ester, which by reaction with Me$_3$Sn$^-$ ions in liquid ammonia under irradiation affords 2,2′-bis(trimethylstannyl)-1,1′-binaphthyl in an overall yield of 66% (Scheme 10.27).

The photoinduced reaction of the Me$_2$Sn^{2-} dianion with 4-MeOC$_6$H$_4$Cl, 4-MeC$_6$H$_4$Cl, 4-NCC$_6$H$_4$Cl, and 1-ClC$_{10}$H$_7$ gives functionalized Ar$_2$SnMe$_2$ in good yields (68–85%) in liquid ammonia. These stannanes are used in homocoupling reactions in the presence of Cu(NO$_3$)$_2$·2.5H$_2$O to prepare almost quantitatively the corresponding biaryls [45].

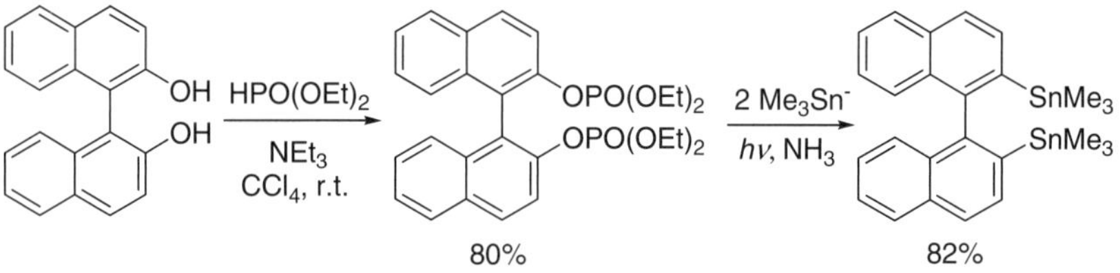

Scheme 10.27

Ar$_2$SnMe$_2$ can also be synthesized in liquid ammonia through a two-step, one-pot procedure. Hence, the treatment of ArSnMe$_3$ with Na metal in liquid ammonia, obtained by a photoinduced S$_{RN}$1 reaction between ArCl and Me$_3$Sn$^-$ ions, yields ArSnMe$_2^-$ ions by a selective Sn–alkyl bond fragmentation. A second consecutive S$_{RN}$1 reaction with the latter anion affords Ar$_2$SnMe$_2$ in 59% isolated yield for Ar = 4-MeOC$_6$H$_4$ [45].

Ph$_3$Sn$^-$ ions are also reactive nucleophiles for the photoinitiated S$_{RN}$1 process. This anion gives substitution products in good yields (62–100%) with ArX (X = Cl, Br, OP(O)(EtO)$_2$) in liquid ammonia [38, 42]. Additionally, the reaction of p-C$_6$H$_4$Cl$_2$ with Ph$_3$Sn$^-$ ions gives 75% yield of disubstitution in liquid ammonia [38].

The organostannyl compounds obtained by S$_{RN}$1 are suitable substrates for cross-coupling reactions catalyzed by Pd(0) to afford biaryl compounds (see Section 10.5.1).

10.4.2
Sulfur Nucleophiles

In contrast to the observed reactivity of phenoxide and aryl alkoxide ions, arene and heteroarene thiolate ions typically couple with aryl radical to generate C–S bonds. The only exception to this regioselective reaction is the addition of 1-naphthalene thiolate ion to p-anisyl radical to render both C- and S-substitutions in 14% and 65% yields, respectively, while with 1-naphthyl radical, 95% of C–S coupling is obtained. In general, PhS$^-$ ions react with ArI in liquid ammonia under photostimulation to afford good yields of ArSPh or heteroaryl-SAr (70–100%). Substitution of the less-reactive ArBr can be achieved under photochemical initiation in DMF, MeCN, or DMSO [1].

The reactivity of sulfur-centered nucleophiles such as thiourea anion [46] and thioacetate anion [17] in photoinduced S$_{RN}$1 reaction has been reported as a one-pot, two-step method for the synthesis of several sulfur-aromatic compounds from moderate to good yields. Without isolation, the ArS$^-$ ions obtained by the aromatic substitution are quenched with MeI to yield ArSMe in a one-pot procedure, together with Ar$_2$S in variable yields, from an S$_{RN}$1 between ArS$^-$ and aryl radicals (Scheme 10.3).

For substrates with EWGs or 2-substituted-ArX, the yields of ArSMe are good (49–87%) (Scheme 10.28). In these reactions, the competitive S$_{RN}$1 process of the ArS$^-$ ions with the aryl radical is retarded since the reactivity of ArS$^-$ ions is diminished [46].

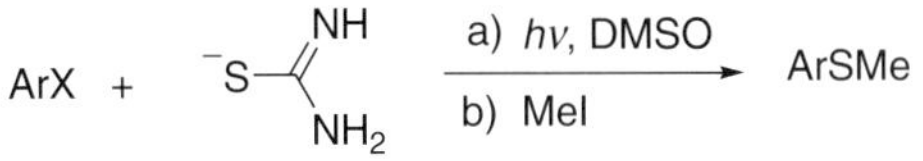

ArX + ⁻S−C(=NH)(NH₂) → a) hv, DMSO b) MeI → ArSMe

4-BrC₆H₄CN	49%
4-IC₆H₄NO₂	77%
4-PhCOC₆H₄Br	81%
2-MeCOC₆H₄Br	64%
3- and 2-Cl Pyridine	58%, 87%
2-Cl Quinoline	64%
2-Cl Pyrazine	83%

Scheme 10.28

By optimizing the reaction conditions, the formation of the Ar_2S can be improved from moderate to good yields (64–83%) [17]. Different chemical transformations are also possible for ArS^- ions formed in the $S_{RN}1$ reaction without isolation, and with this strategy a variety of sulfur compounds have been synthesized (Scheme 10.29) [46].

4-Iodo-1,1,2,2,9,9,10,10-octafluoro[2.2]paracyclophane (**4**) undergoes high-yield nucleophilic substitution with PhS^-, $2\text{-MeC}_6H_4S^-$, $3\text{-MeC}_6H_4S^-$ ions by the photoinduced $S_{RN}1$ mechanism, providing ready access to a new group of paracyclophane derivatives in 88%, 70%, and 91% yields, respectively [24].

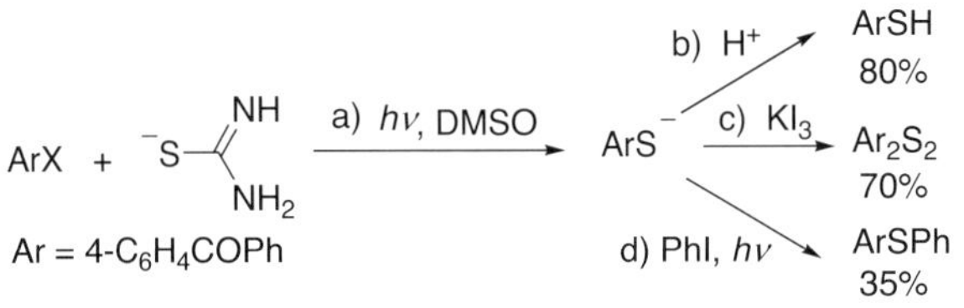

ArX + ⁻S−C(=NH)(NH₂) → a) hv, DMSO → ArS⁻ →
  b) H⁺ → ArSH 80%
  c) KI₃ → Ar₂S₂ 70%
  d) PhI, hv → ArSPh 35%

Ar = 4-C₆H₄COPh

Scheme 10.29

10.5
Synthesis of Bi-, Tri-, and Polyaryls

10.5.1
Consecutive $S_{RN}1$–Pd(0)-Catalyzed Crosscoupling Reactions

Trimethylstannyl- and poly(trimethylstannyl)substituted aryl- and heteroaryl compounds are easily available via $S_{RN}1$ reactions of Me_3Sn^- with appropriate aryl derivatives (see Section 10.4.1) Moreover, these arylstannanes are suitable intermediates for Pd(0)-catalyzed crosscoupling processes with carbon electrophiles (the Stille reaction) leading to C–C bond formation. Thus, a sequence of $S_{RN}1$–Stille reactions has been developed to obtain the polyphenylated compounds shown in Scheme 10.30 [47]. Following the same process, 1,3,5-triphenylbenzene was obtained in 61% isolated yield from 1,3,5-$C_6H_3Cl_3$ in a one-pot method; this was the first report of a trisubstitution by a crosscoupling reaction catalyzed by Pd(0).

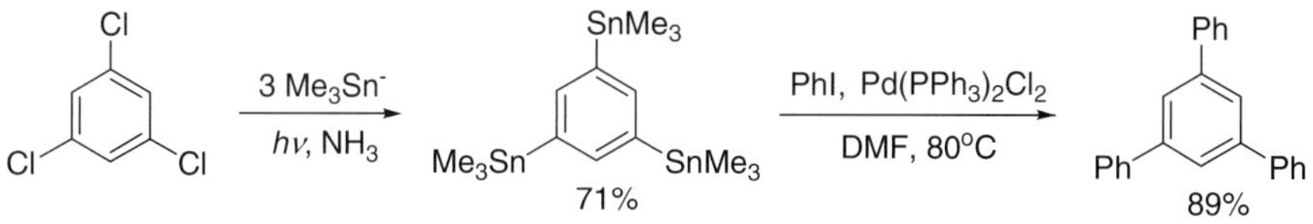

10.5 Synthesis of Bi-, Tri-, and Polyaryls

Scheme 10.30

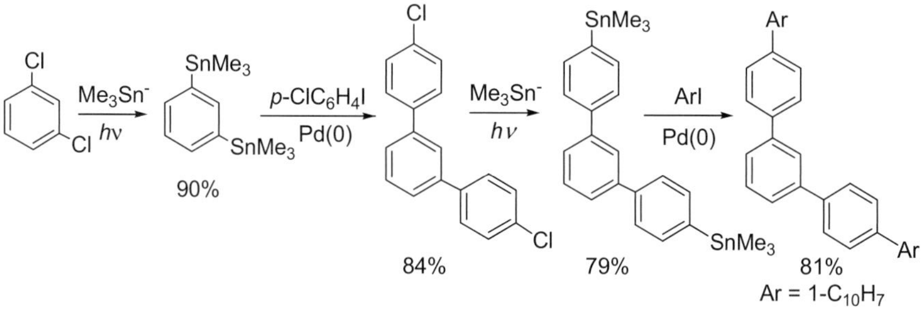

Scheme 10.31

If the palladium-catalyzed reaction is performed with a chloroiodoarene as substrate, the stannane reacts faster by the C—I bond via a chemoselective cross-coupling reaction to give an ArCl as product; this can be further arylated by a consecutive $S_{RN}1$–Stille reaction. These sequential reactions can also be carried out with substrates with two leavings groups to afford products in high yields (Scheme 10.31) [41].

A similar approach, based on the different reactivity of ArBr and ArOP(O)(OEt)$_2$ in Pd(0)-catalyzed crosscoupling reactions, is applied to the synthesis of asymmetric triaryl compounds in acceptable global yields. Mixed biaryl compounds are also obtained in moderate yields through an $S_{RN}1$–Stille sequence (Scheme 10.32). An attractive feature of this methodology is that simple commercial benzenediols, chloro- and methoxy phenols might represent useful starting substrates as their phosphate esters, leading to final products in good overall yields [48].

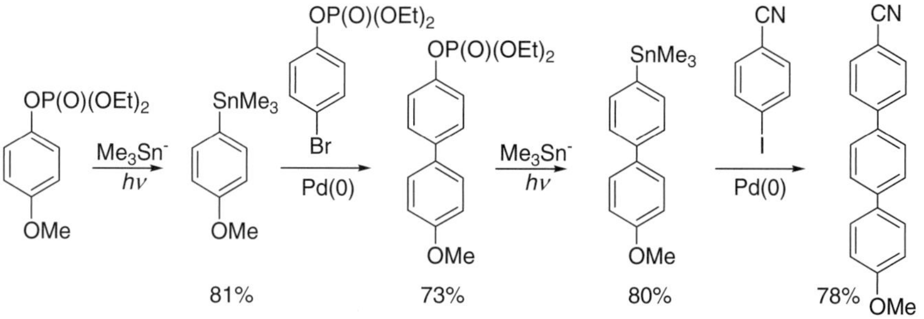

Scheme 10.32

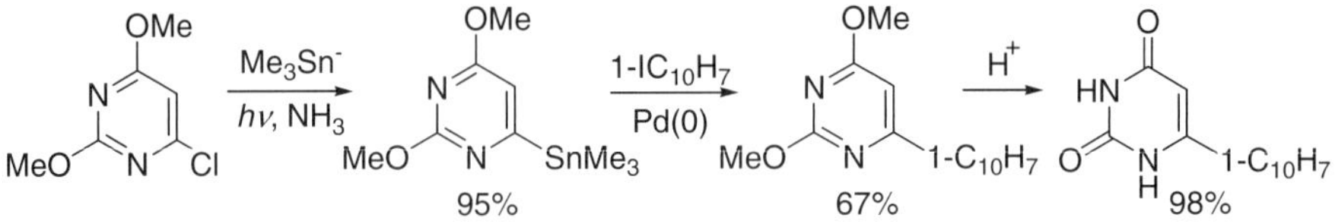

Scheme 10.33

The synthesis of 6-substituted uracils in a three-step reaction ($S_{RN}1$–Stille–hydrolysis) is accomplished in good yields using commercial 6-chloro-2,4-dimethoxypyrimidine as the starting substrate (Scheme 10.33) [49].

When this novel three-step approach was applied to a one-pot reaction, 6-substituted uracils (1-naphthyl, 4-chlorophenyl, 3-chlorophenyl, 2,3,4,5,6-pentafluorophenyl) were formed and isolated (43–57%) as pure products [49].

Arylboronic acids are readily available via transmetalation between $ArSnMe_3$ and borane in THF. Consequently, a variety of arylpolyboronic acids is achievable from the stannanes previously synthesized by the $S_{RN}1$ mechanism from ArCl or ArOH in around 80% yields [50, 51]. These arene di- and tri-boronic acids can be used as starting materials for the synthesis of polycyclic aromatic systems via double or triple Suzuki crosscoupling reactions [50, 51], as shown in Scheme 10.34 [51].

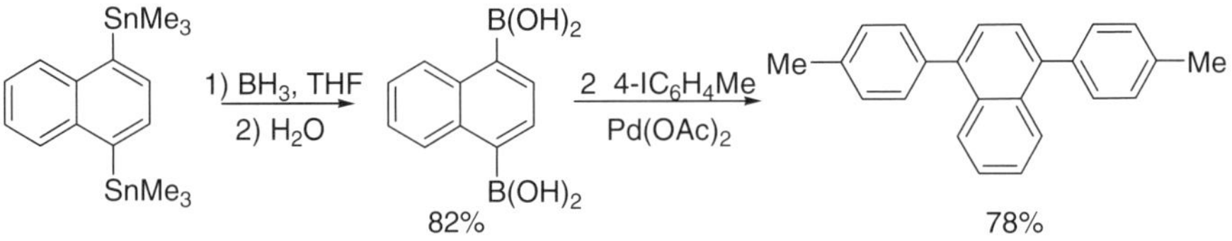

Scheme 10.34

10.5.2
Photo-S_N1 as an Alternative to Metal Catalysis

The combination of $S_{RN}1$ and Stille reactions is, as described, a versatile tool for the preparation of bi, tri, and polyarenes. In general, transition metal-catalyzed reactions involve the coupling of a nucleophilic organometallic derivative ArM (M = SnR_3: Stille reaction; M = MgX: Kumada reaction; M = $B(OH)_2$: Suzuki reaction) with an electrophilic reagent ArX (X = halogen or triflate) under Pd(0) catalysis. In this context, the photo S_N1 process proved to be an environmentally friendly alternative to these reactions, where the light that activates the Ar–X bond acts as a clean reagent in place of the toxic metals. A lower sensitivity to impurities such as residual water, and the use of a less-expensive starting material such as ArCl and aryl esters such as mesylates or phosphates, can be identified as major advantages of these reactions [8]. Through this methodology, the arylation of benzene afforded monosubstituted biphenyls by reaction between the aryl cation in the triplet state with benzene or heteroaromatics (in excess). This reaction occurs with a wide variety of functional groups from EDGs to EWGs; for the former, the direct irradiation of chlorides or ester

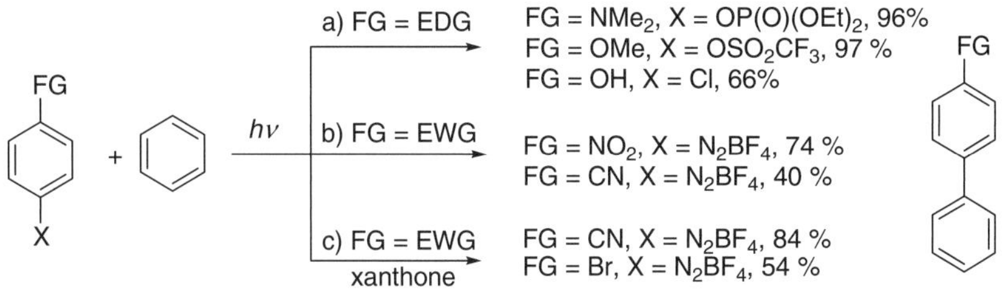

10.5 Synthesis of Bi-, Tri-, and Polyaryls | 337

Scheme 10.35

a) FG = EDG
- FG = NMe$_2$, X = OP(O)(OEt)$_2$, 96%
- FG = OMe, X = OSO$_2$CF$_3$, 97 %
- FG = OH, X = Cl, 66%

b) FG = EWG
- FG = NO$_2$, X = N$_2$BF$_4$, 74 %
- FG = CN, X = N$_2$BF$_4$, 40 %

c) FG = EWG, xanthone
- FG = CN, X = N$_2$BF$_4$, 84 %
- FG = Br, X = N$_2$BF$_4$, 54 %

phosphate derivatives yields biphenyls in almost quantitative yields (Scheme 10.35, pathway a) [26, 27]. When EWGs are present in the molecule, formation of the aryl cation becomes less favorable and a better leaving group is thus required. In order to fulfill this requirement, diazonium salts are employed to yield biphenyls by direct photolysis. Those groups that facilitate ISC generate aryl cations in the triplet state due to the spin conservation along the reaction pathway (Scheme 10.35, pathway b) [13]. Conversely, halide and cyano substituents show a better performance by indirect photolysis when xanthone is used as a triplet energy sensitizer, enhancing the yield of biphenyls (Scheme 10.35, pathway c) [13].

The procedure has been extended to symmetric methylbenzenes, namely p-xylene, 1,3,5-trimethylbenzene (mesitylene) and 1,2,4,5-tetramethylbenzene (durene) [52]. The steric hindrance causes no major limitation to the reaction and a clean monoarylation rather than a polysubstitution is obtained. For instance, a variety of ArX under irradiation in the presence of mesitylene affords the bulky biaryl compounds in satisfactory yields (>50%) (Scheme 10.36) [52].

Cation **6** adds to furan, pyrrole, thiophene and some of their 2,4-dimethyl derivatives bearing a free α-position, in MeCN as solvent. Furan is the least reactive substrate giving 2-arylated product in 32% yield, whereas under the same reaction conditions the sulfur analogue provides 54% of 2-substituted-thiophene **9c** (Scheme 10.37). Pyrroles are the most reactive heterocycles, and 2-arylated products **9a-b** are found with 64% and 75% yields using pyrrole and 2,4-dimethyl-pyrrole, respectively. Only with 2,5-dimethylated furan and pyrrole, will arylation

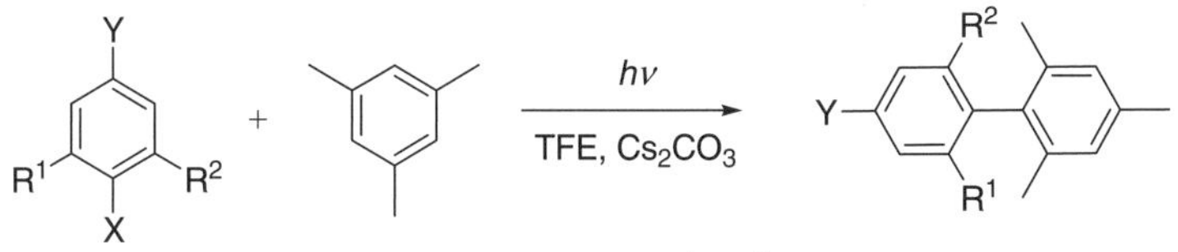

- X = Cl, Y = SMe, R^1 = R^2 = H, 70%
- X = OMs, Y = H, R^1 = OMe, R^2 = H, 77%*
- X = Cl, Y = NH$_2$, R^1 = Me, R^2 = H, 62%
- X = Cl, Y = OMe, R^1 = R^2 = Me, 84%*
- X = Cl, Y = OSiMe$_2$t-Bu, R^1 = R^2 = Me, 77%*

*with acetone as sensitizer

Scheme 10.36

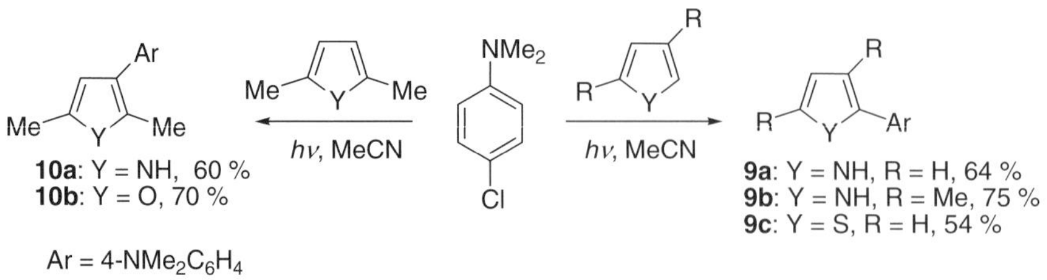
Ar
Me
Y
Me
10a: Y = NH, 60 %
10b: Y = O, 70 %
Me
Y
Me
hν, MeCN
NMe2
Cl
R
R
Y
hν, MeCN
R
R
Y
Ar
9a: Y = NH, R = H, 64 %
9b: Y = NH, R = Me, 75 %
9c: Y = S, R = H, 54 %
Ar = 4-NMe2C6H4

Scheme 10.37

occur with equal efficiency in the β-position, giving the corresponding heterocycle **10** (Scheme 10.37) [53].

Accordingly, whilst five-membered heterocycles exclusively give arylation in α-position toward aryl cations, the β-position is only arylated in 2,5-dimethyl derivatives. Such selectivity for attacking in the α-position is greater than in other electrophilic substitutions, which indicates that triplet phenyl cations are not localized cations and therefore are more selective electrophiles [53].

10.6
Synthesis of Carbocycles and Heterocycles

One of the most important achievements of the $S_{RN}1$ mechanism relates to its synthetic capability to obtain aromatic cyclic compounds in good yields. For this purpose, different approaches are used:

- $S_{RN}1$ substitution followed by a polar ring closure reaction with an *ortho* substituent to the leaving group.
- $S_{RN}1$ substitution with substrates bearing two leaving groups and bidentate or monodentate nucleophiles, followed by a polar or catalyzed reactions.
- Intramolecular $S_{RN}1$ reactions.
- $S_{RN}1$ substitution where a radical cyclization takes place in the propagation step (tandem cyclization-$S_{RN}1$ sequence and *vice versa*).

A cyclization reaction by the S_N1 mechanism is also possible either by the intramolecular addition of the aryl cation onto a tethered double bond, or by an intermolecular tandem formation of an Ar—C and a C—O bond with the concomitant formation of cyclic products.

10.6.1
Carbocycles

The photoinduced reaction of 1,8-diiodonaphthalene with anions $^{-}CH_2COBu$-*t* and $^{-}CH_2COMe$ in liquid ammonia affords substituted dihydrophenalenes **11** from a spontaneous and efficient aldol condensation of the disubstituted product **12** (Scheme 10.38) [54].

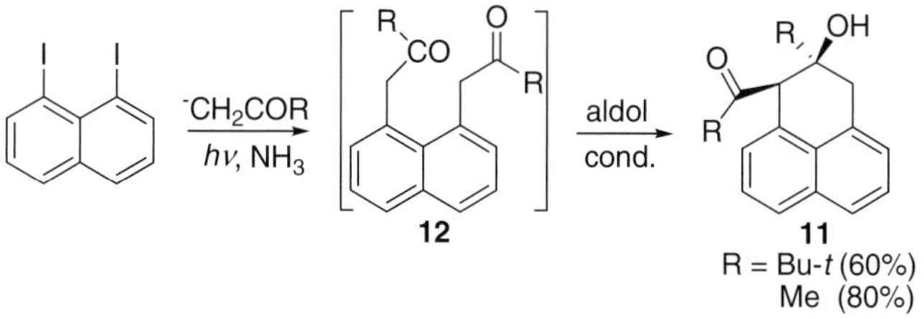

10.6 Synthesis of Carbocycles and Heterocycles | 339

Scheme 10.38

The use of ω-(2-bromophenyl)alkyl-2-oxazolines (**13**) in base-promoted ring-closure reactions provides a good access to 1-phenyl-indane and 1-phenyl-tetralin derivatives (**14**) that contain easily manipulated oxazoline moieties. Satisfactory isolated yields of cyclized products **14** are obtained with lithium diisopropylamide (LDA) in THF at room temperature under irradiation with UV light, even when *quaternary* centers are formed (Scheme 10.39) [55]. The yield of **14** ($n = 1$, R = Ph) increases to 75% when the reaction proceeds under laboratory light for 48 h at room temperature.

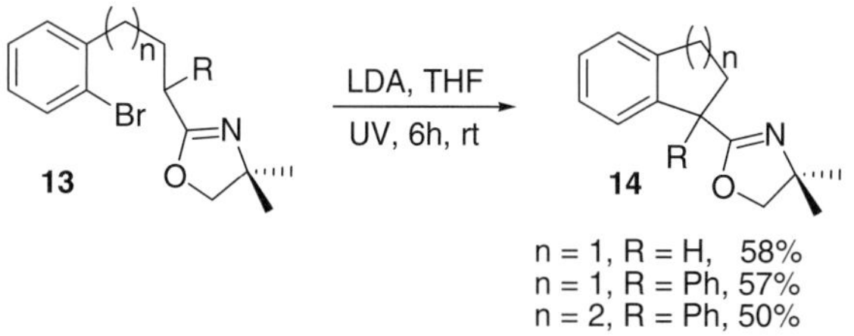

$n = 1$, R = H, 58%
$n = 1$, R = Ph, 57%
$n = 2$, R = Ph, 50%

Scheme 10.39

The diastereoselective ring-closure reaction of 2-bromophenylpropyl precursors containing chiral oxazolines yields diastereoisomeric indanes. However, the maximum selectivity achieved with a 4-isopropyl substituent in the oxazoline (48% d.r.) is modest [55].

The effect of chain length on the radical cyclization reaction of bromoaryl alkyl-linked oxazolines has also been examined (Scheme 10.40). The major products from the precursor with a 5-C atom spacer are derivatives of benzocycloheptane **15** in which the oxazoline group has undergone a novel areneotropic migration from the end of the spacer to the benzo ring. The corresponding 2-C-atom precursor affords a 9-oxazolinophenanthrene derivative **16** by aromatization in the reaction medium of

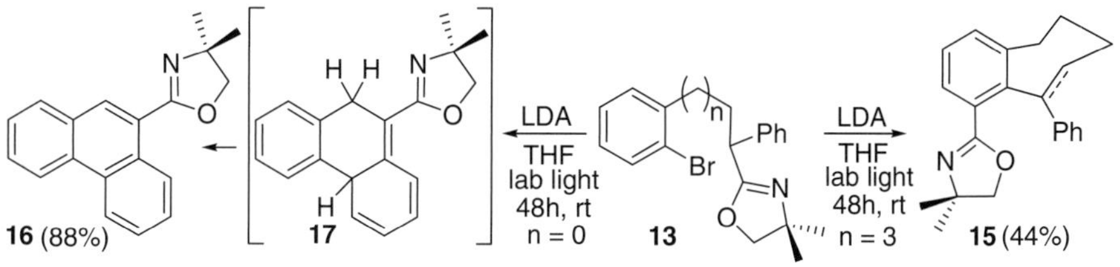

Scheme 10.40

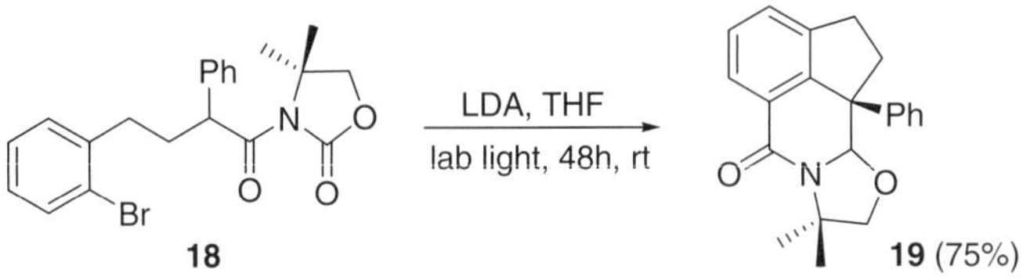

Scheme 10.41

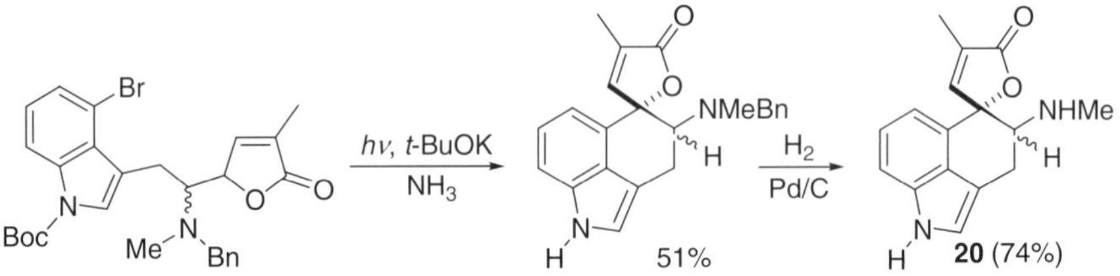

Scheme 10.42

product **17** formed by coupling with a resonance structure of the azaenolate on the phenyl ring [56].

The analogous precursor **18** containing an oxazolidinone ring reacts upon treatment with LDA in THF under laboratory light. Under these conditions, a new fused tetracyclic cyclopentaoxazoloisoquinolinone system **19** is assembled in a multistage rearrangement. The initial steps involve a ring closure reaction to form efficiently a *quaternary* center (Scheme 10.41) [57].

The $S_{RN}1$ mechanism represents an excellent alternative to accomplishing the following intramolecular cyclization reaction (Scheme 10.42), which constitutes one of the steps in the synthesis of rugulovasines A and B (**20**, α-H and β-H), novel structures within the family of ergot alkaloids [58].

The aporphine skeleton **21** can be synthesized through the $S_{RN}1$ mechanism by exploiting the bidentate behavior of phenoxide ions that allows one to obtain the *ortho*-arylation in good yields, as shown in Scheme 10.43 [59]. Tetrahydroisoquinoline precursors that contained nitrogen-based EWGs (i.e., amides, sulfonamides,

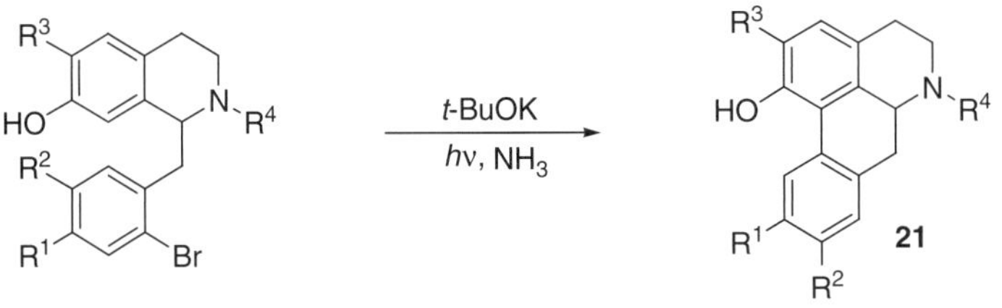

$R^1 = R^2 = H$, $R^3 = OMe$, $R^4 = CO_2Me$, COMe   75%, 82%
$R^1 = F$, $R^2 = H$, $R^3 = OMe$, $R^4 = CO_2Me$ 50%
$R^1 = R^2 = R^3 = OMe$, $R^4 = CO_2Me$ 75%
$R^1 = R^2 = R^3 = H$, $R^4 = SO_2C_6H_4Me$ 54%

Scheme 10.43

carbamates) gave cyclized products, whereas precursors containing basic nitrogens groups (i.e., NH or NMe) either failed to yield cyclized products or rendered aporphine only in low yield. This was most likely due to a hydrogen shift from the N–H or the N–Me groups to the aryl radical intermediates [59]. Finally, this strategy has been extended for the first time to the synthesis of a homoaporphine derivative (40% yield) [59].

10.6.2
Nitrogen Heterocycles

The synthesis of indoles by the photoinduced substitution reaction of o-haloanilines (22) with carbanions derived from aliphatic ketones in liquid ammonia is an important example of the $S_{RN}1$ reaction followed by a spontaneous ring closure in the reaction media in moderate to excellent yield (53%–100%) [1]. Although the enolate anions of aromatic ketones do not react in liquid ammonia with 22, they will cyclize to indoles in DMSO under photoinitiation (Scheme 10.44) [60, 61].

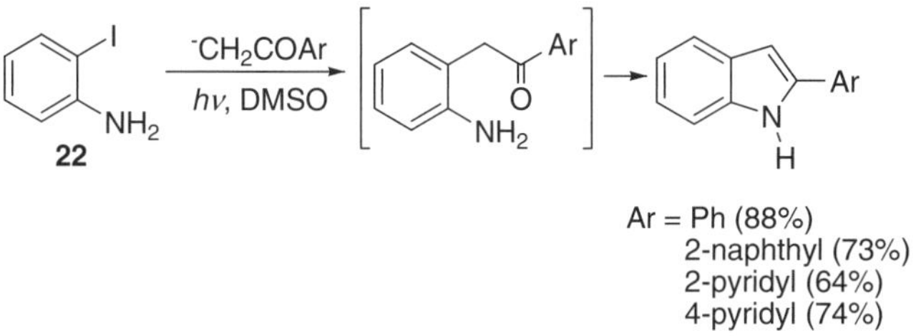

Ar = Ph (88%)
2-naphthyl (73%)
2-pyridyl (64%)
4-pyridyl (74%)

Scheme 10.44

Both, benzo[e]indole and benzo[g]indole can be obtained by the reaction of 1-bromo-2-naphthylamine or 2-bromo-1-naphthylamine with $^-$CH$_2$COPh in DMSO [61], and $^-$CH$_2$COBu-t and $^-$CH$_2$COCH(OMe)$_2$ anions in liquid ammonia in variable yields (46–82%) [62]. In addition to this, the photostimulated reaction of vic aminohalo pyridines with the anion of $^-$CH$_2$COBu-t or $^-$CH$_2$COMe ions leads to azaindoles in high yields (75–95%) [63].

Examples of the syntheses of fused indoles by the $S_{RN}1$ mechanism are depicted in Scheme 10.45. The photoinduced reactions of 22 with the enolate anions of cyclic ketones afford the corresponding indoles from moderate to good yields [61].

The preparation of methoxy carbazoles is also achieved by this procedure. Thus, the photostimulated reaction of 22 with enolate anion 23 in DMSO gives methoxy-5,11-dihydro-6H-benzo[a]carbazoles (24) in 42–61% yields (Scheme 10.46) [64].

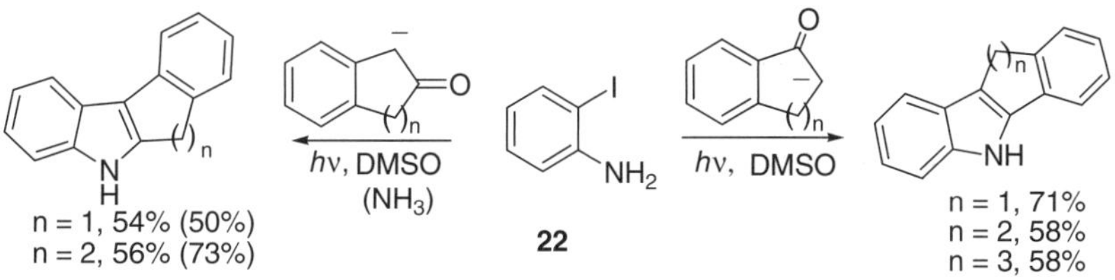

n = 1, 54% (50%)
n = 2, 56% (73%)

n = 1, 71%
n = 2, 58%
n = 3, 58%

Scheme 10.45

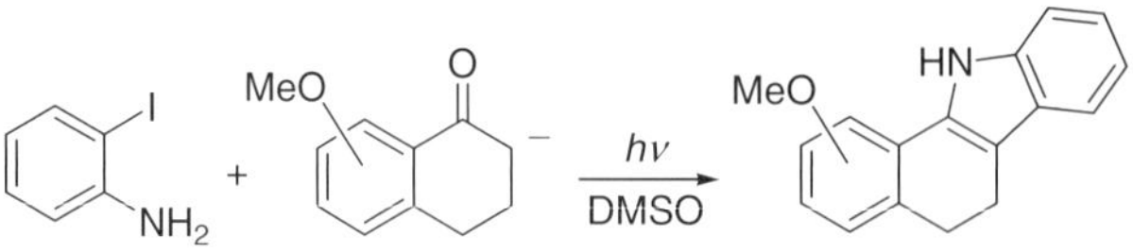

Scheme 10.46

Scheme 10.47

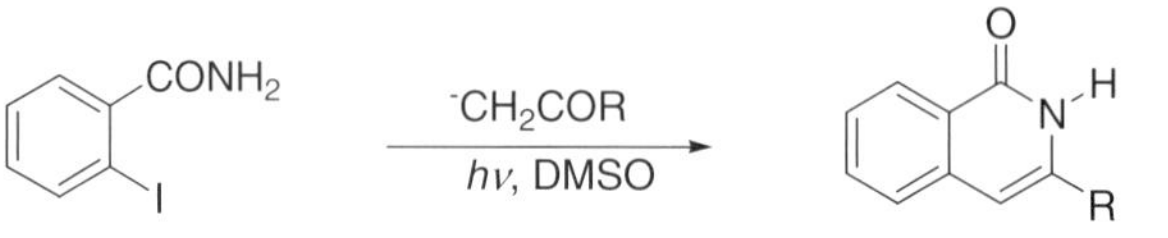

R = Ph — 91%
benzo[d][1,3]dioxol-6-yl — 87%
1-, 2-naphthyl — 83%, 68%
2-, 3-, 4-pyridyl — 81%, 79%, 72%
1-adamantyl — 71%

The synthesis of 3-substituted isoquinolinones and fused isoquinolinones can be performed by the photoinduced $S_{RN}1$ reactions in DMSO of o-iodobenzamide with the enolates of acyclic aromatic and aliphatic ketones and cyclic ketones, respectively. These reactions proceed from moderate to good yields (Schemes 10.47 and 10.48) [65].

The photostimulated reactions with the enolate anions of 1-indanone and α-tetralone lead to the reduction product benzamide (23% and 44%, respectively), together with the target fused isoquinolinones (51% and 42%, respectively) [65]. The formation of benzamide can probably be ascribed to the hydrogen atom abstraction from the cyclic ketone.

The intramolecular ortho-arylation of aryl amide ions with aryl iodides has been recently applied to the synthesis of phenanthridines **25** in excellent yields (Scheme 10.49) [66]. By applying the same procedure, N-(2-iodobenzyl) naphthalen-2-amine and N-(2-chlorobenzyl) naphthalen-1-amine yield benzo[a]- and benzo[c] phenanthridines in 98% and 84% yields, respectively [66].

3-Substituted 2,3-dihydro-1-H-indoles **26** are accessible by the versatile application of a 5-exo ring closure process during the propagation cycle in the $S_{RN}1$ reaction [67]. Following the initial ET and fragmentation of the radical anion of the substrate, the

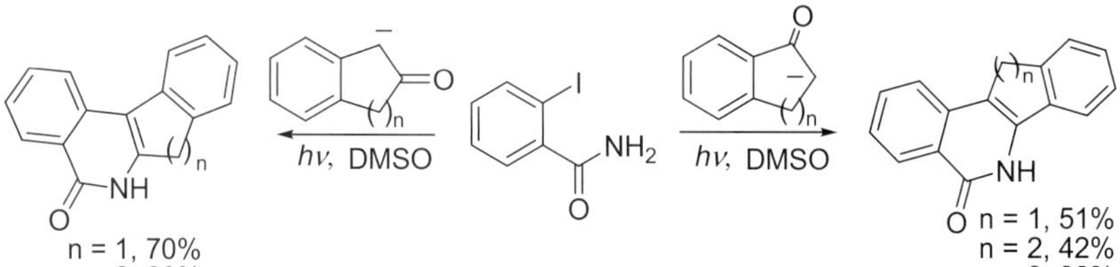

n = 1, 70%
n = 2, 80%

n = 1, 51%
n = 2, 42%
n = 3, 88%

Scheme 10.48

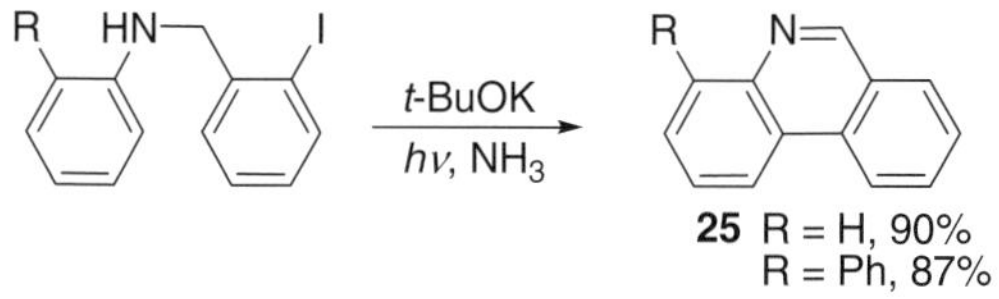

10.6 Synthesis of Carbocycles and Heterocycles

Scheme 10.49

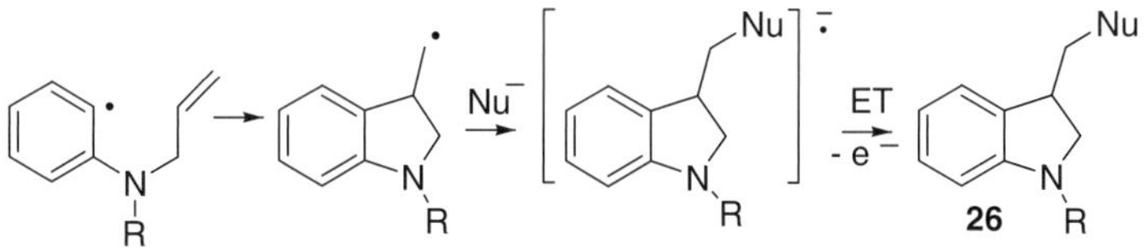

Scheme 10.50

aryl radical intermediate formed cyclizes in a 5-*exo*-trig fashion to yield a new carbon radical. This alkyl radical is able to couple with the nucleophiles, thus continuing propagation cycle of the $S_{RN}1$ (Scheme 10.50).

Hence, N,N-diallyl-(2-chlorophenyl)-amine and N-allyl-N-(2-chlorophenyl)-acetamide react with Me_3Sn^- and Ph_2P^- ions in liquid ammonia under photostimulation to afford cyclized-substituted compounds in good yield (Scheme 10.51) [67].

Nitromethane anion gives the corresponding N-allyl-3-(2-nitro-ethyl)- 2,3-dihydro-1-*H*-indole in 60% yield after photostimulated reaction with the bromide analogue, and in the presence of the enolate ion and acetone as an *entrainment* reagent (Scheme 10.51) [67].

The key step in the synthesis of the azaphenanthrene alkaloid eupoulauramine **27** (56% yield) is an intramolecular $S_{RN}1$ reaction, followed by *in situ* stilbene photocyclization, and further methylation (Scheme 10.52) [68].

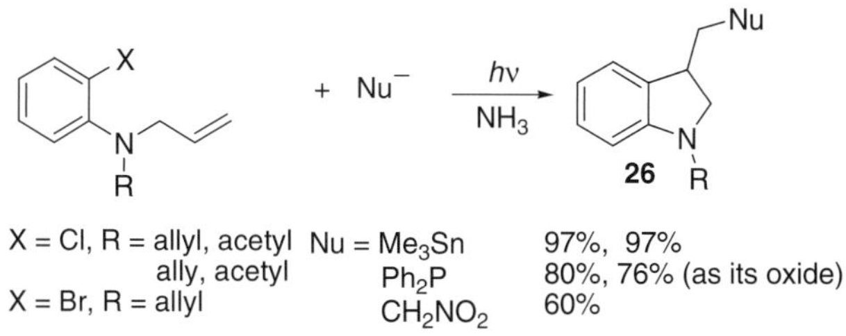

X = Cl, R = allyl, acetyl   Nu = Me$_3$Sn   97%, 97%
            ally, acetyl   Ph$_2$P   80%, 76% (as its oxide)
X = Br, R = allyl   CH$_2$NO$_2$ 60%

Scheme 10.51

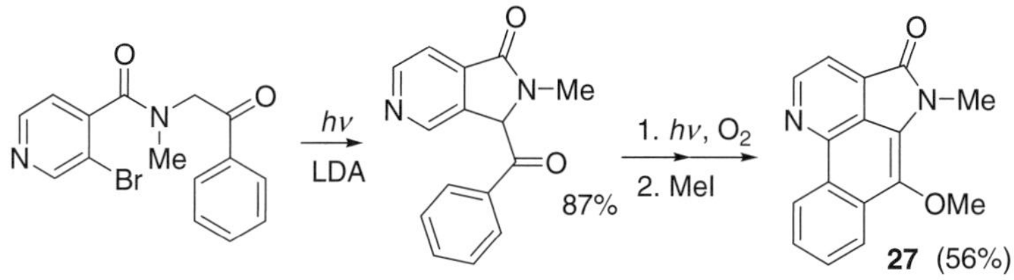

Scheme 10.52

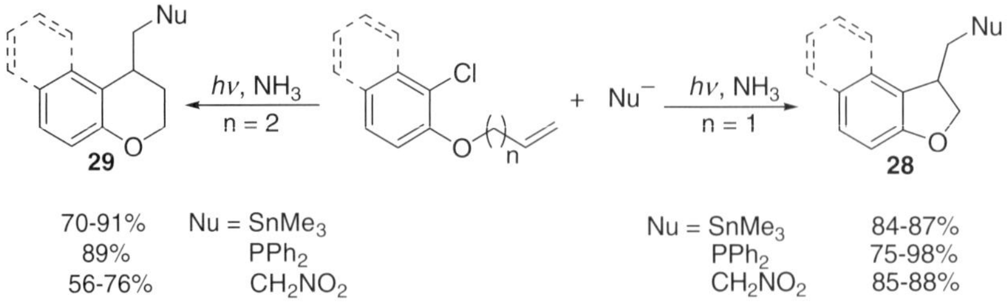

Scheme 10.53

Yields for compound **29**: 70–91% (Nu = SnMe₃), 89% (PPh₂), 56–76% (CH₂NO₂)
Yields for compound **28**: 84–87% (Nu = SnMe₃), 75–98% (PPh₂), 85–88% (CH₂NO₂)

10.6.3
Oxygen Heterocycles

The tandem 5-*exo* [67], and 6-*exo*-trig cyclization-S$_{RN}$1 [69] reactions afford 3-substituted dihydrobenzofuranes and dihydronaphtho[2,1-*b*]furanes (**28**) for the former, and 4-substituted chromanes and benzo[*f*]chromanes (**29**) for the latter. In these reactions, at least one C–C bond is formed and the products obtained are feasible for further synthetic transformation (Scheme 10.53). Another approach to O-heterocycles is provided via S$_{RN}$1 substitutions followed by a copper-catalyzed reaction (see Scheme 10.8).

The synthesis of a variety of dihydrobenzofuranes can be accomplished via the Aryl S$_N$1 mechanism. The acetone-sensitized irradiation of various *o*-chlorophenyl allyl ethers in polar solvents affords the aryl cation, which adds onto the tethered double bond following the 5-*exo*-trig mode leading to (dihydro)benzofurans. The reaction is solvent-dependent, and byproducts are also generally obtained. The best results are found when ethyl acetate is used. Depending on the substrate employed, the cation intermediate can yield products as a result of proton elimination or trapping by chloride ions (Scheme 10.54) [70].

The photo S$_N$1 reactions allow the formation of cyclic products by the consecutive formation of an Ar–C and C–O bonds. The process is based on the intermolecular phenyl cation addition onto olefins bearing O-nucleophiles in the tethered chain. As a consequence, the addition of a phenyl cation to a ω-alkenoic acid, followed by an intramolecular attack by the carboxylic group, forms regioselectively 5-benzyl-γ-lactones. ArCl gives the best yields and, surprisingly, the reaction also occurs with chloro aromatics bearing poor EDGs such as the *n*-butyl group (Scheme 10.55).

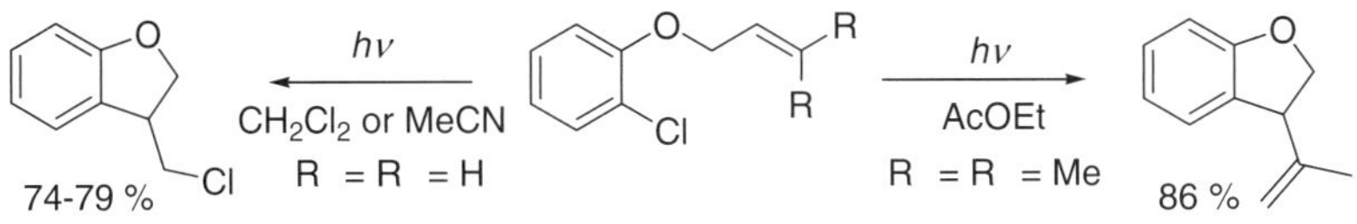

Scheme 10.54

Left: 74–79% (CH₂Cl₂ or MeCN, R = R = H)
Right: 86% (AcOEt, R = R = Me)

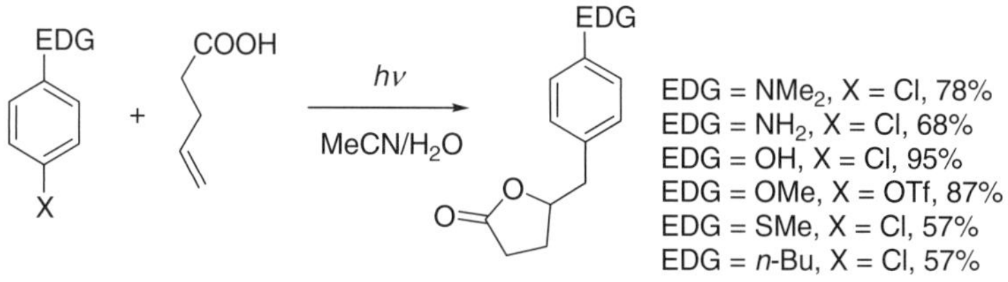

10.6 Synthesis of Carbocycles and Heterocycles

Scheme 10.55

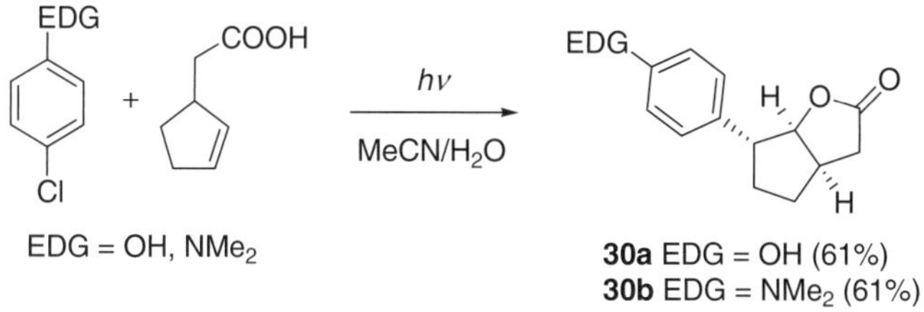

EDG = NMe$_2$, X = Cl, 78%
EDG = NH$_2$, X = Cl, 68%
EDG = OH, X = Cl, 95%
EDG = OMe, X = OTf, 87%
EDG = SMe, X = Cl, 57%
EDG = n-Bu, X = Cl, 57%

Scheme 10.56

EDG = OH, NMe$_2$

30a EDG = OH (61%)
30b EDG = NMe$_2$ (61%)

Furthermore, when a cyclic alkenoic acid is used (e.g., 2-cyclopentenacetic acid), the reaction is fully regioselective and stereoselective, affording hexahydrocyclopentan[b]furan-2-one **30** (Scheme 10.56) [71].

The reaction between phenyl cations and alkenols represents another example of a one-pot, tandem synthesis of cyclic ethers. Accordingly, 2-benzyltetrahydrofurans are obtained by the addition of photogenerated phenyl cations to γ-hydroxyalkenes. For example, 4-penten-1-ol (a terminal alkene) reacts with phenyl chlorides bearing EDGs under irradiation in polar protic solvents, such as TFE or a mixture MeCN/H$_2$O (Scheme 10.57) [72].

With the same methodology, the preparation of 3-phenyl-substituted THFs can be assessed by irradiating ArCl in the presence of a nonterminal alkene. Hence, the photogenerated phenyl cations react with the (E)-3-hexen-1-ol, and this results in the stereoselective formation of *trans*-2-ethyl-3-aryltetrahydrofuran derivative in good to very good yields (52–88%) [72].

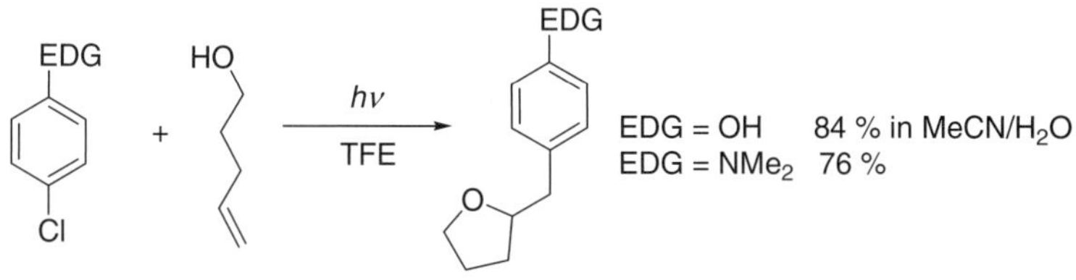

Scheme 10.57

EDG = OH 84 % in MeCN/H$_2$O
EDG = NMe$_2$ 76 %

10.6.4
Sulfur Heterocycles

The photostimulated reaction of 1,8-diiodonaphthalene with p-methyl-benzenethiolate ions in DMSO yields the substituted cyclized product 10-methyl-7-thia-benzo[de]anthracene (**31**) in moderate yield (Scheme 10.58) [54]. The mechanism proposed to explain product **31** involves an intramolecular radical cyclization after monosubstitution in the propagation cycle of the $S_{RN}1$ process.

2-Iodobenzenesulfonamide (**32**) undergoes photostimulated $S_{RN}1$ reactions in liquid ammonia, with the potassium enolates derived from acetone, pinacolone, 3-methyl-2-butanone, cyclopentanone, cyclohexanone and cyclooctanone, to give fair to good yields of 2H-1,2-benzothiazine-1,1-dioxides (Scheme 10.59) [73].

The reductive dehalogenation predominates in the photoinduced reactions of **32** with 3-pentanone, 2-methyl-3-pentanone, and 2,4-dimethyl-3-pentanone. Although, substitution is favored over reduction in all of the reactions of **32** with cyclic ketone enolates, a competing reduction is responsible for the lower yields observed with cyclopentanone and cyclohexanone compared to cyclooctanone (Scheme 10.59) [73].

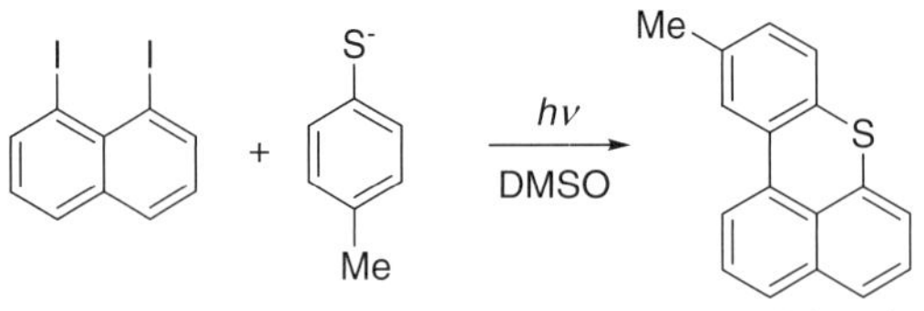

Scheme 10.58

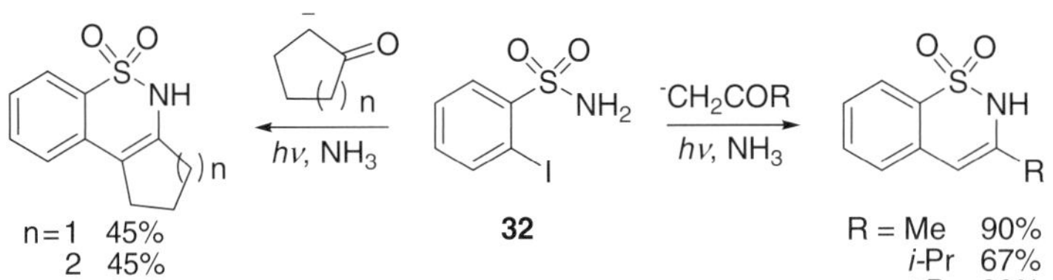

Scheme 10.59

References

1 Rossi, R.A., Pierini, A.B. and Peñéñory, A.B. (2003) Nucleophilic substitution reactions by electron transfer. *Chemical Review*, **103**, 71–167.

2 Rossi, R.A. and Peñéñory, A.B. (2003) The photostimulated $S_{RN}1$ process: reaction of haloarenes with carbanions, in *CRC Handbook of Organic Photochemistry and Photobiology*, 2nd edn (eds W.M. Horspool and F. Lenci), CRC Press Inc., Boca Raton, pp. 47-1–47-24.

3 Rossi, R.A., Pierini, A.B. and Santiago, A.N. (1999) Aromatic substitution by the $S_{RN}1$ reaction, in *Organic Reactions*,

Vol. 54 (eds L.A. Paquette and R. Bittman), John Wiley & Sons, Ltd, Chichester, pp. 1–271.

4 Rossi, R.A. and Peñéñory, A.B. (2006) Strategies in synthetic radical organic chemistry. recent advances on cyclization and S$_{RN}$1 reactions. *Current Organic Synthesis*, **3**, 121–158.

5 Rossi, R.A. (2005) Chapter 16 in *Photoinduced Aromatic Nucleophilic Substitution Reactions. Synthetic Organic Photochemistry* (eds A.G. Griesbeck and J. Mattay), Marcel Dekker, New York, pp. 495–527.

6 Dichiarante, V. and Fagnoni, M. (2008) Aryl cation chemistry as an emerging versatile tool for metal-free arylations. *Synlett*, (6), 787–800.

7 Fagnoni, M. (2006) The photo S$_N$1 reaction Via phenyl cation. *Letters in Organic Chemistry*, **3**, 253–259.

8 Fagnoni, M. and Albini, A. (2005) Arylation reactions: the photo-S$_N$1 path via phenyl cation as sn alternative to metal catalysis. *Account of Chemical Research*, **38**, 713–721.

9 Schmidt, L.C., Argüello, J.E. and Peñéñory, A.B. (2007) Nature of the chain propagation in the photostimulated reaction of 1-bromonaphthalene with sulfur centered nucleophiles. *Journal of Organic Chemistry*, **72**, 2936–2944.

10 Argüello, J.E., Peñéñory, A.B. and Rossi, R.A. (2000) Quantum yields of the initiation step and chain propagation turnovers in S$_{RN}$1 reactions: photostimulated reaction of 1-iodo-2-methyl-2-phenyl propane with carbanions in DMSO. *Journal of Organic Chemistry*, **65**, 7175–7182.

11 Lazzaroni, S., Dondi, D., Fagnoni, M. and Albini, A. (2008) Geometry and energy of substituted phenyl cations. *Journal of Organic Chemistry*, **73**, 206–211.

12 Mella, M., Coppo, P., Guizzardi, B., Fagnoni, M., Freccero, M. and Albini, M. (2001) Photoinduced, ionic meerwein arylation of olefins. *Journal of Organic Chemistry*, **66**, 6344–6352.

13 Milanesi, S., Fagnoni, M. and Albini, A. (2005) (Sensitized) photolysis of diazonium salts as a mild general method for the generation of aryl cations. chemoselectivity of the singlet and triplet 4-substituted phenyl cations. *Journal of Organic Chemistry*, **70**, 603–610.

14 Manet, I., Monti, S., Grabner, G., Protti, S., Dondi, D., Dichiarante, V., Fagnoni, M. and Albini, A. (2008) Revealing phenylium, phenonium, vinylenephenonium, and benzenium ions in solution. *Chemistry – A European Journal*, **14**, 1029–1039.

15 Ferrayoli, C.G., Palacios, S.M. and Alonso, R.A. (1995) Alternative synthesis of 2-Aryl propanoic acids from enolate and aryl halides. *Journal of the Chemical Society, Perkin Transactions* 1, 1635–1638.

16 Wu, B.Q., Zeng, F.W., Zhao, Y. and Wu, G.S. (1992) Stereochemistry of the coupling step in Photo-S$_{RN}$1 reaction. *Chinese Journal of Chemistry*, **10**, 253–261.

17 Schmidt, L.C., Rey, V. and Peñéñory, A.B. (2006) Photoinduced nucleophilic substitution of aryl halides with potassium thioacetate. A one-pot approach to aryl methyl and diaryl sulfides. *European Journal of Organic Chemistry*, 2210–2214.

18 Baumgartner, M.T., Pierini, A.B. and Rossi, R.A. (1999) Reactions of 2-and 3-acetyl-1-methylpyrrole enolate ions with iodoarenes and neopentyl iodides by the S$_{RN}$1 mechanism. *Journal of Organic Chemistry*, **64**, 6487–6489.

19 Baumgartner, M.T., Jiménez, L.B., Pierini, A.B. and Rossi, R.A. (2002) Reactions of o-iodohalobenzenes with carbanions of aromatic ketones. Synthesis of 1-aryl-2-(o-halophenyl)ethanones. *Journal of the Chemical Society, Perkin Transactions* 2, 1092–1097.

20 Beugelmans, R. and Chastanet, J. (1993) S$_{RN}$1 reactions of chlorotrifluoromethyl pyridines with naphtholate. Phenolate and

malonate anions. *Tetrahedron*, **49**, 7883–7890.
21 Nwokogu, G.C., Wong, J.-W., Greenwood, T.D. and Wolfe, J.F. (2000) Photostimulated reactions of phenylacetic acid dianions with aryl halides. Influence of the metallic cation on the regiochemistry of arylation. *Organic Letters*, **2**, 2643–2646.
22 Palacios, S.M., Asis, S.E. and Rossi, R.A. (1993) Synthesis of N,N-dimethyl α-aryl and α,α-diarylacetamides by radical nucleophilic substitution reactions. *Bulletin de la Societe Chimique de France*, **130**, 111–116.
23 Baumgartner, M.T., Lotz, G.A. and Palacios, S.M. (2004) Diastereoselective C-arylation of prochiral enolates by the $S_{RN}1$ recation. *Chirality*, **16**, 212–219.
24 Wu, K., Dolbier, W.R. Jr., Battiste, M.A. and Zhai, Y. (2006) The $S_{RN}1$ chemistry of 4-Iodo-1,1,2,2,9,9,10,10-octafluoro[2.2] paracyclophane. *Mendeleev Communication*, **16** (3), 146–147.
25 Wong, J.W., Natalie, K.J., Nwokogu, G.C., Pisipati, J.S., Flaherty, P.T., Greenwood, T.D. and Wolfe, J.F. (1997) Compatibility of various carbanion nucleophiles with heteroaromatic nucleophilic substitution by the $S_{RN}1$ mechanism. *Journal of Organic Chemistry*, **62**, 6152–6159.
26 Protti, S., Fagnoni, M., Mella, M. and Albini, A. (2004) Aryl cations from aromatic halides. photogeneration and reactivity of 4-hydroxy(methoxy)phenyl cation. *Journal of Organic Chemistry*, **69**, 3465–3473.
27 De Carolis, M., Protti, S., Fagnoni, M. and Albini, A. (2005) Metal-free cross-coupling reactions of aryl sulfonates and phosphates through photoheterolysis of aryl–oxygen bonds. *Angewandte Chemie, International Edition*, **44**, 1232–1236.
28 Protti, S., Fagnoni, M. and Albini, A. (2005) Expeditious synthesis of bioactive allylphenol constituents of the genus *Piper* through a metal-free photoallylation procedure. *Organic Biomolecular Chemistry*, **3**, 2868–2871.
29 Protti, S., Fagnoni, M. and Albini, A. (2005) Photo-cross-coupling reaction of electron-rich aryl chlorides and aryl esters with alkynes: a metal-free alkynylation. *Angewandte Chemie, International Edition*, **44**, 5675–5678.
30 Guizzardi, B., Mella, M., Fagnoni, M. and Albini, A. (2003) Photochemical reaction of N,N-dimethyl-4-chloroaniline with dienes: new synthetic paths via a phenyl cation. *Chemistry – A European Journal*, **9**, 1549–1555.
31 Fraboni, A., Fagnoni, M. and Albini, A. (2003) A novel α-arylation of ketones, aldehydes, and esters via a photoinduced S_N1 reaction through 4-aminophenyl cations. *Journal of Organic Chemistry*, **68**, 4886–4893.
32 Argüello, J.E. and Peñéñory, A.B. (2003) Fluorescent quenching of 2-naphthoxide anion by aliphatic and aromatic halides. mechanism and consequences of electron transfer reactions. *Journal of Organic Chemistry*, **68**, 2362–2368.
33 Beugelmans, R. and Bois-Choussy, M. (1991) $S_{RN}1$-based methodology for synthesis of naphthylquinolines and naphthylisoquinolines. *Journal of Organic Chemistry*, **56**, 2518–2522.
34 Baumgartner, M.T., Tempesti, T.C. and Pierini, A.B. (2003) Steric effects in the synthesis of ortho-substituted 1,1′-binaphthalene derivatives by the $S_{RN}1$ and the Stille reaction. *ARKIVOK*, **X**, 420–433.
35 Tempesti, C.T., Pierini, A.B. and Baumgartner, M.T. (2005) A different route to the synthesis of 9,10-disubstituted phenanthrenes. *Journal of Organic Chemistry*, **70**, 6508–6511.
36 Pierini, A.B., Baumgartner, M.T. and Rossi, R.A. (1987) Photostimulated reactions of haloarenes with 2-naphthylamide ions. A facile synthesis of 1-aryl-2-naphthylamines. *Tetrahedron Letters*, **28**, 4653–4656.

37 Dichiarante, V., Fagnoni, M. and Albini, A. (2006) Convenient synthesis of electron-donating substituted benzonitriles by photolysis of phenyl halides and esters. *Chemical Communications*, 3001–3003.

38 Yammal, C.C., Podesta, J.C. and Rossi, R.A. (1992) Reactions of triorganostannyl ions with haloarenes in liquid ammonia. Competition between halogen-metal exchange and electron-transfer reactions. *Journal of Organic Chemistry*, **57**, 5720–5725.

39 Córsico, E.F. and Rossi, R.A. (2000) Reactions of trimethylstannyl ions with mono-, di- and trichloro-substituted aromatic substrates by the $S_{RN}1$ mechanism. *Synlett*, 227–229.

40 Santiago, A.N., Basso, S.M., Montañez, J.P. and Rossi, R.A. (2006) Trimethylstannylation of mono- and dichloroarenes by the $S_{RN}1$ mechanism in liquid ammonia. *Journal of Physical Organic Chemistry*, **19**, 829–835.

41 Córsico, E.F. and Rossi, R.A. (2002) Sequential photostimulated reactions of trimethylstannyl anions with aromatic compounds followed by palladium-catalyzed cross-coupling processes. *Journal of Organic Chemistry*, **67**, 3311–3316.

42 Chopa, A.B., Lockhart, M.T. and Dorn, V.B. (2002) Phenols as starting materials for the synthesis of arylstannanes via $S_{RN}1$. *Organometallics*, **21**, 1425–1429.

43 Chopa, A.B., Lockhart, M.T. and Silbestri, G. (2001) Synthesis of arylstannanes from arylamines. *Organometallics*, **20**, 3358–3360.

44 Chopa, A.B., Lockhart, M.T. and Silbestri, G. (2002) Synthesis of bis-(trimethylstannyl)aryl compounds via an $S_{RN}1$ mechanism with the intermediacy of monosubstitution products. *Organometallics*, **21**, 5874–5878.

45 Uberman, P.M., Martín, S.E. and Rossi, R.A. (2005) Synthesis of functionalized diaryldimethylstannanes from the Me_2Sn^{2-} dianion by $S_{RN}1$ reactions. *Journal of Organic Chemistry*, **70**, 9063–9066.

46 Argüello, J.E., Schmidt, L.C. and Peñéñory, A.B. (2003) One pot two-step synthesis of aryl sulfur compounds by photoinduced reactions of thiourea anion with aryl halides. *Organic Letters*, **5**, 4133–4136.

47 Córsico, E.F. and Rossi, R.A. (2000) Synthesis of mono-, di-, and tri-phenyl arenes by sequential photostimulated $S_{RN}1$ and Pd(0)-catalyzed cross coupling reactions on aryl halides. *Synlett*, 230–232.

48 Chopa, A.B., Silbestri, G.F. and Lockhart, M.T. (2005) Strategies for the synthesis of bi- and triarylic materials starting from commercially available phenols. *Journal of Organometallic Chemistry*, **690**, 3865–3877.

49 Bardagi, J.I. and Rossi, R.A. (2008) A novel approach to the synthesis of 6-substituted uracils in three-step, one-pot reactions. *Journal of Organic Chemistry*, **73**, 4491–4495.

50 Mandolesi, S.D., Vaillard, S.E., Podesta, J.C. and Rossi, R.A. (2002) Synthesis of benzene- and pyridinediboronic acids via organotin compounds. *Organometallics*, **21**, 4886–4888.

51 Fidelibus, P.M., Silbestri, G.F., Lockhart, M.T., Mandolesi, S.D., Chopa, A.B. and Podest,á J.C. (2007) Three-step synthesis of arylpolyboronic acids from phenols via organotin compounds. *Applied Organometallic Chemistry*, **21**, 682–688.

52 Dichiarante, V., Fagnoni, M. and Albini, A. (2007) Metal-free synthesis of sterically crowded biphenyl by direct Ar-H substitution in alkyl benzenes. *Angewandte Chemie, International Edition*, **46**, 6495–6498.

53 Guizzardi, B., Mella, M., Fagnoni, M. and Albini, A. (2000) Easy photochemical preparation of 2-dimethylamino-phenylfurans-pyrroles and -thiophenes. *Tetrahedron*, **56**, 9383–9389.

54 Norris, R.K. and McMahon, J.A. (2003) The $S_{RN}1$ reaction of 1,8-

diiodonaphthalene. *ARKIVOK*, **X**, 139–155.

55 Roydhouse, M.D. and Walton, J.C. (2005) Radical-carbanion cyclo-coupling in armed aromatics: overriding steric hindrance to ring closure. *Chemical Communications*, 4453–4455.

56 Marshall, L.J., Roydhouse, M.D., Slawin, A.M.Z. and Walton, J.C. (2007) Effect of chain length on radical to carbanion cyclo-coupling of bromoaryl alkyl-linked oxazolines: 1,3-areneotropic migration of oxazolines. *Journal of Organic Chemistry*, **72**, 898–911.

57 Roydhouse, M.D. and Walton, J.C. (2007) Formation of a tetracyclic isoquinoline derivative by rearrangement of a [(bromophenyl)butyryl]oxazolidinone. *European Journal of Organic Chemistry*, 1059–1063.

58 Liras, S., Lynch, C.L., Fryer, A.M., Vu, B.T. and Martin, S.F. (2001) Applications of vinylogous Mannich reactions. total synthesis of the ergot alkaloids rugulovasines A and B and setoclavine. *Journal of the American Chemical Society*, **123**, 5918–5924.

59 Barolo, S.M., Teng, X., Cuny, G.D. and Rossi, R.A. (2006) Syntheses of aporphine and homoaporphine alkaloids by intramolecular *ortho*-arylation of phenols with aryl halides via $S_{RN}1$ reactions in liquid ammonia. *Journal of Organic Chemistry*, **71**, 8493–8499.

60 Baumgartner, M.T., Nazareno, M.A., Murguía, M.C., Pierini, A.B. and Rossi, R.A. (1999) Reaction of *o*-iodoaniline with aromatic ketones in DMSO: synthesis of 2-aryl or 2-hetarylindoles. *Synthesis-Stuttgart*, 2053–2056.

61 Barolo, S.M., Lukach, A.E. and Rossi, R.A. (2003) Syntheses of 2-substituted indoles and fused indoles by photostimulated reactions of *o*-iodoanilines with carbanions by the $S_{RN}1$ mechanism. *Journal of Organic Chemistry*, **68**, 2807–2811.

62 Beugelmans, R. and Chbani, M. (1995) $S_{RN}1$ – synthesis of nitrogen, sulfur and phosphorus heterocyclic compounds linked to naphthalene. *Bulletin de la Societe Chimique de France*, **132**, 729–733.

63 Estel, L., Marsais, F. and Queguiner, G. (1988) Metalation/$S_{RN}1$ coupling in heterocyclic synthesis. A convenient methodology for ring functionalization. *Journal of Organic Chemistry*, **53**, 2740–2744.

64 Barolo, S.M., Rosales, C., Angel Guío, J.E. and Rossi, R.A. (2006) One pot synthesis of substituted dihydroindeno[1,2-*b*]indoles and dihydrobenzo[*a*]carbazoles by photostimulated reactions of *o*-iodoaniline with carbanions by the $S_{RN}1$ mechanism. *Journal of Heterocyclic Chemistry*, **43**, 695–699.

65 Guastavino, J.F., Barolo, S.M. and Rossi, R.A. (2006) One-pot synthesis of 3-substituted isoquinolin-1-(2*H*)-ones and fused isoquinolin-1-(2*H*)-ones by $S_{RN}1$ reactions in DMSO. *European Journal of Organic Chemistry*, 3898–3902.

66 Buden, M.E. and Rossi, R.A. (2007) Syntheses of phenanthridines and benzophenanthridines by intramolecular *ortho*-arylation of aryl amide ions with aryl halides via $S_{RN}1$ reactions. *Tetrahedron Letters*, **48**, 8739–8742.

67 Vaillard, S.E., Postigo, A. and Rossi, R.A. (2002) Syntheses of 3-substituted 2,3-dihydrobenzofuranes 1, 2-dihydronaphtho [2,1-*b*]furanes, and 2,3-dihydro-1*H*-indoles by tandem ring closure-$S_{RN}1$ reactions. *Journal of Organic Chemistry*, **67**, 8500–8,506.

68 Goehring, R.R. (1992) An exceptionally brief synthesis of eupolauramine. *Tetrahedron Letters*, **33**, 6045–6048.

69 Bardagí J.I., Vaillard, S.E. and Rossi, R.A. (2007) Synthesis of 4-substituted chromanes and 4-substituted benzo[*f*] chromanes by tandem 6-*exo*-trig cyclization-$S_{RN}1$reactions. *ARKIVOC*, **IV**, 73–83.

70 Dichiarante, V., Fagnoni, M., Mella, M. and Albini, A. (2006) Intramolecular photoarylation of alkenes by phenyl

cations. *Chemistry – A European Journal*, **12**, 3905–3915.
71 Protti, S., Fagnoni, M. and Albini, A. (2006) Benzyl (Phenyl) γ- and δ-lactones via photoinduced tandem Ar—C, C—O bond formation. *Journal of the American Chemical Society*, **128**, 10670–10671.
72 Protti, S., Dondi, D., Fagnoni, M. and Albini, A. (2008) Photochemical arylation of alkenols: role of intermediates and synthetic significance. *European Journal of Organic Chemistry*, 2240–2247.
73 Layman, W.J. Jr, Greenwood, T.D., Downey, A.L. and Wolfe, J.F. (2005) Synthesis of 2H-1,2-benzothiazine 1,1-dioxides via heteroannulation reactions of 2-iodobenzenesulfonamide. *Journal of Organic Chemistry*, **70**, 9147–9155.

11
Singlet Oxygen as a Reagent in Organic Synthesis
Matibur Zamadar and Alexander Greer

11.1
Introduction

In this chapter we review the synthesis of organic peroxides from singlet molecular oxygen [$^1O_2$ ($^1\Delta_g$)] via dye-sensitized photooxygenations. The chapter comprises four sections which describe the synthesis of dioxetanes from [2 + 2]-cycloadditions, endoperoxides from [2 + 4]-cycloadditions, hydroperoxides from "ene" reactions, and tandem combinations there of (see Scheme 11.1). Examples include endoperoxides and bisperoxides that rearrange to spiro compounds.

Whilst this chapter is not comprehensive, attention is focused on those reports of photooxygenation where peroxide yields approached or exceeded 50%. The isolation of peroxides involves meticulous and potentially dangerous work [1]. Thus, percentage yields are not always reported in the same way, and may relate to the feasibility of purification and safety and, perhaps because of this issue, safety information is often incomplete. Analytical methods such as low-temperature nuclear magnetic resonance (NMR) spectroscopy has allowed the characterization of unstable peroxides in reaction mixtures. Some peroxides have been isolated in traps under vacuum or by low-temperature silica gel chromatography, while others were recrystallized at room temperature [2–4].

Singlet oxygen is conveniently generated by sensitization and, in view of the low energy of this species, dyes are effective and can be used in low concentrations due to the high absorptivity.

$$\text{Dye} + h\nu \rightarrow \text{Dye}^* \quad \text{Dye}^* + {}^3O_2 \rightarrow \text{Dye} + {}^1O_2$$

Suitable light sources for exciting the sensitizer include halogen or mercury lamps equipped with a 400 nm cut-off filter to transmit visible light. The sensitizers methylene blue (MB), *meso*-tetraphenylporphine (TPP), sulfonated aluminum phthalocyanine, and Rose Bengal (RB) are widely used in homogeneous solutions. The main appeal of heterogeneous sensitizers is that they are easy to separate from

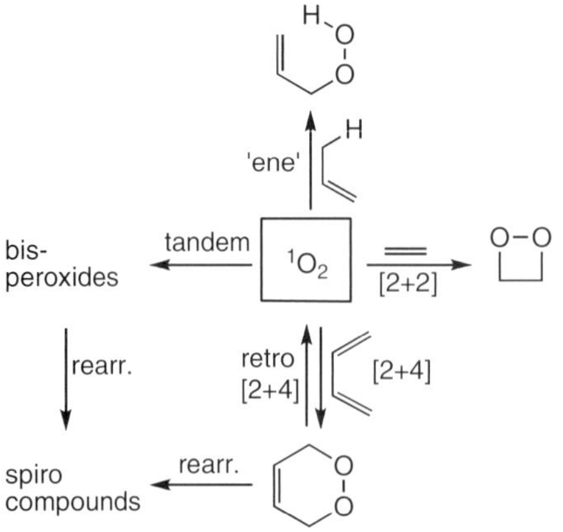

Scheme 11.1

solution after reaction; examples include, polymer-supported RB or *seco*-porphyrazine, MB-doped zeolites, and fullerene-supported silica gel [5, 6].

The lifetime of singlet oxygen is critically dependent on the solvent used [7–9]. The longer lifetime of singlet oxygen in halogenated solvents than organic or aqueous solvents is well known, for example, carbon tetrachloride (59 ms) compared to benzene (30 µs), water (3.5 µs), and deuterium oxide (65 µs). The use of halogenated solvents also comes about in response to the need for low-temperature solvents in view of the instability of peroxide products. The combination of halogenated solvents with low-temperature conditions is seen often in peroxide synthesis with singlet oxygen [10, 11]. Whether intended or not, low-temperature conditions offer another advantage, namely that of enhancing the reaction rates as entropy-controlled reactivity often accompanies singlet oxygen chemistry [12]. Peroxides are formed also through mechanisms that do not involve singlet oxygen, in particular via an electron transfer path (see Section 11.3.3), although preparative examples are limited in number.

11.2
Dioxetanes

11.2.1
Background Information

1,2-Dioxetanes are four-membered ring peroxides which have been synthesized from singlet oxygen–alkene reactions as pure compounds (e.g., **18–33**; Schemes 11.2 and 11.3). The rosette of Scheme 11.3 shows reported examples of dioxetanes formed in good yield from the corresponding alkenes displayed in Scheme 11.2. The sensitized photooxidation of alkenes **1–17** proceeded rapidly with visible light, but stopped after the uptake of one equivalent of oxygen.

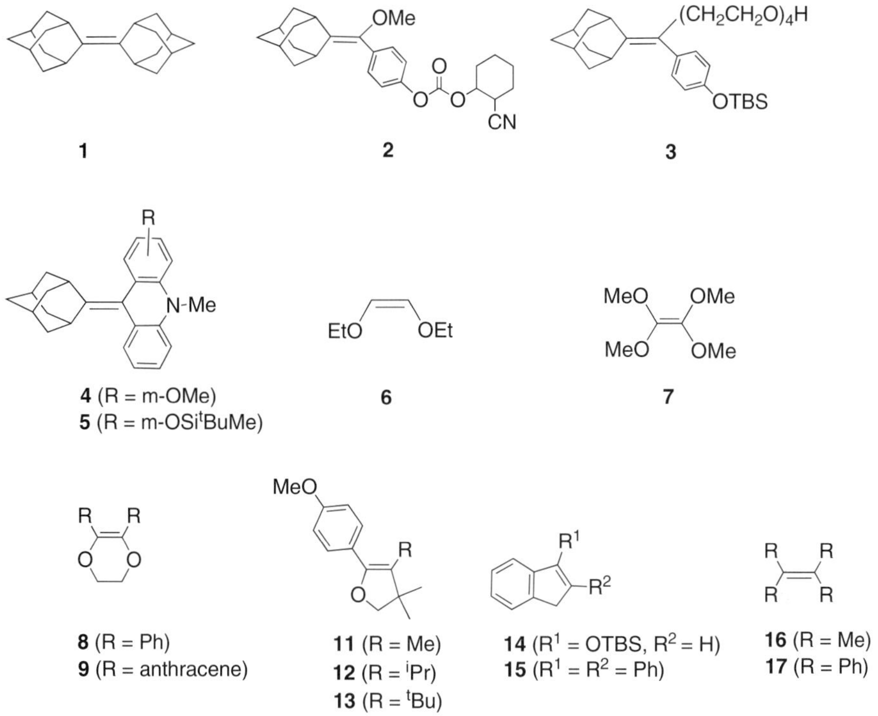
OMe
$(CH_2CH_2O)_4H$
O
O O
CN
OTBS
1
2
3
R
N-Me
MeO OMe
EtO OEt
MeO OMe
4 (R = m-OMe)
5 (R = m-OSi^{t}BuMe)
6
7
MeO
R R
O O
R
O
R^1
R^2
R R
R R
8 (R = Ph)
9 (R = anthracene)
11 (R = Me)
12 (R = ^{i}Pr)
13 (R = ^{t}Bu)
14 (R^1 = OTBS, R^2 = H)
15 (R^1 = R^2 = Ph)
16 (R = Me)
17 (R = Ph)

Scheme 11.2

11.2.2
Adamantyl-Substituted Alkenes

In 1975, adamantylideneadamantane-1,2-dioxetane (**18**) was prepared from an MB-sensitized reaction of biadamantylidene (**1**) with singlet oxygen in CH_2Cl_2 at −78 °C in good yield (66%) after recrystallization [13]. The scope of the singlet oxygen reaction include monospiroadamantyl-dioxetanes, which were also stable towards isolation and characterization [14]. For example, in 1998, a singlet oxygenation of *cis*-2-cyanocyclohexyl carbonate **2** with polymer-supported RB in O_2-saturated CH_2Cl_2 at −78 °C provided dioxetane **19** quantitatively (100% yield) [15]. In 2006, the formation of spiroadamantane dioxetane **20** was reported from a tetraglycol-substituted alkene **3** in environmental water samples containing chromophoric dissolved organic matter (CDOM) as the photosensitizer [16, 17]. The CDOM was not characterized structurally, but was proposed to contain amphiphilic macromolecular species with micelle-like properties. Dioxetane **20** served as a hydrophobic trap-and-trigger chemiluminescent probe to validate singlet oxygen concentrations in natural waters [16, 17]. In 1997, photooxygenations of methoxy- and siloxy-substituted adamantylideneacridanes (**4** and **5**) produced the corresponding dioxetanes **21** and **22** [16]. The reaction was conducted with TPP as the photosensitizer in $CHCl_3$

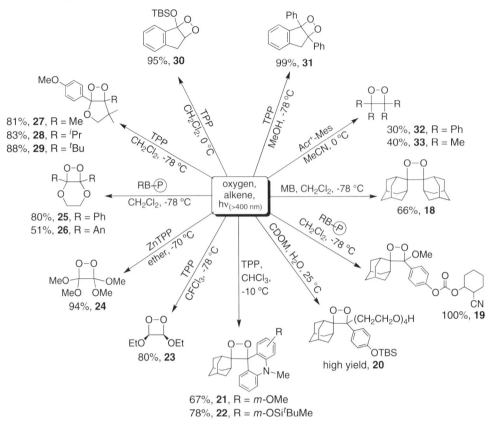

Scheme 11.3

at −10 °C. To avoid decomposition, dioxetanes **21** and **22** were isolated at reduced temperature (−10 °C) during silica-gel chromatography, and were obtained in 67% and 78% yield, respectively.

11.2.3
Alkoxy-Substituted Alkenes

Electron-rich olefins also react with singlet oxygen and form dioxetanes. In 1970, the formation of *cis*-1,2-dioxetane **23** was achieved stereospecifically in greater than 80% yield in a photooxidation reaction with *cis*-diethoxyethylene **6** (Scheme 11.3) [18]. The photolysis of **6** was carried out in O_2-saturated TPP-sensitized $CFCl_3$ at −78 °C, where no *trans*-1,2-dioxetane was obtained. *Trans*-1,2-dioxetane can be prepared itself stereospecifically from the singlet oxygenation of *trans*-diethoxyethylene [18]. In 1970, the irradiation of tetramethoxyethylene (**7**) sensitized by zinc tetraphenylporphine or dinaphthalenethiophene in an O_2-saturated ether solution at −70 °C led to the production of dioxetane **24** [19]. Purification was conducted by evaporating the solvent at −78 °C and then distilling the residue at 25 °C, to give **24** as a liquid in 94% yield.

In 1978, the photooxidation of 2,3-diphenyl-1,4-dioxene (**8**) was reported in CH_2Cl_2 solution at −78 °C [3, 20]. Compound **8**, when reacted with singlet oxygen from the irradiation of polymer-bound RB, afforded the corresponding dioxetane (1,6-diphenyl-2,5,7,8-tetraoxabicyclo[4.2.0]octane, **25**) in 80% yield. Bicyclic dioxetane **25** was insufficiently stable for combustion analysis. On the basis of ^{13}C NMR spectroscopy, the dioxetane structure was assigned [20]. 2,3-Di(2-anthryl)-1,4-dioxene (**9**) was also photooxidized in CH_2Cl_2 solution at −78 °C, and dioxetane **26** obtained in 51% yield [21]. Interestingly, an enhanced chemiluminescence from silica gel was observed with the catalytic decomposition of **26** [21]. In 1997, dioxetanes **27–29** were synthesized in 81–88% yields from singlet oxygenation of the corresponding 4-alkyl-3-dimethyl-2,3-dihydrofurans (**11–13**) in CH_2Cl_2 at −78 °C [22]. This reaction has been conducted with a variety of dihydrofuran derivatives to generate similar dioxetanes. In 1999, an excellent yield of dioxetane **30** was obtained (95%) in the photooxidation of silyl enol ether **14** [23]. Dioxetane **30** was characterized at low temperature, and was unstable at room temperature. ^{13}C NMR signals at δ_{C1} (~113 ppm) and δ_{C2} (~92 ppm) were characteristic of the dioxetane ring, and were the basis for the structure assignment.

Singlet oxygen reactions involving rearrangements of initially formed endoperoxides to dioxetanes have been reported [24, 25]. For example, 2-(2′-anthryl)-1,4-dioxene **9** initially formed an endoperoxide by a [2 + 4]-cycloaddition reaction, and later rearranged to a dioxetane when separated on silica gel with o-xylene [24].

11.2.4
Phenyl- or Methyl-Substituted Alkenes

11.2.4.1 Diphenylindene Photooxidation
In 1979, an interesting effect on the dioxetane/diendoperoxide product ratio was observed upon the addition of methanol to acetone. The addition of methanol increased the dioxetane/diendoperoxide ratio. The photooxidation of 2,3-diphenylindene **15** afforded its dioxetane **31** in methanol at −78 °C [26]. On the basis of liquid chromatography (LC) analysis and NMR spectroscopy, dioxetane **31** formed in high yields in methanol or methanol-acetone (3:7) mixtures. However, the formation also occurred of a diendoperoxide with the solvents Freon-11 and acetone-d_6 (the topic of bisperoxides will be discussed in Section 11.5.2.1). The synthesis of benzofuran dioxetanes has also been accomplished from a TPP-sensitized photooxidation of benzofurans [27, 28].

11.2.4.2 Electron-Transfer Photooxidation
Despite the fact that singlet oxygen oxidizes tetramethylethylene (**16**) via the "ene" pathway, with abstraction of one of the 12 α protons, under electron-transfer photooxidation conditions dioxetane **32** (R = Me) has been generated [29]. The electron-deficient sensitizer 9-mesityl-10-methylacridinium ion (Acr$^+$-Mes) in its excited state Acr$^{\bullet}$-Mes$^{\bullet+}$ was generated, in which the alkene radical cation and

the superoxide ion ($O_2^{\bullet-}$) were proposed, coupling to the dioxetane. Similar results were observed in the electron transfer-photosensitized oxidation of tetraphenylethylene (**17**).

11.2.5
Summary

Reports of photooxidations producing dioxetanes are quite extensive in number [30, 31]. Steric and polar factors have a profound influence in activating alkenes to react with singlet oxygen. Isolable dioxetanes from singlet oxygen have been restricted to adamantyl-, alkoxy-, and phenyl-substituted alkenes. The [2+2]-cycloaddition of singlet oxygen to alkenes that bear α protons has shown little success in dioxetane synthesis, in view of hydroperoxide formation from the singlet oxygen "ene" reaction. However, an electron-transfer photooxidation reaction to dioxetanes offers a means of overcoming the structural restrictions of singlet oxygen [29]; for example, the tetramethylethylene radical cation coupled to $O_2^{\bullet-}$, and formation of the dioxetane, but the broader synthetic utility has not yet been established [29, 32].

Dioxetanes vary in their stability towards decomposition to excited carbonyl products; some require heating to induce decomposition to carbonyl compounds with chemiluminescence (e.g., **18**). Base-catalyzed β-elimination of the *cis*-2-cyano-cyclohexylcarbonate protecting group of dioxetane **19** produced the corresponding dioxetane phenoxide anion a structure that facilitated the molecule's fragmentation into ketones and subsequent chemically initiated electron exchange luminescence (CIEEL). Upon treatment with fluoride ions, dioxetanes **21** and **22** also underwent fragmentation into ketones via CIEEL. The thermolytic stability of dioxetanes has been reviewed [33]. Dioxetanes can catalytically decompose from trace impurities in solvents, such as transition metal ions [34], or can decompose upon silica-gel chromatography work-up (e.g., **26**) [14, 21]. Finally, the treatment of dioxetanes with phosphines led to phosphorane intermediates, and subsequently to phosphine oxides and epoxides [35–37].

11.3
Endoperoxides

11.3.1
Background Information

Endoperoxides (sometimes called 1,4-epiperoxides or dioxapropellanes) are six-membered ring peroxides, which arise from singlet oxygenations of arene and diene compounds and have been synthesized as pure compounds (e.g., **43–52** and **62–72**; Schemes 11.4 to 11.7). Many endoperoxide structures have been reported [38–41]; the rosettes of Schemes 11.5 and 11.7 show examples where endoperoxides were formed in good yield.

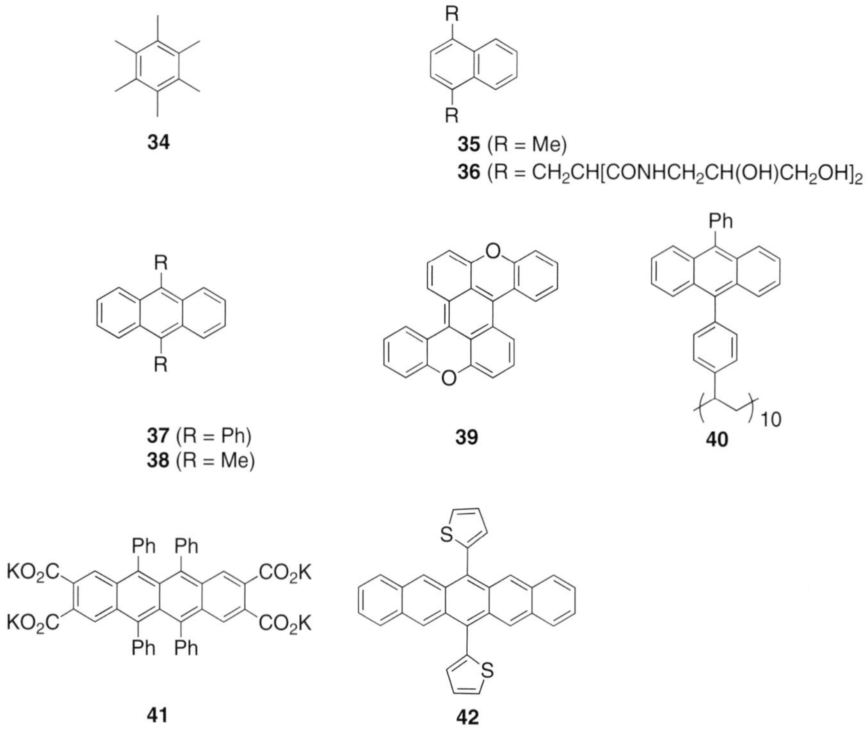

R
R
34
35 (R = Me)
36 (R = CH2CH[CONHCH2CH(OH)CH2OH]2
R
R
37 (R = Ph)
38 (R = Me)
O
O
39
Ph
10
40
Ph Ph
KO2C
KO2C
CO2K
CO2K
Ph Ph
41
S
S
42

Scheme 11.4

11.3.2 Arenes

The rosette of Scheme 11.5 shows the endoperoxidation of the corresponding arene compounds in Scheme 11.4.

11.3.2.1 Benzenes

Although, benzene, toluene, and xylenes do not react chemically with singlet oxygen, hexamethylbenzene (**34**) is sufficiently electron-rich to react with singlet oxygen [42]. Hexamethylbenzene-1,4-endoperoxide (**43**) formed in a MB-sensitized reaction in CD_3NO_2 at 15 °C and, over time, also endoperoxy-hydroperoxide **43′**, arose from a secondary singlet oxygenation in 90% yield (not isolated, quantitated from the NMR spectrum with an added internal standard) (Scheme 11.8). The second equivalent of oxygen added, interestingly, by a singlet oxygen "ene" reaction on the initially formed endoperoxide **43**. (Bisperoxides formed from the addition of two singlet oxygen molecules are detailed in Section 11.5.2.) Heating the primary endoperoxide product **43** at 40 °C for 1 h led to oxygen evolution and the regeneration of hexamethylbenzene **34**. Singlet oxygen also reacted with pentamethylbenzene, but much more slowly than did hexamethylbenzene [42, 43]. The reaction of phenol with singlet oxygen led to *p*-catechol and *p*-quinone products in near-quantitative yield,

360 | *11 Singlet Oxygen as a Reagent in Organic Synthesis*

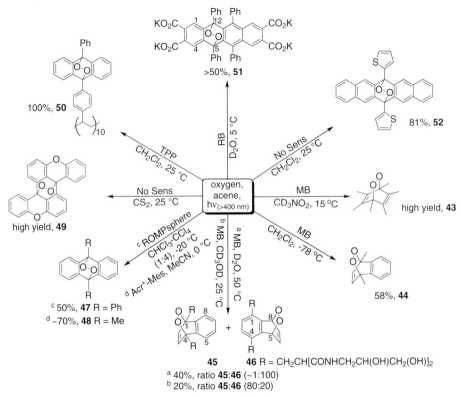

Scheme 11.5

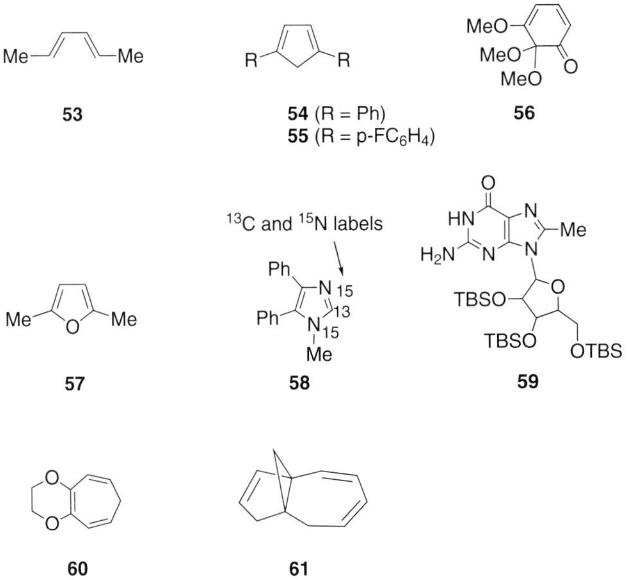

Scheme 11.6

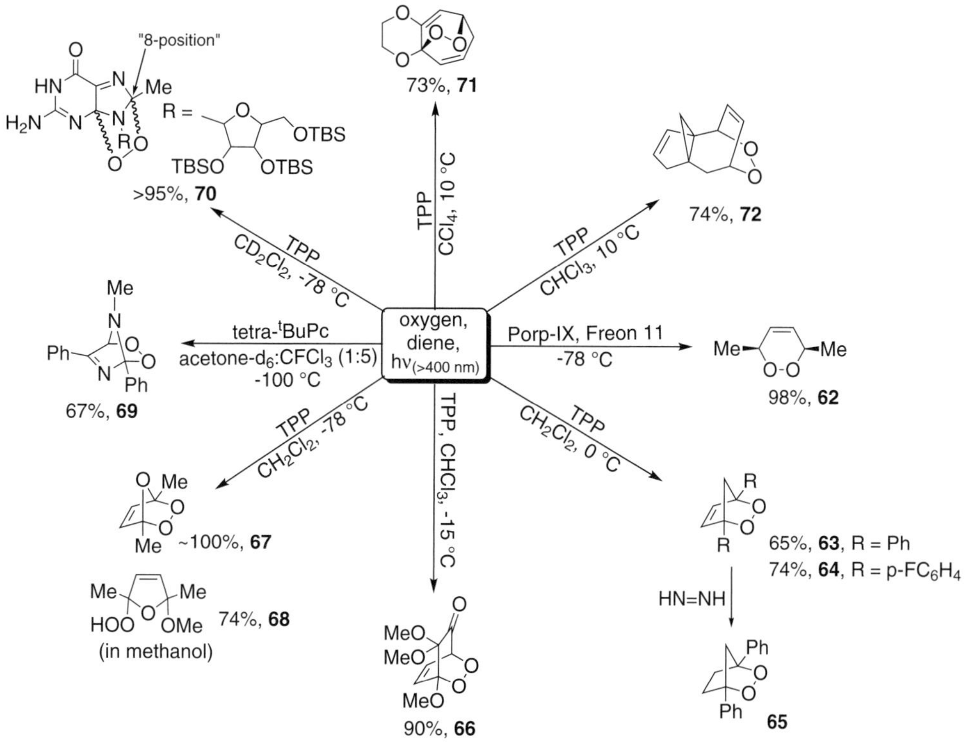

Scheme 11.7

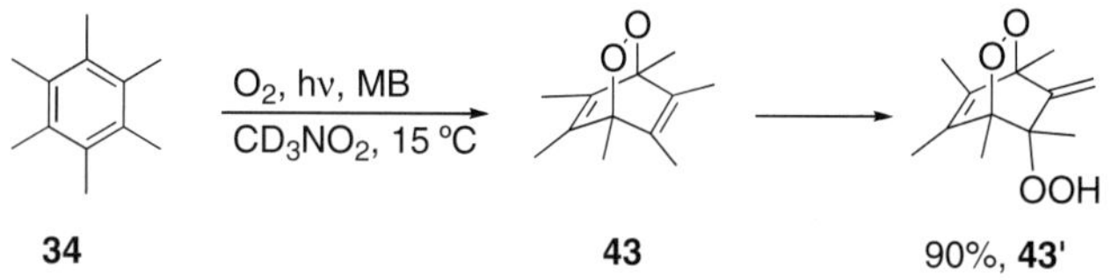

Scheme 11.8

although the proposed endoperoxide intermediate was not directly observed (Scheme 11.9) [44, 45].

11.3.2.2 Naphthalenes

Many naphthalene endoperoxides have been prepared [46–49]. In 1972, the singlet oxygenation of 1,4-dimethylnaphthalene (**35**), in which MB served as sensitizer in O_2-saturated CH_2Cl_2 at $-78\,°C$ gave, regioselectively, 1,4-dimethylnaphthalene-1,4-endoperoxide **44** in 58% yield after column chromatography at $0\,°C$ [47]. Compound **35** reversibly binds singlet oxygen, forms the endoperoxide **44**, and then gradually decomposes back to **35** by dissociation of singlet oxygen [47]. With enriched $^{18}O_2$ gas used in the photooxidation, an ^{18}O-^{18}O labeled naphthalene endoperoxide has also been synthesized [50]. Bulky substituents at the 1,4-positions of naphthalene

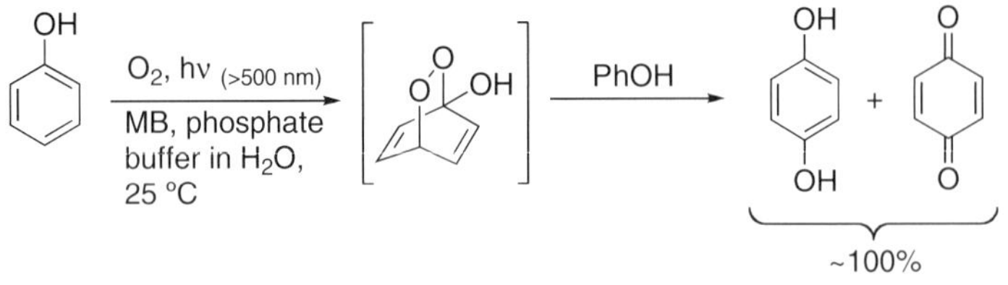

Scheme 11.9

slowed the rate of singlet oxygen addition to **36** with competitive binding of singlet oxygen at the 1,4 and 5,8-positions, **45** and **46**, respectively [51, 52]. The maloamide substituents exerted steric interactions that slowed the binding of $^1O_2$ at the 1,4-positions. Rather, the binding of $^1O_2$ at the 5,8-positions was favored in D_2O at 50 °C (ratio of **45:46** was ~1:100), in which the stability of **46** ($t_{1/2} = 5$ h) was greater than that of **45** ($t_{1/2} = 0.5$ h) at 25 °C due to a higher crowding at the dioxapropellane core of **45**. The maloamide chains are not themselves oxidized; for example, no side-chain hydroperoxides were formed. Solvent effects also profoundly influence the singlet oxygen endoperoxidation [51–53]. The binding of singlet oxygen at the 1,4-positions was favored in CD_3OD at 25 °C (ratio of **45:46** was 80:20) [51]. Studies of the solvent effects of singlet oxygen [2+4]-cycloadditions point to exciplex formation preceding the endoperoxide product in view of the dependence on solvent parameters π^* and σ_H [53].

Naphthalene endoperoxides have been synthesized with methyl groups located at the 1,4-, 1,8-, 1,2,4-, 1,4,5-, 1,2,3,4-positions, and 1,2,3,4,5,6,7,8-positions in good yield after recrystallization of 58%, 88%, 93%, 54%, 83%, and 69%, respectively [48, 49]. The C–O bond formation preferably takes place at aryl sites bonded by the methyl groups. The stabilities of the endoperoxides were greater with increased numbers of methyl groups on naphthalene [at 25 °C in $CDCl_3$, compare 1,2,4-trimethylnaphthalene-1,4-endoperoxide ($t_{1/2} = 70$ h) with 1,2,3,4-tetramethyl-naphthalene-1,4-endoperoxide ($t_{1/2} = $ indefinite)], and where *peri* methyl-methyl interactions increased stability [at 25 °C in $CDCl_3$, compare **44** ($t_{1/2} = 5$ h) with 1,8-dimethylnaphthalene-1,4-endoperoxide ($t_{1/2} = 30$ h)] [48].

11.3.2.3 Anthracenes, Polyacenes, and Carbon Nanotubes

In 2006, the singlet oxygenation of 9,10-diphenylanthracene (**37**) led to 9,10-diphenylanthracene-9,10-endoperoxide (**47**) in O_2-purged $CHCl_3$-CCl_4 (1:4) at −20 °C [54]. The photosensitizer *seco*-porphyrazine was immobilized in a ring-opening metathesis sphere (ROMP-sphere) in a heterogeneous reaction [55]. In 1982, the photooxidation of anthracene derivative **39** in CS_2 at 25 °C gave selectively endoperoxide **49** in high yield [56, 57]. Here, sunlight was used as the light source and the substrate **39** served as its own sensitizer. Oligomeric anthracene endoperoxide (**50**) was prepared from a singlet oxygenation of thin anthracene films of **40** mixed with TPP or MB [58]. Thermolysis of polyendoperoxide **50** led to the release of singlet oxygen molecules and gave back the starting anthracene **40**.

Singlet oxygenation of a tetraphenyltetracene compound bearing four carboxylate groups (**41**) gave endoperoxide **51** in high yield in D_2O [46, 59]. Here, singlet oxygen

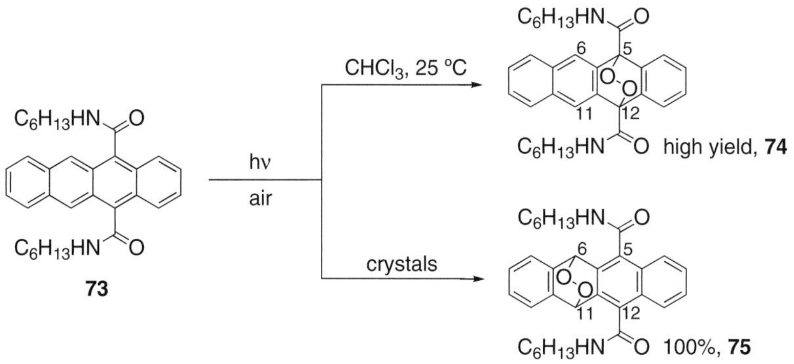

Scheme 11.10

reacted with **41** regioselectivity at one of the two center rings [59]. Photooxidation experiments showed that the parent tetracene (naphthacene, $C_{18}H_{12}$) gave the 5,12-endoperoxide but not the 1,4-endoperoxide [60, 61]. In CDCl$_3$, the ^{1}H NMR absorptions for naphthacene at δ_{H5} (6.15 ppm) and δ_{H6} (7.84 ppm) were characteristic of the 5,12-endoperoxide [62]. In 2008, singlet oxygen was found to add to 5,12-diamide-substituted tetracene **73** crystals at the 6,11-positions (**75**; Scheme 11.10) [62]; however, the singlet oxygen addition in chloroform afforded the 5,12-endoperoxide **74**. The product distribution was apparently distinguished by a H-bond-mediated $^1O_2$ endoperoxidation. The regioselectivity derives from an *anti* diamide conformation in the solid state, which gives the opposite regioselectivity to that in chloroform. Hydroxyl substituents are also known to assist singlet oxygen binding with facial selectivity in endoperoxidation reactions [60].

An investigation into a pentacene–$^1O_2$ reaction was reported in 2007 [63], when 6,13-thienylpentacene-6,13-endoperoxide (**52**) was synthesized in 81% yield in the reaction of 6,13-thienyl-substituted pentacene **42** with singlet oxygen in CH$_2$Cl$_2$ at 25 °C [63]. Here, compound **42** served as its own sensitizer. The endoperoxide structure **52** was assigned on the basis of ^{13}C NMR spectroscopy and X-ray crystallography studies. The X-ray structure displayed intramolecular through space sulfur–oxygen (S—O) interactions [63]. Photooxidation of the parent pentacene $C_{22}H_{14}$ molecule followed similar chemistry, yielding the 6,13-endoperoxide but neither the 1,4- nor 5,14-endoperoxide [64, 65]. Pentacene-6,13-endoperoxide decomposed to pentacenequinone [64, 65]. Reports on higher polyacene–$^1O_2$ reactions are scarce because of ground-state $^3O_2$ reactions which take place preferentially.

In 2007, a report on single-walled carbon nanotubes (SWCNTs) described the reaction with singlet oxygen [66]. Purification was simplified due to the contrasting solubility properties of unfunctionalized and oxyfunctionalized SWCNTs, although it was uncertain whether the SWCNT peroxides were formed from a [2 + 2]-cycloaddition, a [2 + 4]-cycloaddition, or both [66]. Subsequent reports found SWCNTs to be of very low chemical reactivity with $^1O_2$ [67, 68]. Rather, SWCNTs were proposed to physically quench singlet oxygen (to convert $^1O_2$ to $^3O_2$) with rate constants on the order of 10^8 M^{-1} s^{-1}.

11.3.3
Electron-Transfer Photooxidation

In 2004, a report appeared that 9,10-dimethylanthracene-9,10-endoperoxide (**48**) arose from an electron-transfer photooxidation of 9,10-dimethylanthracene (**38**) with the use of the sensitizer Acr$^+$-Mes in O$_2$-saturated CH$_3$CN at 0 °C [32]. Subsequent coupling of the anthracene radical cation and the superoxide ion generated endoperoxide **48**. Endoperoxide **48** was detected during the initial stage of the photooxidation, but was not isolated; over time, the reaction yielded 10-hydroxanthrone, anthraquinone, and H$_2$O$_2$.

11.3.4
Conjugated Dienes

The rosette of Scheme 11.7 shows the endoperoxidation of the corresponding diene compounds in Scheme 11.6. In one case, a solvent addition adduct was formed.

11.3.4.1 Acyclic Dienes
In 1989, the irradiation of (*E,E*)-2,4-hexadiene **53** sensitized by *meso*-porphyrin IX dimethyl ester led to the formation of *cis*-3,6-dimethyl-1,2-dioxene (**62**), which was the major product detected at −78 °C in Freon 11 [69]. Endoperoxide **62** was purified under vacuum at 0.75 mmHg, and collected in a trap (98% isolated yield). Dienes that can adopt a *cisoid* conformation, such as **53** or (*E,E*)-1,4-diphenylbutadiene, were photooxidized by the corresponding endoperoxides in high or quantitative yield in a suprafacial Diels–Alder reaction [60, 70]. Dienes that cannot readily adopt *cisoid* conformations, such as (*E,Z*)-2,4-hexadienes and (*Z,Z*)-2,4-hexadienes, lose their stereochemistry in the singlet oxygen [2 + 4]-cycloaddition [60]. (*E,Z*)- and (*Z,Z*)-dienes give a complex mixture of hydroperoxides and aldehydes, which suggests the intervention of intermediate zwitterions or 1,4-diradicals [71].

11.3.4.2 Cyclopentadienes and Cyclohexadienes
Singlet oxygen reactions have been conducted with many cyclopentadiene and cyclohexadiene compounds as a means of generating bicyclic endoperoxides [72]. The synthesis of an endoperoxide from the parent cyclopentadiene in O$_2$-purged CHCl$_3$-CCl$_4$ (1 : 4) at −20 °C has been reported in 93% yield, in which the photosensitizer was a ROMP-sphere-supported *seco*-porphyrazine [55]. In 2005, endoperoxides **63** and **64** were synthesized, in 65% and 74% yields in CH$_2$Cl$_2$ at 0 °C, from TPP-sensitized photooxygenations of dienes **54** and **55** [73]. Diimide (HN=NH) from potassium azodicarboxylate and acetic acid selectively reduced the carbon–carbon double bond of endoperoxide **63** to give **65** [39, 74]. The reaction of endoperoxide **63** with thiourea leads to *cis*-2-cyclopenten-1,4-diol [39]. In 2008, the singlet oxygenation of diene **56** afforded endoperoxide **66** in good yield (90%), from a TPP-sensitized photooxygenation reaction in CHCl$_3$ at −15 °C [75]. The photooxidation of dienes

similar to **56** in methanol gave 4-hydroxy-2-cyclopentenones via ring contraction reactions upon the addition of thiourea [75].

11.3.4.3 Heterocycles and Cyclohexatriene

Furans have been employed to generate ozonides via photooxidations. The formation of ozonide **67** resulted from the TPP-sensitized photooxygenation of 2,5-dimethylfuran (**57**) in CH_2Cl_2 at $-78\,°C$ [72]. Low-temperature NMR spectra were recorded, and the reaction can presumably be taken to 100% yield, although the endoperoxide was not isolated. The spectral data revealed the existence of the monomeric furan endoperoxide **67**, while low temperatures were maintained to prevent its dimerization [76]. A reaction similar to **57** was observed in the photosensitized oxidation of a $^{13}C,^{15}N$-labeled N-methyl-4,5-diphenylimidazole (**58**) [77]. The sensitizer used here was 2,9,16,23-tetra-*tert*-butyl-29H,31H-phthalocyanine (tetra-^{t}BuPc), while the solvent was acetone-d_6 and fluorotrichloromethane in a 1:5 ratio. Singlet oxygen reacted with **58** via a [2 + 4]-cycloaddition to form the 2,5-endoperoxide **69** (67%, NMR spectrum of the reaction mixture), which was stable at $-100\,°C$; however, upon warming this decomposed to a hydroperoxide. In 1993, diastereomeric endoperoxides **70** were characterized in the low-temperature photooxidation of 2′,3′,5′-*O*-(*tert*-butyldimethylsilyl)-8-methylguanosine (**59**) [78]. The *O*-(*tert*-butyldimethylsilyl) (OTBS) groups provided solubility for low-temperature characterization in CD_2Cl_2, while the 8-methyl group prevented the facile rearrangement that occurs rapidly in guanine peroxides bearing an 8-H substituent. On the basis of $-20\,°C$ ^1H and ^{13}C NMR spectroscopy and low-temperature fast-atom-bombardment (FAB) mass spectrometry, endoperoxides **70** were assigned in greater than 95% NMR yield [78]. Unlike CD_2Cl_2, in methanol the photooxidation of **59** gave endoperoxide **70** in low yields. In methanol, a methoxy-hydroperoxide intermediate (**68**) was formed in 74% from the photooxygenation of **57** [79, 80]. The formation of 1,2- and 1,4-methanol adducts had been observed previously; for example, the photooxidation of 2,5-dimethyl-2,4-hexadiene in methanol produced such solvent adducts [71].

Butenolides arise from the reaction of furans, such as **76**, with singlet oxygen in the presence of alkyl amine bases [81]. Scheme 11.11 shows an intermediate endoperoxide (**77**) converting to diastereomeric γ-hydroxyacrylbutenolides **78** in 88% yield after column chromatography. Singlet oxygenation of **76** was conducted in the presence of Hünig's base in CH_2Cl_2 at $-78\,°C$. The scheme also shows the proposed conversion of intermediate endoperoxide **77** to diastereomeric butenolides **78**. The absence of Hünig's base in the photooxidation of **76** resulted in a complex mixture of compounds.

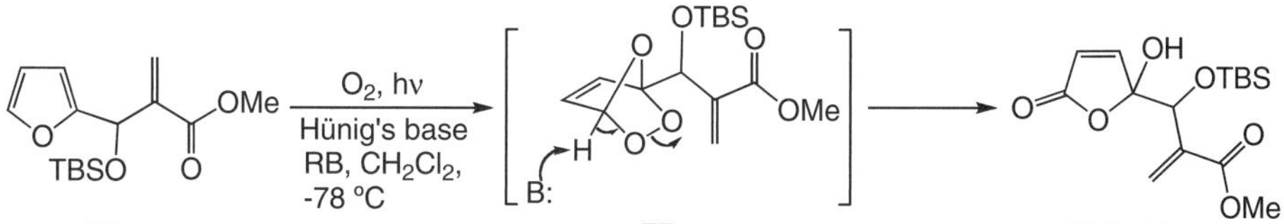

Scheme 11.11

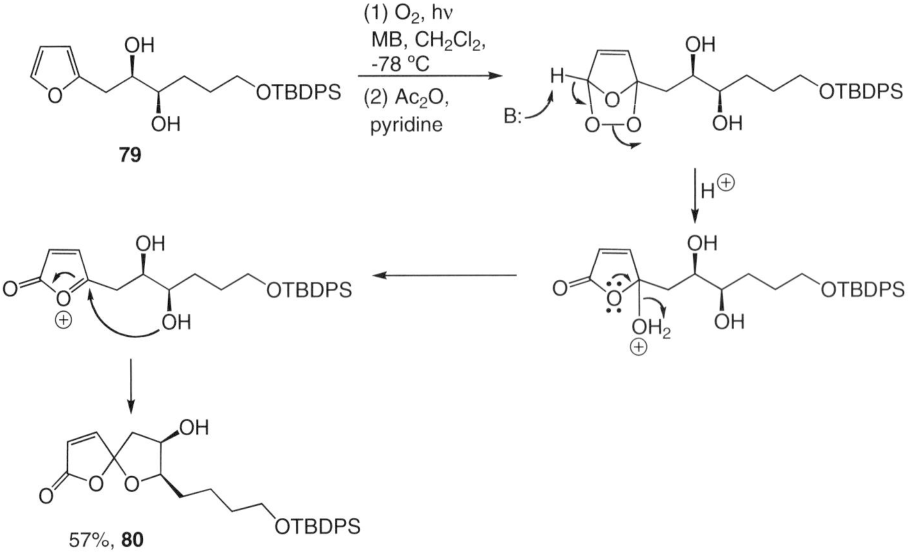

366 | *11 Singlet Oxygen as a Reagent in Organic Synthesis*

Scheme 11.12

57%, **80**

Two examples are presented in which the intramolecular nucleophilic attack of adjacent alcohol groups led to the formation of spiro compounds (Schemes 11.12 and 11.13). In 2008, spirolactone **80** was synthesized in CH_2Cl_2 from an MB-sensitized photooxygenation of furan **79** [82]. A possible mechanism, with an

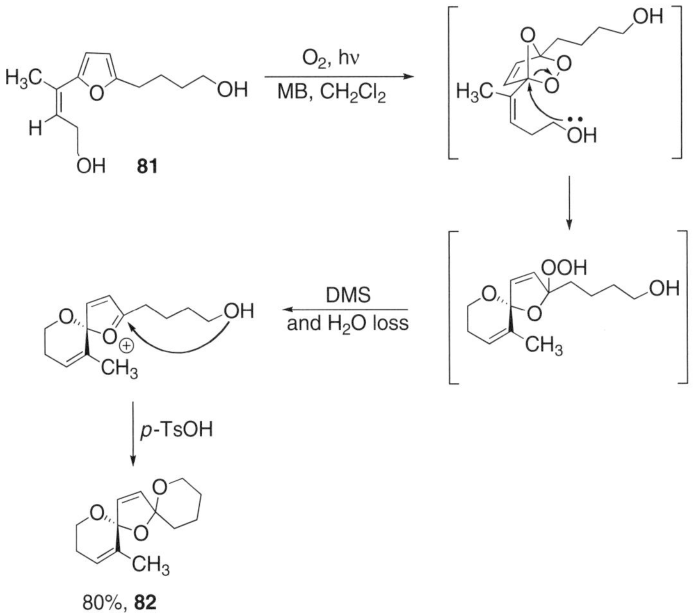

80%, **82**

Scheme 11.13

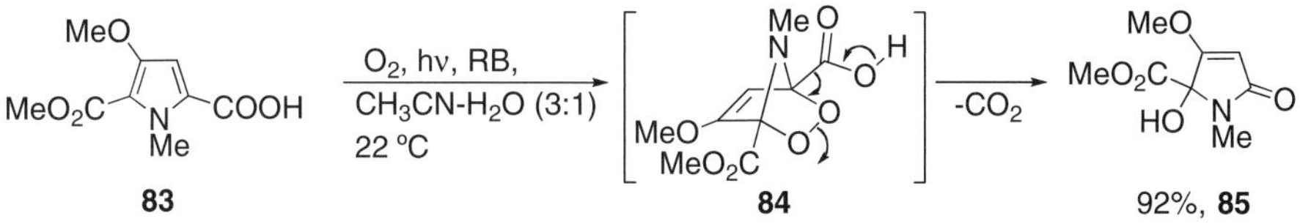

Scheme 11.14

endoperoxide intermediate converting and setting up an intramolecular alcohol attack at the γ-ring site, is shown in Scheme 11.12. Acetic anhydride and pyridine were added to the reaction, which provided **80** in 57% yield over the two steps. In 2006, the singlet oxygenation of 2-substituted furan **81** led to [6,5,6]-bis-spiroketal (**82**) in an MB-sensitized reaction, followed by the addition of dimethylsulfide and *p*-TsOH [83]. In 1991, the singlet oxygenation of 2-carboxy-4-methoxy-5-(methoxycarbonyl)-1-methylpyrrole (**83**), in which RB served as the sensitizer in O_2-saturated CH_3CN-H_2O (3:1) at 22 °C, gave 5-hydroxy-4-methoxy-5-(methoxycarbonyl)-2-methyl-3-pyrrolin-2-one (**85**) in 92% yield after silica-gel chromatography (Scheme 11.14) [84]. The reaction most likely produced the endoperoxide intermediate **84** prior to decarboxylation and formation of pyrrolin-2-one **85** [84]. The reported examples in Schemes 11.11 to 11.14 did not yield isolable endoperoxides, but revealed the potential for endoperoxide decomposition in the presence of a base or internal alcohol.

In 2006, the TPP-sensitized photooxidation of a cycloheptatriene derivative (**60**) led to a tricyclic endoperoxide (**71**) as the major product in O_2-purged CCl_4 at 10 °C (Scheme 11.15) [85]. The treatment of **71** with triethylamine in CH_2Cl_2 at −30 °C led to O−O bond cleavage and the generation of tropolone **86**. On the other hand, heating **71** at 160 °C resulted in the formation of epoxyketal **87**. In 1982, the photooxidation of tricyclo[5.3.1.0]undeca-2,4,9-triene (**61**) in $CHCl_3$ at 10 °C gave selectively endoperoxide **72** in high yield (see Schemes 11.6 and 11.7) [86]. After removal of the solvent, the residue was analyzed chromatographically on silica gel at 0 °C, and the endoperoxide (**72**) obtained in 74% yield as an oil that crystallized in the refrigerator.

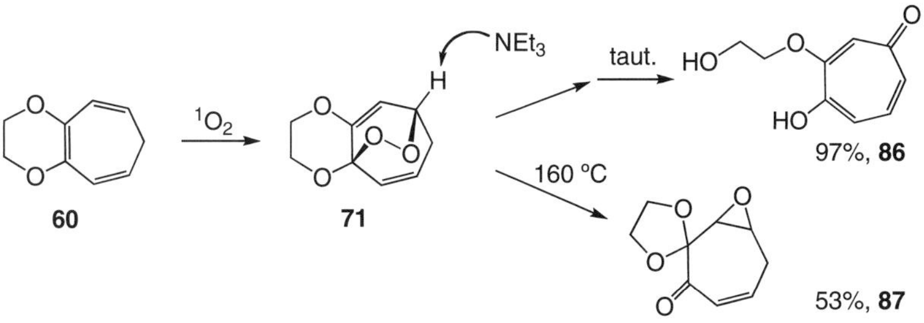

Scheme 11.15

11.3.5
Summary

Reports of photooxidations yielding endoperoxides have been made in several reviews [2, 38–41]. The clean production of endoperoxides has been achieved, and many have been isolated in pure form and in high yields. Although endoperoxides have been created in organic and aqueous media, some are unstable above 0 °C, and their decomposition may lead to the dissociation of oxygen [46–48]. Endoperoxide decomposition may also occur by O–O bond cleavage, which results in the formation of bisepoxides (e.g., spiro[2.4]hepta-4,6-diene endoperoxide) [39] or ketones (an example is the conversion of the natural product kalasinamide via its endoperoxide to the corresponding quinone, marcanine A) [87]. Endoperoxides are not known to rearrange to dioxetanes; for example, vinyl dioxetanes do not rearrange to the corresponding endoperoxides, the latter having been trapped with phosphines. The insertion of phosphines into the O–O bond of unsaturated endoperoxide **61** and other bicyclic endoperoxides has been conducted, and points to a biphilic insertion mechanism with varying degrees of nucleophilic and electrophilic contributions in the transition states to produce the bicyclic dioxaphosphoranes [88]. Finally alcoholic solvents can lead to trapping, while addition of the solvent will yield hydroperoxy alcohols.

11.4
Allylic Hydroperoxides

11.4.1
Background Information

Singlet oxygen chemistry has been employed in the synthesis of allylic hydroperoxides [89, 90]. The $^1O_2$ "ene" reaction of alkenes gives allylic hydroperoxides, where the double bond has shifted; an example was presented in Scheme 11.8 (*vide supra*; Section 11.3.2.1) [42, 43]. Schemes 11.16 to 11.19 demonstrate several reported examples of hydroperoxides formed in good to high yields, where singlet oxygen reacts rapidly with allylic alkenes in solution or confined within zeolites. Zeolites have been used to affect the regiochemical and stereochemical outcome of the hydroperoxide products [91, 92].

11.4.2
Simple Alkenes

An example of a singlet oxygen "ene" reaction was provided in 2006, where hydroperoxide **89** was prepared from C_{60}-supported silica gel in a solvent-free reaction of tetramethylethylene (**88**) at 25 °C (Scheme 11.16) [93]. Purification was conducted by washing the silica gel with chloroform or ethyl acetate, and obtaining **89** after evaporating the solvent. The scope of the solvent-free $^1O_2$ reaction in these studies included mono-alkenes, dienes, and diethylsulfide [93]. Singlet oxygen "ene" reactions of allylic alcohols have also been performed using solid-state photooxygenations in dye-crosslinked polystyrene, to produce diastereomeric hydroperoxides [94, 95].

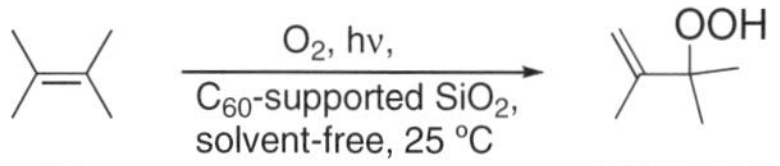

Scheme 11.16

In 2004, excellent yields of hydroperoxides **91** and **92** were obtained (>95%) in the photooxidation of racemic acid **90** (Scheme 11.17) [96]. The diastereoselectivity of the photooxidation favored the *trans*-configured products due to the *anti*-directing effect of the protonated carboxylic acid group. Regioselectivity favored the formation of **92** over **91**. Pyridine helped to accelerate the deprotonation of the CH next to COOH of **90**, affording **92**, and also suppressed the formation of aromatic compounds which otherwise would take place in the absence of pyridine. A great many studies of "ene" reactions have been conducted in organic solvents [97–101] or aqueous media [102–106]; one such report appeared which used a fluorous biphasic solution [107].

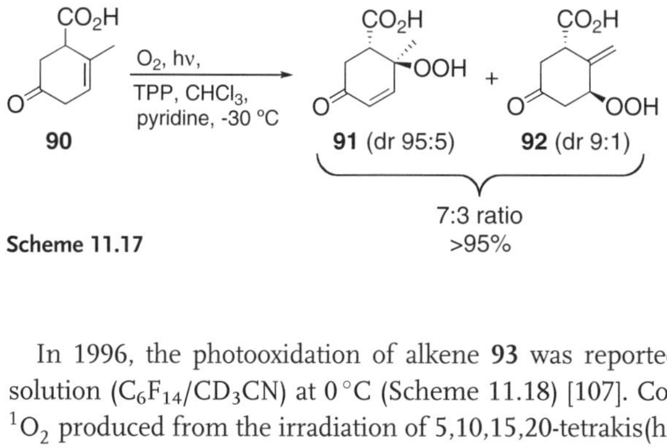

Scheme 11.17

In 1996, the photooxidation of alkene **93** was reported in a fluorous biphasic solution (C_6F_{14}/CD_3CN) at 0 °C (Scheme 11.18) [107]. Compound **93** reacted with $^1O_2$ produced from the irradiation of 5,10,15,20-tetrakis(heptafluoropropyl)porphyrin (TPFPP), and **94** was obtained in 57% yield where a shift in the double bond took place. Acetonitrile was chosen as the cosolvent due to a favorable partition coefficient of the TPFPP sensitizer, which was important. Although, over time, the sensitizers eventually decompose, TPFPP proved to be more chemically stable compared to TPP in the presence of singlet oxygen; this was attributed to the physical segregation of TPFPP from **93** and **94**. A three-phase, solid–gas–liquid experiment also noted an enhanced sensitizer stability when it had become distanced from the reagents [108]. In 2007, a report was made of the photooxidations of alkenes within fluorinated zeolite media which, due to an enhanced lifetime of $^1O_2$, proved useful for "ene" reactions [109a]. The subject of zeolites is discussed in the following subsection [109].

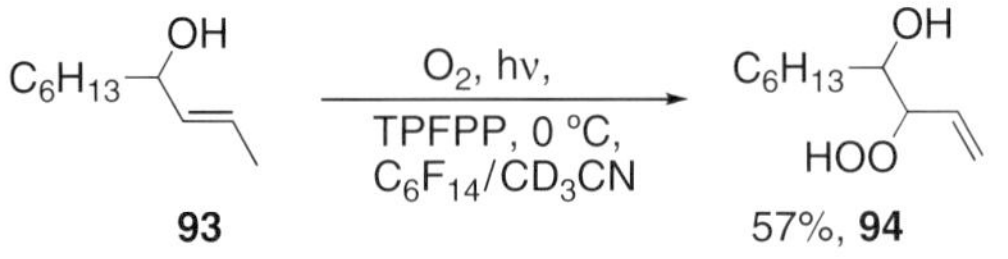

Scheme 11.18

C$_6$H$_{13}$–CH(OH)–CH=CH– (**93**) $\xrightarrow{\text{O}_2,\ h\nu,\ \text{TPFPP, 0 °C, C}_6\text{F}_{14}/\text{CD}_3\text{CN}}$ C$_6$H$_{13}$–CH(OH)–CH(OOH)–CH=CH$_2$ (57%, **94**)

11.4.3
"Ene" Reactions Confined in Zeolites

In recent years, singlet oxygen hydroperoxidations have been conducted in zeolites because different product distributions are observed [91, 92, 110, 111]. Diastereoselection in singlet oxygen "ene" reactions of chiral alkenes has been observed due to zeolite confinement [111]. Trisubtituted alkenes gave different diastereoselectivity in solution than when carried out within zeolite NaY; the latter point was attributed to a synergistic effect of sterics and cation–π interactions. In 1996, a thionin-supported cation-exchanged zeolite reaction was reported for the ^{1}O$_2$ "ene" reaction with 2-methyl-2-pentene (**95**) (Scheme 11.19) [112]. The interior of the cavity helped to control the direction of attack by singlet oxygen on **95**. Singlet oxygen adds −OOH selectively to the 3-position of **95** within the zeolite. In acetonitrile solution, singlet oxygen adds to **95**, yielding the 3- and 2-hydroperoxides (**96** and **97**, respectively).

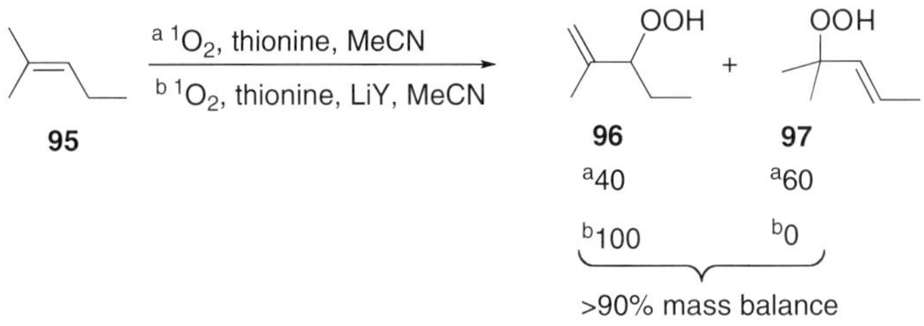

Scheme 11.19

95 $\xrightarrow[\text{b }^1\text{O}_2,\ \text{thionine, LiY, MeCN}]{\text{a }^1\text{O}_2,\ \text{thionine, MeCN}}$ **96** (OOH at 3-position) + **97** (OOH at 2-position)

- **96**: a 40, b 100
- **97**: a 60, b 0

>90% mass balance

11.4.4
Summary

Allylic hydroperoxides have been synthesized as pure compounds in high yield. The sensitized photooxidation of the allylic alkenes proceeded rapidly with visible light, and ceased after uptake of 1 equivalent of oxygen. Various groups have studied singlet oxygen "ene" reactions extensively, and reviews have recently appeared on the subject [89, 90]. Allylic alkenes not only react readily with singlet oxygen to afford hydroperoxide products, but also provide synthetic routes to asymmetric hydroperoxides. Zeolite-mediated asymmetric hydroperoxidations have been of recent interest. In 2003, a theoretical study concluded that the perepoxide is not a minimum on the potential energy surface in reactions of *cis*-2-butene or tetramethylethylene with ^{1}O$_2$ [113].

11.5
Tandem Singlet Oxygen Reactions

11.5.1
Background Information

The bisperoxides described here represent compounds that have taken up 2 equivalents of singlet oxygen. In Sections 11.3.2.1 and 11.3.2.3, details were provided of the formation of a bisperoxide **43'** and a polyendoperoxide **50**. Schemes 11.20 to 11.25 demonstrate reported examples of bisperoxides formed from tandem singlet oxygen reactions in good to high yields. A degree of correlation among furan and guanine photooxidations is mentioned, and in both cases spiro compounds are formed.

11.5.2
Bisperoxides

11.5.2.1 Phenyl-Substituted Alkenes

In 1979, a bisdioxetane (**99**) was prepared from a double [2 + 2]-cycloaddition reaction of two molecules of singlet oxygen with bisdioxene **98**. The reaction used polymer-supported RB in O_2-saturated CH_2Cl_2 at $-78\,°C$ and formed **99** in good yield (70%; Scheme 11.20) [114]. Another example is the tandem [2 + 4]/[2 + 4] singlet oxygenations of 2,3-diphenylindene **15**, which were reported in 1979 (Scheme 11.21). The reaction afforded a diendoperoxide (**100**) in Freon-11 or acetone-d_6 at $-78\,°C$ (100% NMR yield, not isolated) [26]; the initially formed monoendoperoxide was not characterized structurally. The second mole of $^1O_2$ was proposed to add rapidly on the opposite side (on the six-membered ring portion), thereby forming **100**. Interestingly, evidence showed that the photooxidation of **15** in methanol or in methanol–acetone solvent mixtures produced a dioxetane (*vide supra*; Section 11.2.4). In 1972 and 1973, the photooxidation of benzhydrylidene–cyclobutane (**101**) in chloroform and 0.5% ethanol at $-60\,°C$ was reported to lead to diendoperoxide **102** (63% yield), a compound which exploded above $130\,°C$ (Scheme 11.22) [115, 116].

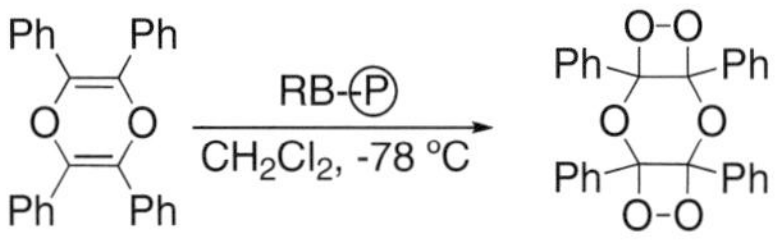

Scheme 11.20

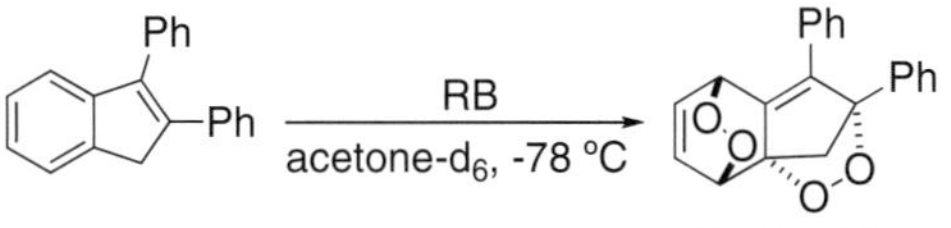

Scheme 11.21

11.5.2.2 Cyclic Alkenes

In 1995, the RB-sensitized photooxygenation of 6,6-diethoxyfulvene (**103**) in acetone-d_6 at −78 °C afforded diethoxyfulvene endoperoxide (**104**) as a primary product (Scheme 11.23) [117]. On the basis of ^{1}H NMR spectroscopy at −50 °C, the initial endoperoxide **104** was assigned, but the dioxetane, **105**, was not observed. Further photooxygenation resulted in the formation of bisperoxides *syn*-**106** and the *anti*-**106** in a 9 : 1 ratio. Peroxides **104** and **106** were not stable and not isolated. Above −30 °C, endoperoxide **104** released oxygen and gave back fulvene **103**, while the bisperoxides **106** decomposed to *cis*-butenedial, diethylcarbonate, and carbon monoxide.

The TPP-sensitized photooxidation of 1,4-cyclohexadiene (**107**) yielded several peroxide products, **109**–**112** (Scheme 11.24) [118–120]. Once the primary "ene" product had formed (**108**), endoperoxy-hydroperoxides **109** and **110** arose by secondary [2 + 4]-cycloadditions in 87% and 9% yield, respectively, while bishydroperoxides **111** and **112** arose by secondary "ene" reactions in less than 4% yield combined. The purification of peroxides **109**–**112** was accomplished by silica-gel chromatography at low temperature; *cis*-hydroperoxide **112** was subsequently recrystallized. In 2006, a hydroperoxy-endoperoxide (**114**) was isolated in 80% chromatographed yield from a TPP-sensitized oxygenation of tetrahydronaphthalene (**113**) in CH$_2$Cl$_2$ at 15 °C (Scheme 11.25) [121]. The primary singlet oxygen "ene" product was not observed, but its formation produced a diene unit capable of adding the second equivalent of singlet oxygen. The configuration of hydroperoxy-endoperoxide **114** was determined indirectly based on the X-ray structure of **116** produced from the bromination of **114** via intermediate **115**.

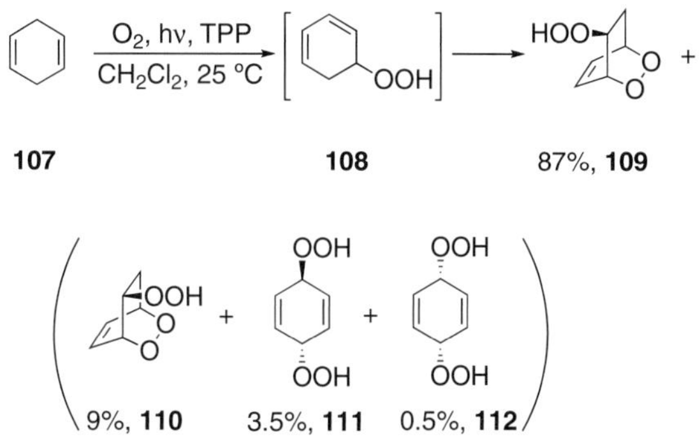

11.5 Tandem Singlet Oxygen Reactions

Scheme 11.24

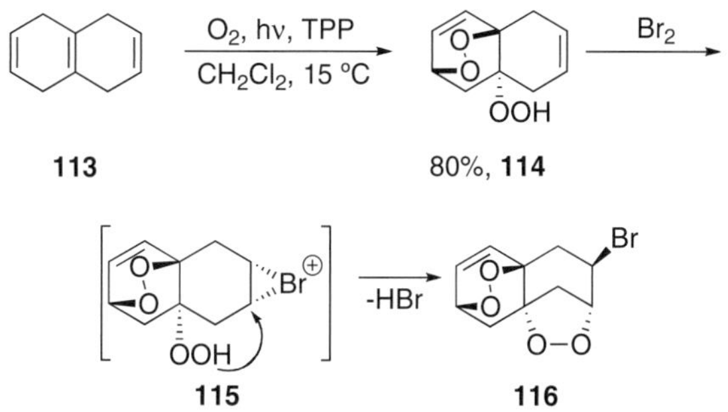

Scheme 11.25

In 2005, 4α-hydroxy peroxyguaienes *anti*-**118** and *syn*-**118** were formed from the photooxidation of diene **117** in CH_2Cl_2 upon reduction of the hydroperoxide group with triphenylphosphine (Scheme 11.26) [122]. The first addition of singlet oxygen was found to be stereoselective. However, the second addition of singlet oxygen was less selective, presumably forming endoperoxy-hydroperoxides prior to reduction by the triphenylphosphine to give mixtures of endoperoxy-alcohols **118** in 56% combined yield.

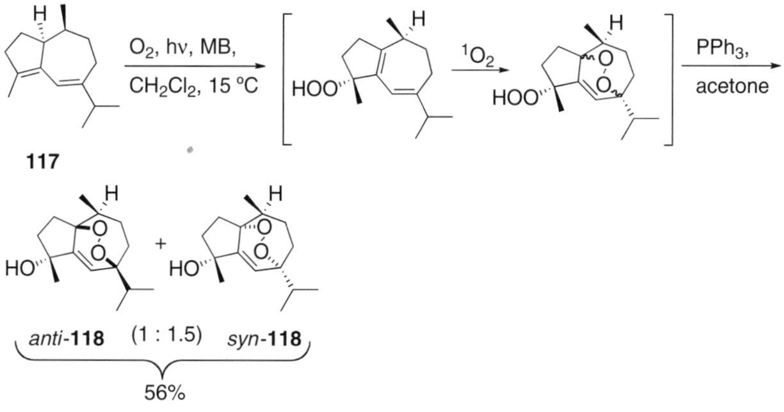

Scheme 11.26

11.5.3
Rearrangement to a Hemiketal Hydroperoxide

In 1998, the singlet oxygenation of biphenylene (**119**) with TPP in O_2-saturated acetone at $-40\,^\circ$C yielded a hemiketal hydroperoxide (**122**) in 56% yield after silica-gel chromatography (Scheme 11.27) [123]. On the basis of ^{1}H NMR spectroscopy, the primary product was assigned to the monoendoperoxide (**120**), which underwent fragmentation into a dicarbonyl product **121**. After further cyclization and tautomerization steps, a furanone derivative added a second singlet oxygen molecule by the "ene" reaction to reach the hemiketal hydroperoxide, **122**.

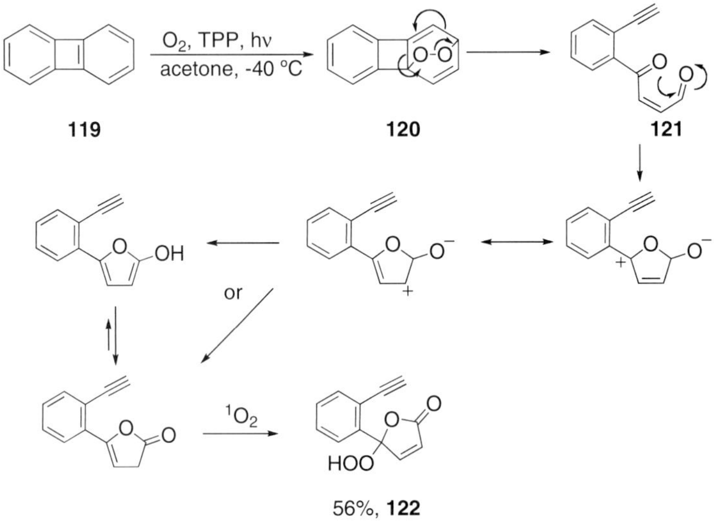

Scheme 11.27

11.5.4
Rearrangements to Spiro Compounds

In 2005, a tandem singlet oxygen reaction with Z- or E-silylated furan **123** was studied, which gave diastereomeric bis-4-hydroxybutenolides **124** as the initial products (Scheme 11.28) [82, 124]. After work-up with mild acid from silica-gel chromatography or p-TsOH, spiroketals **125** were isolated in good yield (80%). The spiroketal moiety of cis- and trans-**125** represents the core structure of the natural product prunolide C. Replacement of the $SiMe_3$ group for H in unsubstituted furans has been shown to be synthetically useful [125]. A mechanism for the singlet oxygenation of 2-methyl-5-trimethylsilylfuran (**126**) proposed silyl migration and methanolysis to give γ-hydroxy-γ-lactone **127** (Scheme 11.29) [125].

In 2007, a tandem singlet oxygen [2 + 4]-cycloaddition and "ene" reaction and ketalization was reported for **128** (Scheme 11.30) [126]. The reaction was conducted with MB as the photosensitizer in $CDCl_3$ at $8\,^\circ$C. The overall yield of 8-epi and

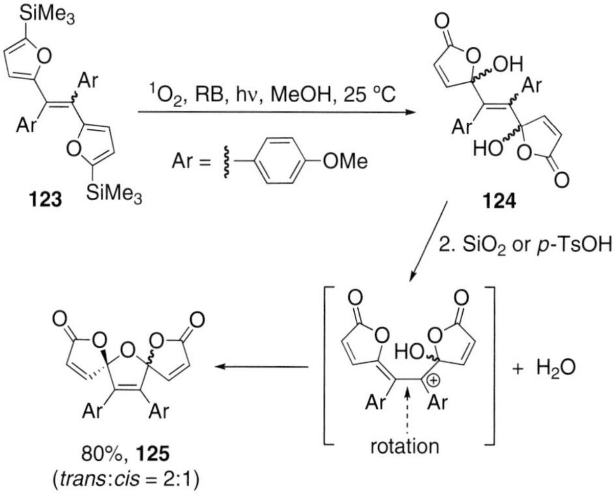

Scheme 11.28

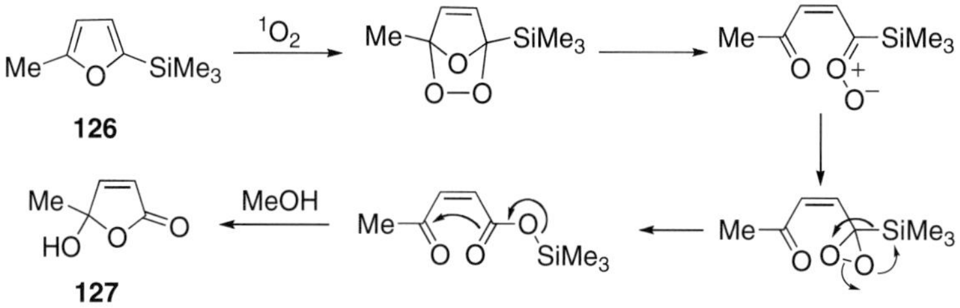

Scheme 11.29

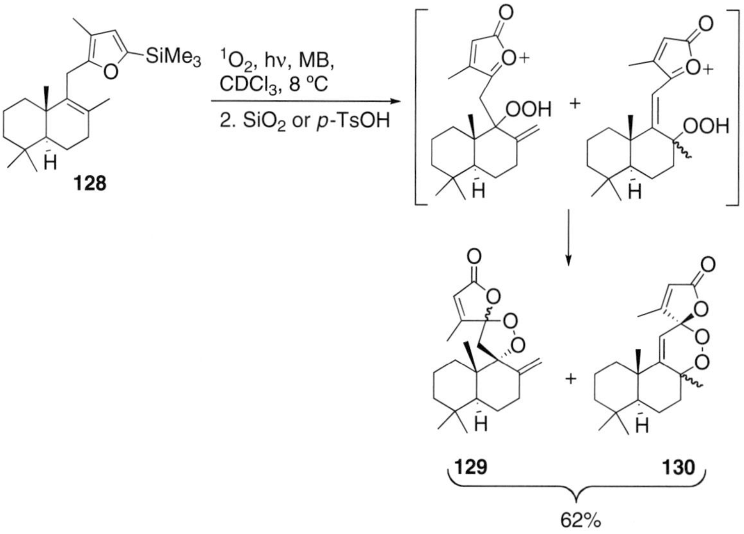

Scheme 11.30

(+)-premnalane was 62% (ratio **129**:**130** was 2:1), and these could be separated using column chromatography. In addition to the NMR spectroscopy evidence for **129** and **130**, X-ray crystallography data was obtained for one of the diastereomers of (+)-premnalane **130**.

Unlike guanosine **59** (as described in Section 11.3.4.3 [78]), the reaction of singlet oxygen with 2′,3′,5′-tris-(O-tert-butyldimethylsilyl)guanosine (**131**) produced spirodiimidohydantoin (**135**) [127]. A facile rearrangement occurred because the guanosine peroxide bore a 8-H substituent. A possible mechanism for the conversion of guanosine **131** to **135** is shown in Scheme 11.31. The secondary photooxidation of 8-oxoG **132** yielded 5-hydroperoxy-8-oxo-7,8-dihydroguanosine (**133**) at low temperature. Reduction of **133** yielded the corresponding alcohol (**134**), which rearranged to a stable final product spirodiimidohydantoin (**135**).

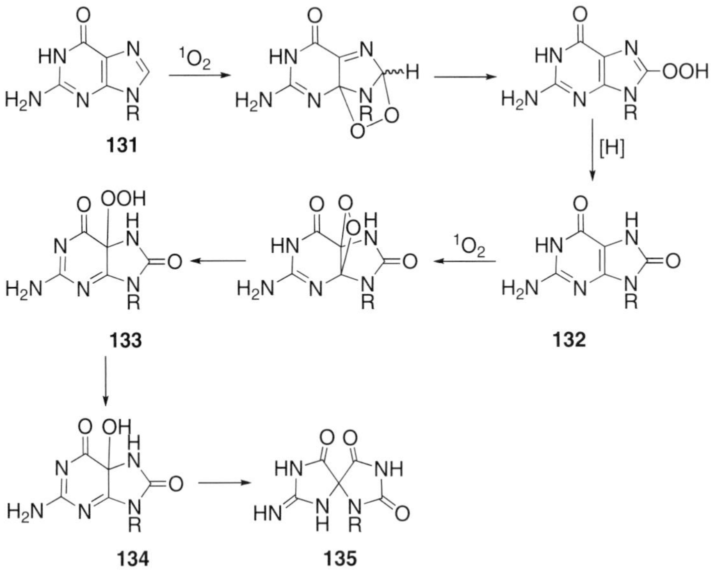

Scheme 11.31

The first report of the structure of the spirodiimidohydantoin nucleoside coming from 8-oxoG was in 2000 [128], in which the oxidation of 8-oxoG by either Ir(IV) or singlet oxygen produced the same spiro compound according to polymerase, ^{13}C-, ^{15}N-, and ^{18}O-labeling, and density functional theory studies [129]. The reaction of guanosine with singlet oxygen [128–130], and the reaction of furans (e.g., **123**) with singlet oxygen [82, 124], showed a similar tendency to form spiro compounds, although they appeared to do so by different mechanisms.

11.5.5
Summary

The reactions in this section proceed through stepwise tandem additions of singlet oxygen. The overoxidation of unsaturated hydrocarbons **15** (in Freon-11 or acetone-d_6),

98, 101, 103, 107, 113, 117, 119, 123, 128, and 131 led to the uptake of 2 equivalents of oxygen. Compounds 15, 98, 101, 103, and 107 led to the formation of bisperoxides. Growing experimental data has been accumulated for singlet oxygen reactions, which give rise to spiro compounds [82, 83, 127–130], or yield bis- or polyperoxides, or polyepoxides [42, 58, 114–120, 131].

11.6
Concluding Remarks

This chapter describes largely singlet oxygen reactions with synthetic potential. Dioxetanes, endoperoxides, hydroperoxides, and bisperoxides were target molecules in dye-sensitized photooxidations. Oxygen, sensitizer, and light were all required to produce the peroxide products. Some polyunsaturated hydrocarbons were subject to singlet oxygen overoxidation reactions, while others were efficient physical quenchers of singlet oxygen. Amine- and metal-containing compounds also efficiently convert $^1O_2$ to $^3O_2$ (physical quenching), and chemically react slowly with singlet oxygen, which may represent a shortcoming for their use as substrates in this chemistry. Although a large body of photooxygenation data are available for peroxides, only a fraction of this is focused on electron-transfer photooxidation chemistry [29, 32, 132–134], and its broader synthetic utility has not yet been established.

Acknowledgments

These studies were supported by a research grant from the National Institutes of Health (GM076168-01).

References

1 (a) Suggested reading: Medard, L.A. (1989) *Accidental Explosions: Types of Explosive Substances*, vol. 2, Ellis Horwood Limited, Chichester; (b) Patnaik, P.A. (1999) *Comprehensive Guide to the Hazardous Properties of Chemical Substances*, 2nd edn, John Wiley & Sons, New York.

2 Wasserman, H.H. and Murray, R.W. (1979) *Singlet Oxygen*, Academic Press, Inc, New York, especially pp. 174–238, pp. 244–283, pp. 287–419, pp. 430–506.

3 Wasserman, H.H. and Ives, J.L. (1981) Singlet oxygen in organic synthesis. *Tetrahedron*, 37 (10), 1825–1852.

4 Hoffmann, N. (2008) Photochemical reactions as key steps in organic synthesis. *Chemical Reviews*, 108 (3), 1052–1103.

5 Wahlen, J., De Vos, D.E., Jacobs, P.A. and Alsters, P.L. (2004) Solid materials as sources for synthetically useful singlet oxygen. *Advanced Synthesis & Catalysis*, 346 (2–3), 152–164.

6 (a) Blossey, E.C., Neckers, D.C., Thayer, A.L. and Schaap, A.P. (1973) Polymer-based sensitizers for photooxidations. *Journal of the American Chemical Society*, 95 (17), 5820–5822; (b) Fuchter, M.J., Hoffman, B.M. and Barrett, A.G.M.

(2006) Ring-opening metathesis polymer sphere-supported *seco*-porphyrazines: efficient and recyclable photooxygenation catalysts. *Journal of Organic Chemistry*, **71** (2), 724–729; (c) Shailaja, J., Sivaguru, J., Robbins, R.J., Ramamurthy, V., Sunoj, R.B. and Chandrasekhar, J. (2000) Singlet oxygen mediated oxidation of olefins within zeolites: selectivity and complexities. *Tetrahedron*, **56** (36), 6927–6943.

7 Schweitzer, C. and Schmidt, R. (2003) Physical mechanisms of generation and deactivation of singlet oxygen. *Chemical Reviews*, **103**, 1685–1757.

8 Wilkinson, F., Helman, W.P. and Ross, A.B. (1995) Rate constants for the decay and reactions of the lowest electronically excited singlet state of molecular oxygen in solution. An expanded and revised compilation. *Journal of Physical Chemistry Reference Data*, **24** (2), 663–1021.

9 (a) Worrall, D.R., Abdel-Shafi, A.A. and Wilkinson, F. (2001) Factors affecting the rate of decay of the first excited singlet state of molecular oxygen $O_2(a^1\Delta_g)$ in supercritical fluid carbon dioxide. *Journal of Physical Chemistry A*, **105**, 1270–1276; (b) Bourne, R.A., Han, X., Chapman, A.O., Arrowsmith, N.J., Kawanami, H., Poliakoff, M. and George, M.W. (2008) *Chemical Communications*, 4457–4459.

10 Matsumoto, M. (1985) Synthesis with singlet oxygen, in *Singlet O_2*, vol. 2 (ed. A.A. Frimer), CRC Press, Boca Raton, FL, pp. 205–274.

11 (a) The full potential of synthetic organic reactions in water has not yet been realized: Grieco, P.A. (ed.) (1998) *Organic Synthesis in Water*, Blackie Academic & Professional Press, London; (b) Lindström, U.M. (2002) Stereoselective organic reactions in water. *Chemical Reviews*, **102** (8), 2751–2772.

12 (a) Gorman, A.A., Lovering, G. and Rodgers, M.A.J. (1979) The entropy-controlled reactivity of singlet oxygen ($^1\Delta_g$) toward furans and indoles in toluene. A variable-temperature study by pulse radiolysis. *Journal of the American Chemical Society*, **101** (11), 3050–3055; (b) Greer, A. (2003) A view of unusual peroxides. *Science*, **302**, 235–236.

13 Schuster, G.B., Turro, N.J., Steinmetzer, H.C., Schaap, A.P., Faler, G. and Adam, W. (1975) Adamantylideneadamantane-1,2-dioxetane. An investigation of the chemiluminescence and decomposition kinetics of an unusually stable 1,2-dioxetane. *Journal of the American Chemical Society*, **97** (24), 7110–7118.

14 (a) Bastos, E.L., Ciscato, L.F.M.L., Weiss, D., Beckert, R. and Badder, W.J. (2006) Comparison of convenient alternative synthetic approaches to 4-[(3-*tert*-butyldimethylsilyloxy) phenyl]-4-methoxyspiro [1,2-dioxetane-3,2′-adamantane]. *Synthesis*, (11), 1781–1786; (b) Adam, W. and Reinhardt, D. (1997) Chemical triggering of dioxetanes derived from 9-adamantylideneacridanes: fluoride- and base-induced chemiluminescence (CIEEL) of siloxy- and acetoxy-substituted dioxetanes. *Journal of the Chemical Society, Perkin Transactions 2*, (8), 1453–1464.

15 Sawa, M., Imaeda, Y., Hiratake, J., Fujii, R., Umeshita, R., Watanabe, M., Kondo, H. and Oda, J. (1998) Toward the antibody-catalyzed chemiluminescence. Design and synthesis of hapten. *Bioorganic & Medicinal Chemistry Letters*, **8** (6), 647–652.

16 Latch, D.E. and McNeill, K. (2006) Microheterogeneity of singlet oxygen distributions in irradiated humic acid solutions. *Science*, **311**, 1743–1747.

17 Grandbois, M., Latch, D.E. and McNeill, K. (2008) Microheterogeneous concentrations of singlet oxygen in natural organic matter isolate solutions. *Environmental Science & Technology*, **42** (24), 9184–9190.

18 Bartlett, P.D. and Schaap, A.P. (1970) Stereospecific formation of 1,2-dioxetanes from *cis*- and *trans*-

diethoxyethylenes by singlet oxygen. *Journal of the American Chemical Society*, **92** (10), 3223–3225.

19 Mazur, S. and Foote, C.S. (1970) Chemistry of singlet oxygen. IX. A stable dioxetane from photooxygenation of tetramethoxyethylene. *Journal of the American Chemical Society*, **92** (10), 3225–3226.

20 Zaklika, K.A., Thayer, A.L. and Schaap, A.P. (1978) Substituent effects on the decomposition of 1,2-dioxetanes. *Journal of the American Chemical Society*, **100** (15), 4916–4918.

21 Zaklika, K.A., Burns, P.A. and Schaap, A.P. (1978) Enhanced chemiluminescence from the silica gel catalyzed decomposition of a 1,2-dioxetane. *Journal of the American Chemical Society*, **100** (1), 318–320.

22 Matsumoto, M., Watanabe, N., Kasuga, N.C., Hamada, F. and Tadokoro, K. (1997) Synthesis of 5-alkyl-1-aryl-4,4-dimethyl-2,6,7-trioxabicyclo[3.2.0]heptanes as a chemiluminescent substrate with remarkable thermal stability. *Tetrahedron Letters*, **38** (16), 2863–2866.

23 Einaga, H., Nojima, M. and Abe, M. (1999) Photooxygenation ($^1O_2$) of silyl enol ethers derived from indan-1-ones: competitive formation of tricyclic 3-siloxy-1,2-dioxetane and α-silylperoxy ketone. *Journal of the Chemical Society, Perkin Transactions 1*, (17), 2507–2512.

24 Schapp, A.P., Burns, P.A. and Zaklika, D.A. (1977) Silica gel-catalyzed rearrangement of an endoperoxide to a 1,2- dioxetane. *Journal of the American Chemical Society*, **99** (4), 1270–1272.

25 Matsumoto, M., Nasu, S., Takeda, M., Murakami, H. and Watanabe, N. (2000) Singlet oxygenation of 4-(4-tert-butyl-3,3-dimethyl-2,3-dihydrofuran-5-yl)-2-pyridone: non-stereospecific 1,4-addition of singlet oxygen to a 1,3-diene system and thermal rearrangement of the resulting 1,4- endoperoxides to stable 1,2-dioxetanes. *Chemical Communications*, (10), 821–822.

26 Boyd, J.D. and Foote, C.S. (1979) Chemistry of singlet oxygen. 32. Unusual products from low-temperature photooxygenation of indenes and *trans*-stilbene. *Journal of the American Chemical Society*, **101** (22), 6758–6759.

27 Berdahl, D.R. and Wasserman, H.H. (1983) The reaction of psoralens and related benzofurans with singlet oxygen. An account of current research. *Israel Journal of Chemistry*, **23** (4), 409–414.

28 Sauter, M. and Adam, W. (1995) Oxyfunctionalization of benzofurans by singlet oxygen, dioxiranes, and peracids: chemical model studies for the DNA-damaging activity of benzofuran dioxetanes (oxidation) and epoxides (alkylation). *Accounts of Chemical Research*, **28** (7), 289–298.

29 Ohkubo, K., Nanjo, T. and Fukuzumi, S. (2005) Efficient photocatalytic oxygenation of aromatic alkene to 1,2-dioxetane with oxygen via electron transfer. *Organic Letters*, **7** (19), 4265–4268.

30 Adam, W. and Trofimov, A.V. (2006) Contemporary trends in dioxetane chemistry, in *Chemistry of Peroxides* (ed. Z. Rappoport), vol. **2** (Pt. 2), Wiley-VCH, Weinheim, pp. 1171–1209.

31 Adam, W., Heil, M., Mosandl, T. and Saha-Moeller, C.R. (1992) Dioxetanes and α-peroxy lactones, four-membered ring cyclic peroxides, in *Organic Peroxides* (ed. W. Ando), John Wiley & Sons, Chichester, UK, pp. 221–254.

32 Kotani, H., Ohkubo, K. and Fukuzumi, S. (2004) Photocatalytic oxygenation of anthracenes and olefins with dioxygen via selective radical coupling using 9-mesityl-10-methylacridinium ion as an effective electron-transfer photocatalyst. *Journal of the American Chemical Society*, **126** (49), 15999–16006.

33 Baumstark, A.L. (1988) Thermolysis of alkyl-1,2-dioxetanes, in *Advances in Oxygenated Processes* (ed. A.L. Baumstark), JAI Press, Greenwich, CT, pp. 31–84.

34 Bartlett, P.D., Baumstark, A.L. and Landis, M.E. (1977) Rearrangement of tetramethyl-1,2- dioxetane by boron trifluoride in aprotic solvents. *Journal of the American Chemical Society*, **99** (6), 1890–1892.

35 Baumstark, A.L. and Vasquez, P.C. (1984) Reaction of tetramethyl-1,2- dioxetane with phosphines: deuterium isotope effects. *Journal of Organic Chemistry*, **49** (5), 793–798.

36 Bartlett, P.D., Baumstark, A.L. and Landis, M.E. (1973) Insertion reaction of triphenylphosphine with tetramethyl-1,2-dioxetane. Deoxygenation of a dioxetane to an epoxide. *Journal of the American Chemical Society*, **95** (19), 6486–6487.

37 Vasquez, P.C., Chen, Y.-X. and Baumstark, A.L. (1995) The reaction of trivalent phosphorus compounds with 1,2-dioxetanes and 1,2-dioxolanes, in *Advances in Oxygenated Processes* (ed. A.L. Baumstark), JAI Press, Greenwich, CT, pp. 203–224.

38 Clennan, E.L. and Foote, C.S. (1992) Endoperoxides, in *Organic Peroxides* (ed. W. Ando), John Wiley & Sons, Chichester, UK, pp. 255–318.

39 Balci, M. (1981) Bicyclic endoperoxides and synthetic applications. *Chemical Reviews*, **81** (1), 91–108.

40 Bartoschek, A., El-Idreesy, T.T., Griesbeck, A.G., Höinck, L.-O., Lex, J., Miara, C. and Neudörfl, J.M. (2005) A family of new 1,2,4-trioxanes by photooxygenation of allylic alcohols in sensitizer-doped polymers and secondary reactions. *Synthesis*, **14**, 2433–2444.

41 Clennan, E.L. (1991) Synthetic and mechanistic aspects of 1,3-diene photooxidation. *Tetrahedron*, **47** (8), 1343–1382.

42 Van den Heuvel, C.J.M., Hofland, A., Steinberg, H. and De Boer, T.J. (1980) The photo-oxidation of hexamethylbenzene and pentamethylbenzene by singlet oxygen. *Recueil des Travaux Chimiques des Pays-Bas*, **99** (9), 275–278.

43 Heuvel, C.J.M., Steinberg, H. and De Boer, T.J. (1977) The photo-oxidation of hexamethylbenzene by singlet oxygen. *Recueil des Travaux Chimiques des Pays-Bas*, **96** (6), 157–159.

44 Briviba, K., Debasagayam, T.P.A., Sies, H. and Steenken, S. (1993) Selective para-hydroxylation of phenol and aniline by singlet molecular oxygen. *Chemical Research in Toxicology*, **6** (4), 548–553.

45 A similar reaction was conducted with 1,5-dihydroxynaphthalene: Shimakoshi, H., Baba, T., Iseki, Y., Endo, A., Adachi, C., Watanabe, M. and Hisaeda, Y. (2008) Photosensitizing properties of the porphycene immobilized in sol–gel-derived silica coating films. *Tetrahedron Letters*, **49** (43), 6198–6201.

46 Aubry, J.-M., Pierlot, C., Rigaudy, J. and Schmidt, R. (2003) Reversible binding of oxygen to aromatic compounds. *Accounts of Chemical Research*, **36** (9), 668–675.

47 Wasserman, H.H. and Larsen, D.L. (1972) Formation of 1,4-endoperoxides from the dye-sensitized photo-oxygenation of alkyl-naphthalenes. *Journal of the Chemical Society, Chemical Communications*, (5), 253–254.

48 Wasserman, H.H., Wiberg, K.B., Larsen, D.L. and Parr, J. (2005) Photooxidation of methylnaphthalenes. *Journal of Organic Chemistry*, **70** (1), 105–109.

49 Hart, H. and Oku, A. (1972) Octamethyl-naphthalene 1,4-Endoperoxide. *Journal of the Chemical Society, Chemical Communications*, (5), 254–255.

50 Martinez, G.R., Ravanat, J.-L., Medeiros, M.H.G., Cadet, J. and Di Mascio, P. (2000) Synthesis of a naphthalene endoperoxide as a source of ^{18}O-labeled singlet oxygen for mechanistic studies. *Journal of the American Chemical Society*, **122** (41), 10212–10213.

51 Pierlot, C. and Aubry, J.-M. (1997) First evidence of the formation of 5,8-endoperoxide from the oxidation of 1,4-disubstituted naphthalene by singlet oxygen. *Journal of the Chemical Society,*

Chemical Communications, (23), 2289–2290.

52 Pierlot, C., Poprawski, J., Marko, J. and Aubry, J.-M. (2000) Effects of oxygenated substituents on the [4 + 2] cycloaddition of singlet oxygen in the photooxygenation of water-soluble naphthyl ethers. *Tetrahedron Letters*, **41** (25), 5063–5067.

53 Aubry, J.-M., Mandard-Cazin, B., Rougee, M. and Bensasson, R.V. (1995) Kinetic studies of singlet oxygen [4 + 2]-cycloadditions with cyclic 1,3-dienes in 28 solvents. *Journal of the American Chemical Society*, **117** (36), 9159–9164.

54 Fuchter, M.J., Hoffman, B.M. and Barrett, A.G.M. (2006) Ring-opening metathesis polymer sphere-supported seco-porphyrazines: efficient and recyclable photooxygenation catalysts. *Journal of Organic Chemistry*, **71** (2), 724–729.

55 Barrett, A.G.M., Hopkins, B.T. and Kibberling, J. (2002) ROMP gel reagents in parallel synthesis. *Chemical Reviews*, **102** (10), 3301–3324.

56 Schmidt, R., Drews, W. and Bauer, H.-D. (1982) Developments of a new photochromic structural principle based on reversible photooxidation. *Journal of Photochemistry*, **18** (4), 365–378.

57 Bauer, H.-D. and Schmidt, R. (2000) New assignment of the electronically excited states of anthracene-9,10-endoperoxide and its derivatives. *Journal of Physical Chemistry A*, **104**, 164–165.

58 Linker, T. and Fudickar, W. (2006) Imaging by sensitized oxygenations of photochromic anthracene films: examination of effects that improve performance and reversibility. *Chemistry – A European Journal*, **12** (36), 9276–9283.

59 Nardello, V. and Aubry, J.-M. (2000) Measurement of photogenerated singlet oxygen in aqueous media. *Methods in Enzymology*, **319**, 50–58.

60 Adam, W. and Prein, M. (1996) π-facial diastereoselectivity in the [4 + 2] cycloaddition of singlet oxygen as a mechanistic probe. *Accounts of Chemical Research*, **29** (6), 275–283.

61 Frizsche, J. (1867) *Comptes Rendues*, **64**, 1035–1037.

62 Liang, Z., Zhao, W., Wang, S., Tang, Q., Lam, S.-C. and Miao, Q. (2008) Unexpected photooxidation of H-bonded tetracene. *Organic Letters*, **10** (10), 2007–2010.

63 Ono, K., Totani, H., Hiei, T., Yoshino, A., Saito, K., Eguchi, K., Tomura, M., Nishida, J.-I. and Yamashita, Y. (2007) Photooxidation and reproduction of pentacene derivatives substituted by aromatic groups. *Tetrahedron*, **63** (39), 9699–9704.

64 Yamada, H., Yamashita, Y., Kikuchi, M., Watanabe, H., Okujima, T., Uno, H., Ogawa, T., Ohara, K. and Ono, N. (2005) Photochemical synthesis of pentacene and its derivatives. *Chemistry – A European Journal*, **11**, 6212–6220.

65 Sparfel, D., Gobert, F. and Rigaudy, J. (1980) Transformations thermiques des photooxydes *méso* des acénes–VI: Cas des photooxydes de pentacenes. *Tetrahedron*, **36** (15), 2225–2235.

66 Hamon, M.A., Stensaas, K.L., Sugar, M.A., Tumminello, K.C. and Allred, A.K. (2007) Reacting soluble single-walled carbon nanotubes with singlet oxygen. *Chemical Physics Letters*, **447** (1–3), 1–4.

67 Lebedkin, S., Kareev, I., Hennrich, F. and Kappes, M.M. (2008) Efficient quenching of singlet oxygen via energy transfer to semiconducting single-walled carbon nanotubes. *Journal of Physical Chemistry C*, **112** (42), 16236–16239.

68 Zhu, Z., Tang, Z., Phillips, J.A., Yang, R., Wang, H. and Tan, W. (2008) Regulation of singlet oxygen generation using single-walled carbon nanotubes. *Journal of the American Chemical Society*, **130** (33), 10856–10857.

69 O'Shea, K. and Foote, C.S. (1989) Quantitative rearrangement of monocyclic endoperoxides to furans catalyzed by cobalt(II). *Journal of Organic Chemistry*, **54** (14), 3475–3477.

70 Rigaudy, J., Capdevielle, P., Combrisson, S. and Maumy, M. (1974) Concerted opening of adducts of bis[1,2-(*E*,*E*)-benzylidene]-*cis*-3,4-diphenylcyclobutane. Reassignment of structures of derived products. *Tetrahedron Letters*, **32**, 2757–2760.

71 Manring, L.E. and Foote, C.S. (1983) Chemistry of singlet oxygen. 44. Mechanism of photooxidation of 2,5-dimethyl-2,4-diene and 2-methyl-2-pentene. *Journal of the American Chemical Society*, **105**, 4710–4717.

72 Clennan, E.L. and Mohrsheikh-Mohammadi, M.E. (1984) Mechanism of endoperoxide formation. 3. Utilization of the Young and Carlsson kinetic techniques. *Journal of the American Chemical Society*, **106** (23), 7112–7118.

73 O'Neill, P.M., Rawe, S.L., Storr, R.C., Ward, S.A. and Posner, G.H. (2005) Lewis acid catalysed rearrangements of unsaturated bicyclic [2,2. n] endoperoxides in the presence of vinyl silanes; access to novel Fenozan BO-7 analogues. *Tetrahedron Letters*, **46** (17), 3029–3032.

74 Coughlin, D.J. and Salomon, R.G. (1977) Synthesis and thermal reactivity of some 2,3-dioxabicyclo[2.2.1]heptane models of prostaglandin endoperoxides. *Journal of the American Chemical Society*, **99** (2), 655–657.

75 Kao, T.-C., Chuang, G.J. and Liao, C.-C. (2008) Photooxygenation of masked o-benzoquinones: an efficient entry into highly functionalized cyclopentenones from 2-methoxyphenols. *Angewandte Chemie, International Edition*, **47** (38), 7325–7327.

76 Gollnick, K. and Griesbeck, A. (1983) [4 + 2]-cycloaddition of singlet oxygen to 2,5-dimethylfuran: Isolation and reactions of the monomeric and dimeric endoperoxides. *Angewandte Chemie, International Edition*, **22** (29), 726–727.

77 Kang, P. and Foote, C.S. (2000) Synthesis of a ^{13}C, ^{15}N labeled imidazole and characterization of the 2,5-endoperoxide and its decomposition. *Tetrahedron Letters*, **41** (49), 9623–9626.

78 Sheu, C. and Foote, C.S. (1993) Endoperoxide formation in a guanosine derivative. *Journal of the American Chemical Society*, **115** (22), 10446–10447.

79 Greer, A. (2006) Christopher Foote's discovery of the role of singlet oxygen [$^{1}O_{2}$ ($^{1}\Delta_{g}$)] in photosensitized oxidation reactions. *Accounts of Chemical Research*, **39**, 797–804.

80 Foote, C.S. and Wexler, S. (1964) Singlet oxygen. A probable intermediate in photosensitized autoxidations. *Journal of the American Chemical Society*, **86**, 3880–3881.

81 Patil, S.N., Stephens, B.E. and Liu, F. (2008) A tandem Baylis–Hillman-singlet oxygen oxidation reaction for facile synthesis of γ-substituted γ-hydroxybutenolides. *Tetrahedron*, **64** (48), 10831–10836.

82 Montagnon, T., Tofi, M. and Vassilikogiannakis, G. (2008) Using singlet oxygen to synthesize polyoxygenated natural products from furans. *Accounts of Chemical Research*, **41** (8), 1001–1011.

83 Georgiou, T., Montagnon, T., Tofi, M. and Vassilikogiannakis, G. (2006) A versatile and general one-pot method for synthesizing bis-spiroketal motifs. *Organic Letters*, **8** (9), 1945–1948.

84 (a) Boger, D.L. and Baldino, C.M. (1991) Singlet oxygen mediated oxidative decarboxylation of pyrrole-2-carboxylic acids. *Journal of Organic Chemistry*, **56**, 6942–6944; (b) de Mayo, P. and Reid, S.T. (1962) The photooxidation of pyrrole: A simple synthesis of maleimide. *Chemistry & Industry*, **35**, 1576–1577.

85 (a) Dastan, A. and Balci, M. (2006) Chemistry of dioxine-annelated cycloheptatriene endoperoxides and their conversion into tropolone derivatives: an unusual non-benzenoid singlet oxygen source. *Tetrahedron*, **67** (17), 4003–4010; (b) Mete, E., Altundas, R., Secen, H.

and Balci, M. (2003) Studies on the mechanism of base-catalyzed decomposition of bicyclic endoperoxides. *Turkish Journal of Chemistry*, **27**, 145–153.

86 Scott, L.T., Erden, I., Brunsvold, W.R., Schultz, T.H., Houk, K.N. and Paddon-Row, M.N. (1982) Competitive [6 + 2], [4 + 2], and [2 + 2] cycloadditions. Experimental classification of two-electron cycloaddends. *Journal of the American Chemical Society*, **104** (13), 3659–3664.

87 Gandy, M.N. and Piggott, M.J. (2008) Synthesis of kalasinamide, a putative plant defense phototoxin. *Journal of Natural Products*, **71**, 866–868.

88 Clennan, E.L. and Heah, P.C. (1984) Bicyclic dioxaphosphorane. 4. A kinetic investigation of the reactions of trivalent phosphorus compounds with bicyclic endoperoxides. *Journal of Organic Chemistry*, **49**, 2284–2286.

89 Stratakis, M. and Orfanopoulos, M. (2000) Regioselectivity in the ene reaction of singlet oxygen with alkenes. *Tetrahedron*, **56** (12), 1595–1615.

90 Alberti, M.N. and Orfanopoulos, M. (2006) Stereoelectronic and solvent effects on the allylic oxyfunctionalization of alkenes with singlet oxygen. *Tetrahedron*, **62** (46), 10660–10675.

91 Lissi, E.A., Lemp, E. and Zanocco, A.L. (2001) Singlet-oxygen reactions: solvent and compartmentalization effects, in *Understanding and Manipulating Excited State Processes*, vol. 8 (eds V. Ramamurthy and K.S. Schanze), Marcel Dekker Inc., pp. 287–316.

92 Clennan, E.L. (2003) Molecular oxygenations in zeolites. *Molecular and Supramolecular Photochemistry*, **9**, 275–308.

93 Hino, T., Anzai, T. and Kuramoto, N. (2006) Visible-light induced solvent-free photooxygenations of organic substrates by using [60]fullerene-linked silica gels as heterogeneous catalysts and as solid-phase reaction fields. *Tetrahedron Letters*, **47** (9), 1429–1432.

94 Griesbeck, A.G., El-Idreesy, T.T. and Bartoschek, A. (2005) Photooxygenation in polymer matrices: En route to highly active antimalarial peroxides. *Pure and Applied Chemistry*, **77** (7), 1059–1074.

95 Bartoschek, A., El-Idreesy, T.T., Griesbeck, A.G., Höink, L.-O., Lex, J., Miara, C. and Neudörfl, J.M. (2005) A family of new 1,2,4-trioxanes by photooxygenation of allylic alcohols in sensitizer-doped polymers and secondary reactions. *Synthesis*, (14), 2433–2444.

96 Fröhlich, L. and Torsten, L. (2004) Simple and diastereoselective synthesis of an A-ring precursor of dihydroxyvitamin D_3 (calcitriol) by photooxygenation. *Synlett*, (15), 2725–2727.

97 Stratakis, M., Orfanopoulos, M. and Foote, C.S. (1998) Reactions of singlet oxygen and N-methyltriazolinediones with β,β-dimethylstyrene. Exceptional syn selectivity in the ene products. *Journal of Organic Chemistry*, **63** (4), 1315–1318.

98 Stratakis, M. and Orfanopoulos, M. (1993) Regioselective formation of cyclic and allylic hydroperoxides. *Synthetic Communications*, **23** (4), 425–430.

99 Griesbeck, A.G., El-Idreesy, T.T., Fiege, M. and Brun, R. (2002) Synthesis of antimalarial 1,2,4-trioxanes via photooxygenation of a chiral allylic alcohol. *Organic Letters*, **4** (24), 4193–4195.

100 Singh, C. and Malik, H. (2005) Protection of the carbonyl group as 1,2,4-trioxane and its regeneration under basic conditions. *Organic Letters*, **7** (25), 5673–5676.

101 Singh, C., Pandey, S., Saxena, G., Srivastava, N. and Sharma, M. (2006) Synthesis of 1,2,4-trioxepanes and 1,2,4-trioxocanes via photooxygenation of homoallylic alcohols. *Journal of Organic Chemistry*, **71** (24), 9057–9061.

102 Stensaas, K.L., Bajaj, A. and Al-Turk, A. (2005) Novel regiochemistry in the aqueous singlet oxygen ene reactions of carboxylic acid salts: a comparison of substrate structure. *Tetrahedron Letters*, **46** (4), 715–718.

103 Stensaas, K.L. and Bajaj, A. (2005) Aqueous solvents dictate the regiochemistry in the photooxidations of α,β-unsaturated carboxylic acid salts. *Synthesis*, (15), 2623–2624.

104 Natarajan, A., Kaanumalle, L.S., Jockusch, S., Gibb, C.L.D., Gibb, B.C., Turro, N.J. and Ramamurthy, V. (2007) Controlling photoreactions with restricted spaces and weak intermolecular forces: exquisite selectivity during oxidation of olefins by singlet oxygen. *Journal of American Chemical Society*, **129** (14), 4132–4133.

105 Greer, A. (2007) Molecular cross-talk. *Nature*, **447**, 273–274.

106 Aebisher, D., Azar, N.S., Zamadar, M., Gandra, N., Gafney, H.D., Gao, R. and Greer, A. (2008) Singlet oxygen chemistry in water: a porous vycor glass supported photosensitizer. *Journal of Physical Chemistry B*, **112** (7), 1913–1917.

107 Di Magno, S.G., Dussault, P.H. and Schultz, J.A. (1996) Fluorous biphasic singlet oxygenation with a perfluoroalkylated photosensitizer. *Journal of American Chemical Society*, **118** (22), 5312–5313.

108 Midden, W.R. and Wang, S.Y. (1983) Singlet oxygen generation for solution kinetics: clean and simple. *Journal of American Chemical Society*, **105**, 4129–4135.

109 (a) Pace, A.P.P., Buscemi, S., Vivona, N. and Clennan, E.L. (2007) Photooxidations of alkenes in fluorinated constrained media: fluoro-organically modified nay as improved reactors for singlet oxygen "ene" reactions. *Journal of Organic Chemistry*, **72** (7), 2644–2646; (b) Pace, A. and Clennan, E.L. (2002) A new experimental protocol for intrazeolite photooxidations. The first product-based estimate of an upper limit for the intrazeolite singlet oxygen lifetime. *Journal of the American Chemical Society*, **124**, (38), 11236–11237; (c) Cojocaru, B., Laferrière, M., Carbonell, E., Parvulescu, V., García, H. and Scaiano, J.C. (2008) Direct time-resolved detection of singlet oxygen in zeolite-based photocatalysts. *Langmuir*, **24** (9), 4478–4481.

110 Clennan, E.L. (2008) Mechanisms of oxygenations in zeolites. *Advances in Physical Organic Chemistry*, **42**, 225–269.

111 Stratakis, M., Raptis, C., Sofikiti, N., Tsangarakis, C., Kosmas, G., Zaravinos, I.-P., Kalaitzakis, D., Stavroulakis, D., Baskakis, C. and Stathoulopoulou, A. (2006) Intrazeolite photooxygenation of chiral alkenes. Control of facial selectivity by confinement and cation–π interactions. *Tetrahedron*, **62** (46), 10623–10632.

112 Li, X. and Ramamurthy, V. (1996) Selective oxidation of olefins within organic dye cation-exchanged zeolites. *Journal of the American Chemical Society*, **118** (43), 10666–10667.

113 Singleton, D.A., Szymanski, M.J., Meyer, M.P., Leach, A.G., Kuwata, K.T., Chen, J.S., Greer, A., Foote, C.S. and Houk, K.N. (2003) Mechanism of ene reactions of singlet oxygen. A Two-Step No-intermediate mechanism. *Journal of the American Chemical Society*, **125**, 1319–1328.

114 Adam, W., Cheng, C.C., Cueto, O., Erden, I. and Zinner, K. (1979) A stable bisdioxetane. *Journal of the American Chemical Society*, **101** (16), 4735–4736.

115 Rio, G., Bricout, D. and Lacombe, L. (1972) Sensitized photooxidation of benzhydrylidenecyclobutane. Addition of singlet oxygen to a conjugated ethylenic-aromatic system. Determination of the structure of the bis-peroxide formed. *Tetrahedron Letters*, **13** (34), 3583–3586.

116 Rio, G., Bricout, D. and Lacombe, L. (1973) Sensitized photooxidation of benzhydrylidenecyclobutane. Addition of singlet oxygen to a conjugated ethylenic-aromatic system. *Tetrahedron*, **29** (22), 3553–3563.

117 Zhang, X., Feng, L. and Foote, C.S. (1995) Sensitized photooxygenation of 6-heteroatom-substituted fulvenes: Primary products and their chemical

transformations. *Journal of Organic Chemistry*, **60** (5), 1333–1338.

118 Salamci, E., Secen, H., Sütbeyaz, Y. and Balci, M. (1997) A concise and convenient synthesis of dl-*proto*-quercitol and dl-*gala*-quercitol *via* ene reaction of singlet oxygen combined with [2 + 4] cycloaddition to cyclohexadiene. *Journal of Organic Chemistry*, **62** (8), 2453–2457.

119 Adam, W., Balci, M. and Kilic, H. (2000) 2,3-Dioxabicyclo[2.2.2]oct-7-en-5-one: synthesis and reactions of the keto endoperoxide of phenol. *Journal of Organic Chemistry*, **65** (19), 5926–5931.

120 Gültekin, M.S., Salamci, E. and Balci, M. (2003) A novel and short synthesis of (1,4/2)-cyclohex-5-ene-triol and its conversion to (±)-proto-quercitol. *Carbohydrate Research*, **338** (16), 1615–1619.

121 Kishali, N.H., Sahin, E. and Kara, Y. (2006) An intramolecular substitution of a hydroperoxy-endoperoxide to a bis-endoperoxide. *Organic Letters*, **8** (9), 1791–1793.

122 Blay, G., Garcia, B., Molina, E. and Pedro, J.R. (2005) Total syntheses of four stereoisomers of 4α-hydroxy-1β,7β-peroxy-10βH-guaia-5-ene. *Organic Letters*, **7** (15), 3291–3294.

123 Adam, W., Balci, M. and Kilic, H. (1998) Oxyfunctionalization of biphenylene by singlet oxygen, hydrogen peroxide, methyltrioxorhenium, and dimethyldioxirane. *Journal of Organic Chemistry*, **63** (23), 8544–8546.

124 Sofikiti, N., Tofi, M., Montagnon, T., Vassilikogiannakis, G. and Stratakis, M. (2005) Synthesis of the spirocyclic core of the prunolides using a singlet oxygen-mediated cascade sequence. *Organic Letters*, **7** (12), 2357–2359.

125 Adam, W. and Rodriguez, A. (1981) Intramolecular silyl migration in the singlet oxygenation of 2-methyl-5-trimethylsilylfuran. *Tetrahedron Letters*, **22** (36), 3505–3508.

126 Margaros, I., Montagon, T. and Vassilikogiannakis, G. (2007) Spiroperoxy lactones from furans in one pot: synthesis of (+)-Premnalane A. *Organic Letters*, **9** (26), 5585–5588.

127 McCallum, J.E.B., Kuniyoshi, C.Y. and Foote, C.S. (2004) Characterization of 5-hydroxy-8-oxo-7,8-dihydroguanosine in the photosensitized oxidation of 8-oxo-7,8-dihydroguanosine and its rearrangement to spiroiminodihydantoin. *Journal of the American Chemical Society*, **126**, 16777–16782.

128 Luo, W., Muller, J.G., Rachlin, E.M. and Burrows, C.J. (2000) Characterization of spiroiminodihydantoin as a product of one-electron oxidation of 8-oxo-7,8-dihydroguanosine. *Organic Letters*, **2** (5), 613–616.

129 (a) Duarte, V., Muller, J.G. and Burrows, C.J. (1999) Insertion of dGMP and dAMP during *in vitro* DNA synthesis opposite an oxidized form of 7,8-dihydro-8-oxoguanine. *Nucleic Acids Research*, **27**, 496–502; (b) Ye, Y., Muller, J.G., Luo, W., Mayne, C.L., Shallop, A.J., Jones, R.A. and Burrows, C.J. (2003) Formation of $^{13}C,.^{15}N$, and ^{18}O-Labeled guanidinohydantoin from guanosine oxidation with singlet oxygen. implications for structure and mechanism. *Journal of the American Chemical Society*, **125**, 13926–13927; (c) Munk, B.H., Burrows, C.J. and Schlegel, H.B. (2008) An exploration of mechanisms for the transformation of 8-oxoguanine to guanidinohydantoin and spiroiminodihydantoin by density functional theory. *Journal of the American Chemical Society*, **130**, 5245–5256.

130 (a) Cadet, J., Douki, T. and Ravanat, J.L. (2008) Oxidatively generated damage to the guanine moiety of DNA: mechanistic aspects and formation in cells. *Accounts of Chemical Research*, **41**, (8), 1075–1083; (b) Gimisis, T. and Cismas, C. (2006) Isolation, characterization, and independent synthesis of guanine oxidation products. *European Journal of Organic Chemistry*, **6**, 1351–1378.

131 Washington, I., Jockosch, S., Itagaki, Y., Turro, N.J. and Nakanishi, K. (2005) Superoxidation of bisretinpoids. *Angewandte Chemie, International Edition*, **44**, 7097–7100.

132 Bonesi, S.M., Fagnoni, M. and Albini, A. (2008) Photosensitized electron transfer oxidation of sulfides: a steady-state study. *European Journal of Organic Chemistry*, **15**, 2612–2620.

133 Baciocchi, E., Del Giacco, T., Elisei, F., Gerini, M.F., Guerra, M., Lapi, A. and Liberali, P. (2003) Electron transfer and singlet oxygen mechanisms in the photooxygenation of dibutyl sulfide and thioanisole in MeCN sensitized by N-Methylquinolinium tetrafluoroborate and 9,10-dicyanoanthracene. The probable involvement of a thiadioxirane intermediate in electron transfer photo-oxygenations. *Journal of the American Chemical Society*, **125** (52), 16444–16454.

134 Wenhui, Zhou and Clennan, Edward L. (1999) Organic reactions in zeolites. 1. photooxidations of sulfides in methylene blue doped zeolite Y. *Journal of the American Chemical Society*, **121** (12), 2915–2916.

12
Synthesis of Heteroaromatics via Rearrangement Reactions

Nicolò Vivona, Silvestre Buscemi, Ivana Pibiri, Antonio Palumbo Piccionello, and Andrea Pace

12.1
Introduction

The varied family of heteroaromatics contains a large fraction of organic compounds, many of which have found important applications in different fields. Many synthetic methodologies have been developed to obtain target systems and, among them, photochemical methods represent valid and elegant alternatives when the final targets are difficult to obtain through *ground-state* chemistry. In this chapter, we describe the photochemical synthesis of heteroaromatic compounds through rearrangement reactions, particularly via ring-transformations of heterocyclic precursors [1–7].

A classification of all photoinduced heterocyclic ring-transformation reactions is quite difficult to achieve due to the large variety of mechanistic pathways, experimental conditions and affecting parameters. In some cases, the photochemical product and the starting substrate have the same heterocyclic nucleus. In this chapter, we will refer to such reactions, characterized by the identity of the starting and final ring, as "ring-degenerate" rearrangements.

In general, photoinduced rearrangements of heterocycles leading to heteroaromatic compounds may involve ring-contraction, internal cyclization, ring-expansion, ring-opening or ring-closure, and ring-fragmentation processes. As for five-membered rings, two processes will be repeatedly quoted throughout this chapter:

- The *ring contraction–ring expansion* route (RCRE), which explains the formal interchange between two adjacent ring-atoms. This results from the photoinduced cleavage of the weakest bond in the ring with formation of a three-membered ring intermediate. The latter is then converted, either thermally or photochemically, into the final five-membered heterocyclic product (see e.g., Scheme 12.22).
- The *internal cyclization–isomerization* route (ICI), which involves the formation of a bicyclic species featuring a single bond between the ring-atoms in the 2- and 5-positions of the original heterocycle. Sigmatropic shifts then occur, leading to different bicyclic intermediates from which the final products will arise (see e.g., Scheme 12.2).

Handbook of Synthetic Photochemistry. Edited by Angelo Albini and Maurizio Fagnoni
Copyright © 2010 WILEY-VCH Verlag GmbH & Co. KGaA, Weinheim
ISBN: 978-3-527-32391-3

In other rearrangements, side-chain substituents can act as an internal reagent, so that the atoms originally belonging to the side chain end up being ring-atoms in the final heterocycle. In some cases, the photoinduced loss of stable molecules, such as N_2, CO_2, or RCN, from the heterocyclic nucleus of the substrate leads to open-chain intermediates which can further cyclize, either spontaneously or in the presence of a catalyst, into the final heterocycle (see e.g., Scheme 12.21). In other cases, an added external reagent can react with the primary photoproduct and is involved in the formation of the target molecule or of its precursors (see e.g., Scheme 12.28, last line).

Although several photochemical reactions have been found to produce a heterocyclic compound, many of these are not of synthetic importance, as they are devoted to the study of the photochemical reactivity of a given substrate, or to mechanistic aspects. The aim of this chapter is to provide the reader with a guide to the synthetic aspects of photoinduced ring-rearrangements which lead to heteroaromatics, mainly five-membered heterocycles containing one, two, or three heteroatoms. In addition, some representative examples of photoinduced ring-transformations in the synthesis of six-membered heteroaromatics and seven-membered heterocyclic targets are reported. The material will be presented according to the target molecule rather than to the class of substrate or type of photoreaction. The literature coverage is not exhaustive; examples showing general synthetic applications have been preferred. In some cases, the availability of the starting material has been considered an important factor in determining the practicality of the photochemical strategy in a synthetic project. In describing the reactions, priority will be given to the experimental details, such as irradiation conditions and isolated yields from preparative-scale reactions, rather than to mechanistic aspects; the latter point will be commented on, if necessary, for a better understanding or to explain particularly interesting reactions. Finally, in the case of more than one similar photochemical synthesis, recent literature or higher-yielding reactions have been quoted.

12.2
Synthesis of Five-Membered Rings with One Heteroatom

12.2.1
Pyrroles

The photochemical synthesis of targeted pyrroles can be achieved from other pyrrole precursors. Some of these reactions consist of a photoinduced migration of groups linked to the ring nitrogen toward a ring carbon atom. For instance, 2-acetylpyrrole **1** can be obtained in 54% yield by irradiation (at 254 nm in dioxane) of *N*-acetyl-pyrrole **2a**. In the case of *N*-acetyl-2,5-dimethylpyrrole **2b**, where both the positions adjacent to the N(1) are occupied, only the 3-acetyl-isomer **3** (36%) was obtained (Scheme 12.1) [8]. This synthetic strategy could assume some significance considering the easy accessibility of a variety of N-substituted pyrroles.

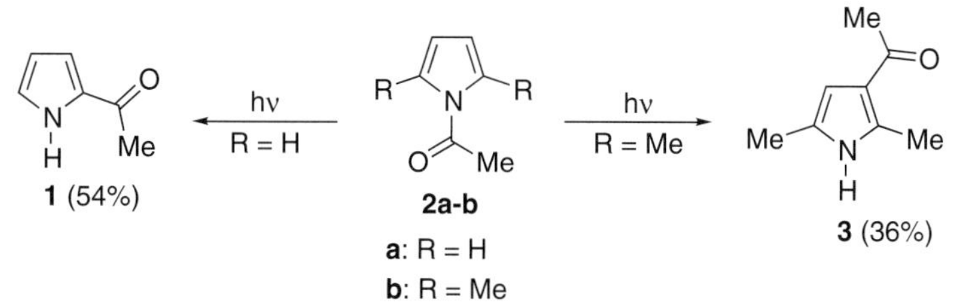

12.2 Synthesis of Five-Membered Rings with One Heteroatom | 389

Scheme 12.1

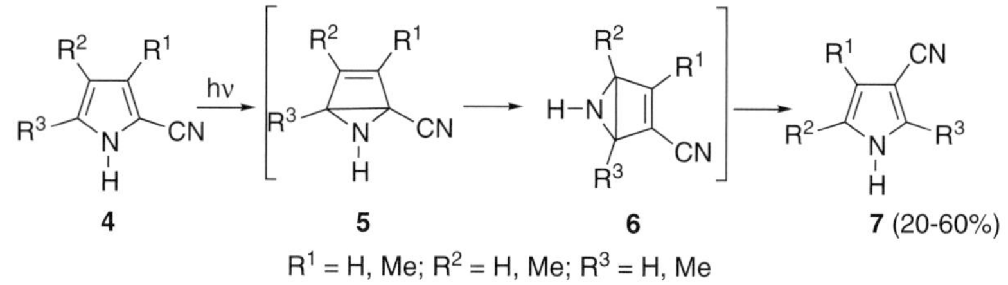

Scheme 12.2

$R^1 = H, Me; R^2 = H, Me; R^3 = H, Me$

On the other hand, the irradiation of 2-cyanopyrrole **4** ($R^1 = R^2 = R^3 = H$), in methanol with a low-pressure Hg lamp, produces the 3-cyanopyrrole **7** ($R^1 = R^2 = R^3 = H$) as a major photoproduct (55%) through the ring-degenerate ICI rearrangement illustrated above (formation of a bond between positions 2 and 5) [9]. Similarly, 3-cyanopyrroles **7** (R^1 or R^2 or $R^3 = Me$) can be obtained by irradiation of the corresponding monomethylated 2-cyanopyrroles (Scheme 12.2) [10].

An important application of photoinduced ICI rearrangements in the pyrrole series regards the synthesis of 3-trimethylsilyl- **8** and 3,4-*bis*(trimethylsilyl)-*N*-methylpyrrole **11**, which are useful synthons for 3-substituted and 3,4-disubstituted pyrroles. The former can be obtained in high yields by irradiation, in pentane with a 450 W Hanovia Hg lamp, of 2-(trimethylsilyl)-*N*-methylpyrrole **9** (R = H). Under similar conditions, the irradiation of compound **9** (R = SiMe₃) afforded a mixture of 2,3- and 3,4-*bis*(trimethylsilyl)-*N*-methylpyrroles **10** and **11** (Scheme 12.3) [11]. Similarly to 2-cyano-substituted pyrroles, this photoisomerization takes advantage of the ability of the silyl group to stabilize an adjacent negative charge that may be present in a polarized structure of the excited state for the internal cyclization.

Pyrroles can be photochemically obtained also from other heterocyclic precursors. For instance, high yields of 1-arylpyrroles **14** are formed by photolysis, at 254 nm in

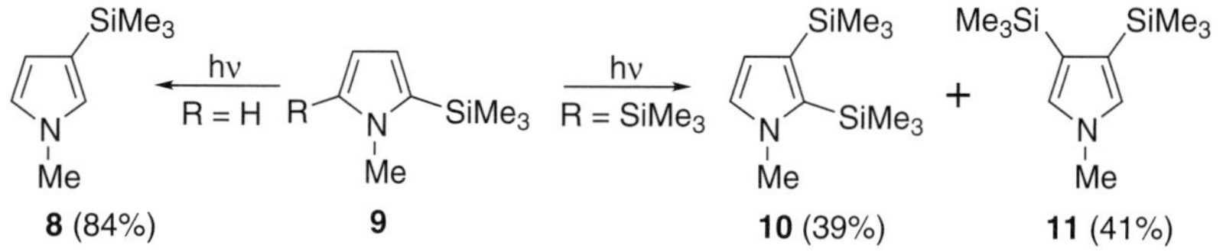

Scheme 12.3

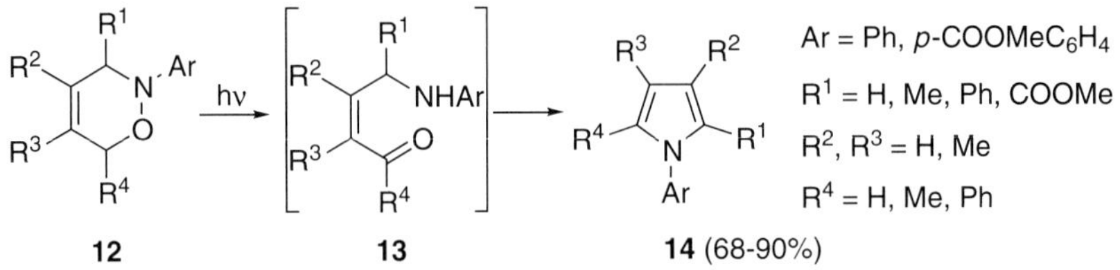

Ar = Ph, p-COOMeC6H4
R1 = H, Me, Ph, COOMe
R2, R3 = H, Me
R4 = H, Me, Ph
12
13
14 (68-90%)

Scheme 12.4

methanol, of dihydro-1,2-oxazines **12**, which are easily obtained by reaction of nitrosoarenes with the appropriate diene (Scheme 12.4) [12, 13].

The reaction occurs through a photoinduced cleavage of the ring O–N bond, followed by ring closure of the open-chain intermediate **13**. Interestingly, rearranged pyrroles **14** (Ar = *p*-COOMe-C_6H_4) lead to the corresponding *N*-unsubstituted pyrroles by Birch reduction [13].

On the other hand, 2-acylpyrroles **17** can be obtained by photoinduced ring contraction of pyridine-*N*-oxides **15**. (Scheme 12.5) [14, 15]. Unfortunately, medium to low yields of target pyrroles are usually obtained, and the use of this photosynthetic strategy should be evaluated case by case. In the formation of a series of 2-formylpyrroles **17** ($R^1 = H$), improvement of the yields, from a small percentage to 32–42%, has been claimed by performing the irradiation of compounds **15** ($R^1 = H$) in water and in the presence of an excess of $CuSO_4$, by using a high-pressure Hg lamp [16]. Other irradiations have been conducted on **15** ($R^1 \neq H$) to synthesize the corresponding 2-acylpyrroles **17** [14, 17, 18].

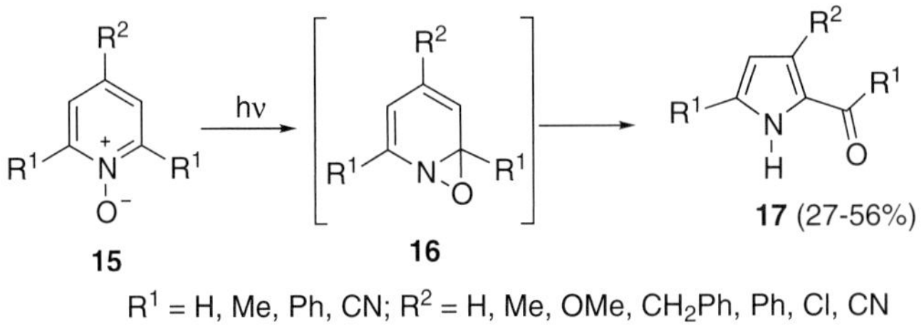

15
16
17 (27-56%)
R1 = H, Me, Ph, CN; R2 = H, Me, OMe, CH2Ph, Ph, Cl, CN

Scheme 12.5

By applying a similar strategy to benzocondensed derivatives, the photoreactivity of quinoline-*N*-oxides has been used efficiently for the synthesis of 3-formylindole targets.

For example, irradiation (in a quartz immersion apparatus with a high-pressure Hg lamp) of compounds **18a** and **18c** in acetonitrile (or of **18b** in methanol) produces 3-formylindoles **20a–c** (37–70%) through the formation of benzoxazepines **19a–c**, followed by ring-contraction (and final decarboxylation in the case of **19b**) [19, 20]. In a typical experiment, the preparative-scale irradiation of **18c**, in acetone (400 W, high-pressure Hg lamp, Pyrex filter, 18 min) allowed the isolation of benzoxazepine **19c** in 94% yield. The latter compound, when irradiated for 8 h in acetonitrile by using the same light source without the Pyrex filter, gave indole **20c** in 70% yield (Scheme 12.6) [19].

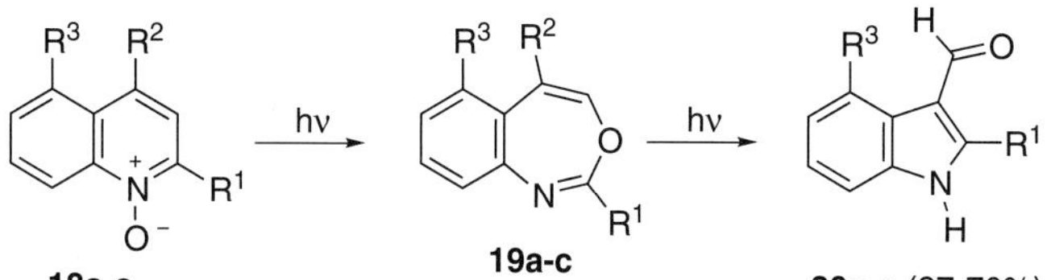

12.2 Synthesis of Five-Membered Rings with One Heteroatom

Scheme 12.6

a: R^1 = Ph, CN; R^2 = R^3 = H
b: R^1 = Ph, Me; R^2 = COOH; R^3 = H
c: R^1 = CN; R^2 = H; R^3 = COOMe

Indole-3-carboxylic acid amides **23a** [21, 22] and esters **23b** [23] can be obtained by irradiation in dichloromethane (DCM) of the 3-diazo-4-oxo-3,4-dihydroquinoline **21** in the presence of alkylamines, dialkylamines, arylamines or alcohols (ZH in Scheme 12.7), through a Wolff–type rearrangement involving a carbenoid species leading to the ketene-like intermediate **22** (Scheme 12.7).

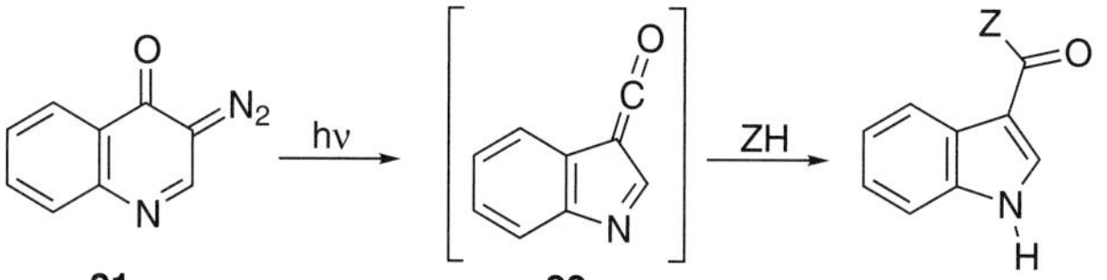

23a: Z = NR^1R^2 (53-70%)
23b: Z = OR (30-60%)

Scheme 12.7

12.2.2
Furans

Photoinduced ring-degenerate transformations in the furan series are scarcely useful for synthetic applications. On the other hand, furans can be obtained photochemically by different heterocyclic precursors such as pyran-4-ones and pyridazine-*N*-oxides. For example, high yields (>95%, analytically determined) of 2-acylfurans **27** are formed (along with variable amounts of pyran-2-one regioisomers depending on the irradiation medium) in the photolysis at 254 nm of 3-alkyl substituted pyran-4-ones **24** in 50% aqueous H_2SO_4, where acidity conditions were proven to be optimal to obtain the highest yield of the furan derivative (Scheme 12.8). Under these

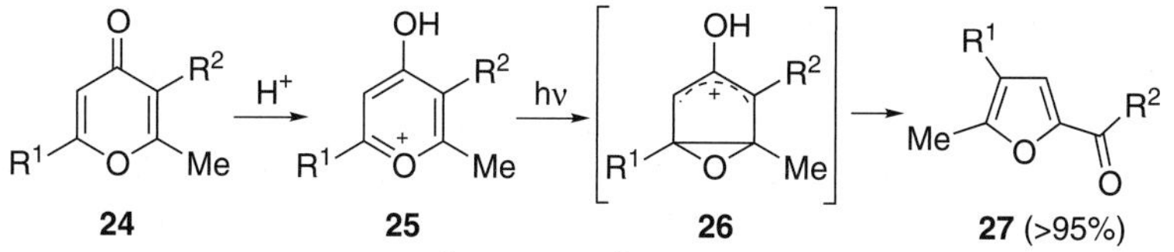

R^1 = H, Me; R^2 = Me, Et

Scheme 12.8

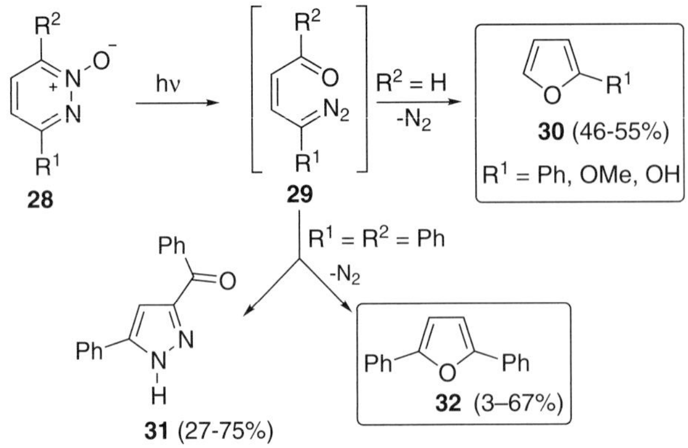

Scheme 12.9

conditions, the pyran-4-ones are present in their 4-hydroxypyrylium cationic form **25**, and the photochemical process consists of an initial internal cyclization to give **26**, which evolves through thermal nucleophilic addition, ring-opening and ring-closure, into finally **27**. After neutralization, the final products are isolated using preparative gas chromatography [24, 25].

2-Substituted furans **30** can be obtained in moderate yields by irradiation, in DCM with a high-pressure Hg lamp, of pyridazine-1-oxides (**28**; $R^2 = H$) opportunely substituted with a group able to stabilize a carbene intermediate. The mechanism involves the open-chain diazo-ketone **29**, which undergoes a photoinduced loss of molecular nitrogen leading to a carbene, which is a precursor of the final furans **30** (Scheme 12.9) [26].

In contrast, in the photolysis of 3,6-diphenylpyridazine-N-oxide **28** ($R^1 = R^2 = Ph$), a mixture of 3-benzoyl-5-phenylpyrazole **31** and 2,5-diphenylfuran **32** is formed. Diazoketone **29** ($R^1 = R^2 = Ph$) undergoes two competing reactions: (i) thermal internal cyclization into the pyrazole **31**; or (ii) photoinduced formation of a carbene leading to the final furan **32** (Scheme 12.9); therefore, the product distribution depends heavily on the reaction conditions. Thus, the formation of pyrazole **31** (75%) is favored over that of furan **32** (3%) by irradiation in acetone in a Rayonet reactor equipped with lamps irradiating at 350 nm. On the other hand, by irradiation with a Hanovia immersion lamp the yield of **31** decreases to 27%, whereas the yield of **32** increases to 67%. Moreover, the exclusive formation of the furan derivative **32** (43%) was observed by irradiation in the Rayonet reactor at low temperatures ($-65\,°C$) [27].

12.2.3
Thiophenes

In contrast to furans, photoinduced ring-degenerate rearrangements are largely represented in the thiophene series. Reported results have been often rationalized in terms of different competing mechanisms such as the *zwitterion-tricyclic* route, which takes advantage of the valence shell expansion of a sulfur atom and involves

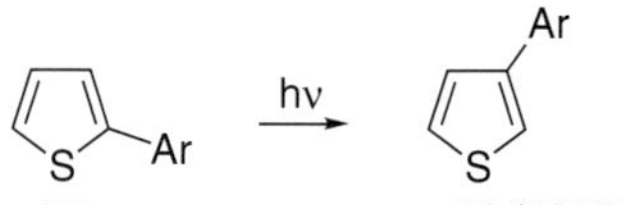

12.3 Synthesis of Five-Membered Rings with Two Heteroatoms

33 → **34** (40%)
Ar = Ph, α-naphthyl, p-tolyl
Scheme 12.10

"benzvalene-like" intermediates, as well as the above-mentioned RCRE and ICI pathways [2, 4].

The photoisomerization of 2-substituted thiophenes into the corresponding 3-substituted derivatives is a common pathway observed for both 2-aryl and 2-alkylthiophenes [4]. However, for synthetic applications, a significant example can be recognized in the photoisomerization of 2-arylthiophenes **33** into the corresponding 3-aryl derivatives **34** by the large-scale irradiation in Et_2O with a high-pressure Hg lamp (Scheme 12.10) [28].

An interesting photochemical synthesis of 2,5-disubstituted-thiophenes **38** stems from the first report of the photoinduced sulfur extrusion from variously 3,6-disubstituted 1,2-dithiins **35a** by exposure to daylight in ethyl acetate or in DCM, or by irradiation at 365 nm in methanol (Scheme 12.11) [29]. Other examples involve naturally occurring thiarubrines **35b** [30], as well as variously substituted synthetic derivatives [31]. This photochemical reactivity has been generalized with an application in the synthesis of selenophene **40** obtained in almost quantitative yield from **39** [32].

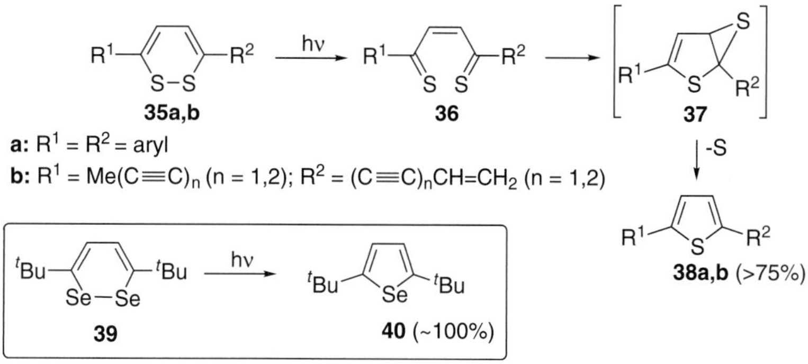

a: $R^1 = R^2$ = aryl
b: $R^1 = Me(C\equiv C)_n$ (n = 1,2); $R^2 = (C\equiv C)_nCH=CH_2$ (n = 1,2)

38a,b (>75%)

40 (~100%)

Scheme 12.11

12.3
Synthesis of Five-Membered Rings with Two Heteroatoms

12.3.1
Pyrazoles

Only a few examples of photoinduced rearrangement reactions for the synthesis of pyrazoles are worthy of mention. For example, N-ethoxycarbonylpyrazoles **43** are formed in the irradiation, in acetone or benzene solutions (high-pressure Hg

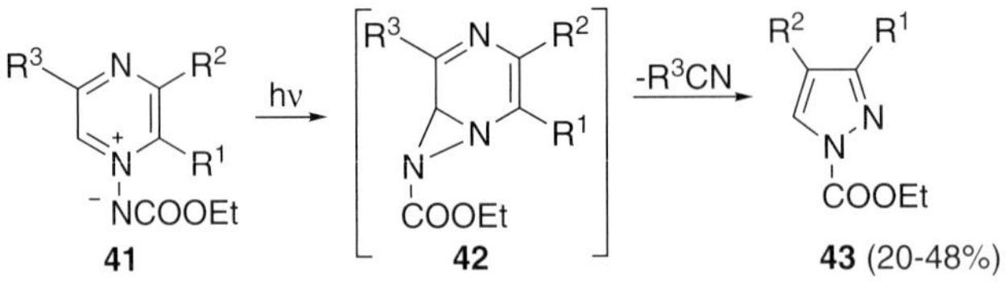

12 Synthesis of Heteroaromatics via Rearrangement Reactions

Scheme 12.12

R^1 = H, Me; R^2 = H, Me, Ph; R^3 = H, Me, Ph

lamp, Pyrex filter), of pyrazinium-N-imides **41** (Scheme 12.12) [33]. This photoreaction is thought to proceed by an initial formation of a diaziridine intermediate **42**, followed by ring-expansion–ring-contraction processes involving the extrusion of a nitrile species and leading to target pyrazoles. A similar ring-contraction into a N-ethoxycarbonylpyrazole is also observed in the irradiation of a pyrimidinium-N-ethoxycarbonyl-imide [33]. Finally, as mentioned above (see Section 12.2), 3-benzoyl-5-phenylpyrazole **31** is formed, together with 2,5-diphenylfuran **32**, from the photoinduced rearrangement of 3,6-diphenylpyridazine-N-oxide (**28**; $R^1 = R^2 =$ Ph) [27].

12.3.2
Imidazoles

Unlike pyrazoles, many imidazoles that have been synthesized by photochemical rearrangement are reported. Some significant examples use imidazoles as precursors, and involve either the migration of the N-linked group, or a ring-degenerate ICI rearrangement. For the first strategy, N-acyl-imidazoles **44a**, when irradiated in tetrahydrofuran (THF), acetonitrile or methanol, with a low- or medium-pressure Hg lamp (depending on the substrate), yield a mixture of 4(or 5)-acyl- **45a** and 2-acylimidazoles **46a**. The reaction has been applied also to the synthesis of 4(or 5)-benzyl-imidazole **45b** and of 2-benzyl-imidazole **46b**, from the irradiation of the N-benzyl derivative **44b** (Scheme 12.13) [34]. Furthermore, 4(or 5)-acetyl-2-alkylimidazoles **45c** can be obtained by preparative-scale irradiation, at 254 nm in THF, of N-acetyl-2-alkylimidazoles **44c** (Scheme 12.13) [35]. This photochemical

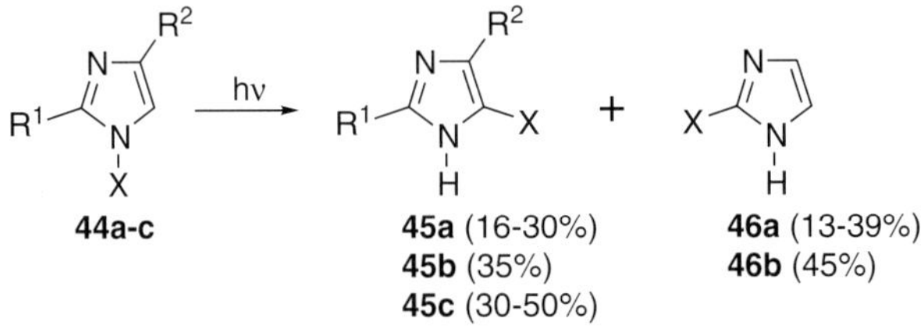

44a-c → **45a** (16-30%), **45b** (35%), **45c** (30-50%) + **46a** (13-39%), **46b** (45%)

a: $R^1 = R^2$ = H; X = COR (R = alkyl, cycloalkyl, alkenyl, Ph)
b: $R^1 = R^2$ = H; X = CH$_2$Ph
c: R^1 = Me, Et, n-Pr, Ph(CH$_2$)$_n$ (n = 1-3); R^2 = H, Me; X = COMe

Scheme 12.13

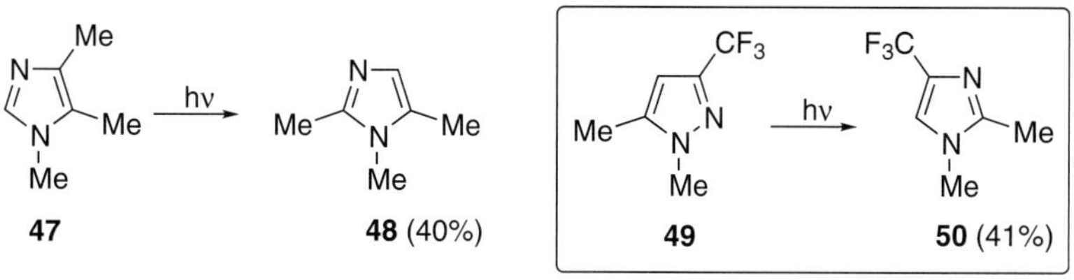

12.3 Synthesis of Five-Membered Rings with Two Heteroatoms

Scheme 12.14

approach appears to be of particular importance due not only to the accessibility of the starting material but also to the difficulties encountered when attempting to introduce acyl groups at the 4(or 5) position of the imidazole nucleus.

An interesting example of preparative-scale photorearrangements proceeding through a ring-degenerate ICI route concerns the formation of 1,2,5-trimethylimidazole **48** by the irradiation of 1,4,5-trimethylimidazole **47** in absolute EtOH with a low pressure-Hg lamp (Scheme 12.14) [36]. Besides imidazoles themselves, other heterocyclic precursors can be used for the photochemical synthesis of heterocycles containing the imidazole nucleus. Photorearrangements of pyrazoles into imidazoles, which are generally obtained in low yields, have been reported; however, not all of these have been explained through the RCRE mechanism [2, 3, 5, 37]. In fact, some experimental details have suggested that both the RCRE and ICI pathways are operative, and that open-chain intermediates are involved in some cases [5, 38]. Consequently, these reactions are not always of synthetic utility. An illustrative and useful example regards the irradiation of the 3-trifluoromethyl substituted pyrazole **49**, which gave only the 4-trifluoromethyl substituted imidazole **50** in moderate isolated yields (41%) through an ICI route (Scheme 12.14) [38]. Mechanistic studies of the photoinduced rearrangements of phenyl-, methyl-, or fluorine-substituted N-methylpyrazoles have been also reported [5].

Variously substituted imidazoles **51** and **53** can be prepared in moderate to good yields by photolysis at 254 nm in different solvents, depending on the substrate, of 1-alkenyl-tetrazoles **52** (Scheme 12.15). The reaction can be considered of general application, and is explained by an intramolecular heterocyclization of the iminonitrene species, arising from fragmentation of the tetrazole ring, involving the alkenyl double bond [39]. Interestingly, irradiation of 1-alkenyl-tetrazoles (**52**; $R^3 = COX$) leads to 4(or 5)-carbonyl-substituted imidazoles **53** (Scheme 12.15) [40].

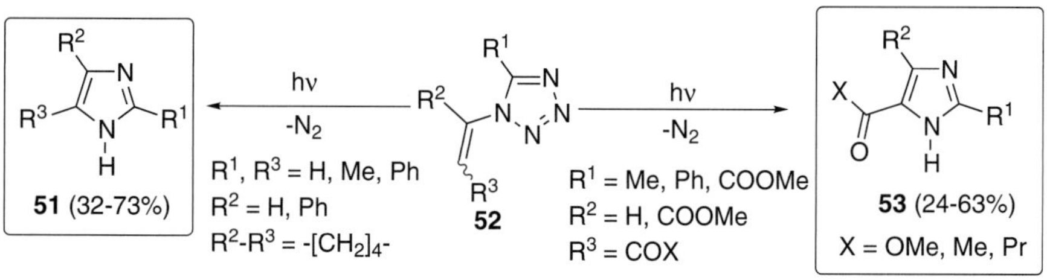

Scheme 12.15

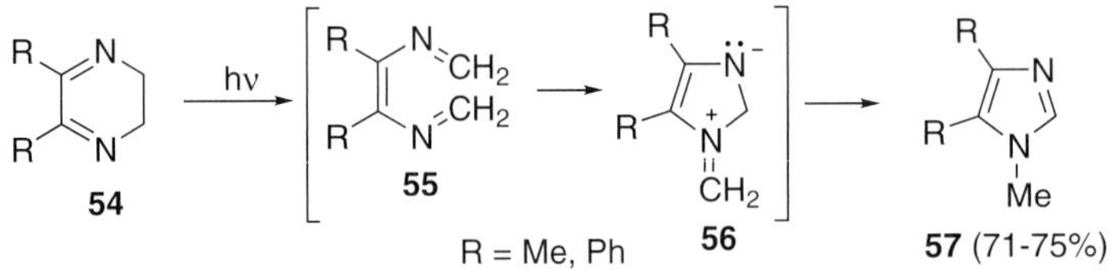

Scheme 12.16

An interesting photochemical approach to 4,5-disubstituted *N*-alkylimidazoles consists of the photolysis of 2,3-dihydro-5,6-disubstituted-pyrazines that can be easily prepared from 1,2-diketones and 1,2-diamino-alkanes. For example, the preparative-scale photolysis, in absolute EtOH with high-pressure Hg lamp (Pyrex filter), of 5,6-dimethyl- or 5,6-diphenyl-2,3-dihydropyrazines **54**, yields the corresponding *N*-methyl-imidazoles **57** in high yields (Scheme 12.16). The reaction mechanism involves the formation of an enediimine intermediate **55**, followed by cyclization and re-aromatization [41].

The pyrazine skeleton has been used in a different manner for the synthesis of imidazoles. Thus, 2,4(or 5)-disubstituted imidazoles **61a** can be obtained, in moderate to high yields, together with the corresponding precursors *N*-cyanoimidazoles **61b**, by irradiation in EtOH (high-pressure Hg lamp, Pyrex filter) of 2-azidopyrazines **58** (Scheme 12.17) [42]. The reaction implies a photochemically generated nitrene species which is converted into **59**. Subsequent ring expansion–ring contraction processes lead to the mixture of final imidazoles **61a** and **61b**.

The synthesis of target imidazoles can be also realized by photoinduced ring-contraction of the pyrimidine nucleus. For example, significant yield of 4(or 5)-acetylimidazoles **64** are isolated from the photolysis, at 254 nm in methanol or benzene, of 4-substituted 2,6-dimethyl-pyrimidine-*N*-oxides **62** (Scheme 12.18) [43].

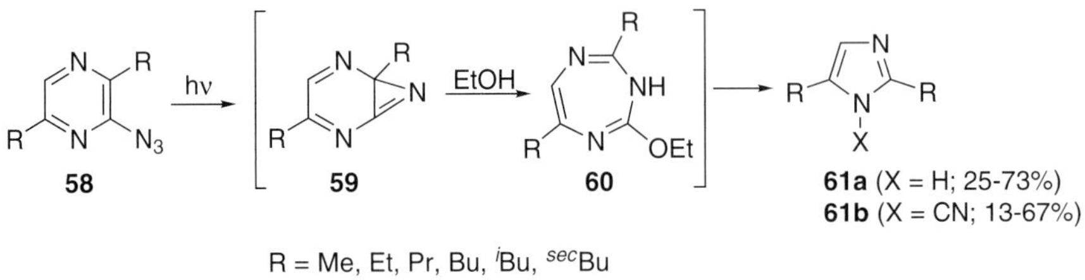

Scheme 12.17

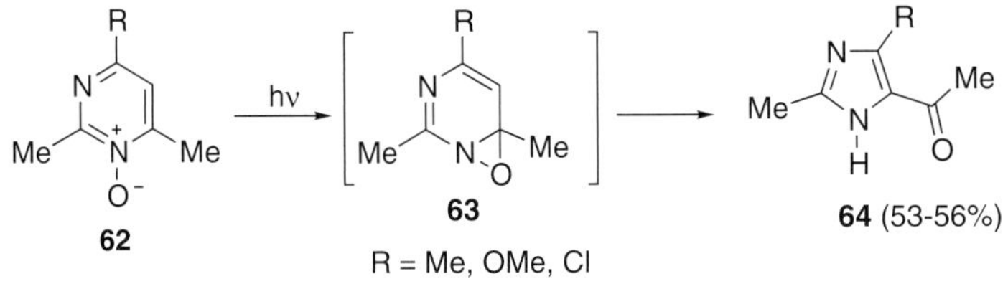

Scheme 12.18

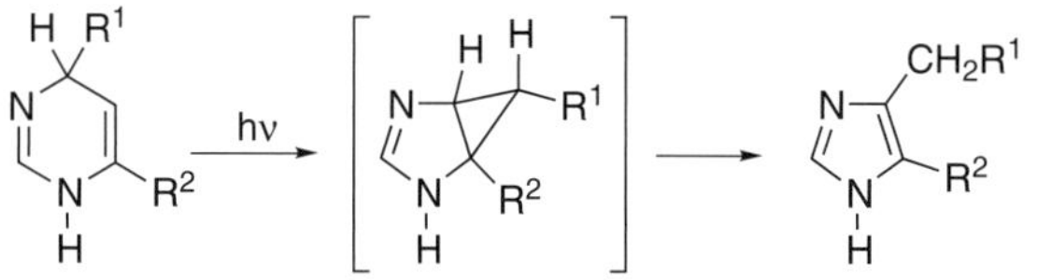

12.3 Synthesis of Five-Membered Rings with Two Heteroatoms

Scheme 12.19

R^1 = 2-thiazolyl, 2-pyridyl, 4-pyridyl; R^2 = H, Ph

Moreover, 4(or 5)-heteroarylmethyl-imidazoles **67** can be obtained by a photo-induced ring-contraction of 4-heteroaryl-substituted 1,4(or 3,4)-dihydropyrimidines **65** easily prepared by reaction of the corresponding pyrimidine with heteroaryllithium (Scheme 12.19). The irradiations were performed in acetone in a Rayonet preparative photoreactor equipped with RUL 300 lamps, and the products isolated in 43–57% yields. The reaction was determined by the anion-stabilizing effect of the heteroaryl group R^1, which affects the thermal development of the bicyclic intermediate **66** [44].

As for benzocondensated derivatives, an historical synthesis of 2-phenylbenzimidazole has been reported through the photochemical rearrangement of 1,5-diphenyltetrazole [45] or of 2,3-diphenyl-1,2,4-oxadiazolin-5-one [46]. Moreover, in analogy to the pyrazole-to-imidazole photorearrangement, N-unsubstituted benzimidazoles are obtained, in yields that are generally low (11–36%), by the photoinduced rearrangement of the corresponding indazoles [3, 37]. On the other hand, high yields of 1-alkyl-benzimidazoles (73–96%) are claimed from preparative-scale irradiations of 2-alkyl-indazoles through an ICI rearrangement [3, 37].

A particular case of a photochemical strategy towards functionalized benzimidazoles regards the synthesis of 2-acylamino derivatives **70**, which can be obtained from irradiations in methanol of 3-(o-amino)phenyl-1,2,4-oxadiazoles **68**. In this reaction, a photolytic cleavage of the ring O–N bond, followed by migration of the aryl substituent, leads to the carbodiimide intermediate **69**, a precursor of the benzimidazole nucleus (Scheme 12.20) [47].

An interesting application of the photochemical reactivity of isoxazolin-5-ones (see also Section 12.3) has been reported for the synthesis of a series of annulated imidazole derivatives of general structure **72**, which are obtained in almost quantitative yield. The methodology consists of an initial N-functionalization of the isoxazolin-5-one with a α-chloro-azaheterocycle, leading to 2-(azahetaryl)isoxazol-5(2H)-ones **71**.

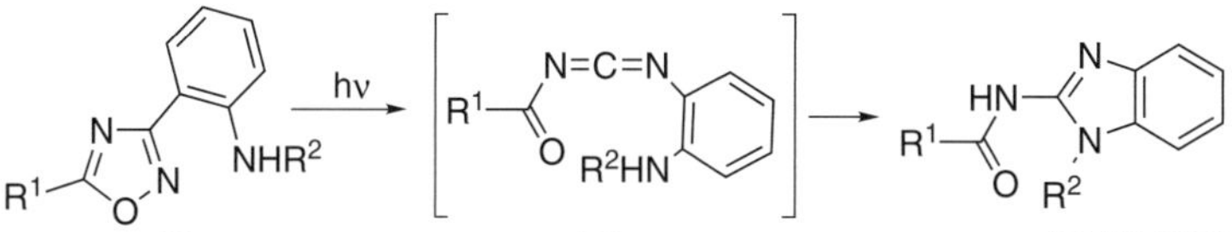

Scheme 12.20

R^1 = Me, Ph; R^2 = H, Me

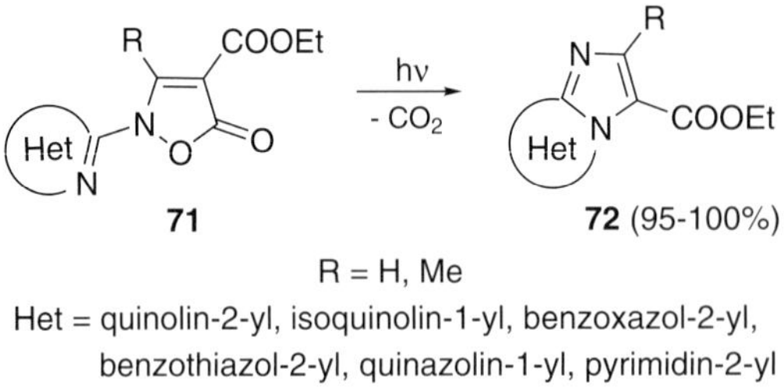
R
COOEt
Het
N
N
O
O
hν
- CO_2
R
N
N
COOEt
Het
71
72 (95-100%)
R = H, Me
Het = quinolin-2-yl, isoquinolin-1-yl, benzoxazol-2-yl, benzothiazol-2-yl, quinazolin-1-yl, pyrimidin-2-yl

Scheme 12.21

Photolysis of these derivatives (at 300 nm in either acetic acid or acetonitrile in the presence of 1 M trifluoroacetic acid) produces a loss of CO_2. Intramolecular cyclization onto the imine moiety then leads to annulated compounds **72** (Scheme 12.21) [48].

12.3.3
Oxazoles

Some aryl-substituted oxazoles have been photochemically obtained through ring-degenerate rearrangements of differently substituted aryloxazoles, via an exchange of annular positions following the ICI route. Although this strategy is applicable to preparative-scale irradiations (2∼3 g), a mixture of photoisomers is obtained (because of the intervening of competing paths) and the yields of target oxazoles are generally low [49]. On the other hand, the photoinduced RCRE rearrangement has been largely applied to obtaining oxazoles from the corresponding isoxazoles, and this approach acquires some significance considering the easy accessibility of the isoxazole ring through classical synthetic methodologies. An historical report describes the irradiation, at 254 nm in Et_2O, of 3,5-diarylisoxazoles **73**, which produces the corresponding 2,5-diaryloxazoles **75** [50]. The reaction involves the azirine intermediate **74** which can be isolated. Further irradiation of the azirines will lead either to the oxazole (product) or to the isoxazole (starting material), depending on the wavelength used (Scheme 12.22) [50].

Since then, several other examples of this photoinduced isomerization (though not always of synthetic utility) have been reported, sometimes as a typical photochemical reactivity of the isoxazole ring [2, 3]. Representative examples include: (i) the synthesis of 2-trifluoromethyloxazole **77** from a preparative-scale irradiation at 254 nm in acetonitrile of the easily accessible 3-trifluoromethylisoxazole **76** [51]; and

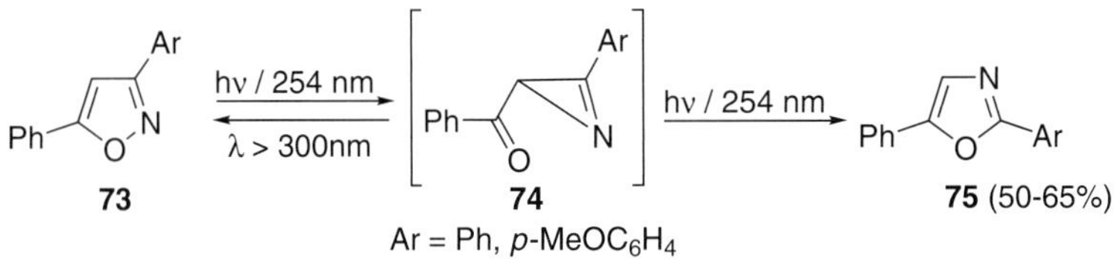
Ar
Ph
O
N
hν / 254 nm
λ > 300nm
Ar
Ph
O
N
hν / 254 nm
N
Ph
O
Ar
73
74
75 (50-65%)
Ar = Ph, p-MeOC6H4

Scheme 12.22

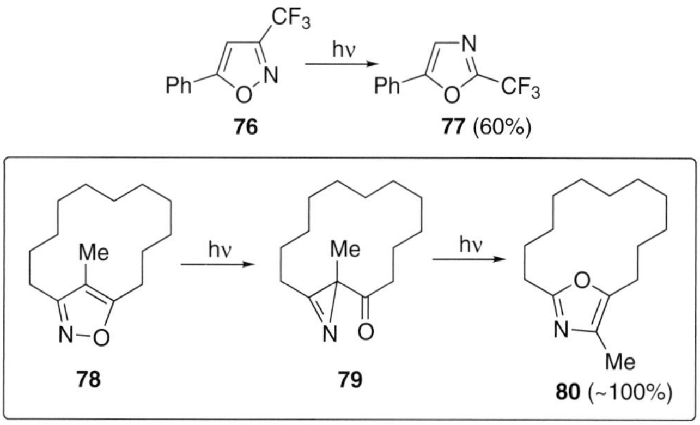

12.3 Synthesis of Five-Membered Rings with Two Heteroatoms

Scheme 12.23

(ii) the synthesis of the oxazolophane **80** obtained in quantitative yields from the irradiation at 254 nm in Et$_2$O of the azirine **79**. The latter was obtained in 40% yield from the irradiation (acetone, 125 W high-pressure Hg lamp, Pyrex filter) of isoxazolophane **78** (Scheme 12.23) [52].

The isoxazole-to-oxazole rearrangement has been also used in the synthesis of benzocondensed derivatives. The irradiation of benzisoxazoles **81** in water (high-pressure Hg lamp) produces almost quantitative yields of the corresponding benzoxazoles **82** (Scheme 12.24) [53]. Benzoxazolin-2-ones **83** are similarly obtained [54]. Furthermore, oxazolo[5,4-b]pyridines **84** and oxazolo[5,4-b]quinoline **85** are obtained by preparative-scale photolysis in ether with a high-pressure Hg lamp of the corresponding isoxazolo[5,4-b]pyridines or isoxazolo[5,4-b]quinoline, respectively (Scheme 12.24) [55].

Another methodology for the synthesis of oxazoles takes advantage of the photochemical reactivity of N-acyl-isoxazolin-5-ones **86**. The irradiation of these compounds is performed on a preparative scale in either acetonitrile (at 254 nm) or in acetone (at 300 nm), and the final oxazoles **87** are obtained in moderate to excellent

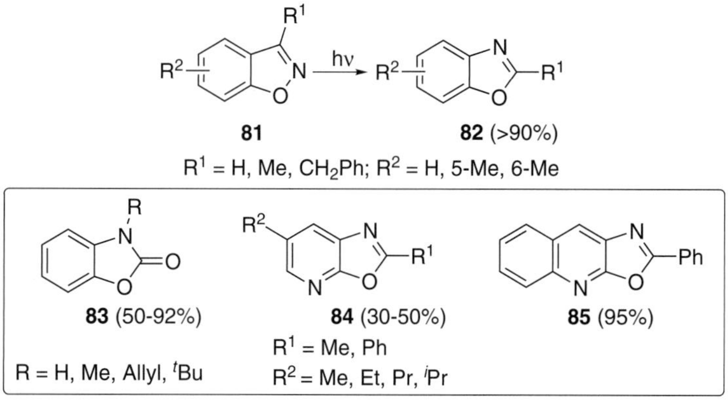

Scheme 12.24

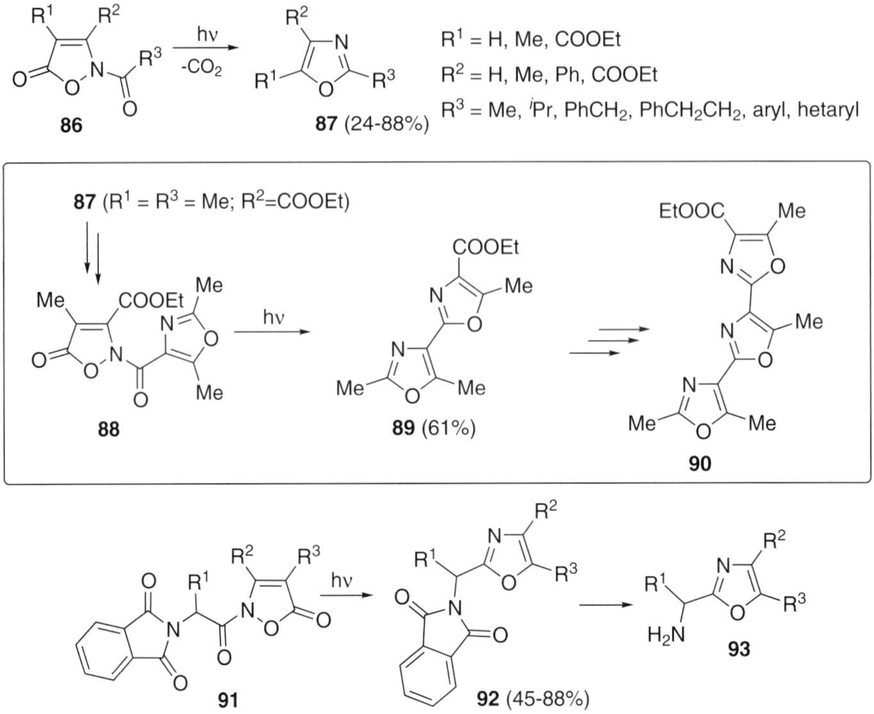

Scheme 12.25

yields, depending on the substrate and the irradiation conditions (Scheme 12.25) [56]. This strategy appears generally applicable to the synthesis of a large variety of 2-substituted targets, which can be obtained by simply using the appropriate acylating reagent on the isoxazolin-5-ones to form **86** that will be subsequently irradiated. Among other interesting examples are the synthesis of *bis*-oxazole **89**, the precursor of *tris*-oxazole **90** [56], and of 2-(1-aminoalkyl)oxazoles **93** (Scheme 12.25) [57, 58]. The latter compounds are obtained without significant racemization at the stereogenic center by a sequence of reactions consisting of N-acylation of the appropriate isoxazolin-5-one with phthalimidoaminoacids, followed by photolysis (in acetone at 300 nm) of the resulting **91**. This key step forms the substituted oxazole **92** that is finally deprotected by hydrazine (Scheme 12.25) [57, 58]. This procedure has also been used for the synthesis of a precursor of natural Almazoles, the oxazole-bearing metabolites from red seaweed [59].

12.3.4
Thiazoles

Examples of the formation of thiazoles through photochemical ring-degenerate rearrangements have been reported [4, 6]. However, in many cases, mixtures of photoisomers are obtained through competing mechanisms, and the product

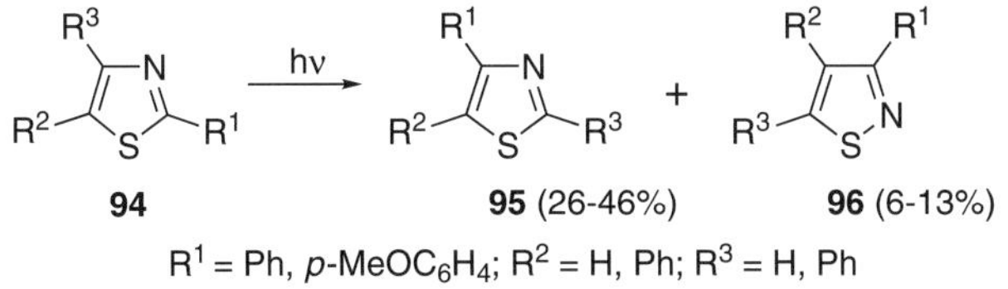

12.3 Synthesis of Five-Membered Rings with Two Heteroatoms | 401

Scheme 12.26

94 → 95 (26–46%) + 96 (6–13%)

R^1 = Ph, p-MeOC$_6$H$_4$; R^2 = H, Ph; R^3 = H, Ph

distribution depends on the substrate and solvent used. Among the reported preparative-scale (∼2–3 g) irradiations of some arylthiazoles [60], representative examples include the irradiation (high-pressure Hg lamp) of thiazoles **94** producing the corresponding thiazole isomers **95**, together with small amounts of isothiazoles **96** (Scheme 12.26) [60].

Other significant examples in synthesis regard the photoinduced ring-degenerate rearrangement of 4-carbomethoxy-2-phenylthiazole **97** into 2-carbomethoxy-4-phenylthiazole **98**, and of 2′-methyl-4-carbomethoxy-2,4′-bithiazole **99** into the corresponding 4,4′-bithiazole **100**, by irradiation in acetonitrile at 302 nm. The high yields reported were determined by analytical high-performance liquid chromatography (HPLC) (Scheme 12.27) [61].

In analogy to the isoxazole-to-oxazole rearrangement, thiazoles can be obtained photochemically from irradiation of the corresponding isothiazoles. In several cases, however, irradiations have been conducted for mechanistic purposes only, and mixtures of photoisomers arising from competing reaction mechanisms are often formed [4, 6].

Similar to the strategy used to obtain oxazoles (see Section 12.3), a fruitful photochemical methodology for the synthesis of the thiazole nucleus takes advantage of the photoreactivity of isoxazolin-5-ones. In a general picture, the photolysis of N-thioacylisoxazol-5(2H)-ones **101**, which are easily obtained in good yields by the thioacylation of isoxazolin-5-ones with suitable thioacylchlorides, produces thiazoles **102** (Scheme 12.28) [62]. The irradiations were performed in acetone at 300 nm (in a Rayonet photoreactor), and the two series of thiazoles **103** (R = PhO, PhS, 4-ClC$_6$H$_4$O) and **104** were reported to be formed in good to high yields; the exception

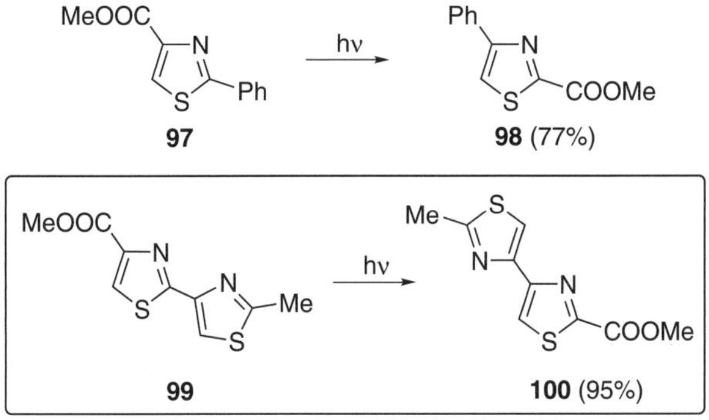

97 → 98 (77%)

99 → 100 (95%)

Scheme 12.27

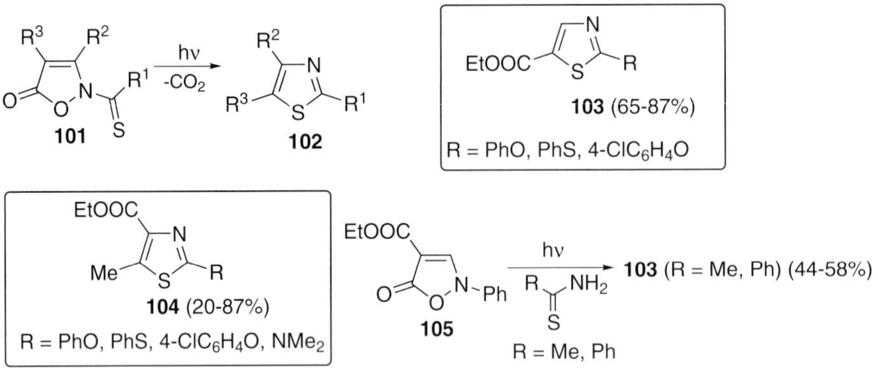

Scheme 12.28

was **104** (R = SPh, NMe$_2$), which was formed in low yields due to the formation of significant amounts of byproducts [62]. The photolysis of **105** in the presence of thioamides (Hanovia medium-pressure Hg lamp) in acetonitrile containing 0.5% trifluoroacetic acid gives thiazoles **103** (R = Me, Ph) in 44–58% isolated yields. In this case, the iminocarbene intermediate reacts with the added thioamides to produce, after acid-catalyzed re-aromatization, the target thiazoles (Scheme 12.28) [63].

12.4
Synthesis of Five-Membered Rings with Three Heteroatoms

12.4.1
Oxadiazoles

A series of 3-amino (or 3-*N*-substituted amino)-1,2,4-oxadiazoles **108** can be obtained in moderate to high yields by the photoinduced rearrangement of 3-acylamino-1,2,5-oxadiazoles (furazans) **106** irradiated in the presence of an excess of ammonia or aliphatic amines [64–66]. The reaction follows a fragmentation–cycloaddition route, with the initial formation of a nitrile and a nitrile-oxide; the latter is attacked by a nitrogen nucleophilic reagent (ZH in Scheme 12.29), and the open-chain interme-

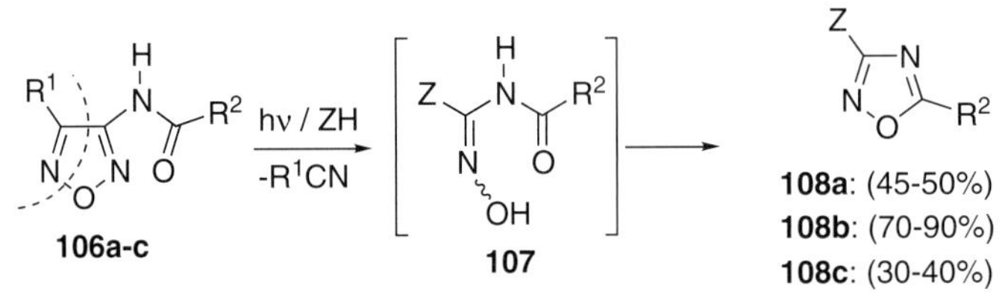

108a: (45-50%)
108b: (70-90%)
108c: (30-40%)

a: R^1 = Me; R^2 = Ph; Z = NH$_2$, MeNH, Me$_2$N, *N*-piperidinyl, *N*-pyrrolidinyl
b: R^1 = Ph; R^2 = Me, Et, Pr, Bu, PhCH$_2$; Z = NH$_2$, MeNH, Me$_2$N, BuNH, *N*-pyrrolidinyl
c: R^1 = Ph; R^2 = C$_3$F$_7$, C$_7$F$_{15}$; Z = NH$_2$, MeNH, PrNH

Scheme 12.29

diate **107** in turn cyclizes intramolecularly to the final oxadiazole **108** (Scheme 12.29). Although the reaction is of general application, the irradiation wavelength should be selected taking into account the substrate, and the reaction carefully monitored as a function of time, since prolonged irradiation may lead to secondary products. Moreover, leaving the photolysate solution after irradiation (in the dark) increases the yields of the final heterocyclized of acylamino-amidoxime **107**.

This methodology has been applied to the synthesis of 3-amino-, 3-N-alkylamino-, or 3-N,N-dialkylamino-5-phenyl-1,2,4-oxadiazoles **108a**, obtained from irradiations in MeOH at 254 nm of 3-benzoylamino-4-methylfurazan **106a** in the presence of an excess of the nitrogen nucleophile [64, 65]. Under similar conditions, when 3-alkanoylamino-4-phenylfurazans **106b** were irradiated at 313 nm, 3-amino- or 3-(N-substituted-amino)-5-alkyl-1,2,4-oxadiazoles **108b** were obtained in higher yields due to their photostability at the used wavelength, which allowed an almost complete photoconversion of the starting material (Scheme 12.29) [66]. The same procedure has been successfully used for the synthesis of 3-amino-(or 3-N-alkylamino)-5-perfluoroalkyl-1,2,4-oxadiazoles **108c** (Scheme 12.29). Unlike the corresponding 5-alkyl congeners, 5-perfluoroalkyl-oxadiazoles **108c** showed some photoreactivity under the employed wavelength. Therefore, irradiations were stopped at a low photoconversion of starting material and lower yields of the final products were obtained. However, if compared with some nonphotochemical procedures, the photochemical approach appears to be a rather general and convenient method to achieve these perfluoroalkyl-substituted oxadiazoles [67].

In analogy with the isoxazole-to-oxazole rearrangement, examples of the photochemical synthesis of 1,3,4-oxadiazoles can be identified in the RCRE rearrangement of some 1,2,4-oxadiazoles. In solution, this rearrangement has been observed for 5-aryl-1,2,4-oxadiazoles bearing a tautomerizable group at the C(3) position (such as NH_2 or NHMe), and appreciable yields of rearranged products were isolated when the photoreactions were performed in MeOH and in the presence of a base [68]. This photochemical approach to 2-amino-1,3,4-oxadiazoles is more significantly applied to the synthesis of perfluoroalkylated derivatives. For instance, 2-amino-5-perfluoroalkyl-1,3,4-oxadiazoles **110** can be obtained by irradiating the corresponding 5-perfluoroalkyl-1,2,4-oxadiazoles **109** (R = H) at 313 nm in methanol containing a 10-fold excess of triethylamine (Scheme 12.30) [69]. In this photoreaction, additional products (12–22%) were the 5-amino-3-perfluoroalkyl oxadiazoles **111** formed by a competing ICI route. A similar irradiation of 3-methylamino-5-perfluoroalkyl-1,2,4-oxadiazoles **109** (R = Me) produced the rearranged 1,3,4-oxadiazoles **112**, together with an amount of 3-methoxy-triazoles **113** (Scheme 12.30) arising from a competing C(3)-N(2) *migration–nucleophilic attack–cyclization* (MNAC) [70] photoreaction involving an *exocyclic* diazirine intermediate and the nucleophilic solvent [71]. Interestingly, in a confined environment such as the cavity of NaY zeolites, the photoinduced RCRE rearrangement has been observed also for 3,5-diphenyl-1,2,4-oxadiazole **114**. The corresponding 2,5-diphenyl-1,3,4-oxadiazole **115** was obtained in about 30% isolated yield after preparative-scale irradiation for 6 h in a quartz tube at 254 nm of a continuously stirred perfluorohexane slurry of NaY loaded with one molecule of substrate per cavity. The yields could be improved up to 90% (monitored

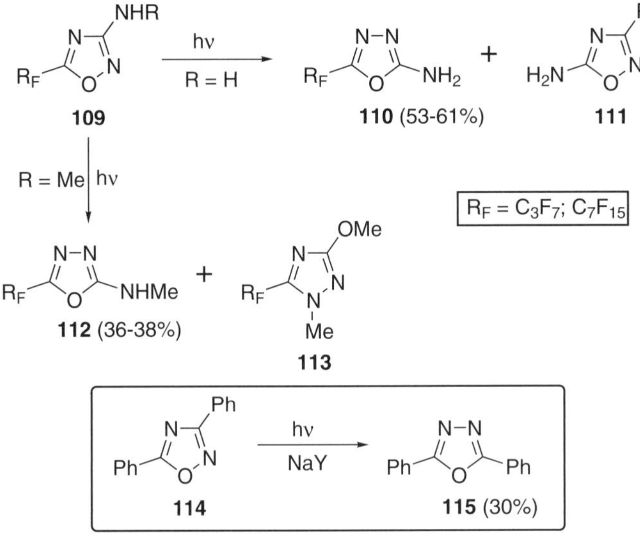

Scheme 12.30

by HPLC) by using a lower loading level of the substrate in the zeolite, which demonstrated the synthetic potential of this photochemical approach (Scheme 12.30) [72].

12.4.2
Triazoles

A series of 1,2,4-triazolin-5-ones **118** are formed from the irradiation of 5-amino-3-phenyl-1,2,4-oxadiazole **116** at 254 nm in methanol, in the presence of ammonia or primary aliphatic amines (Scheme 12.31) [47]. The reaction involves the photolytic species, arising from the ring O−N bond cleavage, and the nitrogen nucleophilic reagent, leading to open-chain intermediates **117** from which the final products are derived. When applied to other 3,5-disubstituted 1,2,4-oxadiazoles, irradiation in the presence of methylamine produced the corresponding 3,5-disubstituted triazoles in lower yields due to the low conversion of the starting material.

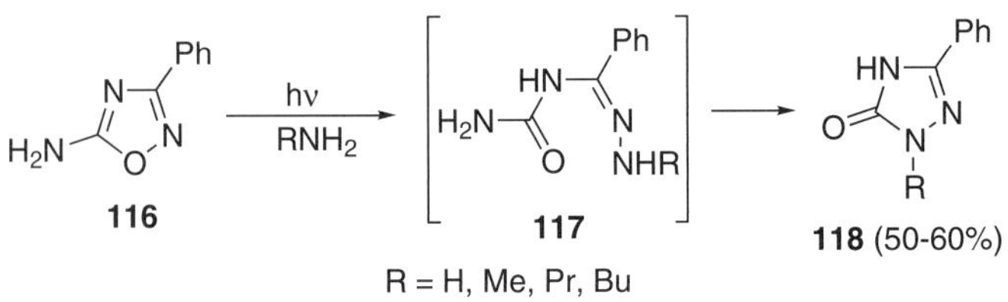

Scheme 12.31

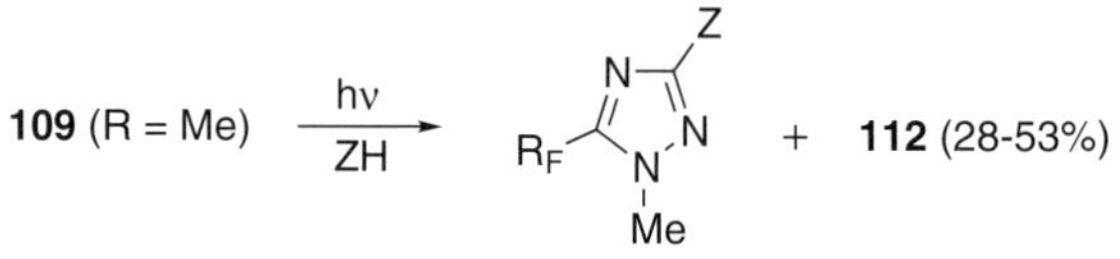

109 (R = Me) $\xrightarrow[ZH]{h\nu}$ [structure with R$_F$, Z, Me, N] + **112** (28-53%)

R$_F$ = C$_3$F$_7$, C$_7$F$_{15}$ **119** (21-27%)
Z = NH$_2$, MeNH, BuNH, N-pyrrolidinyl

Scheme 12.32

An interesting synthesis of targeted 5-perfluoroalkyl-1,2,4-triazoles takes advantage of the aforementioned MNAC [70] photorearrangement of 3-methylamino-1,2,4-oxadiazoles **109** (R = Me) which, when irradiated at 313 nm in methanol and in the presence of various nitrogen nucleophiles (ZH in Scheme 12.32), produces triazoles **119** together with the 1,3,4-oxadiazoles **112** expected from the base-catalyzed RCRE route (Scheme 12.32) [71].

12.4.3
Thiadiazoles

In analogy to the reactivity described above, 1,2,4-oxadiazoles have also been used as photochemical precursors of 1,2,4-thiadiazoles. When 5-amino-3-phenyl-1,2,4-oxadiazole **116** is irradiated at 254 nm in methanol, in the presence of a sulfur nucleophile such as thioureas, significant yields of thiadiazoles **121** were obtained, most likely through the open-chain species **120** (Scheme 12.33) [73].

Moreover, 3,4-disubstituted-1,2,4-thiadiazolin-5-ones **123** are obtained by irradiation (in methanol by using low-pressure Hanovia photoreactor) of the 3,4-disubstituted 1,2,4-oxadiazolin-5-thiones **122**, through a ring-opening–ring-closure sequence consisting of the photoinduced cleavage of the ring O−N bond, and re-cyclization with N−S bond formation (Scheme 12.33) [74].

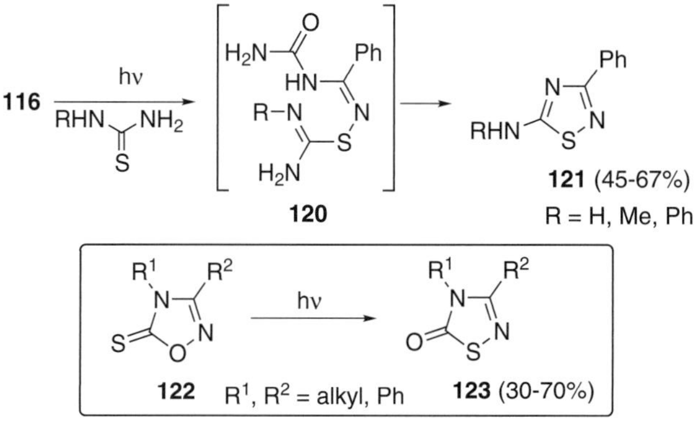

116 $\xrightarrow{h\nu}$ [**120**] → **121** (45-67%), R = H, Me, Ph

122 $\xrightarrow{h\nu}$ **123** (30-70%), R^1, R^2 = alkyl, Ph

Scheme 12.33

12.5
Synthesis of Six-Membered Rings

Several cases of photochemically induced permutations of ring-atoms have been reported for six-membered ring heterocycles. However, these reactions are rarely useful for a synthetic strategy and in most cases such permutations have been investigated only for mechanistic purposes and academic questions [3, 5].

Some synthetically useful examples involve fluorinated pyridazines. Thus, the preparative-scale irradiation of perfluoroalkylpyridazines **124** at 254 nm in the vapor phase produces perfluoroalkylpyrazines **127**. This permutation process is suggested to occur through the bicyclic species **125** and **126**, which have been isolated and characterized (Scheme 12.34) [75, 76]. Another example involves the vapor-phase irradiation of 4,5-difluoro-3,6-*bis*(trifluoromethyl)pyridazine to produce the corresponding perfluoro-2,5-dimethylpyrazine in 58% isolated yield, with 40% of the starting material being recovered [77].

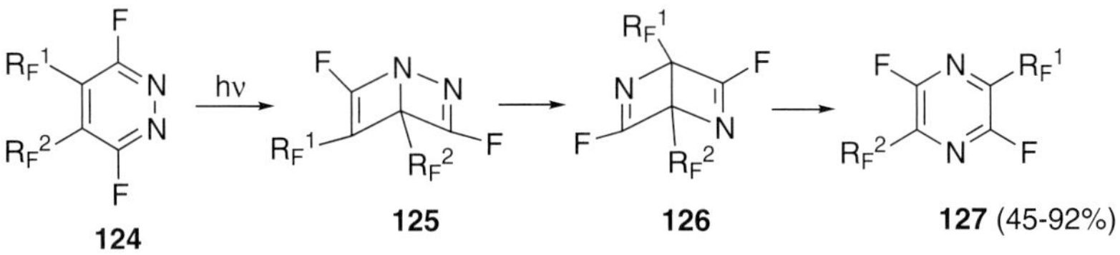

Scheme 12.34

12.6
Synthesis of Seven-Membered Rings

12.6.1
Azepines

It is well known that a significant photochemical procedure for the synthesis of azepines takes advantage of the photorearrangement of arylazides in the presence of nucleophiles such as amines or alcohols. The reaction proceeds through the ring-expansion of photolytically formed singlet arylnitrene, and the azepine formation was found to depend largely on the electronic effects of the substituent on the aryl moiety [78, 79]. To mention a representative example, 2-diethylamino-3*H*-azepine **130** has been obtained in 34% yield from the irradiation of phenylazide **128** (R = H) in diethylamine, by using Hanovia Hg lamps in Pyrex vessels (Scheme 12.35) [80]. Furthermore, alkoxy-azepines **133** were formed from the photolysis of *o*-azidobenzoic acid derivatives **132** in the presence of alcohols [81, 82]. Finally, 5-substituted 3*H*-azepine-2-ones **131** have been obtained via a preparative-scale photolysis of the corresponding arylazides **128** in THF/H$_2$O mixture (1/1) by using 400 W medium-pressure Hg lamps and a Pyrex filter (Scheme 12.35) [83].

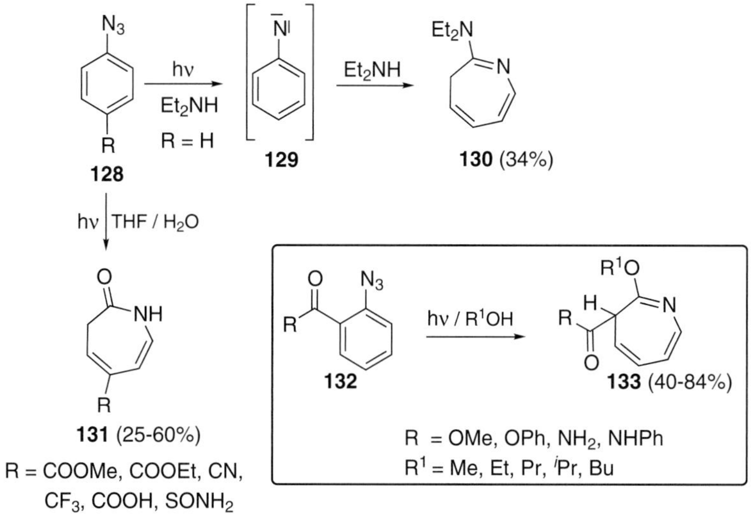
N3
hν
Et2NH
R = H
R
128
NI
129
Et2NH
Et2N
N
130 (34%)
hν THF / H2O
O
NH
R
131 (25-60%)
R = COOMe, COOEt, CN,
CF3, COOH, SONH2
O
N3
R
132
hν / R1OH
R1O
H
N
R
O
133 (40-84%)
R = OMe, OPh, NH2, NHPh
R1 = Me, Et, Pr, iPr, Bu

Scheme 12.35

12.6.2
Diazepines

A photochemical synthesis of 1*H*-1,2-diazepines consists of the photoinduced ring enlargement of 1-iminopyridinium ylides **134** [84–89]. The irradiations are performed on a variety of substituted substrates by using different solvents (benzene, dioxane, acetone, DCM) and a high-pressure Hg lamp in Pyrex vessels, or in a Rayonet apparatus equipped with 350 nm lamps. Diazepines can be isolated in good to excellent yields, depending both on the substrate and on the irradiation solvent. Since 1,2-diazepines are excellent synthons for more complex polycyclic compounds, large-scale processes have been also developed by using dynamic thin-film photoreactors. Subsequently, large batches of final products have been realized, provided that suitable UV lamps are used. With regards to the directing effect of the substituents, 1-iminopyridinium ylides substituted at the C(2) **134b** led to the regiospecific formation of the 3-substituted-1,2-diazepines **136b**, regardless of the electronic properties of the substituent. On the other hand, substituents at C(3) were found to direct the reaction towards the formation of one regioisomer (**136c** from **134c**) or of a mixture of both possible diazepine regioisomers (**136d** and **137** from **134d**), depending on their electronic effects [90]. These representative examples of photoreactions producing 1,2,-diazepines are all illustrated in Scheme 12.36.

The photoinduced ring-enlargement of 2-azidopyridines appears to represent a general procedure for the synthesis of 1,3-diazepines. 2-Dialkylamino-1*H*-1,3-diazepines **141a** and 2-alkoxy-1*H*-1,3-diazepines **141b** are obtained by the preparative-scale photolysis of substituted 2-azidopyridines **138** (Scheme 12.37). The irradiations were generally performed in dioxane, and in the presence of nucleophiles such as

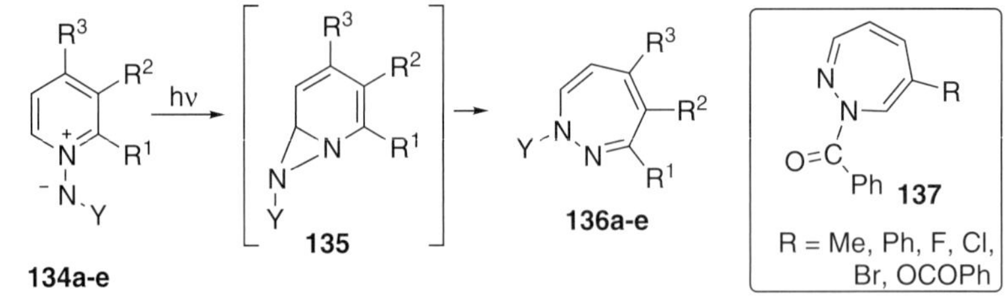

Scheme 12.36

a: Y = COOEt, COOPr; R¹ = R² = R³ = H (90-95%)
b: Y = COOEt; R¹ = Me, CN; R² = R³ = H (80-86%)
c: Y = COOEt, COPh; R¹ = H; R² = COOEt, CN, CONH₂; R³ = H (80-86%)
d: Y = COPh, R¹ = H; R² = Me, Ph, F, Cl, Br, OCOPh; R³ = H (**136d** + **137**: 72-80%)
   (**136d**/**137** = 1.2~2.3)
e: Y = COOEt, COPh; R¹ = R² = H; R³ = Me, Ph, Cl, NMe₂ (65-90%)

Scheme 12.37

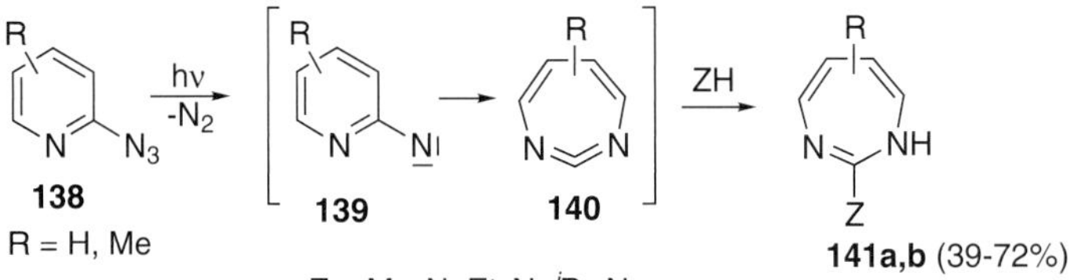

a: Z = Me₂N, Et₂N, ^{i}Pr₂N
b: Z = MeO, EtO, ^{i}PrO

141a,b (39-72%)

alcohols or secondary amines (ZH in Scheme 12.37), by using high-pressure Hg/Xe lamp [91, 92]. The reaction can be considered quite general and efficient, and is explained through the pyridylnitrene **139** and the cyclic carbodiimide **140**. In the case of the alkoxy compounds, the 1-*H* form is generally predominant; this differs in the case of the dialkylamino compounds, where the 5-*H* form is favored [92].

Variously substituted 2-azido-pyridines have been irradiated on a preparative scale in the presence of alcohols or amines for the synthesis of the corresponding variously substituted 1,3-diazepines, in good to high yields. In the presence of halo substituents on the pyridinium moiety, a nucleophilic substitution by the ZH reagent can also occur [93]. Among others, representative examples of obtained target are reported in Figure 12.1 [91, 93].

A similar strategy has been used to synthesize the 1,4-diazepine skeleton, starting from 4-azidopyridines. The preparative-scale irradiation (400 W, high-pressure Hg lamp, Pyrex filter) of azides **148** in MeOH/dioxane (1 : 1) containing NaOMe resulted in the formation of 5-methoxy-6*H*-1,4-diazepines **149** as the more stable tautomer (Scheme 12.38) [94]. It should be noted that compounds **149** are found to be susceptible to decomposition on silica or alumina columns; hence, care must be taken in their purification and storage.

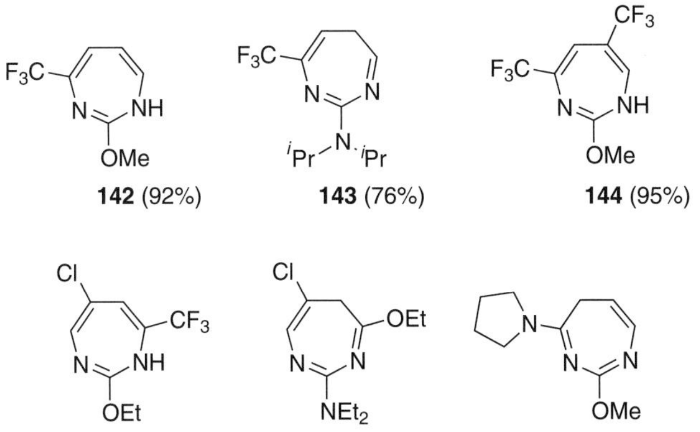

142 (92%)   **143** (76%)   **144** (95%)

145 (57%)   **146** (89%)   **147** (74%)

Figure 12.1 Selected examples of 1,3-diazepines from the irradiation of 2-azido-pyridines.

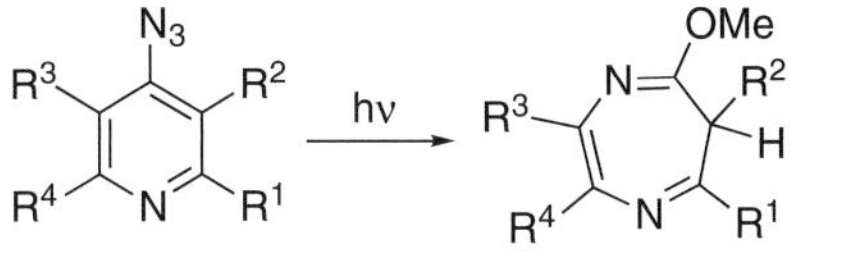

148 → **149** (35–68%)

$R^1, R^2, R^3, R^4 = H, Me$

Scheme 12.38

12.6.3
Oxazepines

The photolysis of some tetraaryl- and pentaaryl-pyridine N-oxides **150** (in benzene solution, in a Rayonet photoreactor by using 3500, 3000, and 2537 Å lamps) allows the isolation of high yields of the corresponding 1,3-oxazepines **151**, the structures of which have been confirmed using X-ray crystallography for **151f** (Scheme 12.39) [95].

As with the benzocondensated series, the production of 3,1-benzoxazepines via the irradiation of quinoline-N-oxides is strongly affected by the solvent, and also by the nature and position of substituents on the pyridine moiety of the quinoline (aprotic solvents and Ph, CN, OMe, or CF_3 groups attached to C(2) position favor formation of

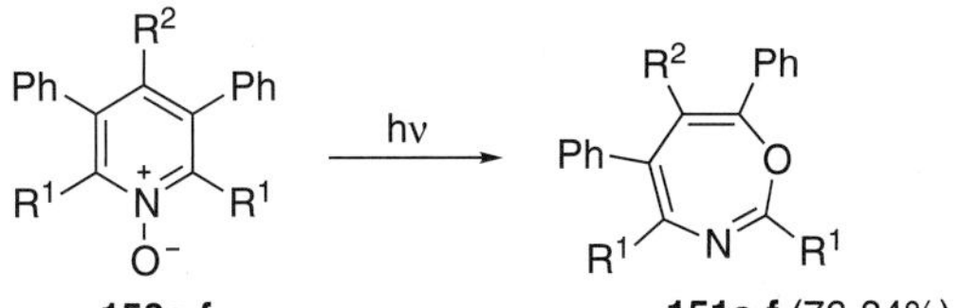

150a–f → **151a–f** (76–84%)

a: R^1 = Ph; R^2 = H
b: R^1 = 4-MeC$_5$H$_4$; R^2 = H
c: R^1 = 4-ClC$_5$H$_4$; R^2 = H
d: R^1 = 4-BrC$_5$H$_4$; R^2 = H
e: R^1 = R^2 = Ph
f: R^1 = Ph; R^2 = 4-BrC$_5$H$_4$

Scheme 12.39

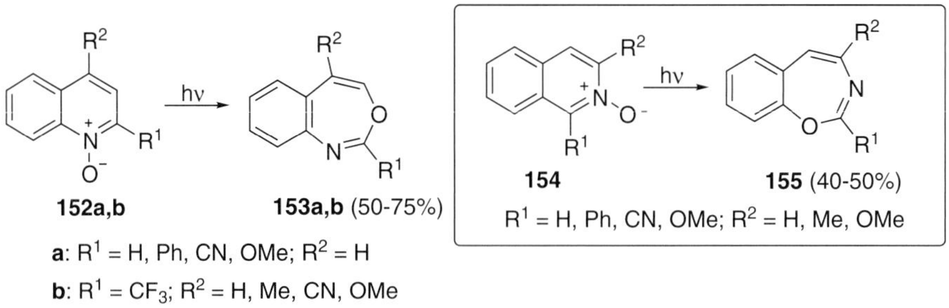

Scheme 12.40

152a,b → 153a,b (50-75%)

a: R^1 = H, Ph, CN, OMe; R^2 = H
b: R^1 = CF_3; R^2 = H, Me, CN, OMe

154 → 155 (40-50%)
R^1 = H, Ph, CN, OMe; R^2 = H, Me, OMe

the seven-membered ring). In some cases, the formed benzoxazepines are reported to be moisture-sensitive, which makes their isolation difficult through normal chromatographic techniques. A large-scale synthesis of the parent 3,1-benzoxazepine 153a (R^1 = H) has been described in detail by the irradiation of quinoline-N-oxide 152a (R^1 = H) in anhydrous benzene, with 50% yields of the pure final product being reported (Scheme 12.40) [96]. Substituted 3,1-benzoxazepines 153a ($R^1 \neq$ H) are also formed in high yields by a preparative-scale irradiation of the corresponding substituted quinoline N-oxides 152a ($R^1 \neq$ H) in different solvents, such as benzene, ethanol or cyclohexane, depending on the substrate [97, 98]. High yields of 3,1-benzoxazepines 153b have also been reported in the irradiation (in ether, by using Hanovia high-pressure Hg lamp with Pyrex filter) of some 2-trifluoromethylsubstituted quinoline-N-oxides 152b (Scheme 12.40) [99].

A similar strategy has been reported for the synthesis of 1,3-benzoxazepines such as 155, which are formed during irradiation of the corresponding isoquinoline N-oxides 154 (Scheme 12.40) [98, 100].

12.7
Concluding Remarks

In concluding this chapter, it is important to emphasize the varying degree to which the photorearrangement of heterocycles can be applied to the synthesis of heteroaromatics. Whilst it is not always possible to distinguish between analytical and preparative-scale irradiations, and accepting that photochemical processes may be highly sensitive to factors such as substrate structure, the nature of the solvent, and the light source, this methodology represents an exclusive means of obtaining a desired heterocyclic target. In preparing this chapter, we have attempted to describe different ways in which advantage may be taken of the photochemical approach to synthesize differently functionalized heteroaromatic compounds.

References

1. Lablache-Combier, A. (1976) Photoisomerization of five membered heterocyclic compounds, in *Photochemistry of Heterocyclic Compounds* (ed. O. Buchardt), John Wiley, New York, pp. 123–206.
2. Padwa, A. (1980) Photochemical rearrangements of five membered ring heterocycles, in *Rearrangements in Ground and Excited States*, Vol. 3 (ed. P. de Mayo), Academic Press, New York, pp. 501–547.
3. Lablache-Combier, A. (1995) Photorearrangement of nitrogen-containing arenes, in *CRC Handbook of Photochemistry and Photobiology* (eds W.M. Horspool and P.-S. Song), CRC Press, Boca Raton, Fl, pp. 1063–1120.
4. Lablache-Combier, A. (1995) Photorearrangement of thioarenes, in *CRC Handbook of Photochemistry and Photobiology* (eds W.M. Horspool, and P.-S. Song), CRC Press, Boca Raton, Fl, pp. 803–823.
5. Pavlik, J.W. (2004) Photoisomerization of some nitrogen-containing heteroaromatic compounds, in *CRC Handbook of Photochemistry and Photobiology*, 2nd edn (eds W.M. Horspool and F. Lenci), CRC Press, Boca Raton, Fl, pp. 97-1–97-22.
6. Pavlik, J.W. (2004) Photochemistry of Thiazoles, Isothiazoles, and 1,2,4-Thiadiazoles, in *CRC Handbook of Photochemistry and Photobiology*, 2nd edn (eds W.M. Horspool and F. Lenci), CRC Press, Boca Raton, Fl, pp. 98-1–98-14.
7. D'Auria, M. (2001) Photochemical isomerization of pentaatomic heterocycles. *Advances in Heterocyclic Chemistry*, **79**, 41–88.
8. Patterson, J.M. and Bruser, D.M. (1973) Photoisomerization of some substituted N-acetylpyrroles. *Tetrahedron Letters*, **14** (31), 2959–2962.
9. Hiraoka, H. (1970) Photoisomerization of 2-cyanopyrrole. *Journal of the Chemical Society (D), Chemical Communications*, (20), 1306.
10. Baltrop, J., Day, A.C., Moxon, P.D. and Ward, R.R. (1975) Permutation patterns and the phototranspositions of 2-cyanopyrroles. *Journal of the Chemical Society, Chemical Communications*, (19), 786–787.
11. Barton, T.J. and Hussmann, G.P. (1985) Photoisomerization of 2-(trimethylsilyl) pyrroles. *Journal of Organic Chemistry*, **50** (26), 5881–5882.
12. Scheiner, P., Chapman, O.L. and Lassila, J.D. (1969) The photolysis of dihydro-1,2-oxazines. *Journal of Organic Chemistry*, **34** (4), 813–816.
13. Givens, R.S., Choo, D.J., Merchant, S.N., Stitt, R.P. and Matuszewski, B. (1982) Photochemistry of 3,6-dihydro-1,2-oxazines: a versatile route to substituted pyrroles. *Tetrahedron Letters*, **23** (13), 1327–1330.
14. Albini, A. and Alpegiani, M. (1984) The photochemistry of the N-oxide function. *Chemical Reviews*, **84** (1), 43–71.
15. Albini, A. and Fagnoni, M. (2004) Photochemistry of N-oxides, in *CRC Handbook in Organic Photochemistry and Photobiology*, 2nd edn (eds W.M. Horspool and F. Lenci), CRC Press, Boca Raton, Fl, pp. 99-1–99-21.
16. Bellamy, F., Martz, P. and Streith, J. (1975) An intriguing copper salt effect upon the photochemistry of pyridine-N-oxides. Specific photoinduced synthesis of 3-substituted 2-formylpyrroles. *Heterocycles*, **3** (5), 395–400.
17. Albini, A., Fasani, E. and Lohse, C. (1988) Photochemistry of pyridine N-oxides. Trapping of an intermediate with amines. *Heterocycles*, **27** (1), 113–124.
18. Ishikawa, M., Kaneko, C., Yokoe, I. and Yamada, S. (1969) Photolysis of 2,6-dicyanopyridine 1-oxides. *Tetrahedron*, **25** (2), 295–300.

19 Kaneko, C., Yamamoto, A. and Hashiba, M. (1979) Ring contraction reactions of methyl quinolyne 1-oxide 5-carboxylates via the corresponding benz[d]-3,1-oxazepines. A facile synthesis of methyl indole 4-carboxylate and its derivatives. *Chemical & Pharmaceutical Bulletin*, **27** (4), 946–952.

20 Kaneko, C., Fujii, H., Kawai, S., Yamamoto, A., Hashiba, K., Kimata, T., Hayashi, R. and Somei, M. (1980) Studies on the N-oxides of π-deficient N-heteroaromatics. XXXIV. A novel synthesis of substituted indoles by photochemical ring contraction of 3,1-benzoxazepines. *Chemical & Pharmaceutical Bulletin*, **28** (4), 1157–1171.

21 Carlock, J.T., Bradshaw, J.S., Stanovnik, B. and Tisler, M. (1977) 3-Diazo-4-oxo-3,4-dihydroquinoline. A novel synthon for indole-3-carboxamides. *Journal of Organic Chemistry*, **42** (11), 1883–1885.

22 Carlock, J.T., Bradshaw, J.S., Stanovnik, B. and Tisler, M. (1977) A noteworthy improvement of the 3-diazo-4-oxo-3,4-dihydroquinoline photosynthesis of indole-3-carboxamides. *Journal of Heterocyclic Chemistry*, **14** (3), 519–520.

23 Stanovnik, B., Tisler, M. and Carlock, J.T. (1976) Heterocycles, CLIX. A novel synthesis of alkyl indole-3-carboxylates. *Synthesis*, (11), 754–755.

24 Pavlik, J.W., Bolin, R.R., Bradford, K.C. and Anderson, W.C. (1977) Photoisomerization of 4-hydroxypyrylium cations. Furyl cation formation. *Journal of the American Chemical Society*, **99** (8), 2816–2818.

25 Pavlik, J.W. and Spada, A.P. (1979) Photo-ring contraction reactions of 4-hydroxypyrylium cations. *Tetrahedron Letters*, **20** (46), 4441–4444.

26 Tsuchiya, T., Arai, H. and Igeta, H. (1973) Photochemistry – IX. Formation of cyclopropenyl ketones and furans from pyridazine N-oxides by irradiation. *Tetrahedron*, **29** (18), 2747–2751.

27 Tomer, K.B., Harrit, N., Rosenthal, I., Buchardt, O., Kumler, P.L. and Creed, D. (1973) Photochemical behaviour of aromatic 1,2-diazine N-oxides. *Journal of the American Chemical Society*, **95** (22), 7402–7406.

28 Winberg, H., van Driel, H., Kellogg, R.M. and Buter, J. (1967) The photochemistry of thiophenes. IV. Observations on the scope of arylthiophene rearrangements. *Journal of the American Chemical Society*, **89** (14), 3487–3494.

29 Schroth, W., Billig, F. and Reinhold, G. (1967) 1,2-Dithiins, a new type of heterocycle. *Angewandte Chemie, International Edition in English*, **6** (8), 698–699.

30 Norton, R.A., Finlayson, A.J. and Towers, G.H. (1985) Two dithiacyclohexadiene polyacetylenes from *Chaenactis douglassi* and *Eriophyllum lanatum*. *Phytochemistry*, **24** (4), 356–357.

31 Schroth, W., Dunger, S., Billig, F., Spitzner, R., Herzschuh, R., Vogt, A., Jende, T., Israel, G., Barche, J. and Strohl, D. (1996) 1,2-Dithines and precursors, XVI: synthesis, structure, and reactivity of non-anellated 1,2-dithiines. *Tetrahedron*, **52** (39), 12677–12698.

32 Block, E., Birringer, M., DeOrazio, R., Fabian, J., Glass, R.S., Guo, C., He, C., Lorance, E., Qian, Q., Schroeder, T.B., Shan, Z., Thiruvazhi, M., Wilson, G.S. and Zhang, X. (2000) Synthesis, properties, oxidation, and electrochemistry of 1,2-dichalcogenins. *Journal of the American Chemical Society*, **122** (21), 5052–5064.

33 Tsuchiya, T., Kurita, J. and Takayama, K. (1980) Studies on diazepines. XIII. Photochemical behaviour of pyrazine, pyrimidine, and pyridazine N-imides. *Chemical & Pharmaceutical Bulletin*, **28** (9), 2676–2681.

34 Iwasaki, S. (1976) Photochemistry of imidazolides. I. The photo-Fries-type rearrangement of N-substituted imidazoles. *Helvetica Chimica Acta*, **59** (8), 2738–2752.

35. LaMattina, J.L., Suleske, R.T. and Taylor, R.L. (1983) Synthesis of 1H-5-acetyl-2-alkylimidazoles. *Journal of Organic Chemistry*, **48** (6), 897–898.
36. Beak, P. and Messer, W.R. (1969) Photorearrangements of some N-methyldiazoles. *Tetrahedron*, **25** (16), 3287–3295.
37. Tiefenthaler, H., Dorscheln, W., Goth, H. and Schmid, H. (1967) Photoisomerisierung von Pyrazolen und Indazolen zu Imidazolen bzw. Benzimidazolen und 2-Amino-benzonitrilen. *Helvetica Chimica Acta*, **50** (8), 2244–2258.
38. Baltrop, J.A., Day, A.C., Mack, A.G., Shahrisa, A. and Wakamatsu, S. (1981) Competing pathways in the phototransposition of pyrazoles. *Journal of the Chemical Society, Chemical Communications*, (12), 604–606.
39. Casey, M., Moody, C.J. and Rees, C.W. (1984) Synthesis of imidazoles from alkenes. *Journal of the Chemical Society, Perkin Transactions 1*, (8), 1933–1941.
40. Casey, M., Moody, C.J., Rees, C.W. and Young, R.G. (1985) Two-step synthesis of imidazoles from activated alkynes. *Journal of the Chemical Society, Perkin Transactions 1*, (4), 741–745.
41. Beak, P. and Miesel, J.L. (1967) The photorearrangement of 2,3-dihydropyrazines. *Journal of the American, Chemical Society*, **89** (10), 2375–2384.
42. Watanabe, T., Nishiyama, J., Hirate, R., Uehara, K., Inoue, M., Matsumoto, K. and Ohta, A. (1983) Synthesis of some alkyl- and arylimidazoles. *Journal of Heterocyclic Chemistry*, **20** (5), 1277–1281.
43. Roeterdink, F., van der Plas, H.C. and Koudijs, A. (1975) Photochemistry of pyrimidine N-oxides (IV). *Recueil des Travaux Chimiques des Pays-Bas*, **94** (1), 16–18.
44. Van der Stoel, R.E., van der Plas, H.C. and Jongejan, H. (1983) Photochemical ring contraction of 4-heteroaryl-substituted 1,4 (or 3,4)-dihydropyrimidines into imidazoles. *Recueil des Travaux Chimiques des Pays-Bas*, **102** (7/8), 364–367.
45. Preston, P.N. (1974) Synthesis, reactions, and spectroscopic properties of benzimidazoles. *Chemical Reviews*, **74** (3), 279–314.
46. Boyer, J.H. and Ellis, P.S. (1979) Fragmentation-rearrangement of Δ^3-oxadiazolin-5- and 3-ones. *Journal of the Chemical Society, Perkin Transactions 1*, (2), 483–487.
47. Buscemi, S., Vivona, N. and Caronna, T. (1996) Photoinduced molecular rearrangements. The photochemistry of some 1,2,4-oxadiazoles in the presence of nitrogen nucleophiles. Formation of 1,2,4-triazoles, indazoles, and benzimidazoles. *Journal of Organic Chemistry*, **61** (24), 8397–8401.
48. Prager, R.H., Singh, Y. and Weber, B. (1994) The chemistry of 5-oxodihydroisoxazoles. VIII. Photolysis of 2-(heterocyclyl)isoxazol-5(2H)-ones. *Australian Journal of Chemistry*, **47** (7), 1249–1262.
49. Maeda, M. and Kojima, M. (1977) Photorearrangements of phenyloxazoles. *Journal of the Chemical Society, Perkin Transactions 1*, (3), 239–247.
50. Singh, B. and Ullman, E.F. (1967) Photochemical transposition of ring atoms in 3,5-diarylisoxazoles. An unusual example of wavelength control in a photochemical reaction of azirines. *Journal of the American Chemical Society*, **89** (26), 6911–6916.
51. Pavlik, J.W., Martin, H.St., Lambert, K.A., Lowell, J.A., Tsefrikas, V.M., Eddins, C.K. and Kebede, N. (2005) Photochemistry of 4- and 5-phenyl substituted isoxazoles. *Journal of Heterocyclic Chemistry*, **42** (2), 273–281.
52. Albanesi, S., Marchesini, A. and Gioia, B. (1979) Photochemical behaviour of the 16-methyl(3,5)[11]isoxazolophane and of the related 14-methyl-15-azabicyclo [12.1.0]pentadec-15(1)-en-13-one. *Tetrahedron Letters*, **20** (21), 1875–1878.

53 Heizelmann, W. and Marky, M. (1974) Photochemie von Benzisoxazolen. *Helvetica Chimica Acta*, **57** (2), 376–382.

54 Darlage, L.J., Kinstle, T.H. and McIntosh, C.L. (1971) Photochemical rearrangements of 1,2-benzisoxazolinones. *Journal of Organic Chemistry*, **36** (8), 1088–1093.

55 Skotsch, C. and Breitmaier, E. (1979) Oxazolo[5,4-b]pyridine durch photoumlagerung von isoxazolo[5,4-b] pyridinen. *Chemische Berichte*, **112** (9), 3282–3285.

56 Prager, R.H., Smith, J.A., Weber, B. and Williams, C.M. (1997) Chemistry of 5-oxodihydroisoxazoles. Part 18. Synthesis of oxazoles by the photolysis and pyrolysis of 2-acyl-5-oxo-2,5-dihydroisoxazoles. *Journal of the Chemical Society, Perkin Transactions 1*, (17), 2665–2672.

57 Cox, M., Prager, R.H., Svensson, C.E. and Taylor, M.R. (2003) The synthesis of some chiral 2-aminoalkyloxazole-5-carboxylates from isoxazol-5(2H)-ones. *Australian Journal of Chemistry*, **56** (9), 897–901.

58 Cox, M., Prager, R.H. and Svensson, C.E. (2003) The synthesis of some chiral 2-aminoalkyloxazole-4-carboxylates from isoxazol-5(2H)-ones. *Australian Journal of Chemistry*, **56** (9), 887–896.

59 Khalafay, J., Prager, R.H. and Williams, C.M. (1999) The synthesis of Almazoles A and B. *Australian Journal of Chemistry*, **52** (1), 31–36.

60 Maeda, M. and Kojima, M. (1978) Mechanism of the photorearrangements of phenylthiazoles. *Journal of the Chemical Society, Perkin Transactions 1*, (7), 685–692.

61 Saito, I., Morii, T., Okumura, Y., Mori, S., Yamaguchi, K. and Matsuura, T. (1986) Ring-selective photorearrangement of bithiazoles. *Tetrahedron Letters*, **27** (52), 6385–6388.

62 Prager, R.H., Taylor, M.R. and Williams, C.M. (1997) The chemistry of 5-oxodihydroisoxazoles. Part 19. The synthesis and photolysis of N-thioacylisoxazol-5(2H)-ones. *Journal of the Chemical Society, Perkin Transactions 1*, (17), 2673–2678.

63 Fong, M., Janowski, W.K., Prager, R.H. and Taylor, M.R. (2004) A convenient synthesis of 2-substituted thiazole-5-carboxylates. *Australian Journal of Chemistry*, **57** (6), 599–604.

64 Buscemi, S., Frenna, V., Caronna, T. and Vivona, N. (1992) Photochemical behaviour of 1,2,5-oxadiazoles. Irradiation of some 3-acylamino-1,2,5-oxadiazoles in the presence of nucleophiles. *Heterocycles*, **34** (12), 2313–2322.

65 Buscemi, S., Vivona, N. and Caronna, T. (1995) Photoinduced molecular rearrangements. Some investigations of the photochemical behaviour of 3-acylamino-1,2,5-oxadiazoles (furazans). *Journal of Organic Chemistry*, **60** (13), 4096–4101.

66 Buscemi, S., Vivona, N. and Caronna, T. (1995) A generalized and efficient synthesis of 3-amino-, 3-(N-alkylamino)-, 3-(N,N-dialkylamino)-5-alkyl-1,2,4-oxadiazoles by irradiation of 3-alkanoylamino-4-phenyl-1,2,5-oxadiazoles (furazans). *Synthesis*, (8), 917–919.

67 Buscemi, S., Pace, A. and Vivona, N. (2000) Fluoro heterocycles. A photochemical methodology for the synthesis of 3-amino- and 3-(N-alkylamino)-5-perfluoroalkyl-1,2,4-oxadiazoles. *Tetrahedron Letters*, **41** (41), 7977–7981.

68 Buscemi, S., Pace, A., Vivona, N. and Caronna, T. (2001) Photoinduced molecular rearrangements. Some comments on the ring photoisomerization of 1,2,4-oxadiazoles into 1,3,4-oxadiazoles. *Journal of Heterocyclic Chemistry*, **38** (3), 777–780.

69 Buscemi, S., Pace, A., Pibiri, I., Vivona, N. and Caronna, T. (2004) Fluorinated heterocyclic compounds. An assay on the photochemistry of some fluorinated 1-oxa-2-azoles: an expedient route to

fluorinated heterocycles. *Journal of Fluorine Chemistry*, **125** (2), 165–173.

70 Pace, A., Buscemi, S., Vivona, N., Silvestri, A. and Barone, G. (2006) Photochemistry of 1,2,4-oxadiazoles. A DFT study on photoinduced competitive rearrangements of 3-amino- and 3-*N*-methylamino-5-perfluoroalkyl-1,2,4-oxadiazoles. *Journal of Organic Chemistry*, **71** (7), 2740–2749.

71 Pace, A., Pibiri, I., Buscemi, S., Vivona, N. and Malpezzi, L. (2004) The photochemistry of fluorinated heterocyclic compounds. An expedient route for the synthesis of fluorinated 1,3,4-oxadiazoles and 1,2,4-triazoles. *Journal of Organic Chemistry*, **69** (12), 4108–4115.

72 Pace, A., Buscemi, S. and Vivona, N. (2005) Heterocyclic rearrangements in constrained media. A Zeolite-directed photorearrangement of 1,2,4-oxadiazoles. *Journal of Organic Chemistry*, **70** (6), 2322–2324.

73 Vivona, N., Buscemi, S., Asta, S. and Caronna, T. (1997) Photoinduced molecular rearrangements. The photochemistry of 1,2,4-oxadiazoles in the presence of sulphur nucleophiles. Synthesis of 1,2,4-thiadiazoles. *Tetrahedron*, **53** (37), 12629–12636.

74 Sumengen, D. and Pelter, A. (1983) The preparation and rearrangements of 3,4-disubstituted 1,2,4-oxadiazoline-5-thiones. *Journal of the Chemical Society, Perkin Transactions 1*, (4), 687–691.

75 Chambers, R.D., MacBride, J.A., Maslakiewicz, H., Srivastava, J.R. and K.C. (1975) Photochemistry of halogenocarbon compounds. Part I. Rearrangement of pyridazines to pyrazines. *Journal of the Chemical Society, Perkin Transactions 1*, (4), 396–400.

76 Chambers, R.D., Maslakiewicz, J.R. and Srivastava, K.C. (1975) Photochemistry of halogenocarbon compounds. Part II. Valence isomers from fluorinated pyridazines. *Journal of the Chemical Society, Perkin Transactions 1*, (12), 1130–1134.

77 Barlow, M.G., Haszeldine, R.N. and Pickett, J.A. (1978) Heterocyclic polyfluoro-compounds. Part 26. Synthesis of 3,6-*bis*-trifluoromethyl-pyridazines and -dihydropyridazines. *Journal of the Chemical Society, Perkin Transactions 1*, (4), 378–380.

78 Iddon, B., Meth-Cohn, O., Scriven, E.F.V., Suschitzky, H. and Gallagher, P.T. (1979) Developments in arylnitrene chemistry: syntheses and mechanisms. *Angewandte Chemie, International Edition in English*, **18** (12), 900–917.

79 Gritsan, N.P. and Platz, M.S. (2006) Kinetics, spectroscopy, and computational chemistry of arylnitrenes. *Chemical Reviews*, **106** (9), 3844–3867.

80 Doering, W., von, E. and Odum, R.A. (1966) Ring enlargement in the photolysis of phenylazide. *Tetrahedron*, **22** (1), 81–93.

81 Smalley, R.K., Strachan, W.A. and Suschitzky, H. (1974) Synthesis of 2-alkoxy-3-alkoxycarbonyl-3*H*-azepines. *Synthesis*, (5), 503–504.

82 Purvis, R., Smalley, R.K., Strachan, W.A. and Suschitzky, H. (1978) The photolysis of o-azidobenzoic acid derivatives: a practicable synthesis of 2-alkoxy-3-alkoxycarbonyl-3*H*-azepines. *Journal of the Chemical Society, Perkin Transactions 1*, (3), 191–195.

83 Lamara, K. and Smalley, R.K. (1991) 3*H*-Azepines and related systems. Part 4. Preparation of 3*H*-azepin-2-ones and 6*H*-azepino[2,1-b]quinazolin-12-ones by photo-induced ring expansion of aryl azides. *Tetrahedron*, **47** (12/13), 2277–2290.

84 Streith, J. and Cassal, J.M. (1968) Photochemical synthesis of ethyl 1*H*-diazepine-1-carboxylate. *Angewandte Chemie, International Edition in English*, **7** (2), 129.

85 Sasaki, T., Kanamatsu, K., Kakehi, A., Ichikawa, I. and Hayakawa, K. (1970) The chemistry of diazepines. The photochemical intramolecular 1,3-dipolar

cycloaddition of substituted 1-ethoxycarbonyliminopyridinium ylides. *Journal of Organic Chemistry*, **35** (2), 426–433.

86 Balasubramanian, A., McIntosh, J.M. and Snieckus, V. (1970) The photoisomerization of 1-iminopyridinium ylides to 1(1*H*), 2-diazepines. *Journal of Organic Chemistry*, **35** (2), 433–438.

87 Streith, J., Luttringer, J.P. and Nastasi, M. (1971) Photochemical synthesis of 1,2-diazepines. V. Synthesis and rearrangements of 1,2-diazepines. *Journal of Organic Chemistry*, **36** (20), 2962–2967.

88 Streith, J. (1977) Photochemistry as a tool in heterocyclic synthesis: from pyridinium *N*-ylides to diazepines and beyond. *Heterocycles*, **6** (9), 1513–1519.

89 Streith, J. (1977) Photochemistry as a tool in heterocyclic synthesis: from pyridinium *N*-ylides to diazepines and beyond. *Heterocycles*, **6** (12), 2021–2043.

90 Fritz, H., Gleiter, R., Nastasi, M., Schuppiser, J.-L. and Streith, J. (1978) Photochemistry of 3-substituted 1-iminopyridinium ylides. Regiospecific versus non-regiospecific photoisomerization patterns. *Helvetica Chimica Acta*, **61** (8), 2887–2898.

91 Reisinger, A. and Wentrup, C. (1996) Synthesis of 1*H*- and 5*H*-1,3-diazepines from azido and tetrazolo-pyridines. *Chemical Communications*, (7), 813–814.

92 Reisinger, A., Koch, R. and Wentrup, C. (1998) Dihydro-1,3-diazepinones and diazabicyclo[3,2,0]heptenones from pyridyl azides. *Journal of the Chemical Society, Perkin Transactions 1*, (15), 2247–2249.

93 Reisinger, A., Bernhardt, P.V. and Wentrup, C. (2004) Synthesis of 1,3-diazepines and ring contraction to cyanopyrroles. *Organic & Biomolecular Chemistry*, **2** (2), 246–256.

94 Sawanishi, H., Tajima, K. and Tsuchiya, T. (1987) Studies on diazepines. XXVI. Syntheses of 6*H*-1,4-diazepines and 1-acyl-1*H*-1,4-diazepines from 4-pyridyl azides. *Chemical & Pharmaceutical Bulletin*, **35** (8), 3171–3181.

95 Buchardt, O., Pedersen, C.L. and Harrit, N. (1972) Photochemical studies. XVIII. Light-induced ring expansion of pyridine *N*-oxides. *Journal of Organic Chemistry*, **37** (23), 3592–3595.

96 Albini, A., Bettinetti, G.F. and Minoli, G. (1983) 1,3-Oxazepines via photoisomerization of heteroaromatic *N*-oxides: 3,1-benzoxazepine, in *Organic Syntheses*, Vol. 61 (ed. R.V. Stevens), John Wiley & Sons, New York, pp. 98–103.

97 Hata, N. and Ono, I. (1976) The primary process of the photochemical isomerization of azanaphthalene *N*-oxides. *Bulletin of the Chemical Society of Japan*, **49** (7), 1794–1798.

98 Albini, A., Fasani, E. and Maggi Dacrema, L. (1980) Photochemistry of methoxy-substituted quinoline and isoquinoline *N*-oxides. *Journal of Chemical Society, Perkin Transactions 1*, (12), 2738–2742.

99 Kaneko, C., Hayashi, S. and Kobayashi, Y. (1974) Photolysis of 2-(trifluoromethyl) quiniline 1-oxides and 1-(trifluoromethyl) isoquinoline 2-oxide. *Chemical & Pharmaceutical Bulletin*, **22** (9), 2147–2154.

100 Buchardt, O., Lohse, C., Duffield, A.M. and Djerassi, C. (1967) The photolysis of 1-phenyl and 1-cyano substituted isoquinoline *N*-oxides to benz[f]-1,3-oxazepines. *Tetrahedron Letters*, **8** (29), 2741–2745.

13
Photolabile Protecting Groups in Organic Synthesis
Christian G. Bochet and Aurélien Blanc

13.1
Introduction

Protecting groups are crucial tools in organic synthesis, and their involvement increases with the length of a synthetic route. However, as their number increases within the same molecule, the problem of their individual and selective removal becomes a critical issue. This is well known as the *orthogonality problem*. Two groups are called orthogonal if they can be removed selectively under specific reaction conditions without mutual interference, and in any given sequence [1]. Unfortunately, the dimensions in the space of orthogonal groups are limited. Such dimensions are traditionally acidic, basic, nucleophilic, reductive, and oxidative conditions. A few more "exotic" specific conditions exist (and are highly desirable), and photolability is clearly one of these. It is attractive because no other reagent than light is required, thus limiting the chances of interference. Of course, the intrinsic photoreactivity of the remaining functional groups of the molecule is a liability, and many of the chapters in this Handbook have covered this topic. Thus, in order to have only reactivity at the desired site – that is, removal of the protecting group – it is necessary to have a high absorbance of that group and a high quantum yield for its cleavage, to ensure that only short irradiation times are required.

As the field of photolabile protecting groups (PPGs) has been reviewed on several occasions [2–8], the aim of this chapter is not to update the existing catalogues of groups. Indeed, when photolability has been selected as a candidate to solve a problem, the plethora of groups – sometimes with contradictory behaviors and reaction conditions – might deter even the most daring synthetic chemist. In an attempt to better guide the experimentalist, the decision was taken to divide the groups into major families based on a specific reaction, and then to list the variations on the groups, detailing their effects on reactivity. The application of photolabile protecting groups to the chemical functions is reported in Appendix 13.6, followed by a list of typical conditions and yields for protection and deprotection procedures.

Handbook of Synthetic Photochemistry. Edited by Angelo Albini and Maurizio Fagnoni
Copyright © 2010 WILEY-VCH Verlag GmbH & Co. KGaA, Weinheim
ISBN: 978-3-527-32391-3

13.2
Photolabile Protecting Groups

13.2.1
Ortho-Nitrobenzyl Alcohol Derivatives

One of the most common photochemical reaction pathways of carbonyl compounds is the formation of a diradicaloid excited state which is able to abstract a hydrogen atom at the γ (or, more rarely, ε) position, followed by either fragmentation or recombination. This process, which is known as the Norrish type II reaction, has a parallel in the photochemistry of nitro groups; the intramolecular hydrogen abstraction of excited *ortho*-nitrotoluene is actually one of the very early synthetic photochemical transformations [9]. It has been exploited in a family of photolabile protecting groups, most prominent among which are derivatives of *ortho*-nitrobenzyl alcohol, as introduced in 1966 by Barltrop *et al.* (Scheme 13.1) [10, 11].

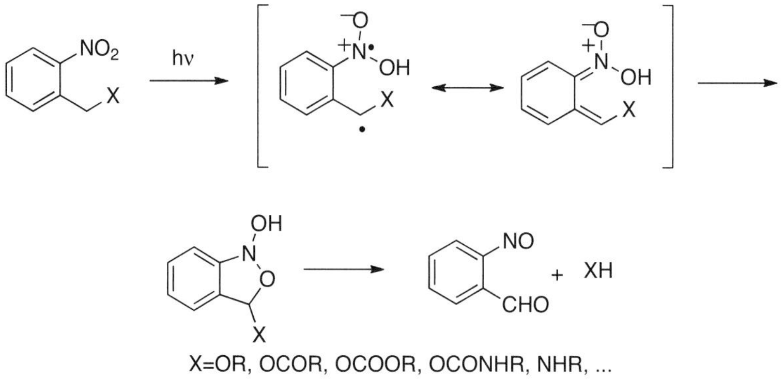

X=OR, OCOR, OCOOR, OCONHR, NHR, ...

Scheme 13.1

Various functional groups can be protected and released in high yields, such as alcohols (as ethers), carboxylic acids (as esters), amides (as amides), and amines (as carbamates). In the latter case, however, two drawbacks must be noted:

- The initially released intermediate is a carbamic acid, which is prone to spontaneous decarboxylation; this step is, however, quite slow, and makes this process inappropriate for fast kinetic experiments (e.g., in neurophysiology).

- The aldehyde side product can react with the freshly released amine, thus limiting the chemical yield. To avoid this reaction one can add an aldehyde scavenger, such as semicarbazide hydrochloride, hydrazine or an aldehyde scavenging resin (e.g., supported hydrazines). An alternative is to alter the structure of the photolabile protecting group.

In order to circumvent these drawbacks, a series of derivatives were proposed (Scheme 13.2). For example, the addition of two methoxy groups to the aromatic ring significantly increases the absorbance at longer wavelength (albeit at the expense of

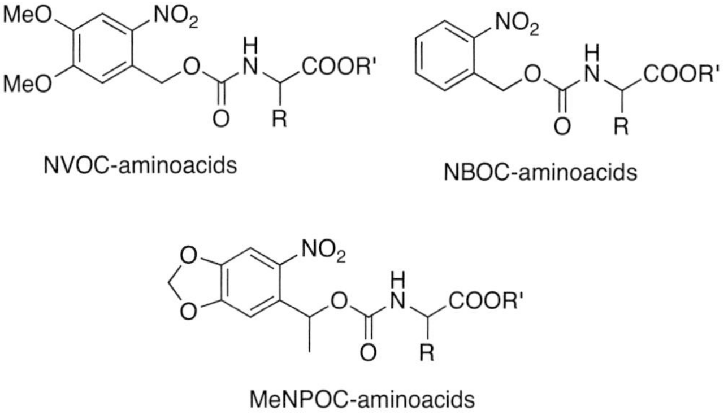

Scheme 13.2

the quantum yield for cleavage), allowing deprotection up to 420 nm. This feature is desirable with potentially photosensitive amino acids such as tryptophan and tyrosine. Another structural modification, aiming this time at minimizing the reaction between the freshly liberated amine and the concomitantly formed nitrosocarbonyl group, is to add an extra methyl group at the benzylic position; the side product is now a ketone, which is much slower at making an imine [12]. It is worth noting that the two methoxy groups were replaced in this example by a methylenedioxy moiety; the electronic impact of this substituent is similar, but not quite identical [13, 14], presumably due to different alignment of the nonbonding oxygen electron pairs.

The usefulness of this family of protecting groups has been proved in countless applications, both in the total synthesis of natural products, where protecting group orthogonality is often a critical issue (e.g., in the total synthesis of Calicheamycin [15] or Aplysiatoxin [16]; Scheme 13.3) or in the light-directed synthesis of oligonucleotides (DNA-chips) [17].

Related PPGs, but operating with a different mechanism, were introduced some years ago; examples are the NPPOC (R = Me), where the side product is a nitrostyrene [18–23], and the NPEOC (R = H) [24].

It has also been used in combination with a nearby strongly absorbing chromophore, playing the role of an antenna in exploiting an efficient energy transfer (Scheme 13.4) [25, 26].

Derivatives of *ortho*-nitrobenzyl alcohol were also used to protect the hydroxy groups of carbohydrates, including in the anomeric position. Thus, the following acetal was quantitatively photolysed to glucose (Scheme 13.5) [27].

A related group was developed by Gravel *et al.*, for the protection of ketones. Thus, a diol derivative was used to form a ketal (under the usual acidic conditions), which was quite stable under acidic conditions (Scheme 13.6) [28, 29] but was, unfortunately, degraded under basic conditions. The introduction of a second aromatic group in the backbone, and provision of a single enantiomer of the diol (thus allowing the protection of chiral ketones without making pairs of diastereoisomers) partially solved the base-stability problem [30].

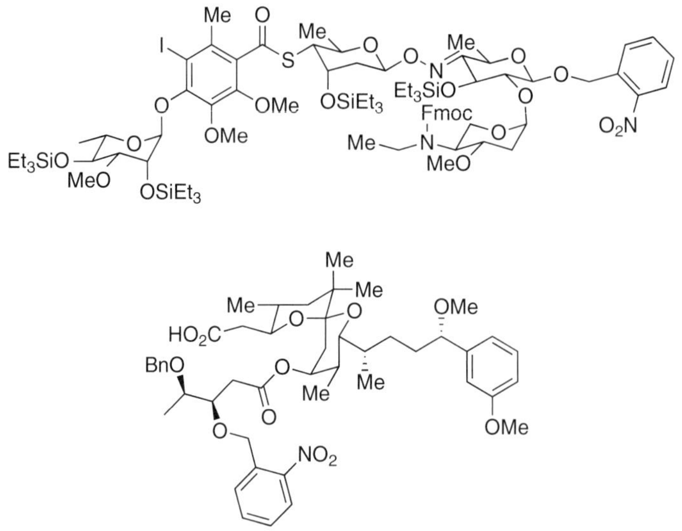

Scheme 13.3

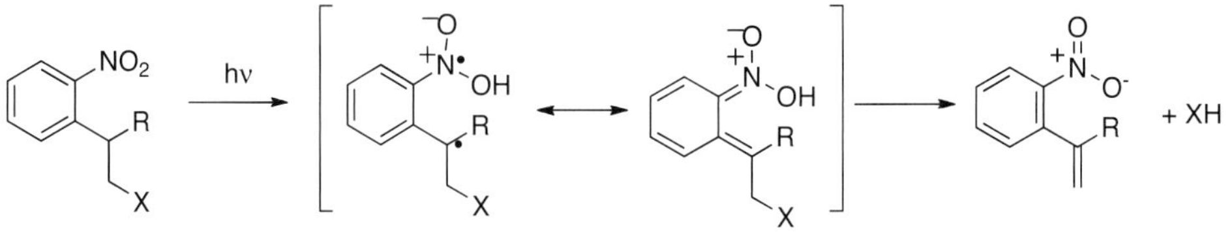

Scheme 13.4

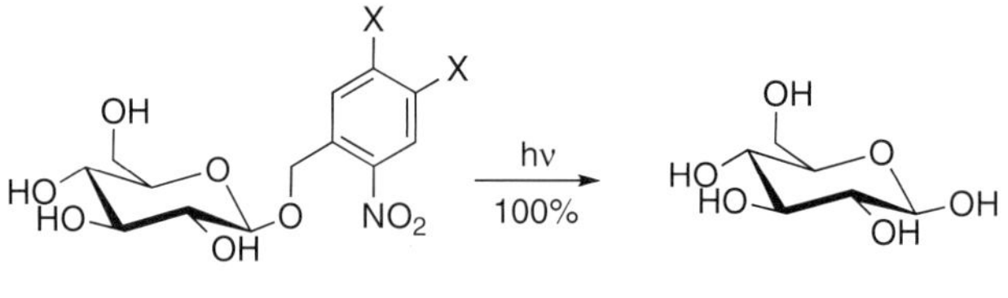

Scheme 13.5

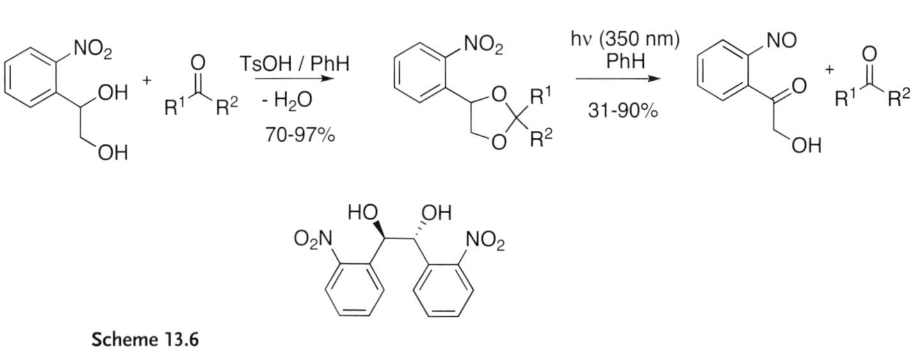

Scheme 13.6

13.2.2
Benzyl Alcohol Derivatives

The photolytic removal of the N-benzyloxycarbonyl group (Cbz) was described at a very early stage in the history of photolabile protecting groups [31, 32]. Relatively high chemical and quantum yields were observed (75%, $\Phi = 0.15$), particularly in the presence of water. The benzyl alcohol side product and a strong pH-dependence suggested a simple heterolysis of the benzylic carbon–oxygen bond at the excited state (Scheme 13.7).

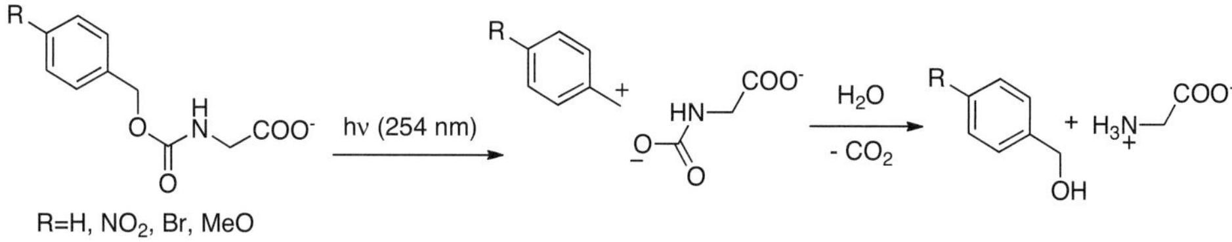

Scheme 13.7

Consistent with this mechanism, the addition of two *meta*-methoxy groups facilitates the reaction, on which was successfully tested a series of amino acids (Scheme 13.8) [33]. One of the main drawback of this group is the need for short-wavelength irradiation, which his frequently incompatible with features on the substrate. The true nature of the mechanism (e.g., homolysis versus heterolysis) has been discussed over the years, using among other arguments the *meta* effect [34–37].

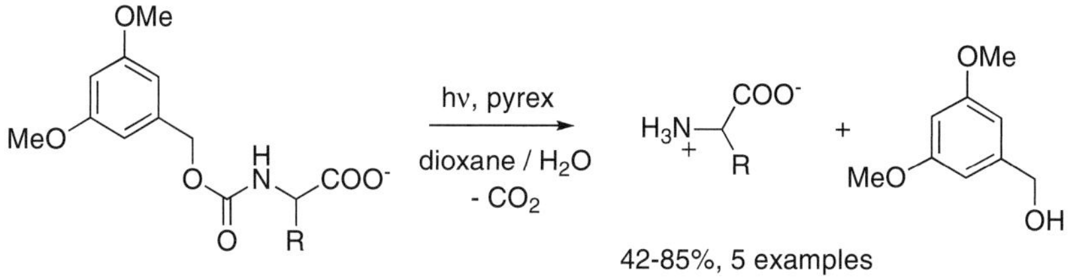

Scheme 13.8

On the other hand, it was found that two additional benzylic methyl groups greatly facilitated the photolysis, as shown by the Ddz variant (Scheme 13.9), where the byproduct was no longer a benzylic alcohol but rather a styrene derivative, albeit prone to $[2\pi + 2\pi]$ dimerization. It is worth noting that this group is also acid-labile, by up to three orders of magnitude more than the Boc group [38, 39].

This reaction was used for the *in situ* generation of a base upon irradiation, a process of significance in imaging applications (Scheme 13.10) [38].

The photochemically promoted rupture of a benzylic carbon–heteroatom bond is exploited in numerous photolabile protecting groups, for example in the 9-phenyl-xanthen-9-yl (also called pixyl group, X = O in Scheme 13.11) [40] and its sulfur-

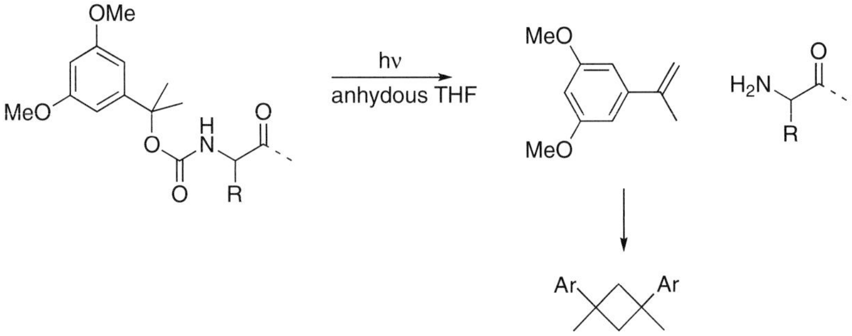

Scheme 13.9

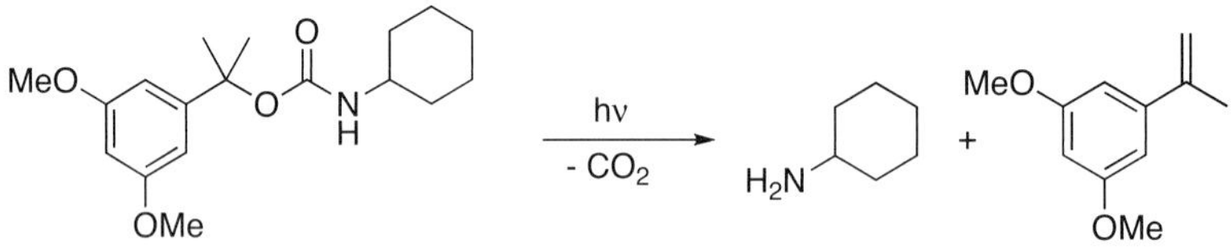

Scheme 13.10

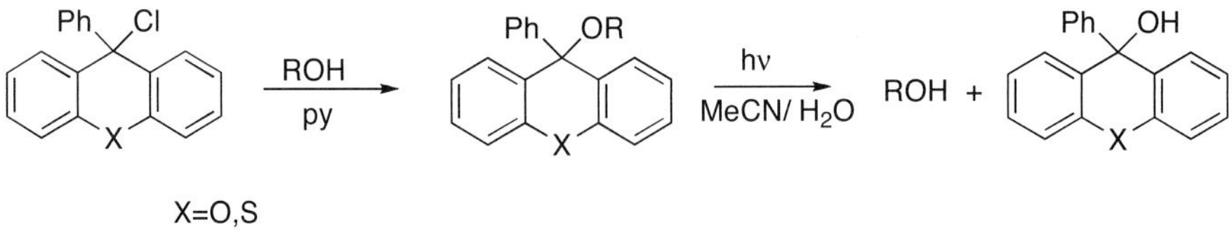

Scheme 13.11

containing analogue (S-pixyl, X = S) [41, 42], which are used in oligonucleotide synthesis.

Based on the same principles, derivatives of anthraquinone (aqmoc) [43], pyrene [44], coumarin [45–53], and phenanthrene have each been used in certain instances (Scheme 13.12) [54].

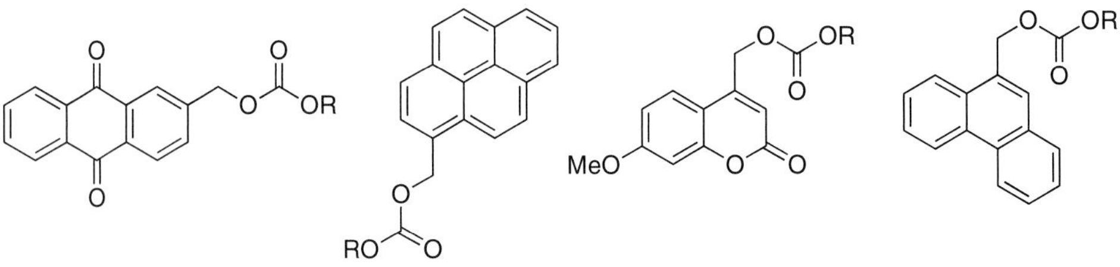

Scheme 13.12

13.2.3
Other Types of Protecting Group

13.2.3.1 Norrish Type II

Norrish type II reactions have been exploited to release functional groups, although these are not untouched by the reaction; consequently, one cannot speak of protecting groups in the true sense of the term. However, this type of reaction has been exploited on several occasions, such as in the release of fragrances (Scheme 13.13). The oxidative release is remarkably robust, and is unaffected by air [55–58].

In such an application, various alkenes were released photochemically, among these being linalyl acetate, which has a pleasant odor (Scheme 13.14) [56–59].

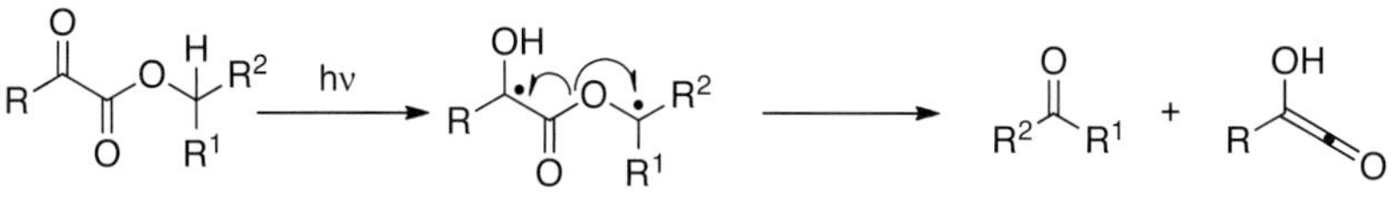

Scheme 13.13

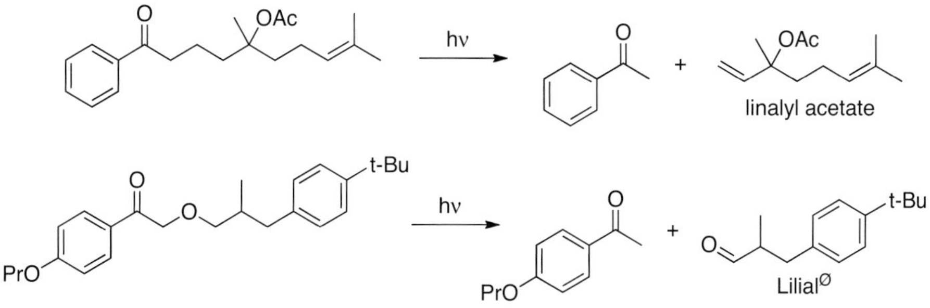

Scheme 13.14

The insensitivity of this reaction towards oxygen makes it very attractive for media where the exclusion of air is not possible, for example, in vivo. As an example, the biologically relevant 4-hydroxy-2-nonenal was released photochemically from an anthraquinone precursor. Interestingly, the phototrigger is reoxidized by molecular oxygen during the reaction (Scheme 13.15) [60].

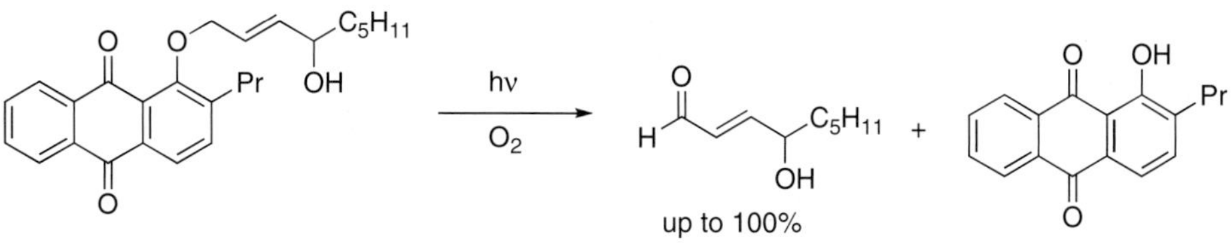

Scheme 13.15

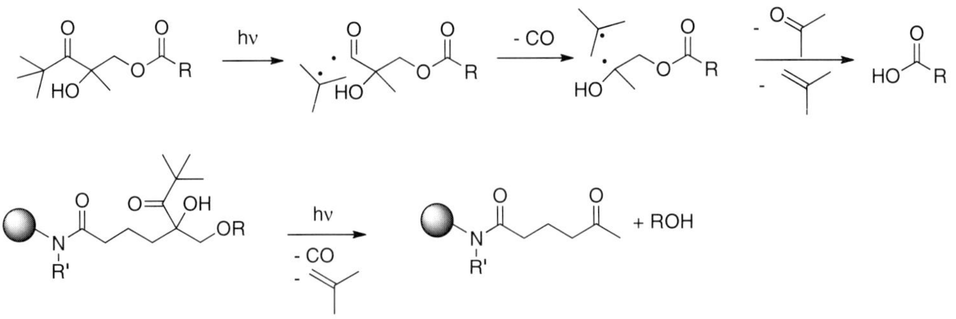

Scheme 13.16

13.2.3.2 Norrish Type I

A PPG based on the Norrish type I fragmentation of *tert*-butylketone derivatives has been introduced by Giese and coworkers [61, 62], and included as part of a linker for solid-phase synthesis. Its tolerance towards other conditions is remarkable (acidic, basic, Pd-catalysis in Suzuki crosscoupling reactions, etc.), and the photolysis releases only volatile or resin-bound side products, which makes the work-up procedure extremely simple (Scheme 13.16). Although the photolysis quantum yield in acetonitrile is very high ($\Phi = 0.56$), the required wavelength is quite short (between 280 and 340 nm). Whilst this might be considered a drawback in some applications, it was actually exploited in combination with other PPGs operating at longer wavelengths (chromatic orthogonality, see below) [63].

13.2.3.3 Thioketals

In a series of different analogues, Kutadeladze proposed an original procedure for protecting ketones, by the addition of the lithium salt of dithiane [64, 65], or with trithiabicyclo[2.2.2]octane (TTBO) derivatives [66]. Photolysis in the presence of a sensitizer returned the ketone and the *ortho*-thioester, which can be reused in another cycle. Interestingly, combinatorial encoding is possible by variations on the *ortho*-thioester backbone (R in Scheme 13.17).

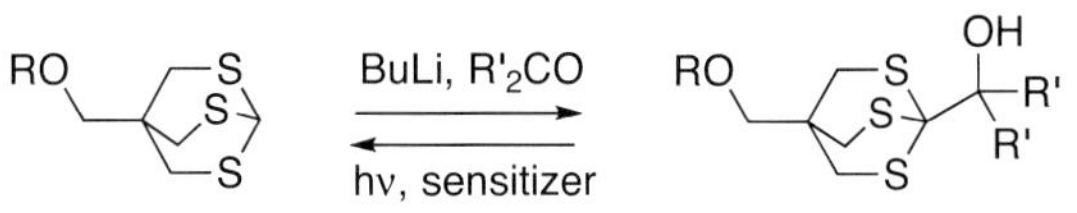

Scheme 13.17

13.2.3.4 Silicon Ethers

As silyl ethers are among the most widely used protecting groups in organic synthesis, a photolabile analogue would be highly desirable. Indeed, such a group was introduced in 1997, namely the tris(trimethylsilyl(silyl)), also called the sisyl group (Scheme 13.18) [67, 68]. This group is reported as being fluoride-resistant under carefully controlled conditions (CsF or KF/18-crown-6), but to be photosensitive at short wavelengths (between 200 and 300 nm). However, despite the group's

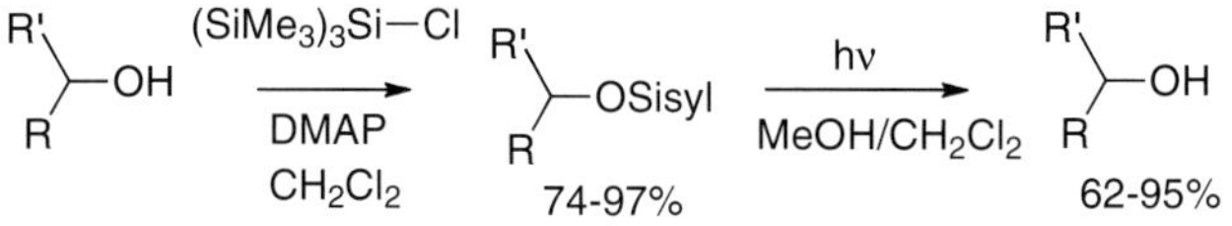

Scheme 13.18

apparent attraction, it has rarely been used in organic synthesis, most likely due to its overall sensitivity towards common conditions in organic synthesis (e.g., BuLi, LiAlH$_4$, TBAF.). However, as the group is very easily introduced under conditions that are compatible with many other functional groups, it still might serve as a good starting group.

Another silicon-based PPG, proposed by Pirrung, was also attractive from the point of view of its easy introduction with alcohol derivatives (Scheme 13.19). The photochemical step is a photoinduced E/Z isomerization (see Section 13.2.4) [69–71].

Although, the photolysis requires quite short wavelengths (254 nm), the reaction is clean and high-yielding (up to 84% of liberated alcohol) in acetonitrile (a different and nonproductive reaction occurs in benzene) with a variety of primary, secondary, and allylic alcohols, as well with 3′- and 5′-protected nucleosides. The chemical stability is very high, approaching that of triisopropylsilyl (TIPS) in some cases (R^2 = Me).

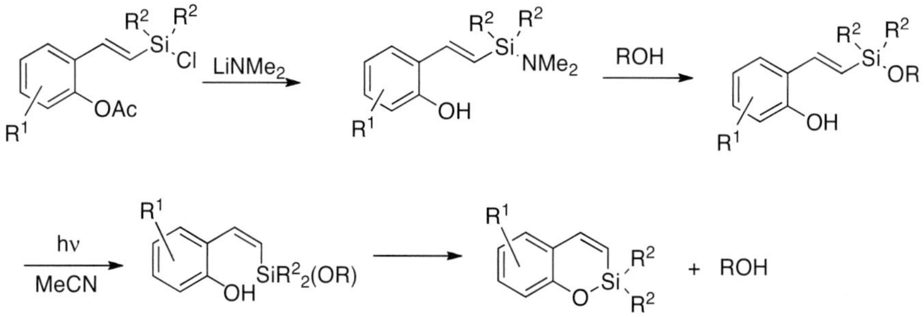

Scheme 13.19

13.2.4
Z/E Photoisomerization

13.2.4.1 Cinnamyl Esters

The well-known cis–trans isomerization of alkenes has been exploited in PPGs in several instances. For example, an enzyme was inactivated by protecting one (or several) of its alcohol functions with a *trans*-hydroxycinammic acid derivative [72–75]. Photoisomerization to the *cis* isomer (using a Pyrex-filtered, medium-pressure mercury-xenon lamp) placed the phenolic OH close to the ester function, thus allowing lactonization and release of the free alcohol (Scheme 13.20); in this way, the enzymatic activity was restored. Although the initial studies highlighted some limitation due to the overlapping spectra of the enzyme and protecting group, the

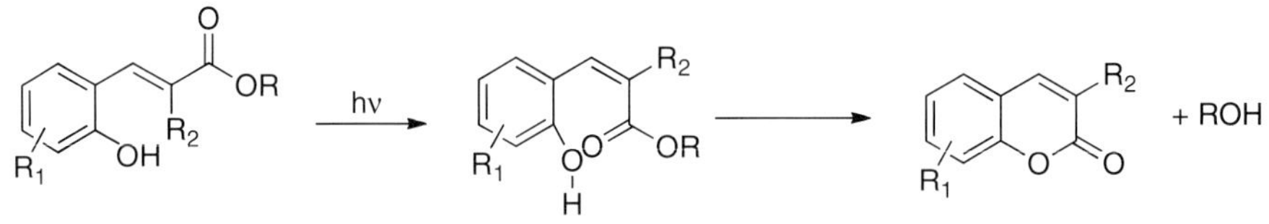

Scheme 13.20

strategy also proved to be useful in other types of applications, such as fragrance release [59].

13.2.5
Phenacyl Derivatives

13.2.5.1 Mechanism

Due to their many photochemical properties, phenacyl esters also represent attractive PPGs, and have been widely used as caging groups in biologically oriented applications (Scheme 13.21). In contrast, the relatively high reactivity of the keto group limits the use of phenacyl esters as protecting groups, and thus has been little used in total syntheses [76–86].

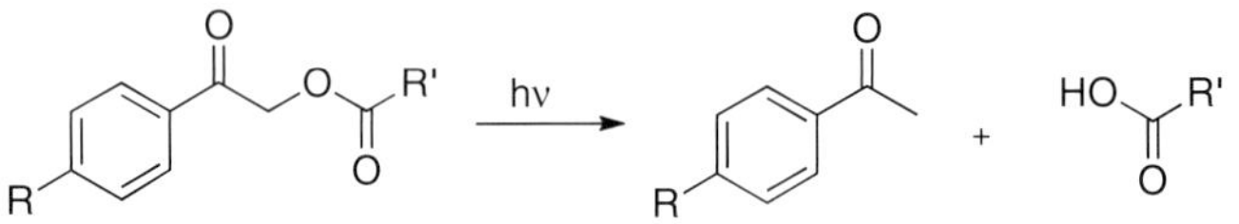

Scheme 13.21

Nevertheless, phenacyl esters have been widely used to release phosphates upon irradiation [80, 81, 87, 88]. Substitution on the aromatic ring can alter their photochemical properties' for example, a hydroxyl group in the para position (R = OH) makes it a fast-release trigger for biological stimulants (Scheme 13.22). The attractive features here are biologically inert byproducts, with little absorption at longer wavelengths (unlike the NVOC derivatives!). However, other substituents (such as an amino group) proved inefficient.

The nature of the solvent is important (small amounts of water can increase the photolysis rate fivefold), but air or triplet quenchers should be excluded, as they create an up to fourfold slowdown.

However, the mechanism depicted in Scheme 13.23 might not be operative for regular esters, because a rapid loss of CO_2 from the carboxyl radical would be

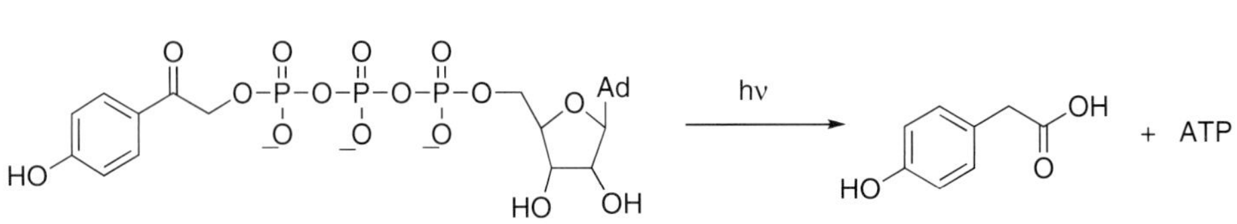

Scheme 13.22

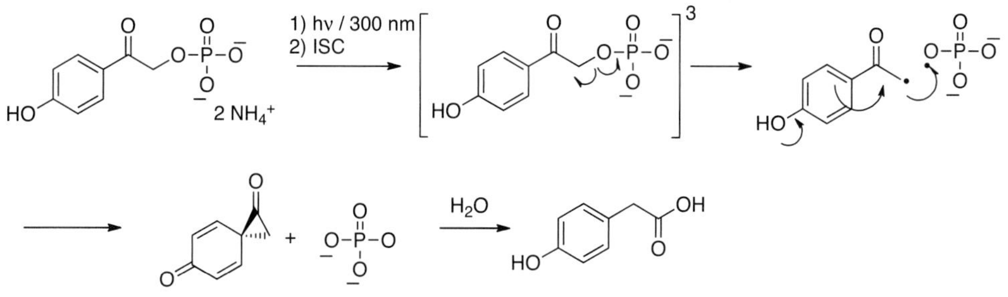

13.2 Photolabile Protecting Groups

Scheme 13.23

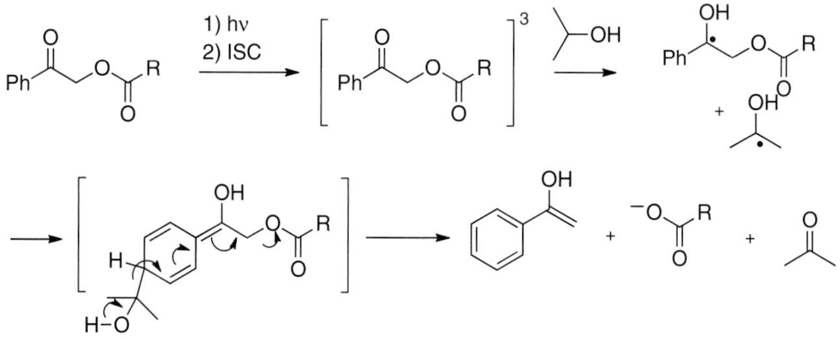

Scheme 13.24

expected. An alternative mechanism in the presence of a hydrogen donor (isopropanol) has been proposed (Scheme 13.24).

Interestingly, phenacyl derivatives bearing an easily abstractable hydrogen might undergo a competitive Norrish type II pathway, and this has been exploited to release fragrant carbonyl compounds (Scheme 13.25) [59].

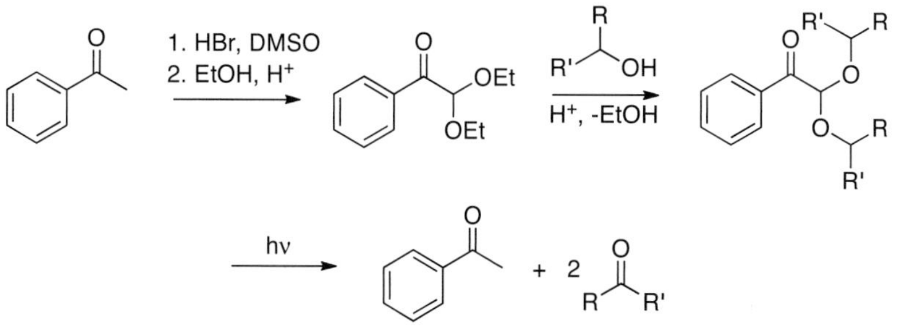

Scheme 13.25

Other phenacyl derivatives were introduced by Klán et al. to release amines (if protected as carbamates) in high chemical yields upon irradiation (up to 97%) [89–93]. The mechanism proceeds through an *ortho*-quinoid aromatic triplet photoenol, which can either cyclize/eliminate or be trapped by the solvent if it is nucleophilic (e.g., MeOH), as shown in Scheme 13.26.

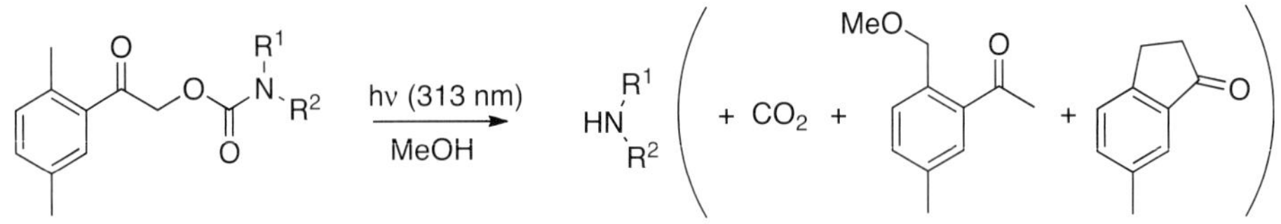

Scheme 13.26

13.2.6
Benzoin Derivatives

13.2.6.1 Mechanism

Benzoin esters were shown to release carboxylic acids in a very high quantum and chemical yields, together with a chemically quite inert side product, a substituted benzofuran (Scheme 13.27) [94]. The exact mechanism remains somewhat controversial, but might involve a radical pathway, a benzylic ionization (see above) [95], or even an intramolecular Paterno–Büchi [2 + 2]-cycloaddition leading to a highly strained "hausane" intermediate. Regardless of the mechanism involved, this reaction most likely originates from a singlet excited state, as evidenced by the lack of quenching by piperylene or naphthalene quenchers [96–100].

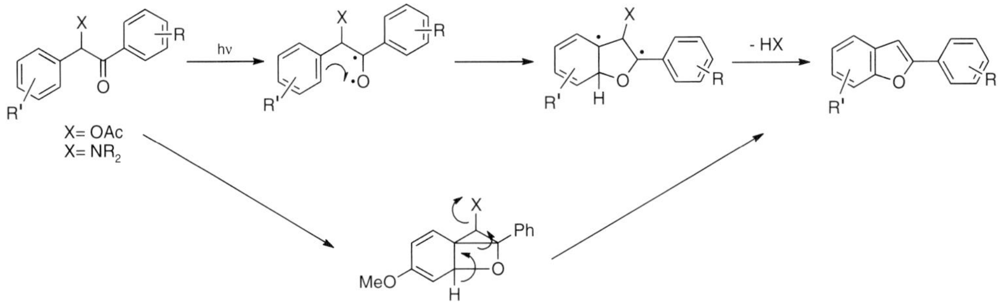

Scheme 13.27

The reaction is of special interest because it has a very high quantum yield, and benzoin esters have little absorbance above 350 nm; this feature allowed the photochemical removal of *other* photolabile protecting groups at longer wavelengths, such as an NVOC group. The term "chromatic orthogonality" has been introduced to denote this dual-wavelength independent removal of PPGs [101, 102]. One limitation of the benzoin PPG is the difficulty of preparing other derivatives than esters, and in particular carbamates [103–107]. In order to increase the chemical robustness and limit the photosensitivity during transformations other than cleavage, an interesting safety-catch approach was shown to be effective, by having first to hydrolyze a thioketal before achieving photosensitivity (Scheme 13.28) [108–110].

It is worthy of mention at this point that the benzoin ester PPG bears a chiral center, which is a potentially problematic feature with chiral substrates. However, this

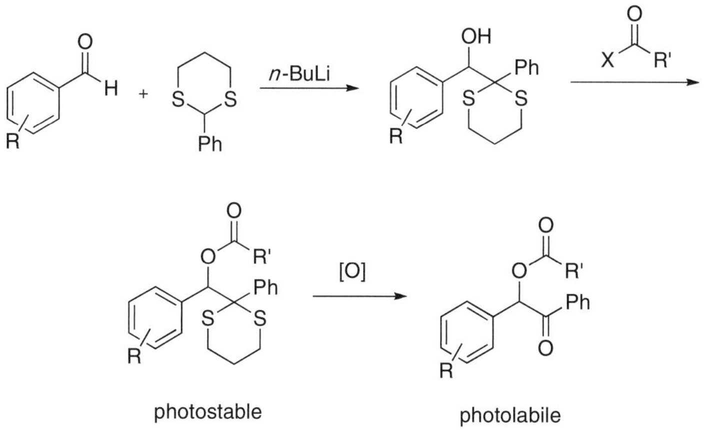
n-BuLi
[O]
photostable
photolabile

Scheme 13.28

problem was overcome by employing an asymmetric synthesis via a cyanohydrin intermediate [96].

13.2.7
Indolines

Although *ortho*-nitroanilines were introduced at an early stage as PPGs, this group was not extensively used as such [111]. In contrast, the cyclic version – the 7-nitroindolines – proved to be much more useful, in particular when it carried a 5-bromo substituent [112]. The reaction was seen to be quite rapid, even at wavelengths up to 400 nm, while the side product – the parent deacylated nitroindoline – was shown to be chemically and photochemically unreactive. Yet, it is typically appropriate for releasing carboxylic acids, as shown by the example of Merrifield in 1977 (Scheme 13.29) [113].

An extremely interesting observation was made when the photolysis was carried out in the presence of a nucleophile (such as an amine), or when methanol was used as a solvent. The isolated product was not the expected carboxylic acid, but rather the amide or the methyl ester. In fact, the reaction involved an acyl transfer, although when water was present the acid was nevertheless observed [114, 115]. Mechanistic studies later revealed that the acyl group underwent a shift from the indoline nitrogen atom to one of the nitro group oxygen atoms, thus forming a highly activated electrophile *in situ* [116].

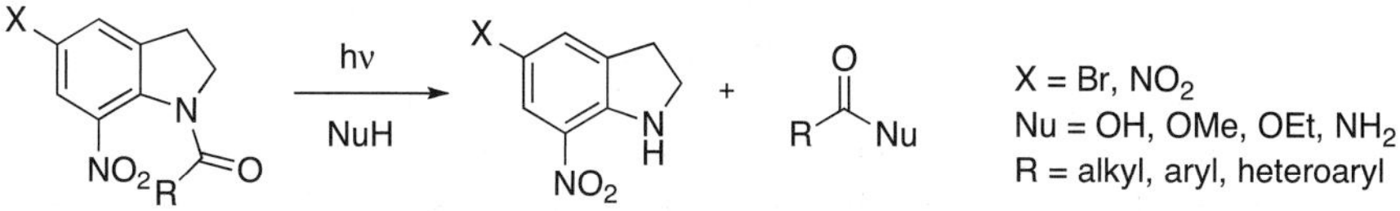
hν
NuH
X = Br, NO2
Nu = OH, OMe, OEt, NH2
R = alkyl, aryl, heteroaryl

Scheme 13.29

Boc-(OBz)-Tyr-Ala-Bni + H-Gly-Phe-Leu-NH$_2$ $\xrightarrow[70\%]{h\nu}$

Boc-(OBz)-Tyr-Ala-Gly-Phe-Leu-NH$_2$

Scheme 13.30

This reaction was exploited by Pass, Amit and Patchornik to directly couple two peptide fragments (Scheme 13.30) [117]. However, one of the main issues to date has been the preparation of such activable amino acids. Two approaches (which, unfortunately, are limited in their scope) are nitration of the parent indoline derivative *after* acylation with the amino acid, or use of the amino acid chloride prepared *in situ*.

Variations of this reaction have since been reported [118], for example replacing the 5-bromo group with another nitro group [119, 120]. This more reactive group allowed the acylation of weaker nucleophiles than amines (such as alcohols), even those that were highly hindered [121]. Whilst this process was used to introduce classical protecting groups onto amines such as Cbz and Fmoc (Scheme 13.31) [119], the introduction of Boc groups, in contrast, proved to be inefficient.

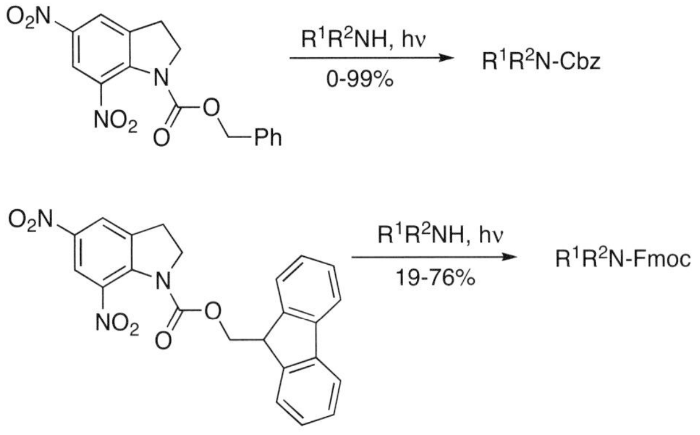

Scheme 13.31

13.3
Chromatic Orthogonality

If some degree of orthogonality were to be provided by PPGs with respect to other classical protecting groups, there would also be the possibility of differentiating two photolabile groups by the wavelength. Although this so-called chromatic orthogonality concept has not yet been generalized, several distinct examples have been identified (Scheme 13.32) [101, 102, 122, 123].

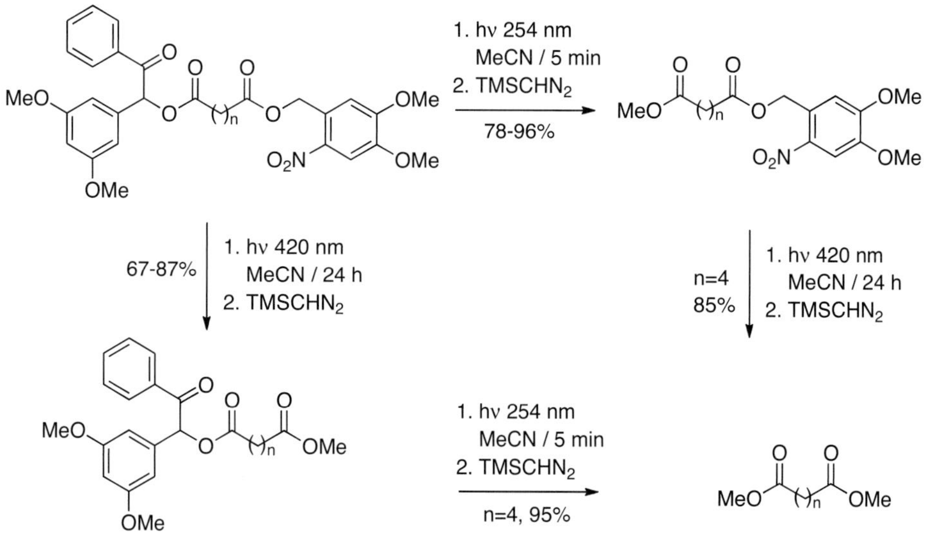

Scheme 13.32

13.4
Two-Photons Absorption

As mentioned earlier, one of the compatibility issues between photolabile protecting groups and other functional groups in a molecule is the use of short wavelengths. This is particularly critical for applications in biological systems, where proteins and nucleic acids are sensitive to wavelengths shorter than 350 nm. An attractive solution to this problem is to use two photons of low energy to cleave the PPGs. It is indeed possible – at least in theory – to excite a chromophore corresponding to an energy gap of, say, 350 nm, by the consecutive absorption of two relatively "benign" 700 nm photons. Whilst this has been verified experimentally, the requirement is for a very intense and focused laser light. An extra "bonus" here is the three-dimensional spatial control (standard photolysis allows control within a plane, whereas the focusing point addresses the third, orthogonal dimension). Chromophores with a large two-photon absorption (2PA) cross-section are still rare, although a few have been described during the past decade (Scheme 13.33) [49, 52, 124–132].

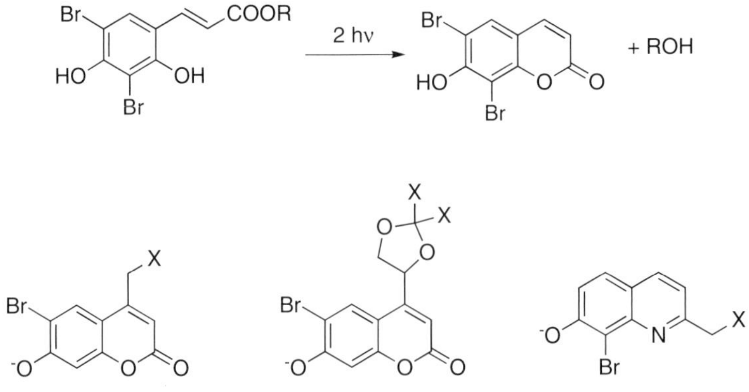

Scheme 13.33

13.5
Concluding Remarks

As noted in the above sections, PPGs constitute a quite large family of protecting groups, and are definitely not limited to the standard *ortho*-nitrobenzylic derivatives. Today, many specific problems in protecting group chemistry can most likely be resolved with recourse to the available arsenal of PPGs, and synthetic organic chemists should not hesitate to use such an approach. Experimental simplicity and high cleavage yields – two essential features of synthesis – are found in many PPGs. Yet, should a photolysis fail, then some simple experimental alterations would usually allow cleavage with a high yield, provided that one is kept aware of the fundamentals of the process, such as the all-too frequent photolysis with a low-pressure mercury lamp in Pyrex glassware!

On the other hand, there is a vast potential for developments in several directions. Among the most obvious, we should perhaps cite the increase in the types of functional group that can be protected (e.g., alkenes and alkynes), the improvement in the efficiency of existing PPGs (in particular for protecting alcohols and thiols), the better tolerance towards chemical conditions and better wavelength differentiation (chemical and chromatic orthogonality), and the higher sensitivity towards longer wavelengths and an increase in the two-photon absorption cross-section. These improvements will, however, occur only if specific needs arise from the users – that is, from the synthetic community. After all, in citing one of the leading reference books describing protecting groups [1], "*Necessity is the mother of invention!*"

13.6 Appendix

The following tables are intended only as a guide for the experimentalist; they are not comprehensive in nature.

13 Photolabile Protecting Groups in Organic Synthesis

13.6 Appendix

Y=	−OR	−OC(O)R	−SR	−NHR	−NHC(O)R	−OC(O)NHR	−OC(O)OR	−OP(O)OR¹OR²	−OCR¹R²O−X
Preparation conditions	SN_2	Esterification	SN_2	Reductive amination	Amidation	Acylation	Carbonoylation	Phosphorylation	Acetalization
Preparative yield	Up to 90%	80%	65%	77%	82%	Not reported	72%	Not reported	67–92%
Photolysis conditions	350 nm	420 nm	366 nm	365 nm	Not reported	>320 nm, di-oxane, CHCl₃, THF, DME, ROH, H₂O	350–400 nm, MeOH	>305 nm	350 nm
	PhMe	MeCN	MeCN + Additives	Phosphate buffer					MeCN or Et₂O
Photolysis yield	Up to 78%	92%	100%	74%	—	100%	45%	77–98%	69–97
Reference	[133]	[102]	[134]	[135]	[136]	[11]	[137]	[138]	[30]

Y=	−OC(O)OR	−OC(O)NHR
Preparation conditions	Carbonoylation	Carbamoylation
Preparative yield	64–81%	78–87%
Photolysis conditions	365 nm	365 nm
	MeOH/H₂O	MeOH + Additives
Photolysis yield	100%	Not reported
Reference	[24]	[19]

436 13 Photolabile Protecting Groups in Organic Synthesis

Y=	–OC(O)NHR (X=H)	–OC(R¹R²)X– (X=O)
Preparation conditions	Carbamoylation	Acetalization
Preparative yield	55–76%	91–99%
Photolysis conditions	254 nm	Pyrex
	Dioxane/H_2O	MeCN/H_2O
Photolysis yield	42–85%	79–93%
Reference	[33]	[139]

Y=	–OR (X=Ph)	–OC(O)NR_2 (X=H)
Preparation conditions	SN_1	Carbamoylation
Preparative yield	65–74%	62–84%
Photolysis conditions	254 or 300 nm	300 nm
	MeCN/H_2O	MeCN/H_2O
Photolysis yield	78–97%	52–90%
Reference	[40]	[140]

13.6 Appendix

Y=	–OR	–NR$_2$
Preparation conditions	Esterification	Amidation
Preparative yield	Not reported	32–99%
Photolysis conditions	365 nm	365 nm
	buffer	MeOH/AcOH (70/1)
Photolysis yield	Not reported	100%
Reference	[72]	[141]

Y=	–OC(O)R	–OC(O)OR	–OC(O)NR^{1}R^2	–OP(O)OR$_2$
Preparation conditions	Esterification	Carbonoylation	Carbamoylation	Phosphorylation
Preparative yield	30–85%	38–88%	42–68%	72–89%
Photolysis conditions	<313 nm	>320 nm	>280 nm	Pyrex
	EtOH or dioxane	MeCN + Sensitizer	MeCN	Dioxane/Et$_3$N
Photolysis yield	33–94%	61–91%	>86%	74–86%
Reference	[78]	[142]	[93]	[80]

438 | 13 Photolabile Protecting Groups in Organic Synthesis

[Structure: MeO-substituted phenyl ketone with Y group, OMe substituent, and phenyl-C(O)- group]

Y=	–OC(O)R	–OC(O)OR	–OC(O)NR$_2$	–OP(O)OR^{1}OR^2
Preparation conditions	Esterification	Carbonoylation	Carbamoylation	Carbonoylation
Preparative yield	87–92%	42–95%	76–90%	45–96%
Photolysis conditions	Pyrex	350 nm	350 nm	350 nm
	PhH	THF	PhH	PhH
Photolysis yield	87–95%	88–98%	56–97%	83–87%
Reference	[94]	[107]	[107]	[96]

[Structure: 4-substituted coumarin with CH$_2$–Y group and R substituent]

Y=	–OC(O)R	–OC(O)OR	–OC(O)NR$_2$	–OP(O)OR^{1}OR^2	–O–(CHR)$_n$–O– (n = 2,4)	–CHOR–CH$_2$OR
Preparation conditions	Esterification	Carbonoylation	Carbamoylation	Phosphorylation	Acetalization	Acetalization
Preparative yield	71–96%	91–100%	40%	23–64%	15–94%	22–57%
Photolysis conditions	350 nm	350 nm	365 or 740 nm (2PA)	333 nm	>348 nm	365 or 740 nm (2PA)
	MeCN		Buffer	Buffer	MeOH/buffer	Buffer
Photolysis yield	80–90%	Not reported	99%	Not reported	≈75%	≈20–50%
Reference	[143]	[48]	[127]	[144]	[145]	[146]

References

1 Kocienski, P. (1994) *Protecting Groups*, Thieme, Stuttgart.
2 Bochet, C.G. (2002) Photolabile protecting groups and linkers. *Journal of the Chemical Society, Perkin Transactions* 1, 125–142.
3 Pillai, V.N.R. (1987) Photolytic deprotection and activation of functional groups, in *Organic Photochemistry*, 9 (ed. A. S Padwa), Marcel Dekker, New York, pp. 225–321.
4 Pillai, V.N.R. (1980) Photoremovable protecting groups in organic synthesis. *Synthesis*, 1, 1–26.
5 Pelliccioli, A.P. and Wirz, J. (2002) Photoremovable protecting groups: reaction mechanisms and applications. *Photochemical & Photobiological Sciences*, (1), 441–458.
6 Goeldner, M. and Givens, R. (2005) *Dynamic Studies in Biology*, John Wiley & Sons, Chichester.
7 Pincock, J.A. (2003) The Photochemistry of Esters of Carboxylic Acids, in *Handbook of Organic Photochemistry and Photobiology*, Vol. II, (eds W.M. Horspool and F. Lenci), CRC Press, Boca Raton, pp. 66.1–66.17.
8 Mayer, G. and Heckel, A. (2006) Biologically active molecules with a "light switch". *Angewandte Chemie, International Edition*, 45, 4900–4921.
9 Ciamician, G. and Silber, P. (1901) Chemical light properties. [Second announcement.]. *Berichte Der Deutschen Chemischen Gesellschaft*, 34, 2040–2046.
10 Barltrop, J.A., Plant, P.J. and Schofield, P. (1966) Photosensitive protective groups. *Journal of the Chemical Society, Chemical Communications*, 822–823.
11 Patchornik, A., Amit, B. and Woodward, R.B. (1970) Photosensitive protecting groups. *Journal of the American Chemical Society*, 92, 6333–6335.
12 McGall, G.H., Barone, A.D., Diggelmann, M., Fodor, S.P.A., Gentalen, E. and Ngo, N. (1997) The efficiency of light-directed synthesis of DNA arrays on glass substrates. *Journal of the American Chemical Society*, 119, 5081–5090.
13 Bley, F., Schaper, K. and Gorner, H. (2008) Photoprocesses of molecules with 2-nitrobenzyl protecting groups and caged organic acids. *Photochemistry and Photobiology*, 84, 162–171.
14 Gorner, H. (2005) Effects of 4,5-dimethoxy groups on the time-resolved photoconversion of 2-nitrobenzyl alcohols and 2-nitrobenzaldehyde into nitroso derivatives. *Photochemical & Photobiological Sciences*, 4, 822–828.
15 Nicolaou, K.C., Hummel, C.W., Nakada, M., Shibayama, K., Pitsinos, E.N., Saimoto, H., Mizuno, Y., Baldenius, K.U. and Smith, A.L. (1993) Total synthesis of calicheamicin-gamma-1(I). 3. The final stages. *Journal of the American Chemical Society*, 115, 7625–7635.
16 Ireland, R.E., Thaisrivongs, S. and Dussault, P.H. (1988) An approach to the total synthesis of aplysiatoxin. *Journal of the American Chemical Society*, 110, 5768–5779.
17 Fodor, S.P.A., Read, J.L., Pirrung, M.C., Stryer, L., Lu, A.T. and Solas, D. (1991) Light-directed, spatially addressable parallel chemical synthesis. *Science (Washington, D.C.)*, 251, 767–773.
18 Pirrung, M.C., Wang, L. and Montague-Smith, M.P. (2001) 3′-Nitrophenylpropyloxycarbonyl (NPPOC) protecting groups for high-fidelity automated 5′ → 3′ photochemical DNA synthesis. *Organic Letters*, 3, 1105–1108.
19 Bhushan, K.R., DeLisi, C. and Laursen, R.A. (2003) Synthesis of photolabile 2-(2-nitrophenyl)propyloxycarbonyl protected amino acids. *Tetrahedron Letters*, 44, 8585–8588.
20 Buehler, S., Lagoja, I., Giegrich, H., Stengele, K.-P. and Pfleiderer, W. (2004) New types of very efficient photolabile protecting groups based upon the [2-(2-nitrophenyl)propoxy]carbonyl (NPPOC)

moiety. *Helvetica Chimica Acta*, **87**, 620–659.

21 Woell, D., Laimgruber, S., Galetskaya, M., Smirnova, J., Pfleiderer, W., Heinz, B., Gilch, P. and Steiner, U.E. (2007) On the mechanism of intramolecular sensitization of photocleavage of the 2-(2-nitrophenyl)propoxycarbonyl (NPPOC) protecting group. *Journal of the American Chemical Society*, **129**, 12148–12158.

22 Bhushan, K.R. (2006) Photolabile peptide nucleic acid monomers: synthesis and photodeprotection. *Synlett*, (13), 2130–2132.

23 Bhushan, K.R. (2006) Light-directed maskless synthesis of peptide arrays using photolabile amino acid monomers. *Organic & Biomolecular Chemistry*, **4**, 1857–1859.

24 Hasan, A., Stengele, K.-P., Giegrich, H., Cornwell, P., Isham, K.R., Sachleben, R. A., Pfleiderer, W. and Foote, R.S. (1997) Photolabile protecting groups for nucleosides: synthesis and photo-deprotection rates. *Tetrahedron*, **53**, 4247–4264.

25 Smirnova, J., Woell, D., Pfleiderer, W. and Steiner, U.E. (2005) Synthesis of caged nucleosides with photoremovable protecting groups linked to intramolecular antennae. *Helvetica Chimica Acta*, **88**, 891–904.

26 Woell, D., Smirnova, J., Pfleiderer, W. and Steiner, U.E. (2006) Highly efficient photolabile protecting groups with intramolecular energy transfer. *Angewandte Chemie, International Edition*, **45**, 2975–2978.

27 Zehavi, U., Amit, B. and Patchornik, A. (1972) Light-sensitive glycosides. I. 6-Nitroveratryl β-D-glucopyranoside and 2-nitrobenzyl β-D-glucopyranoside. *Journal of Organic Chemistry*, **37**, 2281–2285.

28 Gravel, D., Hébert, J. and Thoraval, D. (1983) *o*-Nitrophenylethylene glycol as photoremovable protective group for aldehydes and ketones: syntheses, scope, and limitations. *Canadian Journal of Chemistry*, **61**, 400–410.

29 Hébert, J. and Gravel, D. (1974) *o*-Nitrophenylethylene glycol: a photosensitive protecting group for aldehydes and ketones. *Canadian Journal of Chemistry*, **52**, 187–189.

30 Blanc, A. and Bochet, C.G. (2003) Bis(*o*-nitrophenyl)ethanediol: A practical photolabile protecting group for ketones and aldehydes. *Journal of Organic Chemistry*, **68**, 1138–1141.

31 Barltrop, J.A. and Schofield, P. (1962) Photosensitive protecting groups. *Tetrahedron Letters*, **3**, 697–699.

32 Barltrop, J.A. and Schofield, P. (1965) Organic photochemistry. II. Some photosensitive protecting groups. *Journal of the Chemical Society*, 4758–4765.

33 Chamberlin, J.W. (1966) Use of the 3,5-dimethoxybenzyloxycarbonyl group as a photosensitive N-protecting group. *Journal of Organic Chemistry*, **31**, 1658–1660.

34 Zimmerman, H.E. and Sandel, V.R. (1963) Mechanistic organic photochemistry. II. Solvolytic photochemical reactions. *Journal of the American Chemical Society*, **85**, 915–922.

35 Zimmerman, H.E. (1995) The meta effect in organic photochemistry: mechanistic and exploratory organic photochemistry. *Journal of the American Chemical Society*, **117**, 8988–8991.

36 Pincock, J.A. (1997) Photochemistry of arylmethyl esters in nucleophilic solvents: radical pair and ion pair intermediates. *Accounts of Chemical Research*, **30**, 43–49.

37 Hilborn, J.W., MacKnight, E., Pincock, J. A. and Wedge, P.J. (1994) Photochemistry of substituted benzyl acetates and benzyl pivalates: a reinvestigation of substituent effects. *Journal of the American Chemical Society*, **116**, 3337–3346.

38 Cameron, J.F. and Fréchet, J.M.J. (1990) Base catalysis in imaging materials. 1. Design and synthesis of novel light-sensitive urethanes as photoprecursors of amines. *Journal of Organic Chemistry*, **55**, 5919–5922.

39 Birr, C., Lochinger, W., Stahnke, G. and Lang, P. (1972) The a,a-dimethyl-3,5-dimethoxybenzyloxycarbonyl (Ddz) residue, an N-protecting group labile toward weak acids and irradiation. *Justus Liebigs Annalen der Chemie*, **763**, 162–172.

40 Misetic, A. and Boyd, M.K. (1998) The pixyl (Px) group: a novel photocleavable protecting group for primary alcohols. *Tetrahedron Letters*, **39**, 1653–1656.

41 Coleman, M.P. and Boyd, M.K. (2002) S-Pixyl analogues as photocleavable protecting groups for nucleosides. *Journal of Organic Chemistry*, **67**, 7641–7648.

42 Coleman, M.P. and Boyd, M.K. (1999) The S-pixyl group: an efficient photocleavable protecting group for the 5′ hydroxy function of deoxyribonucleosides. *Tetrahedron Letters*, **40**, 7911–7915.

43 Kemp, D.S. and Reczek, J. (1977) New protective groups for peptide synthesis. III. The Maq ester group. Mild reductive cleavage of 2-acyloxymethyleneanthraquinones. *Tetrahedron Letters*, **18**, 1031–1034.

44 Furuta, T., Torigai, H., Osawa, T. and Iwamura, M. (1993) New photochemically labile protecting group for phosphates. *Chemistry Letters*, 1179–1182.

45 Hagen, V., Bendig, J., Frings, S., Eckardt, T., Helm, S., Reuter, D. and Kaupp, U.B. (2001) Highly efficient and ultrafast phototriggers for cAMP and cGMP by using long-wavelength UV/vis-activation. *Angewandte Chemie, International Edition*, **40**, 1046–1048.

46 Furuta, T., Momotake, A., Sugimoto, M., Hatayama, M., Torigai, H. and Iwamura, M. (1996) Acyloxycoumarinylmethyl-caged cAMP, the photolabile and membrane-permeable derivative of cAMP that effectively stimulates pigment-dispersion response of melanophores. *Biochemical and Biophysical Research Communications*, **228**, 193–198.

47 Furuta, T., Torigai, H., Sugimoto, M. and Iwamura, M. (1995) Photochemical properties of new photolabile cAMP derivatives in a physiological saline solution. *Journal of Organic Chemistry*, **60**, 3953–3956.

48 Suzuki, A.Z., Watanabe, T., Kawamoto, M., Nishiyama, K., Yamashita, H., Ishii, M., Iwamura, M. and Furuta, T. (2003) Coumarin-4-ylmethoxycarbonyls as phototriggers for alcohols and phenols. *Organic Letters*, **5**, 4867–4870.

49 Geissler, D., Antonenko, Y.N., Schmidt, R., Keller, S., Krylova, O.O., Wiesner, B., Bendig, J., Pohl, P. and Hagen, V. (2005) (Coumarin-4-yl)methyl esters as highly efficient, ultrafast phototriggers for protons and their application to acidifying membrane surfaces. *Angewandte Chemie, International Edition*, **44**, 1195–1198.

50 Briand, B., Kotzur, N., Hagen, V. and Beyermann, M. (2008) A new photolabile carboxyl protecting group for native chemical ligation. *Tetrahedron Letters*, **49**, 85–87.

51 Schmidt, R., Geissler, D., Hagen, V. and Bendig, J. (2007) Mechanism of Photocleavage of (Coumarin-4-yl)methyl Esters. *Journal of Physical Chemistry A*, **111**, 5768–5774.

52 Hagen, V., Dekowski, B., Nache, V., Schmidt, R., Geissler, D., Lorenz, D., Eichhorst, J., Keller, S., Kaneko, H., Benndorf, K. and Wiesner, B. (2005) Coumarinylmethyl esters for ultra-fast release of high concentrations of cyclic nucleotides upon one- and two-photon photolysis. *Angewandte Chemie, International Edition*, **44**, 7887–7891.

53 Schmidt, R., Geissler, D., Hagen, V. and Bendig, J. (2005) Kinetics study of the photocleavage of (coumarin-4-yl)methyl esters. *Journal of Physical Chemistry A*, **109**, 5000–5004.

54 Furuta, T., Hirayama, Y. and Iwamura, M. (2001) Anthraquinone-2-ylmethoxycarbonyl (Aqmoc): a new photochemically removable protecting group for alcohols. *Organic Letters*, **3**, 1809–1812.

55 Rochat, S., Minardi, C., De Saint Laumer, J.-Y. and Herrmann, A. (2000) Controlled release of perfumery aldehydes and

ketones by Norrish type-II photofragmentation of a-keto esters in undegassed solution. *Helvetica Chimica Acta*, **83**, 1645–1671.

56 Levrand, B. and Herrmann, A. (2002) Light induced controlled release of fragrances by Norrish type II photofragmentation of alkyl phenyl ketones. *Photochemical & Photobiological Sciences*, **1**, 907–919.

57 Levrand, B. and Herrmann, A. (2007) Controlled light-induced release of volatile aldehydes and ketones by photofragmentation of 2-oxo-(2-phenyl) acetates. *Chimia*, **61**, 661–664.

58 Herrmann, A. (2007) Controlled release of volatiles under mild reaction conditions: from nature to everyday products. *Angewandte Chemie, International Edition*, **46**, 5836–5863.

59 Derrer, S., Flachsmann, F., Plessis, C. and Stang, M. (2007) Applied photochemistry light controlled perfume release. *Chimia*, **61**, 665–669.

60 Brinson, R.G. and Jones, P.B. (2004) Caged trans-4-hydroxy-2-nonenal. *Organic Letters*, **6**, 3767–3770.

61 Peukert, S. and Giese, B. (1998) The pivaloylglycol anchor group: A new platform for a photolabile linker in solid-phase synthesis. *Journal of Organic Chemistry*, **63**, 9045–9051.

62 Glatthar, R. and Giese, B. (2000) A new photocleavable linker in solid-phase chemistry for ether cleavage. *Organic Letters*, **2**, 2315–2317.

63 Kessler, M., Glatthar, R., Giese, B. and Bochet, C.G. (2003) Sequentially photo-cleavable protecting groups in solid-phase synthesis. *Organic Letters*, **5**, 1179–1181.

64 McHale, W.A. and Kutateladze, A.G. (1998) An efficient photo-SET-induced cleavage of dithiane-carbonyl adducts and its relevance to the development of photoremovable protecting groups for ketones and aldehydes. *Journal of Organic Chemistry*, **63**, 9924–9931.

65 Vath, P., Falvey, D.E., Barnhurst, L.A. and Kutateladze, A.G. (2001) Photoinduced C–C bond cleavage in dithiane-carbonyl adducts: a laser flash photolysis study. *Journal of Organic Chemistry*, **66**, 2887–2890.

66 Valiulin, R.A. and Kutateladze, A.G. (2008) 2,6,7-Trithiabicyclo[2.2.2]octanes as promising photolabile tags for combinatorial encoding. *Journal of Organic Chemistry*, **73**, 335–338.

67 Brook, M.A., Gottardo, C., Balduzzi, S. and Mohamed, M. (1997) The sisyl (tris (trimethylsilyl)silyl) group: a fluoride resistant, photolabile alcohol protecting group. *Tetrahedron Letters*, **38**, 6997–7000.

68 Brook, M.A., Balduzzi, S., Mohamed, M. and Gottardo, C. (1999) The photolytic and hydrolytic lability of sisyl (Si(SiMe3)3) ethers, an alcohol protecting group. *Tetrahedron*, **55**, 10027–10040.

69 Pirrung, M.C. and Lee, Y.R. (1993) Photochemically-removable silyl protecting groups. *Journal of Organic Chemistry*, **58**, 6961–6963.

70 Pirrung, M.C., Fallon, L., Zhu, J. and Lee, Y.R. (2001) Photochemically removable silyl protecting groups. *Journal of the American Chemical Society*, **123**, 3638–3643.

71 Pirrung, M.C. and Lee, Y.R. (1993) Tandem photochemical or thermal [1,5]-H, thermal [1,5]-Si migrations of vinyl silanes. *Tetrahedron Letters*, **34**, 8217–8220.

72 Koenigs, P.M., Faust, B.C. and Porter, N.A. (1993) Photochemistry of enzyme-bound cinnamoyl derivatives. *Journal of the American Chemical Society*, **115**, 9371–9379.

73 Turner, A.D., Pizzo, S.V., Rozakis, G.W. and Porter, N.A. (1987) Photochemical activation of acylated a-thrombin. *Journal of the American Chemical Society*, **109**, 1274–1275.

74 Turner, A.D., Pizzo, S.V., Rozakis, G. and Porter, N.A. (1988) Photoreactivation of irreversibly inhibited serine proteinases. *Journal of the American Chemical Society*, **110**, 244–250.

75 Porter, N.A., Thuring, J.W. and Li, H. (1999) Selective inhibition, separation,

and purification of serine proteases: a strategy based on a photoremovable inhibitor. *Journal of the American Chemical Society*, **121**, 7716–7717.

76 Anderson, J.C. and Reese, C.B. (1962) Photo-induced rearrangement involving aryl participation. *Tetrahedron Letters*, **3**, 1–4.

77 Sheehan, J.C. and Daves, G.D. JJrD (1964) Facile alkyl-oxygen ester cleavage. *Journal of Organic Chemistry*, **29**, 2006–2008.

78 Sheehan, J.C. and Umezawa, K. (1973) Phenacyl photosensitive blocking groups. *Journal of Organic Chemistry*, **38**, 3771–3774.

79 Bellof, D. and Mutter, M. (1985) A new phenacyl-type handle for polymer supported peptide synthesis. *Chimia*, **39**, 317–320.

80 Epstein, W.W. and Garrossian, M. (1987) p-Methoxyphenacyl esters as photodeblockable protecting groups for phosphates. *Journal of the Chemical Society, Chemical Communications*, 532–533.

81 Givens, R.S., Athey, P.S., Matuszewski, B., Kueper, L.W. JIII, Xue, J. and Fister, T. (1993) Photochemistry of phosphate esters: α-keto phosphates as a photoprotecting group for caged phosphate. *Journal of the American Chemical Society*, **115**, 6001–6012.

82 Givens, R.S. and Park, F C.-H.D (1996) Hydroxyphenacyl ATP: a new phototrigger. V. *Tetrahedron Letters*, **37**, 6259–6262.

83 Givens, R.S., Jung, A., Park, F C.-H., Weber, J. and Bartlett, W. (1997) New photoactivated protecting groups. 7. p-Hydroxyphenacyl: A phototrigger for excitatory amino acids and peptides. *Journal of the American Chemical Society*, **119**, 8369–8370.

84 Banerjee, A. and Falvey, D.E. (1998) Direct photolysis of phenacyl protecting groups studied by laser flash photolysis: an excited state hydrogen atom abstraction pathway leads to formation of carboxylic acids and acetophenone. *Journal of the American Chemical Society*, **120**, 2965–2966.

85 Givens, R.S., Weber, J.F.W., Conrad, P. G. JII, Orosz, G., Donahue, S.L. and Thayer, S.A. (2000) New phototriggers 9: p-Hydroxyphenacyl as a C-terminal photoremovable protecting group for oligopeptides. *Journal of the American Chemical Society*, **122**, 2687–2697.

86 Specht, A., Loudwig, S., Peng, L. and Goeldner, M. (2002) p-Hydroxyphenacyl bromide as photoremoveable thiol label: a potential phototrigger for thiol-containing biomolecules. *Tetrahedron Letters*, **43**, 8947–8950.

87 Conrad, P.G. JII, Givens, R.S., Hellrung, B., Rajesh, C.S., Ramseier, M. and Wirz, J. (2000) p-Hydroxyphenacyl phototriggers: the reactive excited state of phosphate photorelease. *Journal of the American Chemical Society*, **122**, 9346–9347.

88 Park, C.-H. and Givens, R.S. (1997) New photoactivated protecting groups. 6. p-Hydroxyphenacyl: A phototrigger for chemical and biochemical probes. *Journal of the American Chemical Society*, **119**, 2453–2463.

89 Zabadal, M., Pelliccioli, A.P., Klan, P. and Wirz, J. (2001) 2,5-Dimethyl-phenacyl esters: A photoremovable protecting group for carboxylic acids. *Journal of Physical Chemistry A*, **105**, 10329–10333.

90 Klan, P., Zabadal, M. and Heger, D. (2000) 2,5-Dimethylphenacyl as a new photoreleasable protecting group for carboxylic acids. *Organic Letters*, **2**, 1569–1571.

91 Klan, P., Pelliccioli, A.P., Pospisil, T. and Wirz, J. (2002) 2,5-Dimethylphenacyl esters: A photoremovable protecting group for phosphates and sulfonic acids. *Photochemical & Photobiological Sciences*, **1**, 920–923.

92 Literak, J., Wirz, J. and Klan, P. (2005) 2,5-Dimethylphenacyl carbonates: A photoremovable protecting group for alcohols and phenols. *Photochemical & Photobiological Sciences*, **4**, 43–46.

93 Kammari, L., Plistil, L., Wirz, J. and Klan, P. (2007) 2,5-Dimethylphenacyl carbamate: a photoremovable protecting group for amines and amino acids. *Photochemical & Photobiological Sciences*, **6**, 50–56.

94 Sheehan, J.C., Wilson, R.M. and Oxford, A.W. (1971) The photolysis of methoxy-substituted benzoin esters. A photosensitive protecting group for carboxylic acids. *Journal of the American Chemical Society*, **93**, 7222–7228.

95 Pirrung, M.C., Ye, T., Zhou, Z. and Simon, J.D. (2006) Mechanistic studies on the photochemical deprotection of 3′,5′-dimethoxybenzoin esters. *Photochemistry and Photobiology*, **82**, 1258–1264.

96 Pirrung, M.C. and Shuey, S.W. (1994) Photoremovable protecting groups for phosphorylation of chiral alcohols. Asymmetric synthesis of phosphotriesters of (−)-3′,5′-dimethoxybenzoin. *Journal of Organic Chemistry*, **59**, 3890–3897.

97 Shi, Y., Corrie, J.E.T. and Wan, P. (1997) Mechanism of 3′,5′-dimethoxybenzoin ester photochemistry: heterolytic cleavage intramolecularly assisted by the dimethoxybenzene ring is the primary photochemical step. *Journal of Organic Chemistry*, **62**, 8278–8279.

98 Rajesh, C.S., Givens, R.S. and Wirz, J. (2000) Kinetics and mechanism of phosphate photorelease from benzoin diethyl phosphate: evidence for adiabatic fission to an a-keto cation in the triplet state. *Journal of the American Chemical Society*, **122**, 611–618.

99 Corrie, J.E.T. and Trentham, D.R. (1992) Synthetic, mechanistic and photochemical studies of phosphate esters of substituted benzoins. *Journal of the Chemical Society, Perkin Transactions 1*, 2409–2417.

100 Ma, C., Du, Y., Kwok, W.M. and Phillips, D.L. (2007) Femtosecond transient absorption and nanosecond time-resolved resonance Raman study of the solvent-dependent photo-deprotection reaction of benzoin diethyl phosphate. *Chemistry – A European Journal*, **13**, 2290–2305.

101 Bochet, C.G. (2001) Orthogonal photolysis of protecting groups. *Angewandte Chemie, International Edition*, **40**, 2071–2073.

102 Blanc, A. and Bochet, C.G. (2002) Wavelength-controlled orthogonal photolysis of protecting groups. *Journal of Organic Chemistry*, **67**, 5567–5577.

103 Cameron, J.F., Willson, C.G. and Frechet, J.M.J. (1996) Photogeneration of amines from a-keto carbamates: photochemical studies. *Journal of the American Chemical Society*, **118**, 12925–12937.

104 Cameron, J.F., Willson, C.G. and Fréchet, J.M.J. (1995) New photolabile amino protecting groups: photogeneration of amines from [(3′,5′-dimethoxybenzoinyl)oxy]carbonyl carbamates. *Journal of the Chemical Society, Chemical Communications*, 923–924.

105 Cameron, J.F., Willson, C.G. and Frechet, J.M.J. (1997) Photogeneration of amines from a-keto carbamates: design and preparation of photoactive compounds. *Journal of the Chemical Society, Perkin Transactions 1*, 2429–2442.

106 Papageorgiou, G. and Corrie, J.E.T. (1997) Synthesis and properties of carbamoyl derivatives of photolabile benzoins. *Tetrahedron*, **53**, 3917–3932.

107 Pirrung, M.C. and Huang, F C.-Y.D (1995) Photochemical deprotection of 3′,5′-dimethoxybenzoin (DMB) carbamates derived from secondary amines. *Tetrahedron Letters*, **36**, 5883–5884.

108 Cano, M., Ladlow, M. and Balasubramanian, S. (2002) Practical synthesis of a dithiane-protected 3′,5′-dialkoxybenzoin photolabile safety-catch linker for solid-phase organic synthesis. *Journal of Organic Chemistry*, **67**, 129–135.

109 Lee, H.B. and Balasubramanian, S. (1999) Studies on a dithiane-protected benzoin photolabile safety-catch linker for solid-phase organic synthesis. *Journal of Organic Chemistry*, **64**, 3454–3460.

110 Routledge, A., Abell, C. and Balasubramanian, S. (1997) The use of a dithiane protected benzoin photolabile safety catch linker for solid-phase synthesis. *Tetrahedron Letters*, **38**, 1227–1230.

111 Amit, B. and Patchornik, A. (1973) Photorearrangement of N-substituted o-nitroanilides and nitroveratramides. Potential photosensitive protecting group. *Tetrahedron Letters*, **14**, 2205–2208.

112 Amit, B., Ben-Efraim, D.A. and Patchornik, A. (1976) Light-sensitive amides. The photosolvolysis of substituted 1-acyl-7-nitroindolines. *Journal of the American Chemical Society*, **98**, 843–844.

113 Goissis, G., Erickson, B.W. and Merrifield, R.B. (1977) Synthesis of protected peptide acids and esters by photosolvolysis of 1-peptidyl-5-bromo-7-nitroindolines. Peptides, Proceedings of the 5th American Peptide Symposium, Halsted Press, New York, pp. 559–561.

114 Morrison, J., Wan, P., Corrie, J.E.T. and Papageorgiou, G. (2002) Mechanisms of photorelease of carboxylic acids from 1-acyl-7-nitroindolines in solutions of varying water content. *Photochemical & Photobiological Sciences*, **1**, 960–969.

115 Papageorgiou, G., Ogden, D., Kelly, G. and Corrie, J.E.T. (2005) Synthetic and photochemical studies of substituted 1-acyl-7-nitroindolines. *Photochemical & Photobiological Sciences*, **4**, 887–896.

116 Cohen, A.D., Helgen, C., Bochet, C.G. and Toscano, J.P. (2005) The mechanism of photoinduced acylation of amines by N-acyl-5,7-dinitroindoline as determined by time-resolved infrared spectroscopy. *Organic Letters*, **7**, 2845–2848.

117 Pass, S., Amit, B. and Patchornik, A. (1981) Racemization – free photochemical coupling of peptide segments. *Journal of the American Chemical Society*, **103**, 7674–7675.

118 Nicolaou, K.C., Safina, B.S. and Winssinger, N. (2001) A new photolabile linker for the photoactivation of carboxyl groups. *Synlett*, (Special Issue), 900–903.

119 Helgen, C. and Bochet, C.G. (2003) Photochemical protection of amines with Cbz and fmoc groups. *Journal of Organic Chemistry*, **68**, 2483–2486.

120 Helgen, C. and Bochet, C.G. (2001) Preparation of secondary and tertiary amides under neutral conditions by photochemical acylation of amines. *Synlett*, (12), 1968–1970.

121 Debieux, J.-L., Cosandey, A., Helgen, C. and Bochet, C.G. (2007) Photoacylation of alcohols in neutral medium. *European Journal of Organic Chemistry*, (13), 2073–2077.

122 Bochet, C.G. (2004) Chromatic orthogonality in organic synthesis. *Synlett*, (13), 2268–2274.

123 Blanc, A. and Bochet, C.G. (2007) Isotope effects in photochemistry: Application to chromatic orthogonality. *Organic Letters*, **9**, 2649–2651.

124 Loudwig, S., Specht, A. and Goeldner, M. (2002) Caged compounds, the photolabile precursors of bioactive molecules. Developments and prospects. *Actualité Chimique*, (1), 7–12.

125 Zhao, Y., Zheng, Q., Dakin, K., Xu, K., Martinez, M.L. and Li, W.-H.D (2004) New caged coumarin fluorophores with extraordinary uncaging cross sections suitable for biological imaging applications. *Journal of the American Chemical Society*, **126**, 4653–4663.

126 Matsuzaki, M., Ellis-Davies, G.C.R., Nemoto, T., Miyashita, Y., Iino, M. and Kasai, H. (2001) Dendritic spine geometry is critical for AMPA receptor expression in hippocampal CA1 pyramidal neurons. *Nature Neuroscience*, **4**, 1086–1092.

127 Furuta, T., Wang, S.S.-H., Dantzker, J.L., Dore, T.M., Bybee, W.J., Callaway, E.M., Denk, W. and Tsien, R.Y. (1999) Brominated 7-hydroxycoumarin-4-ylmethyls: photolabile protecting groups with biologically useful cross-sections for two photon photolysis.

128 Pirrung, M.C., Pieper, W.H., Kaliappan, K.P. and Dhananjeyan, M.R. (2003) Combinatorial discovery of two-photon photoremovable protecting groups. *Proceedings of the National Academy of Sciences of the United States, of America*, **100**, 12548–12553.

129 Majjigapu, J.R.R., Kurchan, A.N., Kottani, R., Gustafson, T.P. and Kutateladze, A.G. (2005) Release and report: a new photolabile caging system with a two-photon fluorescence reporting function. *Journal of the American Chemical Society*, **127**, 12458–12459.

130 Kantevari, S., Hoang, C.J., Ogrodnik, J., Egger, M., Niggli, E. and Ellis-Davies, G.C. R. (2006) Synthesis and two-photon photolysis of 6-(ortho-nitroveratryl)-caged IP3 in living cells. *ChemBioChem*, **7**, 174–180.

131 Aujard, I., Benbrahim, C., Gouget, M., Ruel, O., Baudin, F J.-B., Neveu, P. and Jullien, L. (2006) o-Nitrobenzyl photolabile protecting groups with red-shifted absorption: syntheses and uncaging cross-sections for one- and two-photon excitation. *Chemistry – A European Journal*, **12**, 6865–6879.

132 Zhu, Y., Pavlos, C.M., Toscano, J.P. and Dore, T.M. (2006) 8-Bromo-7-hydroxyquinoline as a photoremovable protecting group for physiological use: mechanism and scope. *Journal of the American Chemical Society*, **128**, 4267–4276.

133 Madhavan, N. and Gin, M.S. (2004) Synthesis and photocleavage of a new dimeric bis(o-nitrobenzyl) diether tether. *Chemical Communications*, 2728–2729.

134 Smith, A.B. JIII, Savinov, S.N., Manjappara, U.V. and Chaiken, I.M. (2002) Peptide-small molecule hybrids via orthogonal deprotection-chemoselective conjugation to cysteine-anchored scaffolds. A model study. *Organic Letters*, **4**, 4041–4044.

135 Gareau, Y., Zamboni, R. and Wong, A.W. (1993) Total synthesis of N-Methyl Ltc4 – a novel methodology for the monomethylation of amines. *Journal of Organic Chemistry*, **58**, 1582–1585.

136 Ramesh, D., Wieboldt, R., Billington, A. P., Carpenter, B.K. and Hess, G.P. (1993) Photolabile precursors of biological amides – synthesis and characterization of caged O-nitrobenzyl derivatives of glutamine, asparagine, glycinamide, and gamma-aminobutyramide. *Journal of Organic Chemistry*, **58**, 4599–4605.

137 Hayashi, K., Hashimoto, K., Kusaka, N., Yamazoe, A., Fukaki, H., Tasaka, M. and Nozaki, H. (2006) Caged gene-inducer spatially and temporally controls gene expression and plant development in transgenic Arabidopsis plant. *Bioorganic & Medicinal Chemistry Letters*, **16**, 2470–2474.

138 Rubinstein, M., Amit, B. and Patchornik, A. (1975) Use of a light-sensitive phosphate protecting group for mononucleotide syntheses. *Tetrahedron Letters*, **16**, 1445–1448.

139 Wang, P., Hu, H. and Wang, Y. (2007) Application of the excited state meta effect in photolabile protecting group design. *Organic Letters*, **9**, 2831–2833.

140 Du, H. and Boyd, M.K. (2001) The 9-xanthenylmethyl group: a novel photo-cleavable protecting group for amines. *Tetrahedron Letters*, **42**, 6645–6647.

141 Wang, B. and Zheng, A. (1997) A photo-sensitive protecting group for amines based on coumarin chemistry. *Chemical & Pharmaceutical Bulletin*, **45**, 715–718.

142 Banerjee, A., Lee, K. and Falvey, D.E. (1999) Photoreleasable protecting groups based on electron transfer chemistry. Donor sensitized release of phenacyl groups from alcohols, phosphates and diacids. *Tetrahedron*, **55**, 12699–12710.

143 Piloto, A.M., Rovira, D., Costa, S.P.G. and Goncalves, M.S.T. (2006) Oxobenzo[f] benzopyrans as new fluorescent photolabile protecting groups for the carboxylic function. *Tetrahedron*, **62**, 11955–11962.

144 Schade, B., Hagen, V., Schmidt, R., Herbrich, R., Krause, E., Eckardt, T. and Bendig, J. (1999) Deactivation behavior and excited-state properties of (coumarin-4-yl)methyl derivatives. 1. Photocleavage of (7-methoxycoumarin-4-yl)methyl-caged acids with fluorescence enhancement. *Journal of Organic Chemistry*, **64**, 9109–9117.

145 Lin, W. and Lawrence, D.S. (2002) A strategy for the construction of caged diols using a photolabile protecting group. *Journal of Organic Chemistry*, **67**, 2723–2726.

146 Lu, M., Fedoryak, O.D., Moister, B.R. and Dore, T.M. (2003) Bhc-diol as a photolabile protecting group for aldehydes and ketones. *Organic Letters*, **5**, 2119–2122.

Index

a

acetamide function photochemistry   154
acetone, α-cleavage   54
acetophenone derivatives   146, 149, 189
N-acetyl-2,5-dimethylpyrrole   388
acetylfurans, oxetane formation   224
acid-catalysis in photoreactions   148
acrylanilides reactions   300
acyclic alkenes   174
acyclic dienes   364
acyclic ketones   247
– δ-hydrogen abstraction   247
acyclic N-allyliminium salt   269
acyclic polyenes   267
acyclic vinyl ethers, addition to carbonyls   230
acyl oxyl radical   158
2-acylpyrroles   390
acylsilanes, oxetane formation   232
adamantylideneadamantane-1,2-dioxetane   355
adamantyl-substituted alkenes   355, 356
alanine   252, 254
– photocyclizations   254
aldehydes, oxetane formation   224
alicyclic divinyl-methanes rearrangement   102
alkenes   68, 111, 191, 219
– alkoxy-substituted   356, 357
– alkylated   327
– asymmetric epoxidation   124
– benzene addition to   147
– carbenes addition of   111
– cycloaddition reaction   219
– electron acceptor substituted   69–76
– electron donor substituted   198
– electrophilic, radical addition to   69–76
– hydration   80
– hydroalkoxylation   80

– intramolecular [2+1] reaction with carbenes   111
– intramolecular [2+2] reaction   146, 219
– metal-catalyzed cyclopropanation-supported photochemistry   112
– nucleophilicities   222
– oxetane formation   221
– photocycloadditions   144, 171
– triplet sensitizers for   111
– X–Y addition   83–86
alkenols   196
alkenyldiazo ketones   113
– intramolecular cyclopropanation   113
4-alkenyloxy-2-pyrones   203
α-(alkenyloxy)silyl-substituted diazoacetate   111
– direct photolysis   111
O-alkenyl salicylic acid derivatives, transformation   149
N-alk-4-enyl-substituted maleimides   309
β-alkoxycarbonyl compound   248
3-alkoxycarbonyl-substituted cyclohexenones   178
– trans-products formation   178
β-alkoxy ketones   247, 250
2-alkoxyoxetanes   222, 223
o-alkoxy phenyl ketones, photocyclization   250
4-alkoxypyridones   192
4-alkoxyquinolones   204
allenes   190, 204
– intramolecular [2+2] photocycloaddition   190
(−)-allocoronamic acid synthesis   31
(±)-allocyathin B$_3$ synthesis   182
N-allyl-N-(2-chlorophenyl)-acetamide   343
allyl ethyl ether   196
allylic alcohols   196, 231

Handbook of Synthetic Photochemistry. Edited by Angelo Albini and Maurizio Fagnoni
Copyright © 2010 WILEY-VCH Verlag GmbH & Co. KGaA, Weinheim
ISBN: 978-3-527-32391-3

– bicyclic alcohol   273
– oxetane formation   231
allylic hydroperoxides synthesis   368–370
– from simple alkenes   368–370
– in zeolites   370
allylic silanes   223, 327
allyliminium salts, photochemistry   269
4-allyloxycoumarin   203
alternative energy sources, importance   10
amines addition   72–74
aminoacids (bis) synthesis   48
o-tert-amylbenzophenone   244
angular tricyclic cyclobutene derivative   148
anilides photocyclization   299, 300
anthracene endoperoxide   362
aplysiatoxin, synthesis   419
aquariolide A, synthesis   103, 104
arc
– antimony-xenon   9
– mercury   3–8
– mercury-xenon   9
– sodium   9
– xenon   9
archimedene C120, partial structure   157
arenes
– intermolecular [3+2] cycloaddition with alkenes   119
– intramolecular [3+2] cycloaddition with alkenes   119–122
– oxygenation   359–363
aromatic compounds photocycloaddition   144–150, 311
aromatic ketones   291, 324
– as electron acceptors   291, 292
– in $S_{RN}1$ reactions   324
aromatic nucleophilic substitution, mechanisms   319
ArS_N1 reaction   319–345
$ArS_{RN}1$ reaction   319–345
aryl amide anions   329–331, 342
– intramolecular ortho-arylation   342
aryl amine photoreaction   262
arylation process   319–345
aryl-carbon bond formation   319–345
aryl cations   319–345
arylglyoxylates   228, 229
– oxetane formation   228
aryl halides photoreaction   319–345
aryl phosphates photoreaction   336, 337
α-arylpropionic acids synthesis   325
aryl radical cations   277
aryl sulfonates photoreaction   336, 337, 345
asteltoxin synthesis   228
asymmetric hydroperoxidation   370

asymmetric ring-expanding allylation   187
asymmetric synthesis, of oxetanes   231
aubergenone synthesis   39
avenaciolide synthesis   228
4-aza-2-cycloalkenones photoreaction   183
aza-di-π-methane rearrangement   99, 105, 109, 110
– of β,γ-unsaturated imines   109, 110
azavinylcyclopropane photochemical rearrangement   261
azepines synthesis   406, 407
azepinoindole derivatives synthesis   293
2-azidopyridines photoinduced ring-enlargement   407
aziridine   124, 254, 258
– conversion to L-daunosamine glycoside   124
– photolysis   258
– ring opening to azomethine ylides   258, 278
– synthesis   123–125
azo compounds preparation   33
– aliphatic, nitrogen elimination   33, 34
– Diels–Alder reaction   33, 34
– hydrogenation   33, 34
azoles reaction   78
azomethine ylides   258, 279

b
Barton esters, photolysis   85
Beer's law, negative implications   12
belactosin A precursors synthesis   48, 49
benzaldehyde Paternò Büchi reaction   226, 227
benzannulated perhydroazulenes synthesis   256
benzene   144, 145
– derivatives oxygenation   359–361
– intermolecular [2+2] photocycloaddition   145
benzocyclobutene derivative synthesis   147
benzofuranols synthesis   250
benzofurans synthesis   279
benzoin derivatives photolysis   428–429
benzoindolizidines synthesis   270
benzophenone
– as sensitizer   189, 196
– in Paternò-Büchi reaction   221, 233
benzoxazolin-2-ones synthesis   399
N-benzoylenamines photoreaction   299
1-benzoyl-1-(o-ethylphenyl) cyclopropane photocyclization   246
N-(β-benzoylethyl)-N-tosyl-glycinamides photolysis   250
5-benzoyl-4-methylbicyclo[2.2.2]oct-5-en-2-one, preparation   108, 109

– by Diels–Alder reaction 108
benzvalene-like intermediates 393
benzyl alcohol derivatives as protecting groups 421, 422
N-benzyl β-aziridinylacrylonitrile photoreaction 258
8-benzyl-1-benzoylnaphthalene photoreaction 247
N-benzyloxycarbonyl group, photolytic removal 421
α-(o-benzylphenyl)methyl phenyl ketone photoreaction 246
benzyl(trimethyl)silane radical precursor 71
benzyne 28
bicyclic cyclobutanes synthesis 311
bicyclic 1,3-cyclohexandione derivatives synthesis 147
bicyclic β-lactame synthesis 158
bicyclic oxetanes 224, 228, 229
– exo-selective formation 229
– stereoselective formation 229
bicyclic oxocarbenium ions 277
bicyclic sulfone synthesis 159
bicycloheptane derivatives synthesis 187
bicyclo[3.2.0]heptan-3-one 36
– photodecarbonylation 36
bicyclo[2.2.1]hept-5-en-2-one 107
– direct photolysis 107
– oxadi-π-methane rearrangement 107, 108
bicyclo[n.1.0]alkyl synthesis 117, 118
bicyclo[2.2.2]oct-5,7-dien-2-ones 106
– oxadi-π-methane/di-π-methane rearrangements 106, 107
bicyclo[2.2.2]oct-5-en-2-ones oxadi-π-methane rearrangement 110
[1,10] binaphthalenyl-2,20-diol (BINOL) derivatives synthesis 330
biphenyl as cosensitizer 267, 268
biradicals 242–246, 248, 250, 252, 254, 260, 288, 289, 291, 307, 308
– 1,4-acyl-alkyl 36, 37
– 4,4-(1,n-alkanediyl)bisbenzyl 308
– by photoelimination 25, 36–38, 40, 60
– 1,5 cyclization 242, 246
– cycloelimination reaction 36
– decarbonylation 36
– 1,n intermediates 291
– intermediates in oxetane formation 220–223, 226, 227, 230, 233
– stability rule 221
– transition state model 246
bisallyloxynaphthalene photoreaction 79
bis(amino acids) preparation 47, 48

2,3-bis(p-methoxy phenyl) oxiranes photoreaction 255
bisperoxides synthesis 371–373
– from cyclic alkenes 372, 373
– from phenyl-substituted alkenes 371, 372
bistannylated aromatic compounds synthesis 332
biyouyanagin A synthesis 184
black light lamp, see Wood's lamp
bridged dihalosulfide formation 40
– with 1,5-cyclooctadiene 40
bromoaryl alkyl-linked oxazolines photoreaction 339
butenolides photoreaction 189, 190, 365
N-tert-butoxycarbonyl(Boc)-protected lactams photoreaction 192
tert-butyl ketone photoreaction 36
γ-butyrolactones photodecarboxylation 39, 40
buxapentalactone synthesis 122

c

cage effect 43, 44, 60
– nonstereospecific 44
– stereospecific 44
calicheamycin, synthesis 419
carbanions in $S_{RN}1$ reactions 323–326
carbazoles, from diphenylamines 262
carbenes 111
– in the synthesis of three membered rings 111–114
– intermediates 32, 36
carbocycles synthesis 270, 338–341
– via $ArS_{RN}1$ reaction 338–341
carbohydrate furans 123
– dye-sensitized photooxygenation 123
carbon–carbon bond formation 25, 27, 28, 30, 33, 42, 44, 172, 187, 266, 323–331
– by photoelimination of small molecules 25
– fragmentation 258
– generality 28
– in crystals 25
– unstable molecules synthesis 27–29
– via carbanions 323–326
– via radical addition onto C–C multiple bonds 67
– with cyanide ions 331
– with nucleophiles 326–329
carbon–carbon double bonds, addition to 69–86
– H–C addition 69–76
– H–N addition 76–78
– H–O addition 80–82
– H–P addition 78–80
– H–S addition 82, 83

- X–Y reagents to alkenes   83–86
- halogenation   84
- with C–C bonds formation   84–86
carbon–carbon triple bonds, addition to   86–88
- hydroalkylation reactions   87
- X–Y reagents addition   87
carbon-heteroatom bond formation   332–334
- via sulfur nucleophiles   333, 334
- via tin nucleophiles   332, 333
carbon nanotubes
- oxygenation   362, 363
- single-walled (SWCNTs)   363
carbonyl compounds
- electron-accepting   219
- electron transfer reaction   223
- reaction with alkenes   217–233
- reactions via the singlet state   220, 221
- reactions via the triplet state   243, 247, 250, 252, 254
carbonyl ylides   254–257, 279
carcerands in photochemistry   28
(±)-carnosadine synthesis   31
α-cedrene synthesis   141
C-furanosides synthesis   274
charge-stabilizing solvent   328
charge transfer (CT) complex   321
chemically initiated electron exchage luminescence (CIEEL)   358
chemoselectivity, in oxetane formation   224
chiral auxiliary approach   108
chiral N-(2-benzoylethyl)-N-tosylglycinamide reaction   251
chiral cyclobutane synthesis   188
chiral hexadiine derivatives photoreaction   150
chiral induction, in the synthesis of oxetanes   232
chiral N-tosyl glycine esters reaction   252
chloroacetamides, electron acceptors   292, 293
4-chloroaniline photoreaction   329
cholesterol-ketoprofen dyads, irradiation   307
chromatic orthogonality   424, 428, 430–431
chromium cyclopropanone intermediate formation   156
chromophoric dissolved organic matter, photosensitizer   355
cinnamylanilides photocyclization   311
cinnamyl esters   425, 426
- Z/E photoisomerization   425, 426
circularly polarized light   175
cisoid conformations   364
- (E,Z)/(Z,Z)-2,4-hexadienes   364

α-cleavage reaction   25, 26, 36, 40, 50, 52, 54, 59, 106
colchicine synthesis   152
(−)-complicatic acid synthesis   110
concerted elimination   30
conjugated dienes oxygenations   364–367
- acyclic dienes   364
- cyclohexadienes   364, 365
- cyclohexatriene   365–367
- cyclopentadienes   364, 365
- heterocycles   365–367
copper-catalyzed [2+2] photocycloadditions   139
cosensitizer, 1,3,5-triphenylbenzene   77
coumarins   193, 194
- [2+2]-photocycloadditions   194
critical micelle concentration   59
crown ethers, synthesis   288
crystalline solids   25, 26
crystalline suspension, photoreactions   25, 59
crystal-phase photochemical reactions   15
CuOTf-catalyzed [2+2] photocycloaddition   141–144
- inter/intramolecular   144
- norbornene derivates   143
(R)/(S)-α-cuparenone   57
- enantioselective synthesis   57, 58
1-cyanonaphthalene
- derivative photoreaction   146
- sesnsitizer   276
1-cyano-2-(4-pentenyl)naphthalene derivatives photoreaction   121
9-cyanophenanthrene as photocatalyst   75
α-cyanostilbene oxides photoreactions   255
2-cyano-substituted pyrroles synthesis   389
3-cyano-1-thiocoumarin photoreaction   198
cyclic alkene reaction   174, 175, 226
cyclic 1,3-diketones photoreaction   186
cyclic ketones   56
- ring-contraction product   56
- cis/trans-α,α'-substituents   56
cyclic oligopeptides synthesis   288
cyclic sulfones   41
- photo-elimination products   41, 42
cyclic tertiary amines addition   73
cyclic triene chromium complexes cycloaddition   306
cyclization, in five-membereed ring formation   242, 245, 250–252, 264, 265, 267, 270, 272, 276
- electron transfer   267
- radical   268, 274
- radical anion   272
- selectivity   246

- stilbene-like 295–298
- vinyl-biphenyls 299
cycloaddition reactions
- [2+1] 111
- [2+2] 35, 137–160, 171–173, 264–311
  - carbonyls with alkenes 217,219,223
  - copper catalyzed 139–144
  - in the synthesis of four membered rings 137–160
  - intramolecular 138
  - of aromatic compounds 144–150
  - of 4-aza-2-cyclohexenones 185
  - of coumarins 193, 194
  - of cyclohexenones 177–181
  - of cyclopentenones 173–177
  - of 1,3-dioxin-4-ones 198–200
  - of dioxopyrrolines 184
  - of enones 173–181
  - of Fischer-type carbene complexes 156
  - of 4-heterocycloalkenones 183–185
  - of 2',3'-O-isopropylideneuridine 201
  - of maleic anhydride and derivatives 196, 197
  - of 4-pyrimidinones 200, 201
  - of quinolones 194, 195
  - of p-quinones 181, 182
  - of 1-thiocoumarin 198
  - α,β-unsaturated carbonyls 172
  - of α,β-unsaturated carboxylic acid derivatives 189–201
  - of α,β-unsaturated lactams 192, 203–205
  - of α,β-unsaturated lactones 189–192, 202, 203
  - of vinylogous amides and esters 182
- [2+2+2] 156
- [3+2] 119, 254–258, 279, 311
  - for the synthesis of natural products 122
  - of arenes with alkenes 119–122, 144
- [4+2] 144, 300
  - photoenolization/Diels Alder reaction 301
  - rules 255, 258
  - singlet oxygen as dienophile 364
  - transition metal template controlled 304–306, 309
- [4+4] 138, 302–305
- [5+2] 310
- [6+2] 305
- [6+4] 305
- 1,3-dipolar 265, 278
- five-membered rings formation 254
- four-membered rings formation 101
- large rings formation 302–305
- *meta* 119, 144
- *para* 144
- with singlet oxygen 353, 357, 358, 364, 373, 374, 376
cycloalkenones photoreactions 182
cycloalkyl peroxides, UV irradiation 46
cyclobutane synthesis 33, 137–160, 171–205
- by elimination 33
- from lactones 180
cyclobutanol derivatives 36, 154, 155, 159
- photodecarbonylation 36
- synthesis of, via Yang cyclization 153–155
cyclobutene synthesis 137, 189
γ-cyclodextrin effect on cycloadditions 120, 181
cycloheptatriene derivative 367
- TPP-sensitized photooxidation 367
cyclohexadienes 364, 365, 372
- TPP-sensitized photooxidation 372
1,5-cyclohexadien-3-one, rearrangement 104
cyclohexatriene oxygenation 365–367
cyclohexenones, [2+2] cycloaddition of 177–180
- 4-hetero-2-cyclohexenones, [2+2] cycloaddition of 185–186
N-cyclohexylmaleimide cycloaddition 197
cyclopentadecanone synthesis 47
cyclopentadienes oxygenation 364, 365
cyclopentanol ring system, formation 242–247, 273
- benzofuranols 250
- indanols, synthesis 243–247
- pyrrolidine derivatives, synthesis 250–254
- tetrahydrofuranols, synthesis 247–254
cyclopentanones 52
- α-cleavage 52
- enamine of 263
cyclopentene reaction 178, 195
- endo-products 174
- exo-products, formation 174
- from vinylcyclopropanes 261
cyclopentenone, [2+2] cycloaddition 173–177
- 4-hetero-2-cyclopentenones cycloaddition 183–185
cyclophanes synthesis 28, 29, 40
- double cyclization reactions 28
- precursors 28
- synthesis 29
cyclopropane formation 31, 37, 114, 115, 254
- by elimination 31
- by intramolecular hydrogen abstraction 114–118
- by intramolecular PET 115–117

- β-morpholinopropiophenones irradiation   115
- transition metal-mediated carbene transfer   95
cyclopropylamines, photoreactions   271
cyclopropylaminoacids synthesis   31, 32, 47, 48
cyclopropyl ketones, electron transfer photoreactions   275
cyclopropyl silyl ethers, photoreactions   268
cylindrical cuvettes, *see* spectrophotometric cuvettes
cynnamyl esters, Z/E isomerization   425, 426
α,β-cyperone synthesis   39

d

decarbonylation, of ketones   25, 28, 30, 35, 36, 50, 54, 309
decarboxylation   28–30, 39, 46, 47, 288, 289, 303, 309
de Mayo reaction   85, 182, 187
demecolcine, synthesis of   152
denitrogenation from azo compounds   28, 31–33
9,10-deoxytridachione, photorearrangement   103
diacyl peroxides   44, 46, 47
- decarboxylation products   46, 47, 48
- photolysis reaction   45
trans-dialkenyl cyclohexanone, irradiation   55
1,2-*trans*-dialkylated cyclopentanes synthesis   272
N,N-diallyl-(2-chlorophenyl)-amine cyclization   343
diarylethenes, photochromic compounds   298
diazepines synthesis   407–409
diazoalkane, in situ generation   31, 32
α-diazo carbonyl compounds   111, 112
- intramolecular [2+1] cycloaddition   112
diazoketone photoreaction   392
dibenzofurans synthesis   262
dibenzothiophenes synthesis   262
N,N-dibenzyl-2-benzoylacetamide photolysis   251
1,5-dichloro-9,10-anthraquinone   81
- photoacid generator   81
1,4-dichloro-2-butene cycloaddition   196
9,10-dicyanoanthracene as sensitizer   256
1,4-dicyanobenzene as sensitizer   77, 267, 301
1,1-dicyano-3,3-dimethyl-1,4,5-hexatriene direct photolysis   101
1,4-dicyanonaphthalene as sensitizer   139, 270

3,3-dicyanostilbene oxide cycloaddition   255
1,4-dicyano-2,3,5,6-tetramethyl benzene as sensitizer   267
Diels–Alder reaction, *see* [4+2] cycloaddition
6,6-diethoxyfulvene   372
- sensitized photooxygenation   372
difluoromethyl radicals   74
dihalobenzenes in $S_{RN}1$ reaction   324
dihydrobenzofuranes synthesis   344
9,10-dihydrophenanthrenes synthesis   299
dihydropyridinone photoreactions   185, 233
- 2,3-dihydropyridin-4(1H)-ones   185
- 5,6-dihydro-1H-pyridin-2-ones   204
1,5-dihydropyrrol-2-ones photoreaction   204
dihydrosantonin   39
- direct photodecarboxylation   39
1,8-diiodonaphthalene   338, 346
- photostimulated reaction   346
1,5-diketone derivatives synthesis   183
1,3-diketones photoaddition   85
1,4-dimethoxynapthalene as sensitizer   272
dimethyl acetylendicarboxylate cycloaddition   256
4-(N,N-dimethylamino)benzonitrile synthesis   331
N,N-dimethylaminophenyl cation   329
1,16-dimethyl-dodecahedrane synthesis   243
2,5-dimethyl-furan   365
- TPP-sensitized photooxygenation   365
2,5-dimethyl-2,4-hexadiene arylation   328
1,4-dimethylnaphthalene-1,4-endoperoxide synthesis   361
dimethyloxosulfonium methylide   217
γ-dimethylvalerophenones photoreaction   243
α,ω-diolefins photoreaction   203
dioxapaddlanes *see* endoperoxides
dioxetanes   354–358
- adamantyl-substituted alkenes   355, 356
- alkoxy-substituted alkenes   356, 357
- phenyl/methyl-substituted alkenes   357, 358
1,3-dioxin-4-ones   198–200
- de Mayo-type photochemistry   199
1,3-dioxol derivatives Paternò-Büchi reaction   229
diphenyl amine, photoreaction   262
2,2-diphenyl-3,3-dimethyloxetane synthesis   221
cis/trans-diphenylindanones photoreaction   50, 51
di-π-methane rearrangement   96–104
- acetone-sensitized   97
- of acyclic systems   100–103

- of barrelene 96, 99, *see also* Zimmerman rearrangement
- of benzobarrelene, acetophenone-sensitized irradiation 96, 97
- of dibenzobarrelene 96, 97
- direct irradiation 97, 98
- in natural compounds 103
- in the solid state 98

dipolarophiles 254–256, 258, 278, 279
disproportionation of biradicals 242, 244
N,N-disubstituted amides, carbanions in $S_{RN}1$ reaction 325
4,5-disubstituted N-alkylimidazoles synthesis 396
3,4-disubstituted-1,2,4-thiadiazolin-5-ones synthesis 405
2,5-disubstituted-thiophenes 393
- photochemical synthesis 393

2,4-di-tert-butylacetophenone photoreaction 243
1,3-divinyl-2-cyclopentanol tetronates photoreaction 202

e

(−)-elacanacin, synthesis 181
electrocyclic reactions 150, 261–263, 295
- cyclobutadiene 287, 295, 299
- cyclobutene ring opening 310
- in crystals 150
- with tropolone derivatives 150

electron paramagnetic resonance (EPR) 28
electron-rich olefins photoreactions 82, 194, 356
electron transfer ring opening
- cyclopropanes 260
- epoxides 261, 267–270, 272, 273, 276, 277, 279

electron-transfer photooxygenation 364
electron-transfer sensitizer 293–295, 357
- dicyanoanthracene (DCA) 293, 311
- dicyanonaphthalene (DCN) 293, 311
- 9-mesityl-10-methylacridinium ion 357

enamides photocyclization 299, 300
endoperoxides, synthesis of 358–368
- by electron-transfer photooxidation 364
- from arenes 359–363
- from conjugated dienes 364–367

ene reactions, (with singlet oxygen) 353, 357–359, 368–370, 372, 374
enol silyl ethers as electron donors 294
enolate anion 265, 323
- in $ArS_{RN}1$ reaction 323–325, 329, 341–344, 346

enones
- photoalkylation 71, 75, 80
- photocycloaddition 173–182

epimaalienone synthesis 39
1,4-epiperoxides, *see* endoperoxides
epoxides 123, 217
- ring-expansion reaction 217
- synthesis 123

α,β-epoxyketone photoreaction 256
(±)-erysotrine, synthesis of the precursor 185
erythrolides B synthesis 104
esters, decarboxylation 28
3-ethoxycarbonylcoumarin photoreaction 193
β-ethoxypropiophenone photoreaction 243, 247
ethyl α-(o-benzoylphenoxy)carboxylates 250
- photocyclization 250

α-(o-ethylphenyl) acetophenone photoreaction 245
ethyl propiolate, radical addition to 88
ethyl vinyl ether
- addition 86
- cycloaddition 257

eupoulauramine, synthesis 343
external irradiation set-up 6

f

facial diastereoselectivity 190
falling film photoreactors 13
farnesol, photoreaction 267
Fischer-type carbene complexes 156
- photochemical [2+2] photocycloaddition 156

Fischer-type chromium carbene complexes 158
- photochemical transformations 158

five-membered ring, formation 241, 387
- azomethine ylides generation 258–260
- [3+2]-cycloadditions formation 254
- from vinyl cyclopropanes 260–261
- internal cyclization-isomerization route (ICI) 387
- oxiranes, photofragmentation 254–257
- ring contraction-ring expansion route (RCRE) 387, 398
- photochemical rearrangement 260–261
- with one heteroatom synthesis 388–393
- with three heteroatoms synthesis 402–405
- with two heteroatoms synthesis 393–402

fluorescent lamps 6
fluorobiprophen, synthesis 323
four-membered ring systems 157, 159
- synthesis 137–160, 171–205

fragmentation/elimination in synthesis 26 27 29 36
3(2H)-furanones reactions 184
furans
– photoreactions 224–227, 261, 337
– synthesis 391
2-furylmethanol photoreaction 230

g

geraniol photoreaction 267
(±)-ginkgolide B synthesis 177
glycosyl radicals 88
(−)-grandisol, synthesis 140, 141
Griesbeck model for oxetane formation 226
(−)-guanacastapene synthesis 177

h

haloarenes
– photocyclization 293
– S_N1 reactions 319
– $S_{RN}1$ reactions 319
halogenation reaction 84
Hammond postulate 26
(±)-heliannuol D synthesis 185
hemiketal hydroperoxide synthesis 374
herbertene synthesis 141, 142
heterobicyclo[4.1.0]alkyl ketones synthesis 117, 118
heterocycles
– oxygenation 365–367
– synthesis 387–410
(E,E)-2,4-hexadiene oxygenation 364
hexamethylbenzene-1,4-endoperoxide synthesis 359
hexatriene cyclization 298
high-pressure mercury lamp 4, 173
high-pressure xenon arcs 9
(−)-hirsutene, synthesis 110, 111, 273
(+)-hirsutic acid, synthesis 110
home-made continuous flow microreactors, advantages 14
homobenzoquinones photoreaction 182
homobenzylic ether radical cations 276
hydroacylation reactions 75, 76
hydroalkylation reactions of alkenes and alkynes 69–76, 87
– hydrocarboxylation reactions 75, 76
– hydrofluoromethylation 74
– via alcohols addition 71, 72
– via alkanes addition 70, 71
– via amines addition 72–74
– via ethers addition 71, 72
– via nitriles addition 75
hydroacylation of olefins 75, 76

hydroamination of olefins 76–78
hydroarylation of olefins 76
hydrocarboxylation of olefins 75, 76
hydrofluoromethylation of olefins 74
hydrogen abstraction
– from unactivated substrates 70–72, 75
– in β 115
 – in the synthesis of cyclopropanols 115
– in γ 117, 118, 241, 243–245, 247–250, 253
 – in the formation of three membered rings 117, 118
 – in the formation of four membered rings 153–156
– in δ 153–156, 181, 242, 245
– remote 287, 307, 309
hydrogen bonding in enantioselective synthesis 196, 233
α-hydroxy(alkoxy)alkyl radicals 71
α-(1-hydroxyalkyl)-substituted α,β-unsaturated ketones 116
– direct photolysis 116
4-hydroxycoumarin photocycloaddition 203
4-hydroxy-2-nonenal photorelease 423
3-hydroxyproline esters synthesis 252
4-hydroxyquinolones photocycloaddition 204
hypoglycin A synthesis 48

i

imidazoles synthesis 394–398
imidazoline reaction 156
iminium salts, photochemistry 269
immersion well irradiation apparatus 7
(−)-incarvilline synthesis 176
indanols synthesis 243–245
– E-indanol 246
indoles, from enamines 263
indolines 263, 429, 430
– as photoremovable protecting group 429, 430
– from electron transfer cyclization 277
– from enamines 263
– cis-indoline 263
– stereochemistry 263
indolylketones, from enaminones 264
(±)-ingenol, synthesis of 199, 200
inner filter effect 13
intense light sources, effect 17
intermolecular cyclopentenone [2+2]-photocycloaddition 175
– facial diastereoselectivity 175
intermolecular dipolar cycloadditions 31, 33
intermolecular hydrogen bonding 177
intersystem crossing (ISC) 153, 171, 220, 323
intramolecular [2+1] cycloaddition 111

intramolecular [2+2] cycloaddition   101, 311
– regioselectivity   176
intramolecular hydrogen abstraction
   114–117, 153, 247, 287
2-iodobenzenesulfonamide
   photoreaction   346
1-iodonaphthalene photoreaction   325
4-iodo-1,1,2,2,9,9,10,10-octafluoro[2.2]
   paracyclophane photoreaction   334
ionic liquid as the reaction medium   144
irradiation apparatus   2–11
– amateur version   5
– light sources   9
– low-pressure mercury arcs lamps   3–6
– medium/high-pressure mercury arcs
   lamps   7, 8
isoleucine derivative, Yang cyclization   154
(±)-iso-retronecanol synthesis   73, 270
isothiochroman derivatives synthesis   302
isoxazole-to-oxazole rearrangement   399, 401
isoxazolin-5-ones   397, 400
– photochemical reactivity   397

k
(−)-kainic acid synthesis   192
kalasinamide, reaction with singlet
   oxygen   368
(±)-kelsoene synthesis   175
ketene silyl acetals, arylation   329
ketones   35, 38, 52, 56, 57, 105, 117, 273
– alkenyl, cyclization   274
– α-cleavage   25, 28, 30, 35, 36, 50, 52, 54
– decarbonylation   25, 26, 36, 40, 52, 54, 59
   – enantiospecific   56
   – in the solid   51
– photodecarbonylation   35, 51, 57
– photoinduced electron-transfer   273
ketoprofen-tetrahydrofuran dyad
   photoreaction   307
ketyl radicals, generation   274

l
(+)-laburnine synthesis   73
β-lactams   155
– formation   41
– Yang cyclization   155
γ-lactols synthesis   72
lactones decarboxylation   39
ladderane synthesis   33–35
– from diazenes   33
lamps   3, 4, 5
– black light   6
– coiled   5
– doped   8

– for external irradiation   5
– Hg-lamps   8
– low-pressure mercury arcs   3–6
– mercury resonance   3
– phosphor coated   6
– photochemical syntheses with   4
– quartz   6
– reaction vessel system   3
– tungsten   9
– tungsten-halogen   9
– U-shaped   5
– Wood's   6
lancifodilactone F synthesis   122
lasers light   9
– XeCl excimer laser   9
leaving groups in photochemistry   117, 417
LEDs see light-emitting diodes
ligand to metal charge transfer   139
light-emitting diodes (LEDs)   9, 10, 15, 18
linalyl acetate photorelease   423
(±)-linderol A synthesis   193
(+)-lineatin synthesis   189, 190
longifolene synthesis   31
low-pressure lamp   6
– advantages   6
– electrodeless   6
– Hg arcs lamps   3–6, 15
low-temperature experiments   8
low-temperature matrices   28
lumicolcine synthesis   152
lumicolechecine derivatives synthesis   152
lumi/lumiketone rearrangement   104
lumisantonin synthesis   104
Lycopodium alkaloid synthesis   110, 111

m
macrocyclic synthesis   288, 293, 307–310
(±)-magellanine synthesis   110, 111
malononitrile   75
– anti-Markovnikov photochemical
   addition   75
marcanine A synthesis   368
mazdasantonin synthesis   104
medium-pressure mercury lamp   14
– immersion well   14
(±)-meloscine synthesis   195
(±)-merrilactone A synthesis   175, 190, 232
metal-catalyzed formation of four membered
   rings   156, 157
metal complexes effect on photoreactivity   304
methanoglucamic acid synthesis   31
methoxy carbazoles preparation   341
3-methoxycarbonylcoumarin
   photoreaction   193

4-methoxypyridone photoaddition 204
4-methoxyquinolone photoaddition 204
δ-methoxyvalerophenone photoreaction 242
α-(*N*-methylanilino) styrene 263
– photocyclization-rearrangement 263
3-methyl-1-butene cycloaddition 193
methyl 1,3-dihydroisoindole-2-carboxylate reaction 119
– with cyclopentene 119, 120
2-methyl-furan 225
– Paternò Büchi reaction 225
3-methylisoxazolo[5,4-b]pyridine, reaction 125, 126
N-methylquinolinium hexafluorophosphate as sensitizer 276
migration-nucleophilic attack-cyclization 403
multilamp apparatus 4, 6
– rotating merry-go-round 6

n

naphthaldehydes, Paternò Büchi reaction 226
naphthalene derivatives 146
– endoperoxides 361, 362
– [2+2] photocycloaddition 146
naphthobarrelene, direct irradiation 99
1,4-naphthoquinone photocycloaddition 181
2-naphthoxide ions $S_{RN}1$ reaction 330
β-necrodol synthesis 141, 142
neocarzinostatin synthesis 41
o-nitrobenzyl alcohol derivatives 418–420
nitrogen heterocycles synthesis by $S_{RN}1$ reaction 341–344
nitromethane anion in $S_{RN}1$ reaction 343
NOCAS reaction 84
nonconjugated alkenes 137
– [2+2]-photocycloaddition 137–144
Norrish reaction 153, 307, 418, 423
– general mechanistic scheme 153
– type I group 36, 159, 424
– type II group 154, 418, 423, 427
– Yang reaction, relation to 153
NPEOC photolabile protecting group 419
N_2 photoelimination reaction 35
NPPOC photolabile protecting group 419
nucleophiles in $S_{RN}1/S_N1$ reaction 321, 326–329

o

olefins *see* alkenes
Oppolzer's chiral acryloyl sultam reaction 258
orthogonality of photoremovable protecting groups 417

ozonides formation 365
4-oxa-2-cyclohexenones photochemistry 185
4-oxa-2-cyclopentenones cycloaddition 183
oxadiazoles 402–404
– ring contraction–ring expansion route rearrangement 403
– synthesis of 402–404
oxa-di-π-methane rearrangement 105–111
– in zeolites 107
– of β,γ-unsaturated ketones and aldehydes 105–109
– synthetic applications 110
oxazepines synthesis 409–410
oxazoles 398–400
– photoinduced RCRE rearrangement 398
– synthesis 398–400
oxetanes synthesis 217–233, *see also* Paternò–Büchi reaction
– asymmetric 231
– bicyclic 228, 229
– chiral induction 232
– diastereoselective formation 231
– enantioselective synthesis 232
– endo-selective formation 227
– exo-selective formation 228, 232
– formation 218
– intramolecular cyclization reactions 218
– *O,S*-ketene silyl ether 227
– preparing methods 217
– radical ion pairs 224
– regioselective formation 221–226
– *cis*-selective formation 227
– site-selective syntheses 221–226
– stereoselective formation 219, 221, 226–233
– temperature effect 225, 226
N-oxides photoreactions 390–392, 410
oxiranes, fragmentation 254–257
oxiranes, opening to carbonyl ylides 254
oxygenations 353, 355, 364, 369, 374, 377

p

palladium-catalyzed reaction 335
Paternò–Büchi reaction 219–233, *see also* oxetane synthesis
– benzoin derivatives 428
– mechanism 220
– regioselectivity 221–226
– singlet state 220, 221
– stereoselectivity 226–233
– temperature effect on selectivity 225
– triplet state 221
pentacene-6,13-endoperoxide synthesis 363
(+)-pentacycloanammoxic acid synthesis 175

(±)-pentalenene synthesis 178
pentaprismane synthesis 138
perfluoroalkylethylenes alkylation 72
perfluoroalkylpyridazines photoreaction 406
(−)-perhydrohistrionico toxin synthesis 188
perhydroindolizines synthesis 270
peristaltic pump 13
phenacyl esters as photolabile groups 426, 427
phenanthrene synthesis 296
9-phenanthrol anion, $S_{RN}1$ reaction 330
phenanthroline synthesis 299
2-phenylbenzimidazole, historical synthesis 397
[7]phenylene synthesis 157
phenylglyoxylic acid esters Paternò Büchi reaction 232
phenyl/methyl-substituted alkenes 357, 358
– diphenylindene photooxidation 357
– electron-transfer photooxidation 357, 358
β-phenyl quenching effect in decarbonylation reaction 59
phenylsulfanylcarbene, photosensitive precursor 114
phenyl vinyl sulfones alkylation 70
(2-phosphinoethyl)silyl chelate ligand synthesis 79
phosphor-coated lamps 6, 18
– advantages 6
photocatalysis 3, 69–76
photochemical C–C bond formation via elimination
– CO photoelimination from ketones in solution 35–39
– CO_2 photoelimination from lactones 39
– four-membered ring synthesis 33–35
– N_2 photoelimination reaction 31–35
– polycyclic compounds synthesis 33–35
– sulfur photoelimination from sulfides/ sulfoxides/sulfones 40
– three-membered rings synthesis 31
photochemical electrocyclic reactions 150–153, 261–266
photochemical electron transfer, see photoinduced electron transfer
photochemical experimental methods 1–22
– concentration and scale 11–15
– experimental parameters 11–19
– impurities effect 15
– irradiation apparatus 2–11
– oxygen effect 15
– photochemical steps 19–22
– safety 17
– synthetic planning 17–22
– temperature effect 15
photochemical synthesis
– advantage 16
– chemical efficiency 40
– conditions for 2, 11
– crystal phase 15, 27, 28, 43, 50, 53, 60
– experimental parameters 11
– features 19
– flow microreactor 14
– general schemes 20–22
– large scale 13
– low temperature 7
– parameters 17
– planning 17–19
– practical hints 11, 15, 17
– process intensification, miniaturization 14
– safety precautions 17
– solar light for 10
– synthetic use, limitations to 13
– temperature independence 16
photochromic compounds 298
photocycloadditions see cycloaddition
photodeoxytridachione synthesis 103
photodimerization reactions 201
photoelectron transfer-initiated cyclizations 266–272, 287–295
– aromatic ketones 291, 292
– chloroacetamides 292, 293
– electron-deficient aromatic compounds 293–295
– phthalimides as electron acceptors 287–291
– radical anion 272–276
– radical cation 266–272
– radical cations, intramolecular trapping 276–279
photoelimination reaction 26, 27, 30, 31, 44
– advantages 27
– concerted 30
– disadvantages 27
– enthalpic feasibility 26
photoenolization/Diels–Alder reaction 302
photo-Fries rearrangement 31
photohydroalkylations 69
photoinduced electron transfer (PET) reaction 68, 78, 220, 256, 287–291, 293, 294, 301, 315, 321
– carbonyl derivatives 220,221,223
– in $ArS_{RN}1$ reaction 320, 321, 324, 343
– in the functionalization of olefin 68, 70, 73–75,78, 80
– mediated cyclizations 266–279
photolabile protecting group 417–432
– benzoin derivatives 428–429

- benzyl alcohol derivatives 421, 422
- indolines 429, 430
- o-nitrobenzyl alcohol derivatives 418–420
- phenacyl derivatives 426, 427
photo-NOCAS process 84
photooxygenations 353, 355, 364, 369, 372, 377
photoreactors 2
- falling film 13
- immersion well 7
- miniaturized 14
- multilamp 3
- Rayonet 4
photosensitizer 29, 37, 288,293,302, 311
photostimulated $S_{RN}1$ reactions 322
phthalimides 287
- cyclization 287–292
- [5+2] cycloaddition 310
- electron acceptors 287–291
- N-linked peptides photoreactions 291
(S)-pipecoline synthesis 299
piperidin-2,4-diones photoreactions 203
pixyl group, photolability of 421
polyaryls synthesis 334–338
- alternative to metal catalysis 333–336
- via photo-S_N1 336 reaction 338
polyenes
- cyclization 267
- electrocyclic reactions 150
polynuclear aromatic compounds synthesis 295
- helicenes 295
- phenacenes 295
(+)-premnalane synthesis 376
N-prenylpyridinium perchlorates photoreaction 270
N-prenylquinolinium perchlorate photoreaction 270
(+)-preussin synthesis 227
process intensification through miniaturization 14
4-propanoyl-6-methylpyrimidine, direct irradiation 118
prostratin synthesis 31
protecting groups, photolability of 417–438
prunolide C, synthesis of the core structure 374
(±)-pualownin synthesis 249
(±)-punctaporonin C synthesis 202
pyrazine photoreactions 396
pyrazinobarrelene, acetone-sensitized irradiation 99
pyrazoles 393, 395
- photorearrangements 395

- synthesis 393, 394
pyrazolines, photodenitrogenation 31
Pyrex vessel 188
- filtered light 261, 297
pyridinium salts, photoinduced cyclization 124, 125
pyridone derivatives 150, 192, 201, 303
- electrocylization 150
- photochemical electrocyclization 151
- selective cycloadditions 303
2,4-pyrimidindiones photoreaction 198
4-pyrimidinones photoreaction 198, 200
pyrroles 337, 388, 389
- arylation 338
- ICI rearrangements 389
- synthesis 388–391
pyrrolidine 258, 270
- synthesis 251, 258, 278

q

quadricyclene formation 34
quartz lamps 6
quinocarcin alkaloid synthesis 258
quinoline-N-oxides, photoreactivity 390
quinolizidine synthesis 185
quinolones 194, 201
- photocycloadditions 181
- 2-quinolone photocycloaddition 194
p-quinones, [2+2] cycloaddition of 181, 182

r

radical anion 272, 276, 294
- scavengers 321
radical cations 266, 267, 270, 272, 276, 277, 291
radical pair 43, 44, 220
- photoelimination 25, 27, 36, 40, 43, 44, 60
radical-radical combination 25, 39, 44, 60
radical scavengers 321
- di-*t*-butylnitroxide (DTBN) 321
- 2,2,6,6-tetramethyl-1-piperidinyloxy (TEMPO) 321
Rayonet reactor 4
Rehm-Weller equation
- and Paternò-Büchi reaction 220
remote intramolecular hydrogen abstraction 307–308
re-precipitation method in solid state photochemistry 59
resorcinol derivatives reaction 147
- benzocyclobutene derivatives synthesis 147
ring contraction 25–60, 308–311

ring-expansion-ring-contraction
  processes 394
ring formation
– five-membered 241–279
– four-membered 137–160, 171–205, 217–233
– large 288, 293, 307–310
– six-membered 287–313
– three-membered 31, 95–126
rugulovasines A and B synthesis 340

s

safety in photochemical reactions 17
salen complex catalyst 123
α-santonin photolysis 104
(±)-sceptrin synthesis 196
selenophenes synthesis 261
semibullvalenes synthesis 100
sensitizers
– electron-transfer 302, 303, 311
– in oxygenations 353
– triplet 29, 37, 288,293
seven-membered heteroaromatic ring synthesis 406–410
silicon ethers as photoremovable hydroxyl protecting group 424, 425
β-silyl effect in cycloadditions 119
silyl enol ethers 195, 222, 223
– Paternò Büchi reaction 222
α-silylmethylamine radical cations 270
silyl phosphates in alkylations 80
γ-silylsubstituted cyclopentenone cycloaddition 175
single electron transfer (SET) see photoinduced electron transfer
singlet biradicals 221
singlet oxygen chemistry 353–377
– tandem reactions 371–377
single-walled carbon nanotubes (SWCNTs) 363
Si–Si bond PET activation of 279
sisyl group as photoremovable group 424
six-membered heteroaromatic ring, synthesis 406
small-scale photoreactions 11
– quartz/Pyrex tubes 11
S_N1 see Ar S_N1 reactions
(+)-solanascone, synthesis 180
solanoeclepin A synthesis 190
solar light, photoreactions 10, 155, 362
solid state, photochemical reactions 15, 27, 28, 43, 50, 53, 60
– decarbonylation reaction 50–60
– decarboxylation 44–50

– natural products synthesis 57
– quenching effects 57
– radical intermediates fate, restriction of 43
– radical stability and selectivity 53
– reaction enantiospecificity 56
– reaction scale/experimental conditions 59
– reactivity/stability 42, 43, 51–53
– scope of the reaction 55
spiro compounds synthesis 268, 374–376
spiro[2.2]pentanes synthesis 117, 118
$S_{RN}1$ see $ArS_{RN}1$ reactions
$S_{RN}1$–Stille sequence 335
stereoisomeric tetrahydrofuran-3-ols synthesis 248
stereoselectivity in oxetane formation 226, 227, 230
(±)-sterpurene synthesis 179
cis-stilbenes 295
– like photocyclization 295–298
styrene 112
– cis-selective/enantioselective cyclopropanation 112
styrylalkyl-N-substituted phthalimides 290
– ring photocyclomerization 290
5-substituted 3-H-azepin-2-ones synthesis 406
2-substituted furans synthesis 392
N-substituted phthalimides reactions 287
2-substituted thiophenes 393
– photoisomerization 393
(−)-sulcatine, synthesis 175
sulfides, photoelimination 40
sulfones, photoelimination 40
sulfoxides photoelimination 40
sulfur 40, 290
– heterocycles 346
– nucleophiles 333, 334
– photochemical elimination 40, 41
1,4-sulfur–oxygen interactions 363
sunlight mediated reactions see solar light
synthesis, photochemical steps in 19

t

tandem radical addition/cyclization reactions 291
tandem singlet oxygen reactions 371–377
– in the synthesis of bisperoxides 371–372
– rearrangement to hemiketal hydroperoxide 374
– rearrangement to spiro compounds 374–376
tetrabutylammonium decatungstate, photocatalyst 70–72, 75

tetracyclic cyclopentaoxazoloisoquinolinone system synthesis 340
tetrahydrofuran derivatives synthesis 72, 254, 255, 394
– via alkylation 72
– via carbonyl ylides, cycloaddition of 254, 278
– via electron transfer cyclization 276
tetrahydrofuranols synthesis 247, 250
(−)-tetrahylipstatin synthesis 72
p-tetrahydronaphthoquinones 182
– intermolecular [2+2]-photocycloaddition 182
tetrahydrophthalic anhydride photoreaction 196
5, 10, 15, 20-tetrakis(heptafluoropropyl) porphyrin sensitizer 369
2, 20,4, 40-tetramethoxybenzophenone sensitizer 112
tetramic acids photochemistry 203
tetra/penta/hexa-cyclic azonia-aromatic compounds 297
– oxidative photocyclization 297
1,1,8,8-tetraphenyl-1-7-octadiene 301
– intramolecular cycloaddition 301
tetrazoamine, synthesis 125
tetronates photochemistry 202
tetronic acid amide 202
– intramolecular photocycloaddition 202
tetronic acid esters 202
– photochemistry 202
4-thia-2-cyclohexenones 186
– [2+2]-photocycloaddition reactions 186
thiadiazoles synthesis 405
thiazoles synthesis 400–402
thiobenzofurans, from aryl vinyl sulfides 266
thiocarbonyl derivatives β-hydrogen abstraction 125
1-thiocoumarin photocycloaddition 198
thioketals as photoremovable carbonyl protecting group 424
thiophenes 261, 392
– synthesis of 392, 393
thiophen-2(5H)-one photoreaction 197
4H-thiopyran-1,1-dioxide 102
– direct/triplet-sensitized irradiation 102
– di-π-methane rearrangement 102
three-membered heterocycles photochemical synthesis 123–126
– aziridines 123–126
– epoxides 123
three-membered ring formation 95–126
– by photochemical methods 95
tin nucleophiles, in $ArS_{RN}1$ reaction 332, 333

N-tosyl piperidine photoreaction 253
transition metals template-controlled reactions 304–306
triarylpyrylium (TPP) salts sensitizers 302
triazoles synthesis 404, 405
1,2,4-triazolin-5-ones synthesis 404
tricyclic cyclobutene derivative synthesis 149, 150
tricyclic housane synthesis 33–35
– from diazenes 33
tricyclic ketone reactions 158
tricyclic substrate 199
– intramolecular [2+2]-photocycloaddition 199
tricyclo[3.3.0.02,8]oct-3-ones synthesis 106
4-trifluoromethyl substituted imidazole synthesis 395
trimethylsilylmethylamine radical cation 270
1,3,5-triphenylbenzene synthesis 334
1,1,3-triphenylpropane formation 50, 51
2,4,6-triphenylpyrylium terafluoroborate sensitizer 279, 302
triple [2+2+2] cycloaddition 157
triplet carbenes generation 112
– triplet sensitizer for 112
triquinane (±)-pentalenene synthesis 178
trithiabicyclo[2.2.2]octane (TTBO) derivatives protecting groups 424
trivinyl-methanes 102
– direct/acetophenone-sensitized irradiation, photoproducts 102
– di-π-methane rearrangement 102
tropolone derivatives 152
– 4π electrocyclization 152, 153
two-photon absorption crosssection 431–432

u

β,γ-unsaturated aldehydes 105
– oxa-di-π-methane rearrangement 105–109
α,β-unsaturated carbonyl compounds 171, 272
– β-activation in cyclization reaction 272
– addition to 80, 85
– alkene E/Z isomerization 108
– alkylation 80, 85, 291
– function transformation 156
– oxa-di-π-methane rearrangement 105–109
– β-reductive activation 273
α,β-unsaturated carboxylic acid derivatives
– coumarins 193, 194
– 1,3-dioxin-4-ones 198–200
– endocyclic heteroatom 198–201
– exocyclic heteroatom 201–203
– lactams 192, 203–205

- lactones 189–192, 202, 203
- maleic anhydride 196, 197
- [2+2]-photocycloadditon 189–201
- 4-pyrimidinones 200, 201
- quinolones 194, 195
- sulfur analogues 197, 198
γ,δ-unsaturated carboxylic acid 276
- photoinduced electron-transfer cyclization 276
α,β-unsaturated-γ,δ-epoxynitrile reactions 257
unsaturated hydrocarbons overoxidation 376
α,β-unsaturated ketones *see* α,β-unsaturated carbonyls
α,β-unsaturated lactams photocycloaddition 192
α,β-unsaturated lactones photocycloaddition 189–191
UV spectra and photoreactivity 15

v
l-valine 98
- 4-benzoylphenyl ester chiral sensitizer 98
vinylarene compounds 144
- [2+2] photocycloadditions of 144
vinyl-biphenyls photocyclization 299
N-vinylcarbazole, cyclodimerization 139
vinyl cyclopropanes 260
- photochemical rearrangement 260, 261
vinylene carbonate, Paternò Büchi reactions 229
vinyl ethers Paternò Büchi reactions 222
vinylogous amides 182
- de Mayo reaction 182

- 4-hetero-2-cyclohexenones 185, 186
- 4-hetero-2-cyclopentenones 183–185
- [2+2]-photocycloaddition 182
vinylogous esters photocycloaddition 188

w
Wolff rearrangement 35, 111
Wood's lamp 6
- for photochemical applications 6
Woodward–Hoffmann rules 137, 219

y
Yang cyclization reaction 117, 153, 156, 307
- in the formation of four membered rings 153–156
ylides 264,265
- azomethine 258, 279
- carbonyl 254–257, 279
- push-pull 256

z
zeolites 153
- ene reactions with singlet oxygen 370
- mediated asymmetric hydroperoxidations 370
- oxa-di-π-rearrangement 107
Zimmerman rearrangement 96
zwitterion intermediate in electrocyclic reaction 261, 265
- dehydrogenation 261
- role of 265
zwitterion-tricyclic route in thiophene synthesis 392